Marine Geology and Geoph

T0264354

Editor-in-Chief

John H. Steele

Marine Policy Center, Woods Hole Oceanographic Institution, Woods Hole,
Massachusetts, USA

Editors

Steve A. Thorpe

National Oceanography Centre, University of Southampton,
Southampton, UK
and
School of Ocean Sciences, Bangor University, Menai Bridge, Anglesey, UK

Karl K. Turekian

Yale University, Department of Geology and Geophysics, New Haven,
Connecticut, USA

Subject Area Volumes from the Second Edition

Climate & Oceans edited by Karl K. Turekian
Elements of Physical Oceanography edited by Steve A. Thorpe
Marine Biology edited by John H. Steele
Marine Chemistry & Geochemistry edited by Karl K. Turekian
Marine Ecological Processes edited by John H. Steele
Marine Geology & Geophysics edited by Karl K. Turekian
Marine Policy & Economics guest edited by Porter Hoagland, Marine Policy Center,
Woods Hole Oceanographic Institution, Woods Hole, Massachusetts
Measurement Techniques, Sensors & Platforms edited by Steve A. Thorpe
Ocean Currents edited by Steve A. Thorpe
The Coastal Ocean edited by Karl K. Turekian
The Upper Ocean edited by Steve A. Thorpe

Marine Geology and Geophysics

Editor-in-Chief

John H. Steele
Marine Policy Center, Woods Hole Oceanographic Institution, Woods Hole,
Massachusetts, USA

Editors

Steve A. Thorpe

National Oceanography Centre, University of Southampton,
Southampton, UK
and
School of Ocean Sciences, Bangor University, Menai Bridge, Anglesey, UK

Karl K. Turekian
Yale University, Department of Geology and Geophysics, New Haven,
Connecticut, USA

Subject Area Volumes from the Second Edition

Climate & Oceans edited by Karl K. Turekian
Elements of Physical Oceanography edited by Steve A. Thorpe
Marine Biology edited by John H. Steele
Marine Chemistry & Geochemistry edited by Karl K. Turekian
Marine Ecological Processes edited by John H. Steele
Marine Geology & Geophysics edited by Karl K. Turekian
Marine Policy & Economics guest edited by Porter Hoagland, Marine Policy Center,
Woods Hole Oceanographic Institution, Woods Hole, Massachusetts
Measurement Techniques, Sensors & Platforms edited by Steve A. Thorpe
Ocean Currents edited by Steve A. Thorpe
The Coastal Ocean edited by Karl K. Turekian
The Upper Ocean edited by Steve A. Thorpe

MARINE GEOLOGY & GEOPHYSICS

A DERIVATIVE OF ENCYCLOPEDIA OF OCEAN SCIENCES, 2ND EDITION

Editor

KARL K. TUREKIAN

AMSTERDAM • BOSTON • HEIDELBERG • LONDON • NEW YORK • OXFORD
PARIS • SAN DIEGO • SAN FRANCISCO • SINGAPORE • SYDNEY • TOKYO
Academic Press is an imprint of Elsevier

ELSEVIER

ACADEMIC
PRESS

Academic Press is an imprint of Elsevier
32 Jamestown Road, London NW1 7BY, UK
30 Corporate Drive, Suite 400, Burlington, MA 01803, USA
525 B Street, Suite 1900, San Diego, CA 92101-4495, USA

Material in the work originally appeared in *Encyclopedia of Ocean Sciences* (Elsevier Ltd., 2001) and *Encyclopedia of Ocean Sciences*, 2nd Edition (Elsevier Ltd., 2009), edited by John H. Steele, Steve A. Thorpe and Karl K. Turekian.

The following article is US government work in the public domain and is not subject to copyright:
Satellite Oceanography, History and Introductory Concepts

Notice
No responsibility is assumed by the publisher for any injury and/or damage to persons or property as a matter of products liability, negligence or otherwise, or from any use or operation of any methods, products, instructions or ideas contained in the material herein, Because of rapid advances in the medical sciences, in particular, independent verification of diagnoses and drug dosages should be made

British Library Cataloguing in Publication Data
A catalogue record for this book is available from the British Library

Library of Congress Cataloging-in-Publication Data
A catalog record for this book is available from the Library of Congress

ISBN: 978-0-08-096484-3

For information on all Academic Press publications
visit our website at www.elsevierdirect.com

Printed and bound by CPI Group (UK) Ltd, Croydon, CR0 4YY
Transferred to Digital Print 2012

CONTENTS

MARINE GEOLOGY & GEOPHYSICS: INTRODUCTION

The dimensions of the ocean basins were not known until the explorations that began in the sixteenth century by Portuguese and Spanish explorers soon followed by British and Dutch seafarers. By the eighteenth century knowledge of the ocean basin limits were fairly well understood except for the region around Antarctica so that explorers like Captain James Cook, who in the late eighteenth century, could concern themselves with the details like the islands of the Pacific and the northern Pacific coastal regions as far north as Alaska. The nineteenth century was a time of massive assaults on the ocean and its boundaries including discoveries such as the recognition of Antarctica (although intimated by Tasman) and the major properties of the oceanic islands both volcanic and atolls. Included in the early explorations were those of the HMS Beagle, made famous by Charles Darwin's participation and the U.S. sponsored Wilkes Expedition several years later that saw the parallel explorations of the Pacific with scientists like James Dwight Dana aboard. The extensive, dedicated scientific exploration of the oceans was made at the end of the century by the HMS Challenger.

The topography of the ocean basins was not easily determined prior to the advent of sonic methods but enough bottom sensing data using plumb lines across the ocean were available, after the middle of the nineteenth century, to show that the ocean bottom was very deep and at least in the Atlantic Ocean had a rise in the middle between America and Europe. This information was gleaned during the planning of the transatlantic cable linking the two continents. In the twentieth century the need to explore the topography of the ocean bottom was evident for the continuation of cable laying and maintenance but the quality of data was not available until the development of the use of sound waves via precision depth recorders and sonar. Advances in both these exploitations of sound in exploring the ocean depths came during World War II. Harry Hess of Princeton, while on active duty with the U.S. Navy, made detailed precision depth PDR (Precision Depth Recorder) charts throughout the Pacific theater of operations and in 1946 published the first detailed topographic map of the Pacific Ocean basin identifying submarine seamounts (called guyots) and thereby showing the complexity of the ocean floor. Soon after, Bruce Heezen and Marie Tharp at Columbia compiled all the data from PDR data obtained from the telephone company laying and servicing transatlantic cables to arrive at remarkably accurate maps of the ocean floor including a detailed representation of the Mid-Atlantic ridge. These two representations were to play an important part in the development of the theory of plate tectonics later on.

Gravity studies at sea were initiated by Veining Meinesz of the Netherlands who showed that there were major differences in the gravity field at sea which could be related to bathymetry. This principle was later used in satellite studies of the ocean surface that reflected the topography of the ocean floor. The gravimetrically determined topography mimicked the bathymetric maps interpolated from sparser PDR data as charted by Heezen and Tharpe earlier.

Seismic studies at sea were pursued aggressively after World War II in part using the excess depth charges available after the war ended. Using explosives at first, the ocean crust and the sediment pile could be explored by seismic refraction and reflection studies. Eventually the exploration of the sediment pile was performed using electronic sound emitters.

Although the magnetic field of the Earth had begun to be studied seriously by William Gilbert in England during the reign of Elizabeth I, the continued exploration of the field structure continued on land and at sea early in the nineteenth century through the work of the Rosses (John, the uncle and James, the nephew). The discovery of magnetic polarity changes over time in the 1950s and 60 s provided a basis for the exploration of magnetic intensity anomalies across the ocean floor in this context. This insight allowed the chronology of the magnetic intensity striping of the ocean floor to be understood in the context of ocean floor spreading, one of the earliest constraints on and certification of the developing plate tectonic theory.

Once cores had been obtained from the ocean floor for sedimentological and paleoclimate studies the holes could be outfitted with thermisters and thus enabling the determination of heat flow at sea as well as it was done on land.

The combination of all these methods of mapping and geophysical studies provided the best evidence for the theory of plate tectonics as well as providing the details of the dimensions and time scales of the process.

Karl K. Turekian
Editor

OCEAN BASIN

OCEAN BASIN

HISTORY OF OCEAN SCIENCES

H. M. Rozwadowski, Georgia Institute of Technology, Atlanta, Georgia, USA

Oceanography is the scientific study of the ocean, its inhabitants, and its physical and chemical conditions. Since its emergence as a recognized scientific discipline, oceanography has been characterized less by the intellectual cohesiveness of a traditional academic discipline than by multidisciplinary, often large-scale, investigation of a complex and forbidding environment. The term 'oceanography' was not applied until the 1880s, at which point it still competed with such alternatives as 'thassalography' and 'oceanology'. In many countries, 'oceanography' now encompasses both biological and physical traditions, but this meaning is not universal. In Russia, for instance, it does not include biological sciences; the umbrella term remains 'oceanology'.

Before scientists studied the ocean as a geographic place with an integrated ecosystem, they addressed individual biological, physical, and chemical questions related to the sea. European expansion promoted the study of marine phenomena, initiating a lasting link between commercial interests and oceanography. Institutional interest in the sea by the Royal Society in London flourished briefly from 1660 to 1675, when members, including Robert Hooke and Robert Boyle, discussed marine research, developed instruments to make observations at sea, and conducted experiments on sea water to discover its physical and chemical properties. Sailors and travelers continued to make scattered observations at sea, working from the research plan and equipment established by the Society. In addition, tidal studies were prosecuted consistently from the late seventeenth century to well into the nineteenth.

Following decades of quiescence, a renewal of interest in marine phenomena occurred in the mid-eighteenth century. As part of that century's growth of astronomy, geophysics, chemistry, geology, and meteorology, investigators began making observations at sea as part of their scientific pursuit of other fields. Initially, observers were traveling gentlemen, naturalists, or eclipse-expedition astronomers; later they were scientific explorers. Beginning with the voyages of Captain James Cook in the last third of the century, British exploration became characterized by attention to scientific observations. The amount of energy devoted to marine science depended on the level of interest of expedition leaders or individual members, but, in the last quarter of the century, the volume of experiments and observations made at sea increased. Virtually all ocean investigation during this time focused on temperature and salinity of water, reflecting the growth of chemical sciences. The concept of oceanic circulation sustained by differences in density, which was widely discussed in the early nineteenth century, emerged at this time. The study of waves became more important while marine natural history remained at a modest level.

Early in the nineteenth century the emphasis on physical and chemical studies continued. Curtailed exploration due to the American Revolutionary War and the Napoleonic Wars slowed work in marine science until another period of rapid expansion between 1815 and 1830. Work then focused largely on currents and salinity, reflecting the fact that marine science was conducted on Arctic expeditions which searched for sperm whaling grounds and the Northwest Passage. Navigation and the pursuit of whales prompted Arctic explorers to investigate water temperature and pressure at depths. Enthusiastic individuals, most frequently ships' captains such as William Scorseby, continued to carry on observation programs at sea, but the 1830s saw a decrease of interest in marine science by physical scientists, who turned to rival fields of discovery including meteorology and terrestrial magnetism. At this time, however, British zoologists directed particular attention to marine fauna, embarking on small boats and yachts to collect with modified oyster dredges. Pursuit of new species as well as living relatives of fossilized ones inspired naturalists to reach deeper and deeper into the sea.

Beginning in the 1850s, both British and American hydrographic institutions began deep sea sounding experiments which were guided by the promise of submarine telegraphy, although initiated in support of whaling and navigational concerns. The decades after 1840 saw the gradual awareness by scientists, sailors, entrepreneurs, and governments that the ocean's depths were commercially and intellectually important places to investigate. An unprecedented increase in popular awareness of the sea accompanied this trend. Prefaced by the rage for seaside vacations which was initiated by railroad access to beaches, cultural interest in the ocean was manifested by the popularity of marine natural history collecting and the new maritime novels as well as the vogue for yachting and the personal experiences of growing numbers of ocean travelers and emigrants. Submarine telegraphy and public interest in maritime issues helped scientists argue successfully for government funding of oceanographic voyages. Declining fisheries emphasized the need to know more about the biology of the seas.

Until 1840, Britain led marine science, although important work was conducted by investigators in other countries, especially Scandinavian countries. In Norway, for example, G. O. Sars carried out important research for fisheries and also studied unusual creatures dredged from very deep water. American marine science began to threaten British dominance after mid-century. Matthew Fontaine Maury was the first investigator to compile wind, current, and whale charts to improve navigation and commerce. He also dispatched the first trans-Atlantic sounding voyages in the early 1850s. Under Alexander Dallas Bache's tenure, the US Coast Survey began to include, alongside routine charting work, special studies such as detailed surveys of the Gulf Stream, microscopic examinations of bottom sediments, and dredging cruises with the renowned zoologist Louis Agassiz and his son Alexander. Spencer F. Baird, Secretary of the Smithsonian Institution and United States Fish Commissioner, oversaw American efforts to study marine fauna. In Britain, while physical work was undertaken by the Hydrographic Office and Admiralty exploring expeditions, marine biological science centered around the British Association for the Advancement of Science Dredging Committee from 1839 until the mid-1860s. After that time, a Royal Society–Admiralty partnership took the lead and dispatched a series of summer expeditions. These culminated in the famous four-year circumnavigation of HMS *Challenger* (1872–1876), the first expedition sent out with a mandate to study the world's oceans. The United States, Britain, and other nations as well, quickly followed the example of the Naples Zoological Station and set up coastal marine biological laboratories in the 1880s and 1890s.

In the last quarter of the nineteenth century, many nations sponsored oceanographic voyages modeled after that of *Challenger*. The United States, Russia, Germany, Norway, France, and Italy contributed to the effort to define the limits and contents of the oceans. Late in the nineteenth century, however, the Scandinavian countries promoted a new style of ocean science to replace the great voyage tradition. Mounting concerns about depleted fisheries inspired national efforts in many countries to study the biology of fish species as well as their migration. Sweden initiated the formation of what became in 1902 the International Council for the Exploration of the Seas (ICES), which coordinated research undertaken by eight northern European member nations. Although not a member, the United States also continued active biological research. Victor Henson's discovery of plankton and efforts to quantify its distribution and the subsequent realization of how to use physical oceanography to investigate the movements of fish populations gave oceanographers confidence with which to study the ocean as an undivided system.

World War I disrupted the international community of oceanographers but provided the impetus for developing echo-sounding technology for submarine detection, which had been pioneered for ice detection partly in response to the *Titanic* disaster. By the late 1920s, echo-sounders revolutionized the study of underwater topography and helped scientists to recognize the rift valleys of midocean ridges, showing them to be active, unstable regions. Fishermen took up this technology for locating schools of pelagic fish; later scientists adapted echo-sounding to create fishery-independent surveying tools. Although government funding of oceanographic research dropped back almost to pre-war levels, the late 1920s saw the appearance of the first oceanographic institutions, sponsored mostly by foundations and private individuals. In the United States, the Scripps Institution changed its mission from biological research to oceanography in 1925 and, five years later, the Woods Hole Oceanographic Institution was established on the Atlantic. The availability of ocean-going vessels in the 1930s spurred development of American oceanography on a new scale. Fisheries research, which blossomed in Europe in the 1930s, included important theoretical advances which provided the foundation for later fish population dynamics modeling as well as open ocean research such as Sir Alister Hardy's Continuous Plankton Recorder surveys and Johannes Schmidt's publicly acclaimed search for the mid-Atlantic spawning ground of the European eel. Economic depression affected practical government science, such as fisheries research, as well as private projects, such as several of Schmidt's voyages, which were sponsored by the Carlsberg Foundation. As a consequence, oceanographic work, which was particularly expensive, slowed down dramatically until preparations began for World War II.

As with other sciences, World War II partnerships helped forge a new relationship between governments and oceanography. Oceanographic work during and after the war carried the imprint of wartime government support and policy in both its problem selection and its scale. Physical studies gained and maintained precedence over biological ones. The areas of inquiry promoted by wartime efforts related to submarine and antisubmarine tactics. New research began on underwater acoustics, and ocean floor sediment charts were compiled from existing data. Wave studies also received precedence for their value in predicting surf conditions for landings. After the war, oceanographers and academic institutions, newly accustomed to generous funding, learned to accept and even encourage government support. The foundation of the National Science Foundation, which became the major federal supporter of marine science in the USA, exemplified the new high levels of support for this technology-intensive science. Oceanography became characterized by large-scale, expensive research projects such as the 1960s deep-sea drilling by proponents of the theory of seafloor spreading. Major international projects also became an integral part of ocean sciences, in conjunction with postwar internationalization of science and other sectors. The International Indian Ocean Expedition, for example, targeted the least well-studied of the world's oceans.

By 1950 physical oceanographers were aware of the wanderings of the Gulf Stream and pressed urgently for systematic exploration of ocean variability. Development of deep ocean mooring technology provided the opportunity to study currents and temperatures continuously. This work showed that mesoscale variability was important and was incorporated into general circulation models. Progress in postwar physical oceanography in general proceeded in step with technological improvements, particularly the development of computing power. Another example, that of radioactive tracers and 'experiments' of atomic bomb-testing at sea, led to an intensification of research into biogeochemical cycling from the 1950s. Through programs such as the Geochemical Ocean Sections Study (GEOSECS), tracer use permitted estimates of mixing and circulation times by the late 1960s and led to programs such as the World Ocean Circulation Experiment (WOCE). Circulation and diffusion studies took on renewed importance as public concern about pollution rose. The idea that it was necessary to understand how the ocean functioned in order to avoid inadvertently destroying it fanned development of biological and chemical oceanography from the late 1960s onward. Much effort was funneled into baseline surveying and monitoring programs, initially focusing on potential threats to human health through eating poisoned seafood but soon broadening to investigations of biological effects of

contaminants. The fast growth of mariculture from this period also encouraged these studies.

Although many oceanographers became interested in open ocean research and theoretical modeling in the postwar period, some in Europe resuscitated efforts, begun within ICES in the 1930s, to integrate hydrographic knowledge with fisheries biology. This work encompassed the practical attempt to guide fishermen to catch more fish as well as the scientific project of trying to relate recruitment fluctuations to environmental factors. Parallel efforts in the United States bore fruit in the California Cooperative Oceanic Fisheries Investigations (CalCOFI), a cooperative government and academic program which investigated causes for the decline of the California coastal sardine fishery. CalCOFI, though, was an exceptional project; in general through the 1960s little significant intellectual exchange occurred between biological oceanography and fisheries science. Within biological oceanography, the legacy of E. Steeman-Nielsen's radioactive carbon tracer method led to active work measuring primary productivity, with global estimates by J. H. Ryther and others in the late 1960s.

Discovery of high biological diversity in the deep sea in the late 1960s led to work which was considerably enhanced by deep-diving submersibles, whose use dramatically changed our perception of the ocean's depths. In the area of marine geology and geophysics, the two greatest achievements since World War II have been the development of the theory of plate tectonics and the deciphering of Earth's paleoclimate record from deep-sea sediments. The first of these was set in motion by the Vine and Matthews hypothesis of 1963, which rocked the scientific community but was accepted with remarkable speed. Research submersibles played a central role in proving the theory of seafloor spreading. The 1979 discovery of hydrothermal vents, made in the process of these geophysical investigations, led to the surprise discovery of chemosynthetic ecosystems. Reconstruction of Earth's paleoclimates, which became possible due to the large accumulation of data and the availability of deep sea cores from the ocean drilling projects, was fueled by growing scientific concerns about global climate change. This work began during the 1960s with C. David Keeling's launch of time-series measurements of carbon dioxide, which provided the data documenting the increase of atmospheric carbon dioxide attributable to human activities and prompted inquiry into the magnitude of exchange of carbon dioxide between the atmosphere and oceans. Studies of the ocean's role in global warming and weather production have grown in importance in recent years.

Accompanying the rise of the environmental movement, oceanography's multi-faceted attempt to understand the oceans as integrated biological and physical environments made it a compelling discipline to ecologically and environmentally minded scientists from the 1970s onward. Food web investigations became prominent, including those which used mesoscale enclosures such as one designed and run at Canada's Pacific Biological Station in Nanaimo. Discovery of the microbial character of the pelagic food web added a new dimension to biological oceanography. New understanding of production, including a distinction between recycled nutrients and new production made by Dugdale and Goering in 1967, provided biological oceanography with the mathematical formalism for rigorous, quantitative modeling of ocean productivity and biogeochemical fluxes. Modeling capability encouraged a link to be forged between physical and biological oceanography, because new concepts required that physical processes of mixing and upwelling be integrated into ecosystem models dealing with new production, fish production, or export of organic material from the surface layer. The international Global Ocean Ecosystem Dynamics program and the recent resurgence of fisheries oceanography exemplify relatively successful efforts to bridge trophic levels, from plankton to marine mammals and sea birds, and thereby study the ecosystem as a whole.

See also

Maritime Archaeology.

Further Reading

Deacon M (1971) *Scientists and the Sea, 1650–1900: A Study of Marine Science*. London: Academic Press.

Herdman W (1923) *Founders of Oceanography and their Work: An Introduction to the Science of the Sea*. London: Edward Arnold.

Idyll CP (1969) *Exploring the Ocean World: A History of Oceanography*. New York: Thomas Y. Crowell.

Lee AJ (1992) *The Directorate of Fisheries Research: Its Origins and Development*. London: Ministry of Agriculture, Fisheries, and Food.

Mills EL (1989) *Biological Oceanography: An Early History, 1870–1960*. Ithaca: Cornell University Press.

Schlee S (1973) *The Edge of an Unfamiliar World: A History of Oceanography*. New York: E.P. Dutton.

Smith TD (1994) *Scaling Fisheries: The Science of Measuring the Effects of Fishing, 1855–1955*. Cambridge: Cambridge University Press.

Went AEJ (1972) Seventy years agrowing: a history of the International Council for the Exploration of the Sea, 1902–1972. *Rapport et Procès-Verbaux des Réunions, Conseil Permanent International pour l'Exploration de la Mer* 165: 1–252.

National Research Council 50 (2000) *Years of Ocean Discovery: National Science Foundation, 1950–2000*. Washington, DC: National Academy Press, 2000.

BATHYMETRY

D. Monahan, University of New Hampshire, Durham, NH, USA

Introduction

Bathymetry is the art and science of mapping the surface of the seafloor.

Unlike subaerial land surfaces which are exposed in their entirety and can be captured using a variety of optical instruments, the surface of the seafloor is invisible beyond a few meters depth, or beyond the light beams of human divers and remote vehicles. Mapping it consists of making measurements between it and a reference surface, plotting those measurements in their correct geographic locations, then interpreting the combined measurements into a mapped surface or a grid of calculated depths, both representatives of the actual seafloor.

Most human activity in the marine environment has a spatial or locational component. Geographic coordinate systems are enormously useful, but are more easily perceived by machines than humans. Humans can more quickly and easily visualize maps, with a shoreline (the edge of the seafloor) and depth portrayed by isobaths (contour lines) or color graduation: consequently, published bathymetry maps are used in all branches of marine science. They are the most common base maps, and are readily available in a number of series. Digital grids of depths are also calculated and published. Despite their value and ubiquitousness, bathymetry maps are often not understood and frequently taken for granted. There is a general belief that they are accurate and complete, and can be used unquestioningly. Unfortunately, this is seldom the case.

The surface of the seafloor is a major component of the Earth, covering as it does some 70% of the globe. It may be the place where sediments (and pollutants) accumulate, or may be composed of bare rock, some dating to the early formation of the Earth's crust, some formed only yesterday. It acts as the deeper wave guide for water motion, regulating the flow and velocity of tides and currents, including tsunamis. It is the home of myriad species, some who live their entire lives within it, some who have only sporadic contact with it. Some parts of it support rich biological habitats, some parts are deserts. Elements of it will form international boundaries under the United Nations Convention on the Law of the Sea (UNCLOS). Where it nears the ocean/atmosphere surface, it can become a hazard to navigation or provide protection from storm waves. Despite the seafloor's importance, it is poorly known, being mapped at a resolution that is 1 or 2 orders of magnitude less than some of the planets and the Moon.

Uses of Bathymetry

Bathymetry is used in most human activities in the marine environment, with uses ranging from providing a geographic reference frame to being a major component of a process. Economic and social uses include, but are not limited to:

- sustainable resource management: comprising fisheries, fisheries enhancement, aquaculture, and petroleum and mineral extraction;
- environmental stewardship: comprising habitat monitoring, land-use planning, environmental quality, national heritage, marine protected areas;
- health and safety: comprising sewage sites, dredge spoils disposal, risk reduction, oil spill readiness plans, coastal planning/global warming sea level rise, hazard mitigation (earthquakes, storm surge and tsunami, and seafloor hazards), accident investigation;
- infrastructure: comprising cables and pipeline, offshore exploration and production platforms, coastal structures, dredging, opening Northwest Passage;
- sovereignty: comprising boundaries definition, juridical continental shelf under UNCLOS, seabed surveillance and intervention, contraband and drugs.

In addition, bathymetry contributes to most other marine sciences. Physical oceanographers study how ocean water circulation is controlled by the shape of the seafloor, while to biologists the seafloor is an important component of the natural habitat of bottom-dwelling creatures, so that better resolution of bathymetry leads to better understanding of these phenomena. The shape of the seafloor can tell geologists the history and composition of the rocks and sediments that compose it, give clues to where seafloor hazards like landslides are likely to occur, and trace depositional patterns of material brought from land, including pollutants. Despite these many uses and its importance, the World Ocean is very poorly mapped.

What Bathymetry Is and Is Not

Bathymetric Maps Differ from Navigation Charts

Two types of maps showing 'depth' are in common use. Both bathymetric maps and navigation charts are highly specialized products, which are designed for different purposes and show different things. Since they both show 'depths', they are sometimes substituted one for the other, often due to the fact that in some areas only one of them existed at or near the desired scale.

Navigation charts are carefully constructed instruments designed to provide a basis for charting at sea. To do so, a chart must (1) form the base for the graphical exercise of navigation; (2) provide information on the identification and characteristics of navigational aides; and (3) provide information on the nature and position of navigational hazards, which in the case of bathymetry means providing shoal-biased depths. By contrast, a bathymetry map is not constrained by the need to protect mariners and their vessels. Like its land-based equivalent, the topographic map, it shows the relief of the seafloor, without any constraints imposed by the demands of safety. Every contour is shown in its best-known location; there is no bias toward depicting shallower rather than deeper depths – isolated deeps are included. In fact, the aim of the bathymetric map is to show every 'nook and cranny' of the seafloor in the best possible way permitted by the data available and the horizontal scale of the map.

It may be useful to think in terms of the surfaces that bathymetry maps and navigation charts show. Imagine a layer model, in which the deepest layer is the actual seafloor, the surface that would be seen if the water were removed. It undulates smoothly in places, changing texture to rough abruptly, the creation of the interplay of different geologic forces acting over different timescales. The next lowest layer is the bathymetry map. It is a representation of the lower layer; its creators would like it to conform exactly to the deepest layer. Although in a few areas of very densely spaced measurements it does, mostly the bathymetry layer conforms closely to the real surface in only a few places. There are two prime reasons for this. First, the real surface has only been measured in a few locations, and these are usually insufficient to allow a complete one-to-one correspondence between the bathymetry map surface and the real surface. Second, where the real surface has been measured by single-beam echo sounders,

their beams tend to smooth the real surface, filling in holes and broadening peaks. The upper layer is the navigation chart. Because the navigation chart's role is to help guide mariners away from areas where a ship could run aground, the navigation chart layer touches only the shallowest portions of the real surface, and magnifies the size of dangers to navigation.

Only about 8% of the ocean is shallower than 200 m, and depths that impact navigation are less than half that depth.

Bathymetric Surfaces Differ from Potential Field Surfaces

Maps of potential field (magnetics, gravity) are made from data collected along lines as is bathymetry. The mathematical approaches to producing contours and grids of potential field data have sometimes been applied to bathymetric acoustic measurements, but doing so can lead to erroneous results. Potential fields are usually (much) smoother than seafloor, since potential fields radiate in all directions, declining according to the inverse distance squared law. This means that potential field can only be single-valued, and has limits to its gradients; bathymetry represents a real world surface that in places includes overhangs and can have faces that are absolutely vertical. Assumptions made when processing potential field data may or may not apply to bathymetry.

Bathymetry of a Small Area

It is easy to begin with a model for constructing a bathymetry map or grid of a restricted area, one in which all the measurements are made using the same equipment. Local bathymetry maps are made almost exclusively from echo soundings, carried out by surface ships, or operated near the bottom in remote vehicles. This means that they do not travel very quickly! Essentially, a measurement is made at one location, the echo sounder moved, another measurement taken, the echo sounder moved again, and so on until resources are used up or weather forces termination of activities. From the resulting set of measurements, bathymetry is created.

Acoustics

Most of the data accepted for use in bathymetry come from acoustic measurements. In these measurements, sound is propagated into the water from a transducer, reflected from the seafloor and its arrival recorded at the point of origin. From the travel time, distance is calculated using a measured or assumed speed of sound.

Acoustic systems can be classified into:

1. single-beam echo sounders, with one sound beam oriented vertically, which produces a profile across the seafloor (these were the only method available until *c.* 1985 and were used in all applications; they are being supplanted by (2) and (3));
2. multibeam echo sounders (MBESs) with many beams radiating across the ship; and
3. side-scan sonars with one broad beam on each side; these are sometimes configured to produce measurements of depth, but more often give only a visual depiction.

These instruments are most frequently hull-mounted, and are also shallow-towed, or carried close to the seafloor by deep-towing or remote vehicles for high-resolution mapping.

Depth measurements are converted to data by being combined with navigation and attitude measurements. Navigation determines the sensor's position, while attitude determines where the sensor is pointed, from a combination of heading, speed, roll, pitch, and yaw.

Nonacoustic Methods

There are other ways of mapping the ocean floor besides using acoustics-based instruments. It has long been the dream of some bathymetrists to develop instruments that would not have to operate in the water, so that the progress of ocean mapping would not be tied to the slow speed and limited number of surface ships and their constrained area of operation.

One nonacoustic instrument that is being used in ocean mapping is LIDAR (light detection and ranging), in which a laser beam is projected from an aircraft to the seafloor. While this approach does benefit from the greater speed of aircraft, it is limited to a depth of 50 m or so, and consequently can only be used in shallow coastal waters.

Uncertainty

Bathymetric uncertainty can most easily be minimized before undertaking a single survey by designing the data collection to cover the entire area to be mapped in a cohesive fashion, using the same positioning and sounding instruments. Once a survey has been completed, the reliability with which we can determine uncertainty is completely dependent upon the degree to which the bathymetric data are redundant (repeated measurements at track crossovers). With sufficient crossovers, working only from the measurements themselves, it is possible to express precision of depths as the standard deviation of differences at crossovers.

Scale and Resolution

A map is a picture representing the Earth's continuous surface drawn at a greatly reduced scale. Conventionally the horizontal scale of a map is the ratio of a unit of length on the map to a unit of length on the Earth (e.g., a scale of 1:1 000 000 means that 1 cm on the map surface represents 1 000 000 cm on the Earth's surface). Determining a suitable scale at which to produce a paper map requires balancing a number of general factors like intended use, cost, method of reproduction, and accuracy required (see Table 1). Although the display of digital maps is not constrained by a fixed scale, they cannot portray the seafloor to a higher resolution than the data captured. A grid is a numerical model of the Earth's continuous surface comprising numerical values at discrete intervals. It too has a scale.

Resolution of a bathymetry map or grid specifies the size of the features it can resolve, with smaller indicating more detail. Scale imposes a limit on resolution, so the trick is to ensure that map portrays features down to the resolution limit.

There are a number of stages in the production of a bathymetry map which contribute to the final resolution of the map.

Table 1 Summary of horizontal and vertical resolution requirements for various oceanographic activities by area

Topic	Area	Horizontal resolution	Vertical resolution
Ocean models	Open ocean	10 km	50 m
	Sills	1–5 km	20 m
Tsunami models	Continental shelf <2.5 m deep	50 m	
	Continental shelf 2.5–10 m deep	100 m	
	Continental shelf 10–250 m deep	100–500 m	
	Canyons and ridges	500 m	
	Open ocean	>2 km	
Tides	Open ocean	5 km	10 m
	Continental slope	1 km	
	Continental shelf	100 m	Few meters
Mid-ocean ridges	Rift valleys	100 m	Few meters
Deep seafloor fabric	Abyssal hills	250 m	10 m
Paleoceanography	Continental shelf	500 m	
Subsurface geology	Continental margin	5 km	10–25 m

First, resolution of bottom features by acoustic measurements varies with the beam width of the echo sounder. As the beam spreads out away from the instrument, an increasingly large area of the seafloor (footprint) is ensonified, meaning that larger and larger features are masked as depth increases. (This is why deep tow is finer resolution.) Thus, even within the same map area, measurement resolution will not be the same if depth varies by hundreds of meters.

Second, the spacing and spatial arrangement of the depth measurements determines resolution. This takes a bit of explanation and a unit of measurement is needed. Horizontal resolution may be expressed by considering the surface to be made up of a series of superimposed waves. All Earth surfaces, be they the seafloor or landscapes, are made up of features of many different sizes. For example, mid-ocean ridges are large, and contain smaller mountains and foothills, which can have smaller valleys and ridges on them, which in turn may have ledges, spurs, steps. A 'wavelength' is the horizontal distance from the peak of one wave to the peak of the next (so that mountains might have a wavelength of tens of kilometers while ledges have one measured in tens of meters). Where the spacing between measurements is greater than any of the wavelengths that comprise the seafloor, those wavelengths cannot be captured; consequently, horizontal resolution is limited to twice the measurement spacing. Wavelengths that can be reconstructed from the measurements are closer to the original seafloor when the sampling interval is less than the original wavelength, but only when the sampling interval is less than one half the original wavelength are the regenerated wavelengths of a size similar to the original. Consequently, evenly spaced measurements can only capture features whose half wavelength is less than twice the distance between measurements. This is called the Nyquist frequency.

Spacing between depth measurements takes several forms. Although single-beam echo sounders record continuously along-track, the depths they record are usually stored at discrete intervals, with twice the spacing being the Nyquist frequency in the direction of the track. The spacing between one track and another may be different than the spacing within a track, and where it is the Nyquist frequency will be twice the line spacing. Consequently, there can be different resolutions in different directions within the same map if made from single-beam soundings. However, in the last few years, the introduction and deployment of the MBES is changing bathymetric measurements, having an impact similar to the one that the invention of the telescope had on astronomy. Rather than collecting a profile, MBES measures the (almost) complete surface of the seafloor on both sides of the instrument out to a distance of 2–5 times the water depth. With the proper data processing, it produces maps very similar to the familiar topographic maps on land. Within the area ensonified, MBES produces a grid of depths, whose spacing yields a Nyquist frequency that is much smaller than that normally obtainable by single-beam sounders.

It sometimes may happen that the measurements create wavelengths on the map that do not exist on the seafloor: this is termed aliasing.

All too often overlooked is the fact that maps also have a vertical scale. On bathymetric maps, vertical scale is expressed in the contour interval, since the minimum height of features that the map can resolve is twice the contour interval. For features of larger amplitude than this, contours portray the undulations of the land surface quite well, but smaller features are masked.

Making a Bathymetry Map of a Large Area

Data Assembly

Bathymetry for large areas includes portions of the deep ocean, where there have been very few routine surveys. Instead, there are usually an assortment of tracks from research cruises on which the data were measured from a variety of platforms, using different positioning and sounding systems, using (or not using) different sound velocities, units and plotting methods, collected over a time span of many years. Although much of the data will be held by the National Geophysical Data Center and other data centers worldwide, many are scattered in other data centers and research institutions round the world.

Assembling these data usually includes identification and accumulation, assessment of quality conversion of analog records to digital form, incorporation of digital data sets into an efficient data management system, construction of plots to portray the distribution of observations within the study area, reduction of all observations to a consistent set of processing parameters (velocity of sound in seawater, seasonal variations of sound velocity), the correction of obvious data errors (blunders), and elimination of bad data sets. The data so assembled usually have random orientation, widely variable spacing of tracks and within tracks, widely variable uncertainty in position and depth corrections and some redundancy. Increasingly, the data will consist of a mixture of MBES and single-beam data. Making maps from this data assemblage is more difficult than working from single data sets; hence, this can be

Table 2 Summary of characteristics of the principal types of bathymetric data

Data type	Spot SDG	Single beam E/S profile	Swath/multibeam	Satellite altimetry
Age	Old, aging	1950s to present	Recent	Recent
Distribution	Limited, remote areas	Most of world	Limited areas	World except Poles
Resolution, wavelength	Twice spacing	High along track, 2 × line spacing across	High	See text
Shown on	Small scale	All scales	Large scale	Small scale
Detail	None	Good along track, poor across track	High	Poor
Redundancy	None	Variable with layout	High	High
Coverage	Very low	Good along track, poor across track	High within survey area	High
Interpretation	Sketchy	Needs high level	Automatic	Automatic

Spot or single soundings are included as a base level. Single-beam echo sounding is being replaced by multibeam, a trend which will continue. These characteristics must be kept in mind when working with the different types of data together.

treated as numerical exercise (algorithm) only up to a point and requires interpretation and consideration of other types of data (see **Table 2**).

Interpretation

Mapping of the land topography is largely an automatic process since the land surface can be seen by the human eye and can be photographed and imaged. However, we have never seen the seafloor, over areas larger than tens of square meters, and have bounced sound off only a small percentage of it, so that producing maps and grids is far from automatic. There are, however, two approaches to how to proceed from the incomplete and possibly inefficiently arranged data available. One is based on the assumption that only the depth data should contribute to the creation of the bathymetry, while the other makes use of other related data.

The first approach fits a surface to the data or calculates individual depth values at grid nodes using mathematical techniques. Different workers use different variations of mathematical contouring, triangulating, trend surfaces, and more, with varying and arguable results.

In the second approach, ancillary geological, geophysical, and physical oceanographic information, as well as satellite altimetry (see next section) and sidescan sonar images, may be used in support of the acoustic data. For instance, the center lines of canyons may be established and the mapped surface generated to conform to them. Another example is that of seafloors made of bedrock having different shapes than those that are sediments, and sediments that are accumulating in a still area having different shapes to those emanating from a canyon. In areas of sparse acoustic data, satellite altimetry gives the main direction and extent of larger features.

The introduction of MBES has added another dimension to interpretation. Within the area of seafloor ensonified during a multibeam survey, there is so much data that there is no need to interpret the shape of the seafloor as part of mapping it. Furthermore, the MBES swathes include more information than just depth. They show characteristics of the surface like texture, roughness, graininess, smoothness, nature, as well as backscatter. These can be extrapolated into the areas not surveyed by MBES.

Satellite Altimetry

In addition to its secondary role of guiding the interpretation of scarce acoustic data, satellite altimetry stands on its own as the most consistent and cohesive data set covering the world oceans (except for the Arctic). The beautiful world-scale maps produced by satellite altimetry techniques are widely distributed. Satellite altimetry has the advantages of producing data of uniform quality at regular and close line spacing. Despite having these advantages, satellite altimetry maps cannot be directly substituted for acoustic maps for two reasons. First, they do not measure water depth directly but calculate it from the displacement of the sea's surface due to variations in mass on the seafloor. Because of the inverse distance squared law inherent in potential fields, the bathymetry produced is much smoother than the actual seafloor. Consequently accuracy is diminished, although this can be improved by incorporating echo sounder data into the satellite data. Second, they resolve only long wavelength (12–20 km) features of the seafloor morphology.

Metadata

Since bathymetry data are scarce and valuable, they should be shared by submitting them to a data center. Submitted data are much more valuable when accompanied by metadata, but the preparation of metadata is often viewed as an onerous task. Nevertheless, metadata is useful to describe data sets' attributes to enable accurate searching and allow the data set to be fully understood and used. It also builds a trail to the originator to allow proper acknowledgement of the contribution. Different data centers publish their recommended standards for metadata on their websites: although no common agreed standard exists at the time of writing, data center specialists from several countries are attempting to construct one.

Uncertainty

Uncertainty over large areas is highly variable and difficult to determine, and it is almost always worse in this case than over small areas. Where measurements are very scarce, rather than trying to calculate a numerical value for uncertainty, standard practice has been to show the locations of the ships' tracks and let the user of the map make their own judgments about the reliability. This cannot be applied to grids, however.

Rates of Progress

The rate at which mapping of any natural surface progresses depends on the interrelationship of:

- the size of the surface to be mapped;
- the speed of the platforms that carry those tools, their amount of use, and the areas that they are deployed in;
- the efficiency of the mapping tools that can be applied; and
- the organizational structures in place to assimilate the raw data and transform it into maps.

The seafloor is twice as large as all the land surface on Earth, research ships travel at about the same speed as a rider taking a leisurely bicycle ride, and MBES is the most efficient tool yet devised to map the seafloor as it maps a swath equal in width to 2 or 3 times the water depth. Unfortunately, research ships have never been deployed in a program to systematically ensonify every part of the ocean. If they were, it would take at least 800 ship-years. Rather than a systematic and complete data set, there is a pattern of tracks that is analogous to the pattern of roads and streets on land. If data were collected along roads, in the cities, data would be closely spaced and arranged with some regularity, but in the rural areas data would be sparse and arranged randomly. Maps of world bathymetry are thus very uneven in quality, accuracy, and resolution.

Despite the absence of a systematic survey program, over the next 10 years data will be collected primarily on continental slopes to satisfy UNCLOS Article 76 requirements to map the foot of the slope and the 2500 m contour, tsunami-affected areas, primarily the shallow water run-up zone, and in areas of specialized scientific interest, for example, the Ridge Program (which focuses on mid-ocean ridges). 'Random' tracks that collect data will decrease due to the shift from the 'expeditionary' style of at-sea oceanographic data collection to repetitive measurements of the same point or small area to collect time series. Every track, especially those outside well-surveyed areas, is valuable and it is essential that all data be assembled for the common good.

Need for Cooperation

Because the ocean is so large and the portion of it that any one ship can survey so relatively small, it has long been recognized that only international cooperation will accomplish the mapping of ocean basins. The first international chart of the world ocean was produced in 1903 as the General Bathymetric Chart of the Oceans (GEBCO), a series that has been continually improved and updated since then. Supporting this production is the IHO Data Center for Digital Bathymetry (IHO DCDB) operated by the National Geophysical Data Center in Boulder, CO, USA. The DCDB and a number of national data centers assemble contributions of depth data from all possible sources, perform fundamental quality control, maintain inventories of digital bathymetric data and their metadata, and make the collected data available to the world. The data centers collaborate with one another in establishing standards and sharing data. A path therefore exists for all research cruises to contribute bathymetry data so that existing bathymetry maps and grids can be continuously upgraded. Collecting and sharing of data by all research cruises is the only way to advance in the absence of a systematic data collection program.

Assembly, Interpretation, Distribution, and Updating

Bathymetry maps and grids are produced at scales ranging from local to world-encompassing. The results of a single research cruise may be portrayed as a figure in a published paper or a poster at a conference. Some countries produce national series of bathymetry maps at a variety of scales. Larger regions are often mapped by International Oceanographic Commission's (IOC's) International Bathymetric Charts (IBC) projects. Maps and grids covering the entire World Ocean are produced by the GEBCO. In the maps not produced from a single cruise, maps covering smaller areas are incorporated into those covering larger areas, along with other depth data. Bathymetry normally needs updating as soon as new measurements are made and submitted, and the method production and distribution must be designed to accommodate continual updating. Early approaches of printing successive editions are being replaced with continuous updating of maps and grids using websites.

See also

Geomorphology. Satellite Altimetry.

Further Reading

Carron MJ, Vogt PR, and Jung W-Y (2001) A proposed international long-term project to systematically map the world's ocean floors from beach to trench: GOMaP (Global Ocean Mapping Program). *International Hydrographic Review* 2: 49–55.

Clarke JEH (1996) Are you really getting 'full bottom coverage'? A collection of thoughts and images. http://www.omg.unb.ca/%7Ejhc/coverage_paper.html (modified 8 Nov. 1996).

Clarke JEH (2002) The challenge of technology: Improving sea-floor mapping methodologies (part of "Visualizing the sea floor: Mapping submarine landscapes"). *Symposium on Visualizing and Looking Beyond Earth.* Washington, DC: American Association for the Advancement of Science. http://www.omg.unb.ca/AAAS/UNB_Seafloor_Mapping.html (modified 12 Feb. 2002).

Clarke JH (2000) Present-day methods of depth measurement. In: Cook PJ and Carleton CM (eds.) *Continental Shelf Limits: The Scientific and Legal Interface*, pp. 139–159. Oxford, UK: Oxford University Press.

Committee on National Needs for Coastal Mapping and Charting Ocean Division of Earth and Life Studies (2003) *Voyage into the Unknown*, 213pp. Washington, DC: The National Academies Press.

Cormier M-H, De Moustier C, Hall JK, Mayer L, Monahan D, and Vogt P (2004) The Global Ocean Mapping Project (GOMaP): Promoting international collaboration for a systematic, high-resolution mapping of the world's oceans. *32nd International Geological Conference*, Florence, Italy, 24–27 August.

Dunn RA, Scheirer DS, and Forsyth DW (2001) A detailed comparison of repeated bathymetric surveys along a 300-km-long section of the southern East Pacific Rise. *Journal of Geophysical Research* 106: 463–471.

Dunne D and Sutton G (2006) Integrating marine data into Google Earth. *e-Newsletter Hydro International*, no. 17. http://cmrc.ucc.ie/publications/journals/Reprint_HI_0706_Dunne.pdf (accessed Mar. 2008).

Hare R (2002) Bathymetry error modelling: Approaches, advances and applications. *Canadian Hydrographic Conference*, Toronto, Canada, 28–31 May (CD-ROM).

Holcombe TL and Moore C (2000) Data sources, management and presentation. In: Cook P and Carleton C (eds.) *Sovereign Limits beneath the Oceans: The Scientific and Technological Aspects of the Definition of the Continental Shelf*, pp. 230–249. Oxford, UK: Oxford University Press.

IOC and UNESCO (2002) Improved global bathymetry. *IOC Technical Series: Final Report of SCOR Working Group 107*, p. 111. http://unesdoc.unesco.org/images/0012/001272/127238e.pdf (accessed Mar. 2008).

Marks KM (2002) Acoustic bathymetry for altimetric bathymetry calibration studies, NOAA Laboratory for Satellite Altimetry. http://ibis.grdl.noaa.gov/SAT/Bathy.intro.html (accessed 22 Apr. 2008).

Monahan D (2000) Interpretation of bathymetry. In: Cook PJ and Carleton CM (eds.) *Continental Shelf Limits: The Scientific and Legal Interface*, pp. 160–176. Oxford, UK: Oxford University Press.

Monahan D (2004a) Altimetry applications to continental shelf delineation under the United Nations Convention on the Law of the Sea. *Oceanography* 17: 71–78.

Monahan D (2004b) GEBCO: The second century. Looking towards a general bathymetric chart. *Hydro International* 8: 45–47.

Sandwell DT and Smith WHF (1996) Global Bathymetric Prediction for Ocean Modelling and Marine Geophysics. http://topex.ucsd.edu/marine_topo/text/topo.html (accessed Mar. 2008).

Smith WH (1998) Seafloor tectonic fabric from satellite altimetry. *Annual Review Earth Planetary Science Letters* 26: 697–738.

Wells D, Kleusberg A, and Vanicek P (1996) A seamless vertical-reference surface for acquisition, management and display of ECDIS hydrographic data. *Proceedings of the Canadian Hydrographic Conference '96*, p. 64. Fredericton, NB: Department of Geodesy and Geomatics Engineering, University of New Brunswick.

Relevant Websites

http://www.awi.de
– Alfred Wegener Institute: Bathymetry data and maps of polar regions.

http://www.bodc.ac.uk
– British Oceanographic Data Center: Producers of GEBCO Digital Atlas (GDA), free downloadable grid, information on how to use the GDA.

http://www.ccom.unh.edu
– Center for Coastal and Ocean Mapping, University of

New Hampshire: Education and research in all elements of ocean mapping, with many examples.

http://gebco.net

– General Bathymetric Chart of the Oceans (GEBCO): Free 1 min grid, information and publications on all things bathymetric.

http://mp-www.nrl.navy.mil

– Global Ocean Mapping Project (US Naval Research Laboratory Marine Physics Branch): Mostly inactive site, valuable figures on deep-water data collection.

http://www.ifremer.fr

– IFREMER: French marine data sets.

http://www.ngdc.noaa.gov

– IHO Data Center for Digital Bathymetry (IHO DCDB), NOAA's National Geophysical Data Center (NGDC): Data base of depth data submitted by National Hydrographic offices worldwide; various bathymetry data sets at Marine Geology and Geophysics Division of NGDC; World Data Center (WDC) System: NOAA links to major data centers worldwide, many with bathymetry.

http://www.ices.dk

– International Council for the Exploration of the Sea (ICES): Bathymetry and its use in fisheries.

http://www.imca-int.com

– International Marine Contractors Association: Graphical Guide to the Offshore Survey Industry, an excellent set of interrelated diagrams showing industrial bathymetric data collection.

http://www.omg.unb.ca

– Ocean Mapping Group, University of New Brunswick: Animations of multibeam interaction with the seafloor.

http://www.pangaea.de

– PANGAEA Publishing Network for Geoscientific and Environmental Data: Listing of maps, grids and cruise data sets.

http://ocean-ridge.ldeo.columbia.edu

– RIDGE Multibeam Synthesis Project (ocean floor databases): Data base of multibeam data over the mid-ocean ridges.

http://topex.ucsd.edu

– Satellite Geodesy at the Scripps Institution of Oceanography, University of California San Diego: Measured and estimated seafloor topography, bathymetry from space.

http://www.scor-int.org

– Scientific Committee on Oceanic Research (SCOR): Interrelationships of all oceanography branches.

GEOMORPHOLOGY

C. Woodroffe, University of Wollongong, Wollongong, NSW, Australia

Introduction

Geomorphology is the study of the form of the earth. Coastal geomorphologists study the way that the coastal zone, one of the most dynamic and changeable parts of the earth, evolves, including its profile, plan-form, and the architecture of foreshore, backshore, and nearshore rock and sediment bodies. To understand these it is necessary to examine wave processes and current action, but it may also involve drainage basins that feed to the coast, and the shallow continental shelves which modify oceanographic processes before they impinge upon the shore. Morphodynamics, study of the mutual co-adjustment of form and process, leads to development of conceptual, physical, mathematical, and simulation models, which may help explain the changes that are experienced on the coast.

In order to understand coastal variability from place to place there are a number of boundary conditions which need to be considered, including geophysical and geological factors, oceanographic factors, and climatic constraints.

At the broadest level plate-tectonic setting is important. Coasts on a plate margin where oceanic plate is subducted under continental crust, such as along the western coast of the Americas, are known as collision coasts. These are typically rocky coasts, parallel to the structural grain, characterized by seismic and volcanic activity and are likely to be uplifting. They contrast with trailing-edge coasts where the continental margin sits mid-plate, which are the locus of large sedimentary basins. Smaller basins are typical of marginal sea coasts, behind a tectonically active island arc.

The nature of the material forming the coast is partly a reflection of these broad plate-tectonic factors. Whether the shoreline is rock or unconsolidated sediment is clearly important. Resistant igneous or metamorphic rocks are more likely to give rise to rocky coasts than are those areas composed of broad sedimentary sequences of clays or mudstones. Within sedimentary coasts sandstones may be more resistant than mudstones. Rocky coasts tend to be relatively resistant to change, whereas coasts composed of sandy sediments are relatively easily reshaped, and muddy coasts, in low-energy environments, accrete slowly.

The relative position of the sea with respect to the land has changed, and represents an important boundary condition. The sea may have flooded areas which were previously land (submergence), or formerly submarine areas may now be dry (emergence). The form of the coast may be inherited from previously subaerial landforms. Particularly distinctive are landscapes that have been shaped by glacial processes, thus fiords are glacially eroded valleys, and fields of drumlins deposited by glacial processes form a prominent feature on paraglacial coasts.

The oceanographic factors that shape coasts include waves, tides, and currents. Waves occur as a result of wind transferring energy to the ocean surface. Waves vary in size depending on the strength of the wind, the duration for which it has blown and the fetch over which it acts. Wave trains move out of the area of formation and are then known as swell. The swell and wave energy received at a coast may be a complex assemblage generated by specific storms from several areas of origin. Tides represent a large-wavelength wave formed as a result of gravitational attractions of the sun and moon. Tides occur as a diurnal or semidiurnal fluctuation of the sea surface that may translate into significant tidal currents particularly in narrow straits and estuaries. In addition storm-generated surges and tsunamis may cause elevated water levels with significant geomorphological consequences.

The coast is shaped by these oceanographic processes working on the rocks and unconsolidated sediments of the shoreline. Climate is significant in terms of several factors. First, wind conditions lead to generation of waves and swell, and may blow sand into dunes along the backshore. Climate also influences the rate at which weathering and catchment processes operate. In addition, regional-scale climate factors such as monsoonal wind systems and the El Niño Southern Oscillation phenomenon demonstrate oscillatory behavior and may reshape the coast seasonally or interannually. Gradual climate change may mean that shoreline fluctuations do not revolve around stationary boundary conditions but may exhibit gradual change themselves.

In this respect the impact of perceived global climate change during recent decades as a result of human-modified environmental factors is likely to be

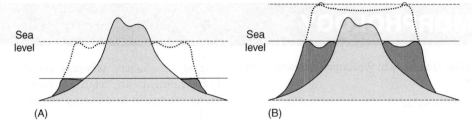

Figure 1 Schematic representation of the original woodcut illustrations by Charles Darwin showing (A) the manner in which he deduced that fringing reefs around a volcanic island would develop into barrier reefs (dotted), and (B) barrier reefs would develop into an atoll, as a result of subsidence and vertical reef growth.

felt in the coastal zone. Of particular concern is anticipated sea-level rise which may have a range of geomorphological effects, as well as other significant socioeconomic impacts.

History

Geomorphology has its origins in the nineteenth century with the results of exploration, and the realization that the surface of the earth had been shaped over a long time through the operation of processes that are largely in operation today (uniformitarianism).

The observations by Charles Darwin during the voyage of the *Beagle* extended this view, particularly his remarkable deduction that fringing reefs might become barrier reefs which in turn might form atolls as a result of gradual subsidence of volcanic islands combined with vertical reef growth (**Figure 1**). In the first part of the twentieth century geomorphology was dominated by the 'geographical cycle' of erosion of William Morris Davis who anticipated landscape

denudation through a series of stages culminating in peneplanation, and subsequent rejuvenation by uplift. This highly conceptual model, across landscapes in geological time, was also applied to the coast by Davis who envisaged progressive erosion of the coast reducing shoreline irregularities with time (**Figure 2**). Such landscape-scale studies were extended by Douglas Johnson who emphasized the role of submergence or emergence as a result of sea-level change (**Figure 3**).

The Davisian view was reassessed in the second half of the twentieth century with a greater emphasis on process geomorphology whereby studies focused on attempting to measure rates of process operation and morphological responses to those processes, reflecting ideas of earlier researchers such as G. K. Gilbert. The concept of the landscape as a system was examined, in which coastal landforms adjusted to equilibrium, perhaps a dynamic equilibrium, in relation to processes at work on them. Studies of

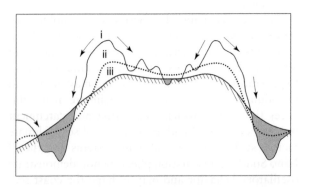

Figure 2 Schematic representation of the planform stages that W. M. Davis conceptualized through which a shoreline would progress from an initial rugged form (1) through maturity (2) to a regularized shoreline (3) that is cliffed (hatched) and infilled with sand (stippled). He envisaged this in parallel to the geographical cycle of erosion by which mountains (like the initial shoreline form, 1) would be reduced to a peneplain (like the solid ultimate shoreline, 3).

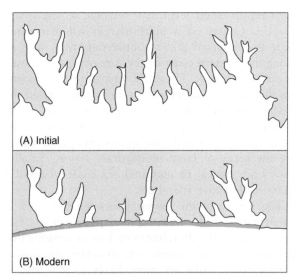

Figure 3 Schematic representation of (A) the initial shoreline form envisaged by Douglas Johnson for an area in Marthas Vineyard, New England, and (B) the modern regularized, form of the shore.

sediment movement and the adjustment of beach shape under different wave conditions were typical.

Scales of Study

Coastal geomorphology now studies landforms and the processes that operate on them at a range of spatial and temporal scales (**Figure 4**). At the smallest scale geomorphology is concerned with an 'instantaneous' timescale where the principles of fluid dynamics apply. It should be possible to determine details of sediment entrainment, complexities of turbulent flow and processes leading to deposition of individual bedform laminae. The laws of physics apply at these scales, though they may operate stochastically. In theory, behavior of an entire embayment could be understood; in practice studies simply cannot be undertaken at that level of detail and broad extrapolations based on important empirical relationships are made.

The next level of study is the 'event' timescale, which may cover a single event such as an individual storm or an aggregation of several lesser events over a year or more. The mechanistic relationships from instantaneous time are scaled up in a deterministic or empirical way to understand the operation of coasts at larger spatial and temporal scales. Thus, stripping of a beach during a single storm, and the more gradual reconstruction to its original state under ensuing calmer periods can be observed. Time taken for reaction to the event, and relaxation back to a more ambient state may be known from surveys of beaches, enabling definition of a 'sweep zone' within which the beach is regularly active.

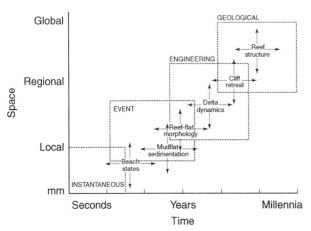

Figure 4 Representation of space and timescales appropriate for the study of coastal geomorphology and schematic representation of some of the examples discussed in the context of instantaneous, event, engineering, and geological scales of enquiry.

At a larger scale of operation, the coastal geomorphologist is interested in the way in which coasts change over timescales that are relevant to societies and at which coastal-zone managers need to plan. This is the 'engineering' timescale, involving several decades. It is perhaps the most difficult timescale on which to understand coastal geomorphology and to anticipate behavior of the shoreline. The largest timescales are geological timescales. Studies over geological timescales are primarily discursive, conceptual models which recognize that boundary conditions, including climate and the rate of operation of oceanographic processes, change. It is clear that sea level itself has changed dramatically over millennia as a result of expansion and contraction of ice sheets during the Quaternary ice ages, and so the position of the coastline has changed substantially between glaciations and interglaciations.

Models of Coastal Evolution

There has been considerable improvement in understanding the long-term development of coasts as a result of significant advances in paleoenvironmental reconstruction and geochronological techniques (especially radiometric dating). Incomplete records of past coastal conditions may be preserved, either as erosional morphology (notches, marine terraces, etc.) or within sedimentary sequences. Although this record is selective, reconstructions of Quaternary paleoenvironments, together with interpretation of geological sequences in older rocks, have enabled the formulation of geomorphological models based on sedimentary evidence. In the case of deltas, where there may be important hydrocarbon reserves, the complementary development of geological models and study of modern deltas has led to better understanding of process and response at longer timescales than those for which observations exist.

Unconsolidated sediments are the key to coastal morphodynamics because coasts change through erosion, transport, and deposition of sediment. Study of coastal systems has led to many insights, particularly in terms of the various pathways through which sediment may move within a coastal compartment or circulation cell. However, it is clear that a completely reductionist approach to coastal geomorphology will not lead to understanding of all components and the way in which they interact. Empirical relationships and the presumed deterministic nature of sediment response to forcing factors remain incomplete and are all too often formulated on the basis of presumed uniform sediment sizes or absence of biotic influence and are ultimately

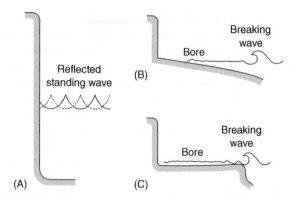

Figure 5 The morphology of cliffs and shore platforms. Plunging cliffs (A) reflect wave energy and are generally not eroded. Shore platforms dissipate wave energy and may be either (B) ramped, or (C) subhorizontal. Change in these systems tends to occur at geological (hard rock) or engineering (soft rock) timescales.

unrealistic. More recently coastal systems have been investigated as nonlinear dynamical systems, because relationships are not linear and behavior is potentially chaotic.

Nonlinear dynamical systems behave in a way that is in part dependent upon antecedent conditions. This is well illustrated by studies of beach state; beach and nearshore sediments are particularly easily reshaped by wave processes, thus a series of characteristics typical of distinct beach states can be recognized (see **Figure 6**). However, the beach is only partly a response to incident wave conditions, its shape being also dependent on the previous shape of the beach which was in the process of adjusting to wave conditions incident at that time.

Coastal systems are inherently unpredictable. However, they may operate within a broad range of conditions with certain states being recurrent. Chaotic systems may tend towards self-organization, for instance patterns of beach cusps characteristic along low-energy beaches which reflect much of the wave energy may adopt a self-organized cuspate morphology in which swash processes, sediment sorting, and form are balanced.

Models may be developed based on patterns of change inferred over geological timescales but consistent with mechanistic processes known to operate over lesser event timescales. Simulation modelling is not intended to reconstruct coastal evolution exactly, but becomes a tool for experimentation and extrapolation within which broad scenarios of change can be modeled and sensitivity to parameterization of variables examined.

Coasts may be divided into rocky coasts, sandy coasts, and muddy coasts, and coral reefs and deltas and estuaries can be differentiated. The

geomorphology of each of these behaves differently. Processes operate at different rates, transitions between different states occur over different timescales and the significance of antecedent conditions varies.

Rocky Coasts

Rocky coasts are characterized by sea cliffs, especially on tectonically active, plate-margin coasts where there are resistant rocks. Cliffed coastlines in resistant rock appear to adopt one of two forms (states): plunging cliffs where a vertical cliff extends below sea level, and shore platforms where a broad bench occurs at sea level in front of a cliff (**Figure 5**). Erosion of cliffs occurs where the erosional force of waves exceeds the resistance of the rocks, and where sufficient time has elapsed.

Plunging cliffs occur where the rock is too resistant to be eroded. The vertical face results in a standing wave which reflects wave energy, so that there is little force to erode the cliff at water level. Waves exert a greater force if they break, or if they are already broken. This can only occur if the water depth is shallow offshore from the cliff face in which case the increased energy from the breaking waves is also able to entrain sediment from the floor (or rock fragments quarried from the foot of the cliff). This process of erosion accelerates through a positive feedback cutting a shore platform.

Shore platforms thus develop in those situations where the erosive force of waves exceeds resistance of the rock. A platform widens as a result of erosion at the foot of the cliff behind the platform. The cliff oversteepens, leading to toppling and fall of detritus onto the rear of the platform, which slows further erosion of the cliff face until that talus has been removed. A series of such negative feedbacks slow the rate at which platforms widen over time, and there is often considerable uniformity of platforms up to a maximum width in any particular lithological setting.

Shore platforms adopt either a gradually sloping ramped form, or a subhorizontal form often with a seaward rampart (**Figure 5**). There has been much discussion as to the relative roles of wave and subaerial processes in the formation of these platforms. Platforms in relatively sheltered locations appear to owe their origin to processes of water-layer leveling (physiochemical processes in pools which persist on platforms at low tide) or wetting and drying and its weakening of the rock. In other cases wave quarrying and abrasion are involved. In many cases both processes may be important. Other platforms may be polygenetic with inheritance from former stands of sea level (reflecting antecedent conditions).

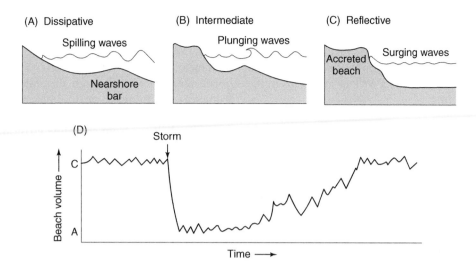

Figure 6 Beach state may be (A) dissipative, where high wave-energy impinges on low-gradient beaches; (B) intermediate (in which several states are possible); or (C) reflective, where low wave-energy surges onto steep or coarse beaches. Change between beach states occurs at event timescales. (D) A beach may react rapidly to a perturbation such as a storm changing from reflective state (C) to more dissipative state (A) but then change much more slowly through intermediate states, readjusting towards state (C) again unless further disrupted.

Coral Reefs

A group of coastlines that are of particular interest to the geomorphologist are those formed by coral reefs. Corals are colonial animals that secrete a limestone exoskeleton that may form the matrix of a reef. Coral reefs flourish in tropical seas in high-energy settings where a significant swell reaches the shoreline. In these circumstances the reef attenuates much of the wave energy such that a relatively quiet-water environment occurs sheltered behind the reef.

Not only is this reef ecosystem of enormous geomorphological significance in terms of the solid reef structure that it forms, but in addition, the carbonate reef material breaks down into calcareous gravels, sands and muds which form the sediments that further modify these coastlines.

On the one hand reefs can be divided into fringing reefs, barrier reefs, and atolls, distinct morphological states that form part of an evolutionary sequence as a result of gradual subsidence of the volcanic basement upon which the reef established (see **Figure 1**). This powerful deduction by Darwin relating to reef structure and operating over geological timescales, has been generally supported by drilling and geochronological studies on mid-plate islands in reef-forming seas. On the other hand, the surface morphology of reefs is extremely dynamic with rapid production of skeletal sediments and their redistribution over event timescales, with the landforms of reefs responding to minor sea-level oscillations, storms, El Niño, coral bleaching, and other perturbations.

Sandy Coasts

Beaches form where sandy (or gravel) material is available forming a sediment wedge at the shoreline. The beach is shaped by incident wave energy, and can undergo modification particularly by formation and migration of nearshore bars which in turn modify the wave-energy spectrum. Various beach states can be recognized across a continuum from beaches which predominantly reflect wave energy, and those which dissipate wave energy across the nearshore zone. Reflective beaches are steep, waves surge up the beach and much of the energy is reflected, and the beach face may develop cusps. Dissipative beaches are much flatter, and waves spill before reaching the shoreline (**Figure 6**).

During a storm, sand is generally eroded from the beach face and deposited in the nearshore, often forming shore-parallel or transverse bars. Waves consequently break on the bars and energy is lost, the form modifying the process in a mutual way. It may be possible to recognize a series of beach states intermediate between reflective and dissipative and any one beach may adopt one or several beach states over time (**Figure 6**). Beach state is clearly modified by incident wave conditions, but the rate of adjustment between states takes time, and a beach is also partly a function of antecedent beach states.

Although erosion and redeposition of sand is the way in which the nearshore adjusts, there may also be long-term storage of sediment. In particular broad, flat beaches may develop dunes behind them.

In other cases a sequence of beach ridges may develop. Geomorphological changes in state over geological timescales represented by beach-ridge plains may relate to variations in supply of sediment to the system (perhaps by rivers) as well as variations in the processes operating.

Deltas and Estuaries

Where rivers bring sediment to the coast, deltas and estuaries can develop. In this case the sediment budget of the shoreline compartment or the receiving basin is augmented and there is generally a positive sediment budget. Deltas are characterized by broad wedges of sediment deposition. Delta morphology tends to reflect the processes that are dominant (**Figure 7**). Where wave energy is low and tides are minor, it is primarily river flows which account for sedimentation patterns. River flow may be likened to a jet, influenced by inertial forces, friction with the basin floor or buoyancy where there are density differences between outflow and the water of the receiving basin.

Wave action tends to smooth the shoreline and a wave-dominated deltaic shoreline will be characterized by shore-parallel bars or ridges. Where river-supplied sediment is relatively low in volume, wave action may form a sandy barrier along the coast and rivers may supply sediment only to lagoons formed behind these barriers. Such is the case for much of the barrier-island shoreline of the eastern shores of the Americas; where barrier islands may be continually reworked landwards during relative sea-level rise. On the other hand if sea level is stable, a stable sand barrier is likely to form closing each

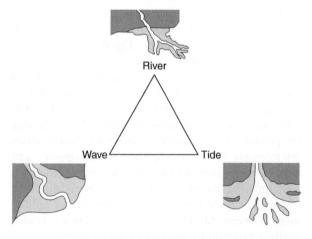

Figure 7 Deltas adopt a variety of forms, but morphology appears to reflect the relative balance of river, wave, and tide action. The broad morphology of the delta tends to be digitate where river-dominated, shore-parallel where wave-dominated, and tapering where tide-dominated.

embayment or creating a barrier estuary as in the case of coastal lagoons in Asia and Australasia.

Estuaries are embayments which are likely to infill incrementally either through the deposition of river-borne sediments or through the influx of sediment from seaward by wave or tidal processes (**Figure 8**). The sediment from seaward may either be derived from the shelf, or from shoreline erosion.

Tidal processes differ from river processes; they are bi-directional, flowing in during flood tide and out during ebb, and the flow is forced by the rising level of the sea. Small embayments are flooded and drained by a tidal prism (the volume of water between low and high tide). Longer estuaries, on the other hand, may have a series of tidal waves which progress up them. Where the tidal range is large this may flow as a tidal bore. Tide-dominated estuarine channels adopt a distinctive tapered form, with width (and depth) decreasing from the mouth upstream (**Figure 8**).

Wave and tidal processes tend to shape deltas and estuaries to varying degrees depending on their relative operation. Many of the deltaic-estuarine processes operate over cycles of change. The hydraulic efficiency of distributaries decreases as they lengthen, until an alternative, shorter course with steeper hydraulic gradient is adopted. The abandoned distributaries of deltas often become tidally dominated with sinuous, tapering tidal creeks dominating what may be a gradually subsiding abandoned delta plain. Wave processes, in wave-dominated settings, smooth and rework the abandoned delta shore, often forming barrier islands.

Muddy Coasts

Muddy coasts are associated with the lowest energy environments. Mud banks may occur in high-energy, wave-exposed settings where large volumes of mud are supplied to the mouth of large rivers. Thus longshore drift north west of the Amazon, and around Bohai Bay downdrift from the Yellow River, enables mud-shoal deposition in open-water settings. Elsewhere mud flats are typical of sheltered settings, within delta interdistributary bays, around coastal lagoons, etc. These muddy environments are likely to be colonized by halophytic vegetation. Mangrove forests occur in tropical settings, whereas salt marsh occurs in higher latitudes, often extending into tropical areas also.

These coastal wetlands further promote retention of fine-grained sediment. The muddy environments are areas of complex hydrodynamics and sedimentation. Sedimentation is likely to occur with a negative feedback such that as the tidal wetlands

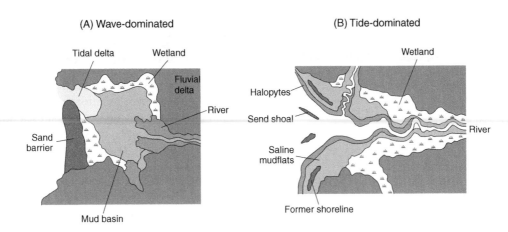

Figure 8 Estuaries are broad embayments which may adopt a wide range of morphologies. They tend to show sand-barrier accumulation where wave-dominated (A), but be prominently tapering where tide-dominated (B). Estuaries are generally sediment sinks with the rates and patterns of infill reflecting the relative dominance of river, wave, and tide processes and sediment sources.

accrete sediment and as the substrate is elevated, they are flooded less frequently and therefore sedimentation decelerates. Boundary conditions, particularly sea level, are likely to vary at rates similar to the rate of sedimentation and prograded coastal plains contain complex sedimentary records of changes in ecological and geomorphological state.

Conclusion

Coastal geomorphology is the study of the evolving form of the shoreline in response to mutually adjusting processes acting upon it. It spans instantaneous and event timescales, over which beaches respond to wave energy, through engineering and geological timescales, over which deltas build seaward or switch distributaries, sea level fluctuates, and cliff morphology evolves. The morphology (state) of the coast changes in response to perturbations, particularly extreme events such as storms, but also thresholds within the system (as when a cliff oversteepens and falls, or a distributary lengthens and then switches). Human action may also represent a perturbation to the system. In each coastal setting, the influence of human modifications is being felt. Thus there are fewer coasts which are not in some way influenced by society. As anthropogenic modification of climate and sea level occurs at a global scale, the human factor increasingly needs to be given prominence in coastal geomorphology.

See also

Beaches, Physical Processes Affecting. Coral Reefs. Rocky Shores. Salt Marshes and Mud Flats.

Further Reading

Boyd R, Dalrymple R, and Zaitlin BA (1992) Classification of clastic coastal depositional environments. *Sedimentary Geology* 80: 139–150.

Carter RWG and Woodroffe CD (eds.) (1994) *Coastal Evolution: Late Quaternary Shoreline Morphodynamics*. Cambridge: Cambridge University Press.

Cowell PJ and Thom BG (1994) Morphodynamics of coastal evolution. In: Carter RWG and Woodroffe CD (eds.) *Coastal Evolution: Late Quaternary Shoreline Morphodynamics*, pp. 33–86. Cambridge: Cambridge University Press.

Darwin C (1842) *The Structure and Distribution of Coral Reefs*. London: Smith Elder.

Davis RA (1985) *Coastal Sedimentary Environments*. New York: Springer-Verlag.

Johnson DW (1919) *Shore Processes and Shoreline Development*. New York: Prentice Hall.

Trenhaile AS (1997) *Coastal Dynamics and Landforms*. Oxford: Clarendon Press.

Wright LD and Thom BG (1977) Coastal depositional landforms: a morphodynamic approach. *Progress in Physical Geography* 1: 412–459.

ICE-INDUCED GOUGING OF THE SEAFLOOR

W. F. Weeks, Portland, OR, USA

Introduction

Inuit hunters have long known that both sea ice and icebergs could interact with the underlying sea floor, in that sea floor sediments could occasionally be seen attached to these icy objects. As early as 1855, no less a scientific luminary than Charles Darwin speculated that gouging icebergs could traverse isobaths, concluding that an iceberg could be driven over great inequalities of [a seafloor] surface easier than could a glacier. Only in 1924 did similar inferences of the possibility of sea ice and icebergs affecting the seafloor again begin to reappear in the scientific literature. However, direct observations of the nature of these interactions were lacking until the early 1970s. The reason for the increased interest in this rather exotic phenomena was initially the discovery of the supergiant oil field on the edge of the Beaufort Sea at Prudhoe Bay, Alaska. Because the initial successful well was located on the coast, there was immediate interest in the possibility of developing offshore fields to the north of Alaska and Canada. It was also apparent that if major oil resources occurred to the north of Alaska and Canada, similar resources might be found on the world's largest continental shelf located to the north of the Russian mainland. Offshore wells are typically tied together by subsea pipelines which take the oil from the individual wells either to a central collection point where tanker pickup is possible or to the coast where the oil can be fed into a pipeline transportation system. If sea ice processes could result in major distrubances of the seafloor, this would clearly become a major consideration in pipeline design. Some time after the Prudhoe Bay find, another large offshore oil discovery was made to the east of Newfoundland. Here the ice-induced gouging problem was caused not by sea ice but by icebergs. In both cases the initial step in treating these perceived problems was to investigate the nature of these ice–seafloor interactions. One needs to know the frequency of gouging events in time and space as well as the widths and depths of the gouges, the water depth range in which this phenomenon occurs, and the effective lifetime of a gouge after its initial formation. Also of importance are the type of subsea soil and the nature and extent of the soil movements below the gouges, as this information is essential in calculating safe burial depths for subsea structures. The following attempts to summarize our current knowledge of this type of naturally occurring phenomenon.

As extensive studies of ice-induced disruptions of the seafloor have been undertaken only during the last 30 years, there is as yet no commonly agreed upon terminology for this phenomenon. In the literature the process has been described as scouring, scoring, plowing and gouging. Here gouging is used because at least in the case of sea ice, it is felt that it more accurately describes the process than do the other terms.

Observational Techniques

A variety of techniques have been used to study the gouging phenomenon. Typically, a fathometer is used to resolve the seafloor relief directly beneath the ship with a precision of better than 10 cm (**Figure 1**) while, at the same time, a sidescan sonar system provides a sonar map of the seafloor on either side of the ship (**Figure 2**). Total sonar swath widths have been typically 200 to 250 m not including a narrow area directly beneath the ship that was not imaged. The simultaneous use of these two different types of records allows one to both measure the depth and width of the gouge (fathometer) at the point where it is crossed by the ship track and also to observe the general orientation and geometry of the gouge track (sonar). Along the Alaskan coast the ships used have been small, allowing them to operate in shallow water. In some cases divers have been used to examine the gouges and, at a few deeper water sites off the Canadian east coast, manned submersibles have been used to gather direct observations of the gouging process.

Results

Arctic Shelves

The most common features gouging the shelves of the Arctic Ocean are the keels of pressure ridges that are made of deformed sea ice. As the ice pack moves under the forces exerted on it by the wind and currents, pressure ridges typically form at floe boundaries as the result of differential movements between the floes. These features can be very large. Pressure

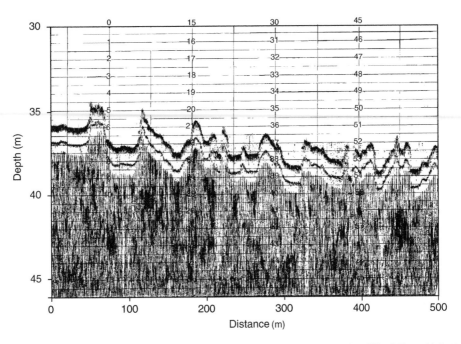

Figure 1 Fathogram of an ice-gouged seafloor. Water depth is 36 m. Record taken 25 km NE of Cape Halkett, Beaufort Sea, offshore Alaska. The multiple reflections from the upper layer of the seafloor are the result of the presence of a thawed active layer in the subsea permafrost. (From Weeks *et al.*, 1983.)

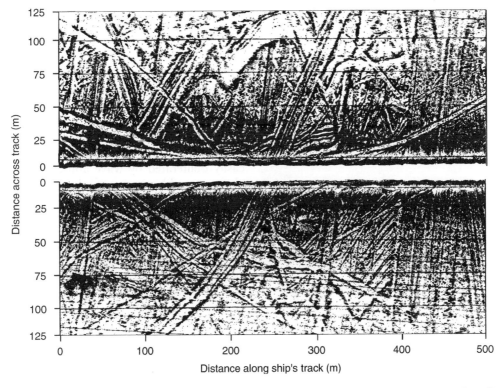

Figure 2 Sonograph of an ice-gouged seafloor. Water depth is 20 m. Record taken 20 km NE of Cape Halkett, Beaufort Sea, offshore Alaska. (From Weeks *et al.*, 1983.)

ridge keels with drafts up to 50 m have been observed in the upward-looking sonar records collected by submarines passing beneath the ice. The distribution of keel depths of these deformed features is approximately a negative exponential, in that there are many shallow ridge keels while deep keels are rare. Deformation can be particularly intense in nearshore regions where the moving ice pack contacts an immovable coast with the formation of large shear ridges that can extend for many kilometers. Such ridges are frequently anchored at shoal areas where large accumulations of highly deformed sea ice can build up. Such grounded ice features, which are referred to using the Russian term stamuki, can have freeboards of as much as 10 m and lateral extents of 10–15 km. As the offshore ice field moves against and along the coast, it exerts force on the sides of any such grounded features, causing them to scrape and plough their way along the seafloor. Considering that surficial sediments along many areas of the Beaufort Sea Coast of Canada and Alaska are fine-grained silts, it is hardly surprising that over a period of time such a process can cause extensive gouging of the seafloor sediments. **Figure 3** is a photograph showing active gouging. The relatively undeformed first-year sea ice in the foreground is moving away from the viewer and pushing against a piece of grounded multiyear sea ice (indicated by its rounded upper surface formed during the previous summer's melt period), pushing it along the coast. The interaction between the first-year and the multiyear ice has resulted in a pileup of broken first-year ice, which when the photograph was taken was higher than the upper surface of the multiyear ice. Also evident is the track cut in the first-year ice as it moves past the

multiyear ice. The fact that both the multiyear ice and the pileup of first year ice are interacting with the seafloor is indicated by the presence of bottom sediment in the deformed first-year ice and on the far side of the multiyear ice.

The maximum water depth in which contemporary sea ice gouging is believed to occur is roughly 50–60 m, corresponding approximately to the draft of the largest pressure ridges. Although gouges occur in water up to 80 m deep, it is generally believed that these deep-water gouges are relicts that formed during periods when sea level was lower than at present.

Not surprisingly the depths of gouges in the seafloor mirrors the keel depth distributions observed in pressure ridge keels in that they are also well approximated by a negative exponential with the character of the falloff as well as the number of gouges varying with water depth. As with ridge keels, shallow gouges are common and deep gouges are rare. As might be expected, there are fewer deep gouges in shallow water as the large ice masses required to produce them have already grounded farther out to sea. For instance, along the Beaufort Coast in water 5 m deep, a 1 m gouge has an exceedance probability of approximately 10^{-4}; that is, 1 gouge in 10 000 will on the average be expected to have a depth equal to or greater than 1 m. In water 30 m deep, a 3.4 m gouge has the same probability of occurrence. Gouges in excess of 3 m deep are not rare and 8 m gouges have been reported in the vicinity of the Mackenzie Delta.

As might be expected, the nature of gouging varies with changes in seafloor sediment types. Along the Beaufort coast where there are two distinct soil types, the gouges in stiff, sandy, clayey silts are typically more frequent and slightly deeper than those found in more sandy sediments. Presumably the gouges in the more sandy material are more easily obliterated by wave and current action. It is also reasonable to assume that the slightly deeper gouges in the silts provide a better picture of the original incision depths. An extreme example of the effect of bottom sediment type on gouging can be found between Sakhalin Island and the eastern Russian mainland. Here the seafloor is very sandy and the currents are very strong. As a result, even though winter observations have conclusively shown that seafloor gouging is common, summer observations reveal that the gouges have been completely erased by infilling.

Gouge tracks in the Beaufort Sea north of the Alaskan and western Canadian coast generally run roughly parallel to the coast, with some regions having an excess of 200 gouges per kilometer.

Figure 3 Photograph of active ice gouging occurring along the coast of the Beaufort Sea. The grounded multiyear ice floe that is being pushed by the first-year ice has a freeboard of ~2 m. The thickness of the first-year ice is ~0.3 m. (Photograph by GFN Cox.)

Histograms of the frequency of occurrence of distances between gouges are also well described by a negative exponential, a result suggesting that the spatial occurrence of gouges may be described as a Poisson process. Gouge occurrence is hardly uniform, however, with significantly higher concentrations of gouges occurring on the seaward sides of shoal areas and barrier islands and fewer gouges on their more protected lee sides. This is clearly shown in **Figure 4** in that the degree of exposure to the moving pack ice decreases in the sequence Jones Islands–Lonely–Harrison Bay–Lagoons. **Figure 5** is a sketch based on divers' observations showing gouging along the Alaskan coast of the Beaufort Sea.

The second features producing gouges on the surface of the outer continental shelf of the Arctic Ocean are ice islands. These features are, in fact, an unusual type of tabular iceberg formed by the gradual breakup of the ice shelves located along the north coast of Ellesmere Island, the northernmost of the Canadian Arctic Islands. Once formed, an ice island can circulate in the Arctic Ocean for many years. For instance, the ice island T-3 drifted in the Arctic Ocean between 1952 and 1979, ultimately completing three circuits of the Beaufort Gyre (the large clockwise oceanic circulation centered in the offshore Beaufort Sea) before exiting the Arctic Ocean through Fram Strait between Greenland and Svalbard (Spitzbergen). The lateral dimensions of ice islands can vary considerably from a few tens of meters to over ten kilometers. Thicknesses, although variable, are typically in the 40–50 m range; ice islands possess freeboards in the same range as exhibited by larger sea ice pressure ridges. In the study of fathometer and sonar data on gouge distributions and patterns, no attempt is usually made to separate gouges made by pressure ridges from gouges made by ice islands, as the ice features that made the gouges are commonly no longer present. However, it is reasonable to assume that many of the very wide, uniform gouges are the result of ice islands interacting with the seafloor.

It is relatively easy to characterize the state of the gouging existing on the seafloor at any given time. It is another matter to answer the question of how deep one must bury a pipeline, a cable, or some other type of fixed structure beneath the seafloor to reduce the chances of it being impacted by moving ice to some acceptable level. To answer this question one needs to know the rates of occurrence of new gouges; these values are in many locations still poorly known, as they require replicate measurements over a period of many years so that new gouges can be counted and the rate of infilling of existing gouges can be estimated. The best available data on gouging occurrence along the Beaufort Coast has been collected by the Canadian and US Geological Surveys and

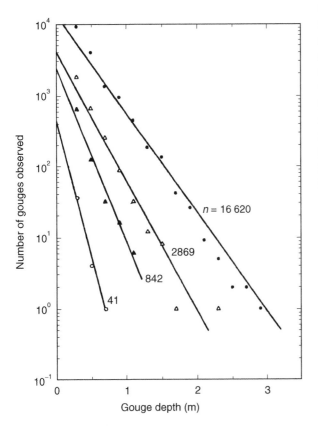

Figure 4 Semilogarithmic plot of the number of gouges observed versus gouge depth for four regions along the Alaskan coast of the Beaufort Sea. ●, Jones Island and east; △, Lonely;, ▲ Harrison Bay; ○, Lagoons. (From Weeks *et al.*, 1983.)

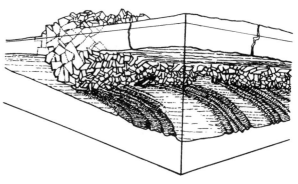

Figure 5 Diver's sketch of active gouging occurring off the coast of the Alaskan Beaufort Sea. Some sense of the scale of this drawing can be gained from the observation that commonly the thickness of the ice blocks in such ridges ranges between 0.3 and 1.0 m. (Drawing by TR Alpha, US Geological Survey.)

indicates that gouging occurs in rather large scale regional events related to severe storms that drive the pack ice inshore – about once every 4–5 years. Because there is no absolute technique for dating the age of existing gouges, there is no current method for determining whether a particular gouge that one observes on the seafloor formed during the last year or sometime during the last 6000 years (after the Alaskan shelf was submerged as the result of rising sea levels at the end of the Pleistocene). It is currently believed that gouges in shallow water formed quite recently. For instance, during the late summer of 1977 when the Beaufort Coast was comparatively ice free, strong wave action during late summer storms obliterated gouges in water less than 13 m deep and caused pronounced infilling of gouges in somewhat deeper water. It is generally estimated that such storms have an average recurrence interval of approximately 25 years. On the other hand, gouges in water deeper than ~60 m are believed to be relatively old in that they are presumed to be below the depth of currently active gouging. The lengths of time represented by the gouges observed in water depths between 20 and 50 m are less well known. Unfortunately, this is the water depth range in which the largest gouges are found. Stochastic models of gouging occurrence, which incorporate approximate simulations of subsea sediment transport, suggest that if seafloor currents are sufficiently strong to exceed the threshhold for sediment movement, gouge infilling will occur comparatively rapidly (within a few years).

To date, two different procedures have been used to estimate the depth of gouges with specified recurrence intervals. In the first case, the available rates and geometric characteristics of new gouges at a particular site are used. If a multiyear high-quality dataset of new gouge data is available for the region of interest, this would clearly appear to be the favored approach. In the other case, information on the draft distribution of pressure ridge keels and pack ice drift rates are combined to calculate the rates of gouging. The problem with using pack ice drift rates and pressure ridge keel depths is that drift rates in the near shore where grounding is occurring are undoubtedly less than in the offshore where the ice can move relatively unimpeded. In addition, it is doubtful that offshore keel depths observed in water sufficiently deep to allow submarine operations provide accurate estimates of keel depths in the near-shore.

One might suggest that offshore engineers should simply bury pipelines and cables at depths below those of known gouges and forget about all these statistical considerations. Although this sounds attractive, it is not a realistic resolution of the problem.

In the first place, for a pipeline to avoid deformation, buckling, and failure it must, depending on the nature of the seafloor sediments, be buried at depths up to two times that of the design gouging event. Considering that gouges in excess of 3 m are not rare at many locations along the Beaufort Coast of Alaska and Canada, this means burial depths >5 m. Burial at such depths is extremely costly and time consuming considering the very short operating season in the offshore Arctic. It is also near the edge of existing subsea trenching technology. Another factor to be considered here is that the deeper the pipeline is buried, the more likely it is that its presence will result in the thawing of subsea permafrost that is known to exist in many areas of the Beaufort and Chukchi Shelves. Such thawing could result in settlement in the vicinity of the pipe, which could threaten the integrity of the pipeline. There are engineering remedies to this problem but, as one might expect, they are expensive.

Canadian East Coast

The primary problem to the east of Newfoundland and off the coast of Labrador is not sea ice but icebergs. Although the majority of these originate from the Greenland Ice Sheet calving to the west into Baffin Bay, some icebergs do originate from the East Greenland coast and from the Canadian Arctic. Considering their great size and deep draft, icebergs are formidable adversaries. Iceberg gouges with lengths >60 km are known and many have lengths >20 km. Gouge depths can exceed 10 m and recent gouging is known to occur at depths of at least 230 m along the Baffin and Labrador Shelves. Icebergs also appear to be able to traverse vertical ranges of bathymetry up to at least 45 m. As with the sea-ice-induced gouges of the Arctic shelves, iceberg gouge depths are exponentially distributed, with small gouges being common and deep gouges being rare. Iceberg gouges can be straight or curved. Pits are also common, which occur when the iceberg draft is suddenly increased through splitting and rolling. The iceberg can then remain fixed to the seafloor while it rocks and twists as a result of the wave and current forces that impinge upon it. At such times pits can be produced that are deeper than the maximum gouge depth otherwise associated with the iceberg.

Although considerable information is available on the statistics of iceberg gouge occurrences, in studies of iceberg groundings more attention has been paid to the energetics of the ice–sediment interactions than to the statistics. Also, more emphasis has been placed on iceberg 'management' in order either to reduce or to remove the threat to a specific offshore operation. For instance, as icebergs are discrete objects as compared to pressure ridges, which are giant piles of ice blocks that are frequently poorly cemented together, icebergs can be deflected away from a production site via the use of tugboats in combination with trajectory modeling. Such schemes can also utilize aerial, satellite, and ship reconnaissance techniques to provide operators with an adequate warning of a potential threat. Surface-based over-the-horizon radar technology has also been developed as an iceberg reconnaissance tool, but at the time of writing it has not been used operationally.

Other Locations

Although the discussion here has primarily focused on the Beaufort Coast of the Arctic Ocean and the offshore regions of Newfoundland and Labrador, this is only because the interest in offshore oil and gas reserves at these locations has resulted in the collection of observational information on the local nature of the ice-induced gouging phenomenon. However, regions where ice-induced gouging is presently active are very large and include the complete continental shelf of the Arctic Ocean, the continental shelves of Greenland, of eastern Canada, and of Svalbard, as well as the continental shelf of the Antarctic continent where typical shelf icebergs are known to have drafts of ~200 m with lateral dimensions as large as 100 km. In addition, relict iceberg gouges exist and as a result affect seafloor topography in regions where icebergs are no longer common or no longer occur, such as the Norwegian Shelf and even the northern slope of Little Bahama Bank and the Straits of Florida. Finally, although 230 m is a reasonable estimate for the maximum depth of active iceberg gouging, relict gouges presumed to have occurred during the Pleistocene have been discovered at depths from 450 m to at least 850 m on the Yermak Plateau located to the northwest of Svalbard in the Arctic Ocean proper.

Conclusions

There is clearly much more that we need to know concerning processes acting on the poorly explored continental shelves of the polar and subpolar regions. The ice-induced gouging phenomenon is clearly one of the more important of these, in that an understanding of this process is essential to both the engineering and the scientific communities. At first glance, icebergs and stamuki zones might appear to be exotic entities that are out of sight somewhere way to the north or south and that can be safely put out of mind. However, as oil production moves to ever more difficult frontier areas such as the offshore Arctic, a quantitative understanding of processes such as ice gouging becomes essential for safe development. Lack of such understanding leads to poor, inefficient designs and the increased possibility of failures. The last thing that either the environmental community or the petroleum industry needs is an offshore oil spill in the high Arctic.

See also

Sub-sea Permafrost.

Further Reading

Barnes PW, Schell DM, and Reimnitz E (eds.) (1984) *The Alaskan Beaufort Sea: Ecosystems and Environments.* Orlando: Academic Press.

Colony R and Thorndike AS (1984) An estimate of the mean field of arctic sea ice motion. *Journal of Geophysical Research* 89(C6): 10623–10629.

Darwin CR (1855) On the power of icebergs to make rectilinear, uniformly-directed grooves across a submarine undulatory surface. *London, Edinburgh, and Dublin Philosophical Magazine and Journal of Science* 10: 96–98.

Goodwin CR, Finley JC and Howard LM (1985) *Ice Scour Bibliography.* Environmental Studies Revolving Funds Rept. No. 010, Ottawa.

Lewis CFM (1977) The frequency and magnitude of drift ice groundings from ice-scour tracks in the Canadian Beaufort Sea. *Proceedings of 4th International Conference on Port Ocean Engineering Under Arctic Conditions.* Newfoundland: Memorial University. vol. 1, 567–576.

Palmer AC, Konuk I, Comfort G, and Been K (1990) Ice gouging and the safety of marine pipelines. *Offshore Technology Conference,* Paper 6371, 235–244.

Reed JC and Sater JE (eds.) (1974) *The Coast and Shelf of the Beaufort Sea.* Arlington, VA: Arctic Institute of North America.

Reimnitz E, Barnes PW, and Alpha TR (1973) *Bottom features and processes related to drifting ice, US Geological Survey Miscellaneous Field Studies,* Map MF-532. Washington, DC: US Geological Survey.

Vogt PR, Crane K, and Sundvor E (1994) Deep Pleistocene iceberg plowmarks on the Yermak Plateau: sidescan and 3.5 kHz evidence for thick calving ice fronts and a possible marine ice sheet in the Arctic Ocean. *Geology* 22: 403–406.

Wadhams P (1988) Sea ice morphology. In: Leppäranta M (ed.) *Physics of Ice-Covered Seas,* pp. 231–287. Helsinki: Helsinki University Printing House.

Weeks WF, Barnes PW, Rearic DM, and Reimnitz E (1983) *Statistical aspects of ice gouging on the Alaskan Shelf of the Beaufort Sea. Cold Regions Research and Engineering*

Laboratory Report. New Hampshire, USA: Hanover. 83–21.

Weeks WF, Tucker WB III, and Niedoroda AW (1985) A numerical simulation of ice gouge formation and infilling on the shelf of the Beaufort Sea. *Proceedings, International Conference on Port Ocean Engineering Under Arctic Conditions, Narssarssuaq, Greenland* 1: 393–407.

Woodworth-Lynas CMT, Simms A, and Rendell CM (1985) Iceberg grounding and scouring on the Labrador continental shelf. *Cold Regions Science and Technology* 10(2): 163–186.

IGNEOUS PROVINCES

M. F. Coffins, University of Texas at Austin,
Austin, TX, USA
O. Eldholm, University of Oslo, Oslo,
Norway

Introduction

Large igneous provinces (LIPs) are massive crustal
emplacements of predominantly Fe- and Mg-rich
(mafic) rock that form by processes other than normal
seafloor spreading. LIP rocks are readily distinguishable from the products of the two other major types of
magmatism, midocean ridge and arc, on the Earth's
surface on the basis of petrologic, geochemical, geo-
chronologic, geophysical, and physical volcanological
data. LIPs occur on both the continents and oceans,
and include continental flood basalts, volcanic passive
margins, oceanic plateaus, submarine ridges, seamounts, and ocean basin flood basalts (**Figure 1,
Table 1**). LIPs and their small-scale analogs, hot spots,
are commonly attributed to decompression melting of
hot, low density mantle material known as mantle
plumes. This type of magmatism currently represent $\sim 10\%$ of the mass and energy flux from the
Earth's deep interior to its crust. The flux may have
been higher in the past, but is episodic over geological
time, in contrast to the relatively steady-state activity
at seafloor spreading centers. Such episodicity reveals
dynamic, nonsteady-state circulation within the

Earth's mantle, and suggests a strong potential for LIP
emplacements to contribute to, if not instigate, major
environmental changes.

Composition, Physical Volcanology, Crustal Structure, and Mantle Roots

LIPs are defined by the characteristics of their dominantly Fe- and Mg-rich (mafic) extrusive rocks;
these most typically consist of subhorizontal, subaerial basalt flows. Individual flows can extend for
hundreds of kilometers, be 10s to 100s of meters
thick, and have volumes as great as 10^4–10^5 km^3. Si-
rich rocks also occur as lavas and intrusive rocks,
and are mostly associated with the initial and late
stages of LIP magmatic activity. Relative to mid-
ocean ridge basalts, LIPs include higher MgO lavas,
basalts with more diverse major element compositions, rocks with more common fractionated
components, both alkalic and tholeiitic differentiates, basalts with predominantly flat light rare earth
element patterns, and lavas erupted in both subaerial
and submarine settings.

As the extrusive component of LIPs is the most
accessible for study, nearly all of our knowledge of
LIPs is derived from lavas forming the uppermost
10% of LIP crust. The extrusive layer may exceed
10 km in thickness. On the basis of geophysical,
predominantly seismic data from LIPs, and from
comparisons with normal oceanic crust, LIP crust
beneath the extrusive layer is believed to consist of
an intrusive layer and a lower crustal body,

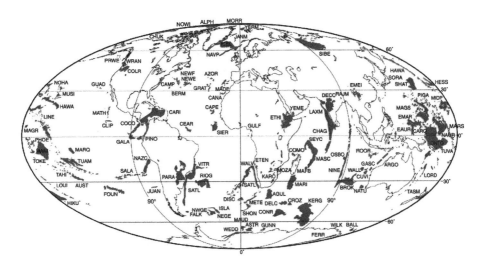

Figure 1 Phanerozoic global LIP distribution (red), with LIPs labeled (**Table 1**).

characterized by compressional wave velocities of 7.0–7.6 km s^{-1}, at the base of the crust (**Figure 2**). Beneath continental crust this body may be considered as a magmatically underplated layer. Seismic wave velocities suggest an intrusive layer that is most likely gabbroic, and a lower crust that is ultramafic. If the LIP forms on pre-existing continental or oceanic crust or along a divergent plate boundary, dikes and sills are probably common in the middle and upper crust. The maximum crustal thickness, including extrusive, intrusive, and the lower crustal body, of an oceanic LIP is ~35 km, determined from seismic and gravity studies of the Ontong Java Plateau (**Figure 1, Table 1**).

Low-velocity zones have been observed recently in the mantle beneath the oceanic Ontong Java Plateau, as well as under the continental Deccan Traps and

Paraná flood basalts (**Figure 1, Table 1**). Interpreted as lithospheric roots or keels, the zones can extend to at least 500–600 km into the mantle. In contrast to high-velocity roots beneath most continental areas, and the absence of lithospheric keels in most oceanic areas, the low-velocity zones beneath LIPs apparently reflect residual chemical and perhaps thermal effects of mantle plume activity. High-buoyancy roots extending well into the mantle beneath oceanic LIPs would suggest a significant role in continental growth via accretion of oceanic LIPs to the edges of continents.

Distribution, Tectonic Setting, and Types

LIPs occur worldwide, in both continental and oceanic crust in purely intraplate settings, and along present and former plate boundaries (**Figure 1, Table 1**), although the tectonic setting of formation is unknown for many features. If a LIP forms at a plate boundary, the entire crustal section is LIP crust (**Figure 2**). Conversely, if one forms in an intraplate setting, the pre-existing crust must be intruded and sandwiched by LIP magmas, albeit to an extent not resolvable by current geological or geophysical techniques.

Continental flood basalts, the most intensively studied LIPs due to their exposure, are erupted from fissures on continental crust (**Figure 1, Table 1**). Most continental flood basalts overlie sedimentary basins that formed via extension, but it is not clear what happened first, the magmatism or the extension. Volcanic passive margins form by excessive magmatism during continental breakup along the trailing, rifted edges of continents. In the deep ocean basins, four types of LIPs are found. Oceanic plateaus, commonly isolated from major continents, are broad, typically flat-topped features generally lying 2000 m or more above the surrounding seafloor. They can form at triple junctions (e.g., Shatsky Rise), midocean ridges (e.g., Iceland), or in intraplate settings (e.g., northern Kerguelen Plateau). Submarine ridges are elongated, steep-sided elevations of the seafloor. Some form along transform plate boundaries, e.g., Ninetyeast Ridge. In the oceanic realm, oceanic plateaus and submarine ridges are the most enigmatic with respect to the tectonic setting in which they are formed. Seamounts, closely related to submarine ridges, are local elevations of the seafloor; they may be discrete, form a linear or random grouping, or be connected along their bases and aligned along a ridge or rise. They commonly form in intraplate regions, e.g., Hawaii. Ocean basin flood

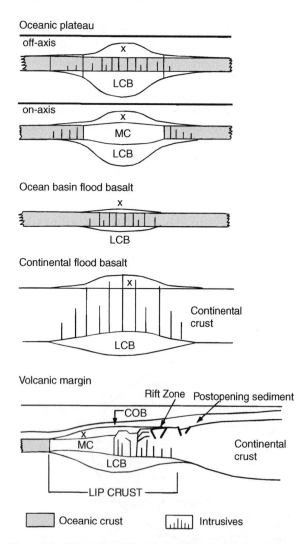

Figure 2 Schematic LIP plate tectonic settings and gross crustal structure. LIP crustal components are: extrusive cover (X), middle crust (MC), and lower crustal body (LCB), continent-ocean boundary (COB), Normal oceanic crust is gray.

Table 1 Large igneous provinces

Large igneous province	Abbreviation (Figure 1)	Type
Agulhas Ridge	AGUL	SR
Alpha-Mendeleyev Ridge	ALPH	SR/OP
Argo Basin	ARGO	VM
Astrid Ridge	ASTR	VM
Austral Seamounts	AUST	SMT
Azores	AZOR	SMT
Balleny Islands	BALL	SMT
Bermuda Rise	BERM	OP
Broken Ridge	BROK	OP
Canary Islands	CANA	SMT
Cape Verde Rise	CAPE	OP
Caribbean Flood Basalt	CARI	OBFB (partly accreted)
Caroline Seamounts	CARO	SMT
Ceara Rise	CEAR	OP
Central Atlantic Magmatic Province (VM only)	CAMP	CFB/VM
Chagos-Laccadive Ridge	CHAG	SR
Chukchi Plateau	CHUK	OP
Clipperton Seamounts	CLIP	SMT
Cocos Ridge	COCO	SR
Columbia River Basalt	COLR	CFB
Comores Archipelago	COMO	SMT
Conrad Rise	CONR	OP
Crozet Plateau	CROZ	OP
Cuvier (Wallaby) Plateau	CUVI	VM
Deccan Traps	DECC	CFB/VM
Del Caño Rise	DELC	OP
Discovery Seamounts	DISC	SMT
Eauripik Rise	EAUR	OP
East Mariana Basin	EMAR	OBFB
Emeishan Basalts	EMEI	CFB
Etendeka	ETEN	CFB
Ethiopian Flood Basalt	ETHI	CFB
Falkland Plateau	FALK	VM
Ferrar Basalts	FERR	CFB
Foundation Seamounts	FOUN	SMT
Galapagos/Carnegie Ridge	GALA	SMT/SR
Gascoyne Margin	GASC	VM
Great Meteor-Atlantis Seamounts	GRAT	SMT
Guadelupe Seamount Chain	GUAD	SMT
Gulf of Guinea	GULF	VM
Gunnerus Ridge	GUNN	VM
Hawaiian-Emperor Seamounts	HAWA	SMT
Hess Rise	HESS	OP
Hikurangi Plateau	HIKU	OP
Iceland/Greenland–Scotland Ridge	ICEL	OP/SR
Islas Orcadas Rise	ISLA	SR
Jan Mayen Ridge	JANM	VM
Juan Fernandez Archipelago	JUAN	SMT
Karoo	KARO	CFB
Kerguelen Plateau	KERG	OP/VM
Laxmi Ridge	LAXM	VM
Line Islands	LINE	SMT
Lord Howe Rise Seamounts	LORD	SMT
Louisville Ridge	LOUI	SMT

continued

Table 1 *continued*

Large igneous province	Abbreviation (Figure 1)	Type
Madagascar Flood Basalts	MAFB	CFB
Madagascar Ridge	MARI	SR/VM?
Madeira Rise	MADE	OP
Magellan Rise	MAGR	OP
Magellan Seamounts	MAGS	SMT
Manihiki Plateau	MANI	OP
Marquesas Islands	MARQ	SMT
Marshall Gilbert Seamounts	MARS	SMT
Mascarene Plateau	MASC	OP
Mathematicians Seamounts	MATH	SMT
Maud Rise	MAUD	OP
Meteor Rise	METE	SR
Mid-Pacific Mountains	MIDP	SMT
Morris Jesup Rise	MORR	VM
Mozambique Basin	MOZA	VM
Musicians Seamounts	MUSI	SMT
Naturaliste Plateau	NATU	VM
Nauru Basin	NAUR	OBFB
Nazca Ridge	NAZC	SR
New England Seamounts	NEWE	SMT
Newfoundland Ridge	NEWF	VM
Ninetyeast Ridge	NINE	SR
North Atlantic Volcanic Province	NAVP	CFB
Northeast Georgia Rise	NEGE	OP
Northwest Georgia Rise	NWGE	OP
Northwest Hawaiian Ridge	NOHA	SR/SMT
Northwind Ridge	NOWI	SR
Ontong Java Plateau	ONTO	OP (partly accreted)
Osborn Knoll	OSBO	OP
Paraná	PARA	CFB
Phoenix Seamounts	PHOE	SMT
Pigafetta Basin	PIGA	OBFB
Piñón Formation (Ecuador)	PINO	OP (accreted)
Pratt-Welker Seamounts	PRWE	SMT
Rajmahal Traps	RAJM	CFB
Rio Grande Rise	RIOG	OP
Roo Rise	ROOR	OP
Sala y Gomez Ridge	SALA	SR
Seychelles Bank	SEYC	VM
Shatsky Rise	SHAT	OP
Shona Ridge	SHON	SR
SiberianTraps	SIBE	CFB
Sierra Leone Rise	SIER	OP
Sorachi Plateau (Japan)	SORA	OP (accreted)
South Atlantic Margins	SATL	VM
Tahiti	TAHI	SMT
Tasmantid Seamounts	TASM	SMT
Tokelau Seamounts	TOKE	SMT
Tuamotu Archipelago	TUAM	SMT
Tuvalu Seamounts	TUVA	SMT
Vitória-Trindade Ridge	VITR	SR/SMT
Wallaby Plateau (Zenith Seamount)	WALL	OP
Walvis Ridge	WALV	SR
Weddell Sea	WEDD	VM
Wrangellia	WRAN	OP (accreted)
Yemen Plateau Basalts	YEME	CFB
Yermak Plateau	YERM	VM
Wilkes Land Margin	WILK	VM

CFB, continental flood basalt; OBFB, ocean basin flood basalt; OP, oceanic plateau; SMT, seamount; SR, submarine ridge basalts, the least studied type of LIP, are extensive submarine flows and sills lying above and postdating normal oceanic crust.

Ages

Age control for all LIPs except continental flood basalts is sparse due to their relative inaccessibility, but the $^{40}Ar/^{39}Ar$ dating technique is having a particularly strong impact on studies of LIP volcanism. Geochronological studies of continental floor basalts (e.g., Siberian, Karoo/Ferrar, Deccan, Columbia River; **Figure 1**) suggest that most LIPs result from mantle plumes which initially transfer huge volumes ($\sim 10^5-10^7$ km^3) of mafic rock into localized regions of the crust over short intervals ($\sim 10^5-10^6$ years), but which subsequently transfer mass at a far lesser rate, albeit over significantly longer intervals (10^7-10^8 years). Transient magmatism during LIP formation is commonly attributed to mantle plume 'heads' reaching the crust following transit through all or part of the Earth's mantle, whereas persistent magmatism is considered to result from steady-state mantle plume 'tails' penetrating the lithosphere which is moving relative to the plume (**Figure 3**). However, not all LIPs have obvious connections to mantle plumes or hot spots, suggesting that more than one source model may be required to explain all LIPs.

LIPs are not distributed uniformly in time. During the past 150 million years for example, many LIPs formed between 50 and 150 million years ago, whereas few have formed during the past 50 million years (**Figure 4**). Such episodicity likely reflects variations in rates of mantle circulation, and this is supported by high rates of seafloor spreading during a portion of the 50–150 million year interval. Thus, although LIPs manifest types of mantle processes distinct from those resulting in seafloor spreading, waxing and waning rates of overall mantle circulation probably affect both sets of processes. A major question that emerges from the global LIP production rate is whether the mantle is circulating less vigorously as the Earth ages.

Large Igneous Provincess and Mantle Dynamics

The formation of various sizes of LIPs in a variety of tectonic settings on both continental and oceanic lithosphere suggests a variety of thermal anomalies in the mantle that give rise to LIPs as well as strong lithospheric control on their formation. Equivalent mantle plumes beneath continental and oceanic lithosphere should produce more magmatism in the latter scenario, as oceanic lithosphere is thinner, allowing more decompression melting. Similarly, equivalent mantle plumes beneath an intraplate region (e.g., Hawaii) and a divergent plate boundary (e.g., Iceland) (**Figure 1, Table 1**) will produce more magmatism in the latter setting, again because decompression melting is enhanced. Recent seismic tomographic images of mantle plumes beneath Iceland and Hawaii show significant differences between the two.

Only recently, seismic tomography has revealed that slabs of subducting lithosphere can penetrate the entire Earth's mantle to the D″ layer at the boundary between the mantle and core at ~2900 km depth

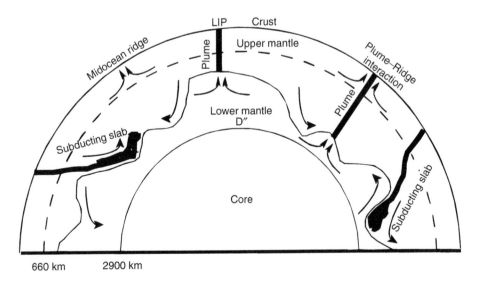

Figure 3 Model of the Earth's interior showing plumes, subducting slabs, and two mantle layers that move in complex patterns, but never mix.

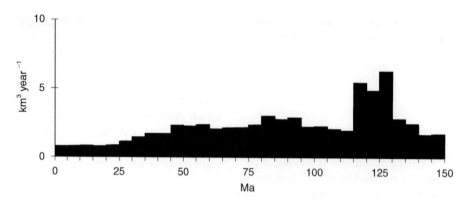

Figure 4   LIP production, corrected for subduction and averaged over a 15 million year running window, since 150 million years ago (Ma).

(**Figure 3**). If we assume that the volume of the Earth's mantle remains roughly constant through geological time, then the mass of crustal material fluxing into the mantle must be balanced by an equivalent mass of material fluxing from the mantle to the crust. Most, if not all of the magmatism associated with the plate tectonic processes of seafloor spreading and subduction is believed, on the basis of geochemistry and seismic tomography, to be derived from the upper mantle (above ~ 660 km depth). It is most reasonable to assume that the lithospheric material that enters the lower mantle is eventually recycled, in some part contributing to plume magmas.

Large Igneous Provincess and the Environment

The formation of LIPs has had documented environmental effects both locally and regionally. The global effects are less well understood, but the formation of some LIPs may have affected the global environment, particularly when conditions were at or near a threshold state. Eruption of enormous volumes of basaltic magma during LIP formation releases volatiles such as CO_2, S, Cl, and F (**Figure 5**). A key factor affecting the magnitude of volatile release is whether eruptions are subaerial or submarine; hydrostatic pressure inhibits vesiculation and degassing of relatively soluble volatile components (H_2O, S, Cl, F) during deep-water submarine eruptions, although low solubility components (CO_2, noble gases) are mostly degassed even at abyssal depths. Investigations of volcanic passive margins and oceanic plateaus have demonstrated widespread and voluminous subaerial basaltic eruptions.

Another important factor in the environmental impact of LIP volcanism is the latitude at which the LIP forms. In most basaltic eruptions, released volatiles remain in the troposphere. However, at high latitudes, the tropopause is relatively low, allowing large mass flux, basaltic fissure eruption plumes to

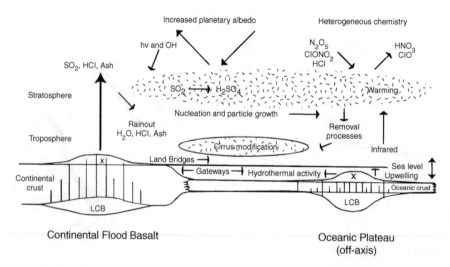

Figure 5   Environmental effects of LIP formation. LIP eruptions can perturb the Earth–ocean–atmosphere system significantly. Note that many oceanic plateaus form at least in part subaerially. LCB, lower crustal body; X, extrusive cover.

transport SO_2 and other volatiles into the stratosphere. Sulfuric acid aerosol particles that form in the stratosphere after such eruptions have a longer residence time and greater global dispersal than if the SO_2 remains in the troposphere; therefore they have greater effects on climate and atmospheric chemistry. The large volume of subaerial basaltic volcanism, over relatively brief geological intervals, at high-latitude LIPs would contribute to potential global environmental effects.

Highly explosive felsic eruptions, such as those documented from volcanic passive margins, an oceanic plateau (Kerguelen) (**Figure 1, Table 1**), and continental flood basalt provinces, can also inject both particulate material and volatiles (SO_2, CO_2) directly into the stratosphere. The total volume of felsic volcanic rocks in LIPs is poorly constrained, but they may account for a small, but not negligible fraction of the volcanic deposits in LIPs. Significant volumes of explosive felsic volcanism would further

contribute to the effects of plume volcanism on the global environment.

Between ~ 145 and ~ 50 million years ago, the global oceans were characterized by variations in chemistry, relatively high temperatures, high relative sea level, episodic deposition of black shales, high production of hydrocarbons, mass extinctions of marine organisms, and radiations of marine flora and fauna (**Figure 6**). Temporal correlations between the intense pulses of igneous activity associated with LIP formation and environmental changes suggest a causal relationship. Perhaps the most dramatic example is the eruption of the Siberian flood basalts (**Figure 1, Table 1**) ~ 250 million years ago, coinciding with the largest extinction of plants and animals in the geological record. Around 90% of all species became extinct at that time. On Iceland, the 1783–84 eruption of Laki provides the only human record of experience with the type of volcanism that constructs LIPs. Although Laki produced a basaltic

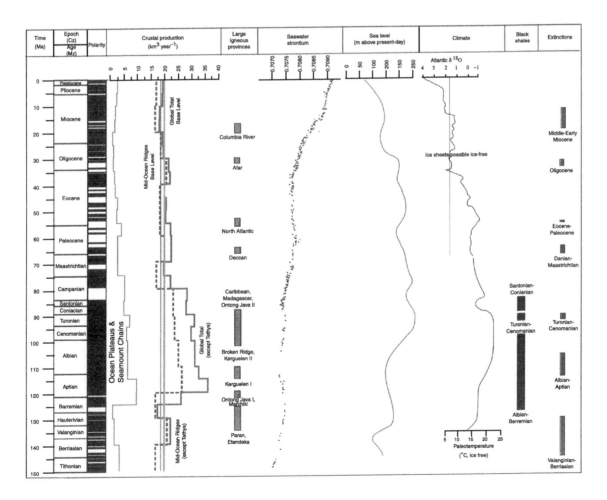

Figure 6 Temporal correlations among geomagnetic polarity, crustal production rates, LIPs, seawater strontium (Sr), sea level, climate, black shales, and extinctions.

lava flow representing ∼1% of the volume of a typical (10^3 km^3) LIP flow, the eruption's environmental impact resulted in the deaths of 75% of Iceland's livestock and 25% of its population from starvation.

Conclusions

Oceanic plateaus, volcanic passive margins, submarine ridges, seamounts, ocean basin flood basalts, and continental flood basalts share geological and geophysical characteristics indicating an origin distinct from igneous rocks formed at midocean ridges and arcs. These characteristics include: (1) broad areal extent ($>10^4$ km^2) of Fe- and Mg-rich lavas; (2) massive transient basaltic volcanism occurring over 10^5–10^6 years; (3) persistent basaltic volcanism from the same source lasting 10^7–10^8 years; (4) lower crustal bodies characterized by compressional wave velocities of 7.0–7.6 km s^{-1}; (5) some component of more Si-rich volcanic rocks; (6) higher MgO lavas, basalts with more diverse major element compositions, rocks with more common fractionated components, both alkalic and tholeiitic differentiates, and basalts with predominantly flat light rare earth element patterns, all relative to midocean ridge basalts; (7) thick (10s–100s of meters) individual basalt flows; (8) long (≤ 750 km) single basalt flows; and (9) lavas erupted in both subaerial and submarine settings.

There is strong evidence that LIPs both manifest a fundamental mode of mantle circulation commonly distinct from that which characterizes plate tectonics, and contribute episodically, at times catastrophically, to global environmental change. Nevertheless, it is important to bear in mind that we have literally only scratched the surface of oceanic, as well as continental LIPs.

See also

Deep-Sea Drilling Results. Geophysical Heat Flow. Gravity. Magnetics. Mid-Ocean Ridge Geochemistry and Petrology. Mid-Ocean Ridge Seismic Structure. Mid-Ocean Ridge Tectonics, Volcanism, and Geomorphology. Propagating Rifts and Microplates. Seamounts and Off-Ridge Volcanism. Seismic Structure

Further Reading

Carlson RW (1991) Physical and chemical evidence on the cause and source characteristics of flood basalt volcanism. *Australian Journal of Earth Sciences* 38: 525–544.

Campbell IH and Griffiths RW (1990) Implications of mantle plume structure for the evolution of flood basalts. *Earth and Planetary Science Letters* 99: 79–93.

Coffin MF and Eldholm O (1994) Large igneous provinces: crustal structure, dimensions, and external consequences. *Reviews of Geophysics* 32: 1–36.

Cox KG (1980) A model for flood basalt vulcanism. *Journal of Petrology* 21: 629–650.

Davies GF (2000) *Dynamic Earth: Plates, Plumes and Mantle Convection*. New York: Cambridge University Press.

Duncan RA and Richards MA (1991) Hotspots, mantle plumes, flood basalts, and true polar wander. *Reviews of Geophysics* 29: 31–50.

Hinsz K (1981) A hypothesis on terrestrial catastrophes: wedges of very thick oceanward dipping layers beneath passive continental margins – their origin and paleoenvironmental significance. *Geologisches Jahrbuch* E22: 3–28.

Macdougall JD (ed.) (1989) *Continental Flood Basalts*. Dordrecht: Kluwer Academic Publishers.

Mahoney JJ and Coffin MF (eds.) (1997) *Large Igneous Provinces: Continental, Oceanic, and Planetary Flood Volcanism*. Washington: American Geophysical Union Geophysical Monograph 100.

Morgan (1981) Hotspot tracks and the opening of the Atlantic and Indian oceans. In: Emiliani C (ed.) *The Oceanic Lithosphere, The Sea* vol. 7, pp. 443–487. New York: John Wiley.

Richards MA, Duncan RA, and Courtillot VE (1989) Flood basalts and hot-spot tracks: plume heads and tails. *Science* 246: 103–107.

Saunders AD, Tarney J, Kerr AC, and Kent RW (1996) The formation and fate of large oceanic igneous provinces. *Lithos* 37: 81–95.

Sigurdsson H, Houghton BF, McNutt SR, Rymer H and Stix J (eds.) (2000) *Encyclopedia of Volcanoes*, San Diego Academic Press: p. 1147.

Sleep NH (1992) Hotspot volcanism and mantle plumes. *Annual Review of Earth and Planetary Science* 20: 19–43.

White R and McKenzie D (1989) Magmatism at rift zones: the generation of volcanic continental margins and flood basalts. *Journal of Geophysical Research* 94: 7685–7729.

White RS and McKenzie D (1995) Mantle plumes and flood basalts. *Journal of Geophysical Research* 100: 17543–17585.

SEAMOUNTS AND OFF-RIDGE VOLCANISM

R. Batiza, Ocean Sciences, National Science
Foundation, VA, USA

Summary

There are three major types of off-axis volcanism
forming the abundant seamounts, islands, ridges,
plateaus, and other volcanic landforms in the world's
oceans. (1) The generally small seamounts that form
near the axes of medium and fast-spreading ridges
but less so at slow-spreading ones. These are most
likely a result of mantle upwelling and melting in a
wide zone below mid-ocean ridges, although off-axis
'mini plumes' cannot be ruled out. (2) The huge
oceanic plateaus and linear volcanic chains that form
from starting plumes and trailing plume conduits
respectively. It is widely believed that mantle plumes
originate in the lower mantle, perhaps near the core–
mantle boundary. (3) Off-ridge volcanism that is not
due to plumes, but which chemically and isotopically
resembles plume volcanism. Emerging data indicate
that much off-axis volcanism previously ascribed to
mantle plumes is not plume-related. Several distinct
types of activity seem to be the result of various
forms of intraplate mantle upwelling, or pervasively
available asthenosphere melt rising in conduits
opened by intraplate stresses, or both. Seamounts,
ridges, and plateaus produced by off-axis volcanism
play important roles in ocean circulation, as biological habitats, and in biogeochemical cycles involving the ocean crust.

Introduction

The seafloor that is produced at mid-ocean ridges is
ideally quite uniform, and except for the regular
abyssal hills and the rugged linear traces of ridge
offsets, it is essentially featureless. In strong contrast,
real ocean crust in the main ocean basins and the
marginal basins and seas is decorated with volcanic
islands, seamounts, ridges, and platforms that range
in size from tiny lava piles only tens of meters high to
vast volcanic outpourings covering huge areas of
seafloor. Volcanoes that are active close to mid-ocean
ridges are related to the ridge processes that build the
ocean crust, whereas those erupting farther away, so-
called off-axis, off-ridge, or intraplate volcanic

features, are the result of processes that are unrelated
to mid-ocean ridges. The largest oceanic volcanic
features, oceanic plateaus and linear chains of islands
and seamounts (known as large igneous provinces or
LIPs) (see Igneous Provinces), are considered to be
the result of rising plumes of hot material that may
originate as deep as the core-mantle boundary. Laboratory models suggest that when first initiated,
plumes consist of large buoyant 'heads' (so-called
starting plumes) trailed by much narrower cylindrical conduits that continue to feed material upward. In these models, the massive starting plume
experiences decompression melting and eruption of
this melt produces large oceanic plateaus. When
starting plumes rise below continents, they produce
huge volcanic outpourings called flood basalts. After
passage of the starting plume, further melting of the
rising cylindrical conduit can build linear island and
seamount chains on the moving, over-riding plate.
Plumes are found within plate interiors and also at
and near mid-ocean ridges, with which they interact.

While volumetrically plumes may be responsible
for most off-ridge volcanism, there are other forms of
off-ridge volcanism that do not appear to be related
to mantle plumes, although chemically and isotopically their magmas are very similar to those of
supposed plumes. These diverse and less voluminous
volcanic features include individual isolated seamounts, *en echelon* volcanic ridges, clustered seamounts, and lava fields. The distinct tectonic settings
in which they occur suggest that their origin is related to stresses induced in moving lithospheric
plates; however, the manner in which melt is produced is uncertain.

This article describes near-ridge seamounts,
plume-related volcanism, and off-axis volcanism that
is not related to mantle plumes. For each of these
three major types, their characteristics are reviewed
briefly and the evidence for their origin and evolution
is discussed. A common theme is the question of
whether volcanism is principally controlled by the
availability of mantle-derived melt, or alternatively,
the extent to which the thermomechanical properties
of ocean lithosphere variably influence the eruption
of this melt. It is clearly more difficult for magma to
penetrate and erupt through thick, cold, and fast-
moving lithosphere than through thin, hot, and slow-
moving lithosphere.

Another common thread is the extent to which
different kinds of off-axis volcanism can be linked
with patterns of mantle flow occurring at various

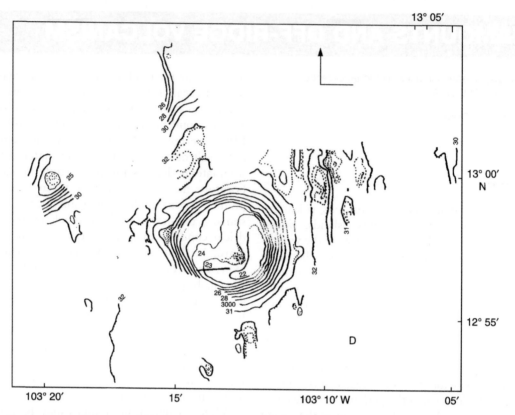

Figure 1 Seabeam map of Seamount 'D' in the eastern equatorial Pacific. Depth contours are in meters (four digits) or hundreds of meters (two digits). The arrow shows the direction of ridge-parallel abyssal hills. Note the relatively flat summit region and the caldera that is breached to the northwest. (Reproduced with permission from Elsevier from Batiza R and Vanko D (1983) Volcanic development of small oceanic central volcanoes on the flanks of the East Pacific Rise inferred from narrow-beam echo-sounder surveys. *Marine Geology* 54: 53–90.)

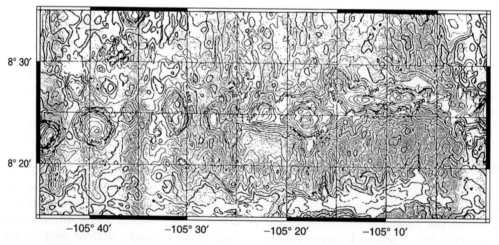

Figure 2 Seabeam map of the seamount chain at 8°20′N (see **Figure 3** for location). Note the diverse seamount shapes. (Reproduced with permission from Schierer DS and Macdonald KC (1995) Near-axis seamounts on the flanks of the East Pacific Rise, 8°N to 17°N. *Journal of Geophysical Research* 100: 2239–2259.)

levels within the Earth's mantle: flow which is linked in fundamental ways to the Earth's heat loss and the dominant plate tectonic processes that control the dynamic outer layer of the Earth. Finally, diverse oceanic volcanic features interact in important ways with ocean currents and biological organisms. Because of volcanic degassing, hydrothermal activity, and slow weathering processes, these volcanic features also affect the chemistry of sea water and influence patterns of sedimentation. the oceanographic effects of seamounts and other off-axis volcanic features are briefly discussed.

Near-ridge Seamounts

The most abundant seamounts on Earth, probably numbering in the millions, are the relatively small, mostly submerged volcanoes that occur on the flanks of mid-ocean ridges. They originate at and grow fairly close to the active mid-ocean ridges, so despite their huge numbers only a small percentage are active at any given time, and because of their small size they contribute only a few percent of material to the ocean crust. Although the existence of abundant seamounts on the ocean floor has been known since the earliest exploration of the ocean, the availability in the early 1980s of multibeam swath mapping sonar systems has made it possible to study large numbers of these seamounts. Several dozen have been studied in detail with deep-sea research submersibles (see Manned Submersibles, Deep Water).

Individual volcanic seamounts vary in size from small dome-shaped lava piles only tens of meters high, to large volcanic edifices several kilometers in height. Commonly, they have steep outer slopes, flat or nearly flat circular summit areas, and collapse features such as calderas and pit craters (**Figure 1**). In general, the smallest volcanoes tend to have the most diverse shapes. They occur as both individual volcanoes and as linear groups consisting of a few to several dozen individual volcanoes (**Figure 2**). In general, those volcanoes comprising chains tend to be larger than the isolated individual ones. Large numbers of seamounts, mostly occurring as linear chains, have been mapped on the flanks of the Juan de Fuca ridge and both the northern (**Figure 3**) and southern (**Figure 4**) East Pacific Rise (EPR).

At the Juan de Fuca ridge and along the southern East Pacific Rise (**Figure 4**), there is a marked asymmetry to the distribution of seamount chains, with most chains present on the Pacific plate. This asymmetry is absent or much less marked along the Pacific-Cocos portion of the northern EPR, and occurs, but with the opposite sense, along the Pacific-Rivera boundary. In contrast with seamount chains, isolated small seamounts near the southern EPR are symmetrically distributed on both flanks of the EPR axis.

Studies at Santa Barbara indicate that near-axis seamounts, whether isolated or in chains, form close to the axis in a zone that is about 0.2–0.3 million years wide and is independent of spreading rate. Many may continue to grow within a wider zone and a much smaller number may continue to be active at even great distances (several hundred kilometers) from the axis. Far from the axis it is difficult to distinguish near-axis seamounts that remain active for very long periods, from near-axis seamounts that are volcanically reactivated, from true intraplate volcanism that was initiated far from the axis. In such cases, the distinction between ridge-related volcanism and true intraplate volcanism can be somewhat blurred.

Near-axis seamounts occur most commonly on the flanks of inflated ridges with large cross-sectional areas and abundant melt supply, and the abundance of large ones (>400 m high) is strongly correlated with spreading rate (**Figure 5**). Abundant seamounts characterize not only modern fast-spreading ridge flanks, but also crust produced at fast spreading rates

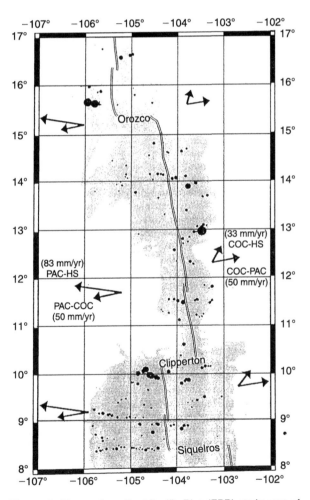

Figure 3 The northern East Pacific Rise (EPR) study area of Schierer and Macdonald (1995). Seamounts >200 m in height are shown as dots, and the double line is the axis of the EPR. The arrows show the magnitude and direction of relative and absolute plate motions. (Reproduced with permission from Schierer DS and Macdonald KC (1995) Near-axis seamounts on the flanks of the East Pacific Rise, 8°N to 17°N. *Journal of Geophysical Research* 100: 2239–2259.)

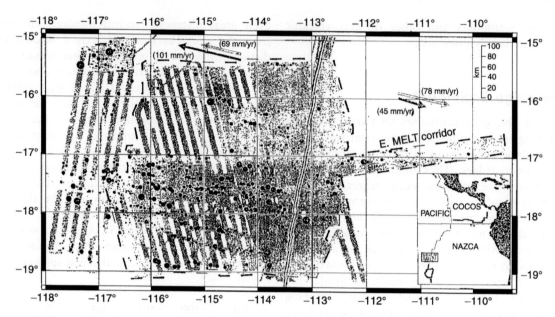

Figure 4 Study area along the southern East Pacific Rise showing the EPR axis (double line) and seamounts as dots. Arrows show the relative (gray) and absolute (black) plate motion vectors. Later studies show much more complete mapping on the Nazca plate to the east of the EPR. Note the very abundant seamount chains present especially on the Pacific plate. (Reproduced with permission from Klewer from Scheirer DS, Macdonald KC, Forsyth DW and Shen Y (1996) Abundant seamounts of the Rano Rahi seamount field near the southern East Pacific Rise, 15° to 19°S. *Marine Geophysical Researches* 18: 14–52.)

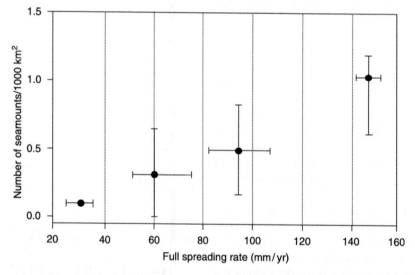

Figure 5 Plot of number of seamounts >400 m in height per 1000 km² versus the full spreading rate for the Mid-Atlantic Ridge, various medium-spreading ridges, and the northern and southern EPR. Faster spreading ridges produce more near-axis seamounts and inflated ridge segments produce more and larger seamounts than segments with smaller cross-sectional area. (Reproduced with permission from Schierer DS and Macdonald KC (1995) Near-axis seamounts on the flanks of the East Pacific Rise, 8°N to 17°N. *Journal of Geophysical Research* 100: 2239–2259.)

in the past, for example, in the Indian Ocean before the collision of India with Asia.

While near-axis seamounts form preferentially at inflated portions of fast-spreading ridges, they also occur near offsets (*see* Mid-Ocean Ridge Tectonics, Volcanism, and Geomorphology) such as overlapping spreading centers (OSCs), where they form closer to the ridge axis. They may also occur on fracture zones, although this is much more common on old versus young ocean crust. At the slow-spreading Mid-Atlantic Ridge (MAR), studies show that small seamounts are very common within the floor of the axial valley. Many or most of these appear to be a manifestation of ridge axis volcanism from both primary volcanic vents as well as off-axis eruptions fed by lava tubes.

Exactly how and why seamounts form near mid-ocean ridge axes is not known, although the composition of their lavas suggests strongly that they have the same mantle sources as volcanics erupted at the axis. Since the zone of melting that feeds the axes of fast-spreading ridges is very wide, extending several hundred kilometers on both sides of the axis, it is possible that near-axis seamounts are simply due to rising axial melt that was ineffectively focused at the axis. This idea explains their chemistry but not why they so commonly form chains. An appealing idea to explain chains is that they are due to mantle heterogeneities akin to 'mini' mantle plumes, which would help explain the occurrence of chains trending in the direction of absolute plate motion and possibly the observed asymmetry of distribution on the flanks of some ridges, as seen at the Juan de Fuca ridge. However, on the Cocos plate, where the absolute and relative motion are very different, most chains are parallel to relative motion, suggesting that perhaps the movement of the lithosphere or convection rolls parallel to relative motion might trigger seamount formation.

However, not all near-axis seamount chains trend parallel or subparallel to relative or absolute plate motion. Lonsdale showed that the oblique trend of the Larson seamounts near the EPR at ~21°N is consistent with its being fed by an asthenospheric melt diapir rising beneath the ridge axis, as envisioned in the model of Schouten and others. A problem with testing this idea further is that, along most of the EPR, the relative and absolute plate motions are quite similar and in this case the Schouten *et al.* trend is not distinct enough from the absolute and relative motion directions to be recognized. In summary, a widely applicable, self-consistent hypothesis to explain all the observations of near-axis seamounts and seamount chains is not yet available.

Finally, on the flanks of the southern EPR (**Figure 4**), numerous chains of near-axis seamounts show an inverse correlation of seamount volume between adjacent chains, suggesting that the magma might originate in plume-like sources in the upper mantle. This is an interesting observation, suggesting that near-axis seamounts are controlled by melt availability. However, the fact that seamounts form in a narrow zone near the axis that corresponds to a lithosphere thickness of 4–8 km and is independent of spreading rate suggests that the lithosphere plays an important role in the origin of near-axis seamounts. In general, the extent to which near-axis seamounts are controlled by magma availability (and whether their sources are distinct from those feeding the axis), or lithospheric vulnerability or both, is presently unknown.

Off-ridge Plume-related Volcanism

At the opposite end of size spectrum from the small near-axis seamounts are the huge oceanic plateaus (**Figure 6**) that occur in all the major ocean basins. As previously discussed, these large igneous provinces (LIPs) are thought to be the result of melting of starting plumes, and it has been proposed that the Pacific plateaus were produced by an immense superplume or group of plumes in Cretaceous time. In addition to these huge plateaus, mantle plumes are thought to produce the long linear island and seamount chains that are so common in the ocean basins (**Figure 7**). Widely held corollaries of the plume hypothesis are that plumes are nearly fixed relative to one another and that they originate in the lower mantle, possibly at the core–mantle boundary. Further, the conventional wisdom is that most of the intraplate volcanism on the planet is due to plumes. The Hawaii-Emperor chain of islands, atolls, seamounts, and drowned islands (guyots) is the classic example of a 'well-behaved' plume, with an orderly and predictable age progression of eruptive ages and bend in direction (**Figure 6**) at 43 Ma when the Pacific plate motion changed from NNW to WNW. Finally, the composition of Hawaiian lavas and those of many other suspected mantle plumes are distinct from the sources that supply mid-ocean ridges, consistent with the hypothesis that plumes sample a different and perhaps deeper region of the Earth's mantle.

Interestingly, in many cases mantle plumes are close to or centered on active mid-ocean ridges, in which case the plume and ridge interact and mixing of mantle sources is observed. Iceland is the classic example of a ridge-centered plume; whereas Galapagos is a good example of plume–ridge interaction. In addition to mixing between plume and ridge mantle sources, plume–ridge interaction can lead to the formation of the second type of hot spot island chain discussed by Morgan, in which case the orientation of the linear chain is not parallel to the absolute plate motion (as for normal plumes), but rather has a trend intermediate between the absolute and relative plate motions. An example of this type of seamount chain may be the Musicians seamounts (**Figure 8**), which consists of a western chain of seamounts oriented NW, with roughly E–W trending ridges progressing eastward. The NW trending chain has the proper orientation for a normal hot spot chain, whereas the E–W ridges appear to have been produced by plume–ridge interaction and are intermediate in trend between the absolute and relative plate motions in the Cretaceous when the Musicians plume interacted with the Pacific-Farallon spreading center.

Off-axis Volcanism not Related to Plumes

There is increasing evidence that plumes may not be the only, or even the most abundant form of intra-plate volcanism within the ocean basins. While studies of non-plume intra-plate volcanism are just

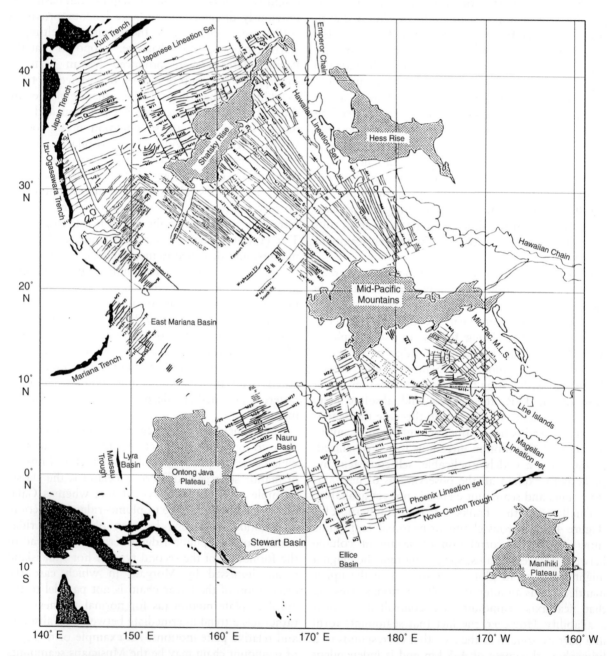

Figure 6 Map of the west and central Pacific showing the major oceanic plateaus of the region. Note also the outline north of the Mid-Pacific Mountains of the Hawaii-Emperor seamount chain with its dogleg just south of the Hess Rise. (Reproduced with permission from Neal CR, Mahoney JJ, Kroenke LW, Duncan RA and Petterson MG (1997) The Ontong Java Plateau. In: *Large Igneous Provinces*, Geophysical Monograph 100, pp. 183–216. Washington, DC: AGU.)

beginning, at least several distinct types of occurrences have been documented. A considerable obstacle to non-plume hypotheses of intraplate volcanism has been the general belief that mantle upwelling is required for melting, as at ridges and plumes, combined with the fact that most models of mantle convection show no upwelling in intraplate regions. One way around this problem is to show that secondary upwelling can occur in intraplate regions, as with Richter and Parson's longitudinal

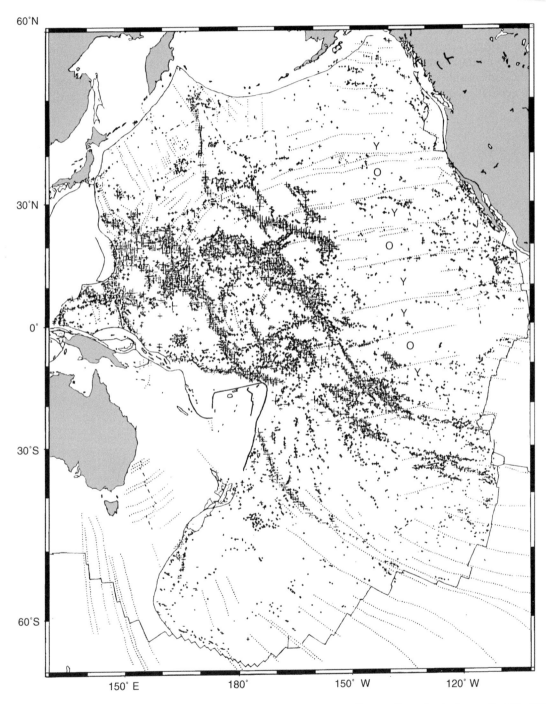

Figure 7 Locations of about 9000 seamounts mapped in the Pacific by satellite gravity methods (crosses), with cross size proportional to the maximum vertical gravity gradient. Note that the western and central Pacific have the most numerous large seamounts. Note that while many seamounts are clustered into linear chains and equant clusters, some are relatively isolated. (Reproduced with permission from Wessel P and Lyons S (1997) Distribution of large Pacific seamounts from Geosat/ERS-1: Implications for the history of intra-plate volcanism. *Journal of Geophysical Research* 102: 22 459–22 475.)

upper mantle convective rolls (called Richter rolls). Another way is to invoke localized upward mantle flow into depressions or recesses in the base of the lithosphere. A final possibility is to invoke diffuse regional mantle upwelling, as might be generated by a weak mantle plume. A completely different way around the problem, discussed by Green and others, is to cause melting not by decompression, but rather by an influx of volatiles, as is thought to occur at convergent margins. In mid-plate settings, volatiles could perhaps migrate upward from the low velocity zone of the asthenosphere. If this occurs, then magmas may generally be present and available below

most ocean lithosphere, and would need only an appropriate pathway for eruption.

Recent surveys have documented the presence on older Pacific seafloor, of long, *en echelon*, linear ridges whose trend is distinct from that of plume traces on the Pacific plate. For example, the Puka Puka ridges (**Figure 9**), stretch for at least several thousand kilometers and their morphology suggests that they are due to eruptions accompanying tensional cracking of the Pacific plate. Interestingly, the lavas of the Puka Puka ridges are chemically similar to those of supposed plumes on the Pacific plate; however, the trend of the ridges and the measured

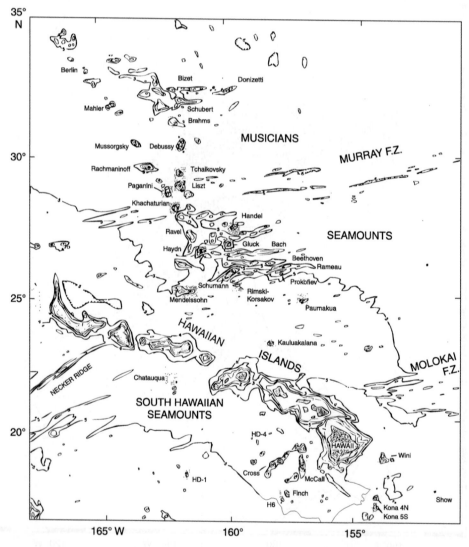

Figure 8 Generalized map of the central Pacific (contours in km) showing a portion of the Hawaiian island chain and the Musicians seamounts. Note that the group comprises a chain of NW trending seamounts including Mahler, Berlin, and Paganini and also E–W trending ridges such as those including Bizet and Donizetti to the north and Bach and Beethoven to the south. (Reproduced with permission from Sager WW and Pringle MS (1987) Paleomagnetic constraints on the origin and evolution of the Musicians and south Hawaiian seamounts, central Pacific Ocean. In: *Seamounts, Islands, and Atolls*, Geophysical Monograph 43, pp. 133–162 Washington, DC: AGU.)

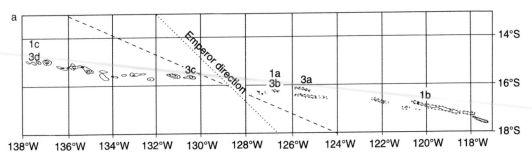

Figure 9 Bathymetric map of part of the Puka Puka Ridges showing only the ridges for clarity. Note their *en echelon* geometry and their trend which is distinct from the Hawaiian chain direction (dashed) and the Emperor chain direction (dotted). (Reproduced with permission from Lynch MA (1999) Linear ridge groups: evidence for tensional cracking in the Pacific Plate. *Journal of Geophysical Research* 104: 29 321–29 333.)

age progression of volcanism indicate that the ridges could not be due to a mantle plume.

Another form of intraplate volcanism not involving mantle plumes is the very common formation of large volcanoes of alkali basalt within the axes of inactive or fossil spreading ridges. Lonsdale has documented their very common existence in fossil ridges of the extinct Pacific–Farallon ridge system, the Mathematician fossil ridge, and the fossil Galapagos Rise (**Figure 10**). In some cases, these volcanoes are large enough to form islands, for example, Guadalupe Island off the coast of Baja California. Samples from these islands, and rarer samples from submerged seamounts, indicate that these lavas also are indistinguishable from supposed plume lavas on the Pacific plate.

A third form of non-plume volcanism that appears to be fairly widespread is associated with flexure caused by loading of the ocean lithosphere. Examples of this type of volcanism include lava fields found on the flexural arch associated with the Hawaiian island chain. The so-called North Arch and South Arch lava fields contain lavas not unlike those of the Hawaiian plume, but they erupted several hundred kilometers from the presumed location of the plume. An example on a smaller scale is Jasper seamount (**Figure 11**), which is surrounded by a ring of seamounts built on its flexural arch. Finally, there is the example of the southern Austral islands and seamounts, which recent studies suggest cannot be explained by a mantle plume, as previously proposed. Instead, it appears that these volcanoes are the result of available melts erupting in response to flexural loading by nearby edifices. In all these cases where samples are available, the lavas are chemically and isotopically similar to supposed plume lavas.

The most incompletely documented occurrences of non-plume volcanism, but potentially important in terms of volumes, are isolated large seamounts and groups of seamounts forming clusters rather than linear chains. About a dozen examples of isolated large intraplate seamounts not due to mantle plumes have been documented in several studies, for example Vesteris seamount (**Figure 12**). However, there are potentially many hundreds or even thousands of such volcanoes in the ocean basins. It is possible that every large volcano that is not clearly associated with a linear chain is a member of this group. Additional examples of large isolated intraplate seamounts are Shimada seamount and Henderson seamount in the eastern Pacific. As shown by the recent studies of Wessel and Kroenke, there are many large seamounts in the Pacific that are not members of linear chains. Further, Chapel and Small have shown that while many of the very largest volcanoes in the Pacific are associated with linear chains, many large volcanoes are clustered in nonlinear groups.

In addition to these forms of non-plume volcanism, recent studies have questioned the plume origin of several linear island chains. Wessel and Kroenke propose that many short island chains without clear age progressions, such as the Cook-Australs, the Marqueses, and the Society Islands, are 'crackspots': sites of extensional volcanism along reactivated zones of weakness induced by intraplate stresses. While these ideas are controversial, they suggest that the Pacific may contain only about five mantle plumes, instead of the several dozen that have previously been proposed. If these suggestions prove to be correct, then much, perhaps even most intraplate volcanism in the oceans will be of the non-plume type, with important implications for mantle convection and mechanisms of advective heat loss. Haase has shown that chemically and isotopically, various types of plume and non-plume intraplate volcanism seem to define a single population that exhibits chemical systematics as a function of the age of the lithosphere affected by intraplate volcanism.

This chemical coherence, along with emerging evidence for the volumetric importance of non-plume volcanism, suggests that much oceanic intraplate volcanism originates in the upper mantle, not in the lower mantle as suggested by the plume hypothesis.

Oceanographic Effects

Seamounts of all types, including large plateaus and platforms, may contribute about 10% or more to the mass of oceanic crust. Since seamounts are volcanic and host active hydrothermal convective systems, seamounts should have a significant effect on element cycles involving sea water and its dynamic interaction with the ocean crust. Likewise, because normal abyssal sedimentation patterns are severely disturbed in the vicinity of seamounts, they exert a significant influence on the average composition of oceanic sediments, including their hydrothermal and biogenic components (see Calcium Carbonates).

Seamounts may also have a significant influence on global ocean circulation patterns because their presence induces much greater mixing than is measured in areas with smooth bottom topography. At a more local scale, seamounts have a great effect on circulation patterns and currents, which in turn have very important effects on seamount biota,

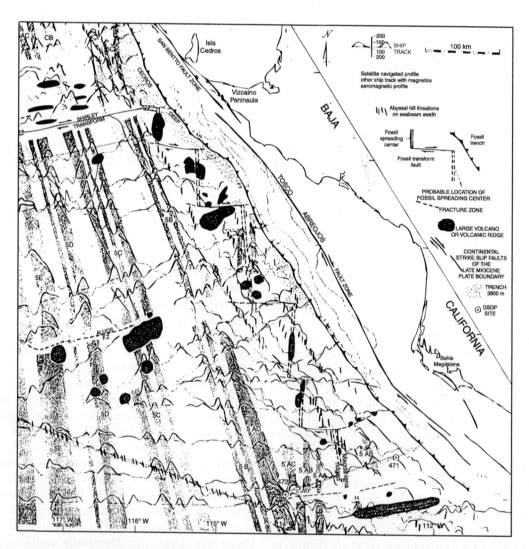

Figure 10 Interpretation of magnetic anomalies off the coast of Baja California showing the locations of probable fossil spreading centers (double dashed lines). Note that many of the fossil spreading centres have large volcanoes or volcanic ridges built in their axes. (Reproduced with permission of the American Association of Petroleum Geologists from Lonsdale P (1991) Structural patterns of the Pacific floor offshore of Peninsular California. In: *The Gulf and Peninsular Province of the Californias*, AAPG Memoir 47, pp. 87–125. Tulsa, OK: AAPG.)

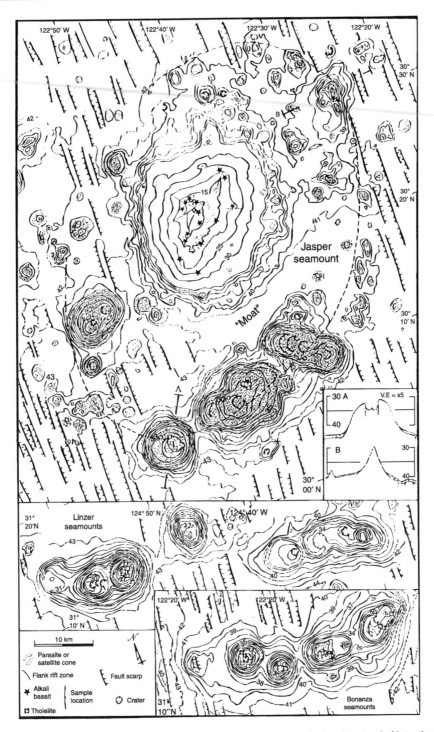

Figure 11 Bathymetric maps of several seamounts (two-digit contours are hundreds of meters). Note the elliptical group of seamounts surrounding Jasper seamount. These are presumed to have erupted on Jasper's flexural arch, similar to seamounts and lava beds elsewhere in the Pacific basin. The Linzer and Bonanza seamounts are additional examples of near-ridge seamounts built on older Pacific crust, although Linzer, like Jasper, may be part of the Fieberling hot spot chain. (Reproduced with permission of the American Association of Petroleum Geologists from Lonsdale P (1991) Structural patterns of the Pacific floor offshore of Peninsular California. In: *The Gulf and Peninsular Province of the Californias*, AAPG Memoir 47, pp. 87–125. Tulsa, OK: AAPG.)

including populations of fishes. In general, seamounts host very diverse and abundant faunas, with important effects on oceanic biology. Thus, while seamounts and off-axis volcanism are interesting on their own, seamounts are also of great interest as obstacles to current flow, biological habitats, and for biogeochemical cycles involving the ocean crust.

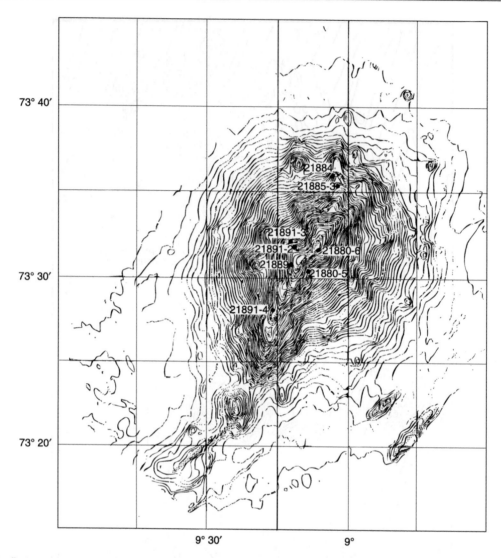

Figure 12 Bathymetric map of Vesteris seamount, an isolated intra-plate seamount in the north Atlantic ocean. Note the volcanic rift zones shown with heavy dark lines. Sample locations and numbers are shown. (Reproduced with permission from Oxford University Press from Haase KM and Devey CW (1994) The petrology and geochemistry of Vesteris seamount, Greenland basin – an intraplate alkaline volcano of non-plume origin. *Journal of Petrology* 35: 295–328.)

See also

Calcium Carbonates. Igneous Provinces. Manned Submersibles, Deep Water. Mid-Ocean Ridge Tectonics, Volcanism, and Geomorphology.

Further Reading

Batiza R, Smith T, and Niu Y (1989) Geologic and petrologic evolution of seamounts near the EPR based on submersible and camera study. *Marine Geophysical Researches* 11: 169–236.

Floyd PA (ed.) (1991) *Oceanic Basalts*. Glasgow: Blackie and Son.

Green DH and Falloon TJ (1998) Pyrolite: A Ringwood concept and its current expression. In: Jackson I (ed.) *The Earth's Mantle*, pp. 311–378. Cambridge: Cambridge University Press.

Haase KM (1996) The relationship between the age of the lithosphere and the composition of oceanic magmas: constraints on partial melting, mantle sources and the thermal structure of the plates. *Earth and Planetary Science Letters* 144: 75–92.

Larson RL (1991) Latest pulse of Earth: Evidence for a mid-Cretaceous superplume. *Geology* 19: 547–550.

Lueck RG and Mudge TD (1997) Topographically induced mixing around a shallow seamount. *Science* 276: 1831–1833.

McNutt MK, Caress DW, Reynolds J, Johrdahl KA, and Duncan RA (1997) Failure of plume theory to explain midplate volcanism in the southern Austral Island. *Nature* 389: 479–482.

Morgan WJ (1978) Rodriquez, Darwin, Amsterdam,..., A second type of hotspot island. *Journal of Geophysical Research* 83: 5355–5360.

Richter FM and Parsons B (1975) On the interaction of two scales of convection in the mantle. *Journal of Geophysical Research* 80: 2529–2541.

Schilling J-G (1991) Fluxes and excess temperatures of mantle plumes inferred from their interaction with migrating mid-ocean ridges. *Nature* 352: 397–403.

Schmidt R and Schmincke H-U (2000) Seamounts and island building. *Encyclopedia of Volcanoes*, pp. 383–402. London: Academic Press.

Schouten H, Klitgord KD, and Whitehead JA (1985) Segmentation of mid-ocean ridges. *Nature.* 317: 225–229.

Smith DK and Cann JR (1999) Constructing the upper crust of the Mid-Atlantic Ridge: A reinterpretation based on the Puna Ridge, Kilauea Volcano. *Journal of Geophysical Research* 104: 25 379–25 399.

Wessel P and Kroenke LW (2000) The Ontong Java Plateau and late Neogene changes in Pacific Plate motion. *Journal of Geophysical Research* 105: 28 255–28 277.

White SM, Macdonald KC, Scheirer DS, and Cormier M-H (1998) Distribution of isolated volcanoes on the flanks of the East Pacific Rise, 15.3°–20°S. *Journal of Geophysical Research* 103: 30 371–30 384.

METHODS OF OCEAN BOTTOM EXPLORATION

SATELLITE ALTIMETRY

R. E. Cheney, Laboratory for Satellite Altimetry, NOAA, Silver Spring, Maryland, USA

Introduction

Students of oceanography are usually surprised to learn that sea level is not very level at all and that the dominant force affecting ocean surface topography is not currents, wind, or tides; rather it is regional variations in the Earth's gravity. Beginning in the 1970s with the advent of satellite radar altimeters, the large-scale shape of the global ocean surface could be observed directly for the first time. What the data revealed came as a shock to most of the oceanographic community, which was more accustomed to observing the sea from ships. Profiles telemetered back from NASA's pioneering altimeter, Geos-3, showed that on horizontal scales of hundreds to thousands of kilometers, the sea surface is extremely complex and bumpy, full of undulating hills and valleys with vertical amplitudes of tens to hundreds of meters. None of this came as a surprise to geodesists and geophysicists who knew that the oceans must conform to these shapes owing to spatial variations in marine gravity. But for the oceanographic community, the concept of sea level was forever changed. During the following two decades, satellite altimetry would provide exciting and revolutionary new insights into a wide range of earth science topics including marine gravity, bathymetry, ocean tides, eddies, and El Niño, not to mention the marine wind and wave fields which can also be derived from the altimeter echo. This chapter briefly addresses the technique of satellite altimetry and provides examples of applications.

Measurement Method

In concept, radar altimetry is among the simplest of remote sensing techniques. Two basic geometric measurements are involved. In the first, the distance between the satellite and the sea surface is determined from the round-trip travel time of microwave pulses emitted downward by the satellite's radar and reflected back from the ocean. For the second measurement, independent tracking systems are used to compute the satellite's three-dimensional position relative to a fixed Earth coordinate system. Combining these two measurements yields profiles of sea surface topography, or sea level, with respect to the reference ellipsoid (a smooth geometric surface which approximates the shape of the Earth).

In practice, the various measurement systems are highly sophisticated and require expertise at the cutting edge of instrument and modeling capabilities. This is because accuracies of a few centimeters must be achieved to properly observe and describe the various oceanographic and geophysical phenomena of interest. **Figure 1** shows a schematic of the Topex/Poseidon (T/P) satellite altimeter system. Launched in 1992 as a joint mission of the American and French Space agencies (and still operating as of 2001), T/P is the most accurate altimeter flown to date. Its microwave radars measure the distance to the sea surface with a precision of 2 cm. Two different frequencies are used to solve for the path delay due to the ionosphere, and a downward-looking microwave radiometer provides measurements of the integrated water vapor content which must also be known. Meteorological models must be used to estimate the attenuation of the radar pulse by the atmosphere, and other models correct for biases created by ocean waves. Three different tracking systems (a laser reflector, a Global Positioning System receiver, and a 'DORIS' Doppler receiver) determine the satellite orbit to within 2 cm in the radial direction. The result of all these measurements is a set of global sea level observations with an absolute accuracy of 3–4 cm at intervals of 1 s, or about 6 km, along the satellite track. The altimeter footprint is exceedingly small – only 2–3 km – so regional maps or 'images' can only be derived by averaging data collected over a week or more.

Gravitational Sea Surface Topography

Sea surface topography associated with spatial variations in the Earth's gravity field has vertical amplitudes 100 times larger than sea level changes generated by tides and ocean currents. To first order, therefore, satellite altimeter data reveal information about marine gravity. Within 1–2% the ocean topography follows a surface of constant gravitational potential energy known as the geoid or the equipotential surface, shown schematically in **Figure 1**. Gravity can be considered to be constant in time for most purposes, even though slight changes do occur as the result of crustal motions,

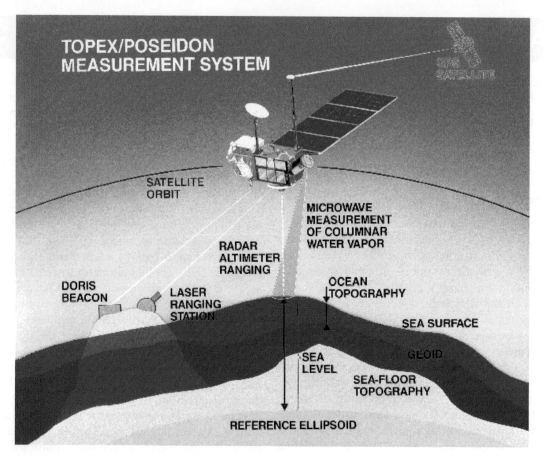

Figure 1 Schematic diagram of satellite radar altimeter system. Range to the sea surface together with independent determination of the satellite orbit yields profiles of sea surface topography (sea level) relative to the Earth's reference ellipsoid. Departures of sea level from the geoid drive surface geostrophic currents.

redistribution of terrestrial ice and water, and other slowly varying phenomena. An illustration of the gravitational component of sea surface topography is provided in **Figure 2**, which shows a T/P altimeter profile collected in December 1999 across the Marianas Trench in the western Pacific. The trench represents a deficit of mass, and therefore a negative gravity anomaly, so that the water is pulled away from the trench axis by positive anomalies on either side. Similarly, seamounts represent positive gravity anomalies and appear at the ocean surface as mounds of water. The sea level signal created by ocean bottom topography ranges from ~ 1 m for seamounts to ~ 10 m for pronounced features like the Marianas Trench, and the peak-to-peak amplitude for the large-scale gravity field is nearly 200 m.

Using altimeter data collected by several different satellites over a period of years, it is possible to create global maps of sea surface topography with extraordinary accuracy and resolution. When these maps are combined with surface gravity measurements, models of the Earth's crust, and bathymetric data collected by ships, it is possible to construct three-dimensional images of the ocean floor – as if all the water were drained away (**Figure 3**). For many oceanic regions, especially in the Southern Hemisphere, these data have provided the first reliable maps of bottom topography. This new data set has many scientific and commercial applications, from numerical ocean modeling, which requires realistic bottom topography, to fisheries, which have been able to take advantage of new fishing grounds over previously uncharted seamounts.

Dynamic Sea Surface Topography

Because of variations in the density of sea water over the globe, the geoid and the mean sea surface are not exactly coincident. Departures of the sea surface with respect to the geoid have amplitudes of about 1 m and constitute what is known as 'dynamic topography'. These sea surface slopes drive the geostrophic circulation: a balance between the surface slope (or surface pressure gradient) and the Coriolis force (created by the Earth's rotation).

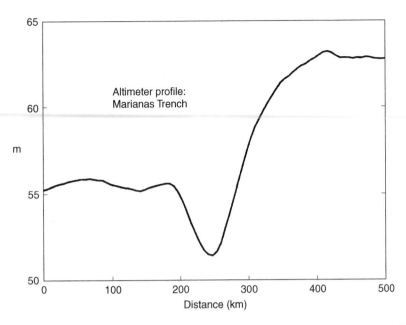

Figure 2 Sea surface topography across the Marianas Trench in the western Pacific measured by Topex/Poseidon. Heights are relative to the Earth's reference ellipsoid.

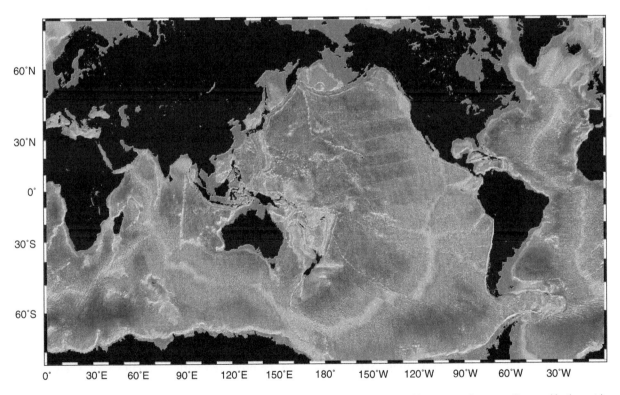

Figure 3 Topography of the ocean bottom determined from a combination of satellite altimetry, gravity anomalies, and bathymetric data collected by ships. (Courtesy of Walter H. F. Smith, NOAA, Silver Spring, MD, USA.)

The illustration in **Figure 4** shows an estimate of the global geostrophic circulation derived by combining a mean altimeter-derived topography with a geoid computed from independent gravity measurements. Variations are between −110 cm (deep blue) and 110 cm (white). The surface flow is along lines of equal dynamic topography (red arrows). In the Northern Hemisphere, the flow is clockwise around

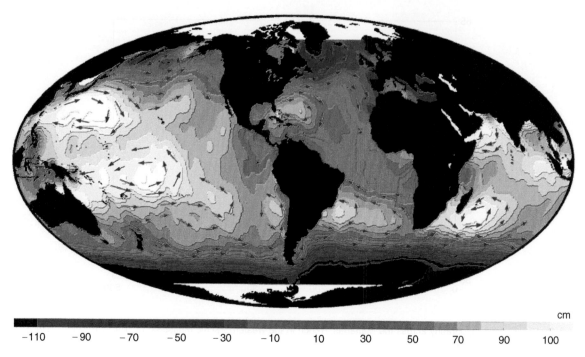

$$cm$$

| −110 | −90 | −70 | −50 | −30 | −10 | 10 | 30 | 50 | 70 | 90 | 100 |

Figure 4 Surface geostrophic circulation determined from a combination of satellite altimetry and a model of the marine gravity field. (Courtesy of Space Oceanography Division, CLS, Toulouse, France.)

the topography highs, while in the Southern Hemisphere, the flow is counter-clockwise around the highs. The map shows all the features of the general circulation such as the ocean gyres and associated western boundary currents (e.g. Gulf Stream, Kuroshio, Brazil/Malvinas Confluence) and the Antarctic Circumpolar Current.

At the time of writing, global geoid models are not sufficiently accurate to reveal significant new information about the surface circulation of the ocean. However, extraordinary gravity fields will soon be available from dedicated satellite missions such as the Challenging Minisatellite Payload (CHAMP: 2000 launch), the Gravity Recovery and Climate Experiment (GRACE: 2002 launch), and the Gravity Field and Steady-state Ocean Circulation Explorer (GOCE: 2005 launch). These satellite missions, sponsored by various agencies in the USA and Europe, will employ accelerometers, gravity gradiometers, and the Global Positioning System to virtually eliminate error in marine geoid models at spatial scales larger than 300 km and will thereby have a dramatic impact on physical oceanography. Not only will it be possible to accurately compute global maps of dynamic topography and geostrophic surface circulation, but the new gravity models will also allow recomputation of orbits for past altimetric satellites back to 1978, permitting studies of long, global sea level time-series. Furthermore,

measurement of the change in gravity as a function of time will provide new information about the global hydrologic cycle and perhaps shed light on the factors contributing to global sea level rise. For example, how much of the rise is due simply to heating and how much to melting of glaciers? Together with complementary geophysical data, satellite gravity data represent a new frontier in studies of the Earth and its fluid envelope.

Sea Level Variability

At any given location in the ocean, sea level rises and falls over time owing to tides, variable geostrophic flow, wind stress, and changes in temperature and salinity. Of these, the tides have the largest signal amplitude, on the order of 1 m in mid-ocean. Satellite altimetry has enabled global tide models to be dramatically improved such that mid-ocean tides can now be predicted with an accuracy of a few cm. In studying ocean dynamics, the contribution of the tides is usually removed using these models so that other dynamic ocean phenomena can be isolated.

The map in **Figure 5** shows the variability of global sea level for the period 1992–98. It is derived from three satellite altimeter data sets: ERS-1, T/P, and ERS-2 (ERS is the European Space Agency Remote Sensing Satellite), from which the tidal signal has been removed. The map is dominated by mesoscale

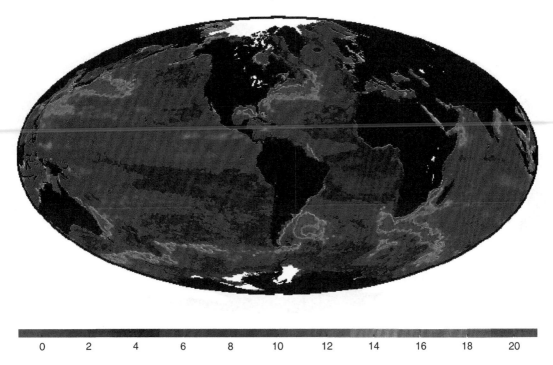

Figure 5 Variability of sea surface topography over the period 1992–98 from three satellite altimeters: Topex/Poseidon, ERS-1, and ERS-2. Highest values (cm) correspond to western boundary currents which meander and generate eddies. (Courtesy of Space Oceanography Division, CLS, Toulouse, France.)

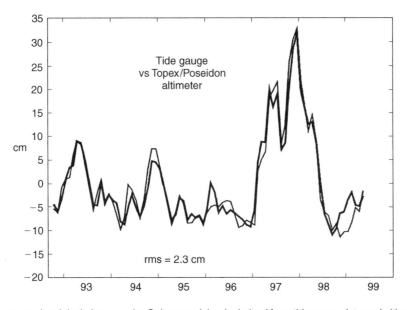

Figure 6 Monthly mean sea level deviation near the Galapagos Islands derived from tide gauge data and altimeter data. The ~2 cm agreement demonstrates the accuracy of altimetry for observing sea level variability. The effect of the 1997–98 El Niño is apparent.

(100–300 km) variability associated with the western boundary currents, where the rms variability can be as high as 30 cm. This is due to a combination of current meandering, eddies, and seasonal heating and cooling. Other bands of relative maxima (10–15 cm) can be seen in the tropics where interannual signals such as El Niño are the dominant contributor. The smallest variability is found in the eastern portions of the major ocean basins where values are <5 cm rms.

To examine a sample of the sea level signal more closely, **Figure 6** shows the record from the region of

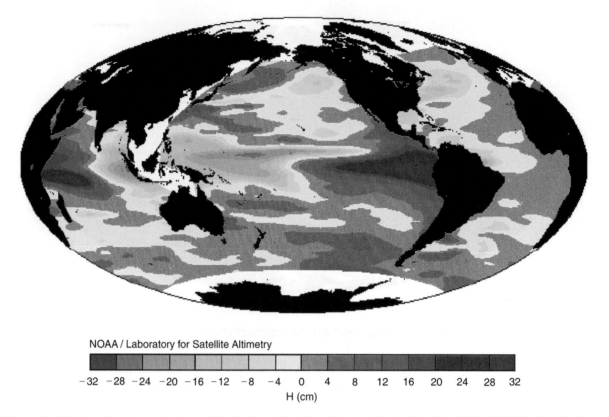

NOAA / Laboratory for Satellite Altimetry

−32 −28 −24 −20 −16 −12 −8 −4 0 4 8 12 16 20 24 28 32

H (cm)

Figure 7 Global sea level anomaly observed by the Topex/Poseidon altimeter at the height of the 1997–98 El Niño. High (low) sea level corresponds to areas of positive (negative) heat anomaly in the ocean's upper layers.

the Galapagos Islands in the eastern equatorial Pacific. The plot includes two time-series: one from the T/P altimeter and the other from an island tide gauge, both averaged over monthly time periods. These independent records agree at the level of 2 cm, an indication of the remarkable reliability of satellite altimetry. The plot also illustrates changes associated with the El Niño event which took place during 1997–98. During El Niño, relaxation of the Pacific trade winds cause a dramatic redistribution of heat in the tropical oceans. In the eastern Pacific, sea level during this event rose to 30 cm above normal by December 1977 and fell by a corresponding amount in the far western Pacific. The global picture of sea level deviations observed by the T/P altimeter at this time is shown in **Figure 7**. Because sea level changes can be interpreted as changes in heat (and to a lesser extent, salinity) in the upper layers, altimetry provides important information for operational ocean models which are used for long-range El Niño forecasts.

Global Sea Level Rise

Tide gauge data collected over the last century indicate that global sea level is rising at about 1.8 mm y^{-1}. Unfortunately, because these data are relatively sparse and contain large interdecadal fluctuations, the observations must be averaged over 50–75 years in order to obtain a stable mean value. It is therefore not possible to say whether sea level rise is accelerating in response to recent global warming. Satellite altimeter data have the advantage of dense, global coverage and may offer new insights on the problem in a relatively short period of time. Based on T/P data collected since 1992, it is thought that 15 years of continuous altimeter measurements may be sufficient to obtain a reliable estimate of the current rate of sea level rise. This will require careful calibration of the end-to-end altimetric system, not to mention cross-calibration of a series of two or three missions (which typically last only 5 years). Furthermore, in order to interpret and fully understand the sea level observations, the various components of the global hydrologic system must be taken into account, for example, polar and glacial ice, ground water, fresh water stored in man-made reservoirs, and the total atmospheric water content. It is a complicated issue, but one which may yield to the increasingly sophisticated observational systems that are being brought to bear on the problem.

Wave Height and Wind Speed

In addition to sea surface topography, altimetry provides indirect measurements of ocean wave height and wind speed (but not wind direction). This is made possible by analysis of the shape and intensity of the reflected radar signal: a calm sea sends the signal back almost perfectly, like a flat mirror, whereas a rough sea scatters and deforms it. Wave height measurements are accurate to about 0.5 m or 10% of the significant wave height, whichever is larger. Wind speed can be measured with an accuracy of about $2 \, \mathrm{m \, s^{-1}}$.

Conclusions

Satellite altimetry is somewhat unique among ocean remote sensing techniques because it provides much more than surface observations. By measuring sea surface topography and its change in time, altimeters provide information on the Earth's gravity field, the shape and structure of the ocean bottom, the integrated heat and salt content of the ocean, and geostrophic ocean currents. Much progress has been made in the development of operational ocean applications, and altimeter data are now routinely assimilated in near-real-time to help forecast El Niño, monitor coastal circulation, and predict hurricane intensity. Although past missions have been flown largely for research purposes, altimetry is rapidly moving into the operational domain and will become a routine component of international satellite systems during the twenty-first century.

See also

Satellite Oceanography, History and Introductory Concepts.

Further Reading

Cheney RE (ed.) (1995) TOPEX/POSEIDON: Scientific Results. *Journal of Geophysical Research* 100: 24 893–25 382

Douglas BC, Kearney MS, and Leatherman SP (eds.) (2001) *Sea Level Rise: History and Consequences.* London: Academic Press.

Fu LL and Cheney RE (1995) Application of satellite altimetry to ocean circulation studies, 1987–1994. *Reviews of Geophysics Suppl:* 213–223.

Fu LL and Cazenave A (eds.) (2001) *Satellite Altimetry and Earth Sciences.* London: Academic Press.

SATELLITE OCEANOGRAPHY, HISTORY AND INTRODUCTORY CONCEPTS

W. S. Wilson, NOAA/NESDIS, Silver Spring, MD, USA
E. J. Lindstrom, NASA Science Mission Directorate, Washington, DC, USA
J. R. Apel[†]**,** Global Ocean Associates, Silver Spring, MD, USA

Published by Elsevier Ltd.

Oceanography from a satellite – the words themselves sound incongruous and, to a generation of scientists accustomed to Nansen bottles and reversing thermometers, the idea may seem absurd.

Gifford C. Ewing (1965)

Introduction: A Story of Two Communities

The history of oceanography from space is a story of the coming together of two communities – satellite remote sensing and traditional oceanography.

For over a century oceanographers have gone to sea in ships, learning how to sample beneath the surface, making detailed observations of the vertical distribution of properties. Gifford Ewing noted that oceanographers had been forced to consider "the class of problems that derive from the vertical distribution of properties at stations widely separated in space and time."

With the introduction of satellite remote sensing in the 1970s, traditional oceanographers were provided with a new tool to collect synoptic observations of conditions at or near the surface of the global ocean. Since that time, there has been dramatic progress; satellites are revolutionizing oceanography. (Appendix 1 provides a brief overview of the principles of satellite remote sensing.)

Yet much remains to be done. Traditional subsurface observations and satellite-derived observations of the sea surface – collected as an integrated set of observations and combined with state-of-the-art models – have the potential to yield estimates of the three-dimensional, time-varying distribution of properties for the global ocean. Neither a satellite nor an *in situ* observing system can do this on its own. Furthermore, if such observations can be collected over the long term, they can provide oceanographers with an observational capability conceptually similar to that which meteorologists use on a daily basis to forecast atmospheric weather.

Our ability to understand and forecast oceanic variability, how the oceans and atmosphere interact, critically depends on an ability to observe the three-dimensional global oceans on a long-term basis. Indeed, the increasing recognition of the role of the ocean in weather and climate variability compels us to implement an integrated, operational satellite and *in situ* observing system for the ocean now – so that it may complement the system which already exists for the atmosphere.

The Early Era

The origins of satellite oceanography can be traced back to World War II – radar, photogrammetry, and the V-2 rocket. By the early 1960s, a few scientists had recognized the possibility of deriving useful

Figure 1 Thermal infrared image of the US southeast coast showing warmer waters of the Gulf Stream and cooler slope waters closer to shore taken in the early 1960s. While the resolution and accuracy of the TV on *Tiros* were not ideal, they were sufficient to convince oceanographers of the potential usefulness of infrared imagery. The advanced very high resolution radiometer (AVHRR) scanner (see text) has improved images considerably. Courtesy of NASA.

[†] Deceased

oceanic information from the existing aerial sensors. These included (1) the polar-orbiting meteorological satellites, especially in the 10–12-μm thermal infrared band; and (2) color photography taken by astronauts in the Mercury, Gemini, and Apollo manned spaceflight programs. Examples of the kinds of data obtained from the National Aeronautics and Space Administration (NASA) flights collected in the 1960s are shown in **Figures 1** and **2**.

Such early imagery held the promise of deriving interesting and useful oceanic information from space, and led to three important conferences on space oceanography during the same time period.

In 1964, NASA sponsored a conference at the Woods Hole Oceanographic Institution (WHOI) to examine the possibilities of conducting scientific research from space. The report from the conference, entitled *Oceanography from Space*, summarized findings to that time; it clearly helped to stimulate a number of NASA projects in ocean observations and sensor development. Moreover, with the exception of the synthetic aperture radar (SAR), all instruments flown through the 1980s used techniques described in

this report. Dr. Ewing has since come to be justifiably regarded as the father of oceanography from space.

A second important step occurred in 1969 when the Williamstown Conference was held at Williams College in Massachusetts. The ensuing Kaula report set forth the possibilities for a space-based geodesy mission to determine the equipotential figure of the Earth using a combination of (1) accurate tracking of satellites and (2) the precision measurement of satellite elevation above the sea surface using radar altimeters. Dr. William Von Arx of WHOI realized the possibilities for determining large-scale oceanic currents with precision altimeters in space. The requirements for measurement precision of 10-cm height error in the elevation of the sea surface with respect to the geoid were articulated. NASA scientists and engineers felt that such accuracy could be achieved in the long run, and the agency initiated the Earth and Ocean Physics Applications Program, the first formal oceans-oriented program to be established within the organization. The required accuracy was not to be realized until 1992 with *TOPEX/ Poseidon*, which was reached only over a 25-year

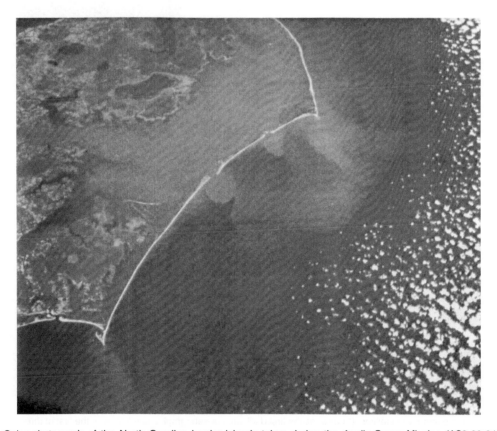

Figure 2 Color photograph of the North Carolina barrier islands taken during the Apollo-Soyuz Mission (AS9-20-3128). Capes Hatteras and Lookout, shoals, sediment- and chlorophyll-bearing flows emanating from the coastal inlets are visible, and to the right, the blue waters of the Gulf Stream. Cloud streets developing offshore the warm current suggest that a recent passage of a cold polar front has occurred, with elevated air–sea evaporative fluxes. Later instruments, such as the coastal zone color scanner (CZCS) on *Nimbus-7* and the SeaWiFS imager have advanced the state of the art considerably. Courtesy of NASA.

period of incremental progress that saw the flights of five US altimetric satellites of steadily increasing capabilities: *Skylab*, *Geos-3*, *Seasat*, *Geosat*, and *TOPEX/Poseidon* (see **Figure 3** for representative satellites).

A third conference, focused on sea surface topography from space, was convened by the National Oceanic and Atmospheric Administration (NOAA), NASA, and the US Navy in Miami in 1972, with 'sea surface topography' being defined as undulations of the ocean surface with scales ranging from

approximately 5000 km down to 1 cm. The conference identified several data requirements in oceanography that could be addressed with space-based radar and radiometers. These included determination of surface currents, Earth and ocean tides, the shape of the marine geoid, wind velocity, wave refraction patterns and spectra, and wave height. The conference established a broad scientific justification for space-based radar and microwave radiometers, and it helped to shape subsequent national programs in space oceanography.

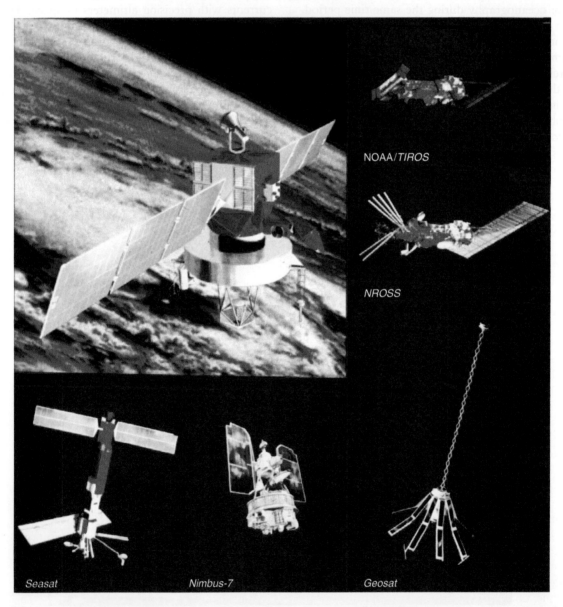

NOAA/TIROS

NROSS

Seasat   Nimbus-7   Geosat

Figure 3 Some representative satellites: (1) *Seasat*, the first dedicated oceanographic satellite, was the first of three major launches in 1978; (2) the *Tiros* series of operational meteorological satellites carried the advanced very high resolution radiometer (AVHRR) surface temperature sensor; *Tiros-N*, the first of this series, was the second major launch in 1978; (3) *Nimbus-7*, carrying the CZCS color scanner, was the third major launch in 1978; (4) *NROSS*, an oceanographic satellite approved as an operational demonstration in 1985, was later cancelled; (5) *Geosat*, an operational altimetric satellite, was launched in 1985; and (6) this early version of *TOPEX* was reconfigured to include the French *Poseidon*; the joint mission *TOPEX/Poseidon* was launched in 1992. Courtesy of NASA.

The First Generation

Two first-generation ocean-viewing satellites, *Skylab* in 1973 and *Geos-3* in 1975, had partially responded to concepts resulting from the first two of these conferences. *Skylab* carried not only several astronauts, but a series of sensors that included the S-193, a radar-altimeter/wind-scatterometer, a long-wavelength microwave radiometer, a visible/infrared scanner, and cameras. S-193, the so-called Rad/Scatt, was advanced by Drs. Richard Moore and Willard Pierson. These scientists held that the scatterometer could return wind velocity measurements whose accuracy, density, and frequency would revolutionize marine meteorology. Later aircraft data gathered by NASA showed that there was merit to their assertions. *Skylab*'s scatterometer was damaged during the opening of the solar cell panels, and as a consequence returned indeterminate results (except for passage over a hurricane), but the altimeter made observations of the geoid anomaly due to the Puerto Rico Trench.

Geos-3 was a small satellite carrying a dual-pulse radar altimeter whose mission was to improve the knowledge of the Earth's marine geoid, and coincidentally to determine the height of ocean waves via the broadening of the short transmitted radar pulse upon reflection from the rough sea surface. Before the end of its 4-year lifetime, *Geos-3* was returning routine wave height measurements to the National Weather Service for inclusion in its Marine Waves Forecast. Altimetry from space had become a clear possibility, with practical uses of the sensor immediately forthcoming. The successes of *Skylab* and *Geos-3* reinforced the case for a second generation of radar-bearing satellites to follow.

The meteorological satellite program also provided measurements of sea surface temperature using far-infrared sensors, such as the visible and infrared scanning radiometer (VISR), which operated at wavelengths near 10 μm, the portion of the terrestrial spectrum wherein thermal radiation at terrestrial temperatures is at its peak, and where coincidentally the atmosphere has a broad passband. The coarse, 5-km resolution of the VISR gave blurred temperature images of the sea, but the promise was clearly there. **Figure 1** is an early 1960s TV image of the southeastern USA taken by the NASA *TIROS* program, showing the Gulf Stream as a dark signal. While doubts were initially held by some oceanographers as to whether such data actually represented the Gulf Stream, nevertheless the repeatability of the phenomenon, the verisimilitude of the positions and temperatures with respect to conventional wisdom, and their own objective judgment finally convinced most workers of the validity of the data. Today, higher-resolution, temperature-calibrated infrared imagery constitutes a valuable data source used frequently by ocean scientists around the world.

During the same period, spacecraft and aircraft programs taking ocean color imagery were delineating the possibilities and difficulties of determining sediment and chlorophyll concentrations remotely. **Figure 2** is a color photograph of the North Carolina barrier islands taken with a hand-held camera, with Cape Hatteras in the center. Shoals and sediment- and chlorophyll-bearing flows emanating from the coastal inlets are visible, and to the right, the blue waters of the Gulf Stream. Cloud streets developing offshore the warm stream suggest a recent passage of a cold polar front and attendant increases in air–sea evaporative fluxes.

The Second Generation

The combination of the early data and advances in scientific understanding that permitted the exploitation of those data resulted in spacecraft sensors explicitly designed to look at the sea surface. Information returned from altimeters and microwave radiometers gave credence and impetus to dedicated microwave spacecraft. Color measurements of the sea made from aircraft had indicated the efficacy of optical sensors for measurement of near-surface chlorophyll concentrations. Infrared radiometers returned useful sea surface temperature measurements. These diverse capabilities came together when, during a 4-month interval in 1978, the USA launched a triad of spacecraft that would profoundly change the way ocean scientists would observe the sea in the future. On 26 June, the first dedicated oceanographic satellite, *Seasat*, was launched; on 13 October, *TIROS-N* was launched immediately after the catastrophic failure of *Seasat* on 10 October; and on 24 October, *Nimbus-7* was lofted. Collectively they carried sensor suites whose capabilities covered virtually all known ways of observing the oceans remotely from space. This second generation of satellites would prove to be extraordinarily successful. They returned data that vindicated their proponents' positions on the measurement capabilities and utility, and they set the direction for almost all subsequent efforts in satellite oceanography.

In spite of its very short life of 99 days, *Seasat* demonstrated the great utility of altimetry by measuring the marine geoid to within a very few meters, by inferring the variability of large-scale ocean surface currents, and by determining wave heights. The wind scatterometer could yield oceanic surface wind velocities equivalent to 20 000 ship observations per day. The scanning multifrequency radiometer also provided wind speed and atmospheric water content data; and the SAR penetrated clouds to show features on the surface of the sea, including surface and internal waves, current boundaries, upwellings, and rainfall patterns. All of these measurements could be extended to basin-wide scales, allowing oceanographers a view of the sea never dreamed of before. *Seasat* stimulated several subsequent generations of ocean-viewing satellites, which outline the chronologies and heritage for the world's ocean-viewing spacecraft. Similarly, the early temperature and color observations have led to successor programs that provide large quantities of quantitative data to oceanographers around the world.

The Third Generation

The second generation of spacecraft would demonstrate that variables of importance to oceanography could be observed from space with scientifically useful accuracy. As such, they would be characterized as successful concept demonstrations. And while both first- and second-generation spacecraft had been exclusively US, international participation in demonstrating the utility of their data would lead to the entry of Canada, the European Space Agency (ESA), France, and Japan into the satellite program during this period. This article focuses on the US effort. Additional background on US third-generation missions covering the period 1980–87 can be found in the series of *Annual Reports for the Oceans Program* (NASA Technical Memoranda 80233, 84467, 85632, 86248, 87565, 88987, and 4025).

Partnership with Oceanography

Up to 1978, the remote sensing community had been the prime driver of oceanography from space and there were overly optimistic expectations. Indeed, the case had not yet been made that these observational techniques were ready to be exploited for ocean science. Consequently, in early 1979, the central task was establishing a partnership with the traditional oceanographic community. This meant involving them in the process of evaluating the performance of *Seasat* and *Nimbus-7*, as well as building an ocean science program at NASA headquarters to complement the ongoing remote sensing effort.

National Oceanographic Satellite System

This partnership with the oceanographic community was lacking in a notable and early false start on the part of NASA, the US Navy, and NOAA – the National Oceanographic Satellite System (NOSS). This was to be an operational system, with a primary and a backup satellite, along with a fully redundant ground data system. NOSS was proposed shortly after the failure of *Seasat*, with a first launch expected in 1986. NASA formed a 'science working group' (SWG) in 1980 under Francis Bretherton to define the potential that NOSS offered the oceanographic community, as well as to recommend sensors to constitute the 25% of its payload allocated for research. However, with oceanographers essentially brought in as junior partners, the job of securing a new start for NOSS fell to the operational community – which it proved unable to do. NOSS was canceled in early 1981. The prevailing and realistic view was that the greater community was not ready to implement such an operational system.

Science Working Groups

During this period, SWGs were formed to look at each promising satellite sensing technique, assess its potential contribution to oceanographic research, and define the requirements for its future flight. The notable early groups were the TOPEX SWG formed in 1980 under Carl Wunsch for altimetry, Satellite Surface Stress SWG in 1981 under James O'Brien for scatterometry, and Satellite Ocean Color SWG in 1981 under John Walsh for color scanners. These SWGs were true partnerships between the remote sensing and oceanographic communities, developing consensus for what would become the third generation of satellites.

Partnership with Field Centers

Up to this time, NASA's Oceans Program had been a collection of relatively autonomous, in-house activities run by NASA field centers. In 1981, an overrun in the space shuttle program forced a significant budget cut at NASA headquarters, including the Oceans Program. This in turn forced a reprioritization and refocusing of NASA programs. This was a blessing in disguise, as it provided an opportunity to initiate a comprehensive, centrally led program – which would ultimately result in significant funding for the oceanographic as well as remote-sensing communities. Outstanding relationships with individuals like Mous Chahine in senior management at the Jet Propulsion Laboratory (JPL) enabled the partnership between NASA headquarters and the two prime ocean-related field centers (JPL and the Goddard Space Flight Center) to flourish.

Partnerships in Implementation

A milestone policy-level meeting occurred on 13 July 1982 when James Beggs, then Administrator of NASA, hosted a meeting of the Ocean Principals Group – an informal group of leaders of the ocean-related agencies. A NASA presentation on opportunities and prospects for oceanography from space was received with much enthusiasm. However, when asked how NASA intended to proceed, Beggs told the group that – while NASA was the sole funding agency for space science and its missions – numerous agencies were involved in and support oceanography. Beggs said that NASA was willing to work with other agencies to implement an ocean satellite program, but that it would not do so on its own. Beggs' statement defined the approach to be pursued in implementing oceanography from space, namely, a joint approach based on partnerships.

Research Strategy for the Decade

As a further step in strengthening its partnership with the oceanographic community, NASA collaborated with the Joint Oceanographic Institutions Incorporated (JOI), a consortium of the oceanographic institutions with a deep-sea-going capability. At the time, JOI was the only organization in a position to represent and speak for the major academic oceanographic institutions. A JOI satellite planning committee (1984) under Jim Baker examined SWG reports, as well as the potential synergy between the variety of oceanic variables which could be measured from space; this led to the idea of understanding the ocean as a system. (From this, it was a small leap to understanding the Earth as a system, the goal of NASA's Earth Observing System (EOS).)

The report of this Committee, *Oceanography from Space: A Research Strategy for the Decade, 1985–1995*, linked altimetry, scatterometry, and ocean color with the major global ocean research programs being planned at that time – the World Ocean Circulation Experiment (WOCE), Tropical Ocean Global Atmosphere (TOGA) program, and Joint Global Ocean Flux Study (JGOFS). This strategy, still being followed today, served as a catalyst to engage the greater community, to identify the most important missions, and to develop an approach for their prioritization. Altimetry, scatterometry, and ocean color emerged from this process as national priorities.

Promotion and Advocacy

The *Research Strategy* also provided a basis for promoting and building an advocacy for the NASA program. If requisite funding was to be secured to pay for proposed missions, it was critical that government policymakers, the Congress, the greater oceanographic community, and the public had a good understanding of oceanography from space and its potential benefits. In response to this need, a set of posters, brochures, folders, and slide sets was designed by Payson Stevens of Internetwork Incorporated and distributed to a mailing list which grew to exceed 3000. These award-winning materials – sharing a common recognizable identity – were both scientifically accurate and esthetically pleasing.

At the same time, dedicated issues of magazines and journals were prepared by the community of involved researchers. The first example was the issue of *Oceanus* (1981), which presented results from the second-generation missions and represented a first step toward educating the greater oceanographic community in a scientifically useful and balanced way about realistic prospects for satellite oceanography.

Implementation Studies

Given the SWG reports taken in the context of the *Research Strategy*, the NASA effort focused on the following sensor systems (listed with each are the various flight opportunities which were studied):

- altimetry – the flight of a dedicated altimeter mission, first *TOPEX* as a NASA mission, and then *TOPEX/Poseidon* jointly with the French Centre Nationale d'Etudes Spatiales (CNES);
- scatterometry – the flight of a NASA scatterometer (NSCAT), first on NOSS, then on the *Navy Remote Ocean Observing Satellite* (NROSS), and finally on the *Advanced Earth Observing Satellite* (ADEOS) of the Japanese National Space Development Agency (NASDA);
- visible radiometry – the flight of a NASA color scanner on a succession of missions (NOSS, *NOAA-H/-I*, *SPOT-3* (Système Pour l'Observation de la Terre), and *Landsat-6*) and finally the purchase of ocean color data from the Sea-viewing Wide Field-of-view Sensor (SeaWiFS) to be flown by the Orbital Sciences Corporation (OSC);
- microwave radiometry – a system to utilize data from the series of special sensor microwave imager (SSMI) radiometers to fly on the Defense Meteorological Satellite Program satellites;
- SAR – a NASA ground station, the Alaska SAR Facility, to enable direct reception of SAR data from the *ERS-1/-2*, *JERS-1*, and *Radarsat* satellites of the ESA, NASDA, and the Canadian Space Agency, respectively.

New Starts

Using the results of the studies listed above, the Oceans Program entered the new start process at NASA headquarters numerous times attempting to secure funds for implementation of elements of the third generation. *TOPEX* was first proposed as a NASA mission in 1980. However, considering limited prospects for success, partnerships were sought and the most promising one was with the French. CNES initially proposed a mission using a *SPOT* bus with a US launch. However, NASA rejected this because *SPOT*, constrained to be Sun-synchronous, would alias solar tidal components. NASA proposed instead a mission using a US bus capable of flying in a non-Sun-synchronous orbit with CNES providing an *Ariane* launch. The NASA proposal was accepted for study in fiscal year (FY) 1983, and a new start was finally secured for the combined *TOPEX/Poseidon* in FY 1987.

In 1982 when the US Navy first proposed *NROSS*, NASA offered to be a partner and provide a scatterometer. The US Navy and NASA obtained new starts for both *NROSS* and NSCAT in FY 1985. However, *NROSS* suffered from a lack of strong support within the navy, experienced a number of delays, and was finally terminated in 1987. Even with this termination, NASA was able to keep NSCAT alive until establishing the partnership with NASDA for its flight on their *ADEOS* mission.

Securing a means to obtain ocean color observations as a follow-on to the coastal zone color scanner (CZCS) was a long and arduous process, finally coming to fruition in 1991 when a contract was signed with the OSC to purchase data from the flight of their SeaWiFS sensor. By that time, a new start had already been secured for NASA's EOS, and ample funds were available in that program for the SeaWiFS data purchase.

Finally, securing support for the Alaska SAR Facility (now the Alaska Satellite Facility to reflect its broader mission) was straightforward; being small in comparison with the cost of flying space hardware, its funding had simply been included in the new start that NSCAT obtained in FY 1985. Also, funding for utilization of SSMI data was small enough to be covered by the Oceans Program itself.

Implementing the Third Generation

With the exception of the US Navy's *Geosat*, these third-generation missions would take a very long time to come into being. As seen in **Table 1**, *TOPEX/Poseidon* was launched in 1992 – 14 years after *Seasat*; NSCAT was launched on *ADEOS* in 1996 – 18 years

after *Seasat*; and SeaWiFS was launched in 1997 – 19 years after *Nimbus-7*. (In addition to the missions mentioned in **Table 1**, the Japanese *ADEOS-1* included the US NSCAT in its sensor complement, and the US *Aqua* included the Japanese advanced microwave scanning radiometer (AMSR); the United States provided a launch for the Canadian *RADARSAT-1*.) In fact, these missions came so late that they had limited overlap with the field phases of the major ocean research programs (WOCE, TOGA, and JGOFS) they were to complement. Why did it take so long?

Understanding and Consensus

First, it took time to develop a physically unambiguous understanding of how well the satellite sensors actually performed, and this involved learning to cope with the data – satellite data rates being orders of magnitude larger than those encountered in traditional oceanography. For example, it was not until 3 years after the launch of *Nimbus-7* that CZCS data could be processed as fast as collected by the satellite. And even with only a 3-month data set from *Seasat*, it took 4 years to produce the first global maps of variables such as those shown in **Figure 4**.

In evaluating the performance of both *Seasat* and *Nimbus-7*, it was necessary to have access to the data. *Seasat* had a free and open data policy; and after a very slow start, the experiment team concept (where team members had a lengthy period of exclusive access to the data) for the *Nimbus-7* CZCS was replaced with that same policy. Given access to the data, delays were due to a combination of sorting out the algorithms for converting the satellite observations into variables of interest, as well as being constrained by having limited access to raw computing power.

In addition, the rationale for the third-generation missions represented a major paradigm shift. While earlier missions had been justified largely as demonstrations of remote sensing concepts, the third-generation missions would be justified on the basis of their potential contribution to oceanography. Hence, the long time it took to understand sensor performance translated into a delay in being able to convince traditional oceanographers that satellites were an important observational tool ready to be exploited for ocean science. As this case was made, it was possible to build consensus across the remote sensing and oceanographic communities.

Space Policy

Having such consensus reflected at the highest levels of government was another matter. The *White House*

Table 1   Some major ocean-related missions

Year	USA	Russia	Japan	Europe	Canada	Other
1968		Kosmos 243				
	Nimbus-3					
1970	Nimbus-4	Kosmos 384				
1972	Nimbus-5					
1974	Skylab					
	Nimbus-6, Geos-3					
1976						
1978	Nimbus-7, Seasat					
		Kosmos 1076				
1980		Kosmos 1151				
1982						
		Kosmos 1500				
1984		Kosmos 1602				
	Geosat					
1986		Kosmos 1776				
		Kosmos 1870	MOS 1A			
1988		OKEAN 1				
1990		OKEAN 2	MOS 1B			
		Almaz-1, OKEAN 3		ERS-1		
1992	Topex/Poseidon[a]		JERS-1			
1994		OKEAN 7				
		OKEAN 8		ERS-2	RADARSAT-1	
1996			ADEOS-1			
	SeaWiFS		TRMM[b]			
1998	GFO					
	Terra, QuikSCAT	OKEAN-O #1				OCEANSAT-1[i]
2000				CHAMP[c]		
		Meteor-3 #1		Jason-1[e]		
2002	Aqua; GRACE[d]			ENVISAT		HY-1A[j]
	WINDSAT		ADEOS-2			
2004	ICESat	SICH-1M[f]				
				CRYOSat		
2006			ALOS	GOCE, MetOp-1		HY-1B[j]
		Meteor-3M #2		SMOS	RADARSAT-2	OCEANSAT-2[i]
2008				OSTM/Jason-2[g]		
	NPP; Aquarius[h]			CryoSat-2		HY-1C, HY-2A[j]
2010			GCOM-W	Sentinel-3, MetOp-2		OCEANSAT-3[i]

[a] US/France TOPEX/Poseidon.
[b] Japan/US TRMM.
[c] German CHAMP.
[d] US/German GRACE.
[e] France/US Jason-1.
[f] Russia/Ukraine Sich-1M.
[g] France/US Jason-2/OSTM.
[h] US/Argentina Aquarius.
[i] India OCEANSAT series.
[j] China HY series.
Updated version of similar data in Wilson WS, Fellous JF, Kawamura H, and Mitnik L (2006) A history of oceanography from space. In: Gower JFR (ed.) *Manual of Remote Sensing, Vol. 6: Remote Sensing of the Environment*, pp. 1–31. Bethesda, MD: American Society for Photogrammetry and Remote Sensing.

Fact Sheet on US Civilian Space Policy of 11 October 1978 states, "… emphasizing space applications … will bring important benefits to our understanding of earth resources, climate, weather, pollution … and provide for the private sector to take an increasing responsibility in remote sensing and other applications." *Landsat* was commercialized in 1979 as part of this space policy. As Robert Stewart explains, "Clearly the mood at the presidential level was that earth remote sensing, including the oceans, was a practical space application more at home outside the scientific community. It took almost a decade to get an

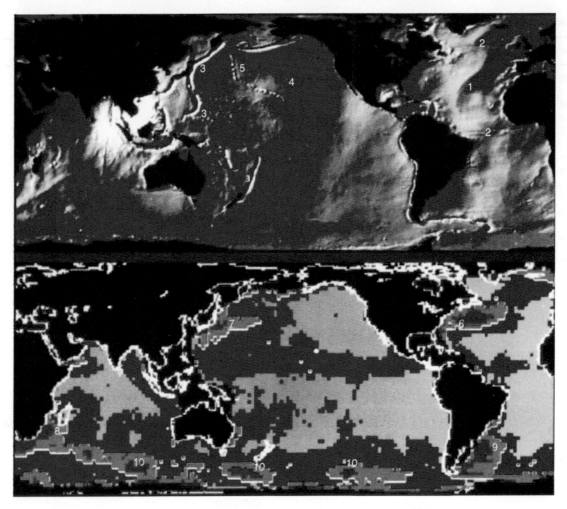

Figure 4 Global sea surface topography c. 1983. This figure shows results computed from the 70 days of *Seasat* altimeter data in 1978. Clearly visible in the mean sea surface topography, the marine geoid (upper panel), are the Mid-Atlantic Ridge (1) and associated fracture zones (2), trenches in the western Pacific (3), the Hawaiian Island chain (4), and the Emperor seamount chain (5). Superimposed on the mean surface is the time-varying sea surface topography, the mesoscale variability (lower panel), associated with the variability of the ocean currents. The largest deviations (10–25 cm), yellow and orange, are associated with the western boundary currents: Gulf Stream (6), Kuroshio (7), Agulhas (8), and Brazil/Falkland Confluence (9); large variations also occur in the West Wind Drift (10). Courtesy of NASA.

understanding at the policy level that scientific needs were also important, and that we did not have the scientific understanding necessary to launch an operational system for climate." The failures of NOSS, and later *NROSS*, were examples of an effort to link remote sensing directly with operational applications without the scientific underpinning.

The view in Europe was not dissimilar; governments felt that cost recovery was a viable financial scheme for ocean satellite missions, that is, the data have commercial value and the user would be willing to pay to help defray the cost of the missions.

Joint Satellite Missions

It is relatively straightforward to plan and implement missions within a single agency, as with NASA's space science program. However, implementing a satellite mission across different organizations, countries, and cultures is both challenging and time-consuming. An enormous amount of time and energy was invested in studies of various flight options, many of which fizzled out, but some were implemented. With the exception of the former Soviet Union, NASA's third-generation missions would be joint with each nation having a space program at that time, as well as with a private company.

The *Geosat* Exception

Geosat was the notable US exception, having been implemented so quickly after the second generation. It was approved in 1981 and launched in 1985 in order to address priority operational needs on the part of the US Navy. During the second half of its mission, data would become available within 1–2 days. As will be discussed below, *Geosat* shared a number of attributes with the meteorological satellites: it had a specific focus; it met priority operational needs for its user; experience was available for understanding and using the observations; and its implementation was done in the context of a single organization.

Challenges Ahead

Scientific Justification

As noted earlier, during the decade of the 1980s, there was a dearth of ocean-related missions in the United States, it being difficult to justify a mission based on its contribution to ocean science. Then later in that decade, NASA conceived of the EOS and was able to make the case that Earth science was sufficient justification for a mission. Also noted earlier, ESA initially had no appropriate framework for Earth science missions, and a project like *ERS-1* was pursued under the assumption that it would help develop commercial and/or operational applications of remote sensing of direct societal benefit. Its successor, *ERS-2*, was justified on the basis of needing continuity of SAR coverage for the land surface, rather than the need to monitor ocean currents. And the *ENVISAT* was initially decided by ESA member states as part of the Columbus program of the International Space Station initiative. The advent of an Earth Explorer program in 1999 represented a change in this situation. As a consequence, new Earth science missions – *GOCE*, *CryoSat*, and *SMOS* – all represent significant steps forward.

This ESA program, together with similar efforts at NASA, are leading to three sets of ground-breaking scientific missions which have the potential to significantly impact oceanography. *GOCE* and *GRACE* will contribute to an improved knowledge of the Earth's gravity field, as well as the mass of water on the surface of the Earth. *CryoSat-2* and *ICESat* will contribute to knowledge of the volume of water locked up in polar and terrestrial ice sheets. Finally, *SMOS* and *Aquarius* will contribute to knowledge of the surface salinity field of the global oceans. Together, these will be key ingredients in addressing the global water cycle.

Data Policy

The variety of missions described above show a mix of data policies, from full and open access without any period of exclusive use (e.g., *TOPEX/Poseidon*, *Jason*, and *QuikSCAT*) to commercial distribution

(e.g., real-time SeaWiFS for nonresearch purposes, *RADARSAT*), along with a variety of intermediate cases (e.g., *ERS*, *ALOS*, and *ENVISAT*). From a scientific perspective, full and open access is the preferred route, in order to obtain the best understanding of how systems perform, to achieve the full potential of the missions for research, and to lay the most solid foundation for an operational system. Full and open access is also a means to facilitate the development of a healthy and competitive private sector to provide value-added services. Further, if the international community is to have an effective observing system for climate, a full and open data policy will be needed, at least for that purpose.

In Situ Observations

Satellites have made an enormous contribution enabling the collection of *in situ* observations from *in situ* platforms distributed over global oceans. The Argos (plural spelling; not to be confused with *Argo* profiling floats) data collection and positioning system has flown on the NOAA series of polar-orbiting operational environmental satellites continuously since 1978. It provides one-way communication from data collection platforms, as well as positioning of those platforms. While an improved Argos capability (including two-way communications) is coming with the launch of *MetOp-1* in 2006, oceanographers are looking at alternatives – Iridium being one example – which offer significant higher data rates, as well as two-way communications.

In addition, it is important to note that the Intergovernmental Oceanographic Commission and the World Meteorological Organization have established the Joint Technical Commission for Oceanography and Marine Meteorology (JCOMM) to bring a focus to the collection, formatting, exchange, and archival of data collected at sea, whether they be oceanic or atmospheric. JCOMM has established a center, JCOMMOPS, to serve as the specific institutional focus to harmonize the national contributions of *Argo* floats, surface-drifting buoys, coastal tide gauges, and fixed and moored buoys. JCOMMOPS will play an important role helping contribute to 'integrated observations' described below.

Integrated Observations

To meet the demands of both the research and the broader user community, it will be necessary to focus, not just on satellites, but 'integrated' observing systems. Such systems involve combinations of satellite and *in situ* systems feeding observations into data-processing systems capable of delivering a comprehensive view of one or more geophysical variables (sea level, surface temperature, winds, etc.).

Three examples help illustrate the nature of integrated observing systems. First, consider global sea level rise. The combination of the *Jason-1* altimeter, its precision orbit determination system, and the suite of precision tide gauges around the globe allow scientists to monitor changes in volume of the oceans. The growing global array of *Argo* profiling floats allows scientists to assess the extent to which those changes in sea level are caused by changes in the temperature and salinity structure of the upper ocean. *ICESat* and *CryoSat* will provide estimates of changes in the volume of ice sheets, helping assess the extent to which their melting contributes to global sea level rise. And *GRACE* and *GOCE* will provide estimates of the changes in the mass of water on the Earth's surface. Together, systems such as these will enable an improved understanding of global sea level rise and, ultimately, a reduction in the wide range of uncertainty in future projections.

Second, global estimates of vector winds at the sea surface are produced from the scatterometer on *QuikSCAT*, a global array of *in situ* surface buoys, and the Seawinds data processing system. Delivery of this product in real time has significant potential to improve marine weather prediction. The third example concerns the *Jason-1* altimeter together with *Argo*. When combined in a sophisticated data assimilation system – using a state-of-the-art ocean model – these data enable the estimation of the physical state of the ocean as it changes through time. This information – the rudimentary 'weather map' depicting the circulation of the oceans – is a critical component of climate models and provides the fundamental context for addressing a broad range of issues in chemical and biological oceanography.

Transition from Research to Operations

The maturing of the discipline of oceanography includes the development of a suite of global oceanographic services being conducted in a manner similar to what exists for weather services. The delivery of these services and their associated informational products will emerge as the result of the successes in ocean science ('research push'), as well as an increasing demand for ocean analyses and forecasts from a variety of sectors ('user pull').

From the research perspective, it is necessary to 'transition' successfully demonstrated ('experimental') observing techniques into regular, long-term, systematic ('operational') observing systems to meet a broad range of user requirements, while maintaining the capability to collect long-term, 'research-quality'

observations. From the operational perspective, it is necessary to implement proven, scientifically sound, cost-effective observing systems – where the uninterrupted supply of real-time data is critical. This is a big challenge to be met by the space systems because of the demand for higher reliability and redundancy, at the same time calling for stringent calibration and accuracy requirements. Meeting these sometimes competing, but quite complementary demands will be the challenge and legacy of the next generation of ocean remote-sensing satellites.

Meteorological Institutional Experience

With the launch in 1960 of the world's first meteorological satellite, the polar-orbiting *Tiros-1* carrying two TV cameras, the value of the resulting imagery to the operational weather services was recognized immediately. The very next year a National Operational Meteorological System was implemented, with NASA to build and launch the satellites and the Weather Bureau to be the operator. The feasibility of using satellite imagery to locate and track tropical storms was soon demonstrated, and by 1969 this capability had become a regular part of operational weather forecasting. In 1985, Richard Hallgren, former Director of the National Weather Service, stated, "the use of satellite information simply permeates every aspect of the [forecast and warning] process and all this in a mere 25 years." In response to these operational needs, there has been a continuing series of more than 50 operational, polar-orbiting satellites in the United States alone!

The first meteorological satellites had a specific focus on synoptic meteorology and weather forecasting. Initial image interpretation was straightforward (i.e., physically unambiguous), and there was a demonstrated value of resulting observations in meeting societal needs. Indeed, since 1960 satellites have ensured that no hurricane has gone undetected. In addition, the coupling between meteorology and remote sensing started very near the beginning. An 'institutional mechanism' for transition from research to operations was established almost immediately. Finally, recognition of this endeavor extended to the highest levels of government, resulting in the financial commitment needed to ensure success.

Oceanographic Institutional Issue

Unlike meteorology where there is a National Weather Service in each country to provide an institutional focus, ocean-observing systems have multiple performer and user institutions whose interests must be reconciled. For oceanography, this is a significant challenge working across 'institutions', where the *in situ* research is in one or more agencies, the space research and development is in another, and operational activities in yet another – with possibly separate civil and military systems. In the United States, the dozen agencies with ocean-related responsibilities are using the National Oceanographic Partnership Program and its Ocean.US Office to provide a focus for reconciling such interests. In the United Kingdom, there is the Interagency Committee on Marine Science and Technology.

In France, there is the Comité des Directeurs d'Organismes sur l'Océanographie, which gathers the heads of seven institutions interested in the development of operational oceanography, including CNES, meteorological service, ocean research institution, French Research Institute for Exploitation of the Sea (IFREMER), and the navy. This group of agencies has worked effectively over the past 20 years to establish a satellite data processing and distribution system (AVISO), the institutional support for a continuing altimetric satellite series (*TOPEX/Poseidon*, *Jason*), the framework for the French contribution to the *Argo* profiling float program (CORIOLIS), and to create a public corporation devoted to ocean modeling and forecasting, using satellite and *in situ* data assimilation in an operational basis since 2001 (Mercator). This partnership could serve as a model in the effort to develop operational oceanography in other countries. Drawing from this experience working together within France, IFREMER is leading the European integrated project, MERSEA, aimed at establishing a basis for a European center for ocean monitoring and forecasting.

Ocean Climate

If we are to adequately address the issue of global climate change, it is essential that we are able to justify the satellite systems required to collect the global observations 'over the long term'. Whether it be global sea level rise or changes in Arctic sea ice cover or sea surface temperature, we must be able to sustain support for the systems needed to produce climate-quality data records, as well as ensure the continuing involvement of the scientific community.

Koblinsky and Smith have outlined the international consensus for ocean climate research needs and identified the associated observational requirements. In addition to their value for research, we are compelled by competing interests to demonstrate the value of such observations in meeting a broad range of societal needs. Climate observations pose

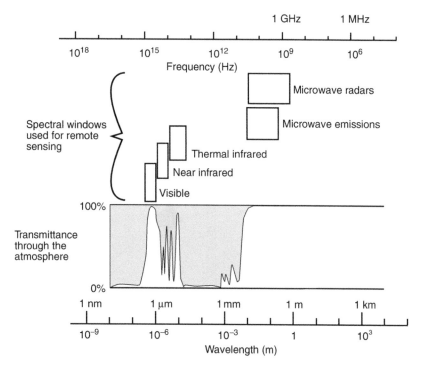

Figure 5 The electromagnetic spectrum showing atmospheric transmitance as a function of frequency and wavelength, along with the spectral windows used for remote sensing. Microwave bands are typically defined by frequency and the visible/infrared by wavelength. Adapted from Robinson IS and Guymer T (1996) Observing oceans from space. In: Summerhayes CP and Thorpe SA (eds.) *Oceanography: An Illustrated Guide*, pp. 69–87. Chichester, UK: Wiley.

challenges, since they require operational discipline to be collected in a long-term systematic manner, yet also require the continuing involvement of the research community to ensure their scientific integrity, and have impacts that may not be known for decades (unlike observations that support weather forecasting whose impact can be assessed within a matter of hours or days). Together, the institutional and observational challenges for ocean climate have been difficult to surmount.

International Integration

The paper by the Ocean Theme Team prepared under the auspices of the Integrated Global Observing Strategy (IGOS) Partnership represents how the space-faring nations are planning for the collection of global ocean observations. IGOS partners include the major global research program sponsors, global observing systems, space agencies, and international organizations.

On 31 July 2003, the First Earth Observations Summit – a high-level meeting involving ministers from over 20 countries – took place in Washington, DC, following a recommendation adopted at the G-8 meeting held in Evian the previous month; this summit proposed to 'plan and implement' a Global Earth Observation System of Systems (GEOSS). Four additional summits have been held, with participation having grown to include 60 nations and 40 international organizations; the GEOSS process provides the political visibility – not only to implement the plans developed within the IGOS Partnership – but to do so in the context of an overall Earth observation framework. This represents a remarkable opportunity to develop an improved understanding of the oceans and their influence on the Earth system, and to contribute to the delivery of improved oceanographic products and services to serve society.

Appendix 1: A Brief Overview of Satellite Remote Sensing

Unlike the severe attenuation in the sea, the atmosphere has 'windows' in which certain electromagnetic (EM) signals are able to propagate. These windows, depicted in **Figure 5**, are defined in terms

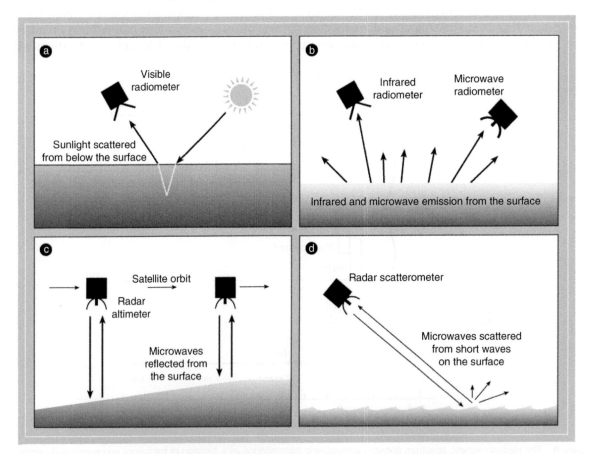

Figure 6 Four techniques for making oceanic observations from satellites: (a) visible radiometry, (b) infrared and microwave radiometry, (c) altimetry, and (d) scatterometry. Adapted from Robinson IS and Guymer T (1996) Observing oceans from space. In: Summerhayes CP and Thorpe SA (eds.) *Oceanography: An Illustrated Guide*, pp. 69–87. Chichester, UK: Wiley.

	Passive sensors (radiometers)			Active sensors (microwave radars)		
Sensor type	Visible	Infrared	Microwave	Altimetry	Scatterometry	SAR
Measured physical variable	Solar radiation backscattered from beneath the sea surface	Infrared emission from the sea surface	Microwave emission from the sea surface	Travel time, shape, and strength of reflected pulse	Strength of return pulse when illuminated from different directions	Strength and phase of return pulse
Applications	Ocean color; chlorophyll; primary production; water clarity; shallow-water bathymetry	Surface temperature; ice cover	Ice cover, age and motion; sea surface temperature; wind speed	Surface topography for geostrophic currents and tides; bathymetry; oceanic geoid; wind and wave conditions	Surface vector winds; ice cover	Surface roughness at fine spatial scales; surface and internal wave patterns; bathymetric patterns; ice cover and motion

Figure 7   Measured physical variables and applications for both passive and active sensors, expressed as a function of sensor type.

of atmospheric transmittance – the percentage of an EM signal which is able to propagate through the atmosphere – expressed as a function of wavelength or frequency.

Given a sensor on board a satellite observing the ocean, it is necessary to understand and remove the effects of the atmosphere (such as scattering and attenuation) as the EM signal propagates through it. For passive sensors (**Figures 6(a)** and **6(b)**), it is then possible to relate the EM signals collected by the sensor to the associated signals at the bottom of the atmosphere, that is, the natural radiation emitted or reflected from the sea surface. Note that passive sensors in the visible band are dependent on the Sun for natural illumination.

Active sensors, microwave radar (**Figures 6(c)** and **6(d)**), provide their own source of illumination and have the capability to penetrate clouds and, to a certain extent, rain. Atmospheric correction must be done to remove effects for a round trip from the satellite to the sea surface.

With atmospheric corrections made, measurements of physical variables are available: emitted radiation for passive sensors, and the strength, phase, and/or travel time for active sensors. **Figure 7** shows typical measured physical variables for both types of sensors in their respective spectral bands, as well as applications or derived variables of interest – ocean color, surface temperature, ice cover, sea level, and surface winds. The companion articles on this topic address various aspects of **Figure 7** in more detail, so only this general overview is given here.

Acknowledgments

The authors would like to acknowledge contributions to this article from Mary Cleave, Murel Cole, William Emery, Michael Freilich, Lee Fu, Rich Gasparovic, Trevor Guymer, Tony Hollingsworth[†], Hiroshi Kawamura, Michele Lefebvre, Leonid Mitnik, Jean-François Minster, Richard Moore, William Patzert, Willard Pierson[†], Jim Purdom, Keith Raney, Payson Stevens, Robert Stewart, Ted Strub, Tasuku Tanaka, William Townsend, Mike Van Woert, and Frank Wentz.

See also

History of Ocean Sciences. Satellite Altimetry.

Further Reading

Apel JR (ed.) (1972) Sea surface topography from space. *NOAA Technical Reports: ERL No. 228, AOML No. 7.* Boulder, CO: NOAA.

Cherny IV and Raizer VY (1998) *Passive Microwave Remote Sensing of Oceans.* Chichester, UK: Praxis.

Committee on Earth Sciences (1995) *Earth Observations from Space: History, Promise, and Reality*. Washington, DC: Space Studies Board, National Research Council.

Ewing GC (1965) Oceanography from space. *Proceedings of a Conference held at Woods Hole, 24–28 August 1964. Woods Hole Oceanographic Institution Ref. No. 65-10*. Woods Hole, MA: WHOI.

Fu L-L, Liu WT, and Abbott MR (1990) Satellite remote sensing of the ocean. In: Le Méhauté B (ed.) *The Sea, Vol. 9: Ocean Engineering Science*, pp. 1193–1236. Cambridge, MA: Harvard University Press.

Guymer TH, Challenor PG, and Srokosz MA (2001) Oceanography from space: Past success, future challenge. In: Deacon M (ed.) *Understanding the Oceans: A Century of Ocean Exploration*, pp. 193–211. London: UCL Press.

JOI Satellite Planning Committee (1984) *Oceanography from Space: A Research Strategy for the Decade, 1985–1995*, parts 1 and 2. Washington, DC: Joint Oceanographic Institutions.

Kaula WM (ed.) (1969) The terrestrial environment: solid-earth and ocean physics. *Proceedings of a Conference held at William College, 11–21 August 1969, NASA CR-1579*. Washington, DC: NASA.

Kawamura H (2000) Era of ocean observations using satellites. *Sokko-Jiho* 67: S1–S9 (in Japanese).

Koblinsky CJ and Smith NR (eds.) (2001) *Observing the Oceans in the 21st Century*. Melbourne: Global Ocean Data Assimilation Experiment and the Bureau of Meteorology.

Masson RA (1991) *Satellite Remote Sensing of Polar Regions*. London: Belhaven Press.

Minster JF and Lefebvre M (1997) *TOPEX/Poseidon* satellite altimetry and the circulation of the oceans. In: Minster JF (ed.) *La Machine Océan*, pp. 111–135 (in French). Paris: Flammarion.

Ocean Theme Team (2001) *An Ocean Theme for the IGOS Partnership*. Washington, DC: NASA. http://www.igospartners.org/docs/theme_reports/IGOS-Oceans-Final-0101.pdf (accessed Mar. 2008).

Purdom JF and Menzel WP (1996) Evolution of satellite observations in the United States and their use in meteorology. In: Fleming JR (ed.) *Historical Essays on Meteorology: 1919–1995*, pp. 99–156. Boston, MA: American Meteorological Society.

Robinson IS and Guymer T (1996) Observing oceans from space. In: Summerhayes CP and Thorpe SA (eds.) *Oceanography: An Illustrated Guide*, pp. 69–87. Chichester, UK: Wiley.

Victorov SV (1996) *Regional Satellite Oceanography*. London: Taylor and Francis.

Wilson WS (ed.) (1981) *Special Issue: Oceanography from Space. Oceanus* 24: 1–76.

Wilson WS, Fellous JF, Kawamura H, and Mitnik L (2006) A history of oceanography from space. In: Gower JFR (ed.) *Manual of Remote Sensing, Vol. 6: Remote Sensing of the Environment*, pp. 1–31. Bethesda, MD: American Society for Photogrammetry and Remote Sensing.

Wilson WS and Withee GW (2003) A question-based approach to the implementation of sustained, systematic observations for the global ocean and climate, using sea level as an example. *MTS Journal* 37: 124–133.

Relevant Websites

http://www.aviso.oceanobs.com
– AVISO.

http://www.coriolis.eu.org
– CORIOLIS.

http://www.eohandbook.com
– Earth Observation Handbook, CEOS.

http://www.igospartners.org
– IGOS.

http://wo.jcommops.org
– JCOMMOPS.

http://www.mercator-ocean.fr
– Mercator Ocean.

SONAR SYSTEMS

A. B. Baggeroer, Massachusetts Institute of
Technology, Cambridge, MA, USA

Introduction and Short History

Sonar (Sound Navigation and Ranging) systems
are the primary method of imaging and communicating within the ocean. Electromagnetic energy does
not propagate very far since it is attenuated by either
absorption or scattering – visibility beyond 100 m is
exceptional. Conversely, sound propagates very well
in the ocean especially at low frequencies; consequently, sonars are by far the most important systems used by both man and marine life within the
ocean for imaging and communication.

Sonars are classified as being either active or passive. In active systems an acoustic pulse, or more
typically a sequence of pulses, is transmitted and a
receiver processes them to form an 'image' or to de-
code a data message if operating as a communication
system. The image can be as simple as the presence of
a discrete echo or as complex as a visual picture. The
receiver may be coincident with the transmitter – a
monostatic system, or separate – a bistatic system.
Both the waveform of the acoustic pulse and the
beamwidths of both the transmitter and receiver are
important and determine the performance of an active
system. One typically associates an active sonar with
the popular perception of sonar systems. Many marine mammals use active sonar for navigation and prey
localization, as well as communication in ways which
we are still attempting to understand. Many of the
signals used by modern sonars have some of the same
features as those of marine mammals.

Passive systems only receive. They sense ambient
sound made by a myriad of sources in the ocean such
as ships, submarines, marine mammals, volcanoes.
These systems have been, and still are, especially
important in anti-submarine warfare (ASW) where
stealth is an important issue, and an active ping
would reveal the location of the source.

The use of sound for detecting underwater objects
was first introduced in a patent by Richardson in
June 1912 for the 'sonic detection of icebergs,' 2
months after the sinking of the *Titanic*.[1] This was

---

[1] Much of this material in the history has been extracted from
Beyer, 1999.

soon followed by the development of the Fessenden
oscillator in 1914 which eventually led to the development of fathometers, an acoustic system for
measuring the depth to the seabed. The French
physicist/chemist Paul Langevin was the first to detect a submarine using sonar in 1918, motivated by
the extensive damage of German U-boats. Between
World Wars I and II both Britain and the US sponsored sonar research, especially on transducers. The
former was conducted under the Antisubmarine
Detection Investigation Committee, or ASDIC as
sonar is still often referred to within the British
military, and the latter was performed at the Naval
Research Laboratory.

The re-emergence of the German U-boat stimulated the modern era of sonars and the physics
of sound propagation in the ocean where major
research programs were chartered in the USA (Columbia, Harvard, Scripps Institution of Oceanography, Woods Hole Oceanographic Institution),
UK, and Russia. A very comprehensive summary was
compiled by the US National Defense Research
Council after World War II, which still remains a
valuable reference (*see* Further Reading Section).

The development of the nuclear submarine, both
as an attack boat (SSN) or as a missile carrier (SSBN)
provided a major emphasis for sonar throughout the
cold war. The USA, UK, Russia, and France all had
substantial research programs on sonar for many
applications, but ASW certainly had a major priority.
The nuclear submarine could deny use of the oceans
but could also unleash massive destruction with
nuclear missiles. With the end of the cold war, ASW
now has a lower priority; however, the submarine
still remains the platform of choice for many countries since modern diesel/electric submarines operating on batteries are extremely hard to detect and
localize. Undoubtedly, the most extensively used
reference was compiled by Urick (1975), which is
frequently referenced as a handbook for sonar
engineers.

While military operations have dominated the
development of sonars, they are now used extensively for both scientific and commercial applications. The use of fathometers and closely related
seismic methods provided much of the important
data validating plate tectonics. There is also a lot
of overlap between geophysical exploration for
hydrocarbons and modern sonars. High resolution and multibeam systems are extensively used
for charting the seabed and its sub-bottom

characteristics, fish finding, current measurements exploiting Doppler, as well as archaeological investigations.

Active Sonar systems

The major components of an active system are indicated in **Figure 1**. A waveform generator forms a pulse or 'ping', which is then modulated, or frequency shifted, to an operating frequency, f_o which may be as low as tens of Hertz for very long-range systems, or as high as 1 MHz, for high resolution short-range imaging sonars. Next, the signal is often 'beamformed' by an array of transducers, that focuses the signal in specific directions either by mechanically rotating the array or by introducing appropriate time delays or phase shifts. The signals are amplified and then converted from an electrical signal to a sound wave by the transmit transducers. Efficient transduction, the conversion of electric power to sound power, and even a modest amount of directivity of the transmitter requires that the transducer have dimensions on the scale of the wavelength of the operating frequency; hence, low frequency transmitters are typically large and not very efficient, whereas high frequency transmitters are smaller and very efficient.

The pinging rate, usually termed the pulse repetition frequency (PRF) is determined by the duration over which strong echos (called 'returns') from the previously transmitted pulse can be expected, so that one return does not overlap and become confused with another. With some systems with well confined response durations, several pulses may be in transit at the same time.

The ocean introduces three important components before it is detected by a receiver.

- There is the desired echo from the target itself. This may be a simple echo, especially if the target is close, but it may also include many multipaths and/or modes as a result of reflections of the ocean surface and bottom as well as paths refracted completely within the ocean itself.
- The ocean is filled with spurious, or unwanted reflectors which produce reverberation. The dominant source of this is the sea bottom, but the sea surface and objects (e.g. fish) can be important as well. Typically, the bottom is characterized in terms of a scattering strength per unit area insonified.
- Finally, the ocean is filled with ambient noise which is created by both natural and man-made sources. At low frequencies, 50–500 Hz, shipping tends to dominate the noise in the Northern Hemisphere, especially near shipping lanes. Wind and wave processes as well as rain can also be important. In specific areas, marine life may be a very important component.

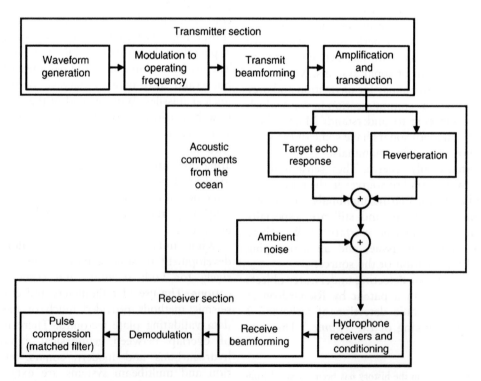

Figure 1 Active sonar system components.

The sonar receiver implements operations similar to the transmitter. Hydrophones convert the acoustic signal to an electric one whereupon it usually undergoes some 'signal conditioning' to amplify it to an appropriate level. In modern sonars the signal is digitized, since most of the subsequent operations are more easily implemented by digital signal processors. Next, a receiver beamformer, which may be quite different from the transmitter, focuses energy arriving from specific directions for spatial imaging. This is also done either by mechanically steering the array or by introducing time delays or phase shifts (if the processing is done in the frequency domain). The signal is then usually demodulated to a low frequency band which simplifies further electronics and signal processing steps. Finally it is 'pulse compressed,' or 'matched filtered,' which is a process that maximizes energy arriving at the travel time that corresponds to the range to a target. The matched filter is simply a correlation operation which seeks the best replica of the transmitted signal among all the signal components introduced by the ocean. In the simplest form of processing a sequence of 'pings' is rastered, i.e. the echo time-series are displayed one after the other, to construct an image. This is typical of a sidescan sonar system. In more sophisticated systems, especially those operating at low frequencies where phase coherence can be preserved or extracted, a sequence of outputs from the pulse compression filter is processed to form images. This is typical of synthetic aperture sonars. In both types of systems display algorithms, sometimes termed 'normalizers,' are important for emphasizing certain features by improving contrast and controlling the dynamic range of the output.

The performance of an active sonar system is captured in the active sonar equation; while imperfect in its details, it is very useful in assessing gross performance. It is expressed in logarithmic units, or decibels which are referenced to a standard level. For a monostatic system it is:

$$SE = SL - 2 * TL + TS - \max(NL, RL) + DI_t$$
$$+ AG_r - DT$$

where[2,3] SE, is the signal excess at the receiver output; SL, is the source level referenced to a pressure level of 1 μPa; TS, is the target strength, which is a function of aspect angle; NL, is the level of the ambient noise at a single hydrophone; RL, is the reverberation which is determined by the area insonified, the scattering strength, and the signal level; DI_t, is the directivity index of the transmitter, which is a measure of the gain compared omnidirectional radiation; AG_r, is the array gain of the receiver in the direction of the target (often this is described as a receiver); directivity index, DI_r; DT, is the direction threshold for a target to be seen on the output display (this can be a complicated function of the complexity of the environment and the sophistication of an operator); TL, is the transmission loss, i.e. the loss in signal energy as it transmits to the target and returns. If the $SE > 0$, then a target is discernible on a display.

The notation $\max(NL, RL)$ distinguishes the two important regimes for an active sonar. When $NL > RL$, the sonar is operating in a noise-limited environment; conversely, when $RL > NL$ the environment is reverberation-limited which is the case in virtually all applications.

Reverberation is the result of unwanted echoes from the sea surface, seafloor, and volume. In the simplest formulation its level for surfaces both bottom and top is usually characterized by a scattering strength per unit area, so the level is given by the product of the resolved area multiplied by the scattering strength (or sum if using a decibel formulation). Often, the Rayleigh parameter $\frac{2\pi\sigma_s}{\lambda} \sin(\phi)$ is used, where σ_s is the rms surface roughness and ϕ is the incident angle as a measure of when surface roughness becomes important, i.e. when the Rayleigh parameter is greater than 1. Similarly with volume scattering, a scattering strength per unit volume is used.

The operating frequency of a sonar is an important parameter in its design and performance. The important design issues are:

- Resolution. There are two aspects of resolution – 'cross-range' resolution and 'in range' resolution. 'Cross-range' resolution is determined by the dimensions of the transmit and receive apertures relative to the wavelength, λ, of the acoustic signal.[4] It is given by $R\lambda/L$ where R is the range and L is the transmitter and/or receiving aperture. Higher resolution requires higher operating frequencies since these result in smaller $R\lambda/L$ ratios.

'In range' resolution is determined by the bandwidth of the signal and is given approximately by

---

[2] There are many versions of the sonar equation and the nomenclature differs among them (see Urick, 1975)

[3] If the system is bistatic wherein the transmitter and receiver are not colocated, then the sonar equation is significantly more complicated (Cox, 1989)

[4] The wavelength in uncomplicated media is given by $\lambda = c/f_o$, where c is the sound speed, nominally 1500 m s^{-1} and f_o is the operating frequency.

$c/2W$, where W is the available bandwidth and c is the sonar speed. Since W is usually proportional to the operating frequency, f_o, one tends to try to use higher frequencies, which limits practical ranges. Also the sonar channel is often very band-limited. As a result in most sonar systems the 'in range' resolution is typically significantly smaller than the 'cross-range' resolution, so care needs to be taken in interpreting images.

- Maximum operating range. Acoustical signals can propagate over very long ranges, but there are a number of phenomena which can both enhance and attenuate the signal power. These include the geometrical spreading, the stratification of the sound speed versus depth, absorption, and scattering processes. The first two are essentially independent of frequency while the latter two have strong dependencies. **Figure 2** indicates the absorption loss in dB km^{-1} of sound at 20° and 35 apt salinity.[5] Essentially, the absorption loss factor increases quadratically with frequency. The two 'knees' in the figure relate to the onset of losses introduced by ionic relaxation phenomena. The net implication is that efforts are made to minimize the operating frequency so that it contributes <10 dB of loss for the desired range, i.e $\alpha(f_o)R < 10$ dB where $\alpha(f_o)R$ is attenuation per unit distance in dB.

The penalty of this, however, is a less directive signal, so it is more difficult to avoid contact with the ocean boundaries and consequent scattering losses, especially the bottom is shallow-water environments. As a result, the actual operating frequency of a sonar is a compromise based on the sound speed profile, the directivity of transmitter, receiver beamformers, and desired operating range and isolation.

- Target cross-section. The physics of sound reflecting from a target can be very complicated and there are few exact solutions, most of which involve long, complicated mathematical functions. In addition, the geometry of the target can introduce a significant aspect dependence. The important scaling number is $2\pi a/\lambda$ where a is a characteristic length scale presented to the incoming sound wave. Typically, if the number is less than unity, the reflected target strength depends upon the fourth power of frequency, so the target strength is quite small; this is the so-called Rayleigh region of scattering. Conversely, if this

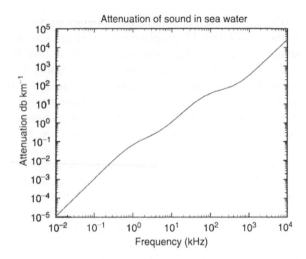

Figure 2 Attenuation of sound vs frequency.

number is larger than unity, the target strength normalized by the presented area is typically between 1 and 10. There are two additional features to consider: (i) large, flat surfaces lead to large returns, often termed specular, and (ii) in a bistatic sonar, the forward scattering, essentially the shadow, is determined almost solely by the intercepted target shape, often called Babinet's principle.

The net effect is a desire for higher operating frequencies in order to stay in the Rayleigh region where there is significant target strength.

Overall high frequencies produce better resolution and higher target strengths. However, at high frequencies acoustic propagation is more complicated and absorption limits the range.

Table 1 indicates the operating frequencies of some typical active sonars.

Sonar System Components

The components in **Figure 1** all have significant impact on the performance of an active sonar. The signal processing issues are complex and there is a large sonar-related literature as well as radar where the issues are similar. The essential problem is to separate a target amidst the reverberation and ambient noise in either of two operating realms – 'noise-limited' and 'reverberation'-limited environments. In a 'noise-limited' environment the ambient background noise limits the performance of a system, so increasing the transmitter output power improves performance. A 'reverberation-limited' environment is one where the noise is composed of mostly unwanted reflections from objects other than the target, so increasing the transmitted power simply increases both the target and reverberation returns simultaneously with no

---

[5] db km^{-1} represents 10 log of the fractional loss in power per kilometer representing an exponential decay versus range.

Table 1 Features of some typical active sonars

Sonar system	Operating frequency range	Wavelength	Nominal range
Long-range, low frequency	50–500 Hz	30–3 m	1000 km
Military ASW sonars	3–4 kHz	0.5–0.75 m	100 km
Bottom-mapping echosounders	3–4 kHz	0.5–0.75 m	vertical
High-resolution fathometers	10–15 kHz	15–10 cm	vertical
Acoustic communications	10–30 kHz	15–5 cm	10 km
Sidescan sonar (long-range)	50–100 kHz	6–1.5 cm	5 km
Sidescan sonar (short range)	500–1000 kHz	3–1.5 mm	100 m
Acoustic localization nets	10–20 kHz	15–5 cm	vertical
Fish-finding sonars	25–200 kHz	6–1 cm	1–5 km
Recreational	100–250 km	15–6 mm	vertical

net gain in signal to noise ping. Most active sonars operate in a reverberation limited environment. Effective design of an active sonar depends upon controlling reverberation through a combination of waveform design and beamforming.

Waveform design There are two basic approaches to waveform design for resolving targets – 'range gating' and 'Doppler gating.' The simplest approach to 'range gating' is a short, high powered pulse. This essentially resolves every reflector and an image is constructed by successive pulses and then the returns are rastered. While this is the simplest waveform it has limitations when operating in environments with high noise and reverberation levels, since the peak power of most sonars, both man-made and marine mammal, is limited. This shortcoming can often be mitigated by exploiting bandwidth (resolution α/β). This has led to a large literature on waveform design with the most popular being frequency modulated (FM) and coded (PRN) signals. With these signals[6] the center frequency is swept, or 'chriped' across a frequency band at the transmitter and correlated, or 'compressed' at the receiver. This class of signals is commonly used by marine mammals including whales and dolphins for target localization.

'Doppler gating' is based on differences in target motion. A moving target imparts a Doppler shift to the reflected signal which is proportional to operating frequency and the ratio v/c, where v is the target speed and c is the sound speed. 'Doppler gating' is particularly useful in some ASW contexts since it is difficult to keep a submarine stationary, thus a properly designed signal, one that resolves Doppler,

can distinguish it against a fixed reverberant background. The ability to resolve Doppler frequency depends upon the duration of a signal, with a dependence of 1/duration, so good 'Doppler gating' waveforms are long.

There has been a lot of research on the topic of optimal waveform design. Ideally, one wants a waveform which can resolve range and Doppler simultaneously which implies long duration, and wide bandwidth. These requirements are difficult to satisfy simultaneously.

Beamforming Both the transmitter and the receiver beamformers provide spatial resolution for the sonar system. The angular resolution in degrees is approximately $\Delta\theta \approx 60\lambda/L$, where λ is the wavelength and L is the aperture length. Since acoustic wavelengths are large when compared with optical wavelengths, the angular resolutions tend to be large especially at low frequencies.[7] Beamformers, often termed array processors in receivers, have been an important research topic for several decades with the advent of digital signal processing which permitted increasingly more sophistication, especially in the realm of adaptive methods. One of the simplest transmit beamformers consists of a line array of transducers each radiating the same signal. The simple receiver and also a line of transducers adds all the signals together. This resolves the paths perpendicular or broadside to the array. If one wants to 'steer' the array, or resolve another direction, the array must be mechanically rotated. This method of beamforming is still used by many systems since it is quite robust. Another simple beamformer is a planar array of transducers.

---

[6] Pseudo Random Noise (PRN) are coded signals which appear to be random noise. Well designed signals have useful mathematical constructs which led to good outputs at the output of the pulse compression, or matched filter, processor.

[7] Sonars with angular resolutions of 1° are generally considered to have high resolution. Compare this with that of the human eye with a nominal diameter of 4 mm and the wavelength of light in the visible region is 0.4 μm leading to a resolution scale of 0.1 ms.

Digital signal processing has led to more sophisticated array processing, especially for receivers. Beamformers which steer beams electronically by introducing delays, or phase shifts, shape beams to control sidelobes, place nulls to control strong reflectors, and reduce jamming are now practical because these features can be practical electronically rather than mechanically.

Examples of Active Sonar Images

This section describes two examples of sonars used for mapping seafloor bathymetry. In the first the sonar is carried on an unmanned underwater vehicle (UUV) close to the seafloor. The operating frequency is 675 kHz and the beam is mechanically steered from port to starboard as well as fore and aft as the UUV proceeds along its track, so the beams are steered forward and directly below the UUV (**Figure 3**). The onset time of the first echo return is the parameter of interest. It is converted to the depth of the seafloor after including the vehicle position, the direction of the beam and possibly refraction effects in the water itself. Usually straight-line acoustic propagation is assumed.

The signals are combined to generate a high resolution map of the seafloor. The processing to achieve this includes editing for spurious responses, registration of the rasters or images from successive transmission using the navigation sensors on the UUV (or more generally any vehicle) and normalization to improve the contrast so that weak features can be detected amidst strong ones.

The second example of an active sonar is a multibeam bathymetric mapper. Most of these systems for deep water operate at a 12 kHz center frequency. The transmit beam is produced by a linear array running fore to aft along the bottom of the ship, thwartships beam, which produces a swath which resolves the seafloor along track (**Figure 4**). The receiver array is oriented port to starboard. The signals from this array are beamformed electronically, so the seafloor is resolved port to starboard within the transmitted swath, since the patch is the product of the transmit and receive beamwidth. This configuration allows two-dimensional resolution with two linear arrays instead of a full planar array. The depths from each of the multibeams are measured by combining the travel time and the ray refraction from the sound speed profile to obtain a depth. Subsequent processing edits anomalous returns and interpolates all the data to generate the contour map. Active sonar systems with additional features have been developed for special applications, but they all use the basic principles described above.

Passive Sonars

Passive sonars that only listen and do not transmit are used in a variety of applications including the military for antisubmarine warfare (ASW), tracking and classification of marine mammals, earthquake detection, and nuclear test ban monitoring.

Since the signals are passive there is no pulse compression, or matched filtering, so a passive sonar design primarily focuses upon the 'short-term' frequency wavenumber spectrum, or the directional spectrum and the power density spectrum and how it evolves in time. The data are nonstationary and inhomogeneous, but many of the processing algorithms

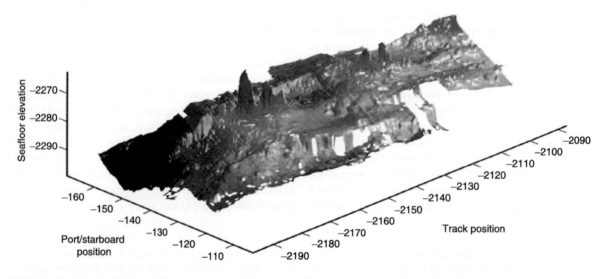

Figure 3 Image with forward-looking and down-looking sonar. (Figure courtesy of Dr Dana Yoerger, Woods Hole Oceanographic Institution.)

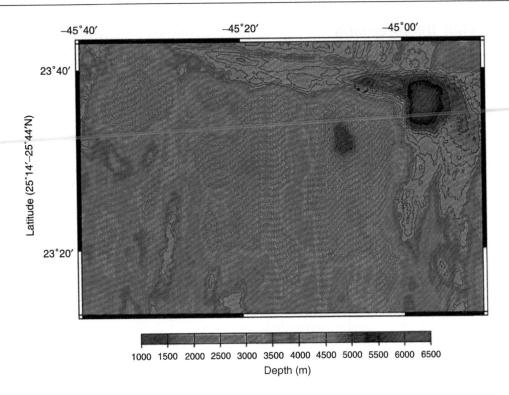

Figure 4 High resolution bathymetric map of the seafloor near the Mid-Atlantic Ridge: (A) contour map; (B) isometric projection (from top-right). (Figures courtesy of Dr Brian Tucholke, Woods Hole Oceanographic Institution.)

are based upon stationary and homogeneous assumptions; hence the term 'short-term.' The performance of a passive system is characterized by the passive sonar equation:

$$SE = SL - TL - NL + AG_r - DT$$

where the terms are essentially the same as for an active sonar. In some applications the arrays are so large that the coherence of the received signal is important, and it is necessary to separate the array gain, AG_r into two terms, or:

$$AG_r = AG_{r,n} - SGD_s$$

where $AG_{r,n}$ is the array gain against the ambient noise and SGD_s is the signal gain degradation due to lack of coherence. $SGD_s = 0$ for a signal that is coherent across the entire array.

Passive Sonar Beamforming

The signals received by the sonar's hydrophone are preconditioned, which might include editing bad data channels, calibration, and filtering. They are then beamformed, either in the time domain by introducing delays to compensate for the travel time across the array, or in the frequency domain. With digital signals the former usually requires upsampling or interpolation of the data to avoid distortion. The latter is accomplished by FFT (fast Fourier transforms), phase shifting to compensate for the delays, and then IFFT (inverse fast Fourier transforming). Frequency domain beamforming allows simpler implementation of adaptive techniques that are useful in cases where the ambient field has many discrete components. Adaptive algorithms form beams with notches, i.e. poor response in the direction of interferers, thereby suppressing them. Many algorithms have been designed to accomplish this, but the MVDR (minimum variance distortion filter – first introduced by Capon) and related algorithms have been used most extensively in practice.

Passive Sonar Display Formats

The output of the beamformer is a time-series for each beam. In certain applications the time-series itself may be of interest, however, in most cases the time-series is further processed to assist in extracting weak signals from the background noise. The parameter for signal processing schemes and display formats includes time (the epoch for data processing, T), angle (azimuth and elevation), and frequency (the spectral content of the data) (**Figure 5**).

Bearing-time Recording

Bearing time processing takes the beam outputs over a specified frequency band and plots the output versus time. Two modes of processing are often used: (i) energy detection, which forms an average of the beam outputs, or (ii) cross-correlation detection, where the array is split at the beamformer and then the two outputs are cross-correlated versus the direction. The processing is often classified according to the width of the band used, passive broadband (PBB) or passive narrowband (PNB).

The data for each time epoch are normalized to improve the contrast for signals of interest and each raster is plotted. Over a sequence of epochs, the directional components in the ambient field which are associated with shipping are observed. By maneuvering the array one can triangulate to obtain a range to each source as well.

Low Frequency Acoustic Recording and Analysis (LOFAR) grams Once a bearing or direction of interest has been determined, spectral analysis of the selected beam is used to produce a LOFAR gram, which is a plot of the signal spectrum for each analysis epoch, T, versus time. By examining the

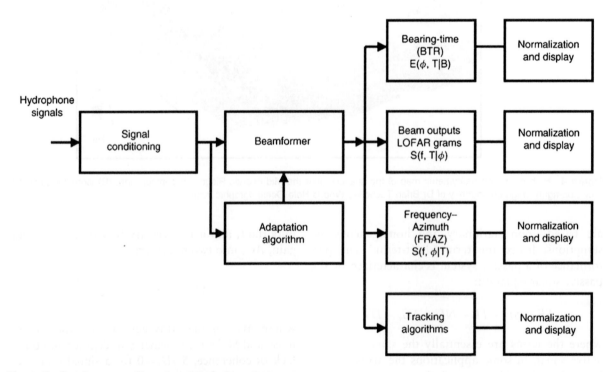

Figure 5 Passive sonar with modes of displaying output.

features of the LOFAR gram, such as frequency, peaks, harmonic signals, and their changes as a function of time, the source of the signal and some of its characteristics, such as speed, can be deduced. As in the case of the bearing-time display, the LOFAR gram is normalized to enhance features of interest.

FRAZ displays Frequency-azimuth, or FRAZ displays plot the spectral content as a function of frequency and azimuth for each epoch, T. Often a number of FRAZ outputs are averaged to improve signal to noise. FRAZ displays allow connection of the spectral content of a single source along a given bearing since the display contains a number of lines for each source at a given azimuth. As with the previous display, normalization algorithms to improve contrast are usually employed.

Trackers The objective of an ASW passive sonar is to detect, classify and track sources of radiating sound. Trackers are used to follow sources through direction and frequency space. There are a number of tracking algorithms build upon various signal models. Some separate the direction and frequency dimensions while others are coupled models. Since the ambient field can often have a number of sources, some targets and some interferers and since there are complicated propagation effects, the design of trackers is difficult. Most involve some form of Kalman filtering and some have integral propagation models.

Advanced beamforming

Most passive sonars are based upon a plane wave model for the ambient field components. Plane wave beamforming is robust, has several computational advantages, and is well understood. However, in many sonars this model is not adequate because the arrays are so long that wavefront curvature becomes an issue or the arrays are used vertical where acoustic propagation introduces multipaths. Advanced beamforming concepts can address these issues.

Wavefront curative becomes important when the target is in the near field of the array (usually given by the Fresnel number $2L^2/\lambda$) which is a consequence of either very long arrays or high frequencies. A quadratic approximation to the curvature is often used and the array is actually designed to focus it at specific range (rather than at infinite range which is the case for plane waves). When focused at short ranges, long-range targets are attenuated. This introduces the focal range as another parameter for displaying the sonar output.

At long ranges and low frequencies or in shallow water acoustic signals have complex multipath or multimode propagation which leads to coherent interference along a vertical array or a very long horizontal array. The appropriate array processing is to determine the full field Green's function for the signal and match the beamforming to it, a technique known as matched field processing (MFP). MFP requires knowledge of the sound speed profile along the propagation path, so its performance depends upon the accuracy of environmental data. It is a computationally intensive process, but has the powerful advantage of being able to resolve both target depth and range, as well as azimuth. MFP is an active subject of passive sonar research.

Further Reading

Baggeroer AB (1978) Sonar signal processing. In: Oppenheim AV (ed.) *Applications of Digital Signal Processing.* Englewood Cliffs NJ: Prentice-Hall.

Baggeroer A, Kuperman WA, and Mikhalevsky PN (1993) An overview of Matched Field Processing. *IEEE Journal of Oceanic Engineering* 18: 401–424.

Beyer RT (1999) *Sounds of Our Times: 200 Years of Acoustics.* New York: Springer-Verlag.

Capon J (1969) High resolution frequency-wavenumber spectrum analysis. *Proceedings of the IEEE* 57: 1408–1418.

Clay C and Medwin H (1998) *Fundamentals of Acoustical Oceanography.* Chestnut Hill, MA: Academic Press.

Cox H (1989) Fundamentals of bistatic active sonar. In: Chan YT (ed.) *Underwater Acoustic Data Processing.* Kluwer Academic Publishers.

National Defense Research Council (1947) *Physics of Sound on the Sea: Parts 1–6.* National Defense Research Council, Division 6, Summary Technical Report.

Urick RJ (1975) *Principles of Underwater Sound for Engineers*, 2nd edn. McGraw-Hill Book Co.

VEHICLES FOR DEEP SEA EXPLORATION

S. E. Humphris, Woods Hole Oceanographic
Institution, Woods Hole, MA, USA

Introduction

Exploring the deep sea has captured the imagination
of humankind ever since Leonardo da Vinci made
drawings of a submarine more than 500 years ago,
and Jules Verne published *20 000 Leagues under the
Sea* in 1875. Since the early twentieth century, people
have been venturing into the ocean in bathyspheres
and bathyscaphs. However, it was not until 1960 that
the dream to go to the bottom of the deepest part of
the ocean was realized, when Jacques Piccard and a
US Navy lieutenant, Don Walsh, descended to the
bottom of the Mariana Trench (10 915 m or 6.8 mil)
in *Trieste* (**Figure 1**). This vehicle consisted of a float
chamber filled with gasoline for buoyancy, and a
separate pressure sphere for the personnel, allowing
for a free dive rather than a tethered one. Containers
filled with iron shot served as ballast to make the
submersible sink. After a 5-h trip to the bottom, and
barely 20 min of observations there, the iron shot was
released and *Trieste* floated back to the surface.

Since that courageous feat almost 50 years ago,
dramatic advances in deep submergence vehicles and
technologies have enabled scientists to routinely
explore the ocean depths. For many years, researchers have towed instruments near the seafloor
to collect various kinds of data (e.g., acoustic, magnetic, and photographic) remotely. With the development of sophisticated acoustic and imaging
systems designed to resolve a wide range of ocean
floor features, towed vehicle systems have become
increasingly complex. Some now use fiber-optic, rather than coaxial, cable as tethers and hence are able
to transmit imagery as well as data in real time.
Examples of deep-towed vehicle systems are included
in **Table 1** and **Figure 2**, and they tend to fall into
two categories. Geophysical systems, such as *SAR*
(IFREMER, France), *TOBI* (National Oceanography
Centre, Southampton, UK), and *Deep Tow 4KS*
(JAMSTEC, Japan), collect sonar imagery, bathymetry, sub-bottom profiles, and magnetics data, as they
are towed tens to hundreds of meters off the bottom.
Imaging systems, such as *TowCam* (WHOI, USA),
Scampi (IFREMER, France), and *Deep Tow 6KC*
(JAMSTEC, Japan), are towed a few meters off the
bottom and provide both video and digital imagery
of the seafloor.

However, since the 1960s, scientists have been
transported to the deep ocean and seafloor in submersibles, or human-occupied vehicles (HOVs), to
make direct observations, collect samples, and deploy instruments. More recently, two other types of
deep submergence vehicles – remotely operated vehicles (ROVs) and autonomous underwater vehicles
(AUVs) – have been developed that promise to
greatly expand our capabilities to map, measure, and
sample in remote and inhospitable parts of the
ocean, and to provide the continual presence necessary to study processes that change over time.

Human-Occupied Vehicles

The deep-sea exploration vehicles most familiar to
the general public are submersibles, or HOVs. This
technology allows a human presence in much of
the world's oceans, with the deepest diving vehicles
capable of reaching 99% of the seafloor.

There exist about 10 submersibles available worldwide for scientific research and exploration that can
dive to depths greater than 1000 m (**Table 2** and
Figure 3). All require a dedicated support ship. These
battery-operated vehicles allow two to four individuals
(pilot(s) and scientist(s)) to descend into the ocean
to make observations and gather data and samples.
The duration of a dive is limited by battery life, human

Figure 1 The bathyscaph *Trieste* hoisted out of the water
in a tropical port, around 1959. Photo was released by the US
Navy Electronics Laboratory, San Diego, California (US Naval
Historical Center Photograph). Photo #NH 96801: US Navy
Bathyscaphe *Trieste* (1958–63).

Table 1 Examples of deep-towed vehicle systems for deep-sea research and exploration (systems that can operate at depths ≥1000 m)

Vehicle	Operating organization	Maximum operating depth (m)	Purpose
TowCam	WHOI, USA	6500	Photo imagery; CTD; volcanic glass samples; water samples
Deep-Tow Survey System	COMRA, China	6000	Sidescan, bathymetry; sub-bottom profiling
DSL-120A	HMRG, USA	6000	Sidescan; bathymetry
IMI-30	HMRG, USA	6000	Sidescan; bathymetry; sub-bottom profiling
Scampi	IFREMER, France	6000	Photo and video imagery
Système Acoustique Remorqué (SAR)	IFREMER, France	6000	Sidescan; sub-bottom profiling; magnetics; bathymetry
SHRIMP	NOC, UK	6000	Photo and video imagery
TOBI	NOC, UK	6000	Sidescan; bathymetry; magnetics
BRIDGET	NOC, UK	6000	Geochemistry
Deep Tow 6KC	JAMSTEC, Japan	6000	Photo and video imagery
Deep Tow 4KC	JAMSTEC, Japan	4000	Photo and video imagery
Deep Tow 4KS	JAMSTEC, Japan	4000	Sidescan; sub-bottom profiling

WHOI, Woods Hole Oceanographic Institution, USA; COMRA, China Ocean Mineral Resources R&D Association; HMRG, Hawai'i Mapping Research Group, USA; IFREMER, French Research Institute for Exploration of the Sea; NOC, National Oceanographic Centre, Southampton, UK; JAMSTEC, Japan Marine Science & Technology Center.

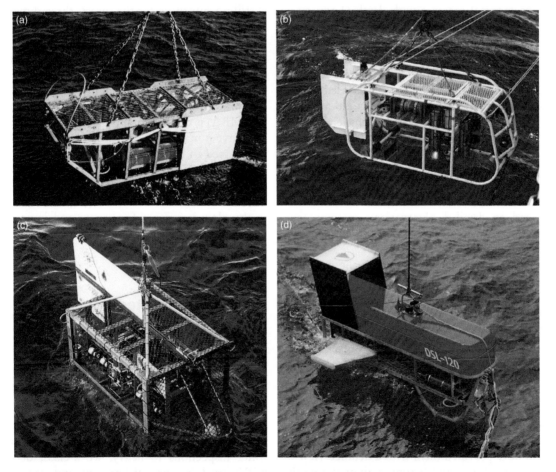

Figure 2 Examples of deep-towed vehicle systems. (a) *SHRIMP*, (b) *Deep Tow*, (c) *Tow Cam*, and (d) *DSL-120A*. (a) Courtesy of David Edge, National Oceanography Centre, UK. (b) © JAMSTEC, Japan, with permission. (c) Photo by Dan Fornari, WHOI, USA. (d) Courtesy of WHOI, USA.

Table 2 HOVs for deep-sea research and exploration (vehicles that can operate at depths ≥1000 m)

Vehicle	Operating organization	Maximum operating depth (m)
HOV (under construction)	COMRA, China	7000
Shinkai 6500	JAMSTEC, Japan	6500
Replacement HOV (in planning stages)	NDSF, WHOI, USA	6500
MIR I and II	P.P. Shirshov Institute of Oceanology, Russia	6000
Nautile	IFREMER, France	6000
Alvin	NDSF, WHOI, USA	4500
Pisces IV	HURL, USA	2170
Pisces V	HURL, USA	2090
Johnson-Sea-Link I and II	HBOI, USA	1000

Abbreviations as in **Table 1**; NDSF, National Deep Submergence Facility; HURL, Hawaii Undersea Research Laboratory; HBOI, Harbor Branch Oceanographic Institution, USA.

endurance, and safety protocols, and typically does not exceed 8–10 h, including transit time to and from the working depth (about 4 h for a seafloor depth of 4000 m). (The Russian *MIR* submersibles are an exception; they operate on a 100-kWh battery that can accommodate dive times in excess of 12 h.) Housed in a personnel sphere (**Figure 4**), the divers are maintained at atmospheric pressure despite the ever-increasing external pressure with depth (1 atm every 10 m). Cameras on pan and tilt mounts with zoom and focus controls are located on the exterior of the vehicles, as well as quartz iodide and/or metal halide lights to illuminate the area. Submersibles are also equipped with robotic arms that can be used to manipulate equipment or pick up samples, and a basket, usually mounted on the front of the vehicle, to transport instruments, equipment, or samples. These vehicles can handle heavy payloads, maintaining neutral buoyancy as their weight changes through a variable ballast control system. All these capabilities, together with their slow speeds (1–2 knots), make submersibles best suited to detailed observations, imaging, and sampling in localized areas, rather than operating in a survey mode.

Many significant discoveries during the past four decades of marine research have resulted from observations and samples taken from submersibles. Through direct observations from submersibles, biologists have discovered many previously unknown animals, and have documented that gelatinous animals (cnidarians, ctenophores, etc.) form a dominant ecological component of mid-water communities. These soft-bodied, fragile animals would have been destroyed by the trawl nets used in earlier days to sample these depths. Submersibles have enabled geologists to explore the global mid-ocean ridge system, and have provided them with a detailed view of the nature of volcanic and tectonic activity during the formation of oceanic crust. Submersibles played an important role in the discovery of hydrothermal vents

and their exotic communities of organisms, and continue to be used extensively for investigation of these extreme deep-sea environments.

HOVs will continue to provide important capabilities for deep-sea research at least for the foreseeable future. Although rapid progress is being made in videography and photography to develop capabilities that match those of the human eye, there is still no substitute for the direct, three-dimensional view that allows divers to make contextual observations and integrate them with the cognitive ability of the human brain. In recognition of this continuing need, there are two submersibles that are under construction or in the planning stages. The China Ocean Mineral Resources R&D Association (COMRA) is constructing their first submersible that will have a maximum operating depth of 7000 m. It is expected to be operational in 2007. In the United States, over 40 years after the submersible *Alvin* was delivered in 1964, the National Deep Submergence Facility at Woods Hole Oceanographic Institution is in the planning stages for a new and improved replacement HOV with an increased operating depth of 6500 m.

Remotely Operated Vehicles

Over the past 20 years, marine scientists have begun to routinely use ROVs to collect deep-sea data and samples. ROVs were originally developed for use in the ocean by the military for remote observations, but were adapted in the mid-1970s by the offshore energy industry to support deep-water operations. There are many ROVs commercially operated today, ranging from small, portable vehicles used for shallow-water inspections to heavy, work-class, deepwater ROVs used by the offshore oil and gas industry in support of subsea cable laying, retrieval, and repair.

Figure 3   Examples of HOVs used to conduct scientific research. (a) *Shinkai 6500*, (b) *Sea Link*, and (c) *Nautile*. (a) © JAMSTEC, Japan, with permission. (b) Courtesy of Harbor Branch Oceanographic Institution, USA. (c) © IFREMER, France, with permission; O. Dugornay.

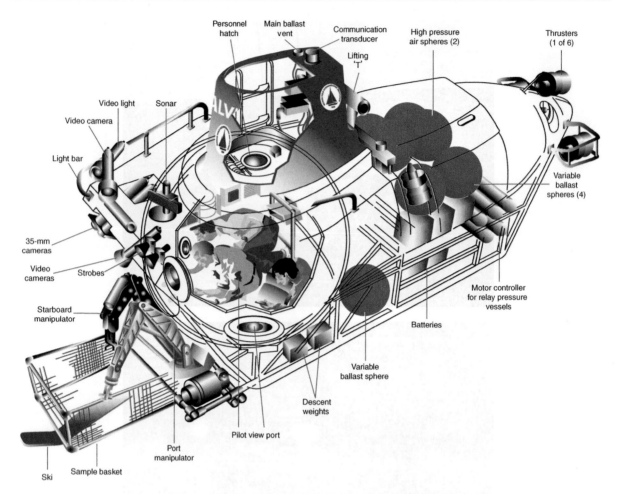

Figure 4 Cutaway illustration of the submersible *Alvin* showing the major components of an HOV. Illustration by E. Paul Oberlander, WHOI, USA.

There are about a dozen ROVs that are available to the international scientific community (**Table 3** and **Figure 5**). While some of these have dedicated support ships, many can operate in the 'flyaway' mode; that is, they can be shipped to, and operated on, a number of different ships. Unlike the HOVs, ROVs are unoccupied, and are tethered to a support ship usually by a fiber-optic cable that has sufficient bandwidth to accommodate a wide variety of oceanographic sensors and imaging tools. The cable provides power and communications from the ship to the ROV, allowing control of the vehicle by a pilot on board the ship. The pilot can also use the manipulator arm(s) to collect samples and perform experiments. The cable transmits images and data from the ROV to the control room on board the ship where monitors display the images of the seafloor or water column in real time. These capabilities, together with their excellent power and lift, allow ROVs to perform many of the same operations as HOVs.

Obvious advantages of using ROVs are that they remove the human risk factor from deep-sea research and exploration and, through the shipboard control room (**Figure 6**), allow a number of scientists and engineers to discuss the incoming data and make collective decisions about the operations. Another distinct advantage is their ability to remain underwater for extended periods of time because power is provided continuously from the ship. This endurance means that scientists can make observations over periods of many days, instead of a few hours a day, and gives them the flexibility to react to unexpected events. The disadvantage of an ROV is that its tether constrains operations because the range of the vehicle with respect to the ship cannot exceed a few hundred meters. Movement of the ship must therefore be carefully coordinated with the movements of the vehicle – this requires a ship equipped with a dynamic positioning system. In addition, the tether is heavy and produces drag on the vehicle, making it less maneuverable and vulnerable to entanglement in rugged terrain. However, with careful tether management, ROVs are well suited to mapping and surveying small areas, as well as to making

Table 3 ROVs for deep-sea research and exploration (vehicles that can operate at depths ≥ 1000 m)

Vehicle	Operating organization	Maximum operating depth (m)
Nereus (hybrid) (under construction)	NDSF, WHOI, USA	11 000
Kaiko 7000	JAMSTEC, Japan	7000
Isis	NOC, UK	6500
Jason II	NDSF, WHOI, USA	6500
ATV	SIO, USA	6000
CV (Wireline Reentry System)	SIO, USA	6000
Victor 6000	IFREMER, France	6000
ROV (on order)	NOAA Office of Ocean Exploration, USA	6000
ROPOS	CSSF, Canada	5000
Tiburon	MBARI, USA	4000
Quest	Research Centre Ocean Margins, Germany	4000
Hercules	Institute for Exploration, USA	4000
Sea Dragon 3500	COMRA, China	3500
Hyper Dolphin	JAMSTEC, Japan	3000
Aglantha	Institute of Marine Research, Norway	2000
Ventana	MBARI, USA	1500
Cherokee	Research Centre Ocean Margins, Germany	1000

Abbreviations as in **Tables 1** and **2**; SIO, Scripps Institution of Oceanography; CSSF, Canadian Scientific Submersible Facility; MBARI, Monterey Bay Aquarium Research Institute; SIO, Scripps Institution of Oceanography.

more detailed observations, imaging, and sampling of specific features.

While many of the ROVs available to the scientific community have a wide range of capabilities, a few are purpose-built. For example, the Wireline Reentry System known as *CV*, and operated by Scripps Institution of Oceanography, is a direct hang-down vehicle designed specifically for precision placement of heavy payloads on the seafloor or in drill holes (**Figure** 7). Unlike conventional, near-neutrally buoyant ROVs, the Wireline Reentry System can handle payloads of a few thousand kilograms, depending on the water depth. It has been used, for example, to install seismometer packages in, and recover instruments packages from, seafloor drill holes in water depths up to 5500 m, as well as to deploy precision acoustic ranging units on the axis of the mid-ocean ridge.

Another ROV being built at Woods Hole Oceanographic Institution for a specific purpose is *Nereus* (**Figure** 8). More correctly referred to as a hybrid remotely operated vehicle, or HROV, because it will be able to switch back and forth to operate as either an AUV or an ROV on the same cruise, *Nereus* will be capable of exploring the deepest parts of the world's oceans, as well as bringing ROV capabilities to ice-covered oceans, such as the Arctic. The HROV will use a lightweight fiber-optic micro-cable, only 1/32 of an inch in diameter, allowing it to operate at great depth without the high-drag and expensive cables typically used with ROV systems. Once the HROV reaches the bottom, it will conduct its

mission while paying out as much as 20 km (about 11 mi) of micro-cable. Once the mission is complete, the HROV will detach from the micro-cable and guide itself to the sea surface for recovery, while the micro-cable is recovered for reuse. In 2008–09, almost 50 years after the dive of the *Trieste*, *Nereus* will dive to the bottom of the Mariana Trench.

Autonomous Underwater Vehicles

Although the concept of AUVs has been around for more than a century, it is only in the last decade or two that AUVs have been applied to deep-sea research and exploration. AUV technology is in a phase of rapid growth and expanding diversity. There are now more than 50 companies or institutions around the world operating AUVs for a variety of purposes. For example, the offshore gas and oil industry uses them for geologic hazards surveys and pipeline inspections, the military uses them for locating mines in harbors among other applications, and AUVs have been used to search for cracks in the aqueducts that supply water to New York City.

There are currently about a dozen AUVs being used specifically for deep-sea exploration (**Table 4** and **Figure** 9), although the numbers continue to increase. These unoccupied, untethered vehicles are preprogrammed and deployed to drift, drive, or glide through the ocean without real-time intervention from human operators. All power is supplied by energy systems carried within the AUV. Data are

Figure 5 Examples of ROVs used for deep-sea research. (a) ROV *Kaiko*, (b) ROV *Jason II*, (c) ROV *Tiburon*, and (d) ROV *Victor 6000*. (a) © JAMSTEC, Japan, with permission. (b) Photo by Tom Bolmer, WHOI, USA. (c) Photo by Todd Walsh © 2006, MBARI, USA, with permission. (d) © IFREMER, France, with permission; M. Bonnefoy.

recorded and are then either transmitted via satellite when the AUV comes to the surface, or are downloaded when the vehicle is recovered. They are generally more portable than HOVs and ROVs and can be deployed off a wide variety of ships. By virtue of their relatively small size, limited capacity for scientific payloads, and autonomous nature, AUVs do not have the range of capabilities of HOVs and ROVs. They are, however, much better suited than HOVs and ROVs to surveying large areas of the ocean that would take years to cover by any other means. They can run missions of many hours or days on their battery power and, with their streamlined shape, can travel many kilometers collecting data of various types depending on which sensors they are carrying.

Hence, AUVs are frequently used to identify regions of interest for further exploration by HOVs and ROVs.

Unlike HOVs and ROVs that are designed with the flexibility to carry different sensors and equipment for different purposes, AUV system design and attributes are driven by the specific research application. Some, such as the autonomous drifters and gliders (essentially drifters with wings and a buoyancy change mechanism that allow the vehicle to change heading, pitch, and roll, and to move horizontally while ascending and descending in the water column), are designed for research in the water column to better understand the circulation of the ocean and its influence on climate. While satellites provide

Figure 6 Portable control van for the ROV *Jason II* constructed from two shipping containers assembled on board the R/V *Knorr*. © Dive and Discover, WHOI, USA.

global coverage of conditions at the sea surface, AUVs are likely to be the only way to continuously access data from the ocean depths. Equipped with oceanographic sensors that measure temperature, salinity, current speed, and phytoplankton abundance, drifters and gliders profile the water column by sinking to a preprogrammed depth, and then rising to the surface where they transmit their data via satellite back to the scientist on shore. By deploying hundreds to thousands of these vehicles, scientists will achieve a long-term presence in the ocean, and will be able to make comprehensive studies of vast oceanic regions.

Other, more sophisticated AUVs are also used to investigate water column characteristics, and ephemeral or localized phenomena, such as algal blooms. The first of the *Dorado Class* of AUVs, operated by Monterey Bay Aquarium Research Institute, was deployed in late 2001 to measure the inflow of water into the Arctic basin through the Fram Strait. *Autosub*, operated by the National Oceanography Centre, Southampton, UK, was deployed to measure flow over the sills in the Strait of Sicily. The *REMUS* (Remote Environmental Monitoring UnitS) class of AUVs is extremely versatile and they have been used on many types of missions. The standard configuration includes an up- and down-looking acoustic Doppler current profiler (ADCP), sidescan sonar, a conductivity–temperature (CT) profiler, and a light scattering sensor. However, many other instruments have been integrated into it for specific missions, including fluorometers, bioluminescence sensors, radiometers, acoustic modems, forward-looking sonar, altimeters, and acoustic Doppler velocimeters. *REMUS* can also carry a video plankton recorder, a plankton pump, video cameras,

Figure 7 The Wireline Reentry System, known as the *CV*, operated by Scripps Institution of Oceanography. This specialized ROV can precisely place heavy payloads on the seafloor and in drill holes. The control vehicle, which weighs about 500 kg in water, is deployed at the end of a 17.3-mm (0.68") electromechanical (coax) or electro-optico-mechanical (three copper conductors, three optical fibers) oceanographic cable. The vehicle consists of a steel frame equipped with two horizontal thrusters mounted orthogonal to each other to control lateral position. The vertical position is controlled by winch operation. Instrumentation includes a compass, pressure gauge, lights, video camera, sonar systems, and electronic interfaces to electrical releases and to a logging probe. Courtesy of Scripps Institution of Oceanography – Marine Physical Laboratory, USA.

electronic still cameras, and, most recently, a towed acoustic array.

Still other AUVs are designed specifically for near-bottom work. They have proved particularly useful for near-bottom surveying and mapping, which can be accomplished autonomously while the support ship simultaneously conducts other, more traditional, operations. One of the earliest vehicles to provide this capability was the *Autonomous Benthic Explorer* (*ABE*) developed at Woods Hole Oceanographic Institution. *ABE* was designed to be extremely stable in pitch and roll and to be reasonably efficient in forward travel. All the buoyancy is built into the two upper pods, while the majority of the weight (the batteries and the main pressure housing) is in the central lower section. The three-hull structure also allows the seven vertical and lateral thrusters to be

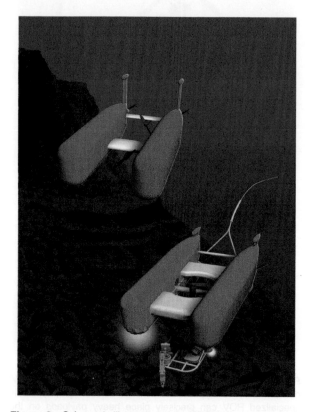

Figure 8 Schematic illustration of the HROV, *Nereus*, currently under construction at Woods Hole Oceanographic Institution, USA, in its autonomous mode (upper) and its ROV mode (lower). Illustration by E. Paul Oberlander, WHOI, USA.

placed between the hulls where they are protected. *ABE* is most efficient traveling forward, but it can also move backward, up or down, left or right, and can hover and turn in place. Equipment that it usually carries includes temperature and salinity sensors, an optical backscatter sensor, a magnetometer to measure near-bottom magnetic fields, and an acoustic altimeter to make bathymetric measurements and for its automated bottom following. *ABE* can dive to depths of 5500 m for 16–34 h, and it uses acoustic transponder navigation to follow preprogrammed track lines automatically. Its capability to maintain a precise course over rugged seafloor terrain gives it the ability to make high-precision seafloor bathymetric maps with features a few tens of centimeters tall and less than a meter long being identifiable.

Other AUVs have been specifically developed for high-resolution optical and acoustic imaging of the seafloor. For example, *SeaBED*, also developed at Woods Hole Oceanographic Institution, was designed specifically to further the growing interests in seafloor optical imaging – specifically, high-resolution color imaging and the processes of photo-mosaicking and three-dimensional image reconstruction. In addition to requiring high-quality sensors, this imposes additional constraints on the ability of the AUV to carry out structured surveys, while closely following the seafloor. The distribution of the four thrusters, coupled with the passive stability inherent in a two-hulled vehicle with a large metacentric height, allows *SeaBED* to survey close to the seafloor, even in very rugged terrain.

In the future, AUVs will play an important role in the development of long-term seafloor observatories.

Table 4   Examples of AUVs for deep-sea research and exploration (vehicles that can operate at depths $\geq$ 1000 m)

Vehicle	Operating organization	Maximum operating depth (m)
Dorado Class	MBARI, USA	6000
CR-01, CR-02	COMRA, China	6000
Sentry	WHOI, USA	6000
REMUS Class	WHOI, USA	6000
Autosub 6000	NOC, UK	6000
Autonomous Benthic Explorer	NDSF, WHOI, USA	5500
Explorer 5000	Research Centre Ocean Margins, Germany	5000
Jaguar/Puma	WHOI, USA	5000
Urashima (hybrid)	JAMSTEC, Japan	3500
Aster x	IFREMER, France	3000
Bluefin AUV	Alfred Wegener Institute, Germany	3000
Bluefin 21 AUV	SIO, USA	3000
Odyssey Class	MIT, USA	3000
SeaBED	WHOI, USA	2000
Autosub 3	National Oceanography Centre, UK	1600
Spray Gliders	WHOI, USA	1500
Seaglider	Univ. of Washington, USA	1000

Abbreviations as in **Tables 1–3**.

Figure 9 Examples of AUVs used in oceanographic research. (a) The *Spray Glider*, (b) *Urashima*, (c) *Autosub*, (d) *SeaBED*, (e) *Dorado Class*, and (f) *ABE*. (a) Photo by Jane Dunworth-Baker, WHOI, USA. (b) © JAMSTEC, Japan, with permission. (c) Courtesy of Gwyn Griffiths, National Oceanography Centre, Southampton, UK. (d) Photo by Tom Kleindinst, WHOI, USA. (e) Photo by Todd Walsh © 2004, MBARI, USA, with permission. (f) Photo by Dan Fornari, WHOI, USA.

Apart from providing the high-resolution maps needed to optimally place geological, chemical, and biological sensors as part of an observatory, AUVs will also operate in a rapid response mode. It is envisaged that deep-sea observatories will include docking stations for AUVs, and there are a number of research groups currently working on developing this technology. When an event – most likely a seismic event – is detected, scientists on shore will be able to program the AUV, via satellite and a cable to a surface buoy, to leave its dock and conduct surveys in the vicinity of the event. The AUV will then return to its dock and return the data to shore for assessment by scientists as to whether further investigation with ships is warranted.

Navigating Deep-Sea Vehicles

Unlike glider and drifter AUVs that can come to the sea surface and determine their positions using a Global Positioning System (GPS), deep-sea vehicles working at the bottom of the ocean have no such reference system because the GPS system's radio frequency signals are blocked by seawater. The technique that has been the standard for three-dimensional acoustic navigation of deep-sea vehicles is long-baseline (LBL) navigation – a technique developed more than 30 years ago. LBL operates on the principle that the distance between an underwater vehicle and a fixed acoustic transponder can be related precisely to the time of flight of an acoustic signal propagating between the vehicle and transponder. Two or more acoustic transponders are dropped over the side of the surface ship and anchored at locations selected to optimize the acoustic range and geometry of planned seafloor operations. Each transponder is a complete subsurface mooring comprised of an anchor, a tether, and a buoyant battery-powered acoustic transponder. The positions of the transponders on the seafloor are determined by using the GPS on board the ship and ranging to them acoustically while the ship circles the point where each transponder was dropped. The positions of the transponders on the seafloor can be determined this way with an accuracy of about 10 m.

Transponders have accurate clocks to measure time very precisely, and they are synchronized with the clocks on the vehicle and on the ship. Each transponder is set to listen for acoustic signals (or pings) transmitted either from the deep-sea vehicle or the ship at a specific frequency. When each transponder hears these acoustic signals, it is programmed to transmit an acoustic signal back to the vehicle and the ship. Each transponder pings at a different frequency, so the ship and the vehicle can discern which transponder sent it. The time of flight of the acoustic signals gives a measure of distance to each transponder, and using simple triangulation, the unique point in three-dimensional space where all distances measured from all the transponders and the ship intersect can be calculated. More recently, conventional LBL navigation has been combined with Doppler navigation data, which measures apparent bottom velocity of the vehicle, for better short-term accuracy.

The Future

The technological breakthroughs in deep-sea vehicle design over the last 40 years have resulted in unprecedented access to the deep ocean. While each type of vehicle has its own advantages and disadvantages, the complementary capacities of all types of deep-submergence vehicles provide synergies that are revolutionizing how scientists conduct research in the deep ocean. They are learning how to exploit those synergies by using a nested survey strategy that employs a combination of tools in sequence for investigations at increasingly finer scales: ship-based swath-mapping systems and towed vehicle systems for reconnaissance over large areas to identify features of interest, followed by more detailed, high-resolution mapping, imagery, and chemical sensing with AUVs, and finally, seafloor observations and experimentation using HOVs and ROVs. A demonstration of the power of such an approach occurred on a cruise to the Galápagos Rift in 2002. The investigative strategy was directed toward ensuring that all potential sites of hydrothermal venting in the rift valley were identified and investigated visually with the HOV *Alvin*. The AUV *ABE* was deployed at night to conduct high-resolution mapping of the seafloor and collect conductivity–temperature–depth (CTD) data in the lower water column to detect sites of venting. Upon its recovery in the morning, micro-bathymetry maps and temperature anomaly maps were quickly generated, compiled with previous data, and then given to the scientists diving in *Alvin* that day for their use in directing the dive. Today, the vehicles are being deployed in various combinations to attack a range of multidisciplinary problems.

Deep-sea vehicles will also play indispensable roles in establishing and servicing long-term seafloor observatories that will be critical for time-series investigations to understand the dynamic processes going on beneath the ocean. AUVs will undertake a variety of mapping and sampling missions while using fixed observatory installations to recharge batteries, offload data, and receive new instructions. They will be used to extend the spatial observational capability of seafloor observatories through surveying activities, and will document horizontal variability in seafloor and water column properties – necessary for establishing the context of point measurements made by fixed instrumentation. HOVs and ROVs will be required to install, service, and repair equipment and instrumentation on the seafloor and in drill holes, as well as collect samples as part of time-series measurements. The additional capabilities that these vehicles will need for service and repair activities will likely build on ROV tools that are currently being used in the commercial undersea cable industry.

Deep-sea vehicles will clearly have a role to play in deep-sea research for the foreseeable future, and they will be at the vanguard of a new era of ocean exploration.

See also

Hydrothermal Vent Deposits. Platforms: Autonomous Underwater Vehicles.

Further Reading

Bachmayer R, Humphris S, Fornari D, et al. (1998) Oceanographic exploration of hydrothermal vent sites on the Mid-Atlantic Ridge at 37°N 32°W using remotely operated vehicles. *Marine Technology Society Journal* 32: 37–47.

Davis RE, Eriksen CE, and Jones CP (2002) Autonomous buoyancy-driven underwater gliders. In: Griffiths G (ed.) *The Technology and Applications of Autonomous Underwater Vehicles*, pp. 37–58. London: Taylor and Francis.

De Moustier C, Spiess FN, Jabson D, et al. (2000) Deep-sea borehole re-entry with fiber optic wireline technology. *Proceedings of the 2000 International Symposium on Underwater Technology*, Tokyo, 23–26 May 2000, pp. 379–384.

Fornari D (2004) Realizing the dreams of da Vinci and Verne. *Oceanus* 42: 20–24.

Fornari DJ, Humphris SE, and Perfit MR (1997) Deep submergence science takes a new approach. *EOS, Transactions of the American Geophysical Union* 78: 402–408.

Fryer P, Fornari DJ, Perfit M, et al. (2002) Being there: The continuing need for human presence in the deep ocean for scientific research and discovery. *EOS, Transactions of the American Geophysical Union* 83(526): 532–533.

Funnell C (2004) *Jane's Underwater Technology 2004–2005*, 800pp, 23rd edn. Alexandria, VA: Jane's Information Group.

National Research Council (2004) *Exploration of the Seas: Voyage into the Unknown*. Washington, DC: National Academies Press.

National Research Council (2004) *Future Needs of Deep Submergence Science*. Washington, DC: National Academies Press.

Reves-Sohn R (2004) Unique vehicles for a unique environment. *Oceanus* 42: 25–27.

Rona P (2001) Deep-diving manned research submersibles. *Marine Technology Society Journal* 33: 13–25.

Rudnick DL, Davis RE, Eriksen CC, Fratantoni DM, and Perry MJ (2004) Underwater gliders for ocean research. *Marine Technology Society Journal* 38: 48–59.

Shank T, Fornari D, Yoerger D, et al. (2003) Deep submergence synergy: *Alvin* and *ABE* explore the Galápagos Rift at 86°W. *EOS, Transactions of the American Geophysical Union* 84(425): 432–433.

Yoerger D, Bradley AM, Walden BB, Singh H, and Bachmayer R (1998) Surveying a subsea lava flow using the *Autonomous Benthic Explorer (ABE)*. *International Journal of Systems Science* 29: 1031–1044.

Relevant Websites

http://auvlab.mit.edu
  – AUV Lab Vehicles, AUV Lab at MIT Sea Grant.
http://www.ropos.com
  – Canadian Scientific Submersible Facility.
http://www.comra.org
  – China Ocean Mineral Resources R&D Association.
http://divediscover.whoi.edu
  – Dive and Discover: Expeditions to the Seafloor.
http://www.soest.hawaii.edu
  – Hawai'i Undersea Research Laboratory (HURL), School of Ocean and Earth Science and Technology.
http://www.ifremer.fr
  – IFREMER Fleet.
http://www.mbari.org
  – Marine Operations: Vessels and Vehicles, Monterey Bay Aquarium Research Institute.
http://www.mpl.ucsd.edu
  – Marine Physical Laboratory, Scripps Institution of Oceanography.
http://www.noc.soton.ac.uk
  – National Oceanography Centre, Southampton.
http://www.jamstec.go.jp
  – Research Vessels, Facilities, and Equipment, JAMSTEC.
http://www.apl.washington.edu
  – *Seaglider*, Applied Physics Laboratory, University of Washington.
http://www.whoi.edu
  – Ships and Technology: National Deep Submergence Facility, Woods Hole Oceanographic Institution.
http://www.rcom.marum.de
  – Technology page, MARUM.

MANNED SUBMERSIBLES, DEEP WATER

H. Hotta, H. Momma, and S. Takagawa,
Japan Marine Science & Technology Center, Japan

Introduction

Deep-ocean underwater investigations are much more difficult to carry out than investigations on land or in outer space. This is because electromagnetic waves, such as light and radio waves, do not penetrate deep into sea water, and they cannot be used for remote sensing and data transmission.

Moreover, deep-sea underwater environments are physically and physiologically too severe for humans to endure the high pressures and low temperatures. First of all, pressure increases by 1 atmosphere for every 10 meters depth because the density of water is 1000 times greater than that of air. Furthermore, as we have no gills we can not breathe under water. Water temperature decreases to 1°C or less in the deep sea and there is almost no ambient light at depth because sunlight can not penetrate through more than a few hundred meters of sea water. These are several of the reasons why we need either manned or unmanned submersibles to work in the deep sea.

A typical manned submersible consists of four major components: a pressure hull, propellers (thrusters), buoyant materials, and observational instruments. The pressure hull is a spherical shell made of high-strength steel or titanium. The typical internal diameter of the hull is approximately 2 m, which allows up to three people to stay at one atmosphere for 8–12 h during underwater operations. In case of emergency, a life-support system enables a stay of three to five days. Several thrusters are usually installed on the body of the submersible to give maneuverability. The buoyant material is syntactic foam, which is made of glass microballoons and an adhesive matrix. Its specific gravity is approximately 0.5 gf ml^{-1}. Observational instruments such as cameras, lights, sonar, CTD (conductivity, temperature, and depth sensors) etc., are also very important for gathering information on the deep-sea environment. It should be mentioned that the power consumption of the lights can reach as much as 15% of the total power consumption of the submersible.

History of Deep Submersibles

The first modern deep diving by humans, to a depth of 923 m, was achieved in 1934 by William Beebe, an American zoologist, and Otis Burton using the bathysphere, which means 'deep sphere'. The bathysphere was a small spherical shell made of cast iron, 135 cm in inside diameter designed for two observers. The bathysphere had an entrance hatch and a small glass view port. As the sphere was lowered by a cable and lacked thrusters, it was impossible to maneuver.

The next advance, using a free-swimming vehicle, occurred after World War II, in 1947. The bathyscaph *FNRS II* was invented by Auguste Piccard, who had been studying cosmic rays using a manned balloon in Switzerland. The principle of the bathyscaph was the same as that of a balloon. Instead of hydrogen or helium gas, gasoline was used as the buoyant material. During descent, air ballast tanks were filled with sea water, and for ascent, iron shot ballast was released. The pressure hull was made of drop-forged iron hemispheres, 2 m in inside diameter and 90 mm in thickness, allowing for two crew members. It was able to maneuver around the seafloor by thrusters driven by electric motors. Later, the second bathyscaph, *Trieste*, was sold to the US Navy, and independently at the same time, the French Navy developed the bathyscaph *FNRS III*, and later *Archimede*. In 1960, the *Trieste* made a dive into the Challenger Deep in the Mariana Trench, to a depth of 10 918 m. This historic dive was conducted by Jacques Piccard, son of Auguste Piccard, and Don Walsh from the US Navy. The bathyscaph was the first generation of deep-diving manned submersibles. It was very big and slow as it needed more than 100 kiloliter capacity gasoline tanks to provide flotation for the 2 m diameter pressure hull.

In 1964, the second generation of deep submersibles began. *Alvin* was funded by the US Navy under the guidance of the Woods Hole Oceanographic Institution (WHOI). At first, its depth capability was only 1800 m. It was small enough to be able to put on board the R/V *Lulu*, which became its support ship. Instead of gasoline flotation, syntactic foam was used. *Alvin* had horizontal and vertical thrusters to maneuver freely in three dimensions. Scientific instruments, including manipulators, cameras, sonar and a navigation system, were installed. Three observation windows were available for the three crew members. In France, the two-person 3000 m-class submersible *Cyana* was built. These two vehicles typified submersibles during the 1960s. At present, the depth capability of the *Alvin* has been increased to 4500 m by replacing the high-strength steel pressure hull with a titanium alloy sphere in 1973. In the 1980s, 6000 m-class submersibles, such as the *Nautile* from France, the *Sea Cliff* of the US Navy, the *Mir I* and *Mir II* from Russia and the *Shinkai 6500* from Japan, were built. They were theoretically able to cover more than 98% of the world's ocean floor.

What will the third generation of deep submersibles be like? Manned submersibles of the third generation, which would be capable of exceeding 10 000 m depth, have not yet been developed at the time of this report. One possibility is a small and highly maneuverable one- or two-person submersible with a transparent acrylic or ceramic pressure hull. Another possibility is a deep submergence laboratory, which would be able to carry several scientists and crew long distances and long durations without the assistance of a mother ship. This would be the realization of the dream like 'Nautilus' in 20 000 Leagues Under The Sea by French novelist Jules Verne. Strong scientific and/or social goals would be needed for such a submersible design to be pursued. And there is a third possibility that the next generation will be evolutionary upgrades of existing second-generation submersibles.

Principles of Modern Submersibles

Descent and Ascent

There are several methods to submerge vehicles into the deep sea. The simplest way is to suspend a sphere by a cable, known as a bathysphere. Mobility, however, is greatly limited. A second method relies on powerful thrusters to adjust vertical position in relatively shallow water. The submersible *Deep Flight* is a high-speed design which uses thrust power coupled with fins for motion control like the wings of a jet fighter. It descends and ascends obliquely in the water column at speeds up to 10 knots. Most submersibles employ a third method that, while using weak thrusters to control attitude and horizontal movement, relies principally on an adjustable buoyancy system for descent and ascent (**Figure 1**).

When on the surface, the submersible's air ballast tank is filled with air creating positive buoyancy, hence it floats. When the dive begins, air is vented from the ballast tank and filled with sea water, thus creating negative buoyancy and sinking the vehicle. As the submersible dives deeper, buoyancy increases modestly due to the increasing water density created by the increasing pressure. Thus the submersible slows slightly as it dives deeper (**Figure 2**).

When the submersible approaches the seafloor (50–100 m in altitude, i.e., height above the bottom), a portion of its ballast (usually lead or some other heavy material) is jettisoned to achieve neutral buoyancy. Perfect neutral buoyancy occurs when the positively buoyant materials (things which tend to float) on the submersible balance the negatively buoyant materials (things which tend to sink). This allows the vehicle to hover weightless in position and move freely about. As perfect neutral buoyancy is difficult to maintain, most submersibles have auxiliary weight-adjusting (trim and ballast) systems. This consists of a sea-water pumping system to draw in or expel water, thus adjusting the buoyancy of the submersible.

Upon completing its mission, the remaining ballast is jettisoned and the submersible now with positive buoyancy begins ascending. When resurfaced, air from a high-pressure bottle is blown into the air ballast tank to give enough draft to the submersible for the recovery operation.

Water Pressure

Water pressure increases by 0.1 MPa per 10 m depth. Thus every component sensitive to pressure must be isolated from intense pressure changes. First and foremost are the passengers which are protected against great ambient pressure by a pressure hull or pressure vessel, maintained at surface pressure. The ambient pressure exerts strong compressional force on the pressure hull which is therefore designed to avoid any tensile stress. The strongest geometric shape against outside pressure relative to volume and hence weight is a sphere, followed by a cylinder (capped at both ends). However, it is not easy to arrange instruments inside a sphere effectively.

In order to increase mobility, it is important to make submersibles small and light. The pressure hull is one of the largest and heaviest components of the submersible. The hull must be as small (and light) as possible, while affording appropriate strength against external pressure. Thus for deep-diving submersibles, a spherical pressure hull is employed whereas shallower vehicles can use a cylindrical shape if so desired.

The material used for the pressure hull is critical. In earlier vehicles, steel was used. Later, titanium alloy was the material of choice. Titanium alloy has very high tensile strength, and is resistant to corrosion and relatively light (specific gravity ~60% that of steel). Recently, the trend in submersible construction is to use nonmetallic materials, such as

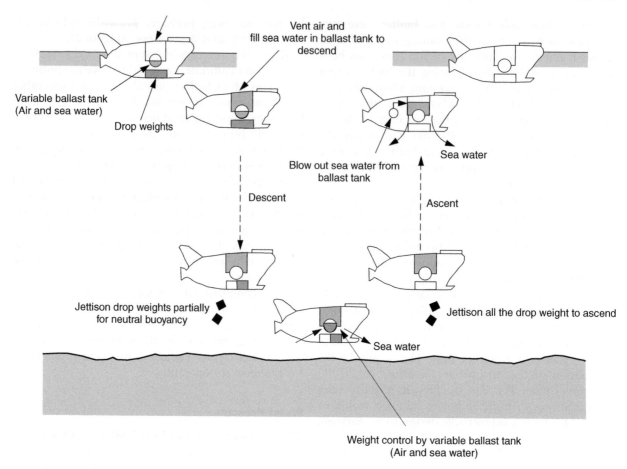

Figure 1 Principle of descent and ascent for a modern deep submersible.

fiber- or graphite-reinforced plastics (FRP or GRP), or ceramics.

Components not sensitive to pressure or saline conditions need no special consideration. Though those devices which require electrical insulation need to be housed in oil-filled compartments called oil-filled pressure compensation systems (**Figure 3**). These systems do not require heavy pressure hulls and thus reduce the weight of the submersible overall. Electric motors, hydraulic systems, batteries, wiring, and power transistors are all housed in pressure compensation systems. Technology is being developed to apply ambient pressure to electronic devices such as integrated circuits (ICs) and large scale ICs (LSIs).

Buoyancy

With the exception of some shallow-water submersibles, the total weight of the essential systems is larger than the total buoyancy. This means that extra buoyancy is needed to balance the excess weight. Wood or foam-rubber cannot be used for this purpose because they shrink under increasing water

pressure. The material providing buoyancy must have a relatively small specific gravity while remaining strong under high-pressure conditions.

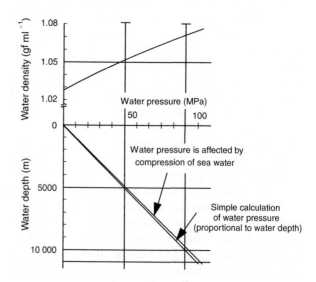

Figure 2 Relations between water depth, pressure and water density at a water temperature of 0°C and salinity of 34.5‰.

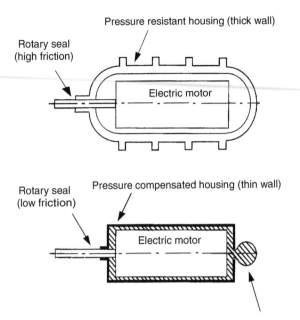

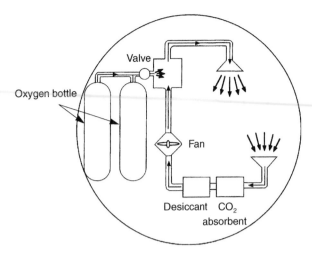

Figure 4 Life support system for a deep submersible.

Figure 3 Pressure resistant and pressure compensated housings for electric motors.

Historically, gasoline was used to provide buoyancy in bathyspheres as it did not lose buoyancy under pressure. However, its specific gravity was too large for practical use – huge volumes are needed to offset the weight. With the invention of syntactic foam, a superior material for deep-diving submersibles became available. Syntactic foam consists of tiny microscopic spheres of glass embedded in resin. These microballoons are 40–200 μm in diameter, and are closely packed with resin filling in the surrounding spaces. Proper selection of the balloons and resin allows the proper pressure tolerance and specific gravity to be created. For example, the syntactic foam used by the *Shinkai 6500* is tolerant up to 130 MPa with a specific gravity of $0.54 \, \text{gf ml}^{-1}$ and the foam used by the ROV *Kaiko* is tolerant up to 160 MPa with a specific gravity of $0.63 \, \text{gf ml}^{-1}$.

Life Support

The pressure hull is a very small space where crew members must stay for up to 20 h, depending on their mission. Since the pressure hull is maintained at ambient pressure, no decompression of the occupants is needed. High-pressure oxygen bottles provide oxygen within the pressure hull, while carbon scrubbers absorb carbon-dioxide (**Figure 4**). Extra life-support is required with varying standards depending on the country (from 32 h to 120 h)

Energy

Energy for deep-diving submersibles is supplied by rechargeable (secondary) batteries. There are several types including: lead acid, nickel cadmium, nickel hydrogen, oxidized silver zinc. Batteries which contain a higher density of energy are preferable to reduce the weight and volume of the submersible. However, such batteries are very expensive. These batteries are housed in oil-filled pressure-compensated systems to reduce weight. Recent developments in fuel cell technology offer the promise of higher density energy coupled with higher efficiency. This has great potential for submersible applications.

Instrumentation

There are many scientific and observational instruments employed on research submersibles. Due to the limited payload, these instruments must be as light and small as possible. For example, bulky camera bodies must be streamlined and aligned with lenses in a compact manner, thus reducing the size and weight of their pressure housing (see **Figure 5**). Furthermore, physical conditions such as extremely cold temperatures or the differential absorption of white light must be considered. Thus, for example, engineers must consider both the compactness of video cameras and their color sensitivity.

Manipulator

Pressure hulls, thrusters, batteries, and buoyancy materials are all essential parts of the modern submersible. One of the most important tools is the manipulator arm. Manipulators extend the arms and hands of the pilot, allowing sample collection and deployment of experimental equipment. Most

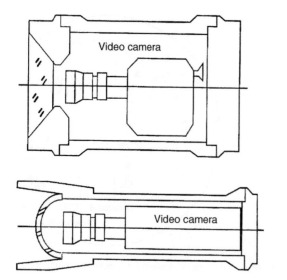

Figure 5 Examples of inside alignments of the pressure case.

manipulators are driven by hydraulic pressure. The most advanced manipulators operate in a master-slave system. The operator handles a master arm (controller) which imitates human arm and hand movements, and the motions are translated to the slave-unit (manipulator) which follows precisely the motions of the master arm. There are usually one or two manipulators on a research submersible.

Navigation

Underwater navigation is one of the most crucial elements of deep-sea submersible researches. Usually, long base line (LBL) or super short base line (SSBL) acoustic navigation systems are used depending on the accuracy required. The positioning error of LBL systems is 5–15 m, and it is approximately 2–5% of slant range for SSBL systems. An advantage of SSBL systems is that seafloor transponders are not necessary, whereas at least two seafloor transponders are necessary for LBL systems. In both systems, absolute or geodetic position is determined by surface navigation systems, such as Differential Global Positioning System (DGPS). Although the position of the submersible is usually reported from the mother ship by voice, the *Shinkai 6500* has an automatic LBL navigation system.

Surface Support (the Mother Ship)

The R/V *Lulu*, the original support ship of the submersible *Alvin*, was retired and replaced by the R/V *Atlantis II* in 1983. In 1997 *Atlantis II* was replaced by the newly commissioned Research Vessel *Atlantis*. The support ship not only supports the diving operation, including launch, recovery, communication, and positioning, but also provides a place to conduct

on-board research during a cruise. Accordingly, research laboratories, computers, instruments for on-board data analysis, multi-narrow beam echo sounders, etc., are necessary. Also, remotely operated vehicles (ROVs) or autonomous underwater vehicles (AUVs), can be operated during the nighttime, or in case the manned submersible cannot be operated because of bad weather.

Deep Submersibles in the World

Alvin (USA)

Alvin (**Figure 6**) was built in 1964 by Litton Industries with funds of the US Navy and operated by WHOI. Its original depth capability was 1800 m with a steel pressure hull. Later, the pressure hull was replaced by titanium alloy to increase depth capability up to 4500 m. *Alvin's* size and weight are: length 7.1 m; width 2.6 m; height 3.7 m and weight 17 tf in air. The outside diameter of the pressure hull is 2.08 m with a wall thickness of 49 mm. It is equipped with three view ports, 120 mm in inside diameter, two manipulators with seven degrees of freedom. The original catamaran support ship, R/V *Lulu*, was replaced by the R/V *Atlantis II*. Launch and recovery take place at the stern A-frame. It has been the leading deep submersible in the world.

Nautile (France)

Nautile (**Figure 7**) was built in 1985 by IFREMER (French Institution for Marine Research and Development) in France. Depth capability of 6000 m was aimed to cover 98% of the world's ocean floor. It is 8 m long, 2.7 m wide, 3.81 m in height and weighs 19.3 tf in air. It is equipped with three view ports, a manipulator, a grabber and a small companion ROV *Robin*. The position of the submersible is directly calculated by interrogating the seafloor transponders. Still video images are transmitted to the support ship through the acoustic link.

Sea Cliff (USA)

Sea Cliff was originally built in 1968 as a 3000 m-class submersible by General Dynamics Corp. for the US Navy as a sister submersible for the *Turtle*. In 1985, the *Sea Cliff* was converted into a 6000 m-class deep submersible. It is 7.9 m long, 3.7 m wide, 3.7 m high and weighs 23 tf in air. (*Sea Cliff* and *Turtle* are currently out of commission.)

Mir I and Mir II (Russia)

The 6000 m-class submersibles *Mir I* and *Mir II* (**Figure 8**) were built in 1987 by Rauma Repola in

Figure 6   US submersible *Alvin*.

Finland for the P.P. Shirshov Institute of Oceanology in Russia. They are 7.8 m long, 3.8 m wide, 3.65 m high and weigh 18.7 tf in air. Inside diameter of the pressure hull, which is made of high-strength steel, is 2.1 m with a wall thickness of 40 mm. Launch and recovery take place by an articulated crane over the side of the support ship, the R/V *Academik Mistilav Keldysh*. If necessary, both *Mir I* and *Mir II* are

Figure 7   French 6000 m-class submersible *Nautile*.

Figure 8   Russian 6000 m-class submersible *Mir I* or *Mir II*.

launched simultaneously to carry out cooperative or independent research. Another characteristic feature of the *Mir* is a powerful secondary battery, 100 kWh of total energy, which allows it to stay more than 20 hours underwater, or to carry out more than 14 h of continuous operation on the bottom.

Shinkai 6500 (Japan)

The *Shinkai 6500* (**Figure 9**) was built in 1989 by Mitsubishi Heavy Industries and operated by the Japan Marine Science Technology Center (JAM-STEC). It is 9.5 m long, 2.71 m wide, 3.21 m high and weighs 25.8 tf in air. The pressure hull is made of titanium alloy, 73.5 mm in thickness, and has an inside diameter of 2 m. It is equipped with three view ports, two manipulators with seven degrees of freedom. Position of the submersible is calculated and displayed in real time by directly interrogating the seafloor transponders. Still color video images are transmitted automatically at 10 s intervals to the support ship, the R/V *Yokosuka*, through the acoustic link during the diving operation. Launch and recovery take place at the stern A-frame of the R/V *Yokosuka*.

Major Contributions of Deep Submersibles

The dive to the Challenger Deep in the Mariana Trench by the bathyscaph *Trieste* in 1960 was one of the most spectacular achievements of the twentieth century. However, the dive was mainly for adventure rather than for science. In 1963, the US nuclear submarine *Thresher* sank in 2500 m of water off Cape Cod in New England. After an extensive search for the submarine, the bathyscaph *Trieste* made dives to inspect the wreck in detail and recover small objects. The operation demonstrated the importance of using deep submersibles and advanced deep ocean technology to increase knowledge of the deep ocean. In 1966, hydrogen bombs were lost with a downed US B-52 bomber off Palomares, Spain. The *Alvin* showed the great utility of deep submersibles by locating and assisting in the recovery of lost objects from the sea.

Between 1973 and 1974, project FAMOUS (French–American Mid-Ocean Undersea Study) was conducted in the Mid-Atlantic Ridge off the Azores using the French bathyscaph *Archimede* and the US submersible *Alvin*. The project was the first systematic and successful use of deep submersibles for science. They discovered and sampled fresh pillow lavas and lava flows at 3000 m deep in the rift valley, where the oceanic crusts were being created, providing visual evidence of Plate Tectonics. In 1977, *Alvin* discovered a hydrothermal vent and vent animals in the East Pacific Rise off the Galapagos Islands at a depth of 2450 m. Discovery of these chemosynthetic animals, which were not dependent on photosynthesis, had a profound impact on biology in the twentieth century.

Manned submersibles now compete with unmanned submersibles, such as ROVs and AUVs. Because of the expense of operation and maintenance, national funding is necessary for manned

Figure 9 Japanese 6000 m-class submersible *Shinkai 6500*.

submersibles. However, ROVs and AUVs can be operated by private companies or institutions. In spite of the costs, the ability of the human observer to rapidly process information to make decisions provides an advantage and justifies continued use of manned submersibles.

See also

Deep Submergence, Science of. Manned Submersibles, Shallow Water. Platforms: Autonomous Underwater Vehicles.

Further Reading

Beebe W (1934) *Half Miles Down*. New York: Harcourt Brace.

Busby RF (1990) *Undersea Vehicles Directory – 1990–91*, 4th edn. Arlington, VA: Busby Associates.

Funnel C (ed.) (1999) *Jane's Underwater Technology*, 2nd edn. UK: Jane's Information Group Limited.

Kaharl VA (1990) *Water Baby – The Story of Alvin*. New York: Oxford University Press.

Piccard A (1956) *Earth Sky and Sea*. New York: Oxford University Press.

Piccard J and Dietz RS (1961) *Seven Miles Down*. New York: G.P. Putnam.

PLATFORMS: AUTONOMOUS UNDERWATER VEHICLES

J. G. Bellingham, Monterey Bay Aquarium Research Institute, Moss Landing, CA, USA

Introduction

Autonomous underwater vehicles (AUVs) are un-tethered mobile platforms used for survey operations by ocean scientists, marine industry, and the military. AUVs are computer-controlled, and may have little or no interaction with a human operator while carrying out a mission. Being untethered, they must also store energy onboard, typically relying on batteries. Motivations for using AUVs include such factors as ability to access otherwise-inaccessible regions, lower cost of operations, improved data quality, and the ability to acquire nearly synoptic observations of processes in the water column. An example of the first is operations under Arctic and Antarctic ice, an environment in which operations of human-occupied vehicles and tethered platforms are either difficult or impossible. Illustrating the next two points, AUVs are becoming the platform of choice for deep-water bathymetric surveying in the offshore oil industry because they are less expensive than towed platforms as well as produce higher-quality data (because they are decoupled from motion of the sea surface). Finally, the use of fleets of AUVs enables the rapid acquisition of distributed data sets over regions as large as $10\,000\,\mathrm{km}^2$.

AUVs are a new class of platform for the ocean sciences, and consequently are evolving rapidly. The Self Propelled Underwater Research Vehicle (SPURV) AUV, built at the University of Washington Applied Physics Laboratory, was first operated in 1967. However, adoption by the ocean sciences community lagged until the late 1990s. Adoption was spurred on by two developments: AUV development teams started supporting science field programs with AUV capabilities, and AUVs that nondevelopers could purchase and operate became available. The first served the purpose of building a user base and demonstrating AUV capabilities. The second enabled scientists to obtain and operate their own vehicles. Today, a wide variety of AUVs are available from commercial manufacturers. An even larger number of companies develop subsystems and sensors for AUVs.

The most common class of AUV in use today is a torpedo-like vehicle with a propeller at its stern, and steerable control surfaces to control turns and vertical motion (see **Figure 1**). These vehicles are used when speed or efficiency of motion is an important consideration. Such torpedo-like vehicles range in weight from a few tens of kilograms to thousands of kilograms. Most typical are vehicles weighing a few hundred kilograms, with an endurance of about a day at a speed of $c.\ 1.5\,\mathrm{m\,s^{-1}}$. Often they have parallel mid-bodies, which allow the vehicle length to be extended without large hydrodynamic consequences. This is useful when it is necessary to add new sensors or batteries to a vehicle. A disadvantage of torpedo-like vehicles is that, like an aircraft, they must maintain forward motion to generate lift over its control surfaces, and thus are not controllable at very low speed through the water.

Gliders are a class of vehicles that use changes in buoyancy rather than a propeller for propulsion. Gliders use their ability to control buoyancy to generate vertical motion. Vertical motion is translated into horizontal motion with lifting surfaces, usually wings mounted in about the middle of the vehicle (see **Figure 1**). Several types of gliders weighing about $50\,\mathrm{kg}$ are in use today. These comparatively small vehicles are designed to move slowly, about $0.25\,\mathrm{m\,s^{-1}}$, and operate sensors consuming a watt or less. By minimizing power consumption, these gliders can operate for periods of months using high-energy-density primary batteries. Disadvantages of this class of system are that they are limited to vertical profiling flight tracks, and can be overwhelmed by ocean currents, especially in the coastal environment or within boundary currents. However, larger gliders in development and testing will operate at higher speeds, and thus not suffer from this limitation.

A final class of AUVs uses multiple thrusters to provide capabilities similar to that of a helicopter or a ship with dynamic positioning (see **Figure 1**). The additional thrusters enable maneuvers such as hovering, translating sideways, and moving vertically. These vehicles are used when maneuverability is needed, for example, when operation near a very rough bottom is a necessity. The disadvantage is that the additional thrusters reduce efficiency for moving large distances or at high speeds.

Figure 1 From top-left corner clockwise: the *Hugin* AUV, the *Spray* glider (courtesy, MBARI), the *ABE* AUV (courtesy, Dana Yoerger, WHOI), and the *Dorado* AUV (courtesy, MBARI). *Hugin* and *Dorado* are examples of propeller-driven vehicles optimized for moving through the water efficiently, and have a torpedo-like configuration. *Spray* is an example of a glider, which is a buoyancy-driven vehicle that has no propeller. *ABE* is a highly maneuverable vehicle capable of hovering or pivoting in place, and of moving straight up or down. Also illustrated are different handling strategies. The large Hugin vehicle is launched and recovered from a ship using a stern ramp. *Spray* is hand-launched and recovered. *ABE* is launched and recovered with a crane. *Dorado* is shown being launched and recovered with a capture mechanism suspended from a J-frame.

While the vehicles described above are representative of the most commonly used systems, a wide range of other vehicles are in development or are in limited use. AUVs that come to the surface and use solar panels to recharge batteries have been demonstrated in seagoing operations. Gliders which extract energy from thermal differences in the ocean for propulsion have also been tested. A hybrid vehicle is being developed for reaching the deepest portion of the ocean. The hybrid vehicle operates as a tethered platform via a disposable fiber optic link for tasks requiring human perception, in other words as a remotely operated vehicle (ROV), but operates as an AUV when that link is severed. These are just a few of the diverse AUVs being developed to answer the needs of ocean science.

Terminology

The military term for an AUV is unmanned underwater vehicle, or UUV. This phrase is ambiguous in that it can also refer to ROVs. While ROVs are unmanned, they are tethered and designed to be operated by human, and thus are not considered autonomous. However, common usage is to employ UUV as a synonym for AUV. Military terminology is relevant as many AUVs in use by the scientific community were developed under navy funding, and the military continues to be the largest single investor in AUV technology. Consequently, the technical literature on AUVs also employs military terminology.

Basics of AUV Performance

Energy is a fundamental limitation for underwater vehicles. Thus, energy efficiency is a fundamental driver for vehicle design and operations. This section outlines the relationship between vehicle speed and endurance, and its dependence on factors such as power consumed by onboard systems.

A simple but useful model for power consumption P of an AUV is as follows:

$$P = P_{\text{prop}} + H \quad \text{where} \quad P_{\text{prop}} = \frac{1}{2}\frac{C_D A \rho v^3}{\eta} \quad [1]$$

Here the total electrical power consumed by the vehicle, P, is equal to the sum of propulsion power, P_{prop} and hotel load, H. Hotel load is simply the power consumed by all subsystems other than propulsion. Propulsion power is a function of the drag coefficient of the vehicle, C_D, the area of the vehicle, A, the density of water, ρ, the speed of the vehicle, v, and the efficiency of the propulsion system, η.

What are the typical values for the coefficients in eqn [1]? Consider an 'example vehicle' which is a 12 3/4″ (0.32 m) diameter torpedo-like AUV. Note that this is a standard for a mid-size class AUV. For such a vehicle, parameter values might be $C_D = 0.2$ (based on frontal area), $A = 0.082\,\text{m}^2$, and $\eta = 0.5$. Hotel load would depend on sensors, but an overall value of 30 W might be representative, although mapping sonars would consume much more power. We use $\rho = 1027\,\text{kg m}^{-3}$. Note that these numbers, except seawater density of course, can differ greatly from vehicle to vehicle. For example,

gliders optimized for low speed and long endurance can operate with hotel loads on the order of a watt or less, but at the cost of operating a few very simple sensors.

The power relationship provides insight into a variety of AUV design considerations. For example, the speed at which the vehicle will consume the least energy per unit distance traveled (the energetically optimum speed) can be computed from observing that power divided by vehicle speed equals energy per unit distance. Finding the minimum of P/v with respect to v yields the optimum speed from an energy conservation perspective:

$$v_{\text{opt}} = \left(\frac{\eta H}{C_D A \rho}\right)^{1/3} \quad [2]$$

For the example vehicle values given above, the optimum vehicle speed is approximately $1\,\text{m s}^{-1}$.

What can we say about vehicle performance at the energetically optimum speed? Substituting eqn [2] into eqn [1] we find that the power consumed at the optimum speed is $(3/2)H$ which for our example vehicle is 45 W. If the total energy capacity of the battery system is E_{cap}, then the maximum range of the vehicle will be:

$$d_{\text{max}} = \frac{2E_{\text{cap}}}{3}\left(\frac{\eta}{C_D A \rho H^2}\right)^{1/3} \quad [3]$$

If our example vehicle carries 10 kg of high-energy-density primary batteries providing a total of 1.3×10^7 J, it will have an endurance of 80 h, and a range of 280 km. The same vehicle with 10 kg of rechargeable batteries, with one-third the energy capacity of the high-energy-density batteries, will have its endurance and range reduced proportionately.

There are caveats to the above discussion. For example, propulsion efficiency, η, is typically a strong function of speed as the electrical motors used tend to have comparatively narrow ranges of efficiency. In practice, a vehicle's propulsion system is optimized for a particular speed and power. Also, vehicles are often operated at higher speeds than the energetically optimum speed given by eqn [2]. For example, operators of an AUV attended by a ship will be more sensitive to minimizing ship costs than to optimizing energy efficiency of the AUV.

AUV Systems and Technology

AUVs are highly integrated devices, containing a variety of mechanical, electrical, and software subsystems. **Figure 2** shows internal and external views of a deep-diving vehicle equipped with mapping

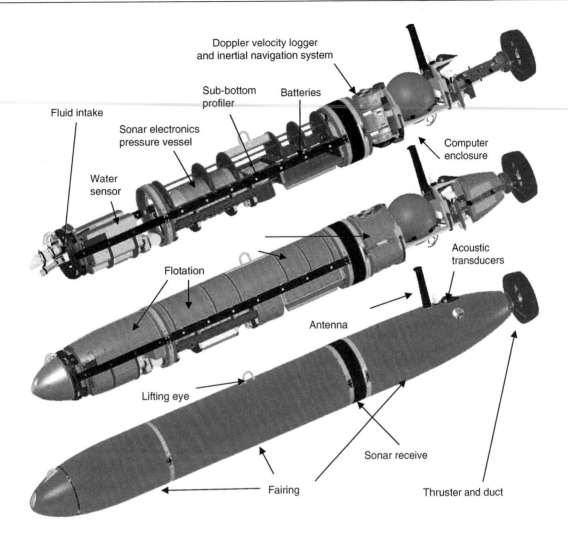

Figure 2 A propeller-driven, modular AUV with labeled subsystems. The top view shows the interior in which an internal mechanical frame supports pressure vessels, internal components, and the propulsion system. This AUV, called a *Dorado*, is a 'flooded' vehicle because the fairings are not watertight, and thus the interior spaces fill with water. Consequently internal components must all be capable of withstanding ambient pressures. Joining rings are visible between the yellow fairing segments on the lower figure. These allow the vehicle to be separated along its axis, allowing replacement or addition of hull sections. This allows reconfiguration of the vehicle with new payloads, and if desired, with new batteries. The propulsion system on this vehicle is a ducted thruster capable of being tilted both vertically and horizontally, to steer the vehicle in the vertical and horizontal planes. Courtesy of Farley Shane, MBARI.

sonars. The anatomy of an AUV typically includes the following subsystems:

- software and computers capable of managing vehicle subsystems to accomplish specific tasks and even complete missions in the absence of human control;
- energy storage to provide power;
- propulsion system;
- a system for controlling vehicle orientation and velocity;
- sensors for measuring vehicle attitude, heading, and depth;
- pressure vessels for housing key electrical components;

- navigation sensors to determine the vehicle position;
- communication devices to allow communication of human operators with the AUV;
- locating devices to allow operators to track the vehicle and locate it for recovery or in the case of emergencies;
- devices for monitoring vehicle health (e.g., leaks or battery failure);
- emergency systems for ensuring vehicle recovery in the event of failure of primary systems.

The mechanical design of an AUV must address issues such as drag, neutral buoyancy, the highly dynamic nature of launch and recovery, and the need to protect many delicate electrical components from seawater. The desire to operate large numbers of sensors for long distances encourages the construction of larger vehicles to hold the necessary equipment and batteries. However, the need for ease of handling and minimizing logistical costs encourages the design of smaller vehicles. Operational demands will also create constraints on vehicle design; for example, launch and recovery factors will impose the need for lift points and discourage external appendages that will be easily broken. The need to service vehicle components imposes a requirement that internal components be easily accessible for servicing and testing, and when necessary, replacement.

In addition, a host of supporting software and hardware are required to operate an AUV. Depending on the nature of the operations, supporting equipment will include:

- software and computers for configuring vehicle mission plans, and for reviewing vehicle data;
- systems for communicating with the vehicle both on deck and when deployed;
- systems for recharging and monitoring vehicle batteries;
- handling gear for transporting, deploying, and recovering AUVs;
- devices for detecting locating devices on the AUV;
- acoustic tracking systems for monitoring the location of the vehicle when in the vicinity of a support vessel.

AUV Mission Software

Functionally, AUV software must address a variety of needs, including: allowing human operators to specify objectives, managing vehicle subsystems to achieve mission objectives, logging data for subsequent review, and ensuring safety of the vehicle in the event of failures or unexpected circumstances. The software must be capable of managing vehicle sensors and control systems to maintain a set heading, speed, and depth. The software might also need to support interacting with a human operator during a mission. In addition to software on the vehicle itself, AUV operators rely on a suite of software applications to configure and validate missions, to maintain vehicle subsystems such as batteries, to review data generated by the vehicle, to prepare mission summaries, and when possible, to track and manage the vehicle while underway. The exponential growth of computational power available for both onboard and off-board computers, as well as the increasingly pervasive nature of the Internet, are supporting a steady increase in software capabilities for AUVs.

Most AUV missions involve sequential tasks such as descending from the surface to a set depth, then transiting to a survey location at a set speed, and then conducting a survey which might involve flying a lawn-mower pattern. The vehicle may be commanded to maintain constant altitude over the bottom if the vehicle is mapping the seafloor. A water-column mission might require the vehicle profile in the vertical plane, moving in a saw-tooth pattern called a yo-yo. The mission will likely include a transit from the end of the survey to a recovery location, with a final ascent to the surface for recovery. During the mission, the vehicle will monitor the performance of onboard subsystems, and in the event of detection of anomalies, like a low battery level, or a failed mission sensor, may abort the mission and return to the recovery point early. A more catastrophic failure might lead to the vehicle shutting down primary systems, and dropping a drop weight so as to float to the surface, and calling for help via satellite or direct radio frequency (RF) communications.

More complex vehicle missions can involve capabilities such as adapting survey operations to obtain better measurements, or managing tasks such as AUV docking. An example of the first might be as simple as a yo-yo mission which cues its vertical inflections from water temperature in order to follow a thermocline. Also in the category of adaptive, but more demanding, surveys, is the capability of following a thermal plume to its source, for example, when an AUV is used to search for hydrothermal vents. The docking of an AUV with an underwater structure encompasses yet a different type of complexity, created by the large number of steps in the process, and the high likelihood that individual steps will fail. For example, docking involves homing on a docking structure, orienting for final approach, engaging the dock, and making physical connections to establish power and communication links. Any one of these steps might fail due to external perturbations; for example, currents or turbulence in the marine environment might cause the vehicle to miss the dock. The vehicle must be able to detect failures and execute a process to recover and try again. Docking is representative of the increasingly complex capabilities AUVs are expected to master with high reliability.

Navigation

The ability for an AUV to determine its location on the Earth is essential for most scientific applications.

However, navigation in the subsea environment is complicated by the opacity of seawater to all but very low frequency electromagnetic radiation, rendering ineffective the use of commonly used technologies such as the Global Positioning System (GPS) and other radio-based navigation techniques. Consequently, navigation underwater relies primarily on various acoustic and dead-reckoning techniques and the occasional excursion to the surface where radio-based methods can be used. There is no single method of underwater navigation that satisfies all operational needs, rather a variety of methods are employed depending on the circumstances.

Dead-reckoning methods integrate a vehicle's velocity in time to obtain an updated location. In order to dead-reckon, the vehicle must know both the direction and speed of its travel. The simplest methods use a magnetic compass to determine direction, and use speed through the water as a proxy for Earth-referenced speed. However, the large number of error sources for magnetic compasses make measurement of heading to better than a degree accuracy technically challenging. Currents pose even more of a problem, as they may be comparable to the vehicle speed in amplitude, yet are not sensed by a water-relative measurement. Dead-reckoning is improved by measuring velocity relative to the seafloor, for example, using a Doppler velocity log (DVL) or a correlation velocity log. A DVL is commonly used by AUVs to measure velocity by measuring the Doppler shift of sound reflected off the seafloor. Correlation velocity logs are more complex in concept, involving measurement of the correlation of two pulses of sounds transmitted by the vehicle, reflected off the seafloor, and received by a hydrophone array. In practice, DVLs are used when a vehicle operates close to the seafloor, perhaps within 200 m, while correlation velocity logs are used when the vehicle is operating in mid-water columns or near the surface in deep water.

Inertial navigation system (INS) technology is well developed, as it is widely used for platforms like aircraft and missiles. However, INS units appropriate for underwater use are expensive enough that they are used only when navigation requirements are stringent, for example, for producing high-accuracy maps. A modern INS includes an array of accelerometers for measuring acceleration on three axes and a laser or fiber optic gyroscope for measuring changes in orientation. Additionally, an INS will include a GPS for initializing the unit's location and orientation, and a computer for acquiring and processing data from INS component sensors. The position reported by an INS will have an error which will grow in time, and thus it is important to constrain INS error with ancillary measurements of velocity and position. For example, combining an INS with a DVL for constraining velocity can result in a system which provides navigation accuracies better than 0.05% of distance traveled.

The two acoustic navigation methodologies most frequently used in AUV operations are ultrashort baseline navigation (USBL) and long baseline (LBL) navigation. A USBL system uses an array of hydrophone separated by a distance comparable to the wavelength of sound to measure the direction of propagation of an acoustic signal. Most often, a USBL system is mounted on a ship, and used to track a vehicle relative to the ship. With knowledge of the ship's location and orientation, the location of the AUV can also be determined. In contrast, LBL navigation acoustically measures the range between the vehicle and an array of widely separated devices of known location. A common LBL approach is to place transponders on the seafloor, and let the vehicle range off the transponders. The process of determining location using ranges from known locations is called spherical navigation, as the vehicle should be located at the intersection of spheres with the measured radius, centered on the respective transponders. An alternative LBL navigation method is to track a vehicle which pings at a preset time to an array of hydrophones at known locations. If the time of the ping is not known, the problem of solving for the vehicle location is called hyperbolic navigation, as only the difference in time of arrival of the ping at the various hydrophones can be determined, and this knowledge constrains the vehicle to be on a hyperbola between the respective receivers. If the time of the ping is known, perhaps triggered at a preselected time by a carefully calibrated clock, then the problem reduces to spherical navigation. In practice, a wide variety of USBL and LBL systems have been implemented for underwater navigation. They must all address the challenges of acoustic propagation in the ocean, which include the absorption of sound by seawater, diffraction by speed of sound variations in the underwater environment, scattering by reflecting surfaces, and acoustic noise generated by physical, geological, biological, and anthropogenic processes.

Other methods of navigation include using geophysical parameters, for example, water depth, to constrain the vehicle location in the context of known maps. These geophysically based navigation methods, similar to terrain contour mapping (TERCOM) navigation used by cruise missiles, depend on having good maps ahead of time. There are software approaches in development that simultaneously build maps and use those same maps for navigation.

These methods are called SLAM for simultaneous localization and mapping.

Using AUVs for Ocean Science

Mapping the Seafloor

AUVs are becoming the platform of choice for high-resolution seafloor maps. Obtaining high-resolution maps requires operating mapping sonars near the seafloor. Alternatives to an AUV include crewed submersibles and tethered platforms. Crewed submersibles are too valuable for routine mapping, and are reserved for other uses which require the presence of humans. Towed vehicles are used for sonar mapping, but have disadvantages as compared with AUVs, especially in deeper water. The principal problem is the high drag of the cable used for a tow sled, which in water depths of several thousand meters will limit speeds to approximately half a meter per second. Even at these slow speeds, a towed platform will stream behind the towing ship, creating several problems. Controlling the position of the towed vehicle over the bottom is very difficult, even when running on a constant heading. When surveying a defined area on the seafloor in a series of passes, the turns between passes may take longer than the actual survey passes themselves, as it is necessary to turn slowly to maintain control of the towed body. If ultrashort baseline acoustic navigation techniques are used to determine the vehicle position, then layback of the towed body behind the ship introduces significant errors as compared with having the ship directly over the sonar platform. For this reason, some commercial use of towed sonar platforms use two ships, one to tow the sonar platform, and one positioned directly over the platform to determine its precise location. Finally, surface motion of the ship will be efficiently coupled to the tow body by the tow cable. Thus, even near the seafloor, the tow body will be subject to sea state experienced by the ship. Consequently, attraction of the use of AUVs includes more economical operations and high data quality. **Figure 3** shows a cost comparison of a commercial deep-water towed survey and an equivalent AUV survey.

Sonar systems used on AUVs for mapping include multibeam sonar, side scan sonar, and sub-bottom profilers. Multibeam sonars, operating at frequencies of hundreds of kilohertz in the case of AUV-mounted systems, allow measurement of range to the seafloor in multiple sonar beams and are used to build up three-dimensional maps such as that in **Figure 4**. Side scan sonars used by AUVs also typically operate at frequencies of hundreds of kilohertz, and are used to image seafloor features. Side scan sonars are particularly useful for finding objects, for example, looking for a shipwreck resting

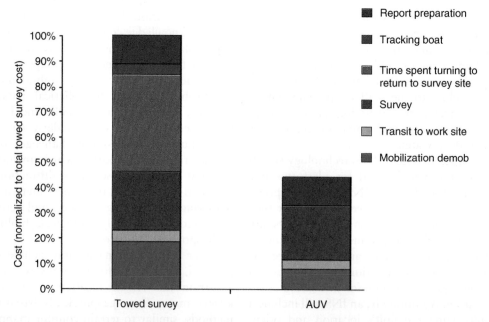

Figure 3 A comparison of the economics of deep survey taken from costs of a survey with a towed vehicle, and projected costs of the same survey with an AUV. The principal cost saving derives from the ability of the AUV to turn much faster than a deep-towed vehicle, reducing the total survey time. Also, the AUV can be acoustically tracked by its mother ship, while a towed vehicle requires a second ship for tracking because the towed vehicle will trail far behind the tow ship. Finally, mobilization and demobilization costs for the AUV can also be lower, although this depends on the size of the AUV employed.

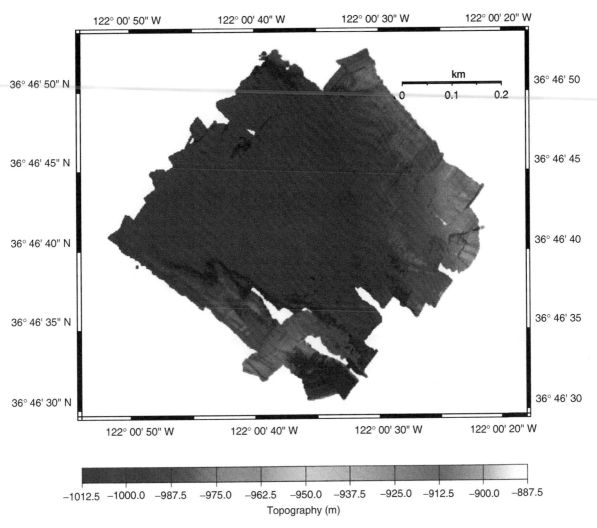

Figure 4   A bathymetric survey produced by an AUV at a depth of about 1000 m. Note the very small size of the survey area and high resolution of the bathymetry. Courtesy of Dave Caress, MBARI.

on the seafloor. Sub-bottom profilers use lower-frequency sound, ranging from 1 kHz to tens of kilohertz in the case of an AUV-mounted system, to penetrate into the seafloor. Depending on the bottom type (e.g., sandy, muddy, or rock), a sub-bottom system might penetrate tens of meters. Cumulatively sonar payloads will consume comparatively large amounts of energy, perhaps hundreds of watts. Mapping also requires high-fidelity navigation, and thus sonar-equipped AUVs will often also use more sophisticated navigation approaches, like inertial navigation. Consequently, mapping AUVs of today are larger, more sophisticated AUVs.

Observing the Water Column

AUVs provide a relatively new tool for observing the physical, chemical, and biological properties of the ocean. The smaller, buoyancy-driven gliders are unique in their combination of mobility and endurance, moving at about a quarter of a meter per second for periods of months. Larger vehicles carry more comprehensive payloads at higher speeds, but for shorter periods. Such vehicles might operate at $1.5\,\mathrm{m\,s^{-1}}$ for a day. A common flight profile is to fly the vehicle on a constant heading, while moving between two depth extremes in a saw-tooth pattern. Often the upper depth extreme will be close to the surface. This strategy allows the production of vertical sections of ocean properties, such as those in **Figure 5**. Variations of this strategy might have the vehicle moving in a lawn-mower or zigzag pattern in the horizontal plane, to develop a full three-dimensional map of ocean properties. **Figure 6** shows a visualization of an internal wave interacting with a phytoplankton layer using such a three-dimensional mapping strategy.

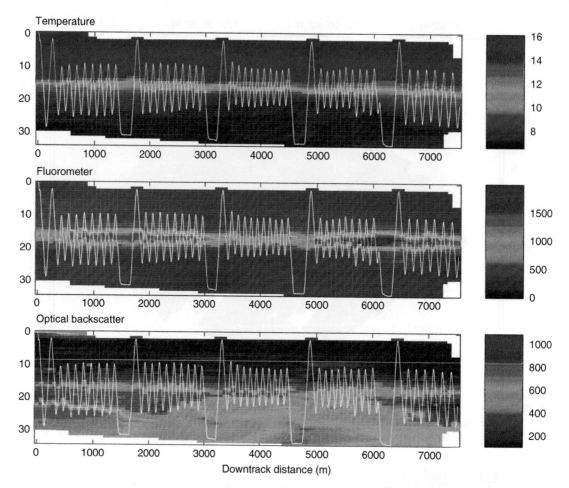

Figure 5 Vertical sections of water properties obtained by an *Odyssey* AUV operating in Massachusetts Bay. The *y*-axis of each figure is depth, in meters, and the *x*-axis is horizontal distance in meters. The top section shows temperature in degrees Celsius, the middle shows chlorophyll fluorescence in arbitrary units, and the bottom shows optical backscatter, also in arbitrary units. The path of the vehicle is shown as a white line, and the interpolated values of the measured property are plotted in color. The vehicle alternated between obtaining high-resolution observations of the thin layer of organisms at the thermocline with full water column profiles.

All AUVs are limited by the availability of sensors. Temperature, salinity, currents, dissolved oxygen, nitrate, optical backscatter properties, and chlorophyll fluorescence are examples of the growing *in situ* sensing capabilities available for AUVs. However, many important properties, for example, pH, dissolved carbon dioxide, and dissolved iron, cannot be measured reliably from a small moving platform. Furthermore, detection of marine organisms is usually accomplished by proxy; for example, chlorophyll fluorescence provides an indicator for phytoplankton abundance. *In situ* methods which directly detect, classify, and quantify marine organism abundance are not available, yet are increasingly important for understanding the structure and dynamics of ocean ecosystems.

Operations in Ice-Covered Oceans

AUVs offer unique operational capabilities for science in ice-covered oceans. Successful under-ice operation has been carried out with AUVs in both the Arctic and Antarctic. Sea ice poses special operational challenges for seagoing ocean scientists. For example, ships with ice-breaking capability can operate in the ice pack, but will typically not be able to hold station, or even assure that tethers and cables deployed over the side will not be severed. AUVs are attractive in that they provide horizontal mobility under ice, and the ability to conduct operations near the seafloor without the complications intrinsic in tether management. Challenges of operating AUVs under ice revolve around the need to assure return of the AUV to the ship for recovery, the process of recovering the AUV through the ice onto the ship, the potential for having an AUV fail and become trapped under ice, and the difficulty of carrying out tasks that would normally be accomplished having an AUV surface (e.g., obtaining a GPS update). Most safety strategies for AUVs in ice-free oceans default to bring

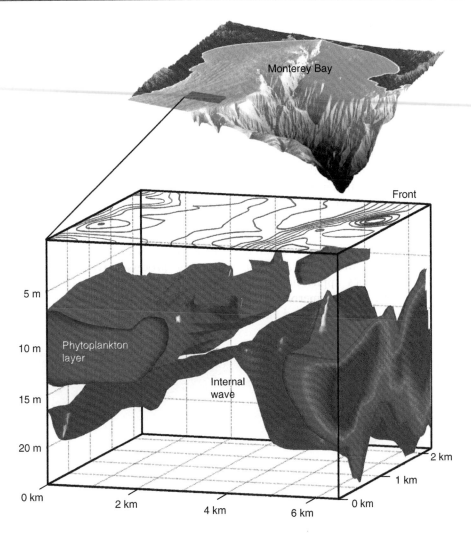

Figure 6 Interaction of layer of phytoplankton with an internal wave in Monterey Bay. Both physical and biological properties were measured by an *Odyssey* AUV, which moved in a horizontal zigzag pattern across the survey volume, while profiling constantly in the vertical plane. The phytoplankton layer, shown in green, was detected by a chlorophyll fluorescence sensor on the AUV. The cyan surface shows deflection of a level of constant density of seawater by a passing internal wave. Courtesy of John Ryan, MBARI.

the vehicle directly to the surface, for example, by dropping a weight. In the Arctic or Antarctic, this strategy could result in the vehicle becoming trapped under very thick ice, making the vehicle harder to find and potentially impossible to recover. The usual surface location devices such as RF beacons, strobes, and combinations of RF communication and satellite navigation will not work. Clearly AUV operations within the ice pack entail higher risk and a more sophisticated vehicle.

Observation Systems, Observatories, and AUVs

An understanding of power consumption of AUVs provides insight to the attractiveness of employing multiple vehicles for certain ocean observation problems. In some circumstances a survey must be accomplished within a set period. In oceanography,

time-constrained surveys are most often encountered when surveying a dynamic process. For example, if the temporal decorrelation of ocean fields associated with upwelling off Monterey Bay is about 48 h, attempts to map the ocean fields need to be accomplished within that time frame. Scales of spatial variability will also determine acceptable separation of observations: for example, decorrelation lengths in Monterey Bay are on the order of 20 km, so observations need to be spaced significantly closer to minimize errors in reconstructing the ocean field. How does this relate to the number of vehicles required to accomplish such a survey? Consider a grid survey of a 100 km × 100 km area with a resolution of 10 km. A single vehicle would have to travel c. 1000 km at a speed of nearly 6 m s^{-1}, traveling in a lawn-mower pattern. Using the example vehicle values from the 'Basics of AUV performance' section, the AUV would consume about 3500 W if it were capable of operating at such a high speed. In contrast, six of the same vehicles operating at their optimum speed would consume a total of 270 W. In other words, the six vehicles would consume 12 times less energy for the complete 48-h survey.

Autonomous mobile platforms are making observation of the interior of the ocean more affordable and more flexible, enabling the practical realization of coupled observation–prediction systems. For example, in late summer 2003, a diverse fleet of AUVs was deployed to observe and predict the evolution of episodic wind-driven upwelling in the environs of Monterey Bay. Over 21 different autonomous robotic systems, three ships, an aircraft, a coastal ocean dynamics application radar (CODAR), drifters, floats, and numerous fixed (moored) observation assets were deployed in the Autonomous Ocean Sampling Network (AOSN) II field program (**Figure 7**). Gliding vehicles, with an endurance of weeks to months, provided a continuous presence with a minimal sensor suite. A few propeller-driven vehicles provided observations of chemical and biological ocean parameters, allowing tracking of ecosystem response to the upwelling process. Observations were fed to two oceanographic models, which provided synoptic realization of ocean fields and predicted future conditions. Among the many lessons are an improved knowledge of the scales of variability of upwelling processes, an understanding of how to scale observation systems to these processes, and insights to strategies for adaptive sampling of comparatively rapidly changing processes with comparatively slow vehicles. These lessons are particularly relevant today, given the present emphasis on developing ocean-observing systems.

Figure 7 Example of a distributed observing system using AUVs. This diagram depicts an AOSN deployment in Monterey Bay.

Further Reading

Allmendinger EE (1990) *Submersible Vehicle Systems Design.* New York: SNAME.

Bradley AM (1992) Low power navigation and control for long range autonomous underwater vehicles. *Proceedings of the Second International Offshore and Polar Conference,* pp. 473–478.

Fossen T (1995) *Guidance and Control of Ocean Vehicles.* New York: Wiley.

Griffiths G (ed.) (2003) Technology and Applications of Autonomous Underwater Vehicles. London: Taylor and Francis.

IEEE (2001) Special Issue: Autonomous Ocean Sampling Networks. *IEEE Journal of Oceanic Engineering* 26(4): 437–446.

Jenkins SA, Humphreys DE, Sherman J, *et al.* (2003) *Underwater glider system study. Scripps Institution of Oceanography Technical Report No. 53.* Arlington, VA: Office of Naval Research.

Rudnick DL and Perry MJ (eds.) (2003) ALPS: Autonomous and Lagrangian Platforms and Sensors, Workshop Report, 64pp. http://www.geo-prose.com/ALPS (accessed Mar. 2008).

Relevant Website

http://www.mbari.org
  – Monterey Bay 2003 Experiment, Autonomous Ocean Sampling Network, MBARI.

DEEP SUBMERGENCE, SCIENCE OF

D. J. Fornari, Woods Hole Oceanographic Institution, Woods Hole, USA

Introduction

The past half-century of oceanographic research has demonstrated that the oceans and seafloor hold the keys to understanding many of the processes responsible for shaping our planet. The Earth's ocean floor contains the most accurate (and complete) record of geologic and tectonic history for the past 200 million years that is available for a planet in our solar system. For the past 30 years, the exploration and study of seafloor terrain throughout the world's oceans using ship-based survey systems and deep submergence platforms has resulted in unraveling plate boundary processes within the paradigm of sea floor spreading; this research has revolutionized the Earth and Oceanographic sciences. This new view of how the Earth works has provided a quantitative context for mineral exploration, land utilization, and earthquake hazard assessment, and provided conceptual models which planetary scientists have used to understand the structure and morphology of other planets in our solar system.

Much of this new knowledge stems from studying the seafloor – its morphology, geophysical structure, and characteristics, and the chemical composition of rocks collected from the ocean floor. Similarly, the discoveries in the late 1970s of deep sea 'black smoker' hydrothermal vents at the midocean ridge (MOR) crest (**Figure 1**) and the chemosynthetic-based animal communities that inhabit the vents have changed the biological sciences, provided a quantitative context for understanding global ocean chemical balances, and suggest modern analogs for the origin of life on Earth and extraterrestrial life processes. Intimately tied to these research themes is the study of the physical oceanography of the global ocean water masses and their chemistry and dynamics, which has resulted in unprecedented perspectives on the processes which drive climate and climate change on our planet. These are but a few of the many examples of how deep submergence research has revolutionized our understanding of our Earth and ocean history, and provide a glimpse at the diversity of scientific frontiers that await exploration in the years to come.

Enabling Deep Submergence Technologies

The events that enabled these breakthroughs was the intensive exploration that typified oceanographic expeditions in the 1950s to 1970s, and focused development of oceanographic technology and instrumentation that facilitated discoveries on many disciplinary levels. Significant among the enabling technologies were satellite communication and global positioning, microchip technology and the widespread development of computers that could be taken into the field, and increasingly sophisticated geophysical and acoustic modeling and imaging techniques. The other key enabling technologies which supplanted traditional mid-twentieth century methods for imaging and sampling the seafloor from the beach to the abyss were submersible vehicles of various types, remote-sensing instruments, and sophisticated acoustic systems designed to resolve a wide spatial and temporal range of ocean floor and oceanographic processes.

Oceanographic science is by nature multidisciplinary. Science carried out using deep submergence vehicles of all types has traditionally involved a wide range of research components because of the time and expense involved with conducting field research on the seafloor using human occupied submersibles or remotely operated vehicles (ROVs), and most recently autonomous underwater vehicles (AUVs) (**Figures 2–6**). Through the use of all these vehicle systems, and most recently with the advent of seafloor observatories, deep submergence science is poised to enter a new millennium where scientists will gain a more detailed understanding of the complex linkages between physical, chemical, biological, and geological processes occurring at and beneath the seafloor in various tectonic settings.

Understanding the temporal dimension of seafloor and sub-seafloor processes will require continued use of deep ocean submersibles and utilization of newly developed ROVs and AUVs for conducting time-series and observatory-based research in the deep ocean and at the seafloor. These approaches will provide new insights into intriguing problems concerning the interrelated processes of crustal generation, evolution, and transport of geochemical fluids in the crust and into the oceans, and origins and proliferation of life both on Earth and beyond.

Since the early twentieth century, people have been venturing into the ocean in a wide range of diving vehicles from bathyscapes to deep diving submersibles.

Figure 1 Hydrothermal vents on the southern East Pacific Rise axis at depths of 2500–2800 m. (A) Titanium fluid sampling bottles aligned along the front rail of *Alvin*'s basket in preparation for fluid sampling. (B) *Alvin*'s manipulator claw preparing to sample the hydrothermal sulfide chimney. (C) View from inside Alvin's forward-looking view port of the temperature probe being inserted into a vent orifice to measure the fluid temperature. (D) Hydrothermal vent after a small chimney was sampled which opened up the orifice through which the fluids are exiting the seafloor. Photos courtesy of Woods Hole Oceanographic Institution – Alvin Group, D. J. Fornari and K. Von Damm and M. Lilley.

Even in ancient times, there was written and graphic evidence of the human spirit seeking the mysteries of the ocean and seafloor. There is unquestionably the continuing need to take the unique human visual and cognitive abilities into the ocean and to the seafloor to make observations and facilitate measurements. For about the past 40 years submersible vehicles of various types have been developed largely to support strategic naval operations of various countries. As a result of that effort, the US deep-diving submersible *Alvin* was constructed. *Alvin* is part of the National Deep Submergence Facility (NDSF) of the University National Oceanographic Laboratories System (UNOLS) operated by the Woods Hole Oceanographic Institution (**Figure 7**). Alvin provides routine scientific and engineering access to depths as great as 4500 m. The US academic research community also has routine,

observational access to the deep ocean and seafloor down to 6000 m depth using the ROV and tethered vehicles of the NDSF (**Figure 7**). These vehicle systems of the NDSF include the ROV *Jason*, and the tethered optical/acoustic mapping systems *Argo II* and DSL-120 sonar (a 120 kHz split-beam sonar system capable of providing 1–2 m pixel resolution back-scatter imagery of the seafloor and phase-bathymetric maps with ∼4 m pixel resolution (**Figure 7**). These fiberoptic-based ROV and mapping systems can work at depths as great as 6000 m.

Alvin has completed over 3600 dives (more than any other submersible of its type), and has participated in making key discoveries such as: imaging, mapping, and sampling the volcanic seafloor on the MOR crest; structural, petrological, and geochemical studies of transforms faults; structural

(A)

(A)

(B)

Figure 2 (A) The submersible *Alvin* being lifted onto the stern of R/V *Atlantis*, its support ship. (B) *Alvin* descending to the seafloor. Photos courtesy of Woods Hole Oceanographic Institution– Alvin Group and R. Catanach.

(B)

Figure 3 (A) The ROV *Jason* being lifted off the stern of a research vessel at the start of a dive. (B) ROV *Jason* recovering amphora on the floor of the Mediterranean Sea. Photos courtesy of Woods Hole Oceanographic Institution – Alvin Group and R. Ballard.

studies of portions of deep-sea trenches off Central America; petrological and geochemical studies of volcanoes in back-arc basins in the western Pacific Ocean; sedimentary and structural studies of submarine canyons, discovering MOR hydrothermal vents; and collecting samples and making time series measurements of biological communities at hydrothermal vents in many MOR settings in the Atlantic and Pacific Oceans. In 1991, scientists in *Alvin* were also the first to witness the vast biological repercussions of submarine eruptions at the MOR axis,

which provided the first hint that a vast subsurface biosphere exists in the crust of the Earth on the ocean floor (**Figure 7**).

Deep Submergence Science Topics

Some of the recent achievements in various fields of deep submergence science include the following.

1. Discoveries of deep ocean hydrothermal communities and hot ($>350°C$) metal-rich vents on many segments of the global mid-ocean ridge (MOR);
2. Documentation of the immediate after-effects of submarine eruptions on the northern East Pacific

Figure 4 Photographs taken from the ROV *Jason* at hydrothermal vents on the Mid-Atlantic Ridge near 37°N on the summit ofLucky Strike Seamount at a depth of 1700 m. All photographs are of a vent named 'Marker d4.' (A) Overall view of vent lookingNorth. White areas are anhydrite and barite deposits, yellowish areas are covered with clumps of vent mussels. (B) Close-up of theside of the vent, *Jason*'s sampling basket is in the foreground. (C) Close-up view of mussels on the side of the hydrothermalchimney; the hydrothermal vent where hot fluids are exiting the mound is at the upper right, where the image is blurry because ofthe shimmering effect of the hot water. Nozzles of a titanium sampling bottle are at the middle-right edge. (D) Insertinga self-recording temperature probe into a beehive chimney. (E) Titanium fluid sampling bottles being held by *Jason*'s manipulatorduring sampling. (F) Close-up of nozzles of sampling bottle inserted into vent orifice during sampling. Photos (D)}(F) are framegrabs of *Jason* video data. Photos courtesy of Woods Hole Oceanographic Institution – ROV Group, D. J. Fornari, S. Humphris, andT. Shank.

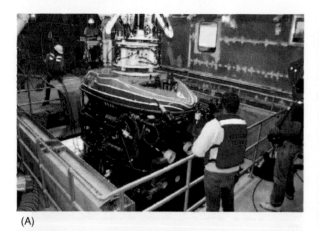

(A)

(B)

Figure 5 (A) ROV *Tiburon* of the Monterey Bay Aquarium Research Institute (MBARI) in the hanger of its support ship R/V *Western Flyer*. This electric-powered ROV can dive to 4000 m. A steel-armored, electro-optical cable connects *Tiburon* to the R/V *Western Flyer* and delivers power to the vehicle. Electric thrusters allow fine maneuvering while minimizing underwater noise and vehicle disturbance. A variable buoyancy control system, together with the syntactic foam pack, enables *Tiburon* to hover inches above the seafloor without creating turbulence, to pick up a rock sample, or maneuver quickly to follow an animal. (B) ROV *Ventana* of the MBARI being launched from its support ship R/V *Pt. Lobos*. This ROV gives researchers the opportunity to make remote observations of the seafloor to depths of 1850 m. The vehicle has two manipulator arms } a seven-function arm with five spatially correspondent joints and another seven-function robot arm with six spatially correspondent joints. Both arms can use a variety of end effectors to suit the type of work being done. *Ventana* is also equipped with a conductivity, temperature and density (CTD) package including a dissolvedoxygen sensor and a transmissometer. Photos courtesy of MBARI.

Rise, Axial Seamount, Gorda Ridge and CoAxial Segment of the Juan de Fuca Ridge;
3. Utilization of Ocean Drilling Program bore holes and specialized vehicle systems (e.g. Scripps Institution's Re-Entry Vehicle) (**Figure 8**) and instrument suites (e.g. CORKs) (**Figure 9**) for a wide range of physical properties, fluid flow and seismological experiments;

Figure 6 The autonomous underwater vehicle (AUV) ABE (Autonomous Benthic Explorer) of the Woods Hole Oceanographic Institution's Deep Submergence Laboratory, which can survey the seafloor completely autonomously to depths up to 4000 m and is especially well suited to working in rugged terrain such as is found on the Mid Ocean Ridge. Photo courtesy of Woods Hole Oceanographic Institution.

4. Discoveries of extensive fluid flow and vent-based biological communities along continental margins and subduction zones;
5. Initial deployment of ocean floor observatories of various types which enable the monitoring and sampling of geological, physical, biological and chemical processes at and beneath the seafloor (**Figures 10** and **11**).

These studies have revolutionized our concepts of deep ocean processes and highlighted the need for more detailed, time-series, multidisciplinary research.

Within the field of biological oceanography, major recent advances have come from the study of the new life forms and chemoautotrophic processes discovered at hydrothermal vents (**Figure 10**). These advances have fundamentally altered

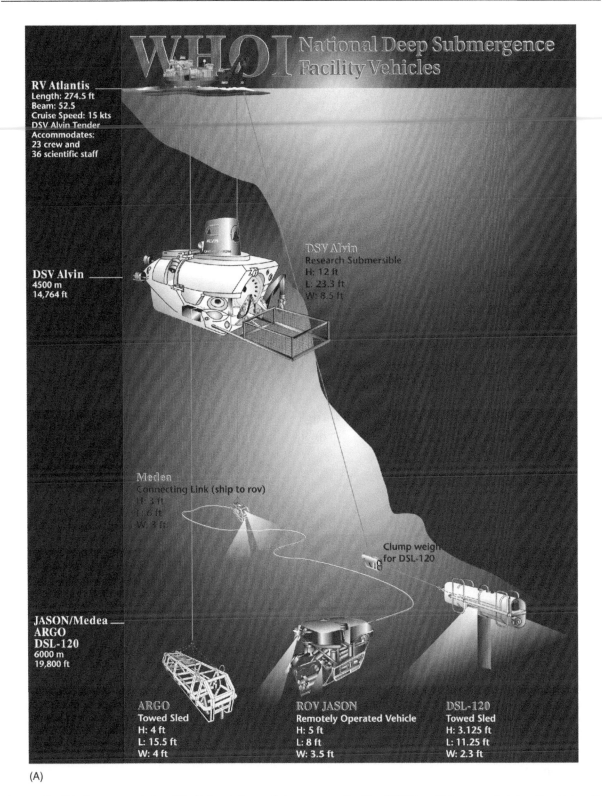

(A)

Figure 7 (A) Summary of the US National Deep Submergence Facility (NDSF) vehicles operated for the University NationalOceanographic Laboratories System (UNOLS) by the Woods Hole Oceanographic Institution (WHOI). (B) Montage of the NDSFvehicles showing examples of the various types of data they collect. The figure also shows the nested quality of the surveysconducted by the various vehicles which allows scientists to explore and map features with dimensions of tens of kilometers (topright multibeam sonar map), to detailed sonar back scatter and bathymetry swaths which have pixel resolution of 1–2 m (DSL-120sonar), which are then further explored with the Argo II imaging system, or sampled using *Alvin* or ROV Jason. Graphics by P.Oberlander, WHOI; photos courtesy of WHOI – Alvin and ROV Group.

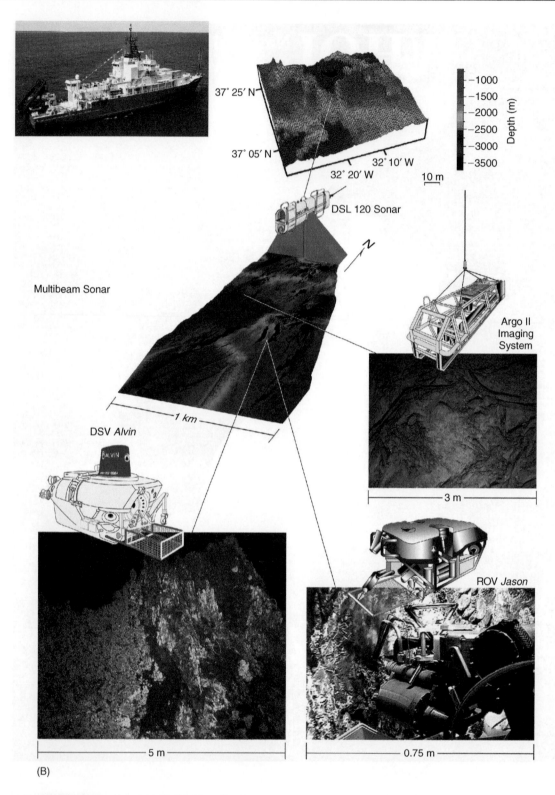

Figure 7   (Continued)

biological classification schemes, extended the known thermal and chemical limits of life, and have pushed the search for origins of life on Earth as well as for new life forms on other planetary bodies.

Recent marine biological studies show that: (1) the biodiversity of every marine community is vastly greater than previously recognized; (2) both sampling statistics and molecular tools indicate that

(A)

Figure 8 (A) The Scripps Institution of Oceanography's Control Vehicle is a specialized ROV that can place instrument strings inside deep-sea boreholes, using a conventional oceanographic vessel capable of dynamic positioning and equipped with a winch carrying 17.3 mm (0.68 in) electromechanical cable with a single coax (RG8-type). The control vehicle is 3.5 m tall and weighs about 500 kg in water (1000 kg in air). It consists of a stainless-steel frame that contains two orthogonal horizontal hydraulic thrusters, a compass, a Paroscientific pressure gauge, four 250 W lights, a video camera, sonar systems, electronic interfaces to electrical releases and to a logging probe, and electronics to control all these sensors and handle data telemetry to and from the ship. Telemetry on the tow cable's single coax is achieved by analog frequency division multiplexing over a frequency band extending from 20 kHz to about 800 kHz. The sonars include a 325 kHz sector-scanning sonar, a 23.5 kHz narrow beam acoustic altimeter, and a 12 kHz sonar for long baseline acoustic navigation. This system was used successfully in the 1998 Ocean Seismic Network Pilot Experiment at ODP Hole 843B, 225 km south-west of the island of Oahu, Hawaii. In 1999, the control vehicle's analog telemetry module was converted into a digital system using fiberoptic technology thus providing bandwidth capabilities in excess of 100 Mbaud. (B) Cartoon showing the configuration of the control vehicle deployed from a research ship as it enters an Ocean Drilling Program bore hole to insert an instrument string. Photo and drawing courtesy of Scripps Institution of Oceanography – Marine Physical Laboratory, F. Spiess, and C. de Moustier.

the large majority of marine species have not been described; (3) the complexity of biological communities is far greater than previously realized; and (4) the response of various communities to both natural and anthropogenic forcing is far more complex than had been understood even a decade ago. Focused studies over long time periods will be required to characterize these communities fully.

The field of marine biology is heading toward a more global time-series approach as a function of recent discoveries largely in the photic zone of the oceans and in the deep ocean at MOR hydrothermal vents. The marine biological, chemical, and physical oceanographic research which will be carried out in the next decade and beyond will certainly have a profound impact on our understanding of the complex food webs in the ocean which control productivity at every level and have direct implications for commercial harvesting of a wide range of resources from the ocean. Meeting the challenge of deciphering the various chemical, biological, and physical influences on these phenomena will require a better understanding and resolution of the causes and consequences of change on scales from hours to millennia. Understanding ocean ecosystems and their constituents will improve dramatically in response to emerging molecular, chemical, optical, and acoustical technologies. Given the relative paucity of information on deep-sea fauna in general, and especially the relatively recent discovery of chemosynthetic ecosystems at MOR hydrothermal vents, this will continue to be a focus for deep ocean biological research in the coming decade and beyond. Time-series and observatory-based research and sampling techniques will be required to answer the myriad of questions regarding the evolution and physiology of these unique biological systems (**Figures 10** and **11**).

Present and future foci for deep submergence science is MOR crests, hydrothermal systems, and the volcano–tectonic processes that create the architecture of the Earth's crust. Geochemists from the marine geology and geophysics community have emphasized the need for studies of: (1) the flux-frequency distribution for ridge-crest hydrothermal activity (heat, fluid, chemistry); (2) the role played by fluid flow in gas hydrate accumulation and determination of how important hydrates are to climate change; (3) slope stability; and (4) determination of how much of a role the microbial community plays in subsurface chemical and physical transformations. Another focus involves subduction zone processes, including: an assessment of the fluxes of fluids and solids through the seismogenic zone; long-term monitoring of changes in

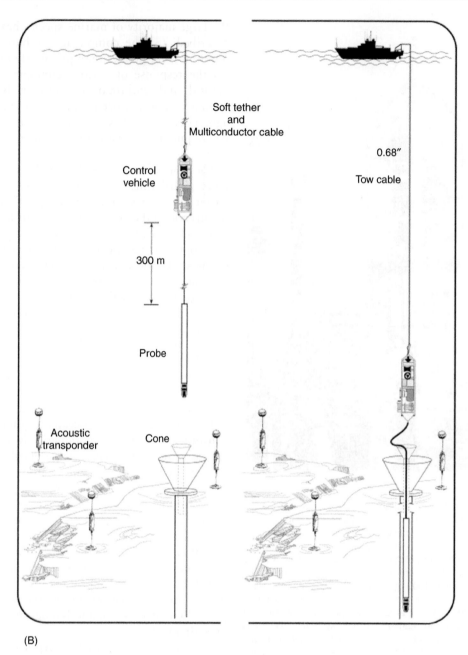

(B)

Figure 8b   (Continued)

seismicity, strain, and fluid flux in the seismogenic zone; and determining the nature of materials in the seismogenic zone.

To answer these types of questions, the marine geology and geophysics community requires systematic studies of temporal evolution of diverse areas on the ocean floor through research that includes mapping, dating, sampling, geophysical investigations, and drilling arrays of crustal holes (**Figures 8, 9** and **11**). The researchers in this field stress that the creation of true seafloor observatories at sites with different tectonic variables, with continuous monitoring of geological, hydrothermal, chemical, and biological activity will be necessary. Whereas traditional geological and geophysical tools will continue to provide some means to address aspects of these problems, it is clear that an array of deep submergence vehicles, *in situ* sensors, and ocean floor observatory systems will be required to address these topics and unravel the variations in the processes that occur over short (seconds/minutes) to decadal timescales. The infrastructural requirements, facility, and development needs required to support the research questions to be asked include: a capability for long-

(A)

(B)

(C)

Figure 9 (A) Diagram of the upper portion of an OceanDrilling Program (ODP) borehole with a Circulation ObviationRetrofit Kit (CORK) assembly. These units serve the samepurpose as a 'cork' which seals a bottle; in the deep-seacase, the bottle is the seafloor which contains fluids that arecirculating in the ocean crust. The CORK allows scientists toaccess the circulating fluids and make controlled hydrologicmeasurements of the pressures and physical properties of thefluids. (B) A CORK observatory on the ocean floor in ODP hole858G off the Pacific north-west coast. (C) A CORK with instrumentsinstalled to measure sub-seafloor fluid circulation processes.Diagram courtesy of Woods Hole Oceanographic Institutionand J. Doucette; photo courtesy of K. Becker andE. Davis.

term seafloor monitoring; effective detection and response capability for a variety of seafloor events (volcanic, seismic, chemical); adequately supported, state-of-the-art seafloor sampling and observational facilities (e.g. submersible, ROVs, and AUVs), and accurate navigation systems, software, and support for shipboard integration of data from mutiscalar and nested surveys (**Figures 7B, 11** and **12**).

As discussed above, the disciplines involved in deep-submergence science are varied and the scales

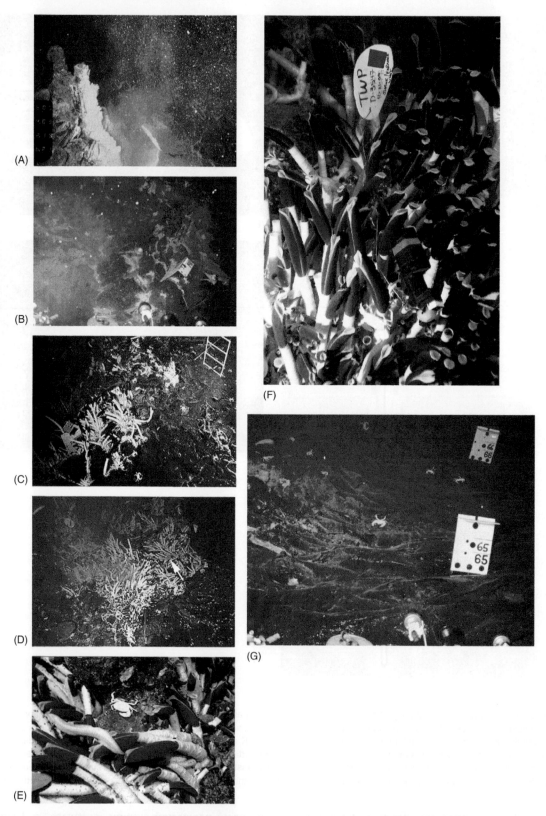

Figure 10 Time-series sequence of photographs taken of the same area of seafloor from the submersible *Alvin* of a hydrothermal vent site on the East Pacific Rise axis near 9°49.8′N at a depth of 2500 m. (A) 'Snow blower' vent spewing white bacterial by-product during the 1991 eruption. (B) Same field of view as (A) about 9 months later. Diffuse venting is still occurring as is bacterial production. White areas in the crevices of the lava flow are juvenile tube worms. (C) Patches of Riftia tube worms colonizing the vent area ~18 months after the March 1991 eruption. (D) Tube worm community has continued to develop, and the venting continues over 5 years after the eruption. (E) Close-up photograph of zoarcid vent fish (middle-left), tube worms, mussels (yellowish oblong individuals) and

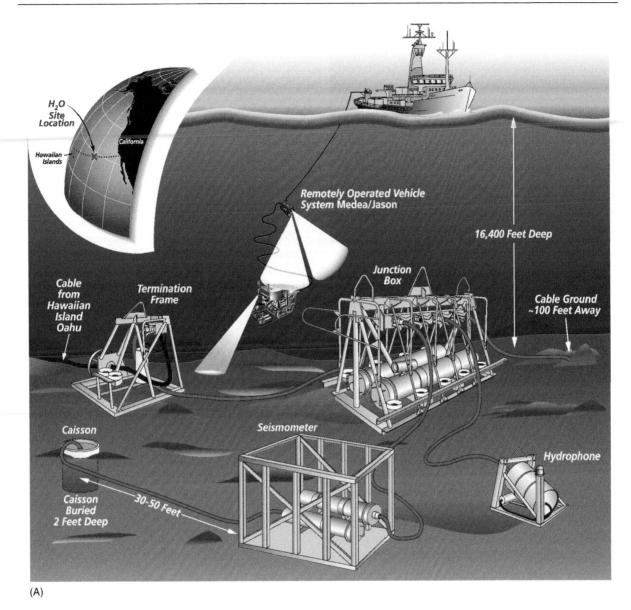

(A)

Figure 11 (A) Diagram of the deployment of the Hawaii-2 Observatory (H2O) one of the first long-term, deep seafloor observatories deployed in the past few years. Scientists used the ROV *Jason* to splice an abandoned submarine telephone cable into a termination frame which acts as an undersea telephone jack. Attached by an umbilical is a junction box, which serves as an electrical outlet for up to six scientific instruments. An ocean bottom seismometer and hydrophone are now functioning at this observatory. (B) The H2O junction box as deployed on the seafloor and photographed by ROV *Jason*. Drawing courtesy of Jayne Doucette, Woods Hole Oceanographic Institution (WHOI), A. Chave, and WHOI–ROV group.

of investigation range many orders of magnitude from molecules and micrometer-sized bacteria to segment-scales of the MOR system (10s to 100s of kilometers long) at depths that range from 1500 m to 6000 m and greater in the deepest trenches. Clearly, the spectrum of scientific problems and environments where they must be investigated require access to the deep ocean floor with a range of safe, reliable, multifaceted, high-resolution vehicles, sensors, and samplers, operated from support ships that have global reach and good station-keeping capabilities in rough weather. Providing the right complement of

bathyurid crab (center). (F) A time-series temperature probe (with black and yellow tape), deployed at a hydrothermal vent (Tube worm pillar) on the East Pacific Rise crest near 9°49.6'N at a depth of 2495 m. The vent is surrounded by a large community of tube worms. (G) Seafloor markers along the Bio-Geo Transect, a series of 210 markers placed on the seafloor in 1992 to monitor the changes in hydrothermal vent biology and seafloor geology over a 1.4 km long section of the East Pacific Rise axis that have occurred after the 1991 volcanic eruption at this site. Photographs courtesy of T. Shank, Woods Hole Oceanographic Institution and R. Lutz, Rutgers University, D. J. Fornari and Woods Hole Oceanographic Institution – Alvin Group.

(B)

Figure 11 (Continued).

deep-submergence vehicles and versatile support ships from which they can operate, and the funding to operate those facilities cost-effectively, is both a requirement and a challenge for satisfying the objectives of deep-sea research in the coming decade and into the twenty-first century.

To meet present and future research and engineering objectives, particularly with a multidisciplinary approach, deep submergence science will require a mix of vehicle systems and infrastructures. As deep submergence science investigations extend into previously unexplored portions of the global seafloor, it is critical that scientists have access to sufficient vehicles with the capability to sample, observe and make time-series measurements in these environments. Submersibles, which provide the cognitive presence of humans and heavy payload capabilities will be critical to future observational, time-series and observatory-based research in the coming decades. Fiberoptic-based ROVs and tethered systems, especially when used in closely timed, nested investigations offer unparalleled maneuverability, mapping and sampling capabilities with long bottom times and without the limitation of human/vehicle endurance. AUVs of various designs will provide unprecedented access to the global ocean, deep ocean and seafloor without dedicated support from a surface ship.

AUVs represent vanguard technology that will revolutionize seafloor and oceanographic measurements and observations in the decades to come. Over approximately the past 5 years scientists at several universities and private laboratories have made enormous advances in the capabilities and field-readiness of AUV systems. One such system is the Autonomous Benthic Explorer (ABE) developed by engineers and scientists at the Woods Hole Oceanographic Institution (**Figure 6**). ABE can survey the

seafloor completely autonomously and is especially well suited to working in rugged terrain such as is found on the MOR. ABE maps the seafloor and the water column near the bottom without any guidance from human operators. It follows programmed track-lines precisely and follows the bottom at heights from 5 to 30 m, depending on the type of survey conducted. ABE's unusual shape allows it to maintain control over a wide range of speeds. Although ABE spends most of its survey time driving forward at constant speed, ABE can slow down or even stop to avoid hitting the seafloor. In practice, ABE has surveyed areas in and around steep scarps and cliffs, and not only survived encounters with the extreme terrain, but also obtained good sensor data throughout the mission.

Recently ABE has been used for several geological and geophysical research programs on the MOR in the north-east and south Pacific which have further proved its reliability as a seafloor survey vehicle, and pointed to its unique characteristics to collect detailed, near-bottom geological and geophysical data, and to ground-truth a wide range of seafloor terrains (**Figure 12**). These new perspectives on seafloor geology and insights into the geophysical properties of the ocean crust have greatly improved our ability to image the deep ocean and seafloor and have already fostered a paradigm shift in field techniques and measurements which will surely result in new perspectives for Earth and oceanographic processes in the coming decades.

Conclusions

One of the most outstanding scientific revelations of the twentieth century is the realization that ocean processes and the creation of the Earth's crust within the oceans may determine the livability of our planet in terms of climate, resources, and hazards. Our discoveries may even enable us to determine how life itself began on Earth and whether it exists on other worlds. The next step is toward discovering the linkages between various phenomena and processes in the oceans and in exploring the interdependencies of these through time. Marine scientists recognize that technological advances in oceanographic sensors and vehicle capabilities are escalating at a increasingly rapid pace, and have created enormous opportunities to achieve a scope of understanding unprecedented even a decade ago. This new knowledge will build on the discoveries in marine sciences over the last several decades, many of which have been made possible only through advances in vehicle and sensor technology. With the rapidly escalating advances in technology,

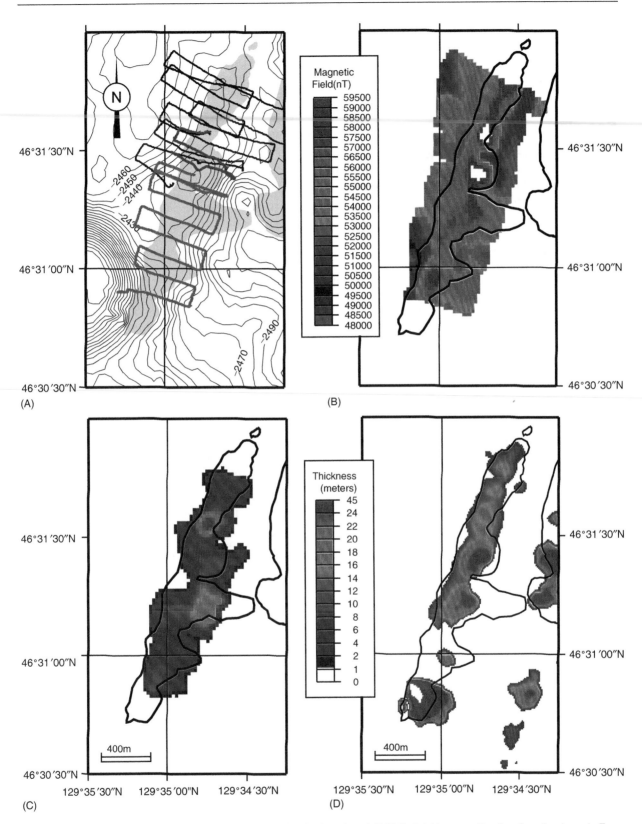

Figure 12 (A) Bathymetry map (contour interval 10 m) showing location of 1993 CoAxial lava eruption (gray) on the Juan de Fuca Ridge off the coast of Washington, and ABE tracklines (each color is a separate dive). (B) Magnetic field map based on ABE tracklines showing strong magnetic field over new lava flow. (C) Computed lava flow thickness assuming an average lava magnetization of 60 A/m compared with (D) lava flow thickness determined from differential swath bathymetry. Figure courtesy of M. Tivey, Woods Hole Oceanographic Institution.

marine scientists agree that the time is ripe to focus efforts on understanding the connections in terms of interdependency of phenomena at work in the world oceans and their variability through time.

See also

Manned Submersibles, Deep Water. Mid-Ocean Ridge Geochemistry and Petrology. Seamounts and Off-Ridge Volcanism. Platforms: Autonomous Underwater Vehicles.

Further Reading

Becker K and Davis EE (2000) Plugging the seafloor with CORKs. *Oceanus* 42(1): 14–16.

Chadwick WW, Embley RW, and Fox C (1995) Seabeam depth changes associated with recent lava flows, coaxial segment, Juan de Fuca Ridge: evidence for multiple eruptions between 1981–1993. *Geophysical Research Letters* 22: 167–170.

Chave AD, Duennebier F, and Butler R (2000) Putting H2O in the ocean. *Oceanus* 42(1): 6–9.

de Moustier C, Spiess FN, Jabson D, *et al.* (2000) Deep-sea borehole re-entry with fiber optic wire line technology. *Proceedings 2000 of the International Symposium on Underwater Technology.* pp. 23–26.

Embley RW and Baker E (1999) Interdisciplinary group explores seafloor eruption with remotely operated vehicle. *Eos Transactions of the American Geophysical Union* 80(19): 213–219 222.

Fornari DJ, Shank T, Von Damm KL, *et al.* (1998) Time-series temperature measurements at high-temperature hydrothermal vents, East Pacific Rise 9°49'–51'N:

monitoring a crustal cracking event. *Earth and Planetary Science Letters* 160: 419–431.

Haymon RH, Fornari DJ, Von Damm KL, *et al.* (1993) Direct submersible observation of a volcanic eruption on the Mid-Ocean Ridge: 1991 eruption of the East Pacific Rise crest at 9°45'–52'N. *Earth and Planetary Science Letters* 119: 85–101.

Humphris SE, Zierenberg RA, Mullineaux L, and Thomson R (1995) *Seafloor Hydrothermal systems: Physical, Chemical, Biological, and Geological Interactions.* American Geophysical Union Monograph, vol. 91, 466 pp.

Ryan WBF (chair) *et al.,* (Committee on seafloor observatories: challenges and opportunities) (2000) *Illuminating the Hidden Planet, the Future of Seafloor Observatory Science.* Washington, DC: Ocean Studies Board, National Research Council, National Academy Press.

Shank TM, Fornari DJ, Von Damm KL, *et al.* (1998) Temporal and spatial patterns of biological community development at nascent deep-sea hydrothermal vents along the East Pacific Rise, 9°49.6'N–9°50.4'N. *Deep Sea Research, II* 45: 465–515.

Tivey MA, Johnson HP, Bradley A, and Yoerger D (1998) Thickness measurements of submarine lava flows determined from near-bottom magnetic field mapping by autonomous underwater vehicle. *Geophysical Research Letters* 25: 805–808.

UNOLS (University National Laboratory System) (1994) *The Global Abyss: An Assessment of Deep Submergence Science in the United States.* Narragansett, RI: UNOLS Office, University of Rhode Island.

Von Damm KL (2000) Chemistry of hydrothermal vent fluids from 9–10°N, East Pacific Rise: 'Time zero' the immediate post-eruptive period. *Journal of Geophysical Research* 105: 11203–11222.

CONVERGENT & DIVERGENT BOUNDARIES

ACCRETIONARY PRISMS

J. C. Moore, University of California at Santa Cruz, Santa Cruz, CA, USA

Introduction

Subduction of oceanic lithosphere along a convergent plate boundary transfers sediments and rocks from the underthrust lithosphere to the overriding plate, producing an accretionary prism. Accretionary prisms develop beneath the inner slopes of the deep ocean trenches that typically mark convergent plate boundaries. The subduction process destabilizes the mantle after about 100 km of underthrusting beneath the upper plate to produce magmas of the volcanic arcs that virtually always occur along convergent plate boundaries (**Figure 1**). As accretionary prisms grow through addition of oceanic material, they become coastal mountain ranges. When a continent collides with a subduction zone, the intervening accretionary prism becomes incorporated into the resultant great mountain belts. Thus, rocks in accretionary prisms sometimes are the only record of ancient vanished ocean basins. Accretionary prisms typically form on the upper plate of subduction zone thrust faults, which host the world's largest earthquakes. Because accretionary prisms incorporate soft sediments at high rates of deformation, they produce some of the world's most complexly deformed rocks,

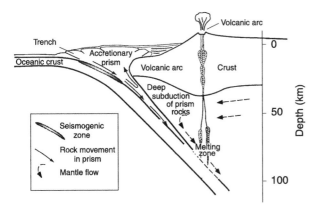

Figure 1 The setting of an accretionary prism in a generalized cross-section of a convergent plate boundary. Although the accretionary prism builds up primarily by scraping off material riding on the oceanic crust, some portions are deeply underthrust and flow back to the surface, while other portions are deeply subducted and participate in the formation of the igneous rocks of the volcanic arc.

commonly called melanges. Sediments offscraped to form accretionary prisms are like sponges that yield fluids as they are squeezed and deformed during prism growth. The fluids affect the mechanics of faults; chemically dissolve, transfer, and deposit material; and support chemosynthetic biological communities. The shape of the accretionary prism is mechanically controlled by the strength of the material comprising the accretionary prism and its internal fluid pressure. At slightly fewer than half of modern convergent plate boundaries, accretionary prisms are not currently forming and the incoming sediment and rock is deeply underthrust. Sometimes the accretionary process is reversed and the upper plates of subduction zones are mechanically abraded or eroded by the underthrust plate, causing subsidence and contraction of the overriding plate.

Origin and Variation of Materials Incorporated in Accretionary Prisms

Accretionary prisms vary in composition depending on the type of material on the subducting oceanic plate. The ideal sequence incoming to a subduction zone consists of oceanic basaltic igneous rocks covered by oceanic or pelagic sediments that change up-section to more rapidly deposited, continentally derived sandstones, shales, and even conglomerates. At a subduction zone starved for sediments, material available for accretionary prism construction may be the igneous rocks of the oceanic plate with thin overlying sedimentary deposits. Alternatively, incoming plates may be sediment-dominated and covered with a kilometers-thick sequence of deposits that are available for accretion. The resulting accretionary prisms may consist of slices of oceanic igneous rocks with minor amounts of interspersed sediments to thick thrust sheets of continentally derived clastic rocks. The Marianas subduction zone of the western Pacific is an example of the former, whereas the Cascadia subduction zone off the northwestern United States and Canada is an example of the latter. Because the sediment-dominated accretionary prisms (**Figure 2**) are more voluminous, they tend to be well recognized in the stratigraphic record, for example, parts of the Franciscan Complex of California. Sediment-starved accretionary prisms are thinner and typically dominated by basaltic and ultramafic igneous rocks. They are harder to recognize in the ancient stratigraphic record.

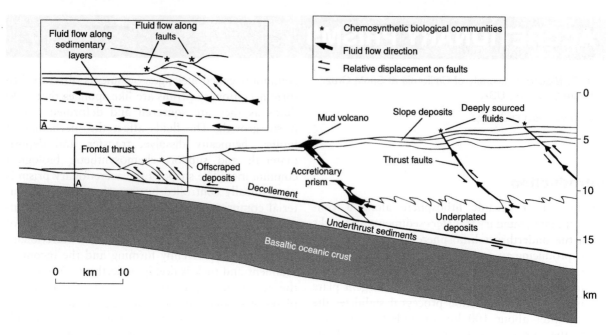

Figure 2 Cross-section of an accretionary prism showing the principal structural elements. Paths of fluid flow from deep sources to surface utilize high-permeability conduits, whether along sedimentary layers or faults.

Factors controlling the amount of sediment and type of sediment available for accretion include the age of the oceanic plate, the types and rates of sediments being deposited along the transport path of the oceanic plate to the subduction zone, and rate of travel or residence time of any plate in a particular sedimentary environment.

Solid Material Transfer in Accretionary Prisms

Sediments and rocks incoming to a subduction zone may be (1) offscraped as a series of thrust sheets at the frontal edge of the accretionary prism; (2) underplated or emplaced at depth along the base of the upper lithospheric plate; or (3) underthrust to great depths to participate in production of volcanic arc magmas or ultimately to be carried into the earth's mantle (**Figures 1** and **2**). In the zone of off-scraping, incoming materials are typically accreted as a series of imbricate thrust sheets extending from the surface to a basal detachment fault or decollement, beneath which all other material is underthrust. The decollement forms in a layer weaker than the adjacent sediment. With continued underthrusting, the weak layer may become stronger because of mineralogical changes or decreasing fluid pressure and step or migrate down through the underthrust plate. This down-stepping process underplates fault-bounded rock packages to the overlying plate (**Figure 2**).

The surfaces of some accretionary prisms and non-accretionary convergent margins (see below) are marked by volcanoes of fluidized mud or serpentine (**Figure 2**). The fluidized mud and serpentine rise through the upper plate of the convergent margin because these materials are of lower density than the surrounding sediments and rocks. Deep under-thrusting of mud and the associated production of natural gas or oil produces mud volcanoes. Serpentine, a low-density rock, is formed by addition of water from underthrust sediments or rocks to mantle rocks. In both cases the low-density rock occurs beneath higher-density rock and buoyancy forces drive the low-density material to the surface.

Incoming sediment that is not offscraped at the front of or underplated beneath the main part of the accretionary prism continues to be underthrust. This underthrust sediment may be underplated beneath volcanic arc basement rocks at any point until it reaches the melting zone (**Figure 1**). Short-lived radioactive isotopes and other chemical tracers in volcanic rocks indicate that sediments less than several million years old are underthrust to the depths of melting (75–100 km) beneath the volcanic arc. Residual material from the melting zone may be deeply underthrust into the earth's mantle.

Continuing sediment accumulation occurs on top of the growing accretionary prism, forming an apron of slope deposits. Locally erosion may remove material from the accretionary prism, redepositing it on the oceanic plate for recycling back into the accretionary prism. In addition to the thrust faults and the decollement associated with accretionary processes, accretionary prisms are cut by thrust, normal, and strike-slip faults that form as the prism adjusts its shape in response to its continuing growth (see Mechanics below).

Accretionary prisms include some of the most complexly deformed and puzzling rocks on the earth. Melanges or 'mixed rocks' and stratally disrupted rocks are included in this category. These rocks are marked by not only stratal discontinuity (**Figure 3A**) but by mixing of incompatible sedimentary and metamorphic environments (**Figure 3B**). These intricately deformed rocks form from hard igneous rocks and sediments that are partially consolidated and lithified in a high-strain and high-strain-rate environment. The extreme variation in strength between the hard rocks and soft sediments and the high strain deformation results in a heterogeneously deformed rock mass. Various return flow processes at depth (**Figure 1**), faulting, and erosion and redeposition of previously accreted rocks contribute to the mixing of rocks derived from differing sedimentary and metamorphic environments (**Figure 3B**).

Accretionary prisms include metamorphic rocks formed under high-pressure, low-temperature conditions. These rocks, called blueschists (**Figure 4**), are diagnostic of this subduction zone metamorphic environment. The high-pressure–low-temperature condition is caused by the rapid underthrusting of old, cold oceanic plates. Burial rates for underthrust rocks at subduction zones exceed $20 \, km \, My^{-1}$. The material is buried and returned to shallow depths more rapidly than it can be warmed by conduction from adjacent warmer parts of the earth. Thus, the low-temperature–high-pressure conditions of the subduction system are imprinted on the rocks and preserved by rapid uplift. The lowest average geothermal gradients through accretionary prisms are less than $10°C \, km^{-1}$, or about a third of the typical gradient through continents. Thermal gradients beneath accretionary prisms may be much higher where young, hot oceanic crust is being subducted. In these cases blueschists are not formed but are replaced by metamorphic rocks characteristic of higher temperature regimes (greenschist and amphibolite).

Overall, the material transfer in accretionary prisms is similar to that in thrust belts in overall form, fault geometry, and mechanics. Thrust belts on land tend to deform more consolidated and lithified sedimentary rocks at slower rates than occur at oceanic convergence zones. Therefore, in thrust belts, both the structural complexity and the effects due to fluid expulsion are less pronounced than in accretionary prisms. A good example of a thrust belt

(A)

(B)

Figure 3 Melanges. (A) Dismembered layers of light-colored sandstone in shale matrix. This type of deformation is typically developed along thrust faults with substantial displacement or along the decollement. (B) Melange consisting of block of basalt with overlying pelagic or open oceanic limestone included in shale matrix. Limestone and shale are incompatible depositional environments that are mixed together in the deformational environment of the accretionary prism. Both photographs are from the 60–65 My old accretionary prism of the Kodiak Islands, Alaska.

Figure 4 Blueschist metamorphic rock from an accretionary prism, indicating high pressures and low temperatures and showing ductile or plastic deformation (fold). This rock was metamorphosed about 200 My ago and occurs in the accretionary prism of the Kodiak Islands of Alaska.

is the Canadian Rocky Mountains just west of Calgary Alberta.

Fluids In and Out of Accretionary Prisms

Accretionary prisms are like sponges, saturated as they begin to form but squeezed virtually dry as their rocks reach maximum depths of burial. About 40% of the sediment section entering the world's subduction zones is composed of water in pore spaces. Additional water resides in pores in the igneous rocks of the crust and is bound to minerals both in the sediments and the oceanic crust.

The rapid burial of incoming sediments and rocks due to incorporation into and underthrusting beneath the accretionary prism increases stress on the rock framework and raises fluid pressure (the sponge is squeezed). Burial also increases temperature, which releases water bound in minerals and converts sedimentary organic matter to oil and natural gas. Sediment microrganisms may also produce natural gas from organic matter. By 150°C, the minerals are substantially dehydrated and most of the organic matter has been converted to hydrocarbons. By about 4 km of burial, the pore volume of sedimentary rocks in the accretionary prism is reduced by about 90%. The high fluid pressure drives fluids out of the accretionary prism through sediment pores and fractures and along faults (**Figure 2**).

Fluid migration out of the accretionary prism affects everything from surface biology, to large-scale structural features, to the fabric of prism rocks. The high fluid pressures that result from the fluid generation process weaken the rocks by reducing the contact stresses on mineral grains and fault surfaces, thereby reducing friction. This pressure-related weakening facilitates long-distance lateral transport on thrust faults (**Figure 2**), such as the decollement, and also controls the shape of the accretionary prism (see Mechanics). At temperatures from 100°C and up, rocks dissolve and re-precipitate (undergo pressure solution) with formation of preferred orientations of minerals and solution seams (fabrics) that record stress orientations. Rocks such as slates commonly form through this process. The constituents of the dissolved carbonate and silicate minerals may be locally precipitated or transported with the fluids and precipitated elsewhere as veins and cements that are common in prism rocks. The fluid-mediated processes of dissolution, transport, and precipitation significantly change the physical properties of the accretionary prism and ultimately affect how it deforms. Fluids expulsed from the surface of accretionary prisms contain dissolved methane and hydrogen sulfide, which are utilized by chemosynthetic organisms that are the basis for cold seep biological communities on the seafloor. In the subsurface, microorganisms both produce and consume various fluid-borne chemical constituents, modifying the physical properties and composition of the host sediments or rocks.

Because fluids alter the physical properties of sediment and rock, they affect the seismic reflection images of the prism interior. Fluid-enriched zones along faults reduce velocity and density and produce strong reflections in the seismic images. Methane near the surface of the accretionary prism freezes to form methane hydrate. Progressively deeper in the prism, the methane hydrate is unstable and is transformed back to free methane. Minor accumulations of free methane gas below the hydrate produce a large change in rock physical properties that is seen as a prominent bottom-simulating reflection in seismic images.

Seismogenesis and Accretionary Prisms

Subduction zone thrust faults produce the largest earthquakes on the Earth because the plate-boundary thrust is in a zone of high strain rate and is inclined shallowly. The shallow inclination provides a large surface area, or seismogenic zone, subject to brittle failure and the production of earthquakes (**Figure 1**). In contrast, more steeply inclined faults, such as the San Andreas Fault, transition down-dip from the region of brittle or seismogenic deformation into the realm of ductile deformation over a shorter distance, therefore limiting the area capable of producing an earthquake. The low thermal gradients characteristic of many subduction zones also extends the brittle–ductile transition to greater depths than in other plate boundary settings and increases the area subject to catastrophic seismogenic failure.

Commonly, the seismogenic zone earthquakes occur beneath accretionary prisms, such as the great Alaskan earthquake of 1964 (magnitude 9.2). The ability of accretionary prisms to build up enough elastic strain to be released in a sudden earthquake event testifies to strength developed during the evolution from soft sediments to hard rocks. In addition to becoming strong and rigid, the materials of the accretionary prism must evolve to fail in a 'velocity weakening' manner, such that there is an acceleration of slip along the fault surface. This accelerating slip produces a discrete seismic event, as opposed to a decelerating creep event that would not produce an earthquake.

Mechanics

In 1983 Davis, Dahlen, and Suppe articulated and formalized the mechanics of accretionary prisms in the widely accepted 'critical Coulomb wedge theory' (**Figure 5**). Virtually all accretionary prisms approximate a wedge in shape, being thinner on the oceanic side and thicker toward the associated volcanic arc. According to Davis *et al.* accretionary wedges resemble piles of dirt (prism materials) being pushed forward by bulldozer (the volcanic arc basement). The stresses driving the prism seaward are that of the arc basement pushing the wedge from the rear and a seaward-directed lateral stress due to unequal gravitational stress resulting from the wedge shape. The latter is similar to the stress causing a pile of sand to fail if it is oversteepened. These driving stresses are resisted by a shear stress along the base of the wedge controlled by the frictional strength of the material, which varies with overburden stress, the fluid pressure along the base, and the material properties. Additionally, the component of the overburden stress acting parallel to the decollement must be overcome to, in essence, lift the prism up the decollement. The wedge is just as thick as it can be in order to move forward; that is, it is at its threshold of failure determined by its frictional strength and internal fluid pressure. If the wedge grows at its leading edge, the area of the decollement resisting motion increases and the prism must thicken at the rear in order to increase the area to which the driving stresses in the arc basement can be applied.

Wedge theory has been largely successful in explaining the shape of accretionary prisms and the observations of fluid pressure, though the latter are not numerous. Because fluid pressure can counteract normal stresses along the decollement, it is a prime parameter controlling the mechanics of accretionary prisms. The generally high fluid pressure along the decollement sharply decreases the frictional resistance there, allowing narrowly tapered prisms to be mechanically stable. The necessity to thicken the landward portion of the prism to keep a stable wedge taper can explain much of the faulting observed in the landward parts of prisms (**Figure 2**). Other mechanical conceptualizations of wedges utilize differing material properties; however, the Coulomb wedge theory, dependent on basic frictional behavior, is most successful at depths up to 10–20 km.

Non-accretionary Convergent Plate Boundaries

According to a compilation by Von Huene and Scholl, somewhat more than half of the world's convergent plate boundaries are forming accretionary prisms now. The remainder of convergent plate boundaries have inner trench slopes underlain by older accretionary prisms or igneous and metamorphic rocks of the continental crustal or volcanic arc origin (**Figure 6**). Some are underlain by igneous and metamorphic rocks of uncertain origin that may represent accreted pieces of seamounts or normal oceanic crust. Recent Ocean Drilling Program results off Costa Rica unequivocally demonstrate that virtually all of the sediment riding on the oceanic plate (Cocos Plate) is underthrust beneath the upper plate (Caribbean Plate) of the subduction zone. This process must also occur at many other convergent margins without accretionary prisms. At a number of other convergent margins, the presence of continental or volcanic arc rocks close to the trench suggests that portions of the forearc may have been tectonically eroded by underthrusting. Moreover, Ocean Drilling Program holes show that rocks currently located in deep water on inner slopes of trenches have subsided from much shallower depths. Presumably this subsidence is due to tectonic erosion of the trench inner slope by the underthrusting plate.

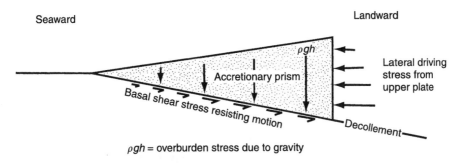

ρgh = overburden stress due to gravity

Figure 5 Diagram showing stresses that control motion of accretionary prisms (and thrust belts). These stresses, integrated over the areas where they act, form a force balance. Forces driving the prism seaward (a push from the rear and internal gravitational forces) are offset by resisting forces (primarily frictional forces and gravitational forces acting along the base of the decollement).

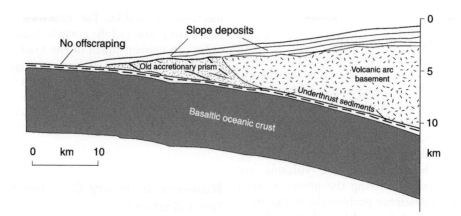

Figure 6   A convergent margin showing a range of features seen worldwide supporting nonaccretion or tectonic erosion. The absence of offscraping indicates no accretion at the front of the margin. At deeper levels, sediment may be being underplated. Locally, sediments and contained fluids are underthrust to great depths and participate in the melting process beneath the volcanic arc (see **Figure 1**). A previous accretionary prism much older than the age of incoming sediment may be being underthrust. Arc basement may be anomalously close to the trench, suggesting erosion of the upper plate. Slope deposits may be of much shallower water origin. These shallow-water deposits suggest substantial subsidence of the margin, which is commonly explained by tectonic erosion of the upper plate.

So, although there are accretionary prisms forming at many convergent margins, the offscraping and underplating at the front of the margin is not a constant process. Accretionary prism formation may be episodic even with continuous subduction. Tectonic erosion may occur. Gaps in the record of accretion are to be expected.

What determines whether convergent plate boundaries form accretionary prisms or tectonically erode the upper plate? The primary control is sediment supply. Where thick sequences of sediment enter the subduction zones, accretionary prisms form. Where the sedimentary sequences are thin, there is less material available to accrete and irregularities in the underthrusting oceanic plate may interact with the overthrusting plate to cause erosion. Protruding fault blocks or seamounts may tectonically abrade the lower surface of the upper plate. Also, as these high areas on the lower plate are underthrust, they may oversteepen and otherwise disturb the upper plate, causing it to fail at the surface by landsliding. The landslides accumulate in the trench, where they are underthrust. Accordingly, the front of the convergent margin is decimated and cycled to deeper levels in the earth.

Conclusions

Accretionary prisms form at the leading edge of convergent plate boundaries by skimming-off sediments and rocks of the lower plate. In detail, the accretion process involves offscraping of rocks and sediments at the front of the prism or underplating (emplacement beneath the prism). This deformational process stacks the sediments into thick vertical piles and shortens them horizontally. Consequently, fluids are expelled and the sediments are progressively transformed to rocks. Because the deformation in accretionary prisms is large and fast, rocks emplaced therein are often severely disrupted and mixed, forming melanges. The rapid rate of underthrusting of the lower plate may carry rocks to great depths before they can heat up, forming a characteristic type of metamorphic rock called a blueschist. The fault surface bounding the base of the accretionary prism, the decollement, is the plate boundary thrust. Because it is shallowly inclined, this fault has a large area undergoing brittle deformation and produces the largest earthquakes on the planet. Mechanically, accretionary prisms resemble a pile of dirt being pushed by a bulldozer. Prisms are pushed from the rear by the volcanic arc basement, with forward motion being resisted by frictional and gravitational forces. Accretion does not occur at the leading edge of all convergent plate boundaries. In these cases, sediments and rocks are underthrust beneath the crustal framework of the upper plate. In some cases accreted rocks have been removed from the upper plate, apparently by tectonic erosion by the lower plate. Thus, an accretionary prism may be a very discontinuous recorder of the incoming sediments and rocks at a convergent plate boundary. Sediments, rocks, and fluids not emplaced in the accretionary prism are carried to great depths and either catalyze subduction zone volcanism or are transported into the mantle.

Acknowledgments

I thank the National Science Foundation for grants supporting my studies of accretionary prisms since

1974 (most recently grant # OCE 9802264). The Ocean Drilling Program provided the opportunity and support to participate in many cruises investigating accretionary prisms. Eli Silver's insight into convergent margin tectonics substantially improved this review. Hilde Schwartz thoughtfully reviewed the final manuscript.

See also

Mid-Ocean Ridge Geochemistry and Petrology. Mid-Ocean Ridge Seismic Structure. Mid-Ocean Ridge Tectonics, Volcanism, and Geomorphology. Seismic Structure.

Further Reading

Bebout GE, Scholl DW, Kirby SH, and Platt JP (1996) *Subduction Top to Bottom*, American Geophysical Union Monograph 96. Washington, DC: American Geophysical Union.

Davis DJ, Suppe J, and Dahlen FA (1983) Mechanics of fold-and-thrust belts and accretionary wedges. *Journal of Geophysical Research* 88: 1153–1172.

Fisher DM (1996) Fabrics and veins in the forearc: a record of cyclic fluid flow at depths of <15 km. In: Bebout GE, Scholl DW, Kirby SH, and Platt JP (eds.) *Subduction Top to Bottom*, American Geophysical Union Monograph 96, pp. 75–89. Washington, DC: American Geophysical Union.

Fryer P, Mottl M, Johnson LE, *et al.* (1995) Serpentine bodies in the forearcs of Western Pacific convergent margins. In: Taylor B and Natland J (eds.) *Active Margins and Marginal Basins of the Western Pacific*, American Geophysical Union Monograph 88, pp. 259–279. Washington, DC: American Geophysical Union.

Hyndman RD (1997) Seismogenic zone of subduction thrust faults. *The Island Arc* 6: 244–260.

Kastner M, Elderfield H, and Martin JB (1991) Fluids in convergent margins: what do we know about their composition, origin, role in diagenesis and importance for oceanic chemical fluxes. *Philosphical Transactions of the Royal Society of London, Series A* 335: 243–259.

Meschede M, Zweigel P, and Kiefer E (1999) Subsidence and extension at a convergent plate margin: evidence for subduction erosion off Costa Rica. *Terra Nova* 11: 112–117.

Moore JC and Vrolijk P (1992) Fluids in accretionary prisms. *Reviews in Geophysics* 30: 113–135.

Moores EM and Twiss RJ (1995) *Tectonics*. New York: WH Freeman.

Morris JD, Leeman WP, and Tera F (1990) The subducted component in island arc lavas; constraints from B-Be isotopes and Be systematics. *Nature* 344: 31–36.

Silver EA (2000) Leg 170 Synthesis of fluid-structural relationships of the Pacific Margin of Costa Rica. In: Silver EA, Kimura G, Blum P and Shipley TH (eds) *Proceedings of the Ocean Drilling Program, Scientific Results* 170 [Online at http://www.odp.tamu.edu/publications/170_SR/VOLUME/CHAPTERS/SR170_04.PDF]

Taira A, Byrne T, and Ashi J (1992) *Photographic Atlas of an Accretionary: Geologic Structures of the Shimanto Belt*. Japan, Tokyo: University of Tokyo Press.

Tarney J, Pickering KT, Knipe RJ, and Dewey JF (1991) The Behaviour and Influence of Fluid in Subduction Zones. *Philosophical Transactions of the The Royal Society of London, Series AE* 335: 225–418.

von Huene R and Scholl DW (1991) Observations at convergent margins concerning sediment subduction, subduction erosion, and the growth of continental crust. *Reviews in Geophysics* 29: 279–316.

OCEAN MARGIN SEDIMENTS

S. L. Goodbred Jr, State University of New York, Stony Brook, NY, USA

Introduction

Ocean margin sediments are largely detrital deposits of terrestrial origin that extend from the shoreline to the foot of the continental rise. Indeed, about 80% of the world's sediment is stored within margin systems, which cover about 14% of the Earth's surface (Table 1; Figure 1). Margin deposits typically consist of sand, silt, and clay-sized particles, the characteristics of which reflect the geology and climate of the adjacent continent. Some of the signals relevant to terrestrial conditions include sediment size, mineralogy, geochemistry, and isotopic signature. Upon entering the marine realm, however, sediments take on new characteristics indicative of coastal ocean processes that include waves, tides, currents, sea level, and biological productivity. Given that these numerous terrestrial and marine processes impart a signature to the sediments, ocean margins preserve an important record of Earth history, providing insights into past atmospheric, terrestrial, and marine conditions. Beyond their significance as environmental recorders ocean margin sediments support major petroleum and mineral resources. Currently, over 50% of the world's oil is recovered along ocean margins, and much of the remaining fraction is held within ancient margin deposits.

In the 1950s, early investigations of ocean margins focused on tectonic structure and how overlying sedimentary sequences developed on timescales of 10^5–10^7 years. Originally aimed at understanding plate tectonics and the nature of ocean–continent boundaries, these large-scale studies continued through the 1960s with specific interest in petroleum resources. Such efforts culminated in the publication of large scientific volumes such as Burk and Drake's *Geology of Continental Margins* (1974). At the time of these summary publications, new models of sedimentary margin systems were already being developed, has also been notably the approach of seismic stratigraphy established at the Exxon Production Research Company. Growing out of this approach was the more general model of sequence stratigraphy, which helped establish that margin strata could be grouped into discrete packages reflecting cycles of sea-level rise and fall. These concepts represent a general approach that has been applied to stratigraphic development and margin evolution in most of the world's ancient and modern sedimentary systems. Sequence stratigraphic data has also been mated with lithologic, magnetic, and biostratigraphic records, allowing researchers to establish the history of sea-level change since the Triassic era (260 million years ago). This historic record significantly advanced out understanding of ocean margin sedimentary records, because sea level is a major control on the distribution and accumulation of margin deposits, as well as an indicator of global climate.

At a much shorter time scale, a great deal of research in recent decades has focused on sediment dynamics along modern margin systems. Aimed at understanding the fate of sediments entering the marine realm, some of the issues driving margin-sediment research included coastal hazards, land loss and development, storm impacts, contaminant fate, and military interests. In part, the field has advanced with the development of new technology such as marine radioisotope geochronology, sonar seafloor mapping, and instruments for remote wave, current, and sediment concentration measures. Recently, research programs have sought to integrate long and short geologic timescales to understand how daily, seasonal, and annual sedimentary beds are ultimately preserved to form thick millennial and longer sedimentary sequences (e.g., US and European STRATA-FORM programs). Other newer initiatives are seeking to take integration a step further by recognizing that ocean margin sediments lie among a continuum of terrestrial and marine influences. Therefore, to better understand this critical accumulation zone, the

Table 1 Area of Earth's major physiographic provinces and the volume of stored sediments. Bracketed are ocean margin components with relative contributions shown

	Area[a] ($10^6 \times km^2$)	Volume[b] ($10^6 \times km^3$)
Land	148.1	45
Interior basins	8.7	35
Shelves	18.4 ⎫	75 ⎫
Slopes	28.7 ⎬ 14%	200 ⎬ 80%
Rises	25.0 ⎭	150 ⎭
Ocean basins	281.2	25
Totals	510.1	530

Data from [a]Burk and Drake (1974) and [b]Kennett (1982).

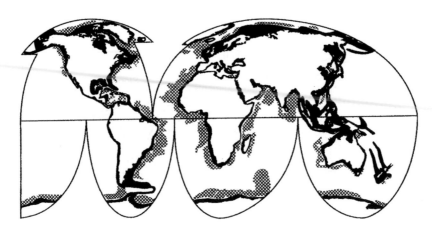

Figure 1 Distribution of major ocean margin deposits, showing extent of continental shelves (black) and rises (hatched). (Modified from Emery (1980).)

terrestrial processes that are responsible for sediment input must be considered in conjunction with the marine processes that are responsible for their redistribution (e.g., US and Japanese MARGINS programs). Such efforts to integrate over various spatial and temporal scales will likely govern the foreseeable future of margin sedimentary research.

Margin Structure

All margins represent past or present tectonic plate boundaries where crusts of varying age, density, and composition meet. In general these boundaries also comprise large-scale basins that trap sediments shed from the adjacent land surface. Most ocean margins specifically denote the boundary between oceanic and continental crusts, where different densities between these adjacent blocks give rise to a steep gradient at the margin. This tectonic structure controls the physiography of the margin as well, which comprises shelf, slope, and rise settings (**Figure 2**). The landward part of the margin, the shelf, overlies the buoyant continental-crust block and is thus the shallowest part of the margin, extending along a low gradient (0.1°) from the coast to the shelf break at 100–200 m water depth. Because of the low gradient and shallow depths, shelf environment are sensitive to sea-level change and isostatic movements (crustal buoyancy) causing the shelf to be periodically flooded and exposed. The seaward edge of the shelf, called the break, is identified by an increase in seafloor gradient (3–6°) and represents a transition between the shelf and continental slope. Roughly overlying the continental and oceanic crusts, sediment accumulation on the slope is partly controlled by on the shelf, because any slope sediments must first bypass this depositional region. Beyond the slope extends the rise, a

low gradient (0.3–1.4°) sedimentary apron that begins in 1500–3500 m water depth. Although the rise may extend hundreds of kilometers into the ocean basin, its sediment source largely derives from sediment originating on the slope and moves as turbidity currents. In addition to the major shelf, slope, and rise features, other regionally significant marginal features may include deltas and canyons, each playing a role in controlling or modifying the dispersal and deposition of sediments.

Tectonically, margins may be broadly divided into divergent (passive) and convergent (active) systems, each with a characteristic structure and physiography that are important to sediment accumulation and transport dynamics. Divergent margins are largely stable crustal boundaries, although the structure and movement of deep basement rocks remain a significant influence on sediment deposition and stratigraphy. Divergent margins are also largely constructive settings in which sediments shed from the continent form thick sedimentary sequences and a wide, low gradient shelf. This loading of sediment onto the margin also drives a slow downwarping of the crust, creating more accommodation space for the accumulation of thick sedimentary sequences. The resulting margin physiography is characterized by a broad shelf, slope, and rise. Margins along the Atlantic Ocean are among the best-studied examples of divergent systems.

In contrast, convergent margins are characterized by the subduction of oceanic crust beneath the adjacent continental block. Active mountain building in convergent settings frequently drives uplift of the margin, limiting the development of thick sedimentary sequences on the shelf. Thus, convergent margins tend to have narrow shelves that bypass most sediment to a steep and well-developed slope. The rise is generally not a significant depocenter because

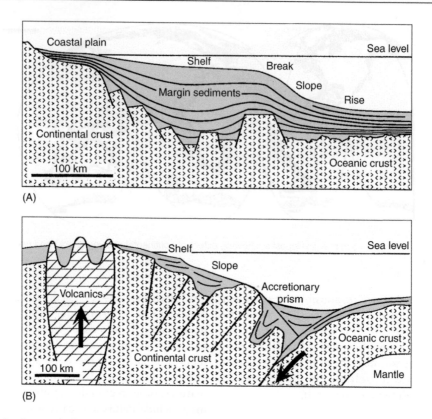

Figure 2 Generalized cross-sections of (A) divergent and (B) convergent-type continental margins. Note the different scales. The divergent margin shows thick sedimentary deposits that have prograded into the ocean basin. Margin deposits in the convergent setting are much thinner, and most deposition occurs below the shelf break in the accretionary prism.

sediment transport is intercepted by trenches and elevated topography that inhibit the long-distance movement of turbidity currents. Seaward of this, though, there is frequently a wedge of deformed marine sediments called an accretionary prism that derives from material scraped from the subducting ocean crust. This situation is often found along the eastern North Pacific margin. Along margins with major trench systems such as those of the western Pacific, the rise and accretionary prism may be altogether absent.

Margin Sediment Sources, Character, and Distribution

Ocean margin sediments generally comprise small particles ranging in size from fine sands (62–125 μm) to silts (4–62 μm) and clays (<4 μm). Larger sediments are occasionally found in ocean margin deposits, including coarse sands to boulders, but these particles are generally either relict (not actively transported), of glacial or ice-rafted origin, or biogenically precipitated (i.e. shell or coral). Overall patterns of grain-size distribution in margin sediments depend on the regional climate, geology, and sediment transport processes operating on the margin.

Three major mineral groups comprise ocean margin sediments, including siliciclastics, carbonates, and evaporites, or any combination of these. Worldwide, however, siliciclastics dominate ocean margin sediments. Siliciclastics are almost exclusively silicon-bearing minerals that derive from the physical and chemical weathering of continental rocks. Because most of the hundreds of rock-forming minerals are weathered before reaching the coast, margin siliciclastics dominantly comprise quartz and four clay species: kaolinite, chlorite, illite, and smectite. There is often a small component (several percent) of remnant feldspars, micas, and heavy minerals (sp. gr. >2.85) as well. Quartz, although an abundant and ubiquitous mineral, is often useful for interpreting margin sediments because different grain sizes reflect varying sources (e.g. eolian input) or energy regimes. For clay minerals, though, relative abundance of the major species in part reflects regional climate, thus making them a useful parameter for interpreting terrestrial signals preserved on the margin (*see* Clay Mineralogy). Furthermore, within the heavy mineral fraction, more unique silicate species can help determine the specific source area (provenance) of margin sediments. This approach is often limited because of regionally common mineralogy, but researchers have begun to use isotope ratios to target more precisely margin sediment source areas. Since ocean margin sediments and deposits are closely tied to continental fluxes, such detailed studies are critical for an integrated understanding of the terrestrial–margin–marine sedimentary system.

Although nearly all siliciclastic margin sediments are derived from the continents, they may be delivered via different mechanisms such as glaciers, rivers, coastal erosion, and wind. Although the latter mechanism is important to deep-sea sedimentation, the others dominate transport to the margin. At high latitudes, glaciers are a major and sometimes sole sediment source. The characteristics of glacial margin sediments is highly variable because of the capacity of ice to transport sediments of all sizes, but in general glacial sediments are coarser and more poorly sorted near the ice front (e.g., till) and become progressively finer and better sorted with distance (e.g., proglacial clays). Examples of glacially dominated modern margins include south-east Alaska, Greenland, and Antarctica. Along most of the world's margins, rivers dominate sediment delivery to the shelf. The grain size of river sediments is typical of margin deposits, mostly including fine sand, silt, and clay-sized particles. However, the range of textures and the total amount of sediment delivered to the margin varies with many factors, including the size, elevation, and climate of the river basin. Because rivers are largely point sources, coastal processes such as waves and tides are important controls on redistributing fluvial sediments once they reach the ocean margin.

In contrast to siliciclastics, carbonate sediments are largely biogenic in origin and are precipitated *in situ* by various corals, algae, bryozoans, mollusks, barnacles, and serpulid worms. Carbonates can be a dominant or significant component of margin sediments where the flux of siliciclastics is not great, often in arid regions or on the outer shelf. Carbonates do not usually form on the slope because of unstable bottom surfaces and depths below the photic zone. In general the production of carbonate sediments is higher in warmer low-latitude waters, but they are found along margins throughout the world. Unlike siliciclastic sediments, carbonates are not widely transported and tend to be concentrated within local regions on the shelf, where the build-up of bioherms and reefs may comprise locally important structural features. Overall, areas of purely carbonate production are rare and limited to ocean margins detached from the continents, such as the Bahamas. In contrast, mixed siliciclastic–carbonate margins are not uncommon and include regions such as the Yucatan and Florida peninsulas, Great Barrier Reef, and Indonesian islands. In very arid regions, margins along smaller marine basins may also support evaporite deposits. These minerals are precipitated directly from the water column to form thick salt deposits. Evaporites are also plastic, low-density sediments that frequently migrate upwards to form salt diapirs, causing major deformation of the overlying margin strata. Evaporite margin deposits are presently forming in the Red Sea and are also an important component of ancient margin sequences around the Mediterranean and Gulf of Mexico.

Along many modern margins, relict and recently eroded sediments comprise the bulk of material that is actively transported on the shelf. An example of a relict margin is the US East coast, where modern fluvial sediment is trapped in coastal estuaries and embayments, thus starving the shelf of significant input. Thus, the shelf there is characterized by a thin sheet of reworked sands (remnant from the Holocene transgression) that overlie much older, partly indurated deposits. Along mountainous high-energy coastlines where rivers are absent or very small, the erosion of headlands can be a major source of sediment to the margin. Often these are also convergent margins with steep shelves that aid in advecting eroded near-shore sediments to the outer shelf and slope. Examples include portions of the eastern Pacific margin and south-eastern Australia (divergent).

Margin Sediment Transport

Rivers are one of the major sources of sediment delivered to the shelf, and thus the fate of river plumes is an important control on the distribution of ocean margin sediments. The initial dispersal of the river plume and the subsequent sediment deposition are controlled by waves, tides, and three basic effluent properties that include the inertia and buoyancy of the plume, and its frictional interaction with the seafloor. In general, wave energy has the effect of keeping sediments close to shore, particularly sands that may be reworked onshore and/or transported alongshore by wave-driven currents. Margins that are wave dominated are typically steep and narrow and found in the high-latitudes, such as southern Australia, southern Africa and the margins of the north and south Pacific (i.e., high wind-stress regions). In contrast to waves, tides force a general onshore–offshore movement of sediment, most significantly along the coast where tidal energy is focused. Tide-dominated margin sediments typically occur on wide shelves and in shallow marginal basins such as the Yellow or North Seas. Compared with waves and tides, the effect of river plume dynamics on sediment dispersal is more varied, particularly as complex feedbacks exist between coastal geology and riverine, wave, and tidal processes.

The fate of sediments along margins is not a simple process of transport, deposition, and burial, but rather involves multiple cycles of erosion, transport, and deposition before being preserved and incorporated into ocean-margin strata (**Figure 3**). Thus, the triad of processes discussed above largely controls the initial phase of dispersal and deposition at the coast and inner shelf. The implication is that sediments undergo a succession of transport steps prior to preservation, and between each step the controlling processes, and thus sediment character, progressively change from riverine to coastal, shelf, and marine-dominated signals. Thus, beyond the coastal zone a different set of mechanisms is responsible for advecting sediments across the margin to the outer shelf, slope, and rise. On the shelf where seafloor gradients are often too low for failure (i.e., mass-wasting events), storms can be a major agent for cross-shelf transport. Two components of storms are involved in this process. First, strong winds generate steep, long-period waves that can resuspend sediment at much greater depths than fair-weather waves (e.g., a storm wave with a 12 s period will begin to affect the seafloor at ∼110 m depth). Second, resuspended sediments may be transported offshore via a near-bottom return flow (set up by wind and wave stresses) or by group-bound infragravity (long period) waves. Although these mechanisms for seaward sediment transport are well recognized, their magnitude, frequency, and distribution are not precisely known.

A second and well-described mode for sediment transport across ocean margins is the suite of gravity-driven mass movements that includes slumps, slides, flows, and currents (**Figure 4**). This suite comprises nearly a continuum of properties from the movement of consolidated sediment blocks as slumps and slides to the superviscous non-Newtonian fluids of debris and sediment flows to the low-density suspended load of turbidity currents. Mass wasting events are common, widespread, and most frequently occur where slopes are oversteepened and/or sediments are underconsolidated (i.e., low shear strength). These last two factors are related since the threshold for oversteepening is partly a function of the sediment's shear strength, with possible failure occurring at gradients <1° in poorly consolidated sediments (e.g., 80–90% water content) and relative stability found on slopes of 20° for well-consolidated material (e.g., 20–30% water content). Many other factors also contribute to mass movements including groundwater pressure, sub-bottom gas production, diapirism, fault planes, and other zones of weakness. In addition to these intrinsic sedimentary characteristics, trigger mechanisms are an important control on mass movements. Triggers may include earthquakes, storms, and wave pumping, each of which may have the effect of inducing shear along a plane of weakness or disrupting the cohesion between sediment grains (i.e., liquefaction). Where trigger mechanisms are frequent and intense, such as along a tectonically active or high-energy margin, mass wasting may occur in deposits that would be stable in a passive, low-energy setting.

Although storm activity can be important on the shelf, mass movements are by far the most dominant

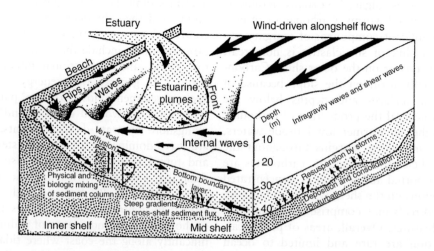

Figure 3 Conceptual diagram illustrating the major physical processes responsible for transporting sediment across the continental shelf. Note the many processes that contribute to the resuspension, erosion, and transport of sediment in this region. (Reproduced with permission from Nittrouer and Wright, 1994.)

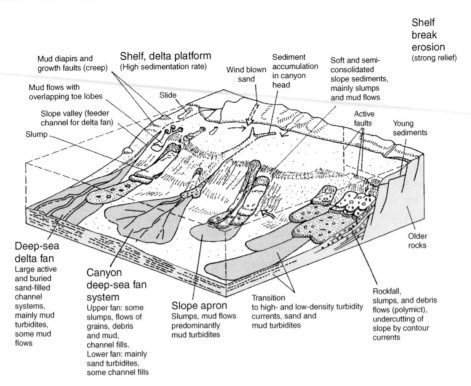

Figure 4 Summary diagram of sedimentary features and transport mechanisms typical of continental margins. Note the variety of mass-wasting events that originate on the slope and extend onto the rise. (Reproduced with permission from Einsele G (1991) Submarine mass flows, deposits and turbidities. In: Einsele G *et al.* (eds) (1991) *Cycles and Events in Stratigraphy*. New York: Springer-Verlag.)

mechanism for transferring sediment to the slope and rise. Indeed, estimates indicate that up to 90% of continental rise deposits derive from turbidity currents. In terms of their occurrence, slumps (movement of consolidated sedimentary blocks) are most common along steep gradients such as the continental slope or along the channels and walls of submarine canyons. Slumps and slides may move over distances of meters to tens of kilometers, and at rates of centimeters per year to centimeters per second. Although slumps and slides do not frequently travel great distances, these energetic events often generate turbidity currents at their front, which can travel hundreds of kilometers across shallow gradients of the continental shelf. Although turbidity currents occur throughout the world, the development of thick and extensive turbidite deposits is largely a function of margin structure, where the trenches and seafloor topography of collision margins inhibit the propagation of these flows (see earlier discussion of **Figure 2**).

Margin Stratigraphy

Strata Formation

Because of the important role that margin sediments play in geochemical cycling (both natural and anthropogenic inputs), there has been a great deal of interest in how seafloor strata are formed and what sort of physical, biological, and chemical impacts sediments undergo before being buried. These processes and controls on strata formation are also of great significance for interpreting the record of Earth's history preserved in ocean-margin deposits. Initially regarded as a one-way sink for organic carbon, dissolved metals, and coastal pollutants, the role of margin sediments has since been shown to be complicated by physical and biological processes that may alter chemical conditions (e.g., redox) and continue re-exposing sediments as deep as several meters to the water column and biogeochemical exchanges. Thus, through mixing processes, seafloor sediments may remain in contact with the water column for 10^1–10^3 years before being permanently buried. In order to quantify the effective importance of biological mixing relative to burial rate, the non-dimensional parameter (G) states that

$$G_b = \frac{D_b/L_b}{A}$$

where D_b represents the rate of biological (diffusive) mixing, L_b is depth of mixing, and A is the sedimentation rate. Field values of G are often 0.1–10, which falls within the range of codominance between mixing and sedimentation. Although the parameter G_b reflects the intensity of biological effects, physical mixing (G_p) may be better represented by T_p/L_p for the upper term, where T_p and L_p are the recurrence interval (period) and depth of physical disturbance, respectively. A diffusive coefficient is not relevant here because physical mixing is primarily caused by resuspension, which results in a complete advective turnover of the affected sediments. The notion that natural mixing intensities fall in the range of codominance is shown by the frequency with which sedimentary sequences are found to be partially mixed (at various scales), displaying juxtaposed laminated and bioturbated strata. Such patterns indicate that mixing is more complex than represented here, but the idea that strata preservation and geochemical availability are a function of sedimentation rate (i.e., burial) versus the rate and depth (i.e., exhumation) of mixing remains a useful, if imperfect, manner in which to consider these processes.

In a spatial sense, the dynamics of strata formation varies systematically across the ocean margin with changes in the controlling processes. Near to shore, the relatively constant processes of waves, tides, and river discharge support the frequent deposition of very thin strata (mm to cm). Although short-term rates of sedimentation are often high, the thin nature of each deposit gives them a low chance of being preserved due to physical and biological mixing. In contrast, mean sedimentation rates on the outer margin are often relatively low, but deposition is dominated by stochastic events such as storms and mass wasting which generally produce thicker deposits. Therefore, mixing between sedimentation events may only affect the upper portion of the deposit and thus preserve lower strata. Another consideration is that the intensity of biological mixing is generally less in offshore regions than near the coast.

Sequences

One of the most difficult problems in sedimentary geology is scaling up from short-term strata formation as discussed above to longer-term preservation and the accumulation of thick (10^2–10^3 m) sediment sequences that comprise ocean margins. Because the majority of sedimentary deposits formed over a particular timescale have a low chance of being preserved, geologists can only model strata formation across two to three orders of magnitude (e.g., incorporation of a 1 m thick deposit into a 100 m thick sequence, or the chance of a seasonal flood layer being preserved for several decades). In reality six or more orders of magnitude are

appropriate, as kilometer thick ancient-margin sequences comprise millimeter scale strata such as tidal deposits, current ripples, and mud layers. Although stratigraphic reconstructions across this large scale remain too complex, geologists do have a good understanding of the incorporation of meter-scale strata into sedimentary packages that are many tens to hundreds of meters thick. Called *sequences*, these stratigraphic units are defined as a series of genetically related strata that are bound by unconformities (surfaces representing erosion or no deposition) (**Figure 5**).

Sequences are the major stratigraphic units that lead to continental margin growth, with deposition generally occurring on the shelf during periods of high sea level and sediments being dispersed to the slope and rise during sea-level lowstands. Forming over 10^4–10^7 years, the major factors in sequence production are: (1) the input of sediments, (2) space in which to store them (accommodation), and (3) some function of cyclicity to produce unconformity-bound groups of deposits. These factors are largely controlled by tectonics, sea level, and sediment supply. In the long term, tectonics is the major control on accommodation, whereby space for sediment storage is created by isostatic subsidence in response to sediment loading and crustal cooling along the margin. At shorter time scales, sea level controls both the availability of accommodation and its location across the margin. Thus, it can be considered that sea level partly controls the formation of sequences, whereas sequence preservation is dependent on tectonic forcings. Sediment supply is perhaps the fundamental control on the size of a sequence, and thus partly affects its chance for preservation. Sediments of riverine and glacial origin are the major sources for sequence production, and changes in these sources with varying climatic conditions is one

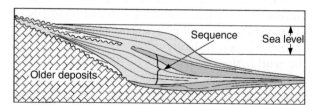

Figure 5 Generalized structure of continental margin sequences formed during cycles of sea-level change. Shown here are two sequences that reflect two periods of sea-level rise and fall. Transgressive deposits (□) are associated with rising seas and the highstand deposits (□) with a high, stable sea level. The lowstand deposits (□) form under falling and low phases of sea level. Reconstruction of these features from continental margins has provided a record of sea-level change over the past 200 million years.

mechanism for generating the unconformity bounding surfaces. Sequence boundaries can also be generated by tectonics and eustatic (global) sea-level change, with the latter being significantly affected by the growth and retreat of continental ice sheets.

Conclusions

Although the study of ocean-margin sediments began more than a century ago, the field has grown most significantly in the past 50 years with the birth of marine geology and oceanography. Typical of geological systems, however, relevant research extends across a range of spatial and temporal scales and incorporates a variety of disciplines including geophysics, geochemistry, and sedimentary dynamics. The future of margin sedimentary research lies with the continued integration of these various scales and disciplines, moving toward the development of coherent models for the production and dispersal of margin sediments and their ultimate incorporation into margin sequences. The impacts of such capability would greatly benefit the world's growing population in the areas of energy and mineral resources, climate change, coastal hazards, sea-level rise, and marine pollution. Ongoing research initiatives are currently advancing the field of ocean-margin sediments toward these goals, and nascent programs are providing a promising lead to the future.

See also

Beaches, Physical Processes Affecting. Calcium Carbonates. Clay Mineralogy. Deep-Sea Sediment Drifts. Geomorphology. Glacial Crustal Rebound, Sea Levels and Shorelines. Mineral Extraction, Authigenic Minerals. Sea Level Variations Over Geologic Time. Sediment Chronologies. Turbulence in the Benthic Boundary Layer.

Further Reading

Aller RC (1998) Mobile deltaic and continental shelf muds as fluidized bed reactors. *Marine Chemistry* 61: 143–155.

Burk CA and Drake CL (eds.) (1974) *The Geology of Continental Margins*. New York: Springer-Verlag.

Emery KO (1980) Continental margins – classification and petroleum prospects. *The American Association of Petroleum Geologists Bulletin* 64: 297–315.

Guinasso NL and Schink DR (1975) Quantitative estimates of biological mixing rates in abyssal sediments. *Journal of Geophysical Research* 80: 3032–3043.

Kennett JP (1982) *Marine Geology*. Englewood Cliffs, NJ: Prentice-Hall.

Nittrouer CA and Wright LD (1995) Transport of particles across continental shelves. *Reviews of Geophysics* 32: 85–113.

Payton CE (ed.) (1977) Seismic stratigraphy – applications to hydrocarbon exploration. *American Association of Petroleum Geologists Memoir* 2.

Sandford LP (1992) New sedimentation, resuspension, and burial. *Limnology and Oceanography* 37: 1164–1178.

Wilgus CW, Hastings BS, Kendall CGStC *et al.* (eds) (1988) Sea-level changes: an integrated approach. *Society of Economic Paleontologists and Mineralogists Special Publication No. 42*.

HYDROTHERMAL VENT DEPOSITS

R. M. Haymon, University of California, CA, USA

Introduction

In April 1979, submersible divers exploring the mid-ocean ridge crest at latitude 21°N on the East Pacific Rise discovered superheated (380±30°C) fluids, blackened by tiny metal-sulfide mineral crystals, spewing from the seafloor through tall mineral conduits. The crystalline conduits at these 'black smoker' hydrothermal vents were made of minerals rich in copper, iron, zinc, and other metals. Since 1979, hundreds of similar hydrothermal deposits have been located along the midocean ridge. It is now clear that deposition of hydrothermal mineral deposits is a common process, and is integrally linked to cracking, magmatism, and cooling of new seafloor as it accretes and spreads away from the ridge (*see* Propagating Rifts and Microplates, Mid-Ocean Ridge Geochemistry and Petrology Seamounts and Off-Ridge Volcanism Mid-Ocean Ridge Seismic Structure).

For thousands of years before mid-ocean ridge hot springs were discovered in the oceans, people mined copper from mineral deposits that were originally formed on oceanic spreading ridges. These fossil deposits are embedded in old fragments of seafloor called 'ophiolites' that have been uplifted and emplaced onto land by fault movements. The copper-rich mineral deposits in the Troodos ophiolite of Cyprus are well-known examples of fossil ocean-ridge deposits that have been mined for at least 2500 years; in fact, the word 'copper' is derived from the Latin word 'cyprium' which means 'from Cyprus.'

The mineral deposits accumulating today at hot springs along the mid-ocean ridge are habitats for a variety of remarkable organisms ranging in size from tiny microbes to large worms. The properties of the mineral deposits are inextricably linked to the organisms that inhabit them. The mineral deposits contain important clues about the physical–chemical environments in which some of these organisms live, and also preserve fossils of some organisms, creating a geologic record of their existence.

Hydrothermal vent deposits are thus a renewable source of metals and a record of the physical, chemical, biological, and geological processes at modern and ancient submarine vents.

Where Deposits Form: Geologic Controls

Less than 2% of the total area of the mid-ocean ridge crest has been studied at a resolution sufficient to reveal the spatial distribution of hydrothermal vents, mineral deposits, and other significant small-scale geologic features. Nevertheless, because study areas have been carefully selected and strategically surveyed, much has been learned about where vents and deposits form, and about the geologic controls on their distribution. The basic requirements for hydrothermal systems include heat to drive fluid circulation, and high-permeability pathways to facilitate fluid flow through crustal rocks. On mid-ocean ridges, vents and deposits are forming at sites where ascending magma intrusions introduce heat into the permeable shallow crust, and at sites where deep cracks provide permeability and fluid access to heat sources at depth.

Fast-spreading Ridges

Near- and on-bottom studies along the fast-spreading East Pacific Rise suggest that most hydrothermal mineral deposits form along the summit of the ridge crest within a narrow 'axial zone' less than 500 m wide. Only a few active sites of mineral deposition have been located outside this zone; however, more exploration of the vast area outside the axial zone is needed to establish unequivocally whether or not mineral deposition is uncommon in this region. The overall spatial distribution of hydrothermal vents and mineral deposits along fast-spreading ridges traces the segmented configuration of cracks and magma sources along the ridge crest (*see* Propagating Rifts and Microplates Mid-Ocean Ridge Geochemistry and Petrology Seamounts and Off-Ridge Volcanism Mid-Ocean Ridge Seismic Structure).

Within the axial zone, mineral deposition is concentrated along the floors and walls of axial troughs created by volcanic collapse and/or faulting along the summit of the ridge crest. The majority of the deposits are located along fissures that have opened above magmatic dike intrusions, and along collapsed lava ponds formed above these fissures by pooling and drainage of erupted lava. Where fault-bounded troughs have formed along the summit of the ridge crest, mineral deposition is focused along the bounding faults and also along fissures and collapsed lava ponds in the trough floor. Hydrothermal vents appear to be most abundant along magmatically inflated segments of fast-spreading ridges; however, the mineral deposits precipitated on the seafloor on magmatically active segments are often buried beneath frequent eruptions of new lava flows. The greatest number of deposits, therefore, are observed on inflated ridge segments that are surfaced by somewhat older flows, i.e., along segments where: (1) much heat is available to power hydrothermal vents; and (2) mineral deposits have had time to develop but have not yet been buried by renewed eruptions.

Intermediate- and Slow-spreading Ridges

Most hydrothermal deposits that have been found on intermediate- and slow-spreading ridge crests are focused along faults, fissures, and volcanic structures within large rift valleys that are several kilometers wide. The fault scarps along the margins of rift valleys are common sites for hydrothermal venting and mineral deposition. Fault intersections are thought to be particularly favorable sites for hydrothermal mineral deposition because they are zones of high permeability that can focus fluid flow. Mineral deposition on rift valley floors is observed along fissures above dike intrusions, along eruptive fissures and volcanic collapse troughs, and on top of volcanic mounds, cones and other constructions. In general at slower-spreading ridges, faults appear to play a greater role in controlling the distribution of hydrothermal vents and mineral deposits than they do at fast-spreading ridges, where magmatic fissures are

clearly a dominant geologic control on where vents and deposits are forming.

Structures, Morphologies, and Sizes of Deposits

A typical hydrothermal mineral deposit on an unsedimented mid-ocean ridge accumulates directly on top of the volcanic flows covering the ridge crest. On sedimented ridges, minerals are deposited within and on top of the sediments. Beneath seafloor mineral deposits are networks of feeder cracks through which fluids travel to the seafloor. Precipitation of hydrothermal minerals in these cracks and in the surrounding rocks or sediments creates a subseafloor zone of mineralization called a 'stockwork'. In hydrothermal systems where fluid flow is weak, unfocused, or where the fluids mix extensively with sea water beneath the seafloor, most of the minerals will precipitate in the stockwork rather than on the seafloor.

Hydrothermal deposits on mid-ocean ridges are composed of: (1) vertical structures, including individual conduits known as 'chimneys' (**Figure 1**) and larger structures of coalesced conduits that are often called 'edifices'; (2) horizontal 'flange' structures that extend outward from chimneys and edifices (**Figure 1**); (3) mounds of accumulated mineral precipitates (**Figure 1**); and (4) horizontal layers of hydrothermal sediments, debris, and encrustations. Chimneys are initially built directly on top of the seabed around focused jets of high-temperature effluents. Chimneys and edifices are physically unstable

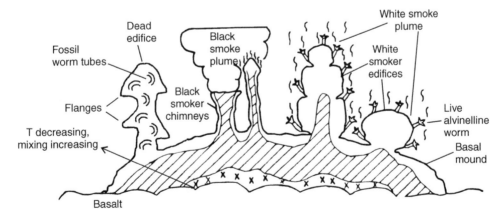

Figure 1 Composite sketch of the mineral structures and zones in hydrothermal mineral deposits on unsedimented ridge crests (modified after Haymon, 1989). Although mound interiors are seldom observed on the seafloor, the simplified sketch of mineral zoning within the mound is predicted by analogy with chimneys and massive sulfide deposits exposed in ophiolites. An outer peripheral zone (unshaded) of anhydrite + amorphous silica + Zn-rich sulfide, dominantly $ZnS + FeS_2$, is replaced in the interior by an inner zone (hatched) of Cu-rich sulfide ($CuFeS_2 + FeS_2$) + minor anhydrite and amorphous silica. The inner zone may be replaced by a basal zone (cross-pattern) of Cu-rich sulfide ($CuFeS_2 + FeS_2$) + quartz. Zones migrate as thermochemical conditions within the mound evolve. Although not shown here, it is expected that zoning around individual fractures cutting through the mound will be superimposed on the simplified zone structure in this sketch.

and often break or collapse into pieces that accumulate into piles of debris. The debris piles are cemented into consolidated mounds by precipitation of minerals from solutions percolating through the piles. New chimneys are constructed on top of the mounds as the mounds grow in size. Hydrothermal plume particles and particulate debris from chimneys settle around the periphery of the mounds to form layers of hydrothermal sediment. Diffuse seepage of fluids also precipitates mineral encrustations on mound surfaces, on volcanic flows and sediments, and on biological substrates, such as microbial mats or the shells and tubes of sessile macrofauna.

The morphologies of chimneys are highly variable and evolve as the chimneys grow, becoming more complex with time. Black smoker chimneys are often colonized by organisms and evolve into 'white smokers' that emit diffuse, diluted vent fluids through a porous carapace of worm tubes (**Figure 1**). Fluid compositions and temperatures, flow dynamics, and biota are all factors that influence the development of chimney morphology. The complexity of the interactions between these factors, and the high degree of spatial–temporal heterogeneity in the physical, chemical, biological and geological conditions influencing chimney growth, account for the diverse morphologies exhibited by chimneys, and present a challenge to researchers attempting to unravel the processes producing these morphologies.

The sizes of hydrothermal mineral deposits on ridges also vary widely. It has been suggested that the largest deposits are accumulating on sedimented ridges, where almost all of the metals in the fluids are deposited within the sediments rather than being dispersed into the oceans by hydrothermal plumes. On unsedimented ridges, the structures deposited on the seabed at fast spreading rates are usually relatively small in dimension (mounds are typically less than a few meters in thickness and less than tens of meters in length, and vertical structures are <15 m high). On intermediate- and slow-spreading ridges, mounds are sometimes much larger (up to tens of meters in thickness, and up to 300 m in length). On the Endeavour Segment of the Juan de Fuca Ridge, vertical structures reach heights of 45 m. The size of a deposit depends on many factors, including: magnitude of the heat source, which influences the duration of venting and mineral deposition; tendency of venting and mineral deposition to recur episodically at a particular site, which depends on the nature of the heat source and plumbing system, and the rate of seafloor spreading; frequency with which deposits are buried beneath lava flows; and the compositions of the vent fluids and minerals. The large deposits found on slower-spreading ridge crests are located on

faults that have moved slowly away from the ridge axis and have experienced repeated episodes of venting and accumulated mineral deposition over thousands of years, without being buried by lava flows. The tall Endeavor Segment edifices are formed because ammonia-enriched fluid compositions favor precipitation of silica in the edifice walls. The silica is strong enough to stabilize these structures so that they do not collapse as they grow taller.

How Do Chimneys Grow?

A relatively simple two-stage inorganic growth model has been advanced to explain the basic characteristics of black smoker chimneys (**Figure 2**). In this model, a chimney wall composed largely of anhydrite (calcium sulfate) precipitates initially from sea water that is heated around discharging jets of hydrothermal fluid. The anhydrite-rich chimney wall precipitated during stage I contains only a small component of metal sulfide mineral particles that crystallize because of rapid chilling of the hydrothermal fluids. In stage II, the anhydrite-rich wall continues to grow upward and to thicken radially, protecting the fluid flowing through the chimney from very rapid chilling and dilution by sea water.

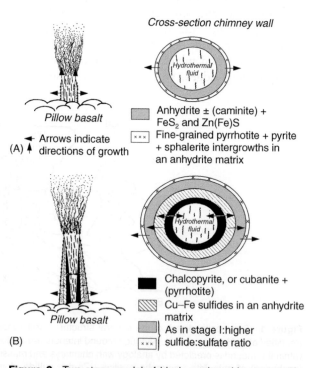

Figure 2 Two-stage model of black smoker chimney growth. (A) Stage I, sulfate-dominated stage; (B) stage II, sulfide replacement stage. During stage II, several different sulfide mineral zonation sequences develop, depending on permeability and thickness of chimney walls, hydrodynamic variables, and hydrothermal fluid composition.

This allows metal sulfide minerals to precipitate into the central conduit of the chimney from the hydrothermal fluid. The hydrothermal fluid percolates outward through the chimney wall, gradually replacing anhydrite and filling voids with metal sulfide minerals. During stage II, the chimney increases in height, girth and wall thickness, and both the calcium sulfate/metal sulfide ratio and permeability of the walls decrease. Equilibration of minerals with pore fluid in the walls occurs continuously along steep, time–variant temperature and chemical gradients between fluids in the central conduit and sea water surrounding the chimney. This equilibration produces sequences of concentric mineral zones across chimney walls that evolve with changes in thermal and chemical gradients and wall permeability.

The model of chimney growth described above is accurate but incomplete, as it does not include the effects on chimney development of fluid phase separation, biological activity, or variations in fluid composition. Augmented models that address these complexities are needed to fully characterize the processes governing chimney growth.

Elemental and Mineral Compositions of Deposits

Ridge crest hydrothermal deposits are composed predominantly of iron-, copper- and zinc sulfide

Table 1 Minerals occurring in ocean ridge hydrothermal mineral deposits

Mineral group/name	Chemical formula
Sulfides/Sulfosalts	
Most abundant	
Sphalerite	$Zn(Fe)S$
Wurtzite	$Zn(Fe)S$
Pyrite	FeS_2
Chalcopyrite	$CuFeS_2$
Less abundant	
Iss-Isocubanite	Variable $CuFe_2S_3$
Marcasite	FeS_2
Melnicovite	FeS_{2-x}
Pyrrhotite	$Fe_{1-x}S$
Bornite–Chalcocite	Cu_5FeS_4–Cu_2S
Covellite	CuS
Digenite	Cu_9S_5
Idaite	$Cu_{5.5}FeS_{6.5}$
Galena	PbS
Jordanite	$Pb_9As_4S_{15}$
Tennantite	$(Cu,Ag)_{10}(Fe,Zn,Cu)_2As_4S_{23}$
Valeriite	$2(Cu,Fe)_2S_23(Mg,Al)(OH)_2$
Sulfates	
Anhydrite	$CaSO_4$
Gypsum	$CaSO_4 \cdot H_2O$
Barite	$BaSO_4$
Caminite	$MgSO_4 \cdot xMg(OH)_2 \cdot (1-2x)H_2O$
Jarosite–Natrojarosite	$(K,Na)Fe_3(SO_4)_2(OH)_6$
Chalcanthite	$CuSO_4 \cdot 5H_2O$
Carbonate	
Magnesite	$MgCO_3$
Calcite	$CaCO_3$
Elements	
Sulfur	S
Oxides/Oxyhydroxides	
Goethite	$FeO(OH)$
Lepidocrocite	$FeO(OH)$
Hematite	Fe_2O_3
Magnetite	Fe_3O_4
Psilomelane	$(Ba,H_2O)_2Mn_5O_{10}$
'Amorphous' Fe-compounds	
'Amorphous' Mn-compounds	
Silicates	
Opaline silica	$SiO_2 \cdot nH_2O$
Quartz	SiO_2
Talc	$Mg_3Si_4O_{10}(OH)_2$
Nontronite	$(Fe,Al,Mg)_2(Si_{3.66}Al_{0.34})O_{10}(OH)_2$
Illite-smectite	
Aluminosilicate colloid	
Hydroxychlorides	
Atacamite	$Cu_2Cl(OH)_3$

Table 2 Ranges of elemental compositions in bulk midocean ridge hydrothermal mineral deposits

Element	Ranges[a]
Cu	0.1–15.0 wt%
Fe	2.0–44.0 wt%
Zn	< 0.1–48.7 wt%
Pb	0.003–0.6 wt%
S	13.0–52.2 wt%
SiO_2	< 0.1–28.0 wt%
Ba	< 0.01–32.5 wt%
Ca	< 0.1–16.5 wt%
Au	< 0.1–4.6 p.p.m.
Ag	3.0–303.0 p.p.m.
As	7.0–918.0 p.p.m.
Sb	2.0–375.0 p.p.m.
Co	< 2.0–3500.0 p.p.m.
Se	< 2.0–224.0 p.p.m.
Ni	< 1.5–226.0 p.p.m.
Cd	< 5–1448 p.p.m.
Mo	1.0–290.0 p.p.m.
Mn	36.0–1847.0 p.p.m.
Sr	2.0–4300.0 p.p.m.

[a]Data sources: Hannington *et al.* (1995) and Haymon (1989).

Morphological and Mineralogical Evolution of Chimneys
on the East Pacific Rise at 9°-10°N

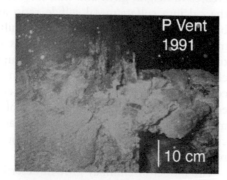

1991 "Proto-chimney"
Anhydrite-dominated
T = 389°-403°C

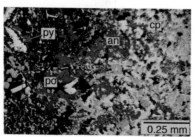

1992-1995
CuFe-sulfide-dominated
T = 340°-392°C

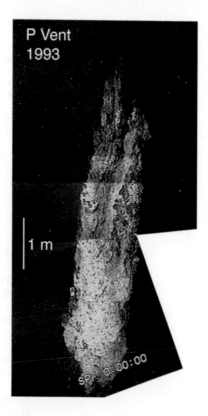

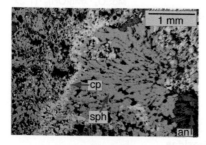

1992-1995
Zn-sulfide-dominated
T = 264°-340°C

KEY
an – anhydrite
cp – chalcopyrite
po – pyrrhotite
py – pyrite

minerals, calcium- and barium-sulfate minerals, iron oxide and iron oxyhydroxide minerals, and silicate minerals (**Table 1**). These minerals precipitate from diverse processes, including: heating of sea water; cooling of hydrothermal fluid; mixing between sea water and hydrothermal fluid; reaction of hydrothermal minerals with fluid, sea water, or fluid–sea water mixtures; reaction between hydrothermal fluid and seafloor rocks and sediments; and reactions that are mediated or catalyzed biologically. This diversity in the processes and environments of mineral precipitation results in the deposition of many different minerals and elements (**Tables 1 and 2**). High concentrations of strategic and precious metals are found in some deposits (**Table 2**). The deposits are potentially valuable, if economic and environmentally safe methods of mining them can be developed.

Chimneys can be classified broadly by composition into four groups: sulfate-rich, copper-rich, zinc-rich and silica-rich structures. Copper-rich chimney compositions are indicative of formation at temperatures above 300°C. Sulfate-rich compositions are characteristic of active and immature chimneys. Many chimneys are mineralogically zoned, with hot interior regions enriched in copper, and cooler exterior zones enriched in iron, zinc, and sulfate (**Figures 1 and 2**). Mounds exhibit a similar gross mineral zoning, and those which are exposed by erosion in ophiolites often have silicified (quartz-rich) interiors (**Figure 2**). Seafloor weathering of deposits after active venting ceases results in dissolution of anhydrite, and oxidation and dissolution of metal-sulfide minerals. Small deposits that are not sealed by silicification or buried by lava flows will not be well preserved in the geologic record (**Figure 3**).

Chimneys as Habitats

Chimney and mound surfaces are substrates populated by microbial colonies and sessile organisms such as vestimentiferan and polychaete worms, limpets, mussels, and clams. It is likely that pore spaces in exterior regions of chimney walls are also inhabited by microbes. All of these organisms that are dependent on chemosynthesis benefit from the seepage of hydrothermal fluid through active mineral structures, and from the thermal and chemical gradients across mineral structures. The structures provide an interface between sea water and hydrothermal fluid that maintains tolerable temperatures for biota, and allows organisms simultaneous access to the chemical constituents in both sea water and hydrothermal fluid. However, organisms attached to active mineral structures must cope with changes in fluid flow across chimney walls (which sometimes occur rapidly), and with ongoing engulfment by mineral precipitation.

Some organisms actively participate in the precipitation of minerals; for example, sulfide-oxidizing microbes mediate the crystallization of native sulfur crystals, and microbes are also thought to participate in the precipitation of marcasite and iron oxide minerals. Additionally, the surfaces of organisms provide favorable sites for nucleation and growth of amorphous silica, metal sulfide and metal oxide crystals, and this facilitates mineral precipitation and fossilization of vent fauna (**Figure 3**).

Fossil Record of Hydrothermal Vent Organisms

Fossil molds and casts of worm tubes, mollusc shells, and microbial filaments have been identified in both modern ridge hydrothermal deposits and in Cretaceous, Jurassic, Devonian, and Silurian deposits. This fossil record establishes the antiquity of vent communities and the long evolutionary history of specific faunal groups. The singular Jurassic fossil assemblage preserved in a small ophiolite-hosted deposit in central California is particularly interesting because it contains fossils of vestimentiferan worms, gastropods and brachiopods, but no clam or mussel fossils. In contrast, modern and Paleozoic faunal assemblages described thus far include clams, mussels and gastropods, but no brachiopods. Does this mean that brachiopods have competed with

Figure 3 On left: a time series of seafloor photographs showing the morphological development of a chimney that grew on top of lava flows erupted in 1991 on the crest of the East Pacific Rise near 9°50.3′N (Haymon et al., 1993). Within a few days-to-weeks after the eruption, anhydrite-rich 'Stage 1 Protochimneys' a few cm high had formed where hot fluids emerged from volcanic outcrops covered with white microbial mats (top left). Eleven months later, the chimney consisted of cylindrical 'Stage 2' anhydrite-sulfide mineral spires approximately one meter in height, and as-yet unpopulated by macrofauna (middle left). Three and a half years after the eruption, the cylindrical conduits had coalesced into a 7 m-high chimneys structure that was covered with inhabited Alvinelline worms tubes (bottom left). On right: photomicrographs of chimney samples from the eruption area that show how the chimneys evolved from Stage 1 (anhydrite-dominated; top right) to Stage 2 (metal-sulfide dominated) mineral compositions (see text). As the fluids passing through the chimneys cooled below ~330°C during Stage 2, the CuFe-sulfide minerals in the chimney walls (middle right) were replaced by Zn- and Fe-sulfide minerals (bottom right).

molluscs for ecological niches at vents, and have moved in and out of the hydrothermal vent environment over time? Fossilization of organisms is a selective process that does not preserve all the fauna that are present at vents. Identification of fossils at the species level is often difficult, especially where microbes are concerned. Notwithstanding, it is important to search for more examples of ancient fossil assemblages and to trace the fossil record of life at hydrothermal vents back as far as possible to shed light on how vent communities have evolved, and whether life on earth might have originated at submarine hydrothermal vents.

Summary

Formation of hydrothermal deposits is an integral aspect of seafloor accretion at mid-ocean ridges. These deposits are valuable for their metals, for the role that they play in fostering hydrothermal vent ecosystems, for the clues that they hold to understanding spatial–temporal variability in hydrothermal vent systems, and as geologic records of how life at hydrothermal vents has evolved. From these deposits we may gain insights about biogeochemical processes at high temperatures and pressures that can be applied to understanding life in inaccessible realms within the earth's crust or on other planetary bodies. We are only beginning to unravel the complexities of ridge hydrothermal vent deposits. Much exploration and interdisciplinary study remains to be done to obtain the valuable information that they contain.

See also

Mid-Ocean Ridge Geochemistry and Petrology. Mid-Ocean Ridge Seismic Structure. Propagating Rifts and Microplates. Seamounts and Off-Ridge Volcanism.

Further Reading

Dilek Y, Moores E, Elthon D, and Nicolas A (eds.) (2000) *Ophiolites and Oceanic Crust: New Insights from Field Studies and the Ocean Drilling Program. Geological Society of America Memoir.* Boulder: Geological Society of America.

Haymon RM (1989) Hydrothermal processes and products on the Galapagos Rift and East Pacific Rise, 1989. In: Winterer EL, Hussong DM, and Decker RW (eds.) *The Geology of North America: The Eastern Pacific Ocean and Hawaii,* vol. N, pp. 125–144. Boulder: Geological Society of America.

Haymon RM (1996) The response of ridge crest hydrothermal systems to segmented, episodic magma supply. In: MacLeod CJ, Tyler P, and Walker CL (eds.) *Tectonic, Magmatic, Hydrothermal, and Biological Segmentation of Mid-Ocean Ridges,* vol. Special Publication 118, pp. 157–168. London: Geological Society.

Humphris SE, Zierenberg RA, Mullineaux LS, and Thomson RE (eds.) (1995) *Seafloor Hydrothermal Systems: Physical, Chemical, Biological, and Geological Interactions, Geophysical Monograph,* vol. 91. Washington, DC: American Geophysical Union.

Little CTS, Herrington RJ, Haymon RM, and Danelian T (1999) Early Jurassic hydrothermal vent community from the Franciscan Complex, San Rafael Mountains, California. *Geology* 27: 167–170.

Tivey MK, Stakes DS, Cook TL, Hannington MD, and Petersen S (1999) A model for growth of steep-sided vent structures on the Endeavour Segment of the Juan de Fuca Ridge: results of a petrological and geochemical study. *Journal of Geophysical Research* 104: 22859–22883.

MID-OCEAN RIDGE GEOCHEMISTRY AND PETROLOGY

M. R. Perfit, Department of Geological Sciences, University of Florida, Gainsville, FL, USA

Introduction

The most volcanically active regions of our planet are concentrated along the axes of the globe, encircling mid-ocean ridges. These undersea mountain ranges, and most of the oceanic crust, result from the complex interplay between magmatic (i.e., eruptions of lavas on the surface and intrusion of magma at depth) and tectonic (i.e., faulting, thrusting, and rifting of the solid portions of the outer layer of the earth) processes. Magmatic and tectonic processes are directly related to the driving forces that cause plate tectonics and seafloor spreading. Exploration of mid-ocean ridges by submersible, remotely operated vehicles (ROV), deep-sea cameras, and other remote sensing devices has provided clear evidence of the effects of recent magmatic activity (e.g., young lavas, hot springs, hydrothermal vents and plumes) along these divergent plate boundaries. Eruptions are rarely observed because of their great depths and remote locations. However, over 60% of Earth's magma flux (approximately $21 \, km^3 \, year^{-1}$) currently occurs along divergent plate margins. Geophysical imaging, detailed mapping, and sampling of mid-ocean ridges and fracture zones between ridge segments followed by laboratory petrologic and geochemical analyses of recovered rocks provide us with a great deal of information about the composition and evolution of the oceanic crust and the processes that generate mid-ocean ridge basalts (MORB).

Mid-ocean ridges are not continuous but rather broken up into various scale segments reflecting breaks in the volcanic plumbing systems that feed the axial zone of magmatism. Recent hypotheses suggest that the shallowest and widest portions of ridge segments correspond to robust areas of magmatism, whereas deep, narrow zones are relatively magma-starved. The unusually elevated segments of some ridges (e.g., south of Iceland, central portion of the Galapagos Rift, Mid-Atlantic Ridge near the Azores) are directly related to the influence of nearby mantle plumes or hot spots that result in voluminous magmatism.

Major differences in the morphology, structure, and scales of magmatism along mid-ocean ridges vary with the rate of spreading. Slowly diverging plate boundaries, which have low volcanic output, are dominated by faulting and tectonism whereas fast-spreading boundaries are controlled more by volcanism. The region along the plate boundary within which volcanic eruptions and high-temperature hydrothermal activity are concentrated is called the neovolcanic zone. The width of the neovolcanic zone, its structure, and the style of volcanism within it, vary considerably with spreading rate. In all cases, the neovolcanic zone on mid-ocean ridges is marked by a roughly linear depression or trough (axial summit collapse trough, ASCT), similar to rift zones in some subaerial volcanoes, but quite different from the circular craters and calderas associated with typical central-vent volcanoes. Not all mid-ocean ridge volcanism occurs along the neovolcanic zone. Relatively small (<1 km high), near-axis seamounts are common within a few tens of kilometers of fast and intermediate spreading ridges. Recent evidence also suggests that significant amounts of volcanism may occur up to 4 km from the axis as off-axis mounds and ridges, or associated with faulting and the formation of abyssal hills.

Lava morphology on slow spreading ridges is dominantly bulbous, pillow lava (**Figure 1A**), which tends to construct hummocks (<50 m high, <500 m diameter), hummocky ridges (1–2 km long), or small circular seamounts (10s–100s of meters high and 100s–1000s of meters in diameter) that commonly coalesce to form axial volcanic ridges (AVR) along the valley floor of the axial rift zone. On fast spreading ridges, lavas are dominantly oblong, lobate flows and fluid sheet flows that vary from remarkably flat and thin (<4 cm) to ropy and jumbled varieties (**Figure 1**). Although the data are somewhat limited, calculated volumes of individual flow units that have been documented on mid-ocean ridges show an inverse exponential relationship to spreading rate, contrary to what might be expected. The largest eruptive units are mounds and cones in the axis of the northern Mid-Atlantic Ridge whereas the smallest units are thin sheet/lobate flows on the East Pacific Rise. Morphologic, petrologic, and structural studies of many ridge segments suggest

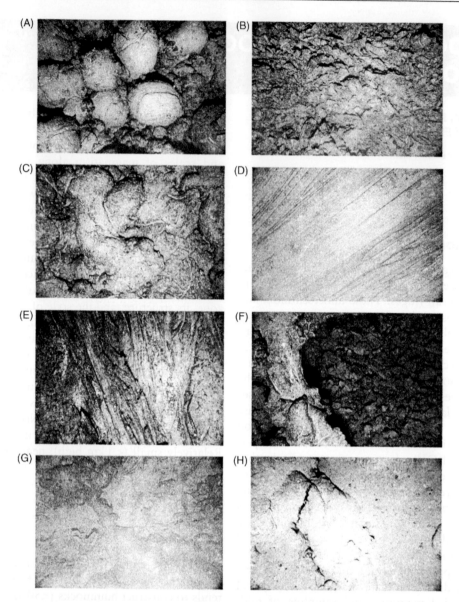

Figure 1 Examples of different morphologies, surface textures and sediment cover on lava flows on the northern East Pacific Rise. Digital images were taken from heights of 5–10 m above the seafloor using the Woods Hole Oceanographic Instution's camera system. The dimensions of the photographs are approximately 4.5 m × 3.0 m. (A) Pillow lava. (B) Hackly or scrambled flow. (C) Lobate lava. (D) Lineated sheet flow. (E) Ropy sheet flow. (F) Collapse structure in lobate flows. (G) A young flow contact on top of older flows. (H) Heavily sediment covered lobate flows with small fissure. Images from Kuras *et al.* 2000.

they evolve through cycles of accretion related to magmatic output followed by amagmatic periods dominated by faulting and extension.

Magma Generation

Primary MORB magmas are generated by partial melting of the upper mantle; believed to be composed of a rock type termed peridotite which is primarily composed of the minerals olivine, pyroxenes (enstatite and diopside), and minor spinel or garnet.

Beneath ridges, mantle moves upward, in part, due to convection in the mantle but possibly more in response to the removal of the lithospheric lid above it, which is spreading laterally. Melting is affected by the decompression of hot, buoyant peridotite that crosses the melting point (solidus curve) for mantle material as it rises to shallow depths (< 100 km), beneath the ridges. Melting continues as the mantle rises as long as the temperature of the peridotite remains above the solidus temperature at a given depth. As the seafloor spreads, basaltic melts formed in a broad region (10s to 100s of kilometers) beneath the ridge accumulate and focus so that they feed a relatively narrow region (a few kilometers) along the axis of the ridge (**Figure 1**).

During ascent from the mantle and cooling in the crust, primary mantle melts are subjected to a variety of physical and chemical processes such as fractional crystallization, magma mixing, crustal assimilation, and thermogravitational diffusion that modify and differentiate the original melt composition. Consequently, primary melts are unlikely to erupt on the seafloor without undergoing some modification. Picritic lavas and magnesian glasses thought to represent likely primary basalts have been recovered from a few ocean floor localities; commonly in transform faults (**Table 1**). MgO contents in these basalts range from ~10 wt% to over 15 wt% and the lavas typically contain significant amounts of olivine crystals. Based on comparisons with high-pressure melting experiments of likely mantle peridotites, the observed range of compositions may reflect variations in source composition and mineralogy (in part controlled by pressure), depth and percentage melting (largely due to temperature differences), and/or types of melting (e.g., batch vs. fractional).

Ocean Floor Volcanism and Construction of the Crust

Oceanic crust formed at spreading ridges is relatively homogeneous in thickness and composition compared to continental crust. On average, oceanic crust is 6–7 km thick and basaltic in composition as compared to the continental crust which averages 35–40 km thick and has a roughly andesitic composition. The entire thickness of the oceanic crust has not been sampled *in situ* and therefore the bulk composition has been estimated based on investigations of ophiolites (fragments of oceanic and back-arc crust that have been thrust up on to the continents), comparisons of the seismic structure of the oceanic crust with laboratory determinations of seismic velocities in known rock types, and samples recovered from the ocean floor by dredging, drilling, submersibles, and remotely operated vehicles.

Rapid cooling of MORB magmas when they come into contact with cold sea water results in the formation of glassy to finely crystalline pillows, lobate flows, or sheet flows (**Figure 1**). These lava flows typically have an ~0.5–1 cm-thick outer rind of glass and a fine-grained, crystalline interior containing only a few percent of millimeter-sized crystals of olivine, plagioclase, and more rarely clinopyroxene in a microscopic matrix of the same minerals. MORB lavas erupt, flow, and accumulate to form the uppermost volcanic layer (Seismic Layer 2A) of ocean crust (**Figure 2**). Magmas that do not reach the seafloor cool more slowly with increasing depth forming intrusive dikes at shallow levels (0.5–3 km) in the crust (layer 2B) and thick bodies of coarsely crystalline gabbros and cumulate ultramafic rocks at the lowest levels (3–7 km) of the crust (layer 3) (**Figure 2**).

Although most magma delivered to a MOR is focused within the neovolcanic zone, defined by the axial summit collapse trough or axial valley, off-axis volcanism and near-axis seamount formation appear to add significant volumes of material to the uppermost crust formed along ridge crests. In some portions of the fast spreading East Pacific Rise, off-axis eruptions appear to be related to syntectonic volcanism and the formation of abyssal hills. Near-axis seamount formation is common along both the East Pacific Rise and medium spreading rate Juan de Fuca Ridge. Even in areas where there are abundant off-axis seamounts they may add only a few percent to the volume of the extrusive crust. More detailed studies of off-axis sections of ridges are needed before accurate estimates of their contribution to the total volume of the oceanic crust can be made.

Oceanic transform faults are supposed to be plate boundaries where crust is neither created nor destroyed, but recent mapping and sampling indicate that magmatism occurs in some transform domains. Volcanism occurs in these locales either at short, intratransform spreading centers or at localized eruptive centers within shear zones or relay zones between the small spreading centers.

Mid-ocean Ridge Basalt Composition

Ocean floor lavas erupted along mid-ocean ridges are low-potassium tholeiites that can range in composition from picrites with high MgO contents to ferrobasalts and FeTi basalts containing lower MgO and high concentrations of FeO and TiO_2, and even to rare, silica-enriched lavas known as icelandites, ferroandesites and rhyodacites (**Table 1**). In most areas, the range of lava compositions, from MgO-rich basalt to FeTi basalt and ultimately to rhyodacite, is generally ascribed to the effects of shallow-level (low-pressure) fractional crystallization in a subaxial magma chamber or lens (**Figure 2**). A pronounced iron-enrichment trend with decreasing magnesium contents (related to decreasing temperature) in suites of genetically related lavas is, in part, what classifies MORB as tholeiitic or part of the tholeiitic magmatic suite (**Figure 3**).

Table 1 Average compositions of normal and enriched types of basalts from mid-ocean ridges and seamounts

Oxide wt%	Normal							Enriched			
	Pacific	Atlantic	Galapagos	Seamounts	Pacific Picritic	Pacific Ferrobasalt	Pacific High-silica	Pacific	Atlantic	Galapagos	Seamounts
SiO_2	50.49	50.64	50.41	50.03	48.80	50.61	55.37	50.10	51.02	49.17	50.19
TiO_2	1.78	1.43	1.54	1.28	0.97	2.36	2.10	1.86	1.46	1.94	1.74
Al_2O_3	14.55	15.17	14.75	15.97	17.12	13.30	12.92	15.69	15.36	16.86	16.71
FeO^*	10.87	10.45	11.19	9.26	8.00	13.61	13.11	9.78	9.56	9.21	8.77
MnO	0.20	0.19	nd	0.15	0.14	0.23	0.21	0.19	0.18		0.15
MgO	7.22	7.53	7.49	8.06	10.28	5.92	3.64	7.00	7.31	6.93	6.80
CaO	11.58	11.62	11.69	12.21	11.93	10.43	8.05	11.17	11.54	10.90	10.67
Na_2O	2.74	2.51	2.28	2.68	2.32	2.74	3.33	3.04	2.52	3.16	3.38
K_2O	0.13	0.11	0.10	0.08	0.03	0.16	0.44	0.43	0.36	0.66	0.75
P_2O_5	0.17	0.14	0.14	0.13	0.07	0.22	0.40	0.24	0.19	0.32	0.33
Sum	99.62	99.61	99.60	99.73	100	99.39	99.40	99.37	99.31	99.14	99.34
K/Ti	7.49	7.70	6.24	6.10	3.0	7.04	13.80	22.26	23.67	32.77	39.05
N =	2303	2148	867	623	10	706	97	304	972	65	197

Analyses done by electron microprobe on natural glasses at the Smithsonian Institution in Washington, D.C. (by W. Melson and T.O'Hearn) except the picritic samples that were analyzed at the USGS in Denver, Co. Enriched MORB in this compilation are any that have K/Ti values greater than 13. High-silica lavas have SiO_2 values between 52 and 64. K/Ti = $(K_2O/TiO_2) \times 100$. N = number of samples used in average. FeO* = total Fe as FeO.

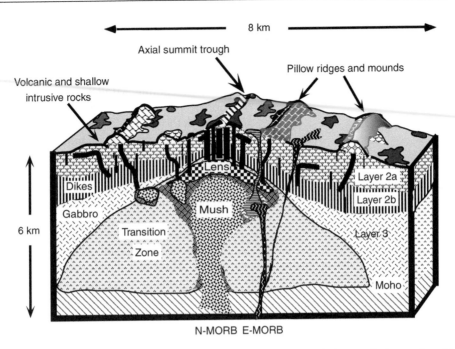

Figure 2 Diagrammatic three-dimensional representation of oceanic crust formed along a fast-spreading ridge showing the seismically determined layers and their known or inferred petrologic composition. Note that although most of the volcanism at mid-ocean ridges appears to be focused within the axial summit trough, a significant amount of off-axis volcanism (often forming pillow mounds or ridges) is believed to occur. Much of the geochemical variability that is observed in MORB probably occurs within the crystal–liquid mush zone and thin magma lens that underlie the ridge crest. The Moho marks the seismic boundary between plutonic rocks that are gabbroic in composition and those that are mostly ultramafic but may have formed by crystal accumulation in the crust.

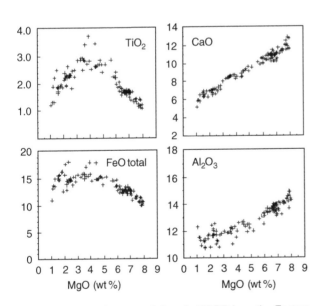

Figure 3 Major element variations in MORB from the Eastern Galapagos Spreading Center showing the chemical trends generated by shallow-level fractional crystallization in the oceanic crust. The rocks range in composition from basalt to ferrobasalt and FeTi basalt to andesite.

Although MORB are petrologically similar to tholeiitic basalts erupted on oceanic islands (OIB), MORB are readily distinguished from OIB based on their comparatively low concentrations of large ion

lithophile elements (including K, Rb, Ba, Cs), light rare earth elements (LREE), volatile elements and other trace elements such as Th, U, Nb, Ta, and Pb that are considered highly incompatible during melting of mantle mineral assemblages. In other words, the most incompatible elements will be the most highly concentrated in partial melts from primitive mantle peridotite. On normalized elemental abundance diagrams and rare earth element plots (**Figure 4**), normal MORB (N-type or N-MORB) exhibit characteristic smooth concave-down patterns reflecting the fact that they were derived from incompatible element-depleted mantle. Isotopic investigations have conclusively shown that values of the radiogenic isotopes of Sr, Nd, Hf and Pb in N-MORB are consistent with their depleted characteristics and indicate incompatible element depletion via one or more episodes of partial melting of upper mantle sources beginning more than 1 billion years ago. Compared to ocean island basalts and lavas erupted in arc or continental settings, MORB comprise a relatively homogeneous and easily distiguishable rock association. Even so, MORB vary from very depleted varieties (D-MORB) to those containing moderately elevated incompatible element abundances and more radiogenic isotopes. These less-depleted MORB are called E-types (E-MORB)

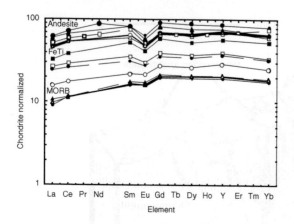

Figure 4 Chondrite-normalized rare earth element (REE) abundances in a suite of cogenetic lavas from the Eastern Galapagos Spreading Center (also shown in **Figures 3** and **6**). Increasing abundances of REE and the size of the negative europium anomaly from MORB to andesite are consistent with evolution of the suite primarily by fractional crystallization. Concave-down patterns are an indication of their 'normal' depleted chemical character (N-MORB).

or P-types, indicative of an 'enriched' or 'plume' component (**Table 1**) typically associated with intraplate 'hot spots'. Transitional varieties are classified as T-MORB. Enriched MORB are volumetrically minor on most normal ridge segments, but can comprise a significant proportion of the crust around regions influenced by plume magmatism such as the Galapagos Islands, the Azores, Tristan, Bouvet, and Iceland.

Mineralogy of Mid-ocean Ridge Basalts

The minerals that crystallize from MORB magmas are not only dependent on the composition of the melt, but also the temperature and pressure during crystallization. Because the majority of MORB magmas have relatively similar major element compositions and probably begin to crystallize within the uppermost mantle and oceanic crust (pressures less than 0.3 GPa), they have similar mineralogy. Textures (including grain size) vary depending on nucleation and crystallization rates. Hence lavas, that are quenched when erupted into sea water, have few phenocrysts in a glassy to cryptocrystalline matrix. Conversely, magmas that cool slowly in subaxial reservoirs or magma chambers form gabbros that are totally crystalline (holocrystalline) and composed of well-formed minerals that can be up to a few centimeters long. Many of the gabbros recovered from the ocean floor do not represent melt compositions but rather reflect the accumulation of crystals and percolation of melt that occurs during convection,

deformation and fractional crystallization in the mush zone hypothesized to exist beneath some mid-ocean ridges (**Figure 2**). These cumulate gabbros are composed of minerals that have settled (or floated) out of cooling MORB magmas and their textures often reflect compaction, magmatic sedimentation, and deformation.

MOR lavas may contain millimeter-sized phenocrysts of the silicate minerals plagioclase (solid solution that ranges from $CaAl_2Si_2O_8$ to $NaAlSi_3O_8$) and olivine (Mg_2SiO_4 to Fe_2SiO_4) and less commonly, clinopyroxene ($Ca[Mg,Fe]Si_2O_6$). Spinel, a Cr-Al rich oxide, is a common accessory phase in more magnesian lavas where it is often enclosed in larger olivine crystals. Olivine is abundant in the most MgO-rich lavas, becomes less abundant in more evolved lavas and is ultimately replaced by pigeonite (a low-Ca pyroxene) in FeO-rich basalts and andesite. Clinopyroxene is only common as a phenocryst phase in relatively evolved lavas. Titanomagnetite, ilmenite and rare apatite are present as microphenocrysts, although not abundantly, in basaltic andesites and andesites.

Intrusive rocks, which cool slowly within the oceanic crust, have similar mineralogy but are holocrystalline and typically much coarser grained. Dikes form fine- to medium-grained diabase containing olivine, plagioclase and clinopyroxene as the major phases, with minor amounts of ilmenite and magnetite. Gabbros vary from medium-grained to very coarse-grained with crystals up to a few centimeters in length. Because of their cumulate nature and extended cooling histories, gabbros often exhibit layering of crystals and have the widest mineralogic variation. Similar to MORB, the least-evolved varieties (troctolites) consist almost entirely of plagioclase and olivine. Some gabbros can be nearly monomineralic such as anorthosites (plagioclase-rich) or contain monomineralic layers (such as olivine that forms layers or lenses of a rock called dunite). The most commonly recovered varieties of gabbro are composed of plagioclase, augite (a clinopyroxene) and hypersthene (orthopyroxene) with minor amounts of olivine, ilmenite and magnetite and, in some cases, hornblende (a hydrous Fe-Mg silicate that forms during the latest stages of crystallization). Highly evolved liquids cool to form ferrogabbros and even rarer silica-rich plutonics known as trondhjemites or plagiogranites.

The descriptions above pertain only to those portions of the oceanic crust that have not been tectonized or chemically altered. Because of the dynamic nature of oceanic ridges and the pervasive hydrothermal circulation related to magmatism, it is common for the basaltic rocks comprising the crust to be chemically altered and metamorphosed. When this occurs, the primary minerals are recrystallized or replaced by a variety of secondary minerals such as smectite, albite, chlorite, epidote, and amphibole that are more stable under lower temperature and more hydrous conditions. MOR basalts, diabases and gabbros are commonly metamorphosed to greenschists and amphibolites. Plutonic rocks and portions of the upper mantle rich in olivine and pyroxene are transformed into serpentinites. Oceanic metamorphic rocks are commonly recovered from transform faults, fracture zones and slowly spreading segments of the MOR where tectonism and faulting facilitate deep penetration of sea water into the crust and upper mantle.

Chemical Variability

Although MORB form a relatively homogeneous population of rock types when compared to lavas erupted at other tectonic localities, there are subtle, yet significant, chemical differences in their chemistry due to variability in source composition, depth and extent of melting, magma mixing, and processes that modify primary magmas in the shallow lithosphere. Chemical differences between MORB exist on all scales, from individual flows erupted along the same ridge segment (e.g., CoAxial Segment of the Juan de Fuca Ridge) to the average composition of basalts from the global ridge system (e.g. Mid-Atlantic Ridge vs. East Pacific Rise). High-density sampling along several MOR segments has shown that quite a diversity of lava compositions can be erupted over short time (10s–100 years) and length scales (100 m to a few kilometers). Slow spreading ridges, which do not have steady-state magma bodies, generally erupt more mafic lavas compared to fast spreading ridges where magmas are more heavily influenced by fractional crystallization in shallow magma bodies. Intermediate rate-spreading centers, where magma lenses may be small and intermittent, show characteristics of both slow- and fast spreading centers. In environments where magma supply is low or mixing is inhibited, such as proximal to transform faults, propagating rift tips and overlapping spreading centers, compositionally diverse and highly differentiated lavas are commonly found (such as the Eastern Galapagos Spreading Center, **Figures 3, 4** and **6**). In these environments, extensive fractional crystallization is a consequence of relatively cooler thermal regimes and the magmatic processes associated with rift propagation.

Local variability in MORB can be divided into two categories: (1) those due to processes that affect an individual parental magma (e.g., fractional crystallization, assimilation) and (2) those created via partial melting and transport in a single melting regime (e.g., melting in a rising diapir). In contrast, global variations reflect regional variations in mantle source chemistry and temperature, as well as the averaging of melts derived from diverse melting regimes (e.g. accumulative polybaric fractional melting). At any given segment of MOR, variations may be due to various combinations of these processes.

Local Variability

Chemical trends defined by suites of related MOR lavas are primarily due to progressive fractional crystallization of variable combinations and proportions of olivine, plagioclase and clinopyroxene as a magma cools. The compositional 'path' that a magma takes is known as its liquid line of descent (LLD). Slightly different trajectories of LLDs (**Figure 5**) are a consequence of the order of crystallization and the different proportions of crystallizing phases that are controlled by initial (and subsequent changing) liquid composition, temperature, and pressure. In some MORB suites, linear elemental trends may be due to mixing of primitive magmas with more evolved magmas that have evolved along an LLD.

Suites of MORB glasses often define distinctive LLDs that match those determined by experimental crystallization of MORB at low to moderate

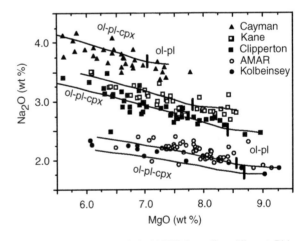

Figure 5 MgO vs. Na$_2$O in MORB from five different Ridge segments (Mid Cayman Rise in the Caribbean; near Kane Fracture Zone on the Mid-Atlantic Ridge, 23°N; AMAR on the Mid-Atlantic Ridge around 37°N; East Pacific Rise near the Clipperton Fracture Zone around 10°N; Kolbeinsey Ridge north of Iceland. Lines are calculated Liquid Lines of Descent (LLDs) from high MgO parents. Bar shows where clinopyroxene joins plagioclase and olivine as a fractionating phase. Na$_8$ is determined by the values of Na$_2$O when the LLD is at MgO of 8 wt%. (Adapted with permission from Langmuir *et al.*, 1992.)

pressures that correspond to depths of ∼1 to 10 km within the oceanic crust and upper mantle. Much of the major element data from fast-spreading ridges like the East Pacific Rise are best explained by low-pressure (∼0.1 GPa) fractional crystallization whereas at slow-spreading ridges like the Mid-Atlantic Ridge data require higher pressure crystallization (∼0.5–1.0 GPa). This is consistent with other evidence suggesting that magmas at fast-spreading ridges evolve in a shallow magma lens or chambers and that magmas at slow-spreading ridges evolve at significantly greater depths; possibly in the mantle lithosphere or at the crust–mantle boundary. Estimated depths of crystallization correlate with increased depths of magma lens or fault rupture depth related to decreasing spreading rate.

Cogenetic lavas (those from the same or similar primary melts) generated by fractional crystallization exhibit up to 10-fold enrichments of incompatible trace elements (e.g., Zr, Nb, Y, Ba, Rb, REE) that covary with indices of fractionation such as decreasing MgO (**Figure 6**) and increasing K_2O concentrations and relatively constant incompatible

trace element ratios irrespective of rock type. In general, the rare earth elements show systematic increases in abundance through the fractionation sequence from MORB to andesite (**Figure 4**) with a slight increase in light rare earth elements relative to the heavy-rare earth elements. The overall enrichments in the trivalent rare earth elements is a consequence of their incompatibility in the crystals separating from the cooling magma. Increasing negative Eu anomalies develop in more fractionated lavas due to the continued removal of plagioclase during crystallization because Eu partially substitutes for Ca in plagioclase which is removed during fractional crystallization.

Global Variability

MORB chemistry of individual ridge segments (local scale) is, in general, controlled by the relative balance between tectonic and magmatic activity, which in turn may determine whether a steady-state magma chamber exists, and for how long. Ultimately, the tectonomagmatic evolution is controlled by temporal variations in input of melt from the mantle. Global correlation of abyssal peridotite and MORB geochemical data suggest that the extent of mantle melting beneath normal ridge segments increases with increasing spreading rate and that both ridge morphology and lava composition are related to spreading rate.

The depths at which primary MORB melts form and equilibrate with surrounding mantle remain controversial (possibly 30 to 100 km), as does the mechanism(s) of flow of magma and solid mantle beneath divergent plate boundaries. The debate is critical for understanding the dynamics of plate spreading and is focused on whether flow is 'passive' plate driven flow or 'active' buoyantly driven solid convection. At present, geological and geophysical observations support passive flow which causes melts from a broad region of upwelling and melting to converge in a narrow zone at ridge crests.

It has also been hypothesized that melting beneath ridges is a dynamic, near-fractional process during which the pressure, temperature, and composition of the upper mantle change. Variations in these parameters as well as in the geometry of the melting region result in the generation of MORB with different chemical characteristics.

Differences in the major element compositions of MORB from different parts of the world's oceans (global scale) have been recognized for some time. In general, it has been shown that N-type MORB from slow-spreading ridges such as the Mid-Atlantic Ridge are more primitive (higher MgO) and have

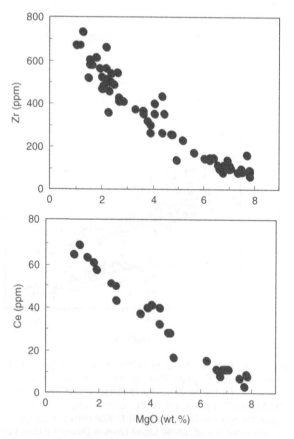

Figure 6 Trace element (Zr and Ce) versus MgO variation diagram showing the systematic enrichments of these highly incompatible elements with increasing fractionation in a suite of cogenetic lavas from the Eastern Galapagos Spreading Center.

greater Na_2O, Al_2O_3 and lower FeO and CaO/Al_2O_3 contents at given MgO values than lavas from medium- and fast-spreading ridges (**Figure 7**). A comparison of ocean floor glass compositions (over 9000) analyzed by electron microprobe at the Smithsonian Institution from major spreading centers and seamounts is presented in **Table 1**. The analyses have been filtered into normal (N-MORB) and enriched (E-MORB) varieties based on their K/Ti ratios (E-MORB $[K_2O/TiO_2] \times 100 > 13$) which reflect enrichment in the highly incompatible elements. These data indicate that on average, MORB are relatively differentiated compared to magmas that might be generated directly from the mantle (compare averages with picritic basalts from the Pacific in **Table 1**). Furthermore, given the variability of glass compositions in each region, N-MORB have quite similar average major element compositions (most elemental concentrations overlap at the 1-sigma level). E-MORB, are more evolved than N-MORB

from comparable regions of the ocean and there are a higher proportion of E-MORB in the Atlantic (31%) compared to the Pacific (12%) and Galapagos Spreading Center region (7%). Unlike the Atlantic where E-MORB are typically associated with inflated portions of the ridge due to the effects of plume–ridge interaction, East Pacific Rise E-MORB are randomly dispersed along-axis and more commonly recovered off-axis. As well as having higher K_2O contents than N-MORB, E-MORB have higher concentrations of P_2O_5, TiO_2, Al_2O_3 and Na_2O and lower concentrations of SiO_2, FeO and CaO. Positive correlations exist between these characteristics, incompatible element enrichments and more radiogenic Sr and Nd isotopes in progressively more enriched MORB.

Direct comparison of elemental abundances between individual MORB (or even groups) is difficult because of the effects of fractional crystallization. Consequently, fundamental differences in chemical characteristics are generally expressed as differences in parameters such as Na_8, Fe_8, Al_8, Si_8 etc. which are the values of these oxides calculated at an MgO content of 8.0 wt% (**Figure 5 and 8**). When using these normalized values, regionally averaged major element data show a strong correlation with ridge depth and possibly, crustal thickness. MORB with high FeO and low Na_2O are sampled from shallow ridge crests with thick crust whereas low FeO– high Na_2O MORB are typically recovered from deep ridges with thin crust (**Figure 8**). This chemical/tectonic correlation gives rise to the so-called 'global array'. Major element melting models indicate there is a strong correlation between the initial depth of melting and the total amount of melt formed. As a consequence, when temperatures are high enough to initiate melting at great depths, the primary MORB melts contain high FeO, low Na_2O and low SiO_2. Conversely, if the geothermal gradient is low, melting is restricted to the uppermost part of the upper mantle, and little melt is generated (hence thinner crust) and the basaltic melts contain low FeO, high Na_2O and relatively high SiO_2.

Although the global systematics appear robust, detailed sampling of individual ridge segments have shown MORB from limited areas commonly exhibit chemical correlations that form a 'local trend' opposite to the chemical correlations observed globally (e.g., FeO and Na_2O show a positive correlation). A local trend may reflect the spectrum of melts formed at different depths beneath one ridge crest rather than the aggregate of all the melt increments.

Although the original hypothesis that global variations in MORB major element chemistry are a consequence of total extents of mantle melting and

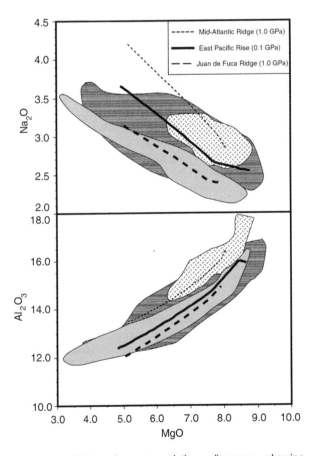

Figure 7 Major element variation diagrams showing compositional ranges from different spreading rate ridges. Generally higher Na_2O and Al_2O_3 concentrations in Mid-Atlantic Ridge (hatchured field) lavas in comparison to MORB from the Juan de Fuca (grey field) and East Pacific Rise (dark field) are shown. Lines show calculated liquid lines of descent at 0.1 and 1.0 GPa for parental magmas from each ridge.

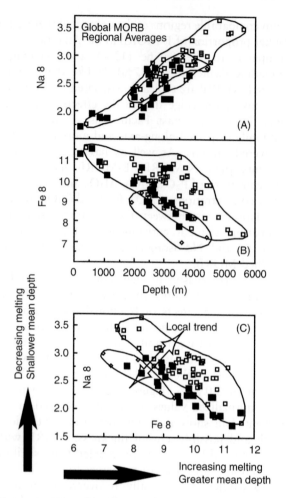

Figure 8 (A) and (B) Global correlations between regional averages of ridge axial depth and the Na_8 and Fe_8 of MORBs. Different groups of MORB are distinguished. □, Normal ridge segments; ◇, ridges behind island arcs; ■, ridges influenced by hot spots. (C) Global trend of Na_8 vs. Fe_8 due to differences in extents and depths of melting. Representative 'Local trend' is common along individual portions of some ridges. (Adapted with permission from Langmuir *et al.*, 1992.)

mean pressure of extraction due to variations in mantle temperature, more recent evidence suggests that heterogeneity in the mantle also plays an important role in defining both global and local chemical trends. In particular, U-series data suggest some MDRB melts equilibrate with highly depleted mantle at shallow depths whereas others equilibrate with less depleted garnet pesidatite at depths greater than ~80 km.

Conclusions

Passive rise of the mantle beneath oceanic spreading centers results in the decompression melting of upwelling peridotite which gives rise to a spectrum of MORB compositions varying from extremely

depleted to moderately enriched varieties. The compositional variability in primary MORB result from combinations of differing source compositions, extents and styles of partial melting, and depths of melt formation. The moderately evolved composition of most MORB primarily reflects the effects of crystal fractionation that occurs as the primary melts ascend from the mantle into the cooler crust. Although MORB are relatively homogeneous compared to basalts from other tectonic environments, they exhibit a range of compositions that provide us with information about the composition of the mantle, the influence of plumes, and dynamic magmatic processes that occur to form the most voluminous part of the Earth's crust.

See also

Mid-Ocean Ridge Seismic Structure. Mid-Ocean Ridge Tectonics, Volcanism, and Geomorphology. Propagating Rifts and Microplates. Seamounts and Off-Ridge Volcanism.

Further Reading

Batiza R and Niu Y (1992) Petrology and magma chamber processes at the East Pacific Rise – 9°30′N. *Journal of Geophysical Research* 97: 6779–6797.

Grove TL, Kinzler RJ, and Bryan WB (1992) Fractionation of Mid-Ocean Ridge Basalt (MORB). In: Phipps-Morgan J, Blackman DK, and Sinton J (eds.) *Mantle Flow and Melt Generation at Mid-ocean Ridges*, Geophys. Monograph 71, pp. 281–310. Washington, DC: American Geophysical Union.

Klein EM and Langmuir CH (1987) Global correlations of ocean ridge basalt chemistry with axial depth and crustal thickness. *Journal of Geophysical Research* 92: 8089–8115.

Kurras GJ, Fomari DJ, Edwards MH, Perfit MR, and Smith MC (2000) Volcanic morphology of the East Pacific Rise Crest 9°49′–52′N:1. Implications for volcanic emplacement processes at fast-spreading mid-ocean ridges. *Marine Geophysical Research* 21: 23–41.

Langmuir CH, Klein EM, and Plank T (1992) Petrological systematics of mid-ocean ridge basalts: constraints on melt generation beneath ocean ridges. In: Phipps-Morgan J, Blackman DK and Sinton J (eds) *Mantle Flow and Melt Generation at Mid-ocean Ridges*, Geophysics Monograph 71, Washington, DC: American Geophysical Union, p. 183–280.

Lundstrom CC, Sampson DE, Perfit MR, Gill J, and Williams Q (1999) Insight, into mid-ocean ridge basalt petrogenesis: U-series disequilibrium from the Siqueiro, Transform, lamont seamounts, and East Pacific Rise *Journal of Geophysical Research* 104: 13035–13048.

Macdonald KC (1998) Linkages between faulting, volcanism, hydrothermal activity and segmentation on fast spreading centers. In: Buck WR, Delaney PT, Karson JA and Lagabrielle Y (eds) *Faulting and Magmatism at Mid-ocean ridges*, American Geophysics Monograph 106, Washington, DC: American Geophysical Union, p. 27–59.

Nicolas A (1990) *The Mid-Ocean Ridges*. Springer Verlag, Berlin.

Niu YL and Batiza R (1997) Trace element evidence from seamounts for recycled oceanic crust in the Eastern Pacific mantle. *Earth Planet. Sci. Lett* 148: 471–483.

Perfit MR and ChadwickWW (1998) Magmatism at mid-ocean ridges: constraints from volcanological and geochemical investigations. In: Buck WR, Delaney PT, Karson JA and Lagabrielle Y (eds), *Faulting and Magmatism at Mid-ocean Ridges*. American Geophysics Monograph 106, Washington, DC: American Geophysics Union, p. 59–115.

Perfit MR and Davidson JP (2000) Plate tectonics and volcanism. In: Sigurdsoon H (ed.) *Encyclopedia of Volcanoes*. San Diego: Academic Press.

Perfit MR, Ridley WI, and Jonasson I (1998) Geologic, petrologic and geochemical relationships between magmatism and massive sulfide mineralization along the eastern Galapagos Spreading Center. *Review in Economic Geology* 8(4): 75–99.

Shen Y and Forsyth DW (1995) Geochemical constraints on initial and final depths of melting beneath mid-ocean ridges. *Journal of Geophysical Research* 100: 2211–2237.

Sigurdsson H (ed.) (2000) *Encyclopedia of Volcanoes*. Academic Press

Sinton JM and Detrick RS (1992) Mid-ocean ridge magma chambers. *Journal of Geophysical Research* 97: 197–216.

Smithsonian Catalog of Basalt Glasses [http://www.nmnh.si.edu/minsci/research/glass/index.htm]

Thompson RN (1987) Phase-equilibria constraints on the genesis and magmatic evolution of oceanic basalts. *Earth Science Review* 24: 161–210.

MID-OCEAN RIDGE TECTONICS, VOLCANISM, AND GEOMORPHOLOGY

K. C. Macdonald, Department of Geological Sciences and Marine Sciences Institute, University of California, Santa Barbara, CA, USA

Introduction

The midocean ridge is the largest mountain chain and the most active system of volcanoes in the solar system. In plate tectonic theory, the ridge is located between plates of the earth's rigid outer shell that are separating at speeds of $\sim$10 to 170 mm y^{-1} (up to 220 mm y^{-1} in the past). The ascent of molten rock from deep in the earth ($\sim$30–60 km) to fill the void between the plates creates new seafloor and a volcanically active ridge. This ridge system wraps around the globe like the seam of a baseball and is approximately 70 000 km long. Yet the ridge itself is only $\sim$5–30 km wide–very small compared to the plates, which can be thousands of kilometers across (**Figure 1**).

Early exploration showed that the gross morphology of spreading centers varies with the rate of plate separation. At slow spreading rates (10–40 mm y^{-1}) a 1–3 km deep rift valley marks the axis, while for fast spreading rates ($>$90 mm y^{-1}) the axis is characterized by an elevation of the seafloor of several hundred meters, called an axial high (**Figure 2**). The rate of magma supply is a second factor that may influence the morphology of midocean ridges. For example, a very high rate of magma supply can produce an axial high even where the spreading rate is slow; the Reykjanes Ridge south of Iceland is a good example. Also, for intermediate spreading rates (40–90 mm y^{-1}) the ridge crest may have either an axial high or rift valley depending on the rate of magma supply. The seafloor deepens from a global average of $\sim$2600 m at the spreading center to $>$5000 m beyond the ridge flanks. The rate of deepening is proportional to the square root of the age of the seafloor because it is caused by the thermal contraction of the lithosphere. Early mapping efforts also showed that the midocean ridge is a discontinuous structure that is offset at right angles to its length at numerous transform faults tens to hundreds of kilometers in length.

Maps are powerful; they inform, excite, and stimulate. Just as the earliest maps of the world in the sixteenth century ushered in a vigorous age of exploration, the first high-resolution, continuous

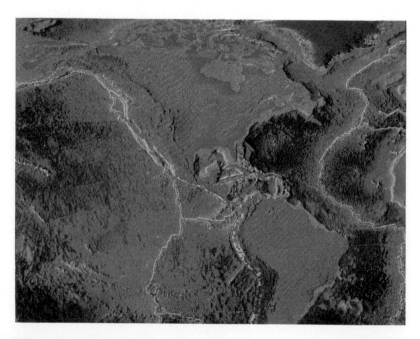

Figure 1 Shaded relief map of the seafloor showing parts of the East Pacific Rise, a fast-spreading center, and the Mid-Atlantic Ridge, a slow-spreading center. Courtesy of National Geophysical Data Center.

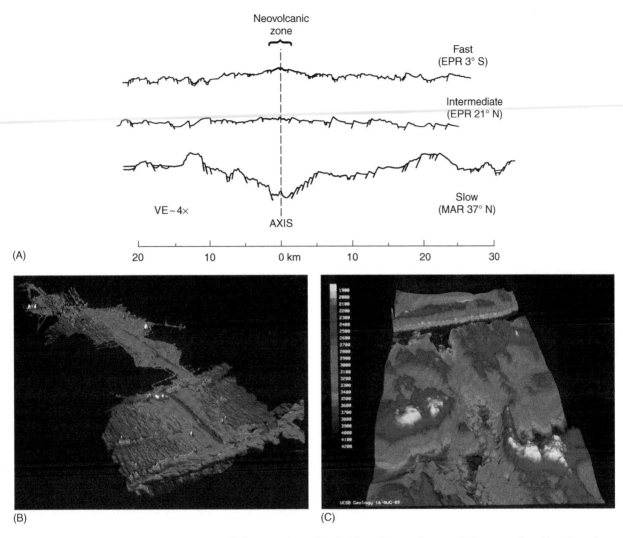

Figure 2 Topography of spreading centers. (A) Cross-sections of typical fast-, intermediate-, and slow-spreading ridges based on high resolution deep-tow profiles. The neovolcanic zone is noted (the zone of active volcanism) and is several kilometers wide; the zone of active faulting extends to the edge of the profiles and is several tens of kilometers wide. After Macdonald *et al.* (1982). (B) Shaded relief map of a 1000 km stretch of the East Pacific Rise extending from 8° to 17°N. Here, the East Pacific Rise is the boundary between the Pacific and Cocos plates, which separate at a 'fast' rate of 110 mm y^{-1}. The map reveals two kinds of discontinuities: large offsets, about 100 km long, known as transform faults and smaller offsets, about 10 km long, called overlapping spreading centers. Colors indicate depths of from 2400 m (pink) to 3500 m (dark blue). (C.) Shaded relief map of the Mid-Atlantic Ridge. Here, the ridge is the plate boundary between the South American and African plates, which are spreading apart at the slow rate of approximately 35 mm y^{-1}. The axis of the ridge is marked by a 1–2 km deep rift valley, which is typical of most slow-spreading ridges. The map reveals a 12 km jog of the rift valley, a second-order discontinuity, and also shows a first-order discontinuity called the Cox transform fault. Colors indicated depths of from 1900 m (pink) to 4200 m (dark blue).

coverage maps of the midocean ridge stimulated investigators from a wide range of fields including petrologists, geochemists, volcanologists, seismologists, tectonicists, and practitioners of marine magnetics and gravity; as well as researchers outside the earth sciences including marine ecologists, chemists, and biochemists. Marine geologists have found that many of the most revealing variations are to be observed by exploring along the axis of the active ridge. This along-strike perspective has revealed the architecture of the global rift system. The ridge axis undulates up and down in a systematic way, defining a fundamental partitioning of the ridge into segments bounded by a variety of discontinuities. These segments behave like giant cracks in the seafloor that can lengthen or shorten, and have episodes of increased volcanic and tectonic activity.

Another important change in perspective came from the discovery of hydrothermal vents by marine geologists and geophysicists. It became clear that in studies of midocean ridge tectonics, volcanism, and hydrothermal activity, the greatest excitement is in the linkages between these different fields. For example, geophysicists searched for hydrothermal activity on mid-ocean ridges for many years by towing arrays of thermisters near the seafloor. However, hydrothermal activity was eventually documented more effectively by photographing the distribution of exotic vent animals. Even now, the best indicators of the recency of volcanic eruptions and the duration of hydrothermal activity emerge from studying the characteristics of benthic faunal communities. For example, during the first deep-sea midocean ridge eruption witnessed from a submersible, divers did not see a slow lumbering cascade of pillow lavas as observed by divers off the coast of Hawaii. What they saw was completely unexpected: white bacterial matting billowing out of the seafloor, creating a scene much like a midwinter blizzard in Iceland, covering all of the freshly erupted, glassy, black lava with a thick blanket of white bacterial 'snow'.

Large-scale Variations in Axial Morphology; Correlations with Magma Supply and Segmentation

The axial depth profile of midocean ridges undulates up and down with a wavelength of tens of kilometers and amplitude of tens to hundreds of meters at fast and intermediate rate ridges. This same pattern is observed for slow-spreading ridges as well, but the wavelength of undulation is shorter and the amplitude is larger (**Figure 3**). In most cases, ridge axis discontinuities (RADs) occur at local maxima along the axial depth profile. These discontinuities include transform faults (first order); overlapping spreading centers (OSCs, second order) and higher-order (third-, fourth-order) discontinuities, which are increasingly short-lived, mobile, and associated with smaller offsets of the ridge (see **Table 1** and **Figure 4**).

A much-debated hypothesis is that the axial depth profile (**Figures 3 and 5**) reflects the magma supply along a ridge segment. According to this idea, the magma supply is enhanced along shallow portions of ridge segments and is relatively starved at segment ends (at discontinuities). In support of this hypothesis is the observation at ridges with an axial high (fast-spreading ridges) that the cross-sectional area or axial volume varies directly with depth (**Figure 6**). Maxima in cross-sectional area ($>2.5\,km^2$) occur at minima along the axial depth profile (generally not near RADS) and are thought to correlate with regions where magma supply is robust. Conversely, small cross-sectional areas ($<1.5\,km^2$) occur at local depth maxima and are interpreted to reflect minima

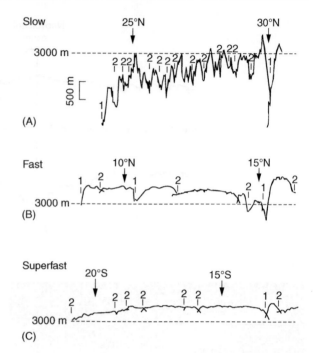

Figure 3 Axial depth profiles for (A) slow-spreading and (B) fast-spreading, and (C) ultrafast-spreading ridges. Discontinuities of orders 1 and 2 typically occur at local depth maxima (discontinuities of orders 3 and 4 are not labeled here). The segments at faster spreading rates are longer and have smoother, lower-amplitude axial depth profiles. These depth variations may reflect the pattern of magma delivery to the ridge.

in the magma supply rate along a given ridge segment. On slow-spreading ridges characterized by an axial rift valley, the cross-sectional area of the valley is at a minimum in the mid-segment regions where the depth is minimum. In addition, there are more volcanoes in the shallow midsegment area, and fewer volcanoes near the segment ends. Studies of crustal magnetization show that very highly magnetized zones occur near segment ends, which is most easily explained by a locally starved magma supply resulting in the eruption of highly fractionated lavas rich in iron.

Multichannel seismic and gravity data support the axial volume/magma supply/segmentation hypothesis (**Figure 6**). A bright reflector, which is phase-reversed in many places, occurs commonly ($>60\%$ of ridge length) beneath the axial region of both the northern and southern portions of the fast- and ultrafast spreading East Pacific Rise (EPR). This reflector has been interpreted to be a thin lens of magma residing at the top of a broader axial magma reservoir. The amount of melt is highly variable along-strike varying from a lens that is primarily crystal mush to one that is close to 100% melt. This 'axial magma chamber' (AMC) reflector is observed where the ridge is shallow and where the axial high has a

Table 1 Characteristics of segmentation. This four-tiered hierarcy of segmentation probably represents a continuum in segmentation

	Order 1	Order 2	Order 3	Order 4
Segments				
Segment length (km)	600 ± 300^{a} $(400 \pm 200)^{b}$	140 ± 90 (50 ± 30)	20 ± 10 $(15 \pm 10?)$	7 ± 5 $(7 \pm 5?)$
Segment longevity (years)	$>5 \times 10^{6}$	$0.5–5 \times 10^{6}$ $(0.5–30 \times 10^{6})$	$\sim 10^{4}–10^{5}$ (?)	$<10^{3}$ (?)
Rate of segment lengthening (long term migration) mm y^{-1}	0–50 $(0–30)$	0–1000 $(0–30)$	Indeterminate: no off-axis trace	Indeterminate: no off-axis trace
Rate of segment lengthening (short term propagation) mm y^{-1}	0–100 (?)	0–1000 $(0–50)$	Indeterminate: no off-axis trace	Indeterminate: no off-axis trace
Discontinuities				
Type	Transform, large propagating rifts	Overlapping spreading centers (oblique shear zones, rift valley jogs)	Overlapping spreading centers (intervolcano gaps), devals	Devals, offsets of axial summit caldera (intravolcano gaps)
Offset (km)	>30	2–30	0.5–2.0	<1
Offset age (years)^{c}	$>0.5 \times 10^{6}$ $(>2 \times 10^{6})$	0.5×10^{6} (2×10^{6})	~ 0	~ 0
Depth anomaly	300–600 $(500–2000)$	100–300 $(300–1000)$	30–100 $(50–300)$	0–50 $(0–100?)$
Off-axis trace	Fracture zone	V-shaped discordant zone	Faint or none	None
High amplitude magnetization?	Yes	Yes	Rarely (?)	No? (?)
Breaks in axial magma chamber?	Always	Yes, except during OSC linkage? (NA)	Yes, except during OSC linkage? (NA)	Rarely
Breaks in axial low-velocity zone?	Yes (NA)	No, but reduction in volume (NA)	Small reduction in volume (NA)	Small reduction in volume? (NA)
Geochemical anomaly?	Yes	Yes	Usually	$\sim 50\%$
Break in high-temperature venting?	Yes	Yes	Yes (NA)	Often (NA)

^{a}Values are ± 1 standard deviation.
^{b}Where information differs for slow- versus fast-spreading ridges (<60 mm y^{-1}), it is placed in parentheses.
^{c}Offset age refers to the age of the seafloor that is juxtaposed to the spreading axis at a discontinuity.
Updated from Macdonald *et al.* (1991).
NA, not applicable; ?, not presently known as poorly constrained.

broad cross-sectional area. Conversely, it is rare where the ridge is deep and narrow, especially near RADs. A reflector may occur beneath RADs during events of propagation and ridge-axis realignment, as may be occurring now on the EPR near 9°N.

There is evidence that major-element geochemistry correlates with axial–cross-sectional area (**Figure 7**). On the EPR 13°–21°S, there is a good correlation between MgO wt% and cross-sectional area (high MgO indicates a higher eruption temperature and perhaps a greater local magmatic budget). The abundance of hydrothermal venting (as measured by light transmission and backscatter in the water column and geochemical tracers) also varies directly with the cross-sectional area of the EPR. It is not often that one sees a correlation between two such different kinds of measurements. It is all the more remarkable considering that the measurements of hydrothermal activity are sensitive to changes on a timescale of days to months, while the cross-sectional area probably reflects a timescale of change measured in tens of thousands of years.

On slow-spreading centers, such as the Mid-Atlantic Ridge (MAR), the picture is less clear.

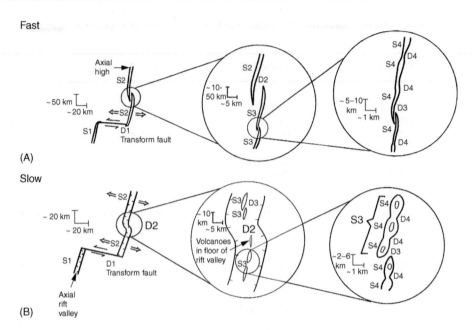

Figure 4 A possible hierarchy of ridge segmentation for (A) fast-spreading and (B) slow-spreading ridges. S1–S4 are ridge segments or order 1–4, and D1–D4 are ridge axis discontinuities of order 1–4. At both fast- and slow-spreading centers, first-order discontinuities are transform faults. Examples of second-order discontinuities are overlapping spreading centers (OSCs) on fast-spreading ridges and oblique shear zones on slow-spreading ridges. Third-order discontinuities are small OSCs on fast-spreading ridges. Fourth-order discontinuities are slight bends or lateral offsets of the axis of less than 1 km on fast-spreading ridges. This four-tiered hierarchy of segmentation is probably a continuum; it has been established, for example, that fourth-order segments and discontinuities can grow to become third-, second-, and even first-order features and vice versa at both slow- and fast-spreading centers. Updated from Macdonald *et al.* (1991).

Seismic and gravity data indicate that the oceanic crust thins significantly near many transform faults, even those with a small offset. This is thought to be the result of highly focused mantle upwelling near mid-segment regions, with very little along axis flow of magma away from the upwelling region. Focused upwelling is inferred from 'bulls-eye'-shaped residual gravity anomalies and by crustal thickness variations documented by seismic refraction and micro-earthquake studies. At slow-spreading centers, melt probably resides in small, isolated, and very short-lived pockets beneath the median valley floor (**Figure 5C**) and beneath elongated axial volcanic ridges. An alternative view is that the observed along-strike variations in topography and crustal thickness can be accounted for by along-strike variations in mechanical thinning of the crust by faulting. There is no conflict between these models, so both focused upwelling and mechanical thinning may occur along each segment.

One might expect the same to hold at fast-spreading centers, i.e., crustal thinning adjacent to OSCs. This does not appear to be the case at 9°N on the EPR, where seismic data suggest a thickening of the crust toward the OSC and a widening of the AMC reflector. There is no indication of crustal

thinning near the Clipperton transform fault either. And yet, as one approaches the 9°N OSC from the north, the axial depth plunges, the axial cross-sectional area decreases, the AMC reflector deepens, average lava age increases, MgO in dredged basalts decreases; hydrothermal activity decreases dramatically, crustal magnetization increases significantly (suggesting eruption of more fractionated basalts in a region of decreased magma supply), crustal fracturing and inferred depth of fracturing increases (indicating a greater ratio of extensional strain to magma supply), and the throw of off-axis normal faults increases (suggesting thicker lithosphere and greater strain) (**Figure 8**). How can these parameters all correlate so well, indicating a decrease in the magmatic budget and an increase in amagmatic extension, while the seismic data suggest crustal thickening off-axis from the OSC and a wider magma lens near the OSC?

One possibility is that mantle upwelling and the axial magmatic budget are enhanced away from RADs even at fast-spreading centers, but that sub-axial flow of magma 'downhill' away from the injection region redistributes magma (**Figure 5**). This along-strike flow and redistribution of magma may be unique to spreading centers with an axial high

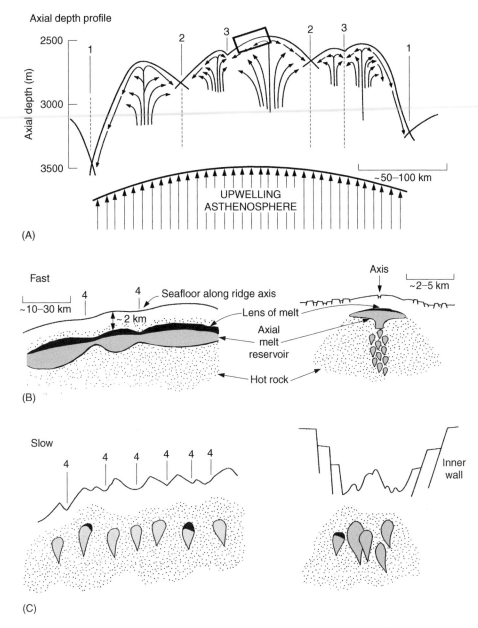

Figure 5 Schematic diagram of how ridge segmentation may be related to mantle upwelling (A), and the distribution of magma supply (B and C). In (A), the depth scale applies only to the axial depth profile; numbers denote discontinuities and segments of orders 1–3. Decompression partial melting in upwelling asthenosphere occurs at depths 30–60 km beneath the ridge. As the melt ascends through a more slowly rising solid residuum, it is partitioned at different levels to feed segments of orders 1–3. Mantle upwelling is hypothesized to be 'sheetlike' in the sense that melt is upwelling along the entire length of the ridge; but the supply of melt is thought to be enhanced beneath shallow parts of the ridge away from major discontinuities. The rectangle is an enlargement to show fine-scale segmentation for (B) a fast-spreading example, and (C) a slow-spreading example. In (B) and (C) along-strike cross-sections showing hypothesized partitioning of the magma supply relative to fourth-order discontinuities (4s) and segments are shown on the left. Across-strike cross-sections for fast- and slow-spreading ridges are shown on the right. Updated from Macdonald *et al.* (1991).

such as the EPR or Reykjanes where the axial region is sufficiently hot at shallow depths to facilitate subaxial flow. It is well documented in Iceland and other volcanic areas analogous to midocean ridges that magma can flow in subsurface chambers and dikes for distances of many tens of kilometers away from the source region before erupting. In this way, thicker crust may occur away from the midsegment injection points, proximal to discontinuities such as OSCs.

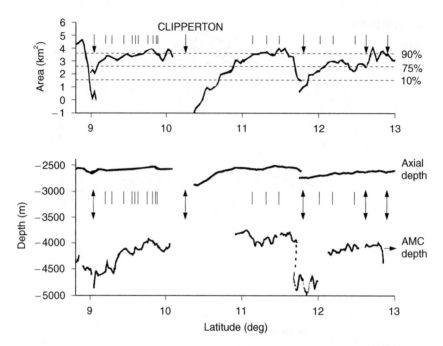

Figure 6 Profiles of the along-axis cross-sectional area, depth, and axial magma chamber (AMC) seismic reflector for the EPR 9°–13°N. The locations of first- and second-order discontinuities are denoted by vertical arrows (first-order discontinuities are named); each occurs at a local minimum of the ridge area profile, and a local maximum in ridge axis depth. Lesser discontinuities are denoted by vertical bars. There is an excellent correlation between ridge axis depth and cross-sectional area; there is a good correlation between cross-sectional area and the existence of an axial magma chamber, but detailed characteristics of the axial magma chamber (depth, width) do not correlate. Updated from Scheirer and Macdonald (1993) and references therein.

Based on studies of the fast-spreading EPR, a 'magma supply' model has been proposed that explains the intriguing correlation between over a dozen structural, geochemical and geophysical variables within a first-, second-, or third-order segment (**Figure 9**). It also addresses the initially puzzling observation that crust is sometimes thinner in the midsegment region where upwelling is supposedly enhanced. Intuitively, one might expect crust to be thickest over the region where upwelling is enhanced as observed on the MAR. However, along-axis redistribution of melt may be the controlling factor on fast-spreading ridges where the subaxial melt region may be well-connected for tens of kilometers. In this model, temporal variations in along-axis melt connectivity may result in thicker crust near midsegment when connectivity is low (most often slow-spreading ridges), and thicker crust closer to the segment ends when connectivity is high (most often, but not always the case at fast-spreading ridges).

The basic concepts of this magma supply model also apply to slow-spreading ridges characterized by an axial rift valey. Mantle melting is enhanced beneath the midsegment regions. However, the axial region is colder (averaged over time) and along-strike redistribution of melt is impeded. Thus, the crust tends to be thickest near the midsegment regions and thinnest near RADs (**Figures 8** and **9**).

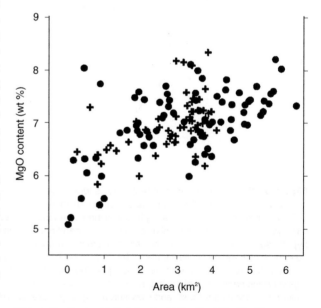

Figure 7 Cross-sectional area of the East Pacific Rise versus MgO content of basalt glass (crosses from EPR 5–14°N, solid circles from 13–23°S). There is a tendency for high MgO contents (interpreted as higher eruption temperatures and perhaps higher magmatic budget) to correlate with larger cross-sectional area. Smaller cross-sectional areas correlate with lower MgO and a greater scatter in MgO content, suggesting magma chambers which are transient and changing. Thus shallow, inflated areas of the ridge tend to erupt hotter lavas. Updated from Scheirer and Macdonald (1993) and references therein.

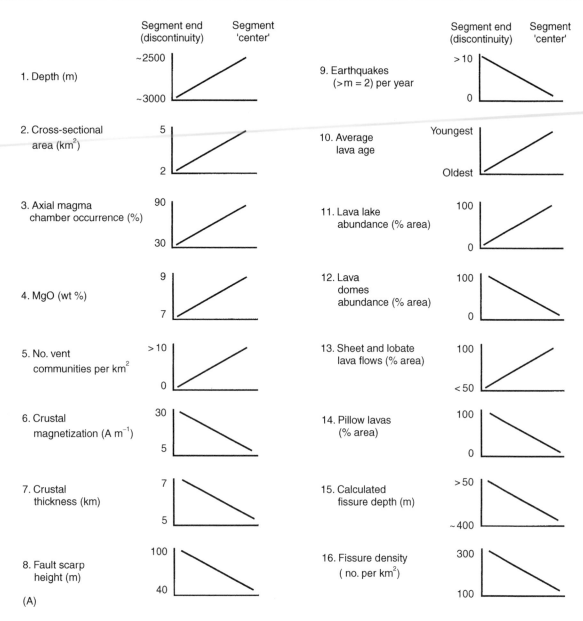

Figure 8   Schematic summary of along-axis variations in spreading center properties from segment end (discontinuity of order 1, 2, or 3) to segment mid-section areas for (A) fast-spreading ridges with axial highs and (B) slow-spreading ridges with axial rift valleys. A large number of parameters correlate well with location within a given segment, indicating that segments are distinct, independent units of crustal accretion and deformation. These variations may reflect a fundamental segmentation of the supply of melt beneath the ridge. (Less than 1% of the ridge has been studied in sufficient detail to create this summary.)

Fine-scale Variations in Ridge Morphology within the Axial Neovolcanic Zone

The axial neovolcanic zone occurs on or near the axis of the axial high on fast-spreading centers, or within the floor of the rift valley on slow-spreading centers (**Figure 5B** and **C**, right). Studies of the widths of the polarity transitions of magnetic anomalies, including *in situ* measurements from the research submersible *Alvin*, document that ~90% of the volcanism that creates the extrusive layer of oceanic crust occurs in a region 1–10 km wide at most spreading centers. Direct qualitative estimates of lava age at spreading centers using submersibles and remotely operated vehicles (ROVs) tend to confirm this, as well as recent high-resolution seismic measurements that show that layer 2A (interpreted

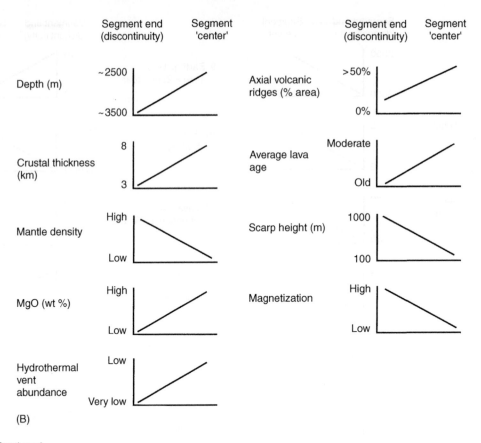

Figure 8   *Continued*

to be the volcanic layer) achieves its full thickness within 1–5 km of the rise axis. However, there are significant exceptions, including small volume off-axis volcanic constructions and voluminous off-axis floods of basaltic sheet flows.

The axial high on fast- and intermediate-spreading centers is usually bisected by an axial summit trough ~ 10–200 m deep that is found along approximately 60–70% of the axis. Along the axial high of fast-spreading ridges, sidescan sonar records show that there is an excellent correlation between the presence of an axial summit trough and an AMC reflector as seen on multichannel seismic records (>90% of ridge length). Neither axial summit troughs nor AMCs occur where the ridge has a very small cross-sectional area.

In rare cases, an axial summit trough is not observed where the cross-sectional area is large. In these locations, volcanic activity is occurring at present or has been within the last decade. For example, on the EPR near 9°45′–52′N, a volcanic eruption documented from the submersible ALVIN was associated with a single major dike intrusion, similar to the 1993 eruption on the Juan de Fuca Ridge. Sidescan sonar records showed that an axial

trough was missing from 9°52′N to 10°02′N, and in subsequent dives it was found that dike intrusion had propagated into this area, producing very recent lava flows and hydrothermal activity complete with bacterial 'snow-storms.' A similar situation has been thoroughly documented at 17°25′–30′S on the EPR where the axial cross-sectional area is large but the axial summit trough is partly filled. Perhaps the axial summit trough has been flooded with lava so recently that magma withdrawal and summit collapse is just occurring now. Thus, the presence of an axial summit trough along the axial high of a fast-spreading ridge is a good indicator of the presence of a subaxial lens of partial melt (AMC); where an axial summit trough is not present but the cross-sectional area is large, this is a good indicator of very recent or current volcanic eruptions; where an axial summit trough is not present and the cross-sectional area is small, this is a good indicator of the absence of a magma lens (AMC).

In contrast to the along-axis continuity of the axial neovolcanic zone on fast-spreading ridges, the neovolcanic zone on slower-spreading ridges is considerably less continuous and there is a great deal of variation from segment to segment. Volcanic

MAGMA SUPPLY MODEL

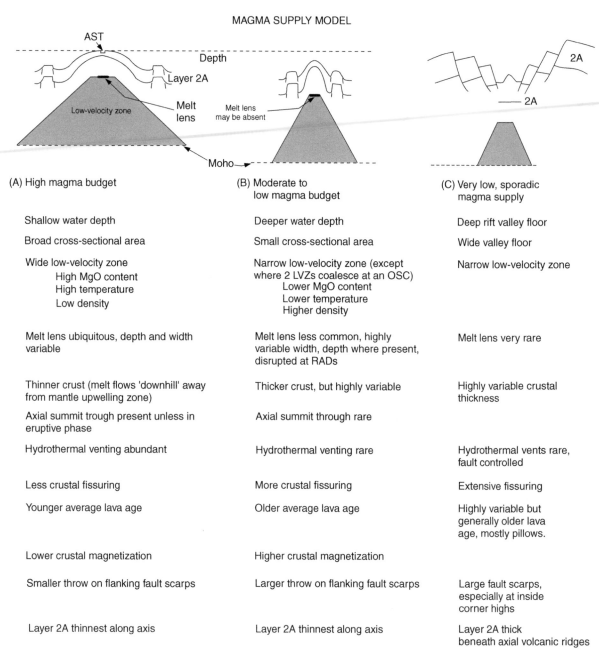

Figure 9 Magma supply model for mid-ocean ridges (see references in Buck *et al.*, 1998). (A) represents a segment with a robust magmatic budget, generally a fast-spreading ridge away from discontinuities or a hotspot dominated ridge with an axial high (AST is the axial summit trough). (B) represents a segment with a moderate magma budget, generally a fast-spreading ridge near a discontinuity or a nonrifted intermediate rate ridge. (C) represents a ridge with a sporadic and diminished magma supply, generally a rifted intermediate to slow rate spreading center (for along-strike variations at a slow ridge, see Fig. 8B).

contructions, called axial volcanic ridges, are most common along the shallow, mid-segment regions of the axial rift valley. Near the ends of segments where the rift valley deepens, widens, and is truncated by transform faults or oblique shear zones, the gaps between axial volcanic ridges become longer. The gaps between axial volcanic ridges are regions of older crust characterized by faulting and a lack of recent volcanism. These gaps may correspond to fine-scale (third- and fourth-order) discontinuities of the ridge.

Another important difference between volcanism on fast- and slow-spreading ridges is that axial volcanic ridges represent a thickening of the volcanic layer atop a lithosphere that may be 5–10 km thick, even on the axis. In contrast, the volcanic layer is

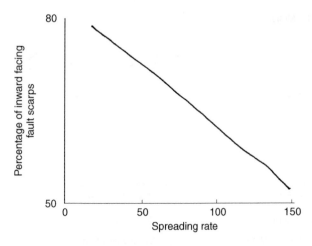

Figure 10 Spreading rate versus percentage of fault scarps that are inward-facing (facing toward the spreading axis versus away from the spreading axis). A significant increase in the percentage of inward-facing scarps occurs at slower spreading rates.

usually thinnest along the axis of the EPR. Thus the axial high on fast-spreading ridges is not a thickened accumulation of lava, while the discontinuous axial volcanic ridges on slow-spreading ridges are.

On both slow- and fast-spreading ridges, pillow and lobate lavas are the most common lava morphology. Based on laboratory studies and observations of terrestrial basaltic eruptions, this means that the lava effusion rates are slow to moderate on most midocean ridges. High volcanic effusion rates, indicated by fossil lava lakes and extensive outcrops of sheet flow lava morphology, are very rare on slow-spreading ridges. High effusion rate eruptions are more common on fast-spreading ridges and are more likely to occur along the shallow, inflated midsegment regions of the rise, in keeping with the magma supply model for ridges discussed earlier. Low effusion rate flows, such as pillow lavas, dominate at segment ends (**Figure 8A**).

Very little is known about eruption frequency. It has been estimated based on some indirect observations that at any given place on a fast-spreading ridge eruptions occur approximately every 5–100 years, and that on slow-spreading ridges it is approximately every 5000–10 000 years. If this is true, then the eruption frequency varies inversely with the spreading rate squared. On intermediate- to fast-spreading centers, if one assumes a typical dike width of ~50 cm and a spreading rate of 5–10 cm y^{-1}, then an eruption could occur ~every 5–10 years. This estimate is in reasonable agreement with the occurrence of megaplumes and eruptions on the well-monitored Juan de Fuca Ridge. However, observations in sheeted dike sequences in Iceland and

ophiolites indicate that only a small percentage of the dikes reach the surface to produce eruptions.

On fast-spreading centers, the axial summit trough is so narrow (30–1000 m) and well-defined in most places that tiny offsets and discontinuities of the rise axis can be detected (**Table 1, Figure 2**). This finest scale of segmentation (fourth-order segments and discontinuities) probably corresponds to individual fissure eruption events similar to the Krafla eruptions in Iceland or the Kilauea east rift zone eruptions in Hawaii. Given a magma chamber depth of 1–2 km, an average dike ascent rate of ~0.1 km h^{-1} and an average lengthening rate of ~1 km h^{-1}, typical diking events would give rise to segments 10–20 km long. This agrees with observations of fourth-order segmentation and the scale of the recent diking event on the Juan de Fuca Ridge and in other volcanic rift zones. The duration of such segments is thought to be very short, ~100–1000 years (too brief in any case to leave even the smallest detectable trace off-axis, **Table 1**). Yet even at this very fine scale, excellent correlations can be seen between average lava age, density of fissuring, the average widths of fissures, and abundance of hydrothermal vents within individual segments. In fact there is even an excellent correlation between ridge cross-sectional area and the abundance of benthic hydrothermal communities (**Figure 8**).

A curious observation on the EPR is that the widest fissures occur in the youngest lava fields. If fissures grow in width with time and increasing extension, one would expect the opposite; the widest fissures should be in the oldest areas. The widest fissures are ~5 m. Using simple fracture mechanics, these fissures probably extend all the way through layer 2A and into the sheeted dike sequence. These have been interpreted as eruptive fissures, and this is where high-temperature vents (>300°C) are concentrated. In contrast to the magma rich, dike-controlled hydrothermal systems that are common on fast-spreading centers, magma-starved hydrothermal systems on slow-spreading ridges tend to be controlled more by the penetration of sea water along faults near the ridge axis. (See Hydrothermal Vent Deposits.)

Faulting

Extension at midocean ridges causes fissuring and normal faulting. The lithosphere is sufficiently thick and strong on slow-spreading centers to support shear failure on the axis, so normal faulting along dipping fault planes can occur on or very close to the axis. These faults produce grabens 1–3 km deep. In

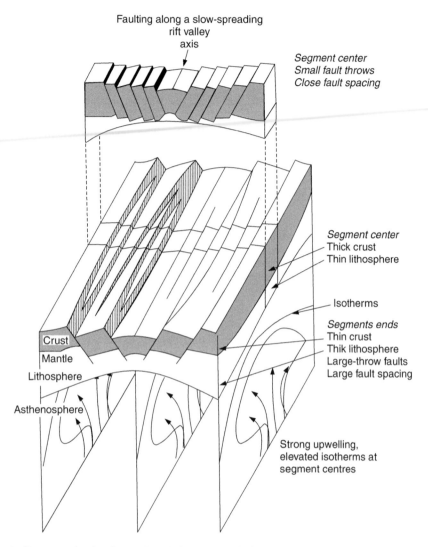

Figure 11   A geological interpretation for along-axis variations in scarp height, and more closely spaced scarps near mid-segment on a slow-spreading center. Cross-section through segment center (top) shows more closely spaced, smaller-throw faults than at the segment ends (bottom). Focused mantle upwelling near the segment center causes this region to be hotter; the lithosphere will be thinner while increased melt supply creates a thicker crust. In contrast to fast-spreading centers, there may be very little melt redistribution along-strike. Near the segment ends, the lithosphere will be thicker and magma supply is less creating thinner crust. Along axis variations in scarp height and spacing reflect these along axis variations in lithospheric thickness. Amagmatic extension across the larger faults near segments ends may also thin the crust, especially at inside corner highs. Modified from Shaw (1992).

contrast, normal faulting along inclined fault planes is not common on fast-spreading centers within ± 2 km of the axis, probably because the lithosphere is too thin and weak to support normal faulting. Instead, the new thin crust fails by simple tensional cracking.

Fault strikes tend to be perpendicular to the least compressive stress; thus they also tend to be perpendicular to the spreading direction. While there is some 'noise' in the fault trends, most of this noise can be accounted for by perturbations to the least compressive stress direction due to shearing in the vicinity of active or fossil ridge axis discontinuities. Once this is accounted for, fault trends faithfully record changes in the direction of opening to within $\pm 3°$ and can be used to study plate motion changes on a finer scale than that provided by seafloor magnetic anomalies. Studies of the cumulative throw of normal faults, seismicity, and fault spacing suggest that most faulting occurs within ± 20–40 km of the axis independent of spreading rate.

There is a spreading rate dependence for the occurrence of inward and outward dipping faults. Most faults dip toward the axis on slow-spreading centers ($\sim 80\%$), but there is a monotonic increase in the occurrence of outward dipping faults with spreading rate (**Figure 10**). Inward and outward facing faults are approximately equally abundant at very fast

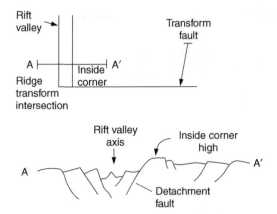

Figure 12 Inside corner high at a slow-spreading ridge transform intersection. Extension is concentrated along a detachment fault for up to 1–2 million years, exposing deep sections of oceanic crust and mantle. The oceanic crust is thinned by this extreme extension; crustal accretion and magmatic activity may also be diminished.

spreading rates. This can be explained by the smaller mean normal stress across a fault plane that dips toward the axis, cutting through thin lithosphere, versus a fault plane that cuts through a much thicker section of lithosphere dipping away from the axis. Given reasonable thermal models, the difference in the thickness of the lithosphere cut by planes dipping toward versus away from the axis (and the mean normal stress across those planes) decreases significantly with spreading rate, making outward dipping faults more likely at fast-spreading rates (**Figures 10** and **11**).

At all spreading rates, important along-strike variations in faulting occur within major (first- and second-order) spreading segments. Fault throws (inferred from scarp heights) decrease in the mid-segment regions away from discontinuities (**Figures 8** and **11**). This may be caused by a combination of thicker crust, thinner lithosphere, greater magma supply and less amagmatic extension away from RADs in the mid-segment region (**Figure 11**). Another possible explanation for along-strike variations in fault throw is along-strike variations in the degree of coupling between the mantle and crust. A ductile lower crust will tend to decouple the upper crust from extensional stresses in the mantle, and the existence of a ductile lower crust will depend on spreading rate, the supply of magma to the ridge, and proximity to major discontinuities.

Estimates of crustal strain due to normal faulting vary from 10–20% on the slow-spreading MAR to ~3–5% on the fast-spreading EPR. This difference may be explained as follows. The rate of magma supply to slow-spreading ridges is relatively low compared with the rate of crustal extension and

faulting, while extension and magma supply rates are in closer balance on fast-spreading ridges. The resulting seismicity is different too. In contrast to slow-spreading ridges where teleseismically detected earthquakes are common, faulting at fast-spreading ridges rarely produces earthquakes of magnitude >4. Nearly all of these events are associated with RADs. The level of seismicity measured at fast-spreading ridges accounts for only a very small percentage of the observed strain due to faulting, whereas fault strain at slow ridges is comparable to the observed seismic moment release. It has been suggested that faults in fast-spreading environments accumulate slip largely by stable sliding (aseismically) owing to the warm temperatures and associated thin brittle layer. At slower spreading rates, faults will extend beyond a frictional stability transition into a field where fault slip occurs unstably (seismically) because of a thicker brittle layer.

Disruption of oceanic crust due to faulting may be particularly extreme on slow-spreading ridges near transform faults (**Figure 12**). Unusually shallow topography occurs on the active transform slip side of ridge transform intersections; this is called the high inside corner. These highs are not volcanoes. Instead they are caused by normal faults which cut deeply and perhaps all the way through oceanic crust. It is thought that crustal extension may occur for 1 to 2 million years on detachment faults with little magmatic activity. This results in extraordinary extension of the crust and exposure of large sections of the deep crust and upper mantle on the seafloor. Corrugated slip surfaces indicating the direction of fault slip are also evident and are called by some investigators, 'megamullions.'

At distances of several tens of kilometers off-axis, topography generated near the spreading center is preserved on the seafloor with little subsequent change until it is subducted, except for the gradual accumulation of pelagic sediments at rates of ~0.5–20 cm per thousand years. The preserved topographic highs and lows are called abyssal hills. At slow-spreading centers characterized by an axial rift valley, back-tilted fault blocks and half-grabens may be the dominant origin of abyssal hills (**Figure 13**), although there is continued controversy over the role of high-angle versus low-angle faults, listric faulting versus planar faulting, and the possible role of punctuated episodes of volcanism versus amagmatic extension. At intermediate-rate spreading centers, abyssal hill structure may vary with the local magmatic budget. Where the budget is starved and the axis is characterized by a rift valley, abyssal hills are generally back-tilted fault blocks. Where the magmatic budget is robust and an axial high is present,

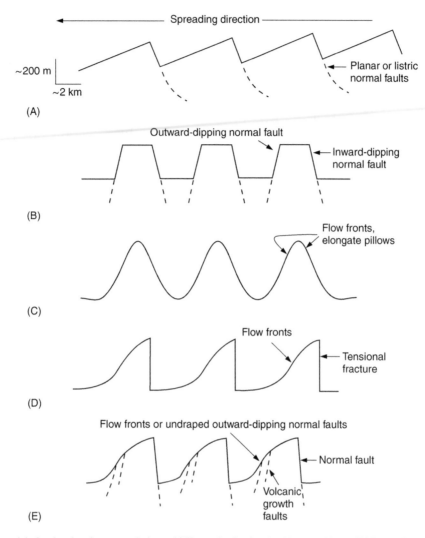

(A)

(B)

(C)

(D)

(E)

Figure 13 Five models for the development of abyssal hills on the flanks of midocean ridges. (A) Back-tilted fault blocks (episodic inward-dipping normal faulting off-axis). (B) Horst and graben (episodic inward- and outward-dipping faulting off-axis). (C) Whole volcanoes (episodic volcanism on-axis). (D) Split volcanoes (episodic volcanism and splitting on-axis). (E) Horsts bounded by inward-dipping normal faults and outward-dipping volcanic growth faults (episodic faulting off-axis and episodic volcanism on or near-axis).

the axial lithosphere is episodically thick enough to support a volcanic construction that may then be rafted away intact or split in two along the spreading axis, resulting in whole-volcano or split-volcano abyssal hills, respectively.

Based on observations made from the submersible ALVIN on the flanks of the EPR, the outward facing slopes of the hills are neither simple outward dipping normal faults, as would be predicted by the horst/graben model, nor are they entirely volcanic-constructional, as would be predicted by the split-volcano model. Instead, the outward facing slopes are 'volcanic growth faults' (**Figure 14**). Outward-facing scarps produced by episodes of normal faulting are buried near the axis by syntectonic lava flows originating along the axial high. Repeated episodes of dip–slip faulting and volcanic burial result in

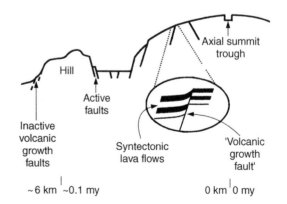

Figure 14 Volcanic growth faults; cross-sectional depiction of the development of volcanic growth faults. Volcanic growth faults are common on fast-spreading centers and explain some of the differences between inward- and outward-facing scarps as well as the morphology and origin of most abyssal hills near fast-spreading centers.

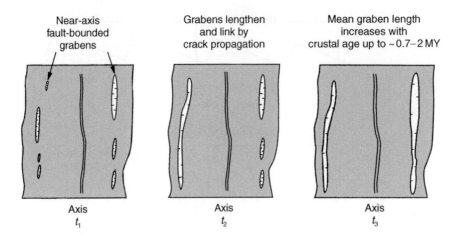

Figure 15   Proposed time sequence of along-strike propagation and linkage of near-axis faults and grabens that define the edges of abyssal hills; time-averaged propagation rates are approximately 20–60 km per million years.

structures resembling growth faults, except that the faults are episodically buried by lava flows rather than being continuously buried by sediment deposition. In contrast, the inward dipping faults act as tectonic dams to lava flows. Thus, the abyssal hills are horsts and the intervening troughs are grabens with the important modification to the horst/graben model that the outward facing slopes are created by volcanic growth faulting rather than traditional normal faulting. Thus, on fast-spreading centers, abyssal hills are asymmetric, bounded by steeply dipping normal faults facing the spreading axis, and bounded by a volcanic growth faults on the opposing side (**Figure 14**). The abyssal hills lengthen at a rate of approximately 60 mm/y for the first ~0.7 my by along strike propagation of individual faults as well as by linkage of neighboring faults (**Figure 15**).

See also

Hydrothermal Vent Deposits. Mid-Ocean Ridge Seismic Structure. Mid-Ocean Ridge Tectonics, Volcanism, and Geomorphology. Propagating Rifts and Microplates.

Further Reading

Buck WR, Delaney PT, Karson JA, and Lagabrielle Y (1998) *Faulting and Magmatism at Mid-Ocean Rides*. AGU Geophysical Monographs 106. Washington, DC: American Geophysical Union.

Humphris SE, Zierenberg RA, Mullineaux LS, and Thompson RE (1995) *Seafloor Hydrothermal Systems: Physical, Chemical, Biological and Geochemical Interactions*. AGU Geophysical Monographs 91. Washington, DC: American Geophysical Union.

Langmuir CH, Bender JF, and Batiza R (1986) Petrological and tectonic segmentation of the East Pacific Rise, 5°30′N–14°30′N. *Nature* 322: 422–429.

Macdonald KC (1982) Mid-ocean ridges: fine scale tectonic, volcanic and hydrothermal processes within the plate boundary zone. *Annual Reviews of Earth and Planetary Science* 10: 155–190.

Macdonald KC and Fox PJ (1990) The mid-ocean ridge. *Scientific American* 262: 72–79.

Macdonald KC, Scheirer DS, and Carbotte SM (1991) Mid-ocean ridges: discontinuities, segments and giant cracks. *Science* 253: 986–994.

Menard H (1986) *Ocean of Truth*. Princeton, NJ: Princeton University Press.

Phipps-Morgan J, Blackman DK, and Sinton J (1992) *Mantle Flow and Melt Generation at Mid-ocean Ridges* AGU Geophysical Monographs 71. Washington, DC: American Geophysical Union.

Shaw PR (1992) Ridge segmentation, faulting and crustal thickness in the Atlantic Ocean. *Nature* 358: 490–493.

Scheirer DS and Macdonald KC (1993) Variation in cross-sectional area of the axial ridge along the East Pacific Rise. Evidence for the magmatic budget of a fast-spreading center. *Journal of Geophysical Research* 98: 7871–7885.

Sinton JM and Detrick RS (1992) Mid-ocean ridge magma chambers. *Journal of Geophysical Research* 97: 197–216.

MID-OCEAN RIDGES: MANTLE CONVECTION AND FORMATION OF THE LITHOSPHERE

G. Ito and R. A. Dunn, University of Hawai'i at Manoa, Honolulu, HI, USA

Introduction

Plate tectonics describes the motion of the outer lithospheric shell of the Earth. It is the surface expression of mantle convection, which is fueled by Earth's radiogenic and primordial heat. Mid-ocean ridges mark the boundaries where oceanic plates separate from one another and thus lie above the upwelling limbs of mantle circulation (**Figure 1**). The upwelling mantle undergoes pressure-release partial melting because the temperature of the mantle solidus decreases with decreasing pressure. Newly formed melt, being less viscous and less dense than the surrounding solid, segregates from the residual mantle matrix and buoyantly rises toward the surface, where it forms new, basaltic, oceanic crust. The crust and mantle cool at the surface by thermal conduction and hydrothermal circulation. This cooling generates a thermal boundary layer, which is rigid to convection and is the newly created edge of the tectonic plate. As the lithosphere moves away from the ridge, it thickens via additional cooling, becomes denser, and sinks deeper into the underlying ductile asthenosphere. This aging process of the plates causes the oceans to double in depth toward continental margins and subduction zones (**Figure 2**), where the oldest parts of plates are eventually thrust downward and returned to the hot underlying mantle from which they came.

Mid-ocean ridges represent one of the most important geological processes shaping the Earth; they produce over two-thirds of the global crust, they are the primary means of geochemical differentiation in the Earth, and they feed vast hydrothermal systems

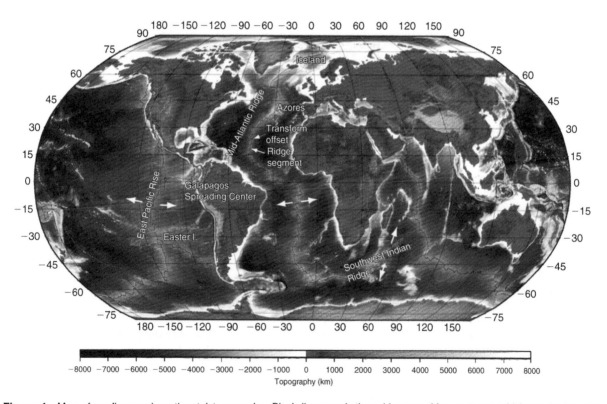

Figure 1 Map of seafloor and continental topography. Black lines mark the mid-ocean ridge systems, which are broken into individual spreading segments separated by large-offset transform faults and smaller nontransform offsets. Mid-ocean ridges encircle the planet with a total length exceeding 50 000 km. Large arrows schematically show the direction of spreading of three ridges discussed in the text.

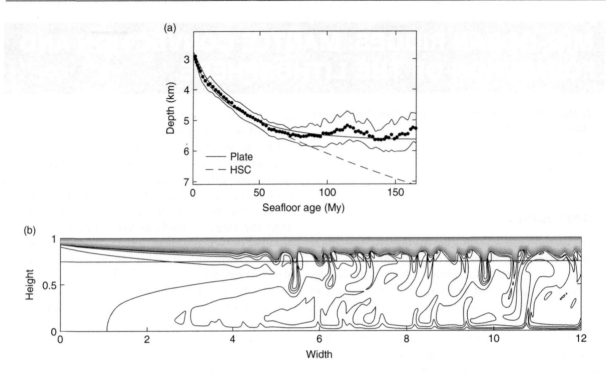

Figure 2 (a) Global average (dots) seafloor depth (after correcting for sedimentation) and standard deviations (light curve) increase with seafloor age. Dashed curve is predicted by assuming that the lithosphere cools and thickens indefinitely, as if it overlies an infinite half space. Solid curve assumes that the lithosphere can cool only to a maximum amount, at which point the lithosphere temperature and thickness remain constant. (b) Temperature contours in a cross section through a 2-D model of mantle convection showing the predicted thickening of the cold thermal boundary layer (i.e., lithosphere, which beneath the oceans reaches a maximum thickness of ~100 km) with distance from a mid-ocean ridge (left side). Midway across the model, small-scale convection occurs which limits the thickening of the plate, a possible cause for the steady depth of the seafloor beyond ~80 Ma. (a) Modified from Stein CA and Stein S (1992) A model for the global variation in oceanic depth and heat flow with lithospheric age. *Nature* 359: 123–129, with permission from Nature Publishing Group. (b) Adapted from Huang J, Zhong S, and van Hunen J (2003) Controls on sublithospheric small-scale convection. *Journal of Geophysical Research* 108 (doi:10.1029/2003JB002456) with permission from American Geophysical Union.

that influence ocean water chemistry and support enormous ecosystems. Around the global ridge system, new lithosphere is formed at rates that differ by more than a factor of 10. Such variability causes large differences in the nature of magmatic, tectonic, and hydrothermal processes. For example, slowly spreading ridges, such as the Mid-Atlantic Ridge (MAR) and Southwest Indian Ridge (SWIR), exhibit heavily faulted axial valleys and large variations in volcanic output with time and space, whereas faster-spreading ridges, such as the East Pacific Rise (EPR), exhibit smooth topographic rises with more uniform magmatism. Such observations and many others can be largely understood in context of two basic processes: asthenospheric dynamics, which modulates deep temperatures and melt production rates; and the balance between heat delivered to the lithosphere, largely by magma migration, versus that lost to the surface by conduction and hydrothermal circulation. Unraveling the nature of these interrelated processes requires the integrated use of geologic mapping, geochemical and petrologic analyses, geophysical sensing, and geodynamic modeling.

Mantle Flow beneath Mid-ocean Ridges

While the mantle beneath mid-ocean ridges is mostly solid rock, it does deform in a ductile sense, very slowly on human timescales but rapidly over geologic time. 'Flow' of the solid mantle is therefore often described using fluid mechanics. The equation of motion for mantle convection comes from momentum equilibrium of a fluid with shear viscosity η and zero Reynolds number (i.e., zero acceleration):

$$\nabla \cdot [\eta(\nabla \mathbf{V} + \nabla \mathbf{V}^T)] - \nabla P + \nabla[(\zeta - 2/3\eta)\nabla \cdot \mathbf{V}] + (1 - \phi)\Delta\rho\mathbf{g} = 0 \qquad [1]$$

The first term describes the net forces associated with matrix shear, where $\mathbf{V}$ is the velocity vector, $\nabla \mathbf{V}$ is the velocity gradient tensor, and $\nabla \mathbf{V}^T$ is its transpose; the second term ∇P is the non-hydrostatic pressure gradient; the third term describes matrix divergence $\nabla \cdot \mathbf{V}$ (with effective bulk viscosity ζ) associated with melt transport; and the last term is the body force, with ϕ being the volume fraction occupied by melt

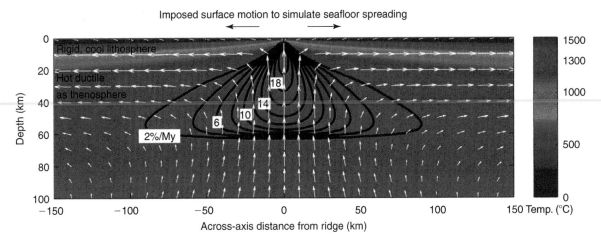

Figure 3 Cross section of a 2-D numerical model that predicts mantle flow (shown by arrows whose lengths are proportional to flow rate), temperatures (shading), and melt production rate beneath a mid-ocean ridge (contoured at intervals of 2 mass %/My). In this particular calculation, asthenospheric flow is driven kinematically by the spreading plates and is not influenced by density variations (i.e., mantle buoyancy is unimportant). Model spreading rate is $6\,\text{mm}\,\text{yr}^{-1}$.

and $\nabla\rho$ being the density contrast between the solid and melt. To first order, the divergence term is negligible, in which case eqn [1] describes a balance between viscous shear stresses, pressure gradients, and buoyancy. Locally beneath mid-ocean ridges, the spreading lithospheric plates act as a kinematic boundary condition to eqn [1] such that seafloor spreading itself drives 'passive' mantle upwelling, which causes decompression melting and ultimately the formation of crust (**Figure 3**). Independent of plate motion, lateral density variations can drive 'active' or 'buoyant' mantle upwelling and further contribute to decompression melting, as we discuss below.

Several lines of evidence indicate that the upwelling is restricted to the upper mantle. The pressures at which key mineralogical transitions occur in the deep upper mantle (i.e., at depths 410 and 660 km) are sensitive to mantle temperature, but global and detailed local seismic studies do not reveal consistent variations in the associated seismic structure in the vicinity of mid-ocean ridges. This finding indicates that any thermal anomaly and buoyant flow beneath the ridge is confined to the upper mantle above the discontinuity at the depth of 410 km. Regional body wave and surface wave studies indicate that it could even be restricted to the upper ~200 km of the mantle. Although some global tomographic images, based on body wave travel times, sometimes show structure beneath mid-ocean ridges extending down to depths of 300–400 km, these studies tend to artificially smear the effects of shallow anomalies below their actual depth extent. Further evidence comes from the directional dependence of seismic wave propagation speeds. This seismic anisotropy is thought to be caused by lattice-preferred orientation of olivine crystals due to mantle flow. Global studies show that at depths of 200–300 km beneath mid-ocean ridges, surface waves involving only horizontal motion (Love waves) tend to propagate slower than surface waves involving vertical motion (Rayleigh waves). This suggests a preferred orientation of olivine consistent with vertical mantle flow at these depths but not much deeper.

Mantle Melting beneath Mid-ocean Ridges

Within the upwelling zone beneath a ridge, the mantle cools adiabatically due to the release of pressure. However, since the temperature at which the mantle begins to melt drops faster with decreasing pressure than the actual temperature, the mantle undergoes pressure-release partial melting (**Figure 3**). In a dry (no dissolved H_2O) mantle, melting is expected to begin at approximately 60-km depth. On the other hand, a small amount of water in the mantle ($\sim 10^2$ ppm) will strongly reduce the mantle solidus such that melting can occur at depths >100 km. Detailed seismic studies observe low-velocity zones extending to depth of 100–200 km, which is consistent with the wet melting scenario.

The thickness of oceanic crust times the rate of seafloor spreading is a good measure of the volume flux (per unit length of ridge axis) of melt delivered from the mantle. Marine seismic studies of the Mohorovičić seismic boundary (or Moho), which is often equated with the transition between the gabbroic lower oceanic crust and the peridotitic upper mantle, find that the depth of the Moho is more or less uniform beneath seafloor formed at spreading rates of $\sim 20\,mm\,yr^{-1}$ and faster (**Figure 4**). This observation indicates that the flux of melt generated in the mantle is, on average, proportional to spreading rate. What then causes such a behavior? Melt flux is proportional to the height of the melting zone as well as the average rate of

upwelling within the melting zone. If the upwelling and melt production rate are proportional to spreading rate, then, all else being equal, this situation explains the invariance of crustal thickness with spreading rate. On the other hand, all else is not likely to be equal: slower spreading tends to lead to a thicker lithospheric boundary layer, a smaller melting zone, and a further reduction in melt production. The cause for the lack of decrease in

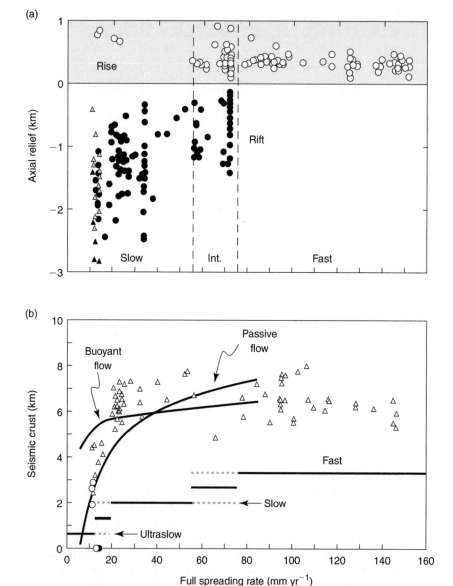

Figure 4   (a) Axial morphology of ridges is predominantly rifted valleys (negative relief) along slow-spreading ridges and axial topographic highs (positive relief) along fast-spreading rates. The ultraslow spreading ridges (triangles) include the Gakkel Ridge in the Arctic and South West Indian Ridge, southeast of Africa. Dashed lines show conventional divisions between slow-, intermediate-, (labeled 'Int.'), and fast-spreading ridges. (b) Seismically determined crustal thicknesses (symbols) are compared to theoretical predictions produced by two types of mantle flow and melting models: the passive flow curve is for a mantle flow model driven kinematically by plate spreading, the buoyant flow curve includes effects of melt buoyancy, which enhances upwelling and melting, particularly beneath slow-spreading ridges. Horizontal bars show a revised classification scheme for spreading rate characteristics. Reprinted by permission from Macmillan Publishers Ltd: *Nature* (Dick HJB, Lin J, and Schouten H (2003) An ultraslow-spreading class of ocean ridge. *Nature* 426: 405–412), copyright (2003).

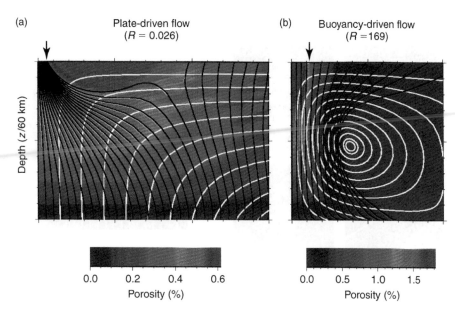

Figure 5 Predictions from 2-D numerical models of mantle flow (white streamlines), melt retention (shading), and pressure-driven melt migration (black streamlines). (a) Fast spreading with high mantle viscosity ($10^{20} - 10^{21}$ Pa s) impairs buoyant flow and leads to large pressure gradients, which draw melt from the broad melting zone toward the ridge axis. (b) Low mantle viscosity ($10^{18} - 10^{19}$ Pa s) allows the low-density, partially molten rock to drive buoyant upwelling, which focuses beneath the ridge axis and allows melt to flow vertically. Black arrows show the predicted width of ridge-axis magmatism, which is still much broader than observed. Figure provided by M. Spiegelman (pers. comm., 2007). Also, see Spiegelman M (1996) Geochemical consequences of melt transport in 2-D: The sensitivity of trace elements to mantle dynamics. *Earth and Planetary Science Letters* 139: 115–132.

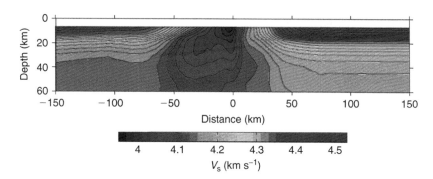

Figure 6 Cross-sectional tomographic image of the upper mantle shear wave velocity structure beneath the southern East Pacific Rise produced from Love wave data. Ridge axis is at $x = 0$ km. At the top of the figure the high-velocity lithospheric lid is clearly evident, as well as its thickening with distance from the ridge. The low-velocity zone beneath the ridge is consistent with the presence of high temperatures and some retained melt. Body wave and Rayleigh wave studies indicate that the low-velocity zone extends even wider below than what is shown here. Adapted from Dunn RA and Forsyth DW (2003) Imaging the transition between the region of mantle melting and the crustal magma chamber beneath the southern East Pacific Rise with short-period Love waves. *Journal of Geophysical Research* 108(B7): 2352 (doi:10.1029/2002JB002217), with permission from American Geophysical Union.

crustal thickness with decreasing spreading rate for rates $> c.$ 20 mm yr^{-1} must involve other processes.

A possible solution could have to do with the likelihood that the relative strengths of the plate-driven (i.e., kinematic) versus buoyancy-driven mantle upwelling change with spreading rate. Let us examine two end-member scenarios for mantle flow and melting. Case 1 considers a situation in which mantle flow is driven entirely kinematically by the separation of the lithospheric plates (**Figures** 3 and 5(a)). This passive flow scenario is predicted if the plate-driven component of flow overwhelms the buoyancy-driven part (i.e., the last term in eqn [1]). The other end-member possibility, case 2, considers a situation in which buoyant flow dominates over the plate-driven part. Lateral density variations, $\Delta\rho$, in the melting zone probably occur due to the presence of small amounts of melt, which has a lower density

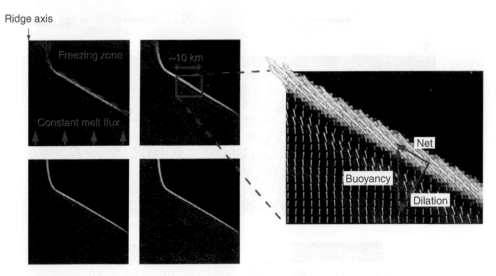

Figure 7 (Left) Cross-sections of a 2-D model of melt migration, with shading showing porosity (black = 0%, white = 3.5%) and at four different times, increasing clockwise from the upper left. A constant melt percolation flux rises through the bottom of the box. A 'freezing boundary' represents the cooler lithosphere which slopes toward the ridge axis (upper left). (Right) Enlarged portion of red box. The freezing boundary halts the rise of melt and diverts melt parallel to it. The net pressure gradient driving the flow is caused by the two components shown by the arrows. The freezing boundary generates porosity waves that propagate away from the boundary. These waves are predicted mathematically to arise from eqns [1], [2], and two others describing conservation of melt and solid mass. Modified from Spiegelman M (1993) Physics of melt extraction: Theory, implications and applications. *Philosophical Transactions of the Royal Society of London, Series A* 342: 23–41, with permission from the Royal Society.

than the solid. A positive feedback can occur between melting and buoyant flow, such that melting increases buoyancy and upwelling, which leads to further melting (**Figure 5(b)**). Returning to the weak dependence on spreading rate for rates >20 mm yr^{-1}; at least at fast-spreading rates, the plate-driven component of flow can be as strong or stronger than any buoyancy component. Thus, numerical models of fast-spreading ridges that do or do not include buoyancy predict similar crustal thicknesses and an insensitivity of crustal thickness to spreading rate (**Figure 4**). However, as spreading rate drops, buoyancy forces can become relatively important, so that – below slow–intermediate-spreading lithosphere – they generate 'fast' mantle upwelling. This fast upwelling is predicted to enhance melt production and compensate for the effects of surface cooling to shrink the melting zone; crustal thickness is therefore maintained as spreading rates decrease toward ~20 mm yr^{-1}.

Observational evidence for the relative importance of plate-driven versus buoyant flow is provided by studies of body and surface wave data along the EPR. Tomographic images produced from these data reveal a broad region of low seismic wave speeds in the upper mantle (**Figure 6**), interpreted to be the region of melt production. To date, there is little indication of a very narrow zone of low wave speeds, such as that predicted for buoyant upwelling zone as depicted by case 2 (**Figure 5(b)**). These findings support the predictions in **Figure 4** that plate-driven flow is strong at fast-spreading ridges, such as the EPR.

At spreading rates less than 20 mm yr^{-1}, however, the melting process appears to change dramatically. Here, melt flux is not proportional to spreading rate (**Figure 4**). At these ultraslow rates, crustal thickness drops rapidly with decreasing spreading rate, suggesting a nonlinear decrease in magma flux. A leading hypothesis suggests that the melt reducing effects of the top-down cooling and corresponding shrinkage of the melting zone overwhelm the melt-enhancing effects of any buoyancy-driven upwelling. Whatever the exact cause is, the large variability in crustal thickness seen at these spreading rates is one example of the large sensitivity of ridge-axis processes to surface cooling at slow or ultraslow spreading rates.

Melt Transport to Ridge Axes

How melt is transported upward from the mantle source to the ridge is another long-standing problem. Seismic studies of oceanic crust indicate that the crust is fully formed within a few of kilometers of the axis of a ridge, requiring either a very narrow melting zone beneath the ridge or some mechanism that focuses melts from a broader melting zone to a narrow region at the ridge axis.

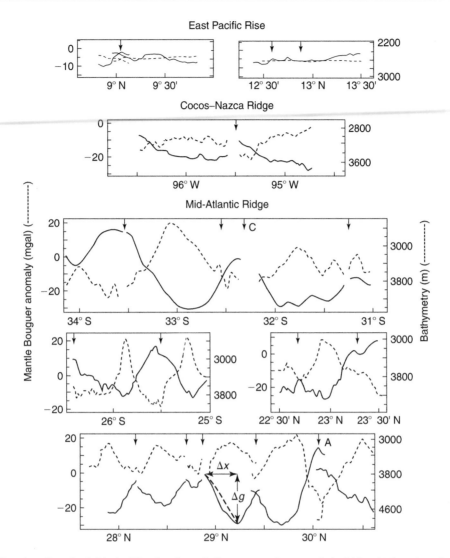

Figure 8   Profiles of seafloor depth (dashed lines) and mantle Bouguer gravity anomaly (solid lines) taken along the axes of various mid-ocean ridges (as indicated in the figure). Arrows mark various ridge-axis discontinuities. Note anticorrelation of gravity and bathymetry along the MAR, indicating shallower bathymetry and thicker crust near the centers of ridge segments. The East Pacific Rise is a fast-spreading ridge, the Cocos–Nazca Ridge (also called the Galapagos Spreading Center) spreads at an intermediate rate, and the MAR is a slow-spreading ridge. Adapted from Lin J and Phipps Morgan J (1992) The spreading rate dependence of three-dimensional mid-ocean ridge gravity structure. *Geophysical Research Letters* 19: 13–16, with permission from American Geophysical Union.

Laboratory experiments and theory show that melt can percolate through the pore space of the matrix (with volume fraction ϕ) in response to matrix pressure gradients ∇P. The Darcy percolation flux is described by

$$\phi(\mathbf{v} - \mathbf{V}) = -(k/\mu)\nabla P \qquad [2]$$

where $(\mathbf{v} - \mathbf{V})$ is the differential velocity of melt $\mathbf{v}$ relative to the matrix $\mathbf{V}$, μ is melt viscosity, and k is the permeability of the porous matrix. Equation [1] shows that ∇P is influenced by both melt buoyancy and matrix shear: melt buoyancy drives vertical percolation while matrix shear can push melt

sideways. In mantle flow case 1, in which plate-driven flow dominates, the zone of melting is predicted to be very wide, requiring large lateral pressure gradients to divert melt 50–100 km sideways toward the ridge (**Figure 5**). For matrix shear to produce such large lateral gradients requires very high asthenospheric viscosities ($10^{20} – 10^{21}$ Pa s). When buoyant flow dominates (case 2), the melting zone is much narrower and the low viscosities ($10^{18} 10^{19}$ Pa s) generate small lateral pressure gradients such that the melt rises mostly vertically to feed the ridge axis. Both of these cases, however, still predict zones of magmatism at the ridge axis that are wider than those typically observed.

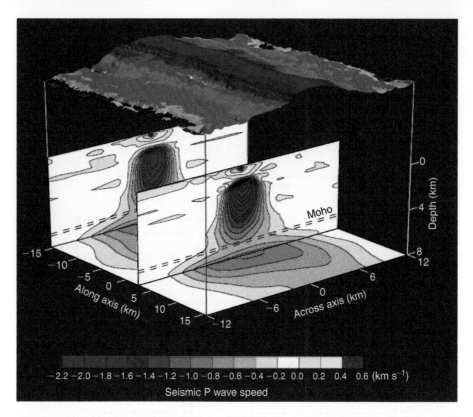

Figure 9 A perspective view of seafloor bathymetry and a seismic tomographic image of the East Pacific Rise (9° N latitude) magmatic system. The image is of the P wave velocity anomaly, relative to an average vertical velocity profile, contoured at $0.2\,km\,s^{-1}$. The vertical planes show the continuity of the crustal magmatic system beneath the ridge (the significant low-velocity region centered beneath the ridge). The deep horizontal plane is located in the mantle just below the crust and shows that the crustal low-velocity region extends downward into the mantle. The mantle velocity anomaly is continuous beneath the ridge, but shows variations in magnitude and location that suggest variations in melt supply. Adapted from Dunn RA, Toomey DR, and Solomon SC (2000) Three-dimensional seismic structure and physical properties of the crust and shallow mantle beneath the East Pacific Rise at 9° 30′ N. *Journal of Geophysical Research* 105: 23537–23555, with permission from the American Geophysical Union.

An additional factor that can help further focus melts toward the ridge is the bottom of the cold lithosphere, which shoals toward the ridge axis. Theoretical studies predict that as melt rises to the base of the lithosphere, it freezes and cannot penetrate the lithosphere (**Figure 7**). But the steady percolation of melt from below causes melt to collect in a high-porosity channel just below the freezing boundary. The net pressure gradient that drives melt percolation is the vector sum of the component that is perpendicular to the freezing front, caused by matrix dilation as melt fills the channel, and the vertical component caused by melt buoyancy. Consequently, melt flows along the freezing front, upward toward the ridge axis.

Regional and Local Variability of the Global Mid-Ocean Ridge System

Major differences in the regional and local structure of mid-ocean ridges are linked to the previously noted processes that influence asthenospheric flow and the heat balance in the lithosphere. One example that highlights a mid-ocean ridge's sensitivity to lithospheric heat balance is the overall shape or morphology of ridges at different spreading rates. At fast-spreading rates, the magmatic (heat) flux is high and this forms a hot crust with thin lithosphere. The mechanical consequences of both a thin lithosphere and relatively frequent magmatic intrusions to accommodate extension generate a relatively smooth, axial topographic ridge standing hundreds of meters above the adjacent seafloor (**Figures 1** and **4**). Slow-spreading ridges, however, have proportionally smaller magma fluxes, cooler crust, and thicker lithosphere. The mechanical effects of a thick lithosphere combined with less-frequent eruptions to accommodate extension cause the axes of slow-spreading ridges to be heavily faulted valleys, as deep as 2 km below the adjacent seafloor. Ridges spreading at intermediate rates show both morphologies, and appear to be sensitive to subtle fluctuations in magma supply. Seismic imaging of the crust reveals that melt

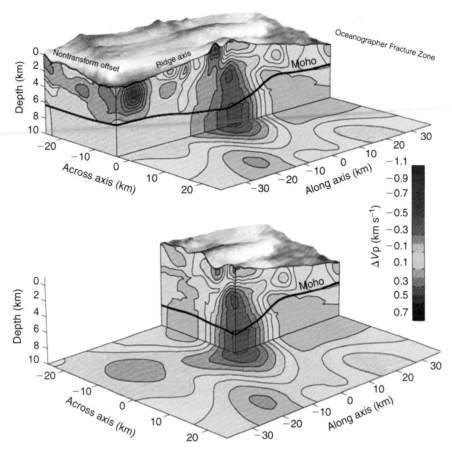

Figure 10 A perspective view of seafloor bathymetry and a tomographic image of the MAR (35° N latitude) magmatic system. The image is of the P wave velocity anomaly, relative to an average vertical velocity profile, contoured at $0.1 \, km \, s^{-1}$. The vertical planes reveal partially molten bodies (the low-velocity regions), which are discontinuous beneath the ridge. The deep horizontal plane is located in the mantle just below the crust and shows that the crustal low-velocity region extends downward into the mantle. Crustal thickness is also greatest at the center of the ridge (black line labeled 'Moho'). The seismic image indicates that as magma rises in the mantle, it becomes focused to the center of the ridge segment where it then feeds into the crust. Adapted from Dunn RA, Lekic V, Detrick RS, and Toomey DR (2005) Three-dimensional seismic structure of the Mid-Atlantic Ridge at 35° N: Focused melt supply and non-uniform plate spreading. *Journal of Geophysical Research* 110: B09101 (doi:10.1029/2004JB003473), with permission from American Geophysical Union.

supply and crustal structure vary with spreading rate (**Figure 4**), geodynamic setting, as well as time.

Another major characteristic is the variability in topography, gravity, and crustal thickness as a function of distance along different mid-ocean ridges. Along individual segments of fast-spreading ridges, topography, gravity, as well as crustal thickness are remarkably uniform, varying by less then ~20% (**Figures 8 and 9**). Such relative uniformity probably indicates a steady magma supply from below as evident from seismic imaging of magma being stored over large distances along fast-spreading ridges (**Figure 9**). Quasi-steady-state crustal magmatic systems have been seismically shown to extend into the underlying mantle.

That said, the variability that is present between and along fast-spreading ridge segments reveals some important processes. Like all mid-ocean ridges,

fast-spreading ridge segments are separated by large-qoffset fracture zones at the largest scale, and also by overlapping spreading centers (OSCs) at an intermediate scale. OSCs are characterized by the overlap of two *en echelon* ridge segments that offset the ridge by several kilometers. In one view, OSCs occur at the boundary between two widely separated, and misaligned, regions of buoyant mantle upwelling. An opposing model states that OSCs are mainly tectonic features created by plate boundary reorganization, below which mantle upwelling is primarily passively driven. A consensus on which hypothesis best explains the observations has not yet been reached.

At still finer scale, segmentation is apparent as minor morphologic deviations from axial linearity (or 'DEVALs') at intervals of 5–25 km. Individual DEVAL-bounded segments of the EPR are associated with higher proportions of melt in the crustal and

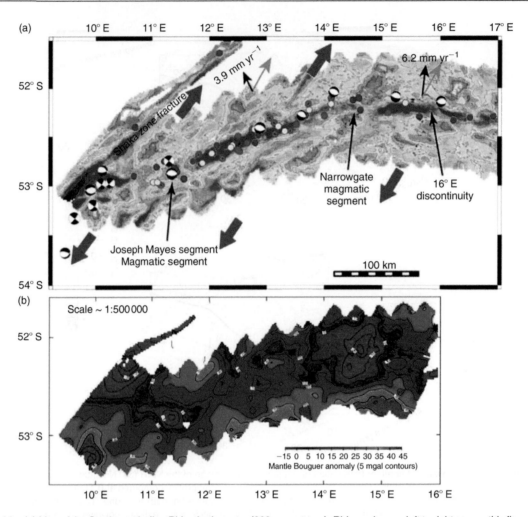

Figure 11 (a) Map of the Southwest Indian Ridge bathymetry (200-m contours). Ridge axis runs left to right across this figure. Large red arrows show relative direction of seafloor spreading (oblique to the ridge axis). Circles with black and white pattern indicate the locations and slip mechanisms of recorded earthquakes. Red and green dots indicate locations where crustal basaltic rocks and mantle peridotite rocks, respectively, were recovered. Significant amounts of mantle peridotite can be found at the seafloor along ultra-slow-spreading ridges. (b) Mantle Bouguer gravity anomaly. The large gravity lows signify thick belts of crust and/or low-density mantle, and correspond to regions where basalts have been predominantly recovered. Reprinted by permission from Macmillan Publishers Ltd: *Nature* (Dick HJB, Lin J, and Schouten H (2003) An ultraslow-spreading class of ocean ridge. *Nature* 426: 405–412), copyright (2003).

upper mantle magmatic system. This suggests that the melt flux from the mantle is locally greater between DEVALs than at the boundaries. The cause is controversial and may even have more than one origin. One possible origin is small-scale mantle diapirism that locally enhances melt production. Such a hypothesis is deduced from deformation fabrics seen in ophiolites, which are sections of oceanic lithosphere that are tectonically thrust onto continents. Alternatively, melt production can be locally enhanced by mantle compositional heterogeneity. Still other possibilities involve shallower processes such as variability in melt transport.

In stark contrast to fast-spreading ridges, slow-spreading ridges show huge variability in topography,

gravity, and crustal thickness along individual spreading segments that are offset by both transform and nontransform boundaries (**Figures 8, 10,** and **11**). The crust is usually thickest near the centers of ridge segments and can decrease by 50% or more toward segment boundaries (**Figure 10**). These and several other observations probably indicate strong along-axis variability in mantle flow and melt production. For example, a recent seismic study reveals a large zone in the middle to lower crust at the center of a slow-spreading ridge segment with very low seismic wave speeds. This finding is consistent with locally elevated temperatures and melt content that extend downward into the uppermost mantle. Although the observations can be explained by several different

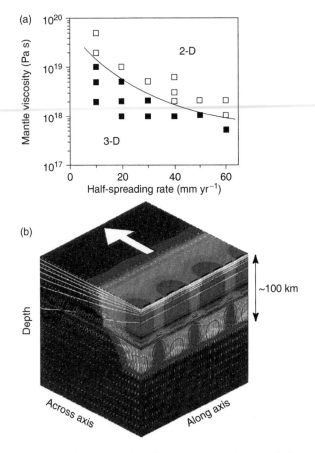

Figure 12 Predictions of a 3-D numerical model of mantle flow and melting. (a) Predicted variability in crustal production along the model ridge is characterized as 3-D or 2-D if the variability is, respectively, larger than or less than an arbitrary threshold. Along-axis variation increases with decreasing mantle viscosity and with decreasing spreading rate. (b) Perspective view showing retained melt (shading, varying from 0% far from the axis to 1.8% at centers of columnar zones), mantle flow (small white arrows of length proportional to flow rate), temperature (white contours), and melt productivity (black contours). The large white arrow depicts a plate spreading slowly at a rate of 12 mm yr^{-1} away from the plane of symmetry at the ridge axis (right vertical plane). Melt retention buoyancy generates convective mantle upwellings in the lower part of the melting zone where viscosities are low (below the red line). In the upper portion of the melting zone (above the red line), viscosity is high, owing to the extraction of water from the solid residue. In this zone, plate-driven mantle flow dominates. Thus, essentially all of the along-axis variability is generated in the lower half of the melting zone. It is the thickness of this lower zone of melting that controls the wavelength of variability. Wavelengths of 50–100 km are typical along the MAR. Adapted from Choblet G and Parmentier EM (2004) Mantle upwelling and melting beneath slow spreading centers: Effects of variable rheology and melt productivity. *Earth and Planetary Science Letters* 184: 589–604, with permission from American Geophysical Union.

mantle flow and melt transport scenarios, a predominant view supports the hypothesis that subridge mantle flow beneath slow-spreading ridges is largely influenced by lateral density variations.

Causes for the major differences in along-axis variability between fast- and slow-spreading ridges have been explored with 3-D numerical models of mantle convection and melting. Models of only plate-driven flow predict that the disruption of the ridge near a segment offset both locally reduces upwelling and enhances lithosphere cooling beneath it, both of which tend to somewhat reduce melt production near the offset. For a given length of segment offset, the size of the along-axis variability is smallest at the fastest spreading rates and increases with decreasing spreading rate. This prediction is broadly consistent with the observations; however, such models still underpredict the dramatic variability observed along many segments of slow-spreading ridges.

Again, a consideration of both plate- and buoyancy-driven flow provides a plausible solution. In the direction parallel to the ridge axis, variations in density can be caused by changes in temperature, retained melt, as well as solid composition due to melt extraction (melting dissolves high-density minerals and extracts high-density elements like iron from the residual solid). All three sources of buoyancy are coupled by the energetics and chemistry of melting and melt transport. As discussed above, models of fast-spreading systems predict plate-driven flow to be most important such that buoyancy causes only subtle along-axis variations in melting (**Figure 12**). As spreading rate decreases, the relative strength of buoyant flow increases as does the predicted variability of melt production. Models that include buoyancy more successfully predict typical amplitudes of variations along slow-spreading ridges.

Even more dramatic melt supply variations are observed at a few locations along ridges, which include the MAR at Iceland and near the Azores Islands, the Galapagos Spreading Center near the Galapagos Archipelago, and the EPR near Easter Island (**Figure 1**). These regions occur where 'hot spots' in the mantle produce so much magmatism that islands are formed. Iceland in fact is a location where a mid-ocean ridge is actually exposed above sea level. These 'hot-spot-influenced' sections of mid-ocean ridges show elevated topography and enhanced crustal thickness over distances of many hundreds of kilometers. The most likely cause for these features are anomalously hot, convective upwellings that rise from depths at least as deep as the base of the upper mantle. Fluid dynamical studies show that plumes of rising mantle can arise from hot thermal boundary layers such as the core mantle boundary. When these hot upwellings eventually rise to the lithosphere, they expand beneath it and can enhance volcanism over large distances (**Figure 13**).

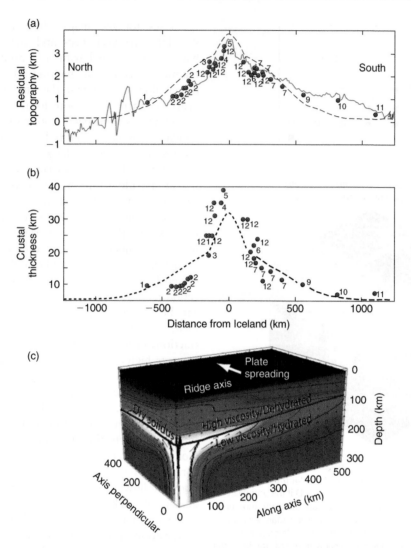

Figure 13 (a) Observed residual topography (solid curve and circles) and (b) crustal thickness of Iceland and the MAR, compared to the predictions of a 3-D model of a hot mantle plume rising beneath the ridge (dashed). (c) Perspective view of potential temperatures (white $>c.$ 1500 °C) within the 3-D model. The vertical cross sections are along (right) and perpendicular (left) to the ridge. Viscosity decreases with temperature and increases at the dry solidus by 10^2 because water is extracted from the solid with partial melting. Thermal buoyancy causes the hot plume material to spread hundreds of kilometers along the MAR away from Iceland. Crustal thickness is predicted to be greatest above the hot plume and to decrease away from Iceland due to decreasing temperatures. Reproduced from Ito G and van Keken PE (2007) Hot spots and melting anomalies. In: Bercovici D (ed.) *Treatise in Geophysics, Vol. 7: Mantle Dynamics.* Amsterdam: Elsevier, with permission from Elsevier.

So far, we have discussed characteristics of the mid-ocean ridge system that are likely to be heavily influenced by differences in heat transport and mantle flow. At the frontier of our understanding of mantle processes is the importance of composition. Studies of seafloor spreading centers at back of the arcs of subduction zones reveal how important mantle composition can be to seafloor creation. For example, the Eastern Lau Spreading Center is characterized by rapid along-strike trends in many observations that are contrary to or unseen along normal mid-ocean ridges. Contrary to the behavior of mid-ocean ridges, as spreading rate increases

along the Eastern Lau back arc spreading system from a slow rate of ∼40 mm yr⁻¹ in the south to an intermediate rate of ∼95 mm yr⁻¹ in the north, the ridge axis changes from an inflated axial high to a faulted axial valley and the evidence for magma storage in the crust disappears. Coincident with this south-to-north variation, the crustal composition changes from andesitic to tholeiitic and isotopic characteristics change from that of the Pacific domain to more like that in the Indian Ocean.

It is hypothesized that many of the along-strike changes along the Eastern Lau Spreading Center are produced by variable geochemical and petrological

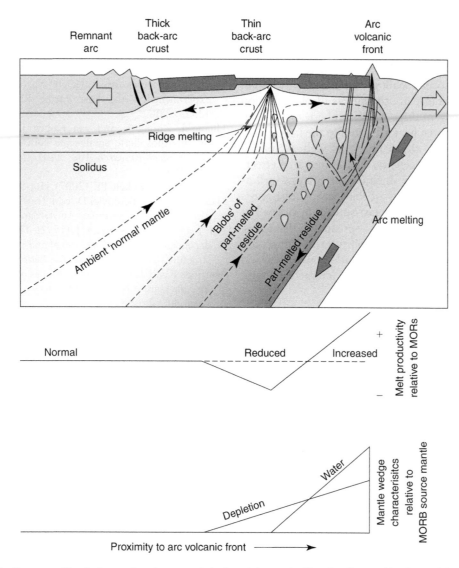

Figure 14 Mantle composition in the wedge above a subducting slab can significantly affect melting beneath back-arc spreading centers. In this scenario, buoyant (partially melted) residual mantle from the arc region is rehydrated by water expelled from the subducting plate, becomes less viscous, and rises into the melting regime of the spreading center, where (because it has already been partially melted) it reduces the total melt production (middle panel). Arc volcanism occurs closer to the subduction zone and originates as fluids percolate from the subducting slab up into the hot mantle wedge and cause melting by reducing melting temperature. If seafloor spreading were closer to this arc melting zone, it would likely form 'thick back-arc crust' and an axial morphology that resembles fast spreading ridges even though spreading here could be slow. Bottom curves schematically show melt depletion and hydration trends in the mantle wedge and their hypothesized effects on the ridge melt productivity with distance from the arc volcanic front. MORS, mid-ocean ridges; MORB, mid-ocean ridge basalt. Adapted from Martinez F and Taylor B (2006) Modes of crustal accretion in back-arc basins: Inferences from the Lau Basin. In: Christie DM, Fisher CR, Lee S-M, and Givens S (eds.) *Geophysical Monograph Series 166: Back-Arc Spreading Systems: Geological, Biological, Chemical, and Physical Interactions*, pp. 5–30 (10.1029/I66GM03). Washington, DC: American Geophysical Union, with permission from American Geophysical Union.

inputs influenced by subduction (**Figure 14**). From south to north, the distance of the ridge from the Tonga arc increases from 30 to 100 km and the depth to the underlying slab increases from 150 to 250 km. To the south, melt production is most likely enhanced by the proximity of the ridge to the arc which causes the ridge to tap arc volcanic melts (slab-hydrated); whereas to the north, melt flux is probably reduced by the absence of arc melts in the ridge melting zone, but in addition, the mantle flow associated with subduction could actually deliver previously melt-depleted residue back to the ridge melting zone. Yet farther to the north, the ridge is sufficiently far away from the slab, such that it taps 'normal' mantle and shows typical characteristics of mid-ocean spreading centers. Similar hypotheses have been formed for other back-arc systems.

Summary

The global mid-ocean ridge system is composed of the divergent plate boundaries of plate tectonics and it is where new ocean seafloor is continually created. Of major importance are the effects of plate motion versus buoyancy to drive asthenospheric upwelling, the balance between heat advected to the lithosphere versus that lost to the seafloor, as well as mantle compositional heterogeneity. Such interacting effects induce variations in the thickness of crust as well as local structural variability of mid-ocean ridge crests that are relatively small at fast-spreading ridges but become more dramatic as spreading decreases. Through examining this variability geoscientists are gaining an understanding of mantle convection and chemical evolution as well as key interactions with the Earth's surface.

See also

Mid-Ocean Ridge Geochemistry and Petrology. Mid-Ocean Ridge Seismic Structure. Mid-Ocean Ridge Seismicity. Mid-Ocean Ridge Tectonics, Volcanism, and Geomorphology. Seamounts and Off-Ridge Volcanism.

Further Reading

Buck WR, Lavier LL, and Poliakov ANB (2005) Modes of faulting at mid-ocean ridges. *Nature* 434: 719–723.

Choblet G and Parmentier EM (2004) Mantle upwelling and melting beneath slow spreading centers: Effects of variable rheology and melt productivity. *Earth and Planetary Science Letters* 184: 589–604.

Dick HJB, Lin J, and Schouten H (2003) An ultraslow-spreading class of ocean ridge. *Nature* 426: 405–412.

Dunn RA and Forsyth DW (2003) Imaging the transition between the region of mantle melting and the crustal magma chamber beneath the southern East Pacific Rise with short-period Love waves. *Journal of Geophysical Research* 108(B7): 2352 (doi:10.1029/2002JB002217).

Dunn RA, Lekic V, Detrick RS, and Toomey DR (2005) Three-dimensional seismic structure of the Mid-Atlantic Ridge at 35°N: Focused melt supply and non-uniform plate spreading. *Journal of Geophysical Research* 110: B09101 (doi:10.1029/2004JB003473).

Dunn RA, Toomey DR, and Solomon SC (2000) Three-dimensional seismic structure and physical properties of the crust and shallow mantle beneath the East Pacific Rise at 9°30′N. *Journal of Geophysical Research* 105: 23537–23555.

Forsyth DW, Webb SC, Dorman LM, and Shen Y (1998) Phase velocities of Rayleigh waves in the MELT experiment on the East Pacific Rise. *Science* 280: 1235–1238.

Huang J, Zhong S, and van Hunen J (2003) Controls on sublithospheric small-scale convection. *Journal of Geophysical Research* 108: 2405 (doi:10.1029/2003 JB002456).

Ito G and van Keken PE (2007) Hot spots and melting anomalies. In: Bercovici D (ed.) *Treatise in Geophysics, Vol. 7: Mantle Dynamics.* Amsterdam: Elsevier.

Lin J and Phipps Morgan J (1992) The spreading rate dependence of three-dimensional mid-ocean ridge gravity structure. *Geophysical Research Letters* 19: 13–16.

Martinez F and Taylor B (2006) Modes of crustal accretion in back-arc basins: Inferences from the Lau Basin. In: Christie DM, Fisher CR, Lee S-M, and Givens S (eds.) *Geophysical Monograph Series 166: Back-Arc Spreading Systems: Geological, Biological, Chemical, and Physical Interactions,* pp. 5–30. Washington, DC: American Geophysical Union (doi:10.1029/l66GM03).

Phipps Morgan J, Blackman DK, and Sinton JM (eds.) (1992) *Geophysical Monograph 71: Mantle Flow and Melt Generation at Mid-Ocean Ridges,* p. 361. Washington, DC: American Geophysical Union.

Phipps Morgan J and Chen YJ (1993) Dependence of ridge-axis morphology on magma supply and spreading rate. *Nature* 364: 706–708.

Shen Y, Sheehan AF, Dueker GD, de Groot-Hedlin C, and Gilbert H (1998) Mantle discontinuity structure beneath the southern East Pacific Rise from P-to-S converted phases. *Science* 280: 1232–1234.

Spiegelman M (1993) Physics of melt extraction: Theory, implications and applications. *Philosophical Transactions of the Royal Society of London, Series A* 342: 23–41.

Spiegelman M (1996) Geochemical consequences of melt transport in 2-D: The sensitivity of trace elements to mantle dynamics. *Earth and Planetary Science Letters* 139: 115–132.

Stein CA and Stein S (1992) A model for the global variation in oceanic depth and heat flow with lithospheric age. *Nature* 359: 123–129.

Relevant Websites

http://www.ridge2000.org
– Ridge 2000 Program.

PROPAGATING RIFTS AND MICROPLATES

Richard Hey, University of Hawaii at Manoa, Honolulu, HI, USA

Introduction

Propagating rifts appear to be the primary mechanism by which Earth's accretional plate boundary geometry is reorganized. Many propagation episodes are caused by or accompany changes in direction of seafloor spreading. Propagating rifts, oriented at a more favorable angle to the new plate motion, gradually break through lithospheric plates. A propagator generally replaces a pre-existing spreading center, causing a sequence of spreading center jumps and leaving a failed rift system in its wake. This results in changes to the classic plate tectonic geometry. There is pervasive shear deformation in the overlap zone between the propagating and failing rifts, much of it accommodated by bookshelf faulting. Rigid plate tectonics breaks down in this zone. When the scale or strength of the overlap zone becomes large enough, it can stop deforming, and instead begin to rotate as a separate microplate between dual active spreading centers. This microplate tectonic behavior generally continues for several million years, until one of the spreading boundaries fails and the microplate is welded to one of the bounding major plates. Active microplates are thus modern analogs for how large-scale (hundreds of kilometers) spreading center jumps occur.

Propagating Rifts

Propagating rifts are extensional plate boundaries that progressively break through mostly rigid lithosphere, transferring lithosphere from one plate to another. If the rifting advances to the seafloor spreading stage, propagating seafloor spreading centers follow, gradually extending through the rifted lithosphere. The orthogonal combination of seafloor spreading and propagation produces a characteristic V-shaped wedge of lithosphere formed at the propagating spreading center, with progressively younger and longer isochrons abutting the 'pseudofaults' that bound this wedge. Although propagation rates as high as 1000 km per million years have been discovered, propagation rates often have similar magnitudes to local spreading rates. **Figure 1** shows several variations of typical mid-ocean ridge propagation geometry, in which a pre-existing 'doomed rift' is replaced by the propagator.

Geometry

Figure 1A shows the discontinuous propagation model, in which periods of seafloor spreading alternate with periods of instantaneous propagation, producing *en echelon* failed rift segments, fossil transform faults and fracture zones, and blocks of progressively younger transferred lithosphere. **Figure 1B** shows the pattern produced if propagation, rift failure, and lithospheric transferral are all continuous. In this idealized model a transform fault migrates continuously with the propagator tip, never existing in one place long enough to form a fracture zone, and thus V-shaped pseudofaults are formed instead of fracture zones. **Figure 1C** shows a geologically more plausible model, in which the spreading rate accelerates from zero to the full rate over some finite time and distance on the propagating spreading center with concomitant decreases on the failing spreading center, so that lithospheric transferral is not instantaneous. Instead of a transform fault, a migrating broad 'non-transform' zone of distributed shear connects the overlapping propagating and failing ridges during the period of transitional spreading. Deformation occurring in this overlap zone is preserved in the zone of transferred lithosphere, cross-hatched in **Figure 1**. This zone is bounded by the failed rifts and inner (proximal) pseudofault. Even more complicated geometries occur in some places on Earth where the doomed rift, instead of failing monotonically as the propagator steadily advances, occasionally itself propagates in the opposite direction. In these 'dueling propagator' systems, both axes curve toward each other. Even in simple propagator systems, the failing rift often curves toward the propagating rift, resembling the smaller scale overlapping spreading center '69' geometry (*see* Mid-Ocean Ridge Tectonics, Volcanism, and Geomorphology), but with about a 1:1 overlap length to width aspect ratio, in contrast to the 3:1 aspect ratio characteristic of overlappers.

Classic plate tectonic geometry holds for the area outside the pseudofaults and zone of transferred lithosphere, but rigid plate tectonics breaks down in the overlap zone where some of the lithosphere formed on the doomed rift is progressively

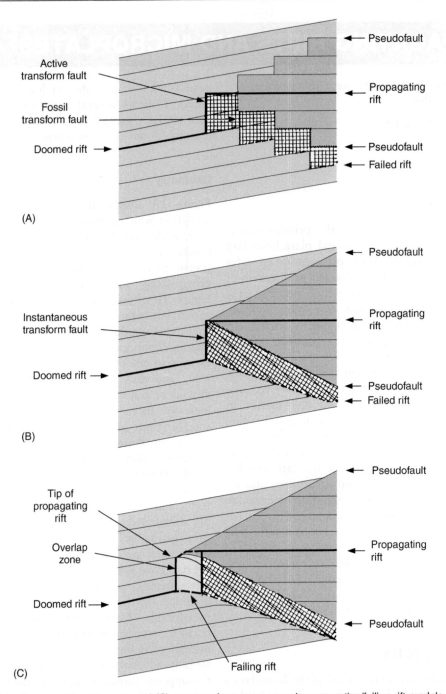

Figure 1 (A) Discontinuous, (B) continuous, and (C) non-transform zone oceanic propagating/failing rift models. Propagating rift lithosphere is marked by dark stipple, normal lithosphere created at the doomed rift is indicated by light stipple, and transferred lithosphere is cross-hatched. Heavy lines show active plate boundaries. In (C), active axes with full spreading rate are shown as heavy lines; active axes with transitional rates are shown as dashed lines. The overlap zone joins these transitional spreading axes. (Reproduced with permission from Hey *et al.*, 1989.)

transferred to the other plate by the rift propagation and resulting migration of the overlap zone. Shear between the overlapping propagating and failing rifts appears to be accommodated by bookshelf faulting, in which, for example, right-lateral plate motion shear produces high angle left-lateral slip, apparently along the pre-existing abyssal hill faults. This produces oblique seafloor fabric, with trends quite different from the ridge-parallel and -perpendicular structures expected on the basis of previous plate tectonic theory. **Figure 2** is a shaded relief map of the type example propagating rift, at 95.5°W along the Cocos-Nazca spreading center. This propagator is breaking westward away from the Galapagos hot

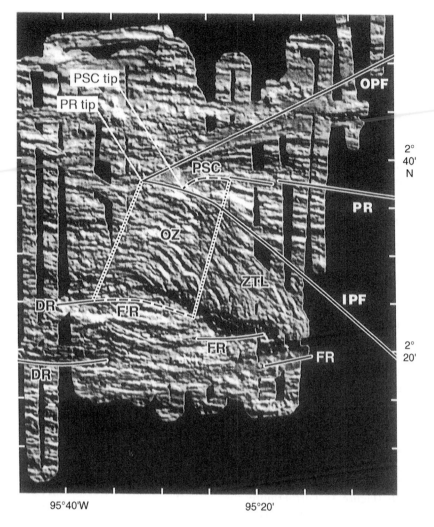

Figure 2 Shaded relief map of digital seabeam swath bathymetry at the Galapagos 95.5°W propagating rift system. The relative plate motion is nearly north–south. Propagation is to the west. The oblique structures in the overlap zone and its wake, the zone of transferred lithosphere, are clearly evident. PR, propagating rift; PSC, propagating spreading center; OPF, IPF, outer and inner pseudofaults; OZ, overlap zone; ZTL, zone of transferred lithosphere; DR, doomed rift; F'R, failing rift; FR, failed rift grabens. (Adapted with permission from Hey *et al.*, 1989.)

spot through 1 million-year-old Cocos lithosphere at a velocity of about 50 km per million years. Well organized seafloor spreading begins about 10 km behind the faulting, fissuring, and extension at the propagating rift tip. This 200 000 year time lag between initial rifting and the rise of asthenosphere through the lithospheric crack to form a steady-state spreading center suggests an asthenospheric viscosity of about 10^{18} Pa-s. The combination of seafloor spreading at about 60 km per million years and propagation produces a V-shaped wedge of young lithosphere surrounded by pre-existing lithosphere. The propagating rift lithosphere is characterized by unusually high amplitude magnetic anomalies and by unusual petrologic diversity, including highly fractionated ferrobasalts. This propagator is replacing a pre existing spreading system about 25 km to the

south, and thus spreading center jumps and failed rifts are being produced. Although propagation is continuous, segmented failed rift grabens seem to form episodically on a timescale of about 200 000 years. This has produced a very systematic pattern of spreading center jumps, in which each jump was younger and slightly longer than the preceding jump. The spreading center orientation is being changed clockwise by about 13°, and more than 10^4 km^3 per million years of Cocos lithosphere is being transferred to the Nazca plate. The active propagating and failing rift axes overlap by about 20 km, and are connected by a broad and anomalously deep zone of distributed shear deformation rather than by a classic transform fault. Most of the seismic activity occurs within this 'non-transform' zone, where the pre-existing abyssal hill fabric originally created on the doomed rift is sheared and tectonically rotated into new oblique trends. Simple equations accurately describe this geometry in terms of ratios of propagation and spreading rates, together with the observed propagating and doomed rift azimuths. For example, for the simplest continuous propagation geometry, if u is the spreading half rate and v is the propagation velocity, the pseudofaults form angles $\tan^{-1}(u/v)$ with the propagator axis, and the isochrons and abyssal hill fabric in the zone of transferred lithosphere have been rotated by an angle $\tan^{-1}(2u/v)$.

Thermal and Mechanical Consequences

The boundaries of the Galapagos high amplitude magnetic anomaly zone, of the ferrobasalt province, and of the spreading center jumps identified from magnetic anomalies, are all essentially coincident with the pseudofaults bounding the propagating rift lithosphere. All of these observations can be explained as mechanical and/or thermal consequences of a new rift and spreading center breaking through cold lithosphere, with increased viscous head loss and diminished magma supply on the propagating spreading center close to the propagator tip. This leads to an unusually deep axial graben, unusually extensive fractional crystallization, and unusually high petrologic diversity.

Basalt glasses erupted along propagating rifts are generally more differentiated than those erupted on normal and failing rifts (**Figure 3**). Furthermore, the Galapagos 95.5°W propagating rift lavas display a striking and unusual compositional diversity, in marked contrast to the narrow compositional range of basalt glasses from the adjacent doomed rift segment. With extremely rare exceptions, ferrobasalts (or FeTi basalts, enriched in iron and titanium) are confined to the wedge of propagating rift lithosphere. The compositional variation of the propagator lavas has been shown to be primarily attributable to shallow-level crystal fractionation of normal mid-ocean ridge basalt parental magmas (MORB, *see* Mid-Ocean Ridge Geochemistry and Petrology), controlled by a balance between the magma cooling and supply rates. The observed variation in erupted lava compositions reflects the evolution of the rift from initial spreading toward the steady-state buffered magma chamber configuration of a normal spreading center. Lava compositions define gradients in mantle source geochemistry roughly centered about the Galapagos hot spot near 91°W. These data indicate that the 95.5°W propagator tip represents a boundary in mantle source geochemistry, implying that this rift propagation can be related to plume-related subaxial asthenospheric flow away from the

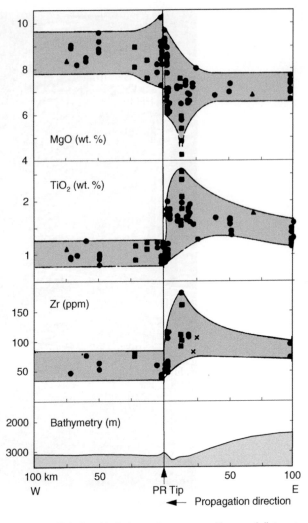

Figure 3   Relationship between lava compositions and distance from the 95.5°W propagating rift tip at the time of eruption. Bathymetric profile shows depth variation along the present-day spreading axis. (Reproduced with permission from Hey *et al.*, 1989.)

hot spot. All six known active Galapagos propagators are propagating away from the Galapagos hot spot.

Causes of Rift Propagation

One important observation is that many rifts and spreading centers propagate down topographic gradients away from hot spots or shallow ridge axis topography. Rift propagation in the Galapagos area is associated with and probably caused by plume-related asthenospheric flow under and away from the Galapagos hot spot. This flow generates gravitational stresses on the shallow spreading center segments near the hot spot that promote crack propagation away from the hot spot. Flow of asthenosphere into these cracks produces new lithosphere at propagating seafloor spreading centers. Regionally high deviatoric tensile stresses associated with regional uplift provide a quantitatively plausible driving mechanism. Crack growth occurs when the stress concentration at the tip, characterized in elastic fracture mechanics by a stress intensity factor, exceeds some threshold value that is a function of the strength of the lithosphere. Propagation can be driven by excess gravity sliding stresses due to the shallow lithosphere near the hot spot, which is almost, but not quite, counterbalanced by the resisting stress intensity contribution due to the local tip depression. The spreading center propagation rate is limited by the viscosity of the asthenosphere flowing into the rift. The overlap/offset ratio of propagating and failing rifts tends to be ~1, close to the ratio at which the stress intensity factor is maximized (see Mid-Ocean Ridge Tectonics, Volcanism, and Geomorphology).

Although many rifts appear to propagate in response to hot spot-related stresses, others appear to propagate because of stresses produced when changes in plate motion occur. A common mechanism in the Northeast Pacific producing such changes and propagation appears to be subduction-related stresses. This probably explains most of the massive reorganizations of the spreading geometry as the Pacific-Farallon ridge neared the Farallon-North America trench, although some propagation away from a hot spot has also occurred in the Juan de Fuca area. Propagation may be produced in many ways over a wide range of scales.

Discussion

Propagating rifts form new extensional plate boundaries and rearrange the geometries of old ones, with the numerous consequences already discussed. These include the formation of anomalous structural, bathymetric, and petrologic provinces; the formation of pseudofaults instead of fracture zones; and the transfer of lithosphere from one plate to another, as well as spreading center reorientations, jumps, and failed rifts.

Propagators that replace pre-existing spreading centers always seem to result in at least slight spreading center reorientation. Whether rifts propagate in response to changes in direction of seafloor spreading, or the spreading direction changes while the rift propagates, rift propagation is the primary mechanism by which many seafloor spreading systems have adjusted to changes in spreading direction. Spreading center jumps are a predictable consequence of rift propagation if there is a failing rift. The largest jumps sometimes involve transient

microplate formation, geometrically similar to the broad overlap zone model of **Figure 1C** but on a much larger scale.

Rift Propagation on the Other Scales

The pervasively deformed zone of transferred lithosphere is characteristic of propagating rift systems in which the offset of the axes ranges from a few kilometers to more than 100 km. Below this offset width, small propagators are sometimes called migrating overlapping spreading centers (see Mid-Ocean Ridge Tectonics, Volcanism, and Geomorphology), the distinction being that this smaller scale reorganization takes place completely within the nonrigid plate boundary zone, before rigid plate tectonics begins, whereas propagators break through rigid lithosphere, thus reorganizing the plate boundary geometry.

Sometimes the overlap zones between propagating and failing rifts are on very large scales, e.g., the 120×120 km pervasively deformed overlap zone between the giant dueling propagators southwest of Easter Island (**Figure 4**). Interestingly, both the Easter microplate just to the north and Juan Fernandez microplate just to the south of this area show evidence for pervasively deformed cores. This is interpreted as evidence for a two-stage evolution, with the deformed core formed during an early rift propagation stage, before later evolution as rapid rotation of a mostly rigid microplate about a pole near the center of the microplate. This suggests that large-scale propagation or duelling propagation could sometimes initiate microplate formation. If an overlap zone becomes too big and strong to deform by pervasive bookshelf faulting, it could instead accommodate the boundary plate motion shear stresses by beginning to rotate as a separate microplate.

Microplates

Microplates are small, mostly rigid areas of lithosphere, located at major plate boundaries but rotating as more or less independent plates. Although small microplates can form in many tectonic settings, and are common in broad continental deformation zones, this article addresses only the type formed along mid-ocean ridges. The two main subtypes – those formed at triple junctions and those formed along ridges away from triple junctions – share many similarities, and plate boundary reorganization by rift propagation is important in both settings. Although it was once thought that stable growing microplates could eventually grow into major oceanic plates, it now appears that these are transient

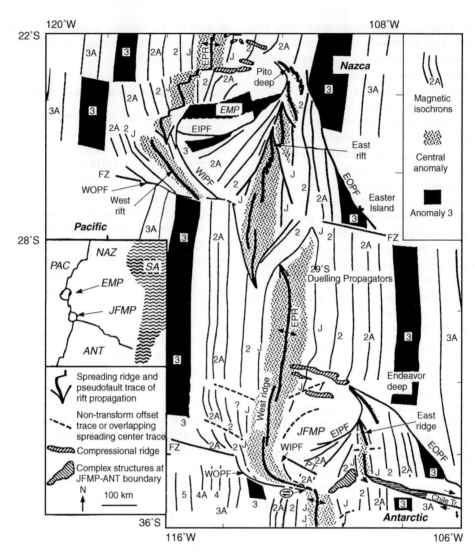

Figure 4 Tectonic boundaries, magnetic isochrons, and structures of Easter (EMP) and Juan Fernandez (JFMP) microplates. EPR, East Pacific Rise; FZ, fracture zone; WOPF, WIPF, EIPF, and EOPF are western outer and inner, and eastern inner and outer, pseudofaults, respectively. (Reproduced with permission from Bird and Naar, 1994.)

phenomena resulting from rift propagation on such a large scale that the overlap zone eventually begins to rotate between dual active spreading centers. The best studied oceanic microplates are the Easter microplate along the Pacific–Nazca ridge, and the Juan Fernandez microplate at the Pacific–Nazca–Antarctica triple junction. Despite their different tectonic settings, they show many striking similarities (**Figure 4**).

Geometry

The scales of the Easter (~ 500 km diameter) and the Juan Fernandez (~ 400 km diameter) microplates are similar. The eastern and western boundaries of both microplates are active spreading centers, propagating north and south respectively. Both microplates began

forming about 5 million years ago, and both East rifts have been propagating into roughly 3 million-year-old Nazca lithosphere. Extremely deep axial valleys occur at their tips, ~ 6000 m at Pito Deep at the northeastern Easter microplate boundary, and ~ 5000 m at Endeavor Deep at the northeastern Juan Fernandez microplate boundary. The northern and southern boundaries are complicated deformation zones, with zones of shear, extension, and significant areas of compression.

Both the Easter and Juan Fernandez microplates have large (~ 100 km $\times 200$ km) complex pervasively deformed cores, which could have formed by bookshelf faulting in overlap zones during an initial large-scale propagating rift stage of evolution. The abyssal hill fabric and magnetic anomalies of the younger seafloor also show a later stage of growth as independent microplates, by rift propagation and seafloor spreading on dual active spreading centers, with deformation during this microplate stage concentrated along the plate boundaries and resulting from the microplate rotation. At present, both microplates are rotating clockwise very rapidly about poles near the microplate centers, spinning like roller-bearings caught between the major bounding plates. The Easter microplate rotation velocity is about 15° per million years and the Juan Fernandez velocity is about 9° per million years.

This roller-bearing analogy has been quantified in an idealized edge-driven model of microplate kinematics (**Figure 5**). If microplate rotation is indeed driven by shear between the microplate and surrounding major plates, the rotation velocity (in radians) is $2u/d$, where $2u$ is the major plate relative velocity, and d is the microplate diameter. This follows because the total spreading on the microplate boundaries must be what the major plate motion would be if the microplate did not exist. The rotation (Euler) poles describing the motion of the microplate relative to the major plates will lie on the microplate boundaries, at the farthest extensions of the rifts,

which must lengthen to stay at the Euler poles because of the microplate rotation.

Evolution

This idealized geometry requires a circular microplate shape, yet also requires seafloor spreading on the dual active ridges, which must constantly change this shape. The more the microplate grows, the more deformation must occur as it rotates, and the less successful the rigid plate model will be. Although it would appear that this inevitable plate growth would soon invalidate the model, numerous episodes of rift propagation helping to maintain the necessary geometry are observed to have occurred at the Easter and Juan Fernandez microplates. These propagators all propagated on the microplate interior side of the failing rifts, thus transferring microplate lithosphere to the major plates, shaving the new microplate growth at the edges and maintaining a circular enough shape for the edge-driven model to be very successful. Although instantaneous spreading rates may be symmetric, time-averaged accration is highly asymmetric, much faster on the major plates than the microplates because of the lithosphere transferred by rift propagation. Nevertheless, some deformation is clearly occurring along the northern boundaries of both microplates, forming large compressional ridges.

According to the edge-driven model, a microplate may stop rotating if one of the bounding ridge axes propagates through to the opposite spreading boundary, eliminating coupling to one of the bounding plates. Dual spreading would no longer occur, spreading would continue on only one bounding ridge, and $10^6-10^7 \, \text{km}^3$ of microplate lithosphere would accrete to one of the neighboring major plates. There is evidence in the older seafloor record that this has happened many times before along the ancestral East Pacific Rise.

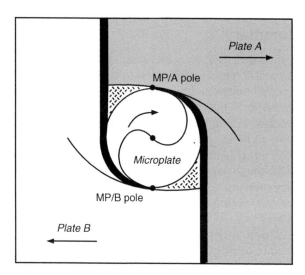

Figure 5 Roller-bearing model of microplates based on a simple, concentrically rotating bearing. The microplate is approximated by a circular plate (white) which is caught between two major plates A and B. The main contacts between microplate and major plates are also the positions of the relative rotation poles (dots). Dark shading shows major spreading centers, overlapping about the microplate. Cross-hatched corners are areas of compression. Medium curved lines are predicted pseudofaults, and arrows show relative motions. This schematic model assumes growth from an infinitesimal point to a present circular shape – the model can be extended to take account of growth from a finite width, eccentric motions, and growth of the microplate. (Reproduced with permission from Searle *et al.*, 1993.)

Causes of Microplate Formation

The occurrence of large, pervasively deformed cores in both microplates, probably formed during early propagating rift stages of evolution, suggests that these oceanic microplates may have originated as giant propagator or duelling propagator systems which formed overlap zones so big that their mechanical behavior had to change. This propagation could have been normal rift propagation, or from another documented type where a new propagator appears to start from near the center of a long transform fault, perhaps on a small intra-transform spreading center.

Both microplates appear to have originated at large left-stepping offsets of the East Pacific Rise. All large right-stepping offsets along the Pacific-Nazca spreading center are transform faults, while all large left-stepping offsets are microplates or the possible duelling propagator proto-microplate between the Easter and Juan Fernandez microplates (**Figure 6**). The Galapagos microplate at the Pacific-Cocos-Nazca triple junction also fits this pattern. This suggests that a recent clockwise change in Pacific-Nazca plate motion could have been an important factor triggering the formation of these microplates, although the right-stepping Wilkes transform also shows evidence for previous boundary reorganization, and the smaller scale 21°S duelling propagators are also right-stepping.

The dominant East rift of the Easter microplate, the dominant West rift of the Juan Fernandez microplate, and the dominant West rift of the duelling propagators between the microplates, are all propagating away from the Easter mantle plume (or the intersection of this plume with the ridge axis), suggesting that microplate formation as well as rift propagation can be driven by plume-related forces.

Earth's fastest active seafloor spreading occurs in this area, and every major mid-ocean ridge segment presently spreading faster than 142 km per million years is reorganizing by duelling propagators or microplates (**Figure 6**). The combination of thin lithosphere produced at these 'superfast' seafloor spreading rates, as well as the unusually hot asthenosphere produced by the Easter mantle plume, would reduce the forces resisting propagation and thus make these plate boundary reorganizations easier, perhaps explaining their common occurrence in this area.

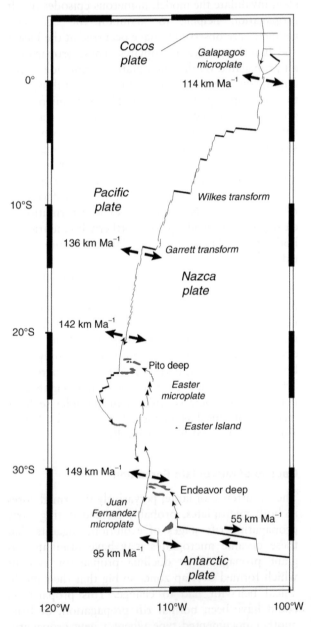

Figure 6 Plate tectonic geometry and relative plate motions along the southern East Pacific Rise. Light lines are ridges, those with arrows are propagating. Heavy straight lines are transform faults. (Adapted with permission from Hey *et al.*, 1995.)

Summary

Rift propagation on scales ranging from overlapping spreading centers with a few kilometers offset up to several hundreds of kilometers at microplate tectonic scales, and indeed all the way up to several thousands of kilometers at continental rifting scales, appears to be the primary mechanism by which Earth's accretional plate boundary geometry is reorganized.

Although conceptually simple, the propagating rift hypothesis has important implications for plate tectonic evolution. It explains the existence of several classes of structures, including pseudofaults, failed rifts, and zones of transferred lithosphere, that are oblique to ridges and transform faults and thus previously seemed incompatible with plate tectonic theory. These are all quantitatively predictable consequences of rift propagation. It explains why passive continental margins are not parallel to the oldest seafloor isochrons, but instead are pseudofaults, bounding lithosphere created on propagating spreading centers and indicating the direction of the continental breakup propagators. It explains the large-scale reorganization of many seafloor spreading systems, including both the origination and termination of many fracture zones, as well as the formation of some transient microplates which appear to be the modern analogs of large-scale spreading center jumps. This hypothesis provides a mechanistic explanation for the way in which many (if not all) spreading center jumps occur and why they occur in systematic patterns. It explains how spreading centers reorient when the direction of seafloor spreading changes, and the origin of large areas of petrologically diverse seafloor, including the major abyssal ferrobasalt provinces. The common occurrence of rift propagation over a wide range of spreading rates and tectonic environments indicates that it represents an efficient mechanism of adjustment of extensional plate boundaries to the forces driving plate motions.

See also

Mid-Ocean Ridge Geochemistry and Petrology. Mid-Ocean Ridge Tectonics, Volcanism, and Geomorphology.

Further Reading

Bird RT and Naar DF (1994) Intratransform origins of mid-ocean ridge microplates. *Geology* 22: 987–990.

Hey RN, Naar DF, Kleinrock MC, *et al.* (1985) Microplate tectonics along a superfast seafloor spreading system near Easter Island. *Nature* 317: 320–325.

Hey RN, Sinton JM, Duennebier FK, and Duennebier FK (1989) Propagating rifts and spreading centers. In: Winterer EL, Hussong DM, and Decker RW (eds.) *Decade of North American Geology: The Eastern Pacific Ocean and Hawaii*, pp. 161–176. Boulder, CO: Geological Society of America.

Hey RN, Johnson PD, Martinez F, *et al.* (1995) Plate boundary reorganization at a large-offset, rapidly propagating rift. *Nature* 378: 167–170.

Kleinrock MC and Hey RN (1989) Migrating transform zone and lithospheric transfer at the Galapagos 95.5°W propagator. *Journal of Geophysical Research* 94: 13859–13878.

Larson RL, Searle MC, Kleinrock MC, *et al.* (1992) Roller-bearing tectonic evolution of the Juan Fernandez microplate. *Nature* 356: 517–576.

Naar DF and Hey RN (1991) Tectonic evolution of the Easter microplate. *Journal of Geophysical Research* 96: 7961–7993.

Phipps Morgan J and Parmentier EM (1985) Causes and rate limiting mechanisms of ridge propagation: a fraction mechanics model. *Journal of Geophysical Research* 90: 8603–8612.

Schouten H, Klitgord KD, and Gallo DG (1993) Edge-driven microplate kinematics. *Journal of Geophysical Research* 98: 6689–6701.

Searle RC, Bird RT, Rusby RI, and Naar DF (1993) The development of two oceanic microplates: Easter and Juan Fernandez microplates, East Pacific Rise. *Journal of the Geological Society* 150: 965–976.

VOLCANIC HELIUM

J. E. Lupton, Hatfield Marine Science Center, Newport, OR, USA

Introduction

Volcanic activity along the global mid-ocean ridge system and at active seamounts introduces a helium-rich signal into the ocean basins that can be used to trace patterns of ocean circulation and mixing. Helium is extracted from oceanic volcanic rocks by circulating sea water and then injected into the ocean as helium dissolved in submarine hydrothermal vent fluids. Hydrothermal venting produces plumes in the ocean that are highly enriched in a variety of tracers, including heat, helium, manganese, iron, methane, and hydrogen. Among these, volcanic helium is a particularly useful tracer because it has such a high concentration in hydrothermal fluids relative to the background values of helium in sea water, and because it is stable and conservative, i.e., helium does not decay radioactively and is not affected by any chemical or biological processes. By making careful measurements of the relative abundance of helium isotopes, it is possible to trace hydrothermal helium plumes for thousands of kilometers from the source regions.

There are two stable isotopes of helium, ^{3}He and ^{4}He, which vary in their ratio by over three orders of magnitude in terrestrial samples. The Earth's atmosphere is well mixed with respect to helium and contains helium with a uniform isotopic composition of ^{3}He/^{4}He $= 1.39 \times 10^{-6}$. Atmospheric helium is a convenient standard for helium isotope determinations, and terrestrial ^{3}He/^{4}He ratios are usually normalized to the air ratio and expressed in units of R/R_A, where $R = ^3$He/^{4}He and $R_A = (^3$He/^{4}He$)_{air}$. In contrast to atmospheric helium ($R/R_A = 1$), the radiogenic helium produced by α-decay of U and Th series isotopes has a much lower ratio of R/R_A 0.1, while the volcanic helium that is derived from the Earth's mantle is highly enriched in ^{3}He ($R/R_A = 5-30$). Thus volcanic helium has an isotopic composition distinct from other sources such as atmospheric helium or the helium produced by radioactive decay. This ^{3}He-rich mantle helium is sometimes called 'primordial' helium, since it is thought to be the remnant of a primitive component trapped in the Earth's interior since the time of its formation. This trapped component probably had ^{3}He/^{4}He $= 1 \times 10^{-4}$ or 100 R_A, similar to the helium found trapped in meteorites or in the solar wind, but has been modified to $R = 30R_A$ by dilution with radiogenic helium since the time the Earth was formed. Although there is a wide variety of volcanic sources in the oceans, including subduction zone volcanoes and hot spot volcanoes, most of the oceanic volcanic helium is derived from activity along the global mid-ocean ridge system. While the ^{3}He/^{4}He ratio of mantle helium shows a wide range of variation, the helium from mid-ocean ridges falls in a much narrower range of $R/R_A = 7-9$.

In order of decreasing importance, the most abundant forms of helium in sea water are dissolved atmospheric helium, volcanic helium, and to a lesser degree radiogenic helium from sediments. There is also an input of pure ^{3}He into the oceans from tritium (^{3}H), the radioactive isotope of hydrogen, which decays to ^{3}He with a half-life of 12.4 years. Because tritium is generally found only in the upper ocean, ^{3}He from tritium decay (tritiogenic helium) is only significant at depths less than about 1000 m.

Although there are only two isotopes of helium, it is still possible to clearly distinguish submarine volcanic helium from the other components because of its high ^{3}He/^{4}He ratio and because volcanic helium is introduced at mid-depth rather than at the ocean surface or on the abyssal plain.

Units

For samples highly enriched in helium such as volcanic rocks and hydrothermal vent fluids, the helium isotope ratio is usually expressed in the R/R_A notation described above. However, for the relatively small variations observed in sea water samples, the ^{3}He/^{4}He variations are usually expressed as $\delta(^3$He$)$, which is the percentage deviation from the ratio in air, defined as in eqn [1].

$$\delta(^3\text{He}) = 100[(R/R_A) - 1] \qquad [1]$$

Here again $R = ^3$He/^{4}He and $R_A = (^3$He/^{4}He$)_{air}$. Thus $R/R_A = 1.50$ is equivalent to $\delta(^3$He$) = 50\%$.

History and Background

The first attempt to detect nonatmospheric helium in the oceans was made by Suess and Wänke in 1965, who predicted that the deep oceans should contain excess ^{4}He due to U and Th decay in sediments and in the ocean crust. Although they were correct about the existence of radiogenic helium in the oceans, their measurements were of insufficient precision to detect any ^{4}He enrichment above the dissolved air component. It is now known that the input of ^{3}He-rich volcanic helium has a greater effect on both the ^{3}He/^{4}He ratio and the ^{4}He concentration in sea water than does the input of radiogenic helium.

Mantle or volcanic helium was first detected on the Earth as an excess in the ^{3}He/^{4}He ratio in deep Pacific waters. Although this oceanic ^{3}He excess is derived from the helium residing in oceanic volcanic rocks, it was not until about five years later that mantle helium was directly measured in the volcanic rocks themselves. Clarke *et al.* in 1969 reported a 21% excess in the ^{3}He concentration at mid-depth above that expected for air-saturated water, and correctly attributed this excess to a flux of primordial helium leaking from the Earth's interior into the oceans and in turn into the atmosphere (see **Figure 1**). Using a box model for oceanic helium, they were able to estimate the global ^{3}He flux from the oceans into the atmosphere at 2 atoms ^{3}He cm^{-2}, a number that is still in reasonable agreement with more recent flux estimates of 4–5 atoms ^{3}He cm^{-2}.

The discovery of excess ^{3}He in the oceans from localized sources distributed along the global mid-ocean ridge system led immediately to the use of this tracer for oceanographic studies. The Geochemical Ocean Sections Study (GEOSECS), which began in 1972, provided the first maps of the global distribution of helium in the oceans. Since then, several

other oceanographic programs, including the World Ocean Circulation Experiment (WOCE), have added to our knowledge of the global helium distribution.

To illustrate the presence of volcanic helium in the oceans, a typical helium profile in the north Pacific Ocean is shown in **Figure 2**. The figure shows the vertical variation in the ^{3}He/^{4}He ratio expressed as $\delta(^3\mathrm{He})$ in %, and the ^{4}He concentration in nmol kg^{-1}. The values expected for air-saturated water (dashed lines) are shown for comparison. For the calculation of air-saturated values it is assumed that each water parcel equilibrated with the atmosphere at the potential temperature of the sample. This profile exhibits a broad maximum in the deep water, reaching a value of $\delta(^3\mathrm{He}) = 25.0\%$ at $\sim 1850\,\mathrm{m}$ depth. Although this station is located at a distance of over $1500\,\mathrm{km}$ from the nearest active spreading center, the profile still exhibits a clear excess in ^{3}He/^{4}He in the 1500–3500 m depth range due to input of volcanic helium from the mid-ocean ridge system. The secondary maximum in the $\delta(^3He)$ profile at $\sim 350\,\mathrm{m}$ depth is due to excess ^{3}He produced by tritium decay. That this peak is tritiogenic helium is evident because the peak in $\delta(^3\mathrm{He})$ at 350 m depth is absent from the ^{4}He profile, indicating input of pure ^{3}He as would be expected for tritium decay. At the ocean surface $\delta(^3\mathrm{He}) = -1.4\%$, which is very close to the expected value of $\delta(^3\mathrm{He}) = -1.35\%$ for water in equilibrium with air (^{3}He is slightly less soluble in water than ^{4}He).

The absolute ^{4}He concentration (**Figure 2B**) also increases with depth, but not as dramatically as the ^{3}He/^{4}He ratio. Part of the ^{4}He increase is due to the

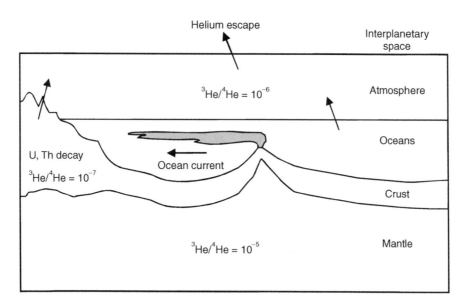

Figure 1 A schematic of the terrestrial helium budget, indicating the flux of helium from the Earth's mantle into the oceans, and in turn into the atmosphere.

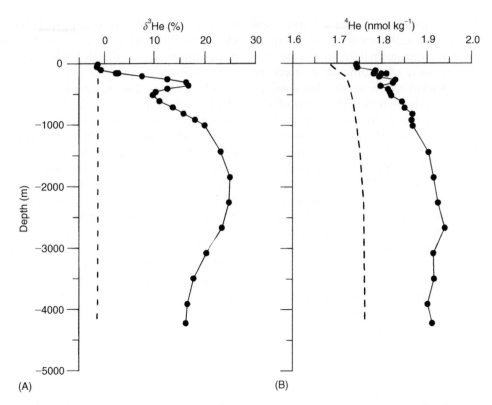

Figure 2 A typical helium profile collected at 28.5°N, 121.6°W in the north Pacific Ocean. (A) The ^{3}He/^{4}He ratio expressed as $\delta(^3\text{He})\%$ plotted versus depth. The sharp peak at $\sim$350 m depth is due to tritium decay, while the broad maximum centered at $\sim$2000 m depth is due to volcanic helium introduced along the mid-ocean ridge system. The dashed line represents the $\delta(^3\text{He})$ for sea water in equilibrium with air. (B) The ^{4}He concentration plotted versus depth for the same samples. The dashed line represents the ^{4}He concentration expected for sea water in equilibrium with air.

higher solubility of helium in the colder deep waters, as shown by the expected solubility values for air-saturated water (dashed line). However, much of the ^{4}He excess above solubility equilibrium is due to the finite amount of ^{4}He present in the volcanic helium signal. At $\sim$2500 m depth, the profile has ^{4}He $= 1.92$ nmol kg^{-1}, about 10% higher than the value of 1.75 nmol kg^{-1} for air-saturated water at those conditions.

The distinct isotopic signature of oceanic volcanic helium can be seen by plotting the ^{3}He concentration versus the ^{4}He concentration as shown in **Figure 3**. In this plot the slope of the trends corresponds to the isotopic ratio of the end-member helium that has been added to the water samples. The thin solid line corresponds to the atmospheric ratio (^{3}He/^{4}He $= 1.39 \times 10^{-6}$ or $R/R_A = 1$), and addition of air would cause the values to migrate along this line. As expected, the range of equilibrium solubility values falls directly on the atmospheric line. Although the measured samples (filled circles) near the ocean surface also fall on this line, the deeper samples fall off the atmospheric trend, defining a much steeper slope. This steeper slope is direct evidence that the helium

that has been added to the deep ocean has a higher ^{3}He/^{4}He ratio than air.

Mid-ocean Ridge Helium

The input of volcanic helium has affected the helium content of all the major ocean basins, although the magnitude of this effect varies greatly. To a large degree, the amount of the excess volcanic helium in each of the ocean basins is controlled by the relative strength of the hydrothermal input, which is in turn roughly proportional to the spreading rate of the ridges. In the Pacific Ocean, where the fastest ridge-crest spreading rates are found, the ^{3}He/^{4}He values at mid-depth average $\delta(^3\text{He}) = 20\%$ for the entire Pacific basin (**Figure 4**). The Indian Ocean, which has ridges spreading at intermediate rates, has $\delta(^3He)$ values averaging about 10–15%. Finally, the Atlantic Ocean, which is bisected by the slow-spreading Mid-Atlantic Ridge, has the lowest ^{3}He enrichments, averaging $\delta(^3\text{He}) = 0–5\%$ (**Figure 4**).

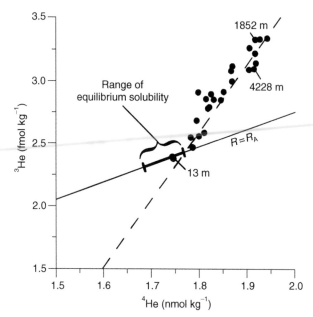

Figure 3 The ^{3}He concentration (in fmol kg^{-1} or 10^{-15} mol kg^{-1}) plotted versus ^{4}He concentration (in nmol kg^{-1} or 10^{-9} mol kg^{-1}) for the samples shown in **Figure 2**. In this plot the slope of any trend corresponds to the isotopic ratio of the end-member helium that has been added to the water samples. The depths in meters of three representative samples are indicated. The thick solid line represents the range of equilibrium solubility values expected for air saturated water (Weiss, 1970; 1971). As expected, the equilibrium solubility values fall on the thin solid line, which is the mixing relation expected for air helium ($R = R_A$). The steep slope of the dashed line, which is a best fit to the sea water samples, indicates that helium with an elevated ^{3}He/^{4}He ratio ($R > R_A$) has been added.

It has been recognized for several decades that the distribution of mantle ^{3}He has great potential for delineating the patterns of circulation and mixing of deep and intermediate water masses. This potential is probably greatest in the Pacific Ocean, because of the strong ^{3}He signal in that ocean. The helium field at mid-depth in the Pacific has been mapped in considerable detail (**Figure 5**). This work has identified several distinct helium plumes emanating from active hydrothermal systems distributed along the mid-ocean ridges. In the eastern equatorial Pacific, two jets of helium-rich water originate at latitude 10°N and at 14°S on the crest of the East Pacific Rise (EPR) and protrude westward into the interior of the basin. Between these two helium jets there is a minimum in the ^{3}He signal on the Equator. This distinct pattern in the helium distribution requires westward transport at mid-depth in the core of these helium plumes, and suggests eastward transport on the Equator (see dashed arrows in **Figure 5**). A separate helium plume is present in the far northeast Pacific produced by input on the Juan de Fuca and Gorda Ridges (JdFR). Although this helium signal is weaker than the helium plumes from the EPR, the JdFR helium is still traceable as a distinct plume that trends south-west into the interior of the north Pacific basin. Farther south at ~20°N, a low-^{3}He tongue penetrates from the west, implying eastward transport at this latitude. Thus the helium field defines a cyclonic (clockwise) circulation pattern at ~2000 m depth in the northeast Pacific.

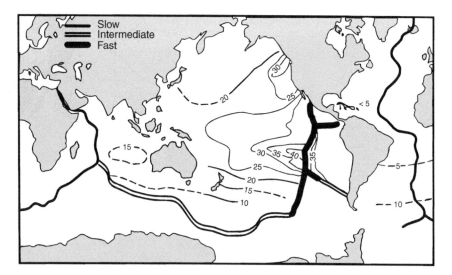

Figure 4 Map of $\delta(^3\text{He})\%$ at mid-depth in the world ocean. The location of the mid-ocean ridges is shown, and the relative spreading rate is indicated by the width of the lines.

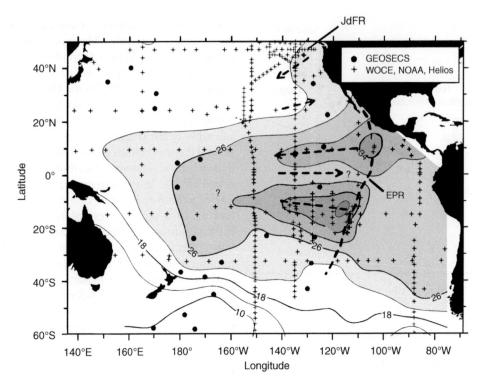

Figure 5 Map of $\delta(^3He)\%$ contoured on a surface at 2500 m depth in the Pacific. Contour interval is 4%. The major helium sources lie along the East Pacific Rise (EPR) and the Juan de Fuca Ridge (JdFR) systems. The dashed arrows indicate areas where the helium plumes define regional circulation patterns. Data along WOCE lines P4 and P6 are from Jenkins (unpublished data). All other data are from Lupton (1998).

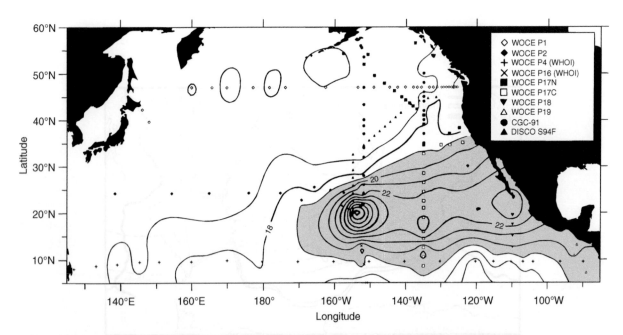

Figure 6 Map of $\delta(^3He)\%$ contoured on a surface at 1100 m depth in the north Pacific, showing the broad lateral extent of a helium plume emanating from Loihi Seamount on the south-eastern flank of the Island of Hawaii. As indicated in the key, data are from several different expeditions.

Hot Spot Helium

In addition to the volcanism along the global mid-ocean ridge system, the oceans are also affected by hot spot volcanoes. Over 100 hot spots have been identified on the Earth's surface, and many of them are located within the ocean basins. One of the best-known examples is the Hawaiian hot spot, which over time has generated the Hawaiian Islands and the Hawaiian-Emperor seamount chain. Unlike mid-ocean ridges, which are submarine, many hot spot volcanoes are subaerial and do not necessarily have direct input into the oceans. Furthermore, hot spot volcanoes have not been explored extensively for their volcanic and hydrothermal activity. Nevertheless, there are several known examples of submarine hydrothermal input at hot spots. Macdonald Seamount in the south Pacific has active vents on its

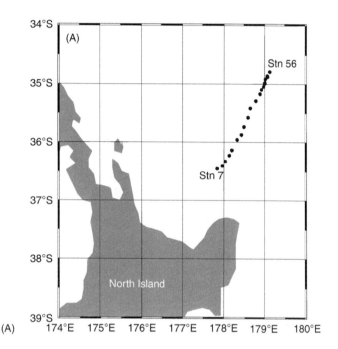

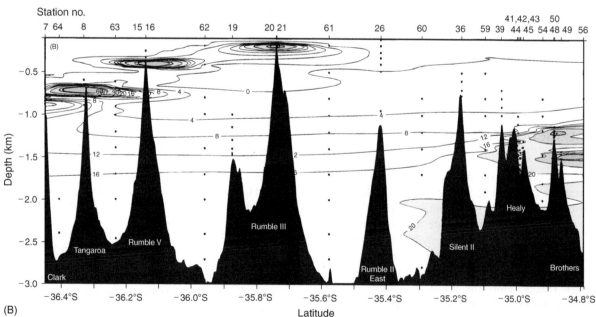

Figure 7 (A) Map showing location of hydrographic stations along the southern end of the Kermadec Arc northeast of New Zealand. (B) $\delta(^3\mathrm{He})\%$ contoured in section view along the southern end of the Kermadec Arc, showing ^{3}He-rich water-column plumes emanating from several of these subduction zone volcanoes. From de Ronde et al. (2000).

summit that inject volatiles into the overlying water column at a depth of $\sim 130\,m$. Loihi Seamount, situated on the south-eastern flank of the island of Hawaii, also has active ventsnear its summit at a depth of $\sim 1000\,m$. Loihi is of considerable importance because it is thought to be the current locus of the Hawaiian hot spot. Loihi lavas and hydrothermal fluids contain helium with a very primitive signature of $R/R_A = 25-30$, indicating a deep mantle origin.

It has been known for some time that hydrothermal venting on Loihi Seamount produces water column plumes that can bedetected with tracers such as temperature, manganese, iron, and methane. However, these tracers are not useful for far-field studies of the Loihi plume because they are either rapidly removed from the water column or arepresent in low concentrations. Because helium is a stable, conservative tracerthat is highly enriched in Loihi vent fluids, the helium signal from Loihi is detectable at considerable distances from the Hawaiian Islands. As shown in **Figure 6**, a map of $\delta(^3He)$ on a surface at $1100\,m$ depth reveals a ^{3}He-rich plume that extends eastward from the Hawaiian Islands for several thousand kilometers, reaching the coast of Mexico at its greatest extent. This far-field plume produced by the Hawaiian hot spot clearly defines an eastward transport at $\sim 1000\,m$ depth in this region ofthe north Pacific. Furthermore, because the end-member helium introduced at Loihi has a ^{3}He/^{4}He ratio three times higher than mid-ocean ridge helium, it should be possible to distinguish the Loihihelium from mid-ocean ridge helium with accurate measurements of ^{3}He and ^{4}He concentrations. The ability to distinguish hot spot helium from mid-ocean ridge helium has been demonstrated for the Loihi helium plume near Hawaii but not yet in the far-field.

Subduction Zone Helium

Submarine volcanism also occurs alongconvergent margins, particularly in regions where two oceanic plates are converging. However, very little is known about the incidence of submarine hydrothermal activity associated with this type of volcanism. Studies of subaerial volcanoes at convergent margins have shown that these volcanoes emit mantle helium with an isotopic ratio of $R/R_A = 3-7$, lower than in mid-ocean ridges. Thus the volcanic helium from subduction zones represents a third type of mantle helium that isisotopically distinct from mid-ocean ridge and hot spot helium.

One clear example of oceanic helium plumesfrom subduction zone volcanism is shown in **Figure 7**, which shows the results of a survey along the southern end of the Kermadec Arc northeast of New Zealand. The Kermadec Arc consists of a series of discrete volcanoes generated by the subduction of the Pacific plate beneath the Australian plate. At the southern end of the arcthese volcanoes are submarine, while farther north some of them are subaerial volcanoes, including Curtis Island, Macauley Island, and Raoul Island. The survey shown in **Figure 7** consisted of a series of hydrographic casts along the arc, and many of the casts were lowered directly over the summits of these arc volcanoes. The section shown in **Figure 7B** shows a series of ^{3}He-rich found at a variety of depths between $150\,m$ for Rumble III volcano down to $1400\,m$ for Brothers volcano. A plot of ^{3}He versus ^{4}He concentration for these samples (notshown), indicated an average ^{3}He/^{4}He ratio of $R/R_A = 6$, in agreement with previous studies of helium from subaerial subduction zone volcanoes. Although the lateral extent of the helium plumes from the Kermadec Arc is not known, this survey confirms that subduction zone volcanoes do produce helium plumes that can be used to trace ocean currents. Furthermore, these subduction zone plumes are potentially quite valuable for tracer studies, since they occur at a wide variety of depths and are generally much shallower than plumes produced at mid-ocean ridges (**Figure 7B**).

See also

Hydrothermal Vent Deposits. Mid-Ocean Ridge Geochemistry and Petrology. Mid-Ocean Ridge Tectonics, Volcanism, and Geomorphology. Propagating Rifts and Microplates. Seamounts and Off-Ridge Volcanism.

Further Reading

Clarke WB, Beg MA, and Craig H (1969) Excess ^{3}He in the sea: evidence for terrestrial primordial helium. *Earth and Planetary Science Letters* 6: 213–220.

Craig H, Clarke WB, and Beg MA (1975) Excess ^{3}He in deep water on the East Pacific Rise. *Earth and Planetary Science Letters* 26: 125–132.

Craig H and Lupton JE (1981) Helium-3 and mantle volatiles in the ocean and the oceanic crust. In: Emiliani C (ed.) The Sea, vol. 7, pp. 391–428. New York: Wiley.

Krylov A Ya, Mamyrin BA, Khabarin L, Maxina TI, and Silin Yu I (1974) Helium isotopes in ocean floor bedrock. *Geokhimiya* 8: 1220–1225.

Lupton JE (1983) Terrestrial inert gases: isotope tracer studies and clues to primordial components in the mantle. *Annual Review of Earth and Planetary Science* 11: 371–414.

Lupton JE (1995) Hydrothermal plumes: near and far field. In: Humphris S et al. (eds.) *Seafloor Hydrothermal Systems, Physical, Chemical, Biological, and Geological Interactions*, Geophysical Monograph Series, vol. 91, pp. 317–346. Washington, DC: American Geophysical Union.

Lupton JE (1998) Hydrothermal helium plumes in the Pacific Ocean. *Journal of Geophysical Research* 103: 15855–15868.

Lupton JE and Craig H (1975) Excess ^{3}He in oceanic basalts: evidence for terrestrial primordial helium. *Earth and Planetary Science Letters* 26: 133–139.

Suess HE and Wänke H (1965) On the possibility of a helium flux through the ocean floor. *Progress in Oceanography* 3: 347–353.

Weiss RF (1970) Helium isotope effect in solution in water and seawater. *Science* 168: 247–248.

Weiss RF (1971) Solubility of helium and neon in water and seawater. *Journal of Chemical Engineering Data* 16: 235–241.

GEOPHYSICAL MEASUREMENTS

GEOPHYSICAL HEAT FLOW

C. A. Stein, University of Illinois at Chicago, Chicago, IL, USA
R. P. Von Herzen, Woods Hole Oceanographic Institution, Woods Hole, MA, USA

Introduction

Heat flow is a directly measurable parameter at Earth's surface that provides important constraints on the thermal state of its interior. When combined with other data, it has proved useful for models of Earth's origin, structure, composition, and convective motion of at least the upper mantle. Because oceanic crust is mostly thin (about 5–8 km) and deficient in heat-producing radioactive elements compared to continental crust, marine heat flow measurements are particularly illuminating for the cooling and dynamics of the oceanic lithosphere. Indeed, they have provided an important constraint in the development and refinement of the plate tectonics paradigm over the past three decades.

Initially influenced by concepts of Sir Edward Bullard, the first successful marine heat flow measurements obtained shortly after the end of World War II were among the first applications of remote electronics in the deep sea. Other than a general understanding of seafloor topography, little information on the nature of the Earth beneath the oceans was then available, so theories to explain the measurements were largely unconstrained. Probably the most important initial finding was that the mean oceanic heat flux was not much different from the mean of the available continental values, contrary to conventional expectations of that era. However, the entire field of marine geophysics (especially seismic reflection and refraction, and geomagnetism) was also being transformed by instrumentation development, and the availability of ships as seagoing platforms gave rise to a 'golden age' of ocean exploration between about 1950 and 1970. The large-scale (>1000 km) variations of marine heat flux were shown to be consistent with the plate tectonic paradigm that explained the mean depth and age of much of the ocean floor. Heat flow is now useful for studies of seafloor tectonics and other marine geophysical problems.

Marine heat flow measurements now number more than 15 000, and with the exception of high latitude regions, are distributed over most of Earth's main ocean basins (**Figure 1**). Values range over about two orders of magnitude (about 10–1000 mW m^{-2}), with a general systematic spatial distribution at large (>1000 km) scales but also a large variability at small (<50 km) scales caused by pervasive hydrothermal circulation in mostly young (<50–70 million years (My)) ocean crust. The spatial distribution and magnitudes of marine heat flow values may be used to deduce the geometry and intensity of such circulation.

Measurements and Techniques

Instrumentation and Development

Techniques for marine measurements were developed with the realization that the temperature of the deep (>2 km) ocean is mostly horizontally stratified, and relatively constant over long time scales (>100 years). In this situation, the heat flux from the Earth's interior is reflected in a relatively uniform temperature gradient with depth beneath a flat seafloor. Heat flow is the product of the magnitude of this temperature gradient and the thermal conductivity of the material over which it is measured. Gradients are commonly measured using a vertically oriented probe with temperature sensors surmounted by a weight that is lowered from a ship to penetrate the relatively soft sediments that cover most of the seafloor. Thermal conductivity is either measured with *in situ* sensors during the seafloor penetration by transient heating experiments or aboard ship on sediment cores.

For instrumental simplicity, the initial heat flow equipment used only two thermal sensors at either end of a 3–4 m long probe with *in situ* analog recording techniques on paper or film. Subsequently the number of sensors has been increased, due to a desire to measure non-linear temperature gradients and thermal conductivity that may vary with depth. The wide range of measured thermal gradients and the developments in solid-state electronics have made digital recording the present standard. 'Pogo' measurements, i.e., multiple penetrations on the same station, combined with nearly real-time acoustic telemetry of raw data have proven useful for investigation of small-scale (<10–20 km) marine heat flow variability, the origins of which are still not fully understood.

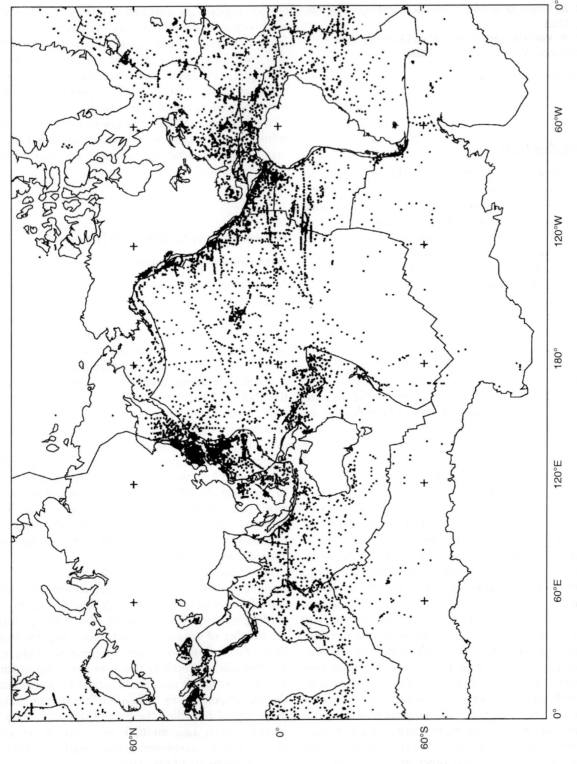

Figure 1 Locations of marine heat flow measurements (dots). Thin lines show the location of the major plate boundaries.

It was necessary for the 'Bullard' probe to remain undisturbed in the seafloor for most of one hour to approach thermal equilibrium with the sediments because of its relatively large diameter (about 2–3 cm). This probe has been largely supplanted by either much smaller (about 3 mm diameter) individual probes mounted in outrigged fashion on a larger ('Ewing') probe to attain deep (5–6 m) penetration, or a 'violin-bow (Lister)' probe consisting of a small (about 1 cm) sensor string mounted parallel to, but separate from, the main strength member (**Figure 2**). Both are capable of *in situ* thermal conductivity measurements, utilizing either a constant heat source or a calibrated pulse (approximating a delta function) after the gradient measurement. The measurement time in the seafloor is 15–20 min. Depending mostly on the desired spacing during pogo operations, the mean time between measurement is 1–2 h. Even with

acoustic data telemetering, battery life can be 2–3 days, thereby allowing many measurements during a single lowering. Real-time ship location accuracy now approaches a few meters with differential Global Positioning System navigation, although the uncertainty of the probe location during pogo operations in normal ocean depths (4–5 km) is typically 200 m or more. Hence 1–2 km is a useful minimum spacing between pogo penetrations unless seafloor acoustic transponder navigation is employed for higher accuracy (about 10 m) navigation. Instrumentation is also available to determine temperatures to depths up to about 600 m below the seafloor during deep-sea drilling to establish the uniformity of heat flow to greater depths, and may be the only method for reliable measurements in shallow seafloor (< 1 km) regions where ocean temperatures are more variable. Geothermal probes have also been developed for use

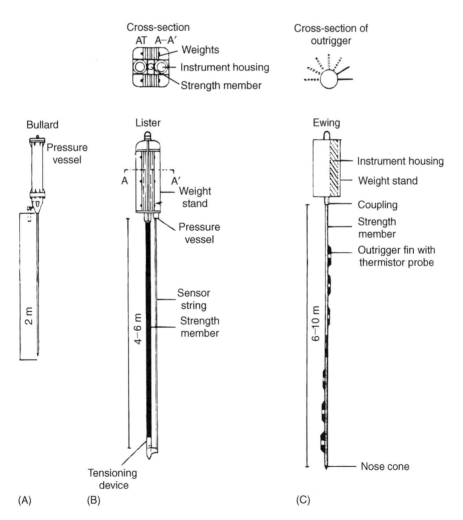

Figure 2 Diagrams of the three most commonly used marine heat flow probes. (Reproduced with permission from Louden KE and Wright JA (1989) Marine heat flow data: a new compilation of observations and brief review of its analysis. In: Wright JA and Londen KE (eds) *Handbook of Seafloor Heat Flow*, pp. 2–67, Boca Raton: CRC Press.)

with manned submersibles, allowing visual control of measurements in regions with large lateral gradients of heat flux or over specific geological features. A development is underway to install geothermal instrumentation on an autonomous underwater vehicle (ABE) for measurements in regions that are remote or difficult for normal ship operations.

Environmental Corrections

Although most measurements on the deep seafloor with the usual instrumentation do not require corrections, they may be needed for some regions with unusual environmental parameters. As mentioned above, shallow water measurements may be subject to temporal bottom water temperature (BWT) variability, causing non-linear gradients in the seafloor. Using heat conduction theory, corrections may be applied if BWT variability is monitored for a sufficient period (months to years if possible) before the geothermal measurements. Conversely, nonlinear temperature–depth profiles can be inverted to yield BWT history assuming that the initial gradient was linear. This inverse procedure is non-unique, although closely spaced measured temperatures to a sufficient depth below the seafloor and well-determined thermal conductivity can reduce uncertainties. Continental margins and some deep-sea trenches are examples of regions where non-linear gradients have been measured, usually correlated with strong and variable deep currents that are probably focused by the seafloor topography. Topography causes lateral heat flow variability even with uniform and constant BWT, because the seafloor topography distorts isotherms that would otherwise be horizontal below a flat surface. Seafloor sedimentation reduces the heat flow measured because recently deposited sediments modify the seafloor boundary condition to which the equilibrium gradient must adjust. Corrections usually become significant when sedimentation rates exceed a few tens of meters per million years and/or the total sediment thickness exceeds 1 km.

Vertical pore water flow in sediments may either enhance (for upward flow) or reduce (for downward flow) temperature gradients because the water advects some of the heat otherwise conducted upwards. The effects become significant for flow rates greater than a few centimeters per year, and nonlinear gradients may be expected if upward flow rates exceed about 10 cm year^{-1}. It is unusual for rates on normal deep seafloor to exceed the latter value because the fluid permeability of pelagic sediments is too low, but coarser and sometimes rapidly deposited sediments

of continental margins may support relatively rapid flows.

Range of Measured Parameters

As discussed later, the plate tectonic paradigm predicts that the highest thermal gradients and heat fluxes should be found in the youngest seafloor, and the lowest values in the oldest. Although this is generally true when measurements are averaged over regions of various seafloor ages, considerable variability occurs over the youngest seafloor as a result of vigorous hydrothermal circulation. The cooling of hot and permeable upper ocean crust supports seawater convection in the crust over lateral circulation scales of at least several to a few tens of kilometers; vertical scales probably do not exceed a few kilometers. The highest gradients and heat fluxes (up to 100°C m^{-1} and 100 W m^{-2}, respectively) are measured in sediments near seafloor vents where fluids upwell, and the lowest (<0.005°C m^{-1} and 0.005 W m^{-2}, respectively) where fluids downwell. Mean upper basement fluid velocities may be several meters per year, and much greater in conduits that support vigorous seafloor venting. The detailed pore water flow and permeability structure for any system of seafloor hydrothermal circulation have not yet been investigated in detail and are probably very complex.

After the seafloor ages to 50–70 My, hydrothermal circulation appears to decrease to a level where modulation of the surface heat flux is small to negligible (**Figure 3**), probably dependent on the thickness and uniformity of sediment cover. For the oldest seafloor (100–180 My), heat flux is relatively uniform at about 50 mW m$^{-2} \pm 10\%$. This value probably reflects the secular cooling of the upper oceanic mantle.

Thermal conductivity of marine sediments generally varies less than a factor of two, from about 0.7 to <1.4 W m^{-1} K^{-1}. The lowest values are associated with red clays or siliceous oozes, which have up to 70% or more of water by weight and the highest with carbonate oozes or coarse sediments near continental margins with a high proportion of quartz. Since water has a low thermal conductivity (0.6 W m^{-1} K^{-1}), sediments with a large percentage of water have low thermal conductivity. Since calcium carbonate and quartz have high thermal conductivity (about 3 W m^{-1} K^{-1} and 4 W m^{-1} K^{-1}, respectively), sediments with a large percentage of either of these minerals have high thermal conductivity. Conductivity variations with depth may be caused by: (1) turbidity flows that sort grain sizes

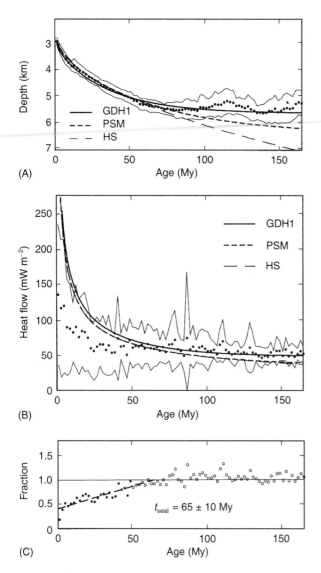

Figure 3 Data and models for ocean depth (A) and heat flow (B) as a function of age. (Reproduced with permission from Stein and Stein, 1992.) The data are averaged in 2 My bins, and one standard deviation about the mean value for each is shown by the envelope. Shown are the plate model of Parsons and Sclater (1977) (PSM), a cooling halfspace model with the same thermal parameters (HS), and the GDH1 plate model from Stein and Stein (1992). (C) Heat flow fraction (observed heat flow/GDH1 prediction) with age averaged in 2 My bins. The discrepancy for ages <50–70 My presumably indicates the fraction of the heat transported by hydrothermal flow. The fractions for ages <50 My (closed circles), which were not used in deriving GDH1, are fit by a least squares line. The sealing age, where the line reaches the fractional value of one, is 65 ± 10 My.

thermal conductivity varies rapidly with depth the temperature–depth profile may not be linear. If thermal conductivity has been measured at closely spaced depth intervals, then the heat flow can be calculated using a method introduced by Bullard. This approach assumes the absence of significant heat sources or sinks and one-dimensional, steady-state, conductive heat flow. Thus there is a linear relationship between temperature and the thermal resistance of the sediments (the sum from the surface to the depth of the temperature measurement of the inverse of thermal conductivity times the sediment thickness for that given thermal conductivity). The slope determined from a least-squares fit of this 'Bullard plot' gives the appropriate conductive heat flux.

Thermal Models

Data for Thermal Modeling

Oceanic lithosphere forms at midocean ridges, where hot magma upwells, and then cools to form plates as the material moves away from the spreading center. As the plate cools, heat flow decreases and the seafloor deepens (**Figure 3**). However, only shallow (less than 1 km) measurements of lithospheric temperatures are possible. Hence, the two primary data sets used to constrain models for the variation in lithospheric temperature with age are seafloor depths and heat flow. The depth, corrected for sediment load, depends on the temperature integrated over the lithospheric thickness. The heat flow is proportional to the temperature gradient. Initially seafloor depths rapidly increase, with the average increase relative to the ridge crest proportional to the square root of the crustal age. However, for ages greater than about 50–70 My, the average increase in depth is slower and the curve is said to 'flatten.' Mean heat flow also decreases rapidly away from the ridge crest, with values approximately proportional to the inverse of the square-root of the age, but after about 50 My this curve also 'flattens.'

Halfspace and Plate Models

Two different mathematical models are often used to describe the thermal evolution of the oceanic lithosphere, the halfspace (or boundary layer) and plate models. For the halfspace model, the predicted lithospheric thickness increases proportionally with the square root of age. Hence depth and heat flow vary as the square root of age and the reciprocal of the square root of age, respectively. However, the halfspace model cannot explain the 'flattening' of the curves. Alternatively, the plate model represents the lithosphere as a layer with a fixed constant

into layers with different proportions of pore water; (2) ice rafting in higher latitudes that may intersperse higher conductivity sediments derived from the continents and lower conductivity marine pelagic sediments; and (3) variability in the proportion of calcium carbonate in sediments near the equator deposited over climate (e.g., glacial) cycles. When temperature at its base. Initially, the cooling for the plate model is the same as the halfspace model, but for older ages the influence of the lower boundary results in slower cooling, approximately predicting the observed flattening of the depths and heat flow at older ages. The plate base is assumed to represent a depth at which additional heat is supplied from the mantle below to prevent the halfspace cooling at older ages. However, the model does not directly describe how this heat is added.

Reference Models

Thermal models are solutions to the inverse problem of finding the temperature as a function of age that best fits the depth and heat flow. The data are used to estimate the primary model parameters (typically plate thickness, basal temperature, and thermal expansion coefficient), subject to other parameters generally specified *a priori*. The global average depth/heat flow variation with age, calculated for a specific model, data, and parameters is used as a reference model to represent 'normal' oceanic lithosphere. 'Anomalies', deviations from the reference model, are then investigated to see if they reflect a significant difference in thermal (or other) processes from the global average. It is important to realize that anomalies (defined as those observed minus the model-predicted values) for various reference models may differ significantly, and hence lead to different tectonic inferences.

Until recently, a 125 km-thick plate model by Parsons and Sclater (denoted here PSM) was commonly used. Subsequently, as more data became available, it was noted that PSM systematically overpredicts depths and underpredicts heat flow for lithosphere older than 70–100 My, causing widespread 'anomalies.' A later joint inversion of the depth and heat flow data by Stein and Stein found that these 'anomalies' are reduced significantly by a plate model termed GDH1. GDH1 has a thinner lithosphere (95 ± 10 km), and a basal temperature of $1450 \pm 100°$C, consistent with the PSM estimate ($1350 \pm 275°$C). GDH1 predicts heat flow in mW m^{-2} as a function of age, t, in millions of years equal to $510 \ t^{-1/2}$ for ages less than about 55 My, and $48 + 96 \exp(-0.0278 \ t)$ for older ages. Inversion of the same data, while prescribing a basal temperature of $1350°$C (a typically assumed temperature for upwelling magma at the ridge based on results from experimental petrology), also yields a thin (100 km) plate, with a somewhat higher value of lithospheric conductivity and hence makes very similar predictions. As the quality of the observed data and our understanding of the physics of processes affecting lithospheric thermal evolution improve, new and better reference models will be developed.

Application to Lithospheric Processes

Using a reference model for the expected heat flow, regions can be examined to observe if the measured heat flow differs from that predicted and, if so, to study the causes of the discrepancy. Two primary discrepancies are assumed to reflect hydrothermal circulation and midplate swells.

Hydrothermal Circulation

Heat flow measurements for crust of ages 0–65 My are generally lower than the predictions of all commonly used reference models (**Figure 3**). Because the heat flow measurements primarily reflect conductive heat flow, this difference has been attributed to hydrothermal water flow in the crust and sediments transporting some of the heat assumed in thermal models to be transferred by conduction. The missing heat transported by convection must appear somewhere, either as high conductive heat flow or as advective discharge to the sea. It is estimated that of the predicted global oceanic heat flux of 32×10^{12} W,

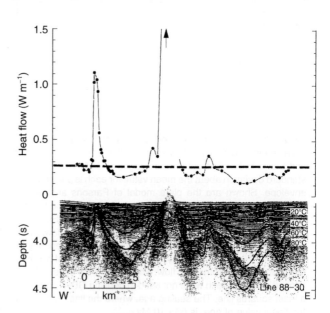

Figure 4 Portion of the heat flow data from the FlankFlux survey on the Juan de Fuca Ridge sites where bare rock penetrates the sediment column have dramatically higher heat flow than the surrounding areas. Away from these outcrops, heat flow varies inversely with depth to basement rock. Expected heat flow for its age is shown with the dashed line. (Reproduced with permission from Davis EE, Chapman DS, Mottl MJ *et al.* (1992) FlankFlux: an experiment to study the nature of hydrothermal circulation in young oceanic crust. *Canadian Journal of Earth Sciences* 29: 925–952.)

approximately one third occurs by hydrothermal flow.

Characteristic heat flow patterns in young crust appear associated with water flow. Surveys at sites like the Galapagos Spreading Center show high heat flow associated with presumed upwelling zones at basement highs or fault scarps, and low heat flow associated with presumably down-flowing water at basement lows. Heat flow over the highs may be two to ten (or more) times greater than at nearby sites where water may be recharging the system (**Figure 4**). Closely spaced surveys in young crust show high scatter, presumably in part due to the local variations in sediment distribution, basement relief, and crustal permeability.

The most spectacular evidence for hydrothermal circulation are vents near the ridge crest where hydrothermally altered water at temperatures up to 400°C exits the seafloor. Hot rock or magma at shallow depths provides a large heat source for the flow. The high sulfide content of the hot fluids supports unusual local biota from bacteria to tube worms. Vent areas show complex flow patterns with very high heat flow close to low heat flow. Although most vigorous vents are near spreading centers, on older crust isolated crystalline crust outcrops surrounded by a well-sedimented area can also vent measurable amounts of hydrothermal fluid (e.g., the FlankFlux survey area on the Juan de Fuca plate).

The persistence of the heat flow discrepancy well away from ridges indicates that hydrothermal heat loss occurs in older crust. The 'sealing age,' defined as that beyond which measured heat flow approximately equals that predicted, is presumed to indicate the near cessation of heat transfer by hydrothermal circulation. The sealing age for the entire global data set is about 65 ± 10 My. Hence, although there are presumably local variations, water flow in older crust seems not, in general, to transport significant amounts of heat. However, some heat flow surveys (e.g., in the north-west Atlantic Ocean on 80 My crust and the Maderia Abyssal Plain on 90 My crust) indicate that hydrothermal circulation may continue, even if relatively little heat is lost by convection into the sea water. The reasons for the 'sealing' are not well understood. An earlier view was that 100–200 m of sediment would be sufficiently impermeable to 'seal' off the crust from the sea, so that heat flow at these sites would yield a 'reliable' value, i.e., that predicted by conduction-only thermal models. However, many such sites have lower than expected values. A simple way to reconcile these observations with the present understanding of seafloor hydrology to assume that if water cannot flow vertically through thick sediments at a particular site, it may flow laterally to a fault or basement outcrop, and then be manifested as either high conductive heat flow or hot water exiting to the sea, such as observed for the FlankFlux area. It has been suggested that the porosity and permeability of the crust decrease with increasing age, thus significantly reducing the water flow. However, recent studies of permeability and seismic velocity suggest that most of the rapid change occurs within the first 5–15 My and that older crust may still retain some relatively permeable pathways.

Hydrothermal circulation has profound implications for the chemistry of the oceanic crust and sea water, because sea water reacts with the crust, giving rise to hydrothermal fluid of significantly different composition. The primary geochemical effects are thought to result from the high-temperature water flow observed at ridge axes. Circulation of sea water through hot rock removes magnesium and sulfate from sea water and enriches calcium, potassium, silica, iron, manganese, and other elements within the hydrothermal solution. Estimating the volume of water flowing through the crust depends on the heat capacity of the water, the heat lost by convection, and the assumed temperature of the water, the latter having the largest uncertainties. Although the heat flow anomaly is greatest in young lithosphere, only about 30% of the hydrothermal heat loss and 7% of the water flow occurs in crust younger than 1 My. This effect should be significant for ocean chemistry because a water volume equivalent to that of the total ocean is estimated to circulate through oceanic crust about once every 0.5–5 My.

Hot Spots

Oceanic midplate swells are identified by seafloor depths shallower than expected for their lithospheric age. Thus, models of the processes giving rise to these regions rely on assessments of how their heat flow and other properties differ from unperturbed lithosphere. The origin of these swells is generally thought to be related to upwelling mantle plumes (hot spots) that result in uplift and volcanism. Two basic types of models have been proposed for their origin. In one, the swell is a thermal effect due to the hot spot thinning and heating the lithosphere at depth. In the second, the uplift is primarily due to the dynamic effects of the upwelling plume, which may largely reflect thermal buoyancy forces within the upwelling mantle. The thermal models predict significantly larger heat flow anomalies than do the dynamic models. Detailed heat flow measurements have been made for Hawaii, Cape Verde, Reunion, Bermuda, and Crozet swells. Relative to the PSM reference curve, large heat flow anomalies are suggested.

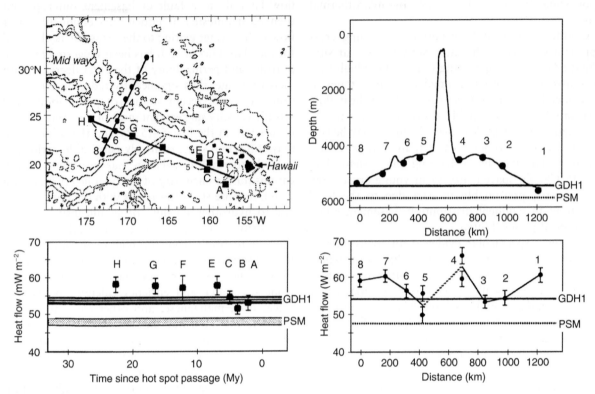

Figure 5 Heat flow data for transects along (lower left) and across (lower right) the Hawaiian Swell at the locations shown (upper left). The heat flow, though anomalously high with respect to the Parsons and Sclater (PSM) model, is at most slightly above that expected for GDH1. The predicted heat flow values are for 100 My (lower right) and 95–110 My (lower left). Figure modified from Von Herzen *et al.* (1982, 1989). (Reproduced with permission from Stein CA and Stein S (1993) Constraints of Pacific midplate swells from global depth-age and heat flow-age models. *The Mesozoic Pacific, American Geophysical Union Monograph* 77: 53–76.)

However, relative to both measured heat flow from off-swell lithosphere and the GDH1 reference curve, smaller heat flow anomalies (less than 5–8 mW m^{-2}) are deduced, supporting the idea of a primary dynamic mechanism for hotspots (**Figure 5**).

Application to Marine Margin Studies

Subduction Zones

Many extensive marine heat flow surveys have been done near the Japanese subduction zones. On average, observed heat flow from the trench axis to the forearc area is lower than that characteristic of the crust seaward of the trench. However, heat flow is higher and more variable over the volcanic arc and back arc region compared to the area seaward of the trench. Recent surveys for subduction zones, including Barbados, Nankai, and Cascadia, show that heat flow is highly variable. Within accretionary prisms, high values are often associated with upward advection of pore fluids, typically found along faults and the bottom decollement. Although they have not been

investigated extensively, the active nonaccretionary forearcs seem to have the lowest mean heat fluxes (20–30 mW m^{-2}). Relatively low heat flow values over the subducting lithosphere are the thermal consequence of the subduction of one plate beneath another. Overall, heat flow depends on the age, rate, geometry, and thermal structure of the subducting plate and the sediment thickness and deformation history of the region. Heat flow in marginal basins behind the volcanic arcs is often high. Many back arc basins with high heat flow appear to have formed by back-arc spreading processes similar to those at midocean ridges, within the last 50 My.

Passive Continental Margins

Passive continental margins form after continental crust is rifted and seafloor spreading occurs. Because rifting heats the lithosphere, heat flow data can be used to constrain models of rifting. Subsequent to rifting, margins slowly subside as cooling occurs. Simple models for this process suggest that subsequent to rifting the additional heat will almost completely dissipate within 100 My. Although the amount of radioactive heating from continental crust may introduce some variability, older passive margins typically have heat flows similar to old oceanic crust, whereas young margins, such as in the Red Sea, have high heat flow.

Gas Hydrates

Gas hydrates, solid composites of biogenically derived gasses (mainly methane) combined with water ice, may be present in marine sediments, especially at continental margins, under the correct pressure and temperature conditions. Gas hydrates are detected by direct sampling, and inferred from seismic reflection data when a strong bottom-simulating reflector (BSR) is produced by the seismic velocity contrast between the gas hydrate and the sediment below. Given a depth (and thus pressure) of a bottom-simulating reflector, its temperature can be predicted from known relationships and a heat flow calculated. Alternatively, heat flow data can be used to estimate the bottom-simulating reflector temperature. The volume of hydrocarbons contained in marine gas hydrates is large. Changes in eustatic sea level (hence pressure) or bottom seawater temperatures could result in their release and thus increase greenhouse gasses and affect global climate.

Further Reading

Humphris S, Mullineaux L, Zierenberg R, and Thomson R (eds.) (1995) *Seafloor hydrothermal systems, physical, chemical, biological, and geological interactions, Geophysical Monographs*, 91, 425–445.

Hyndman RD, Langseth MG, and Von Herzen RP (1987) Deep Sea Drilling Project geothermal measurements: a review. *Review of Geophysics* 25: 1563–1582.

Langseth MG, Jr and Von Herzen RP (1971) Heat Sow through the Soor of the world oceans. In: Maxwell AE (ed.) *The Sea*, vol. IV, part 1, pp. 299–352. New York: Wiley-Interscience.

Lowell RP, Rona PA, and Von Herzen RP (1995) Seafloor hydrothermal systems. *Journal of Geophysical Research* 100: 327–352.

Parsons B and Sclater JG (1977) An analysis of the variation of ocean floor bathymetry and heat flow with age. *Journal of Geophysical Research* 82: 803–827.

Pollack HN, Hurter SJ, *et al.* (1993) Heat flow from the earth's interior: analysis of the global data set. *Review of Geophysics* 31: 267–280.

Stein CA and Stein S (1992) A model for the global variation in oceanic depth and heat flow with lithospheric age. *Nature* 359: 123–129.

Von Herzen RP (1987) Measurement of oceanic heat flow. In: Sammis C and Henyey T (eds.) *Methods of Experimental Physics – Geophysics*, vol. 24, Part B, pp. 227–263. London: Academic Press.

Wright JA and Louden KE (eds.) (1989) *CRC Handbook of Seafloor Heat Flow*. Boca Raton: CRC Press.

GRAVITY

M. McNutt, MBARI, Moss Landing, CA, USA

Introduction

The gravity field varies over the oceans on account of lateral variations in density beneath the ocean surface. The most prominent anomalies arise from undulations on density interfaces, such as occur at the water–rock interface at the seafloor or at the crust–mantle interface, also known as the Moho discontinuity. Because marine gravity is relatively easy to measure, it serves as a remote sensing tool for exploring the earth beneath the oceans. The interpretation of marine gravity anomalies in terms of the Earth's structure is highly nonunique, however, and thus requires simultaneous consideration of other geophysically observed quantities. The most useful auxiliary measurements include depth of the ocean from echo sounders, the shape of buried reflectors from marine seismic reflection data, and/or the density of ocean rocks as determined from dredge samples or inferred from seismic velocities.

Depending on the spatial wavelength of the observed variation in the gravity field, marine gravity observations are applied to the solution of a number of important problems in earth structure and dynamics. At the very longest wavelengths of 1000 to 10 000 km, the marine gravity field is usually combined with anomalies over land to infer the dynamics of the entire planet. At medium wavelengths of several tens to hundreds of kilometers, the gravity field contains important information on the thermal and mechanical properties of the lithospheric plates and on the thickness of their sedimentary cover. At even shorter wavelengths, the field reflects local irregularities in density, such as produced by seafloor bathymetric features, magma chambers, and buried ore bodies. On account of the large number of potential contributions to the marine gravity field, modern methods of analysis include spectral filtering to remove signals outside of the waveband of interest and interpretation within the context of models that obey the laws of phyics.

Units

Gravity is an acceleration. The acceleration of gravity on the earth's surface is about $9.81 \, \text{m s}^{-2}$.

Gravity anomalies (observed gravitational acceleration minus an expected value) are typically much smaller, about 0.5% of the total field. The SI-compatible unit for gravity anomaly is the gravity unit (gu): $1 \, \text{gu} = 10^{-6} \, \text{m s}^{-2}$. However, the older c.g.s unit for gravity anomaly, the milligal (mGal), is still in very wide use: $1 \, \text{mGal} = 10 \, \text{gu}$. Typical small-scale variations in gravity over the ocean range from a few tens to a few hundreds of gravity units. Lateral variations in gravitational acceleration (gravity gradients) are measured in Eötvös units (E): $1 \, \text{E} = 10^{-9} \, \text{s}^{-2}$. Another quantity useful in gravity interpretation is the density of earth materials, measured in kg m^{-3}. In the marine realm, relevant densities range from about $1000 \, \text{kg m}^{-3}$ for water to more than $3300 \, \text{kg m}^{-3}$ for mantle rocks.

A close relative of the marine gravity field is the marine geoid. The geoid, measured in units of height, is the elevation of the sea level equipotential surface. Geoid anomalies are measured in meters and are the departure of the true equipotential surface from that predicted for an idealized spheroidal Earth whose density structure varies only with radius. Geoid anomalies range from 0 to more than $\pm 100 \, \text{m}$. The direction of the force of gravity is everywhere perpendicular to the geoid surface, and the magnitude of the gravitational attraction is the vertical derivative of the geopotential U (eqn [1]).

$$g = -\frac{\partial U}{\partial z} \qquad [1]$$

Geoid height N is related to the same equipotential U via Brun's formula (eqn [2]).

$$N = -\frac{U}{g_0} \qquad [2]$$

in which g_0 is the acceleration of gravity on the spheroid.

For a ship sailing on the sea surface (the equipotential), it is easier to measure gravity. From a satellite in free-fall orbit high over the Earth's surface, radar altimeters can measure with centimeter precision variations in the elevation of sea level, an excellent approximation to the true geoid that would follow the surface of a motionless ocean. Regardless of whether geoid or gravity is the quantity measured directly, simple formulas in the wave-number domain allow gravity to be computed from geoid and vice versa. Given the same equipotential, the gravity representation emphasizes the power in the high-frequency (short-wavelength) part of the spectrum, whereas the geoid representation emphasizes the longer wavelengths. Therefore, for investigations of high-frequency phenomena, gravity is generally the quantity interpreted even if geoid is what was measured. The opposite is true for long-wavelength phenomena.

Measurement of Marine Gravity

Marine gravity measurements can be and have been acquired with several different sorts of sensors and from a variety of platforms, including ships, submarines, airplanes, and satellites. The ideal combination of sensor system and platform depends upon the needed accuracy, spatial coverage, and available time and funds.

Gravimeters

The design for most marine gravimeters is borrowed from their terrestrial counterparts and are either absolute or relative in their measurements. Absolute gravimeters measure the full acceleration of gravity g at the survey site along the direction of the local vertical. Modern marine absolute gravimeters measure precisely the vertical position z of a falling mass (e.g., a corner cube reflector) as a function of time t in a vacuum cylinder using laser interferometry and an atomic clock. The acceleration is then calculated as the second derivative of the position of the falling mass as a function of time (eqn [3]).

$$g = \frac{\mathrm{d}^2 z(t)}{\mathrm{d}t^2} \qquad [3]$$

Absolute gravimeters tend to be larger, more difficult to deploy, costlier to build, and more expensive to run than relative gravimeters, and thus are only used when relative gravimeters are inadequate for the problem being addressed.

Most gravity measurements at sea are relative measurements, Δg: the instrument measures the difference between gravity at the study site and at another site where absolute gravity is known (e.g., the dock where the expedition originates). Modern relative gravimeters are based on Hooke's law for the force F required to extend a spring a distance x (eqn [4]), where k_s is the spring constant, calculated by extending the spring under a known force.

$$F = -k_s x = mg \qquad [4]$$

If a mass m is suspended from this spring at a site where gravity g is known (e.g., by deploying an absolute gravimeter at that base station), then gravity

at other locations can be calculated by observing how much more or less that same spring is stretched at other locations by the same mass. Although such systems are relatively inexpensive to build and easy to deploy, they suffer from drift: in effect, the spring constant changes with time because no physical spring is perfectly elastic. To first order, the drift can be corrected by returning to the same or another base station with the same instrument, and assuming that the drift was linear with time in between. The accuracy of this linear drift assumption improves with more frequent visits to the base station, but this is usually impractical for marine surveys. Through clever design, the latest generation of marine gravimeters has greatly reduced the drift problem as compared with earlier instruments.

The measurement of the gravity gradient tensor was widely used early in the twentieth century for oil exploration, but fell into disfavor in the 1930s as scalar gravimeters became more reliable and easy to use. Gravity gradiometry at sea is currently making a comeback as the result of declassification of military gradiometer technology developed for use in submarines during the Cold War. Gravity gradiometers measure the three-dimensional gradient in the gravity vector using six pairs of aligned gravimeters, with accuracies reaching better than 1 Eötvös. In comparison with measurements of gravity, the gravity gradient has more sensitivity to variations at short wavelenghts (~5 km or less), making it useful for delimiting shallow structures buried beneath the seafloor.

Geoid anomalies can be directly measured from orbiting satellites carrying radar altimeters. The altimeters measure the travel time of a radar pulse from the satellite to the ocean surface, from which it is reflected and bounced back to the satellite. Tracking stations on Earth solve for the position of the satellite with respect to the center of the Earth. These two types of information are then combined to calculate the height of the sea level equipotential surface above the center of the Earth. Because the solid land surface does not follow an equipotential, altimeters cannot be used to constrain the terrestrial geoid. Furthermore, it is difficult to extract geoid from ocean areas covered by sea ice. However, in the near future, laser altimeters deployed from satellites hold the promise of extracting geoid information even over ocean surfaces marred with sea ice, on account of their enhanced resolution.

Platforms

Marine gravity data can be acquired either from moving or from stationary platforms. Because the gravity field from variations in the depth of the sea floor is such a large component of the observed signal, most marine gravity surveys have relied on ships or submarines that enable the simultaneous acquisition of depth observations. However, airborne gravity measurements have been acquired successfully over ice-covered areas of the polar oceans, and orbiting satellites have measured the marine geoid from space.

A major challenge in acquiring gravity data from a floating platform at the sea surface is in separating the acceleration of the platform in the dynamic ocean from the acceleration of gravity. This problem is overcome by mounting marine gravimeters deployed from ships on inertially stabilized tables. These tables employ gyroscopes to maintain a constant attitude despite the pitching and rolling of the ship beneath the table. The nongravitational acceleration is somewhat mitigated by mounting marine gravimeters deep in the hold and as close to the ship's center of motion as possible. Special damping mechanisms also prevent the spring in the gravimeter from responding to extremely high-frequency changes in the force on the suspended mass.

Instruments deployed in submersibles resting on the bottom of the ocean or in instrument packages lowered to the bottom of the ocean do not suffer from the dynamic accelerations of the moving ocean surface, but bottom currents can also be an important source of noise in submarine gravimetry. Installing instruments in boreholes is the most effective way to counter this problem, but it is also an expensive solution.

Reduction of Marine Data

A number of standard corrections must be applied to the raw gravity data (either g or Δg) prior to interpretation. In addition to any drift correction, as mentioned above for relative gravity measurements, a latitude correction is immediately applied to account for the large change in gravity between the poles and the Equator caused by Earth's rotation. Near the Equator, the centrifugal acceleration from the Earth's spin is large, and gravity is about 50 000 gu less, on average, than at the poles. Because this effect is 5000 times larger than typical regional gravity signals of interest, it must be removed from the data using a standard formula for the variation of gravity g_0 on a spheroid of revolution best fitting the shape of the Earth (eqn [5]; $\Theta =$ latitude).

$$g_0(\Theta) = 9.7803185\left(1 + 5.278895 \times 10^{-3}\sin^2\Theta \right. \\ \left. + 2.3462 \times 10^{-5}\sin^4\Theta\right)\text{ms}^{-2} \quad [5]$$

A second correction that must be made if the gravity is measured from a moving vehicle, such as a ship or airplane, accounts for the effect on gravity of the motion of the vehicle with respect to the Earth's spin. A ship steaming to the east is, in effect, rotating faster than the Earth. The centrifugal effect of this increased rate of rotation causes gravity to be less than it would be if the ship were stationary. The opposite effect occurs for a ship steaming to the west. This term, called the Eötvös correction, is largest near the Equator and involves only the east–west component v_{EW} of the ship's velocity vector (eqn [6]), in which ω is the angular velocity of the Earth's rotation.

$$g_{EOT} = 2\omega v_{EW}\cos\Theta \quad [6]$$

The free air gravity correction, which accounts for the elevation of the measurement above the Earth's sea level equipotential surface, is obviously not needed if the measurement is made on the sea surface. The free air correction g_{FA} is required if the measurement is made from a submersible or an airplane: eqn [7], where h is elevation above sea level in meters.

$$g_{FA} = 3.1h \quad \text{gu} \quad [7]$$

This correction is added to the observation if the sensor is deployed above the Earth's surface, and subtracted for stations below sea level.

For land surveys, the Bouguer correction accounts for the extra mass of the topography between the observation and sea level. For its marine equivalent, it adds in the extra gravitational attraction that would be present if rock rather than water existed between sea level and the bottom of the ocean. Except in areas of rugged bathymetry, the Bouguer correction g_B is calculated using the slab formula (eqn [8]).

$$g_B = -2\pi\Delta\rho Gz \quad [8]$$

Here $\Delta\rho$ is the density difference between oceanic crust and sea water, G is Newton's constant, and z is the depth of the sea floor. This correction is seldom used because it produces very large positive gravity anomalies. Furthermore, there are more accurate corrections for the effect of bathymetry that do not make the unrealistic assumption that the expected state for the oceans should be that the entire depth is filled with crustal rocks displacing the water. The Bouguer correction is necessary, however, when gravity measurements are made from a submarine, in order to combine those data with more conventional observations from the sea surface. In this case, the Bouguer correction is applied twice: once to remove the upward attraction of the layer of water above the submarine, and once more to add in that layer's gravitation field below the sensor.

Satellite measurements of sea surface height go through a different processing sequence to recover marine geoid anomalies. The most important step is in calculating precise orbits. Information from tracking stations is supplemented with a 'crossover analysis' that removes long-wavelength bias in orbit elevation by forcing the height valves to agree wherever orbits cross. Corrections are then made for known physical oceanographic effects such as tides, and wave action is averaged out. The height of the sea level geoid above the Earth's center, assuming the standard spheroid, is subtracted from the data to create geoid anomalies.

History

A principal impediment to the acquisition of useful gravity observations at sea was the difficulty in separating the desired acceleration of gravity from the acceleration of the platform floating on the surface of the moving ocean. For this reason, the first successful gravity measurements to be acquired at sea were taken from a submarine by the Dutch pioneer, Vening Meinesz, in 1923. He used a pendulum gravimeter, which was the state of the art for measuring absolute gravity at that time. By accurately timing the period, T, of the swinging pendulum, the acceleration of gravity, g, can be recovered according to eqn [9], in which l is the length of the pendulum arm.

$$T = 2\pi\sqrt{\frac{l}{g}} \qquad [9]$$

By 1959, five thousand gravity measurements had been acquired from submarines globally. These measurements were instrumental in revealing the large gravity anomalies associated with the great trenches along the western margin of the Pacific. However, these gravity observations were very time-consuming to acquire because of the long integration times needed to achieve a high-precision estimate of the pendulum's period, and could not be adapted for use on a surface ship.

Gravity measurements at sea became routine and reliable in the late 1950s with the development of gyroscopically stabilized platforms and heavily damped mass-and-spring systems constrained to move only vertically. The new platforms compensated for the pitch and roll of the ship such that simple mass-and-spring gravimeters could collect

time series of variations in gravity over the oceans from vessels under way. Without any need to stop the ship on station, a time series of gravity measurements could be obtained at only small incremental cost to ship operations. With the advent of the new instrumentation, the catalogue of marine gravity values has grown in the past 40 years to more than 2.5 million measurements.

A new era of precision in marine gravity began with the advent of the Global Positioning System (GPS) in the late 1980s. Prior to this time, the largest source of uncertainty in marine gravity lay in the Eötvös correction. Older navigation systems (dead reckoning, celestial, and even the TRANSIT satellite system) were too imprecise in the absolute position of the ship and too infrequently available to allow accurate velocity estimation from minute to minute, especially if the ship was maneuvering. Typically, gravity data had to be discarded for an hour or so near the time of any change in course. The high positioning accuracy and frequency of GPS fixes now allows such precise calculation of the Eötvös correction that it is no longer the limiting factor in the accuracy of marine gravity data.

A breakthough in determining the global marine gravity field was achieved with the launching of the GEOS-3 (1975–1977) and Seasat (1978) satellites, which carried radar altimeters. Altimeters were deployed for the purpose of measuring dynamic sea surface elevation associated with physical oceanographic effects. The Seasat satellite carried a new, high-precision altimeter that characterized the variations in sea surface elevation with unprecedented detail. The satellite failed prematurely, but not before it returned a wealth of data on the marine gravity field from its observations of the marine geoid. The geoid variations at mid- and short-wavelength were so large that the dynamic oceanographic effects motivating the mission could be considered a much smaller noise term. The success of the Seasat mission led to the launch of Geosat, which measured the geoid at even higher precision and resolution. Unfortunately, most of that data remained classified by the US military until the results from a similar European mission were about to be released into the public domain. The declassification of the Geosat data in 1995 fueled a major revolution in our understanding of the deep seafloor (**Figure 1**).

The latest developments in marine gravity stem from the desire to detect the shortest spatial wavelengths of gravity variations by taking gravimeters to the bottom of the ocean. Gravity is one example of a potential field, and as such the amplitude, A, of the signal of interest decays with distance, z, between source and detector as in eqn [10], where k is the

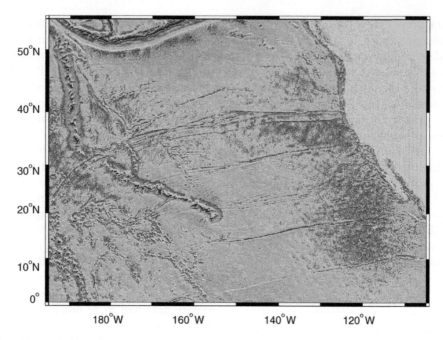

Figure 1   Gravity field over the North Pacific. This view is based on satellite altimetry data from the Geosat and other missions. (Data from Sandwell and Smith (1997).)

modulus of the spatial wavenumber, the reciprocal of the spatial wavelength.

$$A \sim e^{-2\pi k z} \qquad [10]$$

For sensors located on a ship at the sea surface in average ocean depths of 4.5 km, it is extremely difficult to detect short-wavelength variations in gravity of a few kilometers or less. Even lowering the gravimeter to the cruising depth of most submarines (a few hundred meters) does little to overcome the upward attenuation of the signal from localized sources on and beneath the seafloor. The solution to this problem recently has been to take gravimeters to the bottom of the ocean, either in a deep-diving submersible such as *Alvin*, or as an instrument package lowered on a cable. Most gravity measurements at sea are relative measurements. However, recent advances in instrumentation now allow absolute gravimeters to be deployed on the bottom of the ocean, avoiding the problem of instrument drift that adds error to relative gravity measurements. However, noise associated with the short baselines required for operation in the deep sea remains problematic.

Interpretation of Marine Gravity

Short-Wavelength Anomalies

The shortest-wavelength gravity anomalies over the oceans (less than a few tens of kilometers) are the least ambiguous to interpret since they invariably are of shallow origin. The upward continuation factor guarantees that any spatially localized anomalies with deep sources will be undetectable at the ocean surface. Near-bottom gravity measurements are able to improve somewhat the detection of concentrated density anomalies buried at deeper levels, but most are assumed to lie within the oceanic crust.

One of the most useful applications of short-wavelength gravity anomalies has been to predict ocean bathymetry (**Figure 2**). Radar altimeters deployed on the Seasat and Geosat missions measured with centimeter accuracy the height of the underlying sea surface, an excellent approximation to the marine geoid, over all ice-free marine regions. The accuracy and spatial coverage was far better than had been provided from more than a century of marine surveys from ships. At short wavelengths, undulations of the rock–water interface are the largest contribution to the short-wavelength portion of the geoid spectrum, which opened up the possibility of predicting ocean depth from the excellent geoid data. For example, an undersea volcano, or seamount, represents a mass excess over the water it displaces. The extra mass locally raises the equipotential surface, such that positive geoid anomalies are seen over volcanoes and ridges while geoid lows are seen over narrow deeps and trenches. The prediction of bathymetry from marine geoid or gravity data is tricky: the highest frequencies in the bathymetry cannot be estimated because of the upward attenuation

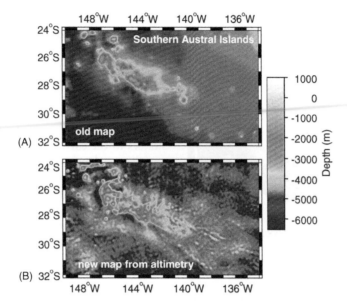

Figure 2 Example of bathymetric prediction from gravity anomalies in a largely unexplored region of the South Pacific. (A) The best available bathymetry from sparse echo soundings available in the early 1990s. (B) A diagram shows a dramatic improvement in definition of the bathymetry when satellite gravity observations are used to constrain the short-wavelength component of the bathymetry. (Adapted from McNutt and Bonneville (1996).)

problem, and the longer wavelengths are canceled out in the geoid by their isostatic compensation (see following section). These longer wavelengths in the bathymetry must be introduced into the solution using traditional echo soundings from sparse ship tracks. Nevertheless, the best map we currently have of the depth of the global ocean is courtesy of satellite altimetry.

Mid-Wavelength Anomalies

The mid-wavelength part of the gravity spectrum (tens to hundreds of kilometers) is dominated by the effects of isostatic compensation. Isostasy is the process by which the Earth supports variations in topography or bathymetry in order to bring about a condition of hydrostatic equilibrium at depth. The definition of isostasy can be extended to include both static and dynamic compensation mechanisms, but at these wavelengths the static mechanisms are most important. There are a number of different types of isostatic compensation at work in the oceans, and the details of the gravity field can be used to distinguish them and to estimate the thermomechanical behavior of oceanic plates.

One of the simplest mechanisms for isostatic compensation is Airy isostasy: the oceanic crust is thickened beneath areas of shallow bathymetry. The thick crustal roots displace denser mantle material, such that the elevated features float on the mantle much like icebergs float in the ocean. Of the various methods of isostatic compensation, this mechanism predicts the smallest gravity anomalies over a given feature. From analysis of marine gravity, we now know that this sort of compensation mechanism is only found where the oceanic crust is extremely weak, such as on very young lithosphere near a midocean ridge. For example, large plateaus formed when hotspots intersect midocean ridges are largely supported by Airy-type isostasy. Elsewhere the oceanic lithosphere is strong enough to exhibit some lateral strength in supporting superimposed volcanoes and other surface loads.

An extremely common form for support of bathymetric features in the oceans is elastic flexure. Oceanic lithosphere has sufficient strength to bend elastically, thus distributing the weight of a topographic feature over an area broader than that of the feature itself (**Figure 3**). Analysis of marine gravity has been instrumental in establishing that the elastic strength of the oceanic lithosphere increases with increasing age. Young lithosphere near the midocean ridge is quite weak, in some cases hardly distinguished from Airy-type isostasy. The oldest oceanic lithosphere displays an effective thickness equivalent to that of a perfectly elastic plate 40 km thick. The fact that this thickness is less than that of the commonly accepted value for the thickness of the mechanical plate that drifts over the asthenosphere indicates that the base of the oceanic lithosphere is not capable of sustaining large deviatoric stresses (of the order of 100 MPa or more) over million-year timescales.

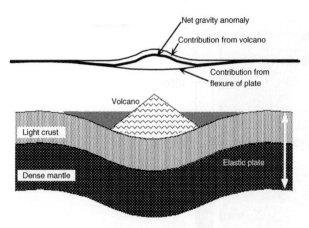

Figure 3 Cartoon showing how the seafloor is warped as an elastic plate under the weight of a small volcano. The gravity anomaly that would be detected by a ship sailing along the sea surface over this feature is the net difference between the positive gravity perturbation from the extra mass of the volcano and the negative gravity perturbation produced when the elastically flexed light crust replaces denser mantle. (Adapted from McNutt and Bonneville (1996).)

Another very important method of isostatic compensation in the ocean is Pratt isostasy. This method of support supposes that the height of a vertical column of bathymetry is inversely proportional to its density. Low-density columns can be higher because they are lighter, whereas heavy columns must be short in order to produce the same integrated mass at some assumed depth of compensation. In the oceans, variations in the temperature of the lithosphere produce elevation changes in the manner of Pratt isostasy. For example, ridges stand 4 km above the deep ocean basins because the underlying lithosphere is hotter when the plate is young. The bathymetric swells around young hotspot volcanoes may also be supported by Pratt-type isostasy, although some combination of crustal thickening and dynamic isostasy may be operating as well. Again, gravity and geoid anomalies have been principal constraints in arguing for the mechanism of support for bathymetric swells.

Long-Wavelength Anomalies

At wavelengths from 1000 to several thousand kilometers, gravity anomalies are usually derived from satellite observations and interpreted using equations appropriate for a spherical earth. Geoid is interpreted more commonly than gravity directly, as it emphasizes the longer wavelengths in the geopotential field. Isostatic compensation for smaller-scale bathymetric features, such as seamounts, can usually be ignored in that the gravity anomaly from bathymetry is canceled out by that from its compensation when spatially averaged over longer wavelengths.

The principal signal at these wavelengths arises from the subduction of lithospheric slabs and other sorts of convective overturn within the mantle. Three sorts of gravitational contributions must be considered: (1) the direct effect of mass anomalies within the mantle, either buoyant risers or dense sinkers which drive convection; (2) the warping of the surface caused by viscous coupling of the risers or sinkers to the earth's surface; and (3) the warping of any deeper density discontinuities (such as the core–mantle boundary) also caused by viscous coupling.

In the 1980s, estimates of the locations and densities of mass anomalies in the mantle responsible for the first contribution above began to become available courtesy of seismic tomography. Travel times of earthquake waves constrained the locations of seismically fast and slow regions in the mantle. By assuming that the seismic velocity variations were caused by temperature differences between hot, rising material and cold, sinking material, it was possible to convert velocity to density using standard relations. Knowing the locations of the mass anomalies driving convection inside the Earth led to a breakthrough in understanding the long-wavelength gravity and geoid fields.

The amount of deformation on density interfaces above and below the mass anomalies inferred from tomography (contributions (2) and (3) above) depends upon the viscosity structure of Earth's mantle. Coupling is more efficient with a more viscous mantle, whereas a weaker mantle is able to soften the transmission of the viscous stresses from the risers and sinkers. Therefore, one of the principal uses of marine gravity anomalies at long wavelengths has been to calibrate the viscosity structure of the oceanic upper mantle. This interpretation must be constrained by estimates of the dynamic surface topography over the oceans, which is actually easier to estimate than over the continents because of the relatively uniform thickness of oceanic crust.

A fairly common result from this sort of analysis is that the oceanic upper mantle must be relatively inviscid. The geoid shows that there are large mass anomalies within the mantle driving convection that are poorly coupled to variations in the depth of the seafloor. If the upper mantle were more viscous, there should be a stronger positive correlation between marine geoid and depth of the seafloor at long wavelengths.

See also

Manned Submersibles, Deep Water. Satellite Altimetry. Satellite Oceanography, History and Introductory Concepts.

Further Reading

Bell RE, Anderson R, and Pratson L (1997) Gravity gradiometry resurfaces. *The Leading Edge* 16: 55–59.

Garland GD (1965) *The Earth's Shape and Gravity*. New York: Pergamon Press.

McNutt MK and Bonneville A (1996) Mapping the seafloor from space. *Endeavour* 20: 157–161.

McNutt MK and Bonneville A (2000) Chemical origin for the Marquesas swell. *Geochemistry, Geophysics and Geosystems* 1.

Sandwell DT and Smith WHF (1997) Marine gravity for Geosat and ERS-1 altimetry. *Journal of Geophysical Research* 102: 10039–10054.

Smith WHF and Sandwell DT (1997) Global seafloor topography from satellite altimetry and ship depth soundings. *Science* 277: 1956–1962.

Turcotte DL and Schubert G (1982) *Geodynamics: Applications of Continuum Physics to Geological Problems*. New York: Wiley.

Watts AB, Bodine JH, and Ribe NM (1980) Observations of flexure and the geological evolution of the Pacific Ocean basin. *Nature* 283: 532–537.

Wessel P and Watts AB (1988) On the accuracy of marine gravity measurements. *Journal of Geophysical Research* 93: 393–413.

Zumberg MA, Hildebrand JA, Stevenson JM, *et al.* (1991) Submarine measurement of the Newtonian gravitational constant. *Physical Review Letters* 67: 3051–3054.

Zumberg MA, Ridgeway JR, and Hildebrand JA (1997) A towed marine gravity meter for near-bottom surveys. *Geophysics* 62: 1386–1393.

SEISMIC STRUCTURE

A. Harding, University of California, San Diego, CA, USA

Introduction

Seismic exploration of the oceans began in earnest in the 1950s. The early seismic experiments were refraction in nature using explosives as sources. The principal data were first arrival, P-wave travel times, which were analyzed to produce primarily one-dimensional models of compressional velocity as a function of depth. Within a decade, the results of these experiments had convincingly demonstrated that the crust beneath the ocean crust was much thinner than continental crust. Moreover, the structure of the deep ocean was unexpectedly uniform, particularly when compared with the continents. In light of this uniformity it made sense to talk of average or 'normal' oceanic crust. The first compilations described the average seismic structure in terms of constant velocity layers, with the igneous crust being divided into an upper layer 2 and an underlying layer 3.

Today, the scale and scope of seismic experiments is much greater, routinely resulting in two- and three-dimensional images of the oceanic crust. Experiments can use arrays of ocean bottom seismographs and/or multichannel streamers to record a wide range of reflection and refraction signals. The source is typically an airgun array, which is much more repeatable than explosives and produces much more densely sampled seismic sections. Seismic models of the oceanic crust are now typically continuous functions of both the horizontal and vertical position, but are still principally P-wave or compressional models, because S-waves can only be produced indirectly through mode conversion in active source experiments.

In spite of their greater resolving power, modern experiments are still too limited in their geographic scope to act as a general database for looking at many of the questions concerning oceanic seismic structure. The main vehicle for looking at the general seismic structure of the oceans is still the catalog of one-dimensional P-wave velocity models built up over approximately 40 years of experiments. The original simple layer terminology, with slight elaboration, is by now firmly entrenched as the means of describing the principal seismic features of the oceanic crust; despite the fact that the representation of the underlying velocity structure has changed significantly over time. The next section discusses the evolution of the velocity model and the layer description. Subsequent sections discuss the interpretation of seismic structure in terms of geologic structure; the seismic structure of anomalous crust; and the relationship of seismic structure to such influences as spreading rate and age.

Normal Oceanic Crust

Table 1 reproduces one of the first definitions of average or 'normal' oceanic crust by Raitt (1963). Even in this era before plate tectonics, Raitt excluded from consideration any areas such as oceanic plateaus that he thought atypical of the deep ocean. Today, compilations count as normal crust formed at midocean ridges away from fracture zones. The early refraction experiments typically consisted of a small set of widely spaced instruments. They were analyzed using the slope-intercept method in which a set of straight lines was fitted to first arrival travel times. This type of analysis naturally leads to stair-step or 'layer-cake' models consisting of a stack of uniform velocity layers separated by steps in velocity. Although their limitations as a description of the earth were recognized, these models provided a simple and convenient means of comparing geographically diverse data sets. Raitt divided the oceanic crust into three layers and included a fourth to represent the upper mantle. The top layer (layer 1), was a variable thickness sedimentary layer. Below this came the two layers that together comprised the igneous oceanic crust, a thinner more variable velocity layer (layer 2), and a thicker, more uniform velocity layer (layer 3). Layer 3 is the most characteristically oceanic of the layers. Arrivals from this layer are the most prominent arrivals in typical refraction profiles. The uniformity of high velocities within this layer mark layer 3 as being compositionally distinct from continental crust. At the base of layer 3 is the Mohorovic discontinuity or Moho, identifiable as such because the velocities of layer 4 were comparable to those seen in the upper mantle beneath continents.

As refraction data sets with better spatial sampling became available, the systematic errors inherent in fitting a few straight lines to the first arrival travel times became more noticeable. This was especially

Table 1 Traditional and modern summaries of average oceanic crystal structure

Parameter	Traditional[a]		Modern[b]		
	Velocity (km s^{-1})	Thickness (km)	Velocity (km s^{-1})	Thickness (km)	Representative gradient (s^{-1})
Layer 2 (igneous crust)	5.07 ± 0.63	1.71 ± 0.75	2.5–6.6	2.11 ± 0.55	1–3
Layer 3 (igneous crust)	6.69 ± 0.26	4.86 ± 1.42	6.6–7.6	4.97 ± 0.90	0.1–0.2
Layer 4 (upper mantle)	8.13 ± 0.24		>7.6		
Total igneous crust		6.57 ± 1.61		7.08 ± 0.78	

[a] From *Raitt (1963)*.
[b] Modified from *White et al.* (1992).

true for layer 2 first arrivals, which appear over a relatively short-range window, but have noticeable curvature because of the wide range of layer 2 velocities. The initial resolution of this problem was to fit more lines to the data and divide layer 2 into smaller, constant velocity sublayers termed 2A, 2B, and 2C. However, there was a more fundamental problem: the layer-cake models were not consistent with the waveform and amplitude behavior of the data. This flaw became apparent when, instead of just using travel times, the entire recorded wavefield began to be modeled using synthetic seismograms. Waveform modeling led to a recasting of the one-dimensional model in terms of smoothly varying velocities, constant velocity gradients, or finely layered stair-steps. Large velocity steps or interfaces are now included in the models only if they are consistent with the amplitude behavior. The stair-step representation is a tacit admission that there is a limit to the resolution of finite bandwidth data. A stair-step model is indistinguishable from a continuous gradient provided the layering is finer than the vertical resolution of data, which for refraction data is some significant fraction of a wavelength. Today, purely travel time analysis based upon densely sampled primary and secondary arrival times and accumulated knowledge can yield accurate models, but seismogram modeling is still required to achieve the best resolution.

The change in the style of the velocity models is illustrated in **Figure 1**, which shows models for a recent Pacific data set. A change in gradient rather than a jump in velocity marks the boundary between layer 2 and 3 in most modern models. Paradoxically, the jump in velocity at the Moho – present in the traditional layer-cake model – is not required by the first arrival times, but is required to fit the secondary arrival times and amplitude behavior of the data. The example also illustrates another general problem with layer-cake models, which is that they systematically underestimate both layer and total crustal thickness.

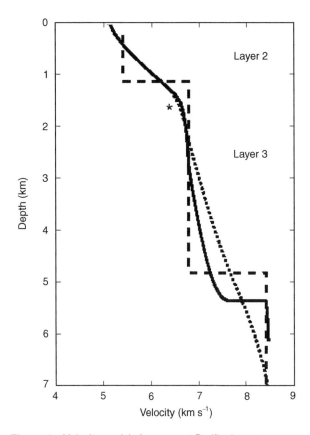

Figure 1 Velocity models for a recent Pacific data set.

Table 1 also presents a more modern summary of average oceanic crustal structure. All models included in this compilation were the result of synthetic seismogram analysis. A range of velocities and typical gradients now characterizes layers 2 and 3. Layer 2 is a region of rapidly increasing velocity at the top of the crust, with typical gradients from 1–3 s^{-1}, while layer 3 is a thicker region of more uniform velocity with gradients between 0 and 0.2 s^{-1}. The layer thicknesses are little changed from the Raitt compilation but are systematically thicker and thus total crustal thickness is also larger.

A virtue of the layer description is that it captures the main features of the seismic data and results without being too precise. So, while the style of velocity model has moved away from one that is strictly layered, the description in terms of layers persists both for historical continuity and linguistic convenience. Without a simple layer-cake model, defining the layer 2/3 transition can be somewhat problematic. Ideally the velocity function will show a small velocity jump or resolvable inflection at the boundary, but often the layer 2/3 boundary is taken as being either at a change in the general velocity gradient of the model or at a particular velocity just below representative layer 3 velocities, sometimes both. The chosen velocity is typically somewhere between 6.5 and 6.7 km s^{-1}.

The use of layers 2 and 3 is almost universally and consistently applied when summarizing the basic features of both seismic data sets and models. More problematic is the use of layers 2A, 2B, and 2C to describe subintervals layer 2. While the use of these layers is widespread in the literature, their application is more variable and has evolved in conjunction with changes in model style and resolution. As a result caution is needed when comparing models from disparate experiments, particularly across tectonic and geographic regions. Today, many authors subdivide layer 2 into 2A and 2B only. Layer 2A is widely recognized as a well defined surficial layer in young oceanic crust near ridges, being associated with velocities <3 km s^{-1} and a transition to velocities >4 km s^{-1} at its base. In this division, layer 2B is simply the lower part of layer 2.

Interpretation of Seismic Structure

The nature of the relationship between the seismic and geologic structure of the oceanic crust is the subject of ongoing debate. The adoption of plate tectonics and seafloor spreading provided a framework for understanding the initially surprising uniformity and simplicity of ocean seismic structure. It also gave rise to the hope that there would be a correspondingly simple and universal interpretation of the seismic layering in terms of geologic structure. However, this expectation has receded as the complexity of the seismic models has increased and our understanding of the diverse magmatic, tectonic, and hydrothermal processes shaping crustal structure, both spatially and temporally, has improved. Fundamentally, there is no unique, unambiguous interpretation of seismic velocity in terms of rock type or geologic structure. Having coincident P- and S-wave velocity models plus other geophysical data can

considerably reduce this ambiguity but ultimately cannot eliminate it. Ideally, reference drill holes through the full oceanic crustal section would be used to calibrate seismic and other geophysical results. These would allow the dominant processes, controlling, for example, the layer 2/3 boundary or the nature of layer 2A to be identified in different tectonic settings. Unfortunately, to date only a limited number of drill holes have penetrated a significant depth into the oceanic crust, and none have penetrated a full crustal section. As a result there has been only limited opportunity for direct comparison between seismic and *in situ* structure. When interpreting seismic results, we must still rely heavily on inferences that draw upon a number of less direct sources including seafloor observations, analogy with ophiolites, and laboratory measurements on dredged rock samples.

The simplest, most straightforward interpretation of seismic velocity is to assume that velocity is dependent on composition and that different velocities indicate different rock types. This reasoning has guided the traditional interpretation of the Moho boundary. In seismic models, the characteristic signature of the Moho is an increase in velocity of between ~0.5 and 1.0 km s^{-1} to velocities >7.6 km s^{-1}. The increase may occur either as a simple interface (at the resolution of the data) or as a transition region up to a kilometer thick. In reflection data, the Moho is often observed as a low frequency, ~10 Hz, quasi-continuous event. In general, layer 3 is considered to be predominantly gabbros, and the most common interpretation of the Moho is as the boundary between a mafic gabbroic crust and an ultramafic upper mantle. The observed Moho structures reflect either a simple contact or a transition zone of interleaved mafic and ultramafic material. These interpretations are supported by observations within ophiolites, reference sections of oceanic lithosphere obducted onto land. However, partially serpentinized, ultramafic peridotites can have P- and S-wave velocities that are for practical purposes indistinguishable from gabbros, when the degree of alteration is between 20 and 40%. For lesser degrees of alteration, the serpentinized rocks will have velocities that are intermediate between gabbros and peridotite and are thus distinguishable. Although widespread serpentinization is unlikely, particularly at fast and intermediate spreading rates, it may be locally important at segment boundaries at slow spreading rates where faulting and extension expose peridotites to pervasive alteration by deeply circulating sea water.

A further complication in interpreting the Moho is the fact that the igneous crust may contain cumulate ultramafic rocks such as dunite, which crystallized as sills from a melt. Seismically, these are indistinguishable from the residual upper mantle harburgites that yielded the crustal melts. Thus a distinction is sometimes made between Moho defined seismically, and a petrologic Moho separating cumulate rocks from source rocks.

There is very little intrinsic difference in composition or velocity to the basaltic rocks – pillow lavas and sheet flows, sheeted dikes, and gabbros – that typically constitute the upper part of the oceanic crust. Certainly not enough to explain the large range of layer 2 velocities or the difference between layers 2 and 3. Instead the seismic character is attributed primarily to the cracks, fractures, and fissures that permeate the upper crust. These reduce the effective stiffness of the rock matrix, and hence the velocity of the upper crust at seismic wavelengths. The relationship between velocity and the size, shape, orientation, and distribution of cracks is a complex nonlinear one that affects P- and S-wave velocities differently. However, at least informally, it is often crack volume or porosity that is taken as the primary control. The large velocity gradients within layer 2 are then seen as being the result of a progressive closure of cracks or reduction in porosity with depth, as confining pressure increases. Once most cracks are closed, velocities are only weakly pressure-dependent and can have the low velocity gradients characteristic of layer 3. An analogous velocity behavior, but with a smaller velocity range is observed in individual rock samples subject to increasing confining pressure.

From this perspective, there can only be a structural or lithologic interpretation of the seismic layers if there is a structural dependence to the crack distribution. Support for such an interpretation comes from composite velocity profiles through ophiolites, constructed using laboratory measurements on hand samples at suitable confining pressures. These profiles showed a broad agreement with oceanic results and led to the standard ophiolite interpretation of seismic structure in which layer 2 is equated with the extrusive section of pillow lavas and sheeted dikes and layer 3 with the intrusive gabbroic section. Moreover, the relative and absolute thicknesses of the extrusive and intrusive sections in ophiolites are comparable to those of the oceanic seismic layers. At a more detailed level, measurements on ophiolites often show a reduction in porosity at the transition from pillow basalts to sheeted dikes: an observation used to bolster the inference that layer 2A is equivalent to pillow lava section in young oceanic crust.

As the only available complete exposures of oceanic crustal sections, ophiolites have had a historically influential role in guiding the interpretation of seismic layering. However, a number of cautions are in order. First, there are inherent uncertainties in extrapolating seismic velocities measured on hand samples to the larger scales and lower frequencies characteristic of seismic experiments. Second, the seismic velocities of ophiolites could have been modified during the obduction process. Finally, most ophiolites are thought to have been produced in back arcs or marginal basin settings and thus while valuable structural analogs may not be representative of the ocean basins as a whole. Ultimately, the ophiolite model can only be used as a guide, albeit an important one, for interpreting oceanic structure, seismic or otherwise and conclusions drawn from it must be weighed against other constraints.

The traditional ophiolite model is a convenient and widely used shorthand for describing seismic structure, that is useful provided that too much is not asked or expected of it. The porosity interpretation of upper crustal velocities gives the basic layer 2/layer 3 division of seismic models a sort of universality that transcends, within bounds, changes in the underlying lithologic structure, and emphasizes the need for taking tectonic setting into consideration when interpreting seismic structure. Any process that either resets or significantly modifies this crack/porosity distribution of the upper crust will imprint itself on the seismic structure. For example, near fracture zones, fracturing and faulting are usually inferred to be dominant controls on layer 2 structure. At Deep Sea Drilling Project Hole 504B, a deep penetration hole in 5.9 million year old crust formed at intermediate spreading rates, the base of seismic layer 2 is found to lie within the sheeted dike section. It is thought that progressive filling of cracks by hydrothermal alteration processes has, over time, raised the depth of the layer 2/3 boundary.

Anomalous Crust

Definitions of normal seismic structure focus on oceanic crust formed at midocean ridges and specifically exclude structures such as fracture zones or oceanic plateaus as anomalous. Particularly at slow spreading rates, fracture zones and the traces of segment boundaries are part of the warp and weft of ocean fabric, making up about 20% of the seafloor. These have been most extensively studied in the northern Atlantic, where the ridge is segmented on scales of 20–100 km, and most segment boundaries are associated with attenuated crust. The degree of crustal thinning shows no simple dependence of segment offset, although the large-offset fracture zones may indeed contain the most extreme structure. Fracture zones and segment traces typically exhibit an inner and outer region of influence. In the outer region, there is gradual thinning of the crust towards the trace extending over a distance of perhaps 20 km. This region is marked by a deepening of the seafloor and a simultaneous shoaling of the Moho. Within the outer region, seismic structure is a thinned but recognizable version of normal crust, and the most extreme structure is associated with the inner region. Here, within a ~10 km wide zone, the crust may be <3 km thick and in one-dimension may appear to be all layer 2 down to the Moho. Looked at in cross-section, there is a coalescence of a gradually thickening layer 2 with a Moho transition region, and the elimination of layer 3. The structure at segment boundaries can be explained as a combination of reduced magma supply, much of it feeding laterally from the segment centers, and pervasive faulting, possibly low angle in nature.

Velocities intermediate between crust and mantle values, $7.1–7.4$ km s^{-1}, are observed beneath small offset transforms and nontransform offsets below about 4 km depth, most likely indicating partial serpentinization of the mantle by deeply circulating water. The upper limit of serpentinization is difficult to determine seismically because of the overlap in velocity between gabbros and altered peridotites at high degrees of alteration. But, from seafloor observations, at least some serpentinites must lie at shallower depths.

In the fast spread Pacific, fracture zones – which are spaced at intervals of a few hundred kilometers – affect only a relatively small fraction of the total crust. In addition, seismic studies suggest that the crustal structure of fracture zones is essentially a slightly thinner version of normal crust with a well-defined layer 3. In addition there can be some thickening and slowing of layer 2 in the vicinity of the transform associated with upper crustal faulting.

The term oceanic large igneous provinces (LIPs) provides a convenient umbrella under which to group such features as oceanic plateaus, aseismic ridges, seamount groups, and volcanic passive margins. They are massive emplacements of mostly mafic extrusive and intrusive material whose origin lies outside the basic framework of seafloor spreading. Together they account for much of the anomalous structure apparent in maps of seafloor bathymetry. At present, the rate of LIP emplacement, including the continents, is estimated to be equal to about 5–10% of midocean ridge production. However, during the formation of the largest LIPs, such as the Ontong Java plateau, off-axis volcanism was a significant fraction of midocean ridge rates. Many LIPs can trace their origin to either transient or persistent (hot-spot) mantle plumes; as such they provide a valuable window into the dynamics of the mantle. Where plumes interact with ridges, they can significantly affect the resulting crustal structure. The most notable example of this, at present, is the influence of the Iceland hot-spot on spreading along the Reykjanes ridge, where the crust is about 10 km thick and includes an approximately 7 km thick layer 3.

Two general features of LIPs seismic structure are a thickened crust, up to 25 km thick, and a high velocity, lower crustal body, reaching up to 7.6 km s^{-1}. However, in detail, the seismic structure depends on the style and setting of their emplacement, including whether the emplacement was submarine or subareal, intraplate or plate boundary. For example, the Kerguelen-Heard Plateau (a province of the larger Kerguelen LIP) is estimated to have 19–21 km thick igneous crust, the majority of which is a 17 km thick layer 3 with velocities between 6.6 and 7.4 km s^{-1}. The plateau is inferred to have formed by seafloor spreading in the vicinity of a hot-spot similar to Iceland. If this is the case, the greater thickness and higher velocities of layer 3 can be attributed to greater than normal extents of partial melting within the upwelling mantle. An example of an intraplate setting is the formation of the Marquesas Island hot-spot. Seismic data reveal that in addition to the extrusive volcanism responsible for the islands, significant intrusive emplacement has created a crustal root beneath the previously existing oceanic crust. Combined, the total crust is up to 17 km thick. The crustal root, with velocities between 7.3 and 7.75 km s^{-1}, may be purely intrusive or a mixture of intrusive rocks with preexisting mantle peridotites.

Systematic Features of the Oceanic Crust

For the most part, seismic investigations of the oceanic crust tend to focus on specific geologic problems. As a consequence, the catalog of published seismic results has sampling biases that make it less than ideal for looking at certain more general questions. There are for example a relatively large number of good measurements of young Pacific crust and old Atlantic crust, but fewer on old Pacific crust and only a handful of measurements on crust formed at ultra-slow spreading rates. Older data sets analyzed by the slope-intercept method are often discounted unless they are the only data available for a particular region. Nevertheless there are a number of systematic features of oceanic crust that can be discerned from compilations of seismic results.

Spreading Rate Dependence of Average Crustal Thickness

Although the style of crustal accretion varies considerably between slow and fast spreading ridges, the average thickness of the crust produced including fracture zones is remarkably uniform at 7 ± 1 km for full spreading rates between 20 and 150 mm a^{-1}. This result indicates that the rate of crustal production is linearly related to spreading rate over this range. Crustal thicknesses are more variable at slower spreading rates, reflecting the more focused magma supply and greater tectonic extension. At ultra-slow spreading rates below 20 mm a^{-1}, there is a measurable and rapid decrease in average crustal thickness. This reduction is expected theoretically, as conductive heat loss inhibits melt production in the upwelling mantle.

Age Dependence of Crustal Structure

The clearest and strongest aging signal in the oceanic crust is the approximate doubling of surficial velocities with age from about 2.5 km s^{-1} at the ridge axis to 5 km s^{-1} off-axis. This increase in velocity was first reported in the mid-1970s based on compilations of surface sonobuoy data. Originally, the velocity signal was interpreted as being associated with a thinning of layer 2A over a period of 20–40 Ma. However, the same data can equally well be explained as simply the increase in velocity of a constant thickness layer, and a compilation of modern seismic data sets indicates that layer 2A velocities increase much more rapidly, almost doubling in < 10 Ma. While both of these inferences are supported by individual flowline profiles extending out from the ridge axis, the distribution bias of modern seismic data sets to the ridge axes makes it hard to assess the robustness of this result.

The increase in layer 2A velocity with age is due to hydrothermal alteration sealing cracks within the upper crust. There need not be a correspondingly large decrease in porosity, as alteration that preferentially seals the small aspect ratio cracks will produce a large velocity increase for a small porosity reduction. Given this mechanism, similar, albeit smaller increases in layer 2B velocities might be expected. Such an increase is not apparent in present compilations, although a small systematic change would be masked by the intrinsic variability of layer 2B and the variability induced by different analysis methods. There is though some indication of systematic change with layer 2B from analysis of ratios of P- and S-wave velocity and as noted in the previous section alteration is thought to have raised the layer 2/3 boundary at Hole 504B.

Anisotropic Structure

Two types of anisotropic structure are frequently reported for the oceanic crust and upper mantle. The P-wave velocities of the upper mantle are found to be faster in the fossil spreading direction, than in the original ridge parallel direction, with the difference being around 7%. This is due to the preferential alignment of the fast a-axis of olivine crystals in the direction of spreading as mantle upwells beneath the midocean ridge.

The other region of the crust that exhibits anisotropy is the extrusive upper crust, which has a fast P-wave propagation direction parallel to ridge axis at all spreading rates. The peak-to-peak magnitude of the anisotropy averages ~10%. Like the velocity structure, this shallow anisotropy is generally ascribed to the crack distribution within the upper crust. Extensional forces in the spreading direction are thought to produce thin cracks and fissures that preferentially align parallel to the ridge axis.

See also

Mid-Ocean Ridge Tectonics, Volcanism, and Geomorphology. Seismology Sensors.

Further Reading

Carlson RL (1998) Seismic velocities in the uppermost oceanic crust: Age dependence and the fate of layer 2A. *Journal of Geophysical Research* 103: 7069–7077.

Fowler CMR (1990) *The Solid Earth*. Cambridge: Cambridge University Press.

Horen H, Zamora M, and Dubuisson G (1996) Seismic waves velocities and anisotropy in serpentinized peridotites from Xigaze ophiolite: Abundance of serpentine in slow spreading ridges. *Geophysical Research Letters* 23: 9–12.

Raitt RW (1963) The crustal rocks. In: Hill MN (ed.) *The Sea*, vol. 3. New York: Interscience.

Spudich P and Orcutt J (1980) A new look at the seismic velocity structure of the oceanic crust. *Reviews in Geophysics* 18: 627–645.

White RS, McKenzie D, and O'Nions RK (1992) Oceanic crustal thickness from seismic measurements and rare earth element inversions. *Journal of Geophysical Research* 97: 19 683–19 715.

MID-OCEAN RIDGE SEISMIC STRUCTURE

S. M. Carbotte, Lamont-Doherty Earth Observatory of Columbia University, Palisades, NY, USA

Introduction

New crust is created at mid-ocean ridges as the oceanic plates separate and mantle material upwells and melts in response through pressure-release melting. Mantle melts rise to the surface and freeze through a variety of processes to form an internally stratified basaltic crust. Seismic methods permit direct imaging of structures within the crust that result from these magmatic processes and are powerful tools for understanding crustal accretion at ridges. Studies carried out since the mid 1980s have focused on three crustal structures; the uppermost crust formed by eruption of lavas, the magma chamber from which the crust is formed, and the Moho, which marks the crust-to-mantle boundary. Each of these three structures and their main characteristics at different mid-ocean ridges will be described here and implications of these observations for how oceanic crust is created will be summarized. The final section will focus on how crustal structure changes at ridges spreading at different rates, and the prevailing models to account for these variations.

Seismic techniques employ sound to create cross-sectional views beneath the seafloor, analogous to how X-rays and sonograms are used to image inside human bodies. These methods fall into two categories; reflection studies, which are based on the reflection of near-vertical seismic waves from interfaces where large contrasts in acoustic properties are present; and refraction studies, which exploit the characteristics of seismic energy that travels horizontally as head waves through rock layers. Reflection methods provide continuous images of crustal boundaries and permit efficient mapping of small-scale variations over large regions. Locating these boundaries at their correct depth within the crust requires knowledge of the seismic velocity of crustal rocks, which is poorly constrained from reflection data. Refraction techniques provide detailed information on crustal velocity structure but typically result in relatively sparse measurements that represent large spatial averages. Hence the types of information obtained from reflection and refraction methods are highly complementary and these data are often collected and interpreted together.

Much of what we know about the seismic structure of ridges has come from studies of the East Pacific Rise. This is a fast-spreading ridge within the eastern Pacific that extends from the Gulf of California to south of Easter Island. Along this ridge, seafloor topography is relatively smooth and seismic studies have been very successful at imaging the internal structure of the crust. Comparatively little is known from other ridges, in part because fewer experiments have been carried out and in part because, with the rougher topography, imaging is more difficult.

Seismic Layer 2A

Early Studies

Seismic layer 2A was first identified in the early 1970s from analysis of refraction data at the Reykjanes Ridge south of Iceland. This layer of low compressional- or P-wave velocities ($<3.5\,\mathrm{km\,s^{-1}}$), which comprises the shallowest portion of the oceanic crust (**Figure 1**) was attributed to extrusive rocks with high porosities due to volcanically generated voids and extensive crustal fracturing. In the late 1980s a bright event corresponding with the base of seismic layer 2A was imaged for the first time using multichannel seismic reflection data. This event is not a true reflection but rather is a refracted arrival resulting from turning waves within a steep velocity gradient zone that marks the base of seismic layer 2A. Within this gradient zone P-wave velocity rapidly increases to velocities typical of seismic layer 2B ($>5.0\,\mathrm{km\,s^{-1}}$) over a depth interval of $\sim100{-}300\,\mathrm{m}$ (**Figure 1A**). The 2A event is seen in the far offset traces of reflection data collected with long receiver arrays ($>2\,\mathrm{km}$) and has been successfully stacked, providing essentially continuous images of the base of layer 2A at mid-ocean ridges.

The Geological Significance of the Layer 2A/2B Transition: Is it a Lithological Transition from Extrusives to Dikes or a Porosity Boundary within the Extrusives?

In most recent studies, layer 2A near the ridge axis is assumed to correspond with extrusive rocks and the base of layer 2A with a lithological transition to the sheeted dike section of oceanic crust. The primary evidence cited for this lithological interpretation comes from studies at Hess Deep in the equatorial

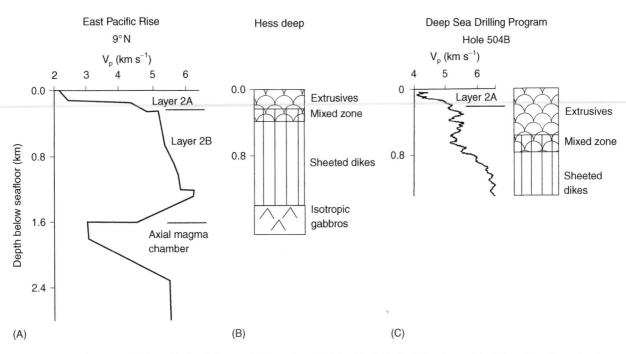

Figure 1 (A) Seismic velocity with depth for newly formed crust at the East Pacific Rise. Layer 2A, 2B, and the low velocities associated with the axial magma chamber (AMC) are identified. (Data from Vera EE, Buhl P, Mutter JC *et al.* (1990), *Journal of Geophysical Research* 95: 15529–15556). (B) Lithological cross section for the upper crust at Hess Deep derived from submersible observations. (Synthesis is from Francheteau J, Armijo R, Cheminee JL *et al.* (1992) *Earth and Planetary Sciences Letters* 111: 109–121. (C) Comparison of P-wave velocities from *in situ* sonic logging within Deep Sea Drilling Hole 504B and the lithological units observed within the hole. (From Becker K *et al.* (1988) *Proceedings of ODP, Initial reports, Part A, v.111, Ocean Drilling Program.* College Station, TX).

eastern Pacific. In this area, observations of fault exposures made from manned submersibles show that the extrusive rocks are ~300–400 m thick, similar to the thickness of layer 2A measured near the crest of the East Pacific Rise (compare **Figures 1A** and **1B**).

Other researchers have suggested that the base of layer 2A may correspond with a porosity boundary within the extrusive section associated with perhaps a fracture front or hydrothermal alteration. This interpretation is based primarily on observations from a deep crustal hole located off the coast of Costa Rica, which was drilled as part of the Deep Sea Drilling Program (DSDP). Within this hole (504B) a velocity transition zone is found that is located entirely within the extrusive section (**Figure 1C**). Here, a thin high-porosity section of rubbly basalts and breccia with P-wave velocities of ~4.2 km s^{-1} overlies a thick lower-porosity section of extrusives with higher P-wave velocities (5.2 km s^{-1}) (**Figure 1**). However, the relevance of these observations for the geological significance of ridge crest velocity structure is questionable. Crust at DSDP hole 504B is 5.9 My old and it is well established that the seismic velocity of the shallow crust increases with age owing to crustal alteration (see below). Indeed, the velocities within the shallowest extrusives at DSDP 504B (~4 km s^{-1}) are much higher than observed at

the ridge crest (2.5–3 km s^{-1}), indicating that significant crustal alteration has occurred (compare **Figures 1A** and **1C**).

Conclusive evidence regarding the geological nature of seismic layer 2A will likely require drilling or observations of faulted exposures of crust at or near the ridge crest made where seismic observations are also available. At present, the bulk of the existing sparse information favors the lithological interpretation and layer 2A is commonly used as a proxy for the extrusive crust. If this interpretation is correct, mapping the layer 2A/2B boundary provides direct constraints on the eruption and dike injection processes that form the uppermost part of oceanic crust.

Characteristics of Layer 2A at Mid-ocean Ridges

Along the crest of the East Pacific Rise, layer 2A is typically 150–250 m thick (**Figures 2** and **3**). Only minor variations in the thickness of this layer are observed along the ridge crest except near transform faults and other ridge offsets where this layer thickens.

Across the ridge axis, layer 2A approximately doubles or triples in thickness over a zone ~2–6 km wide indicating extensive accumulation of extrusives within this wide region (**Figures 4** and **5**). This accumulation may occur through lava flows that travel up to several kilometers from their eruption sites at the axis, either over the seafloor or perhaps transported through subsurface lava tubes. Volcanic eruptions that originate off-axis may also contribute to building the extrusive pile. On the flanks of the East Pacific Rise the base of layer 2A roughly follows the undulating abyssal hill relief of the seafloor (**Figure 6**). Layer 2A is offset at the major faults that bound the abyssal hills. Superimposed on this undulating relief are smaller-scale variations in 2A

thickness (50–100 m) that may reflect local build-up of lavas through ponding at and draping of seafloor faults (**Figures 5** and **6**).

Layer 2A is thicker and more variable in thickness (200–550 m) along the axis of the intermediate spreading Juan de Fuca Ridge, located in the northeast Pacific (see **Figure 9**). At this ridge, the sparse existing data suggest that layer 2A does not systematically thicken away from the ridge axis, and it appears that lavas accumulate within a narrower

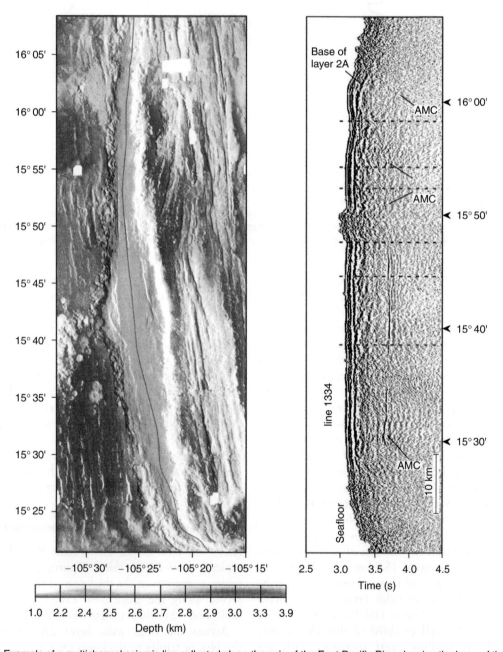

Figure 2 Example of a multichannel seismic line collected along the axis of the East Pacific Rise showing the base of the extrusive crust (layer 2A) and the reflection from the top of the axial magma chamber (AMC). Right-hand panel shows the bathymetry of the ridge axis with the location of the seismic profile in black line. The dashed lines on the seismic section mark the locations of very small offsets that are observed in the narrow depression along the axis where most active volcanism is concentrated.

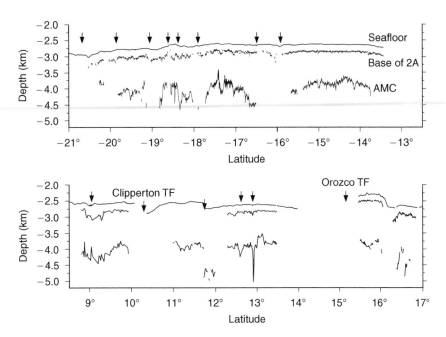

Figure 3 Cross section along the axis of the southern (top panel) and northern (bottom panel) East Pacific Rise showing depth to seafloor, the base of the extrusive crust, and the axial magma chamber reflection. This compilation includes results from all multichannel reflection data available along this ridge. Labeled arrows show the locations of transform faults. Other arrows mark the locations of smaller discontinuities of the ridge axis known as overlapping spreading centers. (Top panel, data from Hooft EE, Detrick RS and Kent GM (1997) *Journal of Geophysical Research* 102: 27319–27340. Bottom panel, data from Kent GM, Harding AJ, and Orcutt JA (1993) *Journal of Geophysical Research* 98: 13945–13696; Detrick RS, Buhl P, Vera E *et al.* (1987) *Nature* 326: 35–41; Babcock JM, Harding AJ, Kent GM and Orcutt JA (1998) *Journal of Geophysical Research* 103: 30451–30467; Carbotte SM, Ponce-Correa G and Solomon A (2000) *Journal of Geophysical Research* 105: 2737–2759).

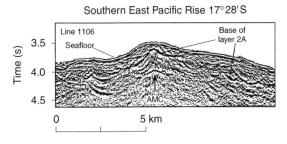

Figure 4 Example of a multichannel seismic profile shot across the ridge axis. Figure shows the magma chamber reflection and the event from the base of layer 2A. (Data from Carbotte SM, Mutter JC and Wu L (1997) *Journal of Geophysical Research* 102: 10165–10184).

zone than at the East Pacific Rise. Along the slow spreading Mid-Atlantic Ridge, the extrusive section is built largely within the floor of the median valley. At all the ridges that have been surveyed to date, extrusive layer thicknesses measured beyond the axial region are very similar (350–600 m).

Evolution of Layer 2A with Age

Global compilations carried out in the mid-1970s showed that the velocity of layer 2A increases as the crust ages away from the ridge axis and that layer 2A

velocities gradually increase to levels typical of layer 2B by 20–40 Ma. Recent compilations of modern seismic data suggest that layer 2A velocities increase abruptly and at quite young crustal ages (<5 Ma), rather than the gradual change evident in the early data. This increase in the velocity of layer 2A is the primary known change in the seismic structure of oceanic crust with age. This change is commonly attributed to precipitation of low-temperature alteration minerals within cracks and voids in the extrusive section during hydrothermal circulation of sea water through the crust. The detailed geochemical and physical processes associated with how hydrothermal precipitation of minerals affects seismic velocities is poorly understood. However, infilling of voids with alteration minerals is believed to increase the mechanical competency of the crust sufficiently to account for the evolution of layer 2A with age.

Axial Magma Chamber

Early Studies

Drawing on observations of the crustal structure of ophiolites (sections of oceanic or oceanic-like crust exposed on land) and the geochemistry of seafloor basalts, geologists long believed that mid-ocean ridges were underlain by large, essentially molten, magma reservoirs. However, until the last 15 years, few actual constraints on the dimensions of magma chambers at ridges were available. Early seismic studies on the East Pacific Rise detected a zone of lower seismic velocity beneath the ridge axis as expected for a region containing melt. A bright reflector was also found indicating the presence of a sharp interface with high acoustic impedance contrast within the upper crust. In the mid 1980s an extensive seismic reflection and refraction experiment was carried out on the northern East Pacific Rise by researchers at the University of Rhode Island, Lamont-Doherty Earth Observatory, and Scripps Institution of Oceanography. This study imaged a bright subhorizontal reflector located 1–2 km below seafloor along much of the ridge. In several locations this reflector was found to be phase-reversed relative to the seafloor reflection, indicating it resulted from an interface with an abrupt drop in seismic velocity. Based on its reversed phase and high amplitude, this event is now recognized as a reflection from a largely molten region located at the top of what is commonly referred to as an axial magma chamber.

Seismic refraction and tomography experiments show that this reflector overlies a broader zone that extends to the base of the crust within which seismic velocities are reduced relative to normal crust. At shallow depths this low-velocity zone is ~2 km wide. It broadens and deepens beneath the ridge flanks and is ~10 km wide at the base of the crust. Because of the relatively small velocity anomaly associated with much of this low-velocity zone (<1 km s^{-1}), this region is interpreted to be hot, largely solidified rock, and crystal mush containing only a few per cent partial melt.

The Characteristics of the Axial Magma Chamber at Mid-ocean Ridges

Several seismic reflection studies have now been carried out along the fast-spreading East Pacific Rise imaging over 1400 km of ridge crest (**Figure 3**). A reflection from the magma chamber roof is detected beneath ~60% of the surveyed region and can be traced continuously in places for tens of kilometers. This reflector is found at a depth of 1–2 km below seafloor and deepens and disappears toward major offsets of the ridge axis, including transform faults and overlapping spreading centers. Most volcanic activity along the East Pacific Rise is concentrated within a narrow depression, <1 km wide, which is interrupted by small steps or offsets that may be the

boundaries between individual dike swarms. In many places, the magma chamber reflector does not disappear beneath these offsets (**Figure 2**). However, changes in the depth and width of the reflector are often seen. Seismic tomography studies centered at 9°30'N on the East Pacific Rise shows that a broader region of low velocities within the crust pinches and narrows beneath two small offsets. These results suggest that segmentation of the axial magma chamber may be associated with the full range of ridge crest offsets observed on the seafloor.

Migration of seismic profiles shot perpendicular to the ridge axis reveals that the magma chamber reflection arises from a narrow feature that is typically less than 1 km in width (e.g., **Figure 4**). Refraction data and waveform studies of the magma chamber reflection suggest that it arises from a thin body of magma a few hundred to perhaps a few tens of meters thick, leading to the notion of a magma lens or sill. Initial studies assumed that this lens contained pure melt. However, recent research suggests that much of the magma lens may have a significant crystal content ($>25\%$) with regions of pure melt limited to pockets only a few kilometers or less in length along the axis.

Possible magma lens reflections, similar to those imaged beneath the East Pacific Rise, have also been imaged along the intermediate spreading Juan de Fuca Ridge and Costa Rica Rift and at the back-arc spreading center in the Lau Basin. In these areas, reflectors 1–2.5 km wide and at 2.5–3 km depth are detected (see **Figures 10 and 11**). Diffractions from the edges of these reflectors are shallower than diffractions due to seafloor topography, indicating that these events clearly lie within the crust. However, there is some debate whether these reflections correspond with magma bodies. Refraction studies along the northern Juan de Fuca show no evidence for a low-velocity zone coincident with the intracrustal reflection. In addition, the Juan de Fuca and Costa Rica Rift data are too noisy to allow determination of the polarity of the event. Hence we cannot rule out the possibility that the shallow crustal reflections observed at these ridges are due to an abrupt velocity increase within the crust, perhaps associated with a frozen magma lens or a cracking front, rather than a velocity decrease associated with the presence of melt.

Along the slow spreading Mid-Atlantic Ridge, evidence for magma lenses has been found in one location along the Reykjanes Ridge. Here an intracrustal reflection at a depth of ~2–5 km is observed, similar to the depths of magma lens events observed beneath portions of the intermediate spreading ridges. The absence of magma lens reflections in seismic

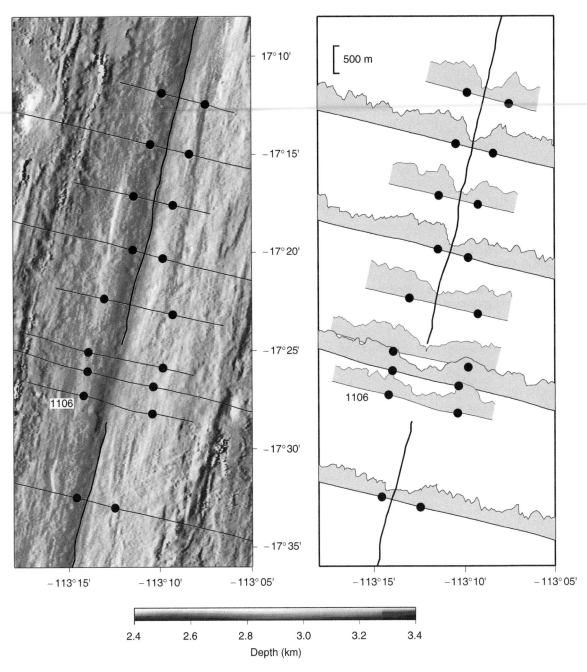

Figure 5 Illustration of the thickening of the seismically inferred extrusive crust (layer 2A) across the axis of the southern East Pacific Rise. Left-hand panel: Bathymetry map of the region with the location of cross-axis seismic lines shown in light line. The bold black line shows the location of the narrow depression along the ridge axis where most volcanic activity occurs. The black dots show the width of the region over which the seismically inferred extrusives accumulate as interpreted from the data shown in the right-hand panel. Right-hand panel: Thickness of the extrusive crust inferred from the seismic data along each cross-axis line. Black dots mark the location where 2A reaches its maximum thickness away from the axis. Seismic line 1106 shown in **Figure 4** is labeled. (Data from Carbotte SM, Mutter JC and Wu L (1997) *Journal of Geophysical Research* 102: 10165–10184).

data collected elsewhere along the Mid-Atlantic Ridge could be due to the imaging problems associated with the very rough topography of the seafloor typical at this ridge. However, there is also evidence from refraction data and seismicity studies that large, steady-state magma bodies are not present beneath this ridge. Microearthquake data show that earthquakes can occur to depths of 8 km beneath parts of the Mid-Atlantic Ridge, indicating that the entire crustal section is sufficiently cool for brittle failure. In other areas, slightly reduced velocities within the crust have been identified, indicating warmer

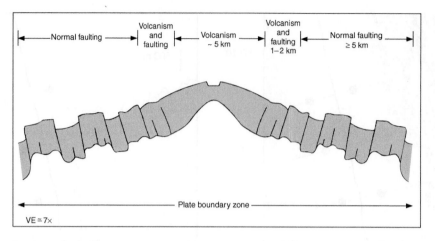

Figure 6 Schematic representation of the accumulation of the extrusive crust on the fast-spreading East Pacific Rise. The extrusive layer gradual increases in thickness within a wide zone centered on the narrow depression that marks the innermost axis. Normal faults begin to develop at the edges of this volcanic zone but may be buried by the occasional lava flow that reaches this distance. Beyond this zone, large-scale normal faulting occurs that gives rise to the fault-bounded abyssal hills and troughs found on the ridge flanks. (Reproduced from Carbotte SM, Mutter JC, and Wu L (1997), *Journal of Geophysical Research* 102: 10165–10184).

temperatures and possibly the presence of small pockets of melt.

The prevailing model for magma chambers beneath ridges (**Figure 7**), incorporates the geophysical constraints on chamber dimensions described above as well as geochemical constraints on magma chamber processes. At fast-spreading ridges (**Figure 7A**), the magma chamber is composed of the narrow and thin melt-rich magma lens that overlies a broader crystal mush zone and surrounding region of hot but solidified rock. The dike injection events and volcanic eruptions that build the upper crust are assumed to tap the magma lens. The lower crust is formed from the crystal residuum within the magma lens and from the broader crystal mush zone. At slow-spreading ridges (**Figure 7B**) a short-lived dike-like crystal mush zone without a steady-state magma lens is envisioned. At these ridges volcanic eruptions occur and the crystal mush zone is replenished during periodic magma injection events from the mantle.

Moho

The base of the crust is marked by the Mohorovcic Discontinuity, or Moho, where P-wave velocities increase from values typical of lower crustal rocks (6.8–$7.0\,\mathrm{km\,s^{-1}}$) to mantle velocities ($>8.0\,\mathrm{km\,s^{-1}}$). The change in P-wave velocity is often sufficiently abrupt that a subhorizontal Moho reflection is observed from which the base of the crust can be mapped. Depth to seismic Moho provides our best estimates of crustal thickness and is used to study how total crustal production varies in different ridge settings.

Characteristics of Moho at Mid-ocean Ridges

Reflection Moho is often imaged in data collected at the East Pacific Rise (**Figure 8**). In places it can be traced below the region of lower crustal velocities found at the ridge and occasionally beneath the magma lens reflection itself. Depths to seismic Moho indicate average crustal thicknesses of 6–7 km. There is no evidence for thickening away from the ridge crest, indicating that the crust acquires its full thickness within a narrow zone at the axis. The Moho reflection has three characteristic appearances on the East Pacific Rise: as a single, a diffuse, or a shingled event. These variations presumably reflect changes in the structure and composition of the crust-to-mantle transition such as are observed in ophiolites, where the base of the crust can vary from a wide band of alternating lenses of mafic and ultramafic rocks to an abrupt and simple transition zone.

At the Mid-Atlantic Ridge, the base of the crust is not marked by a strong Moho reflection such as is imaged at the East Pacific Rise. Here an indistinct boundary is found that is absent in many places. Hence most of our information on crustal thickness at this ridge has come from seismic refraction studies. Detailed refraction surveys are available within a number of locations that show a clear pattern of thinner crust (by 1–4 km) toward transform faults and smaller ridge offsets. These results are interpreted to reflect focused mantle upwelling and greater crustal production within the central regions of ridge segments away from ridge offsets.

Significant variations in crustal thickness are also observed along the East Pacific Rise. However, in the

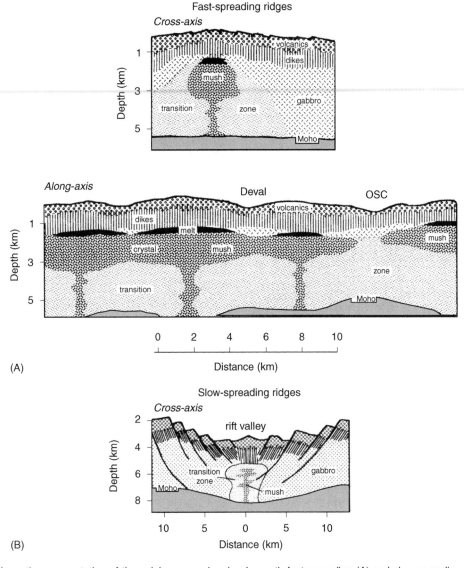

Figure 7 Schematic representation of the axial magma chamber beneath fast-spreading (A) and slow-spreading (B) ridges. At a fast-spreading ridge a thin zone of predominantly melt (black region) is located at 1–2 km below seafloor that grades downward into a partially solidified crystal mush zone. This region is in turn surrounded by a transition zone of solidified but hot rock. Along the ridge axis, the 'melt' sill and crystal mush zone narrows and may disappear at the locations of ridge discontinuities (labeled Deval and OSC in along-axis profile). At a slow-spreading ridge a steady-state melt region is not present. Here a dike-like mush zone forms small sill-like intrusive bodies that crystallize to form the oceanic crust. (Reproduced from Sinton JA and Detrick RS (1991) *Journal of Geophysical Research* 97: 197–216.)

region with the best data constraints (9°–10°N) the spatial relationships are the opposite of those observed on the Mid-Atlantic Ridge. Within this region, crust is ∼2 km thinner, not thicker, within the central portion of the segment where a range of ridge crest observations indicate that active crustal accretion is focused. At this fast-spreading ridge the presence of a steady-state magma chamber and broad region of hot rock (**Figure 7**) may permit efficient redistribution of magma away from regions of focused delivery from the mantle. The absence of a steady-state magma chamber beneath the slow-spreading Mid-Atlantic Ridge may prohibit significant along-axis transport of magma such that at this ridge thicker crust accumulates at the site of focused melt delivery.

Variations in Crustal Structure with Spreading Rate

Spreading rate has long been recognized as a fundamental variable in the crustal accretion process. At slow-spreading ridges new crust is formed within a pronounced topographic depression, whereas at fast spreading ridges a smooth and broad topographic high is found. Gravity anomalies indicate significant variations in crustal and mantle properties along the axis of slow-spreading ridges, whereas they are subdued and quite uniform at fast-spreading ridges. Magnetic anomalies are more complex and often difficult to identify at slow-spreading ridges and seafloor basalts are typically more primitive. These observations indicate that spreading rate plays an important role in how magma is segregated from the mantle and delivered to form new oceanic crust. Seismic techniques provide direct constraints on the significance of spreading rate for the distribution of magma within the crust, total crustal product-ion, and the internal stratification of the crust re-sulting from the magmatic processes of crustal creation.

Some aspects of the seismic structure of ridges are surprisingly similar at all spreading rates. The thickness of the extrusive layer away from the ridge axis is similar (~ 350–$600\,\mathrm{m}$), indicating that the total volume of extrusives produced by seafloor spreading is independent of spreading rate. Average crustal thickness is also comparable at all spreading ridges (6–$7\,\mathrm{km}$) and total crustal production does not appear to depend on spreading rate.

However, the characteristics of crustal magma lenses and the pattern of accumulation of the extrusive layer are different at fast and slow ridges. Throughout the fast-spreading range (85–$150\,\mathrm{mm}$ y^{-1}) the extrusive section is thin ($\sim 200\,\mathrm{m}$) at the ridge axis (**Figure 9**)and accumulates away from the axis within a zone 2–6 km wide. At these rates, magma lenses are imaged beneath much of the axis and have similar widths (**Figure 10**) and

are located at similar depths within the crust (**Figure 11**).

At slower-spreading ridges (<70–$80\,\mathrm{mm\,y^{-1}}$) magma lenses appear to be present only intermit-tently. Where they are observed they lie at a deeper level within the crust ($>2.5\,\mathrm{km}$) and form a second distinct depth population (**Figure 11**). The extrusive layer is thicker along the axis and does not system-atically thicken away from the ridge. At these ridges the extrusive section appears to acquire its full thickness within the innermost axial zone. These differences in the accumulation of extrusives at fast-spreading and slow-spreading ridges could reflect differences in lava and eruption parameters (e.g., eruptive volumes, lava flow viscosity, and morph-ology) that govern flow thicknesses and the distances lavas may travel from their eruption sites.

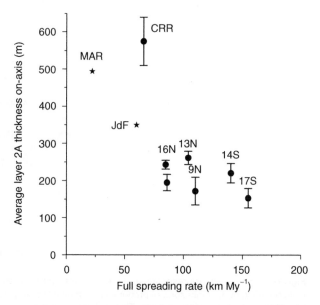

Figure 9 Thickness of the extrusive crust at the ridge axis versus spreading rate. For data obtained from detailed reflection surveys, average thickness is shown with black dots and standard deviations. East Pacific Rise data are labeled by survey location and are from (16N) Carbotte SM, Ponce-Correa G and Solomon A (2000) *Journal of Geophysical Research* 105: 2737–2759; (13N) Babcock JM, Harding AJ, Kent GM and Orcutt JA (1998) *Journal of Geophysical Research* 103: 30451–30467; (9N) Harding AJ, Kent GM and Orcutt JA (1993) *Journal of Geophysical Research* 98: 13925–13944; (14S) Kent GM, Harding AJ, Orcutt JA *et al.* (1994) *Journal of Geophysical Research* 99: 9097–9116; (17S) Carbotte SM, Mutter JC and Wu L (1997). *Journal of Geophysical Research* 102; 10165–10184. Costa Rica Rift (CRR) data are from Buck RW, Carbotte SM, Mutter CZ (1997) *Geology* 25: 935–938. Data from other ridges are derived from other seismic methods and are shown in stars. Data for the Mid-Atlantic Ridge (MAR) are from Hussenoeder SA, Detrick RS and Kent GM (1997), *EOS Transactions AGU*, F692. Data for Juan de Fuca Ridge (JdF) are from McDonald MA, Webb SC, Hildebrand JA, Cornuelle BD and Fox CG (1994) *Journal of Geophysical Research* 99: 4857–4873.

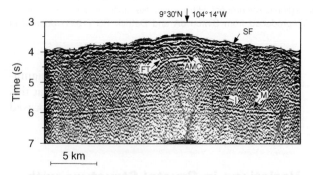

Figure 8 Multichannel seismic line crossing the East Pacific Rise at 9°30′N showing the Moho reflection (labeled M). The seafloor (SF) magma chamber reflection (AMC) and other intra-crustal reflections (FT, I) are labeled. (Reproduced from Barth GA and Mutter JC (1996) *Journal of Geophysical Research* 101: 17951–17975.)

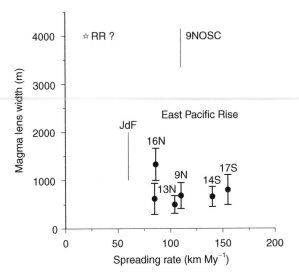

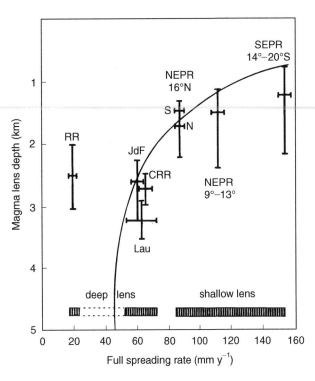

Figure 10 Width of magma lens reflections beneath ridges versus spreading rate. Average widths (black dots) and standard deviations are shown for regions where detailed reflection surveys have been carried out. East Pacific Rise data are labeled by survey location and are from (16N) Carbotte SM, Ponce-Correa G and Solomon A (2000) *Journal of Geophysical Research* 105: 2737–2759; (13N) Babcock JM, Harding AJ, Kent GM and Orcutt JA (1998) *Journal of Geophysical Research* 103: 30451–30467; (9N and 9NOSC) Kent GM, Harding AJ and Orcutt JA (1993) *Journal of Geophysical Research* 98: 13945–13970, 13971–13996 (the data labeled 9NOSC correspond with an unusually wide lens mapped near an overlapping spreading center); (14S) and (17S) from compilation of Hooft EE, Detrick RS and Kent GM (1997) *Journal of Geophysical Research* 102: 27319–27340. An estimate of lens width from wide-angle seismic data along the Reykjanes Ridge (RR) is shown in open star: Sinha MC, Navin DA, MacGregor LM *et al.* (1997) *Philosophical Transactions of the Royal Society of London* 355(1723): 233–253. The Juan de Fuca (JdF) estimate is from Morton JL, Sleep NH, Normark WR and Tompkins DH (1987) *Journal of Geophysical Research* 92: 11315–11326.

What Controls the Depth at which Magma Chambers Reside at Ridges?

Two main hypotheses have been put forward to explain the depths at which magma chambers are found at ridges. One hypothesis is based on the concept of a level of neutral buoyancy for magma within oceanic crust. This model predicts that magma will rise until it reaches a level where the density of the surrounding country rock equals that of the magma. However, at ridges, magma lenses lie at considerably greater depths than the neutral buoyancy level predicted for magma if its density is equivalent to that of lavas erupted onto the seafloor ($2700 \, \mathrm{kg \, m^{-3}}$). Either the average density of magma is greater, or mechanisms other than neutral buoyancy control magma lens depth.

The prevailing hypothesis is that magma chamber depth is controlled by spreading rate-dependent variations in the thermal structure of the ridge. In

Figure 11 Average depth of magma lens reflections beneath ridges versus spreading rate. The curved line shows the depth to the 1200°C isotherm calculated from the ridge thermal model of Phipps Morgan J and Chen YJ (1993) *Journal of Geophysical Research* 98: 6283–6297. Data from different ridges are labeled Reykjanes Ridge (RR), Juan de Fuca Ridge (JdF), Costa Rica Rift (CRR), Lau Basin (Lau), northern and southern East Pacific Rise (NEPR and SEPR, respectively). (Reproduced from Carbotte SM, Mutter CZ, Mutter J and Ponce-Correa G (1998) *Geology* 26: 455–458.)

this model, a mechanical boundary such as a freezing horizon or the brittle–ductile transition acts to prevent magma from rising to its level of neutral buoyancy. Both of these horizons will be controlled by the thermal structure of the ridge axis. Compelling support for this hypothesis was provided by the inverse relation between spreading rate and depth to low-velocity zones at ridges apparent in early datasets. Numerical models of ridge thermal structure have been developed that predict systematic changes in the depth to the 1200°C isotherm (proxy for basaltic melts) with spreading rates that match the first-order depth trends for magma lenses. This model predicts a minor increase in lens depth within the fast-spreading rate range and an abrupt transition to deeper lenses at intermediate spreading rates, consistent with the present dataset (**Figure 11**).

However, there is no evidence for the systematic deepening of magma lenses within the intermediate to slow spreading range that is predicted by the numerical models (**Figure 11**). Instead, where magma lenses have been observed at these ridges, they cluster at 2.5–3 km depth. At these spreading rates there may be large local variations in the supply of magma from the mantle to the axis that control ridge thermal structure and give rise to shallower magma lenses than predicted from spreading rate alone. In light of recent observations, the role of neutral buoyancy may need to be reconsidered. If the magma lens is not a region of 100% melt, magma densities may be considerably higher than used in previous neutral buoyancy calculations. The magma lens may indeed lie at its correct neutrality depth, and the observed variation in magma lens depth may reflect changes in the density of the melt and crystal aggregate found within the lens.

See also

Mid-Ocean Ridge Geochemistry and Petrology. Mid-Ocean Ridge Tectonics, Volcanism, and Geomorphology. Seamounts and Off-Ridge Volcanism. Seismic Structure.

Further Reading

Buck WR, Delaney PT, Karson JA, and Lagabrielle Y (eds.) (1998) *Faulting and Magmatism at Mid-Ocean Ridges,* Geophysical Monograph 106. Washington, DC: American Geophysical Union.

Detrick RS, Buhl P, Vera E, *et al.* (1987) Multichannel seismic imaging of a crustal magma chamber along the East Pacific Rise. *Nature* 326: 35–41.

Jacobson RS (1992) Impact of crustal evolution on changes of the seismic properties of the uppermost oceanic crust. *Reviews of Geophysics* 30: 23–42.

Phipps Morgan J and Chen YJ (1993) The genesis of oceanic crust: magma injection, hydrothermal circulation, and crustal flow. *Journal of Geophysical Research* 98: 6283–6297.

Sinton JA and Detrick RS (1991) Mid-ocean ridge magma chambers. *Journal of Geophysical Research* 97: 197–216.

Solomon SC and Toomey DR (1992) The structure of mid-ocean ridges. *Annual Review of Earth and Planetary Sciences* 20: 329–364.

MID-OCEAN RIDGE SEISMICITY

D. R. Bohnenstiehl, North Carolina State University, Raleigh, NC, USA

R. P. Dziak, Oregon State University/National Oceanic and Atmospheric Administration, Hatfield Marine Science Center, Newport, OR, USA

Introduction

The mid-ocean ridge that divides this planet's ocean basins represents a rift system where new seafloor is emplaced, cooled, and deformed. These processes facilitate hydrothermal circulation and the exchange of elements between the solid Earth and ocean, support complex ecosystems in the absence of sunlight, and form one of the most active and longest belts of seismicity on the planet (**Figure 1**).

This article outlines the tools used to study earthquakes in the oceanic ridge and transform environments, discusses the underlying mechanisms that cause or influence seismicity in these settings, and explores the impacts of earthquakes on submarine hydrothermal systems. As shown below, earthquakes can be used to track a number of important physical processes, including tectonic faulting, subsurface diking, seafloor eruptions, and hydrothermal cracking. As such, studies of earthquake patterns in space and time have become

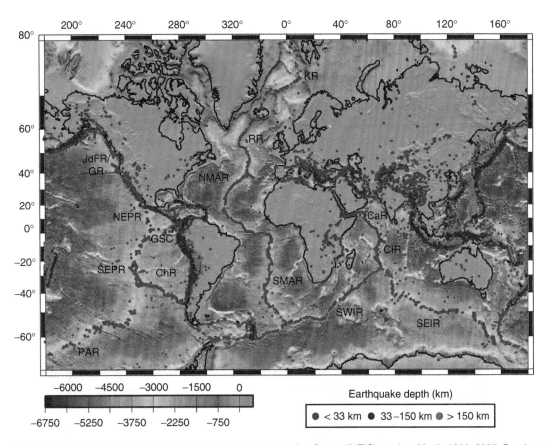

Figure 1 Global map of seismicity from National Earthquake Information Center (NEIC) catalog, M≥5, 1980–2005. Depths <33 km (red), 33–150 km (blue), and >150 km (green). Mid-ocean ridges and oceanic transforms are defined by narrow bands of shallow hypocenter earthquakes. Spreading centers and approximate full spreading rates: Southern East Pacific Rise (SEPR, ~140 mm yr^{-1}), Northern East Pacific Rise (NEPR, ~110 mm yr^{-1}) Pacific–Antarctic Ridge (PAR, ~65 mm yr^{-1}), Galapagos Spreading Center (GSC, ~45–60 mm yr^{-1}), Chile Rise (ChR, ~50 mm yr^{-1}), Northern Mid-Atlantic Ridge (NMAR, 25 mm yr^{-1}), Southern Mid-Atlantic Ridge (SMAR, ~30 mm yr^{-1}), Carlsberg Ridge (CaR, ~30 mm yr^{-1}), Central Indian Ridge (CIR, ~35 mm yr^{-1}), Southwest Indian Ridge (SWIR, ~15 mm yr^{-1}), Southeast Indian Ridge (SEIR, ~70 mm yr^{-1}), Kolbeinsey/Mohns Ridges (KR, ~15–20 mm yr^{-1}), Reykjanes Ridge (RR, ~20 mm yr^{-1}), Juan de Fuca and Gorda Ridges (JdFR/GR, ~60 mm yr^{-1}). Earthquake data from http://earthquake.usgs.gov.

fundamental in efforts to understand the seafloor spreading system within an integrated framework.

Methods of Monitoring Seismicity

Seismicity in the oceanic ridge-transform environment is monitored using a combination of hydroacoustic technologies, which record water-borne acoustic phases associated with submarine earthquakes, and traditional seismic sensors, which record ground motion induced by compressional and shear waves propagating through the solid Earth (**Figure 2**). These technologies should be viewed as complementary, as each presents certain advantages and limitations.

Seismometers

There are hundreds of seismometers deployed on the continents and islands across the globe. These instruments, operated primarily by governments and universities, form networks of sensors that can be used to monitor seismicity, as well as clandestine nuclear tests. However, for spreading centers that lie thousands of kilometers from the nearest seismic station, our ability to detect and locate earthquakes remains limited. Within the most remote ocean basins, global seismic networks consistently detect only earthquakes larger than roughly magnitude (M) 4.5–5, or ruptures having an approximate physical scale of ≥ 1 km. The signals generated by smaller events typically attenuate to below background noise levels as they traverse long solid-Earth paths between the source region and land-based sensors.

Seismometers may also be deployed on or beneath the seafloor in containers that protect the instrumentation from the extreme pressure of the deep ocean (**Figure 2**). Seafloor instruments are commonly known as ocean bottom seismometers (OBSs). They typically are deployed from a surface ship, allowed to free-fall into position, and later located using acoustic ranging techniques. Upon retrieval, an acoustic switch is remotely triggered, causing the instrument package to release a set of anchor weights and rise buoyantly to the surface in the vicinity of a waiting ship.

The detection capabilities of an OBS array depend on the number and distribution of stations. Deployment of a dozen or more instruments for year-long or multiyear observations are becoming increasingly common with recent improvements in technology. OBSs deployed at very local scales, arrays with apertures of only a few kilometers, may be used to monitor small cracking events (often with $M < 0$) in the vicinity of hydrothermal systems. Deployments of larger-aperture arrays, tens of kilometers across, are typically used to study volcanic and tectonic processes within a spreading segment and may provide a record of many earthquakes that would otherwise go undetected by land-based seismometers.

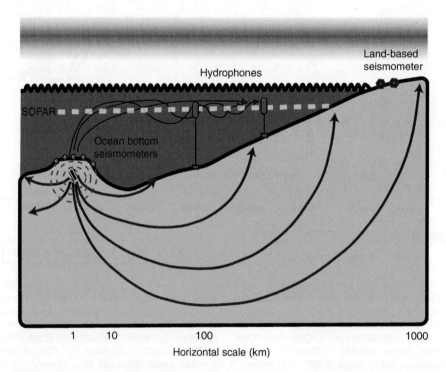

Figure 2　Technologies used to monitor seismicity within the mid-ocean ridge and oceanic transform environments. Horizontal scale indicates network aperture and approximate distances at which earthquakes are commonly monitored using the various methods.

Hydrophones

Due to the combined dependence of ocean sound speed on pressure and temperature, much of the global ocean exhibits a low-velocity region known as the sound fixing and ranging (SOFAR) channel. Seismically generated acoustic energy may become trapped in the SOFAR, where it propagates laterally via a series of upward- and downward-turning refractions having little interaction with the seafloor or sea surface. The attenuation due to geometric spreading is cylindrical (R^{-1}) for SOFAR-guided waves, making transmission significantly more efficient relative to solid-Earth phases that undergo spherical spreading loss (R^{-2}).

Water-borne, earthquake-generated acoustic signals were first observed on near-shore seismic stations. As these signals arrived after the faster propagating primary (P) and secondary (S) solid-Earth phases, they were termed tertiary (T). Today T phases are monitored most commonly by hydrophones deployed directly within the SOFAR.

The first systematic effort to use hydrophone data to produce a continuous catalog of mid-ocean ridge earthquakes began in the early 1990s when NOAA/PMEL gained access to the US Navy's Sound Surveillance System (SOSUS), a permanent network of bottom-mounted hydrophone arrays within the Northeast Pacific. This enabled the real-time detection of T waves generated by seafloor earthquakes and reduced the detection threshold for seismicity at the Juan de Fuca and Gorda Ridges by almost 2 orders of magnitude. The geometry of the SOSUS network also provided a better azimuthal distribution of stations than could be accomplished using land-based seismometers. The station geometry, combined with the existence of a well-defined velocity model of the oceans, also yielded significant improvements in location accuracy.

Early successes using SOSUS facilitated the development of moored autonomous underwater hydrophones (AUHs) that could be used to monitor global ridge segments. In this design, the hydrophone sensor and instrument package are suspended within the SOFAR channel using a seafloor tether and foam flotation (**Figure 2**). These instruments have been deployed successfully along mid-ocean ridge spreading centers in the Atlantic and Pacific Oceans, as well as back-arc spreading systems of the Marianas (W. Pacific) and Bransfield Strait (Antarctica). A typical deployment consists of only six to seven instruments that monitor $\sim 20°$ along axis and provide a consistent record of earthquakes with $M \geq 2.5-3.0$.

Although regional hydrophone arrays can provide improvements in detection and location capabilities,

relative to land-based seismic stations, it is not presently possible to extract the focal mechanism directly from the T waveform. Similarly, only relative depth estimates are acquired through measurement of the T wave rise time, defined as the time between the onset of the signal and its amplitude peak.

Global to Regional Tectonic Patterns

A view of global seismic patterns (**Figure 1**) shows ocean basin earthquakes to be narrowly focused along the spreading axis. Such events have shallow hypocenters and focal mechanisms that dominantly indicate normal faulting on moderately dipping ($\sim 45°$) ridge-parallel structures. Shallow-hypocenter earthquakes also cluster tightly along the oceanic transforms that accommodate the differential motion between offset ridge segments, with the sense of motion along these conservative plate boundaries being dominantly strike-slip on subvertical structures. These patterns reflect a stress regime arising from the motion of the plates and the geometry of their boundaries (**Figure 3**).

Transform and Segment Boundary Seismicity

In the late 1960s, focal mechanism studies of oceanic transform earthquakes provided key evidence in support of plate tectonic theory. Early physiographic maps of the oceans had outlined the mid-ocean ridge system and identified places where this submarine mountain range appeared to be offset laterally. One might infer from these observations that the two ridge segments were once aligned and that their offset represents the cumulative displacement along the connecting fault system. In contrast, the then emerging theory of plate tectonics required the

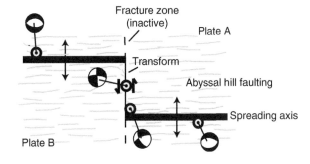

Figure 3 Schematic showing the style of faulting and seismicity associated with an idealized spreading center and transform plate boundary. Double black line indicates the axis of the oceanic spreading center. Thin gray lines show normal faults along the rift flanks. Arrows indicate the direction of relative motion between the plates. Focal mechanisms shown with shaded regions indicating compressive first motions.

opposite sense of motion along these structures in order to accommodate the process of seafloor spreading along two ridge segments that maintain a near-constant offset through time. As Sykes showed in 1967, the first-motion direction of P waves radiated during transform earthquakes clearly supports the latter model.

Today the orientation and slip direction of earthquakes are determined routinely for global events larger than M 5–5.5. These data, combined with detailed morphological observations, provide a more complete picture of the structure and seismicity of oceanic transforms and higher-order segment boundaries.

Many transforms are segmented, stepping in an *en echelon* fashion, with pull-apart basins and more commonly intra-transform spreading centers defining the segmentation. As an earthquake propagates, offsets of sufficient size may arrest the rupture, limiting its spatial extent. The maximum size of a transform earthquake is observed to be a function of slip rate, with the largest transform earthquakes occurring at slow spreading rates ($<40\,\mathrm{mm\,yr^{-1}}$) and having sizes of roughly M 7.5. At spreading rates $>100\,\mathrm{mm\,yr^{-1}}$, the maximum observed magnitude drops to below M 6.5. This reflects the thermal structure of oceanic lithosphere, with a shallower depth of seismogenic rupture at faster spreading rates, and potentially a greater tendency for transform segmentation.

Earthquakes with an oblique sense of slip and orientation are sometimes observed at inside corners, the region where the active transform segment meets the ridge axis (**Figure 3**). Such events may represent slip on curvilinear abyssal hill faults that rotate up to 15–20° as they approach the transform. In other cases, inside corner earthquakes may be better depicted as a compound rupture, where strike-slip motion along the transform and dip-slip motion along an orthogonal abyssal hill fault occur simultaneously. In these cases, the radiation pattern cannot be well described by the double-couple force model used to represent the vast majority of global earthquakes.

Transform seismicity is restricted principally to the region between the two active spreading centers, as no relative motion is required across the fault beyond this region. Fracture-zone scars, however, may be traced across ocean basins and continue to represent zones of weakness. In some areas, intra-plate stresses are sufficient to reactivate these structures, infrequently generating large earthquakes.

Oceanic transforms exhibit similar kinematics to continental transforms, such as the San Andreas Fault in the United States or the Anatolian Fault in Turkey. Oceanic transforms, however, show one major difference – an inventory of earthquakes along oceanic transforms documents a dramatic 'slip deficit', with the majority of all transform motion occurring aseismically. A prediction of the moment release rate along a transform can be obtained as $\sum M_o/t = v\mu lw$, where v is the relative plate velocity, μ is the shear modulus of the rock, l is the length of the transform, w is the down-dip width of the rupture, and t is the time period considered. M_o is the static seismic moment, which for an individual earthquake is the product of fault rupture area (lw), mean slip, and μ. The width w is generally taken to correspond to the ~ 600–$650°$ isotherm, as constrained by slip inversion results and laboratory-based deformation studies (**Figure 4**).

When the sum of the observed seismic moment on a transform is compared with the expected moment, their ratio (or seismic coupling coefficient, α) is typically ~ 1 for continental transforms, but much less than 1 along oceanic transforms. The average α for all oceanic transforms is roughly 0.25, indicating that three-fourths of all transform motion occur aseismically. Studies investigating the velocity dependence of α yield conflicting results. Such assessments are complicated by the presence of significant inter-transform variability in α even at a given spreading rate, uncertainty in the depth of seismic faulting, and the somewhat limited timescale of the observations. The most recent and detailed studies, however, suggest little-to-no systematic relationship with slip rate, or perhaps slightly lower α at fast-spreading rates.

Aseismic fault motions on the oceanic transforms are likely taken up by slow, creeping events. These slow ruptures are inefficient at producing seismic waves and so are termed quiet or silent earthquakes. It recently has been suggested that silent earthquakes may trigger large seismic quakes in neighboring portions of the transform that are more prone to stick-slip behavior. In this view, some seismic events on the transforms may be viewed as aftershocks of these silent earthquakes.

At higher-order segment boundaries, propagating rifts and overlapping spreading centers accommodate ridge segmentation through a mechanism known as bookshelf faulting (**Figure 5**). Here the seafloor between the overlapping rift tips is deformed by simple shear, resulting in the rotation of the initially ridge-parallel seafloor fabric. The style of deformation is reminiscent of a stack of books on a shelf tipping over, with slippage between the 'books' occurring along the preexisting abyssal fault systems. Focal mechanisms indicate a strike-slip sense of motion, with one set of nodal planes trending

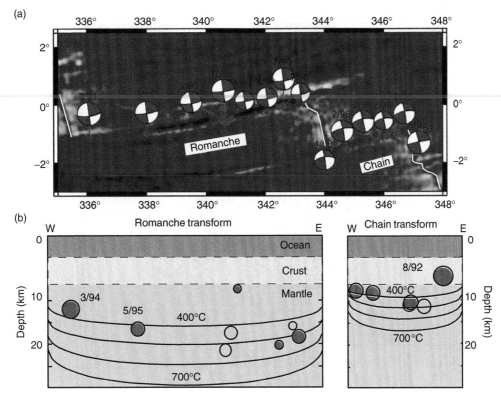

Figure 4 (a) Map of the equatorial Mid-Atlantic Ridge showing the Romanche and Chain transform faults. The focal mechanisms obtained in the broad-band body-wave modeling are joined to their NEIC locations (red circles). (b) The centroid depths of the earthquakes are plotted as circles with symbol size proportional to magnitude. Solid symbols represent earthquakes with the best-resolved parameters, and open symbols those with fewer data or poorer fits. All the depths are accurate to within 3 km. Isotherms are calculated by using a half-space cooling model and averaging both sides of the transform. Slip inversions of the 1994 and 1995 Romanche earthquakes suggest that the former ruptured from near the surface to 20-km depth, and the latter from 10- to 25-km depth. This is consistent with the ~600° isotherm controlling the maximum depth of seismic slip on these transforms. The 1992 Chain earthquake is the only event with a centroid within the crust, but it was large enough to have ruptured into the upper mantle. Reprinted by permission from Macmillan Publishers Ltd: *Nature* (Abercrombie R and Ekström G (2001) Earthquake slip on oceanic transform faults. *Nature* 410: 74–77), copyright (2001).

parallel to an abyssal fabric that becomes progressively rotated within the deformation zone. This mode of deformation is most common at fast to intermediate spreading rates, with well-studied examples along the southern East Pacific Rise ($>120\,\mathrm{mm\,yr^{-1}}$), Galapagos ($\sim60\,\mathrm{mm\,yr^{-1}}$), and Lau Basin spreading centers ($\sim100\,\mathrm{mm\,yr^{-1}}$). A spectacular slow-spreading example, however, can be observed within Iceland's southern seismic zone.

Spreading-Center Earthquakes

When the oceanic ridge systems are monitored using global seismic stations ($M\geq\sim4.5$) or regional hydrophone arrays ($M\geq\sim2.5\text{--}3$) a sharp contrast in the frequency of mid-ocean ridge earthquakes is observed as function of spreading rate (**Figures 1** and **6**). Along fast-spreading ($>80\,\mathrm{mm\,yr^{-1}}$) axial highs, the ridge displays a dearth of small- to moderate-magnitude earthquakes. In contrast, such events are abundant along the rift valleys of a slow-spreading ($<40\,\mathrm{mm\,yr^{-1}}$) ridge system. These first-order patterns largely reflect the variable thickness of the brittle layer, which controls the predicted moment release, and seismic coupling coefficient of the rift-bounding normal fault systems.

The predicted rate of seismic moment release can be estimated in a manner analogous to that described for oceanic transforms: $\sum M_o/t = v\mu lw/\sin\theta\cos\theta$, where θ is the fault dip, v is the full spreading rate, l is the length of the plate boundary, and w is the thickness of the seismogenic lithosphere. A comparison with the observed moment release at a slow-spreading rift zone indicates that only 10–20% of the plate divergence is accommodated by seismic processes. This estimate agrees well with the amount of brittle strain accommodated by the normal fault populations of the abyssal plains. Hence, rift-bounding normal fault systems at slow-spreading centers display a high-level of seismic coupling, with

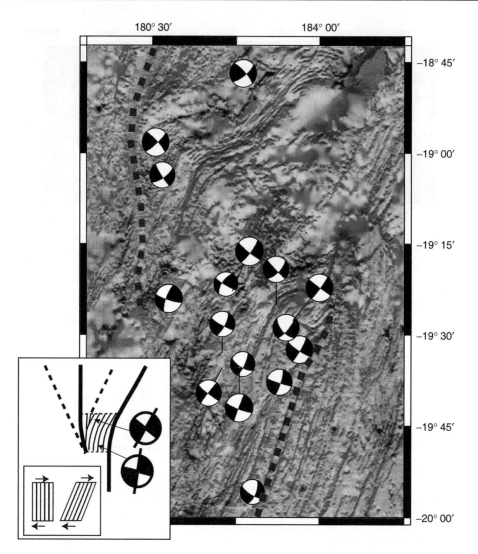

Figure 5 Central Lau Basin propagator with focal mechanics solutions. The Central Lau Basin Spreading Center is propagating to the south as spreading ceases along the northern tip of the Eastern Lau Spreading Center. Inset shows idealized case of bookshelf faulting, with progressive clockwise rotation of the abyssal fabric as the left-offset-propagating rift advances to the south. Focal mechanisms adapted from Wetzel *et al.* (1993) and the Global Centroid Moment Tensor Project (http://www.globalcmt.org). From Wetzel LR, Wiens DA, and Kleinrock MC (1993) Evidence from earthquakes for bookshelf faulting at large non-transform ridge offsets. *Nature* 362: 235–237.

the ratio of seismic fault slip to expected fault slip being close to 1. However, the emplacement of dikes along or near the rift axis takes up the majority (80–90%) of the plate separation.

At fast-spreading ridges, commonly less than 1% of the predicted moment release is observed seismically along the rift. While most of the plate separation is accommodated by diking, as it was at slower rates, faulting studies indicated the fast-spreading lithosphere undergoes an extension of $\sim$4–8%. The observed moment release is insufficient to account for this. Fast-spreading normal fault systems, therefore, appear to have low α and must accrue displacement during aseismic slip events or by

abundant microseismic activity too small to be detected by the hydrophones or global seismic stations.

At intermediate spreading rates, the density of seismic events shows a first-order correspondence with the morphology of the ridge, with small- to moderate-magnitude earthquakes being abundant along rifted-spreading centers and comparatively rare at intermediate-rate axial highs (**Figure 7**). Intermediate-spreading-rate ridges can therefore assume both the morphologic and seismic characteristics of the fast- and slow-spreading end members.

The maximum hypocentral depth of a rift zone earthquake decreases with increasing spreading rate, reflecting the temperature limits of brittle faulting

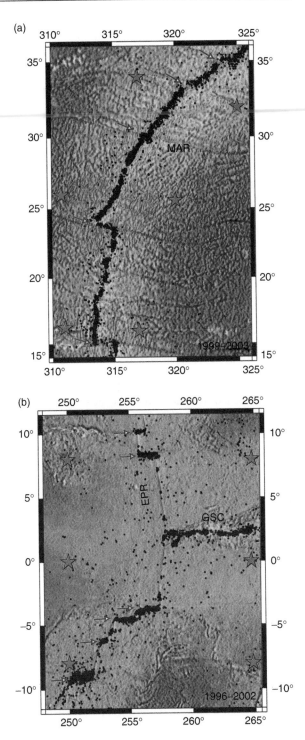

Figure 6 Mid-ocean ridge seismicity viewed from regional hydrophone arrays. (a) Mid-Atlantic Ridge ($\sim 25 \, \text{mm yr}^{-1}$), (b) East Pacific Rise ($\sim 110 \, \text{mm yr}^{-1}$) and Galapagos Spreading Center ($\sim 45–60 \, \text{mm yr}^{-1}$). Events located using four or more hydrophones are shown. $M \geq 2.5 – 3$ earthquakes are consistently detected by these arrays. Yellow stars indicate position of hydrophones in each array. Red arrows mark first-order (transform) offsets of the ridge axis. Data from http://www.pmel.noaa.gov.

and steeper thermal gradients at faster spreading rates (**Figure 8(a)**). OBS studies define additional intra- and inter-segment patterns. Within some slow-spreading segments, the along-axis depth of microseismicity shallows near the center of a segment and deepens near its ends. This reflects the presence of a hotter lithosphere near segment center and is consistent with a three-dimensional (3-D) pattern of upwelling and melt focusing at slow-spreading rates. Studies on the Mid-Atlantic Ridge also show a positive correlation between the relief of the median valley and the maximum depth of microseismicity detected using OBSs (**Figure 8(b)**). This is consistent with stretching models that indicate greater fault offsets in regions of thicker brittle lithosphere.

Magmatism and Diking

Prior to an eruption, mid-ocean ridge melts accumulate in crustal-level chambers. At fast-spreading rates axial magma lenses are ubiquitous, being found beneath $>60\%$ of the axis that has been surveyed using multichannel reflection techniques. At slow-spreading rates, crustal-level melt is absent beneath much of the ridge axis and melt bodies may be ephemeral or localized beneath axial volcanoes. At intermediate rates, the distribution of melt can range between the two extremes, with some axial highs showing nearly continuous along-axis magma chambers reminiscent of fast-spreading centers.

The accumulation of melt within the crustal chamber elevates its pressure relative to the surrounding host rock. This inflation deforms the surrounding crust and triggers earthquakes within the vicinity of the chamber. For magma to leave the chamber, the pressure (P) within must exceed the sum of the minimum confining stresses (σ_3) at the chamber boundary and the tensile strength (T) of the rock. The dike's orientation will be orthogonal to the least compression stress direction and therefore it should be subvertical and aligned parallel to the axis of spreading.

Although cracking in the vicinity of the dike tip creates many small earthquakes, most events of sufficient size to be detected on global seismic or regional hydrophone arrays are thought to occur on preexisting fault surfaces. In the region above the dike, a narrow zone of ridge-normal extension exists where seismicity is localized (**Figure 9**). A broader zone of extension exists beyond the along-axis tips of the intrusion, where seismicity will be triggered in front of a laterally propagating dike. As the dike passes, the lithosphere will be compressed at depth within the region adjacent to the dike.

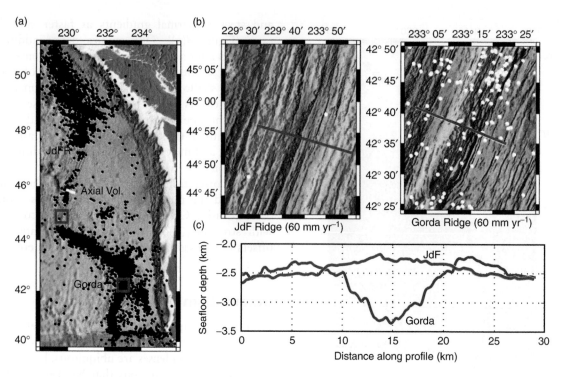

Figure 7 (a) SOSUS-detected earthquakes along the Gorda and Juan de Fuca Ridges in the Northeast Pacific, 1993–2005. Earthquakes located using four or more hydrophones are shown. $M \geq 2.5$ earthquakes are consistently detected by SOSUS. (b) Enlarged view of portions of the Gorda and Juan de Fuca Ridge crests. (c) Cross-axis profiles from these regions.

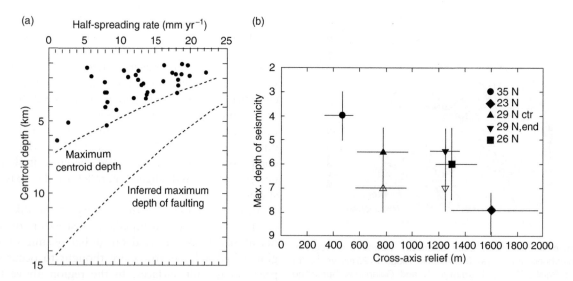

Figure 8 (a) Centroid depth of mid-ocean ridge earthquakes obtained from body waveform inversion of teleseismic arrivals vs. half-spreading rate. Maximum depth of seismic faulting is inferred to be less than twice the maximum centroid depth. (b) Relationship between cross-axis relief and maximum depth of seismicity. Maximum depth of seismicity was inferred from the focal depths of inner valley floor earthquakes, as recorded by OBS studies. (a) Reprinted by permission from Macmillan Publishers Ltd: *Nature* (Solomon SC, Huang PY, and Meinke L (1988) The seismic moment budget of slowly spreading ridges. *Nature* 334: 58–61), copyright (1988). (b) Reproduced from Barclay AH, Toomey DR, and Solomon SC (2001) Microearthquake characteristics and crustal VP/VS structure at the Mid-Atlantic Ridge, 35° N. *Journal of Geophysical Research* 106: 2017–2034 (doi:10.1029/2000JB900371), with permission from American Geophysical Union.

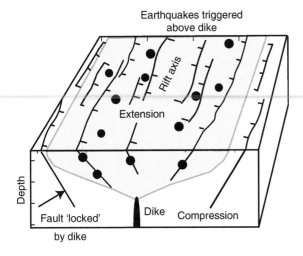

Figure 9 Cartoon showing regions of compression (white) and extension (shaded) associated with a vertically propagating dike. Thin black lines indicate normal faults, with ticks on the down thrown block. Seismicity (black dots) is promoted within the extensional region and inhibited elsewhere. The width of the failure zone scales with the depth to the top of the dike.

Much of what we know about mid-ocean ridge dike injection comes from SOSUS observations of intermediate-rate-spreading centers in the NE Pacific. On 26 June 1993, soon after monitoring began, an intense episode of volcanic seismicity was recorded in real time along the CoAxial Segment of the Juan de Fuca Ridge (JdFR), marking the first observation of such an event (**Figure 10**). The seismic swarm started near $46°\ 15'$ N and migrated northward during the next 40 h to $\sim 46°\ 36'$ N, where majority of earthquake activity occurred during the following 3 weeks. Earthquakes propagated NNE at a velocity of $0.3 \pm 0.1\ \mathrm{m\,s}^{-1}$. The reservoir that acted as the dike's source likely resided beneath, or to the south of, the initial swarm of earthquakes. The character and propagation velocity of this earthquake swarm were very similar to dike injections observed at Krafla and Kilauea Volcanoes. Seismicity went undetected by land-based seismic networks, suggesting earthquakes $M \leq 4.0$.

Since the CoAxial activity, six additional magmatic episodes have been observed hydroacoustically on the Northeast Pacific spreading centers, and several distinctive characteristics have been identified: (1)

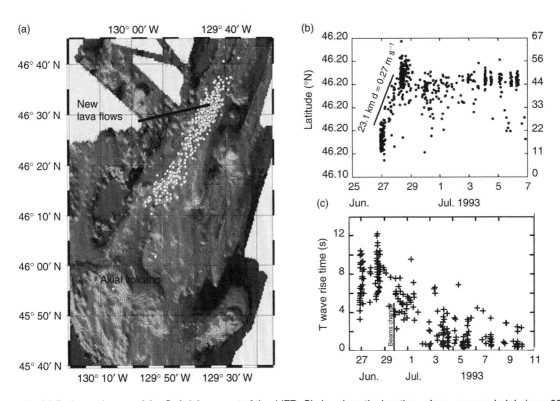

Figure 10 (a) Bathymetric map of the CoAxial segment of the JdFR. Circles show the locations of swarm recorded during a 23-day period of intense activity. (b) Along-segment position of epicenters vs. time during the swarm. (c) T wave rise time vs. time during the swarm. Events show a decrease in rise time consistent with the shallowing of hypocenters as the dike approaches the seafloor, erupting fresh lava in the north. Reproduced from Dziak RP, Fox CG, and Schreiner AE (1995) The June–July 1993 seismo-acoustic event at CoAxial Segment, Juan de Fuca Ridge: Evidence for a lateral dike injection. *Geophysical Research Letters* 22: 135–138, with permission from American Geophysical Union.

the sequences are swarms, lacking a dominant event, with a temporal history that cannot be described by a power-law decay in event rate (i.e., Omori's law); (2) the total number of earthquakes during swarms exceeds 10% (>350 events/week) of the variance of long-term background JdFR–Gorda Ridge seismicity; (3) swarms have several episodes of intense activity that reach 50–100 earthquakes/hour; (4) swarms last from several (>5) days to several weeks; (5) earthquakes may migrate up to tens of kilometers along-axis following the lateral injection of magma through the crust; and (6) swarms may be accompanied by continuous, broad-band energy (3–30 Hz) interpreted as 'intrusion tremor', resulting from magma breaking through the crust.

Autonomous hydrophone recordings from the north-central Mid-Atlantic Ridge ($\sim 25\,\mathrm{mm\,yr^{-1}}$) indicate that diking events are less frequent, consistent with the lower rates of plate separation. During more than 5 years of monitoring, only one probable volcanogenic swarm was detected along $\sim 2500\,\mathrm{km}$ of ridge axis. This activity was within the Lucky Strike segment (37° N), somewhat outside of the AUH array. Earthquake locations could be determined for 147 hydrophone-detected events, with 33 of sufficient size to be located by land-based seismometers ($M\ 3.6$–5.0). In terms of total moment release, this earthquake sequence was one of largest on the Mid-Atlantic Ridge in the last three decades. The activity displayed a swarm-like temporal

behavior and was accompanied by intrusion tremor; however, it was of short duration (<2 days) relative to previous episodes detected on the Northeast Pacific spreading centers.

By far the largest episode of volcanogenic seismicity recorded on a mid-ocean ridge system occurred within the Arctic Ocean basin on the ultra-slow-spreading (<5 mm yr^{-1}) Gakkel Ridge in 1999. Seismic activity began in mid-January and continued vigorously for 3 months, with a reduced rate of activity continuing for 4 additional months (**Figure 11**). In total, 252 events were large enough ($M > 4$) to be recorded on global seismic networks (no hydroacoustic monitoring was in place). Subsequent geophysical studies of the Gakkel indicate the presence of a large, recently erupted flow and a volcanic peak directly in the area of seismic activity, with several large central volcanoes spectacled throughout an otherwise magma starved and heavily sedimented rift system. The duration of the 1999 activity suggests intrusion event(s) spanning several months and indicates a significant volume of magma beneath these central volcanoes.

Because of their small magnitudes, earthquakes generated by diking events along fast-spreading ridge crests may be difficult to detect using hydroacoustic or global seismic techniques. Despite the greater rate of plate separation, during an 8-year period of hydroacoustic monitoring on the East Pacific Rise (10° S–10° N), only a handful of short-duration

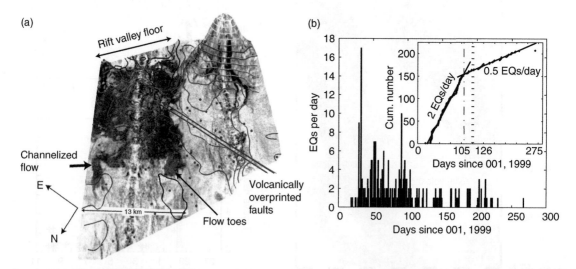

Figure 11 Seismovolcanic activity on the ultra-slow-spreading Gakkel Ridge near 85° N, 85° E. (a) Three-dimensional view from west (bottom) to east (top). The dark, reflective terrain is centered about a close-contoured high having a maximum vertical relief of 500 m. Red circles show the locations of epicenters, Jan.–Sep. 1999. (b) Histogram showing progression of the swarm through time, with each bar representing the number of events per day. Inset figure shows the cumulative number of events through time. There is a clear decrease in the rate of activity on 15 April (dot-dash line). Dashed line shows 6 May, when the USS *Hawkbill* passed over the area collecting the side scan imagery shown in (a). (a) Reprinted by permission from Macmillan Publishers Ltd: *Nature* (Edwards MH, Kurras GJ, Tolstoy M, Bohnenstieh DR, Coakley BJ, and Cochran JR (2001) Evidence of recent volcanic activity on the ultra-slow spreading Gakkel Ridge. *Nature* 409: 808–812), copyright (2001).

(hours to days) swarms containing a few tens to hundreds of locatable earthquakes were detected. They were primarily located along intra-transform spreading centers and near segment ends, where a somewhat colder, thicker lithosphere might be expected to support larger ruptures.

In January of 2006, a submarine diking event and seafloor eruption were captured for the first time by a network of OBSs. The eruption was located within the well-studied 9°50′ N region of the East Pacific Rise. As part of a multidisciplinary study, a network of up to 12 OBSs was deployed at this site in October 2003. Analysis has shown a gradual ramp-up in activity during a more than 2-year interval prior to the eruption, with activity culminating in an intense period of seismicity that lasted ~6 h on 22 January 2006. Although many of the OBS instruments were destroyed by the eruption, the event was also recorded by regional AUHs, and the joint analysis of these data is expected to further illuminate the process of dike injection at a fast-spreading ridge.

It should be emphasized that these somewhat limited observations may not be characteristic of all eruptive activity at mid-ocean ridges. Although ridge scientists often refer to a volcano-tectonic cycle, given a sufficiently long period of observation it is likely that mid-ocean ridge eruptions, like their continental equivalents, will display a power-law scaling with respect to size and duration, with smaller episodes occurring more frequently than larger ones. The scaling exponent and maximum eruption size, however, should be expected to vary as a function of spreading rate, reflecting variability in the dynamics of each system and the size of its crustal reservoir. Likewise, eruption frequency is likely to show a long-term clustering, rather than the periodic behavior sometimes alluded to.

field, strain rates from cooling are estimated to be on the order of $10^{-6}\,yr^{-1}$, about 3 orders of magnitude higher than tectonic strain rates.

In 1995, a vent-scale OBS study lasting 105 days was conducted at 9°50′ N on the East Pacific Rise, where a series of temperature probes also were deployed within the high-temperature vent systems (**Figure 12**). This study recorded a swarm of 162 microearthquakes during a period of about 3 h. Hypocenters defined a subvertical column at a depth of 0.7–1.1 km (**Figure 12**). Four days later, temperature sensors within a high temperature (Bio9) vent began to record increasing fluid temperature of ~1 °C/day, with this trend continuing for 7 days. Vent temperature then returned to background levels during the next ~120 days. Comparisons with vent temperatures during more than 3 years of monitoring showed this postseismic fluctuation to be the largest recorded, making a strong argument that the earthquakes had perturbed the hydrothermal system. It was suggested that fluids rapidly penetrated the new fractures and extracted heat from the fresh rock.

A decade later, multiyear OBS monitoring, in conjunction with *in situ* temperature and chemical and biological studies, has been reestablished in the 9°50′ N region. The results of this ongoing work, combined with similar vent-scale OBS studies conducted at intermediate- and slow-spreading vent sites, show microseismicity to be ubiquitous in hydrothermal areas. Moreover, they suggest a more complex relationship between seismic activity and vent hydrology than could have been envisioned based on the 1995 study, with temperature and flow responses exhibiting variable amplitude, sign, and phase, and many swarms producing no detectable perturbation at the vent sites.

Hydrothermal Seismicity

Within the mid-ocean ridge hydrothermal regime, earthquake activity may be triggered in response to the removal of heat, which leads to thermal contraction and subsequent cracking, or via hydrofracture when trapped pockets of fluid are heated. Such events are typically small, with rupture diameters of 1 m to tens of meters, and can only be recorded using local OBS arrays deployed within the high-temperature vent fields. Within the shallow crust the predicted mode of failure is for purely tensional (mode I) or mixed-mode fracture, rather than the shear failure observed for the larger-magnitude spreading-center earthquakes. Given estimates of the hydrothermal heat flux within a vent

Tidal Triggering

Solar and lunar tidal forces exert short period stress variations that act on the Earth. Beneath the oceans and in coastal areas, stress changes influencing earthquake occurrence reflect the combined effect of the direct Earth tide and indirect loading of the Earth by the ocean tides. Together these changes are on the order of $10^3 - 10^4\,Pa$, much less than the average stress drop of an earthquake or the strength of crustal rocks. Many studies have examined the role of tides in triggering seismicity around the globe. With the exception of some terrestrial volcanic areas, earthquakes and tides are generally not correlated or weakly correlated during periods when tidal fluctuations are at their largest amplitude.

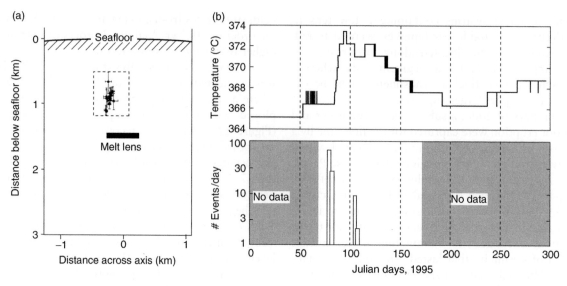

Figure 12   (a) Ridge normal cross section of relocated microearthquakes associated with a 1995 swarm beneath the 9° 50′ N hydrothermal site on the East Pacific Rise. (b) Correlation between the microearthquake swarm and the fluid exit temperature at vent Bio9. Reproduced from Sohn RA, Hildebrand JA, and Webb SC (1999) A microearthquake survey of the high-temperature vent fields on the volcanically active East Pacific Rise (9° 50′ N). *Journal of Geophysical Research* 104: 25367–25378 (doi:10.1029/ 1999JB900263), with permission from American Geophysical Union.

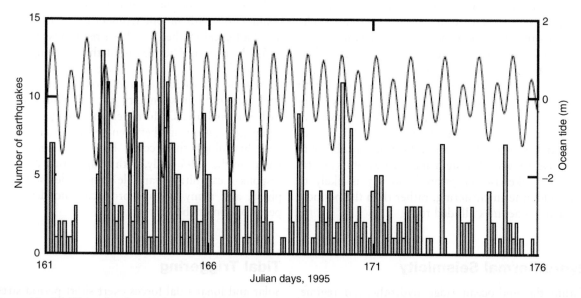

Figure 13   (a) Fifteen-day time histogram of the earthquake count in 2-h bins with the ocean tide time series superimposed. Earthquakes are triggered at periods of low ocean tide, when the seafloor is unloading and stress within the crust is most extensional. Data collected during an ocean bottom seismic experiment near 48° N, 231° E on the Endeavour Segment of the intermediate-rate JdFR. Reproduced from Wilcock W (2001) Tidal triggering of microearthquakes on the Juan de Fuca Ridge. *Geophysical Research Letters* 28: 3999–4002 (doi:10.1029/2001GL013370), with permission from American Geophysical Union.

Similarly, analysis of hydrophone-derived and global seismic catalogs does not show a robust correlation between tidal phase and the occurrence times of small- to moderate-size mid-ocean ridge earthquakes. However, microseismicity ($M < 2$) within axial hydrothermal systems commonly shows a strong correlation, with earthquakes occurring during periods of peak extensional stress. In some locations, such as the JdFR in the Northeast Pacific Ocean, the amplitude of the ocean tides is large and these peak extensional periods correspond to times of low tide, when the seafloor is unloaded as the height of the overlying water column reaches a minimum (**Figure 13**). In other areas, such as the

$9°\ 50'$ N region of the East Pacific Rise, tidal changes in the height of the sea surface are minimal and the Earth tide stresses have been shown to dominate.

One explanation for these observations comes from laboratory studies of rock failure, which indicate that earthquake triggering in response to a periodic loading is dependent on the period of the oscillation relative to the time it takes a slip event to nucleate. Earthquake nucleation time, which is inversely proportional to stressing rate, is estimated to be on the order of a year or more for typical tectonic settings. Consequently, these systems are largely insensitive to tidal stress changes of semi-diurnal frequency. In volcano-hydrothermal systems, however, cooling- and magma-induced stress changes elevate the stressing rate by several orders of magnitude. Therefore, earthquake nucleation times become comparable to tidal periods, and microseismicity in this setting becomes susceptible to tidal influence.

Impacts on Hydrothermal Systems

Earthquakes may disturb the hydrology of marine hydrothermal systems through several mechanisms. As described previously, microearthquakes occurring beneath a vent field can alter the local permeability, creating new fluid pathways that allow the migration of cold water into the reaction zone or redirect the escape of buoyant, heated waters. Since the precipitation of minerals within the up- and downflow zones is predicted to reduce permeability over time, earthquakes events may be critical in maintaining the longevity of hydrothermal systems.

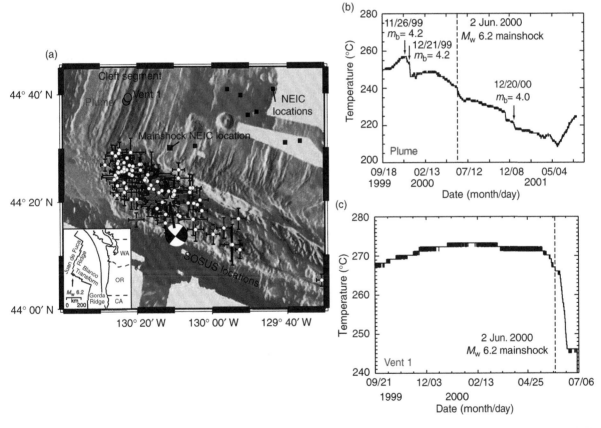

Figure 14 Map showing acoustically derived locations of foreshocks (red dots), mainshock (yellow star), and aftershocks (white dots) of 1–7 Jun. 2000 earthquake sequence on the Western Blanco Transform. Error bars on SOSUS earthquake locations represent one standard error. Note difference in location relative to NEIC seismic epicenters (black squares), which are scattered across the plate interior. Locations of Vent 1 and Plume hydrothermal vent sites along southern JdFR are also shown. (b) Twenty-three-month temperature record at Plume hydrothermal site. Fluid at Plume vent decreased 4.9 °C during the 18 h immediately following mainshock. Three additional temperature drops appear to be associated with much smaller ($M\sim4.0-4.2$) earthquakes along the western Blanco (blue dots in (a)). (b) Eleven-month temperature record at Vent 1 hydrothermal site shows a possible delayed response to the June 2000, M 6.2 earthquake. After 7 months of steady readings (272 ± 3 °C), probe temperature decreased slightly prior to the event, but then dropped markedly by 18 °C begining 7 days after the event. Reproduced from Dziak RP, Chadwick WW, Jr., Fox CG, and Embley RW (2003) Hydrothermal temperature changes at the southern Juan de Fuca Ridge associated with M_W 6.2 Blanco Transform earthquake. *Geology* 31: 119–122, with permission from the Geological Society of America.

Earthquakes also induce static stress changes, with amplitudes that decay with distance from the earthquake source and become insignificant at ranges beyond 1–2 rupture lengths. These changes trigger secondary earthquakes (aftershocks) and dilate (or compress) the surrounding lithosphere. At more remote distances, up to several tens of rupture length from the epicenter, the transient stresses associated with passing seismic waves also may impact the hydrothermal system. In these cases, earthquake-induced ground shaking is thought to jar open some hydrothermal conduits and seal others.

Observations in the marine environment indicate that seismicity may create fluctuations in vent flow, temperature, and chemistry with variable characteristics. For example, drops in temperature at vent sites on the southern JdFR have been observed to correlate with earthquakes on the Western Blanco Transform at distances >30 km (**Figure 14**). The sensitivity, phase, and amplitude of these changes, however, vary significantly even between closely spaced sites (**Figures 14(a)** and **14(b)**). Given the distances involved, temperature changes correlated with moderate-size ($M \sim 4.0$) earthquakes on the Blanco Transform likely reflect the systems response to dynamic stress transients, while much larger earthquakes ($M > 6$) also induced significant static changes.

Importantly, earthquake-induced changes within mid-ocean ridge hydrothermal systems can dramatically affect the biological communities that derive their energy from chemosynthetic processes. For example, the sub-seafloor microbial community is extremely sensitive to temperature, oxygen content, pH, and other environmental variables. It thrives in subsurface zones where the hot hydrothermal fluid mixes with entrained seawater. Changes in crustal fluid temperatures of only a few degrees, or a minor alteration of crustal permeability, can cause the prevailing microbial species to weaken and encourage new species to thrive and become dominant. Similarly, changes in the effluent thermal flux of hydrothermal vents can enhance heat output changing the size and temperature of the thermal boundary layer, a zone just above the axial floor that supports abundant and diverse macrofaunal communities.

Still more dramatic impacts may be associated with the emplacement of magma at depth or the eruption of lava onto the seafloor. The heat supplied by these systems feeds massive bacteria blooms (floc) and may generate event megaplumes within the water column, huge volumes of hydrothermal fluid enriched in reduced chemicals that rise up to 1 km above the seafloor.

See also

Mid-Ocean Ridge Geochemistry and Petrology. Mid-Ocean Ridge Seismic Structure. Mid-Ocean Ridge Tectonics, Volcanism, and Geomorphology. Seamounts and Off-Ridge Volcanism.

Further Reading

Abercrombie R and Ekström G (2001) Earthquake slip on oceanic transform faults. *Nature* 410: 74–77.

Barclay AH, Toomey DR, and Solomon SC (2001) Microearthquake characteristics and crustal VP/VS structure at the Mid-Atlantic Ridge, 35° N. *Journal of Geophysical Research* 106: 2017–2034 (doi:10.1029/2000JB900371).

Bird P, Kagan YY, and Jackson DD (2002) Plate tectonics and earthquake potential of spreading ridges and oceanic transform faults. In: Stein S and Freymueller JT (eds.) *Geodynamics Series 30: Plate Boundary Zones*, pp. 203–218. Washington, DC: American Geophysical Union.

Boettcher MS and Jordan TH (2004) Earthquake scaling relations for mid-ocean ridge transform faults. *Journal of Geophysical Research* 109: B12302 (doi:10.1029/2004JB003110).

Dziak RP, Chadwick WW, Jr., Fox CG, and Embley RW (2003) Hydrothermal temperature changes at the southern Juan de Fuca Ridge associated with M_W 6.2 Blanco Transform earthquake. *Geology* 31: 119–122.

Dziak RP, Fox CG, and Schreiner AE (1995) The June–July 1993 seismo-acoustic event at CoAxial Segment, Juan de Fuca Ridge: Evidence for a lateral dike injection. *Geophysical Research Letters* 22: 135–138.

Edwards MH, Kurras GJ, Tolstoy M, Bohnenstieh DR, Coakley BJ, and Cochran JR (2001) Evidence of recent volcanic activity on the ultra-slow spreading Gakkel Ridge. *Nature* 409: 808–812.

Fox CG, Matsumoto H, and Lau T-KA (2001) Monitoring Pacific Ocean seismicity from an autonomous hydrophone array. *Journal of Geophysical Research* 106: 4183–4206.

Sohn RA, Hildebrand JA, and Webb SC (1999) A microearthquake survey of the high-temperature vent fields on the volcanically active East Pacific Rise (9° 50′ N). *Journal of Geophysical Research* 104: 25367–25378 (doi:10.1029/1999JB900263).

Solomon SC, Huang PY, and Meinke L (1988) The seismic moment budget of slowly spreading ridges. *Nature* 334: 58–61.

Sykes LR (1967) Mechanism of earthquakes and nature of faulting on the mid-oceanic ridges. *Journal of Geophysical Research* 72: 2131–2153.

Tolstoy M, Cowen JP, Baker ET, et al. (2006) A seafloor spreading event captured by seismometers. *Science* 314: 1920–1922 (doi:10.1126/science.1137082).

Wetzel LR, Wiens DA, and Kleinrock MC (1993) Evidence from earthquakes for bookshelf faulting at large non-transform ridge offsets. *Nature* 362: 235–237.

Wilcock W (2001) Tidal triggering of microearthquakes on the Juan de Fuca Ridge. *Geophysical Research Letters* 28: 3999–4002 (doi:10.1029/2001GL013370).

Relevant Websites

http://www.globalcmt.org
 – Global CMT Web Page.

http://www.pmel.noaa.gov
 – National Oceanic and Atmospheric Administration's Acoustic Monitoring Program.

http://earthquake.usgs.gov
 – US Geological Survey Earthquake Hazards Program.

SEISMOLOGY SENSORS

L. M. Dorman, University of California, San Diego,
La Jolla, CA, USA

Introduction

A glance at the globe shows that the Earth's surface is largely water-covered. The logical consequence of this is that seismic studies based on land seismic stations alone will be severely biased because of two factors. The existence of large expanses of ocean distant from land means that many small earthquakes underneath the ocean will remain unobserved. The difference in seismic velocity structure between continent and ocean intruduces a bias in locations, with oceanic earthquakes which are located using only stations on one side of the event being pulled tens of kilometers landward. Additionally, the depths of shallow subduction zone events, which are covered by water, will be very poorly determined. Thus seafloor seismic stations are necessary both for completeness of coverage as well as for precise location of events which are tectonically important. This paper summarizes the status of seafloor seismic instrumentation.

The alternative methods for providing coverage are temporary (pop-up) instruments and permanently connected systems. The high costs of seafloor cabling has thus far precluded dedicated cables of significant length for seismic purposes, although efforts have been made to use existing, disused wires. Accordingly, the main emphasis of this report will be temporary instruments.

Large ongoing programs to investigate oceanic spreading centers (RIDGE) and subductions (MARGINS) have provided impetus for the upgrading of seismic capabilities in oceanic areas.

The past few years has seen a blossoming of ocean bottom seismograph (OBS) instrumentation, both in number and in their capabilities. Active experimental programs are in place in the USA, Europe, and Japan. Increases in the reliability of electronics and in the capacity of storage devices has allowed the development of instruments which are much more reliable and useful. Major construction programs in Japan and the USA are producing hundreds of instruments, a number which allows imaging experiments which have been heretofore associated with the petroleum exploration industry. This contrasts sharply with the severely underdetermined experiments which have characterized earthquake seismology. One change over the past decade has been the disappearance of analog recording systems.

A side effect of this rapid change is that a review such as this provides a snapshot of the technology, rather than a long-lasting reference. The technical details reported below are for instruments at two stages of development: existing instruments (UTIG, SIO/ONR, SIO/IGPP-SP, GEOMAR, LDEO-BB) and instruments still in design and construction (WHOI-SP, WHOI-BB SIO/IGPP-BB) (see **Tables 1** and **2**; **Figures 1–6**). The latter construction project has the acronym 'OBSIP' for OBS instrument pool, and sports a polished, professionally designed web site at http://www.obsip.org.

OBS designs are roughly divided into two categories which for brevity will be called 'short-period' (SP) and 'broadband' (BB). The distinction blurs at times because some instruments of both classes use a common recording system, a possibility which emerges when a high data-rate digitizer has the

Figure 1 The UTIG OBS, a particularly 'clean' mechanical design, which has been in use for many years, with evolving electronics. The anchor is 1.2 m on each side. (Photograph by Gail Christeson, UTIG.)

Table 1 Characteristics of short period ocean bottom seismometers

Parameter	UTIG	WHOI-SP	SIO-IGPP-SP	GEOMAR-SP	JAMSTEC-ORI[a]
Contact/website	Nakamura/http://www.ig.utexas.edu/research/projects/obs/obs.html	Detrick/Collins http://www.obsip.org	Orcutt/Babcock http://www.obsip.org	Flüh/Bialas http://www.geomar.de	–
Seismic sensor(s)	3-component 4.5 Hz Mark Products L-15B or Oyo GS-11D	2 Hz Mark Products L-22 4.5 Hz or L-28 4.5 Hz, vertical component	2 Hz Mark Products L-22, vertical component	optionally uses Webb BB sensor	4.5 Hz vertical
Frequency response	4.5–100 Hz	2–X Hz	2–X Hz	0.05–30 Hz/ 0.01–X Hz	4.5–100 Hz
Nominal sensitivity	2.5 nm s^{-1}	–	–	–	–
Hydrophone	OAS E-2PD crystal	Hightech crystal	Hightech HYI-90-U	OAS E4SD or Cox-Webb DPG	–
Frequency Response	3–100 Hz	5–X Hz	50 mHz-X	0.05–30 Hz	–
Digitizer type, dynamic range, sample rates	14 bits + gain-ranging, 126 dB,112 dB re electronic noise	Quanterra Q330 24-bit,126 dB, 1–200 Hz	Cirrus/Crystal CS5321-CS5322 24-bit, 124–130 dB	oversampling, 120 dB, 25–200 Hz	–
Recording medium and capacity[b]	Disk, semi-continuous	Disk, data download through pressure case	9 Gbyte disk, data download through pressure case	1 Gbyte DAT or 2 Gbyte semiconductor memory	–
Clock type and drift[c]	10 ms	Seascan Precision Timebase < 0.5 ms d^{-1}	Seascan Precision Timebase < 0.5 ms/d^{-1}	Seascan Precision Timebase < 0.5 ms d^{-1}	–
Endurance	8 weeks–6 months	90 days, alkaline/ 1 year, lithium	80–180 days at 250/31.25 Hz sampling 5 days on NiCad rechargeable cells	300 days	–
Power consumption	550 mW	–	420 mW at 31.25 Hz, 1.6 W at 250 Hz for 2 channels	230–250 mW	–
Release type	Burnwire release, acoustically controlled, acoustic release, two backup timers	Burnwire, Edgetech acoustics	Double burnwire, Edgetech acoustics	Acoustic release with back-up timer	–
Mechanical configuration, launch/ recovery weights	Single 43 cm diameter glass sphere, 85 kg/35 kg	63 kg/43 kg	110 kg/80 kg	Vertical cylinders, 175 kg/125 kg	Single 43 cm diameter glass sphere
Number available	37–39[d]	15 now, 40 more under construction	14 now, 74 more under construction	27 OBH + 11 OBS	100?
Total			51		

[a]Information on these instruments is incomplete.
[b]2 gigabytes is about 22 days of data sampling four channels at 128 Hz or 176 days sampling four channels at 16 Hz.
[c]1 ms d^{-1} is about 1 × 10 E-8.
[d]Includes instruments of the same design operated by IRD (formerly ORSTOM) and National Taiwan Ocean University.

Table 2 Characteristics of broadband ocean bottom seismometerrs

Parameter	SIO/ONR[a]	WHOI-BB	LDEO-BB[e]	SIO/IGPP-BB	ORI-BB[f]
Manager	Dorman/ Sauter http:// www-mpl. ucsd.edu/obs	Detrick/ Collins http:// www.obsip.org	Webb/ http:// www.obsip.org	Orcutt/Babcock http://www. obsip.org	–
Seismic sensor	1 Hz Mark Products L4C-3D or PMD 2123	Guralp CMG-3ESP	1 Hz Mark Products L4C-3D	Kinemetrics	PMD 2023
Frequency response	0.033–32 Hz[b]	0.033–50 Hz[b]	0.005–30 Hz[b]	0.02–50 Hz	0.033–50 Hz
Hydrophone	Cox-Webb DPG	HighTech	Cox-Webb DPG for low frequency hydrophone for high frequency	High Tech HYI-90-U or Cox-Webb DPG	–
Frequency response	0.001–5 Hz	0.033–X Hz	0.001–60 Hz	0.05–15 kHz or 0.01–32 Hz	–
Digitizer type, Dynamic range, sample rates	16-bit + gain-ranging, 126 dB, X-128 sps	Quanterra QA330 24-bit, 135 dB	nominal 24-bit ~ 135 dB, 1, 20, 40, 100, 200 sps	24-bit Cirrus/ Crystal CS 5321–CS-5322, 130 dB	20-bit
Recording medium and capacity[c]	Disk, 9–27 Gbyte	Disk, 2 Gbyte	Disk, 18 Gb, 72 Gb planned	Disk, 9 Gbyte	4 × 6.4 Gbyte disks
Clock drift[d]	< 1 ms d^{-1}	< 1 ms d^{-1}	0.5 ms d^{-1}	< 1 ms d^{-1}	–
Endurance	6–12 months	6–12 + months	Up to 15 months	6–12 months	9 months?
Power consumption	400 mW	1.5 W at 20 sps	–	–	~ 600 mW
Release type	Two EG&G 8242 acoustic releases	One EG&G 8242 acoustic relase with back-up acoustic burnwire release	Acoustically controlled burnwire?	EG&G acoustically controlled burnwire	Acoustically controlled burn-plate
Mechanical configuration, launch/ recovery weights	Fiberglass frame, aluminum pressure cases, glass flotation (ONR OBS-style)	ONR OBS-style, floating above anchor, 570 kg/ 472 kg	Aluminum pressure cases, plastic plate frame, 215/145 kg	178/138 kg	50 cm diameter, titanium sphere
Number available	14	25 under construction	64 under construction	15–20 (using recording package from SP instrument)	15
Total			~ 133		

[a]These instruments incorporate a fluid flowmeter/sampler in the instrument frame (see Tryon et al. 2001).
[b]Seismometers are free from spurious resonances below 20 Hz.
[c]2 gigabytes is about 22 days of data sampling four channels at 128 Hz or 176 days sampling four channels at 16 Hz.
[d]1 ms is about 1×10^{-8}.
[e]1 See Webb et al., 2001.
[f]1 See Shiobara et al., 2000.

capability of operation in a low-power, high endurance mode.

Short Period (SP) Instruments

The SP instruments (**Table 1**) are light in weight and easy to deploy, typically use 4.5 Hz geophones, commonly only the vertical component, and/or hydrophones, may have somewhat limited recording capacity and endurance, and are typically used in active-source seismic experiments and for microearthquake studies with durations of a week to a few months. Two types (UTIG and JAMSTEC) are single-sphere instruments.

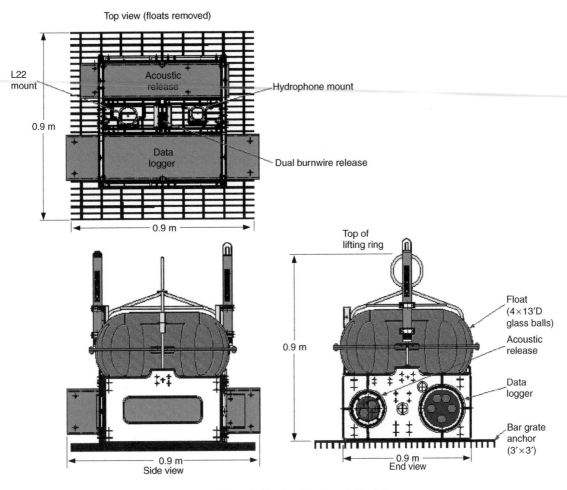

Figure 2 The IGPP-SP instrument. (Figure from Babcock, Harding, Kent, and Orcutt.)

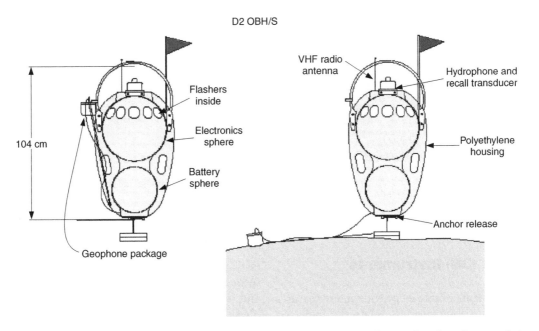

Figure 3 The WHOI-SP instrument. The change of orientation between seafloor and surface modes allows the acoustic transducer an unobstructed view of the surface while on the seafloor and permits acoustic ranging while the instrument is on the surface. (Figure by Beecher Wooding and John Collins.)

Figure 4   The GEOMAR OBH/S. The shipping/storage container is equipped with an overhead rail so that it serves as an instrument dispenser. The OBS version is shown here. (Photograph by Michael Tryon, UCSD.)

Broadband (BB) Instruments

The BB instruments (**Table 2**) provide many features of land seismic observatories, relatively high dynamic range, excellent clock stability $-<1\,\mathrm{ms\,d}^{-1}$ drift. This class of instruments can be equipped with hydrophones useful down to a millihertz. The BB instruments are designed in two parts, the main section contains the recording package, and release and recovery aids, while the sensor package is physically separated from the main section. This configuration allows isolation from mechanical noise and and permits tuning of the mechanical resonance of the sensor–seafloor system.

Figure 5 The SIO-ONR OBSs (Jacobson *et al.* 1991, Sauter *et al,* 1990) being launched in Antarctica. The anchor serves as a collector for the CAT fluid fluxmeter (Tryon *et al.* 2001). The plumbing for the flowmeter is in the light-colored box at the right-hand end of the instrument. (Photograph by Michael Tryon, UCSD.)

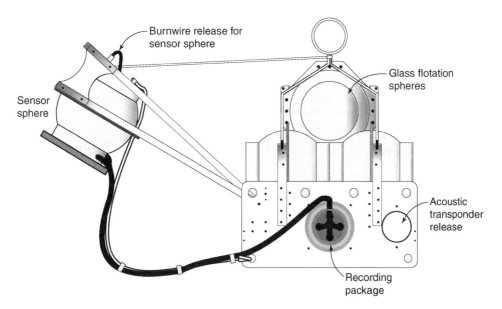

Figure 6 The LDGO-BB OBS. This is based on the Webb design in use during the past few years. The earlier version established a reputation for high reliability and was the lowest noise OBS and lightest of its time period. The main drawback of the earlier version was its limited (16-bit) dynamic range. (Figure from S. Webb, LDEO.)

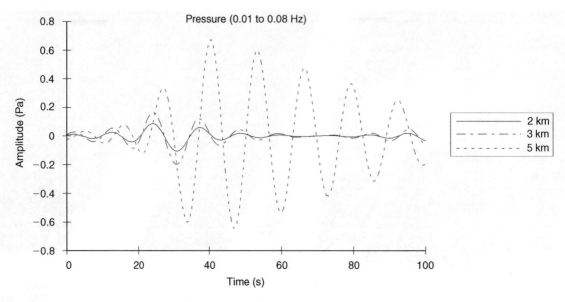

Figure 7   Synthetic seismograms of pressure at three ocean depths.

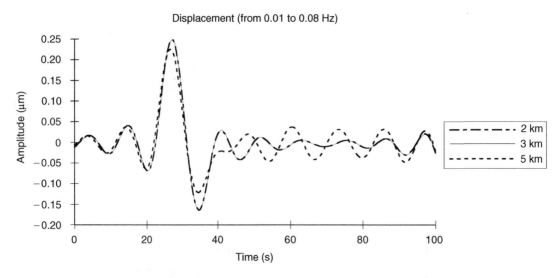

Figure 8   Synthetic seismograms of vertical motion at the same depths as **Figure 7**; note the reduction in the effects of the water column reverberations. (From Lewis and Dorwan, 1998).

Sensor Considerations

Emplacement of a sensor on the seafloor is almost always suboptimal in comparison with land stations. Instruments dropped from the sea surface can land tens of meters from the desired location. The seafloor material is almost always softer than the surficial sediments (these materials have shear velocities as low as a few tens of meters per second. The sensor is thus almost always poorly coupled to the seafloor and sensor resonances can occur within the frequency range of interest for short period sensors. Fortunately, these resonances have little effect on

lower frequencies. However, lower frequency sensors are affected, since a soft foundation permits tilt either in response to sediment deformation by the weight of the sensor or in response to water currents. The existing Webb instruments combat this problem by periodic releveling of the sensor gimbals. The PMD sensors have an advantage here in that the mass element is a fluid and the horizontal components self-level to within 5°.

Why not use hydrophones then? These make leveling unnecessary and are more robust mechanically. In terms of sensitivity, they are comparable to seismometers. The disadvantage of hydrophones lies in the physics of reverberation in the water layer. A pulse incident from below is reflected from the sea surface completely, and when it encounters the seafloor it is reflected to a significant degree. Since the seafloor has a higher acoustic impedance than water, the reflected pressure pulse has the same sign as the incident pulse and the signal is large. However, the seafloor motion associated with pressure pulses traveling in opposite directions is opposite in sign, so cancellation occurs. Unfortunately, the frequency range in which these reverberations are troublesome is in the low noise region. **Figures 7** and **8** show synthetic seismograms of pressure and vertical motion illustrating this effect.

See also

Mid-Ocean Ridge Seismic Structure. Seismic Structure.

Further Reading

Barash TW, Doll CG, Collins JA, Sutton GH, and Solomon SC (1994) Quantitative evaluation of a passively leveled ocean bottom seismometer. *Marine Geophysical Researches* 16: 347–363.

Dorman LM (1997) Propagation in marine sediments. In: Crocker MJ (ed.) *Encyclopedia of Acoustics*, pp. 409–416. New York: John Wiley.

Dorman LM, Schreiner AE, and Bibee LD (1991) The effects of sediment structure on sea floor noise. In: Hovem J, *et al.* (eds.) *Proceedings of Conference on Shear Waves in Marine Sediments*, pp. 239–245. Dordrecht: Kluwer Academic Publishers.

Duennebier FK and Sutton GH (1995) Fidelity of ocean bottom seismic observations. *Marine Geophysical Researches* 17: 535–555.

Jacobson RS, Dorman LM, Purdy GM, Schultz A, and Solomon S (1991) *Ocean Bottom Seismometer Facilities Available*, EOS, Transactions AGU, 72, pp. 506, 515.

Sauter AW, Hallinan J, Currier R *et al.* (1990) *Proceedings of the MTS Conference on Marine Instrumentation*, pp. 99–104.

Tryon M, Brown K, Dorman L, and Sauter A (2001) A new benthic aqueous flux meter for very low to moderate discharge rates. *Deep Sea Research* (in press).

Webb SC (1998) Broadband seismology and noise under the ocean. *Reviews in Geophysics* 36: 105–142.

TSUNAMI

P. L.-F. Liu, Cornell University, Ithaca, NY, USA

Introduction

Tsunami is a Japanese word that is made of two characters: *tsu* and *nami*. The character *tsu* means harbor, while the character *nami* means wave. Therefore, the original word *tsunami* describes large wave oscillations inside a harbor during a 'tsunami' event. In the past, tsunami is often referred to as 'tidal wave', which is a misnomer. Tides, featuring the rising and falling of water level in the ocean in a daily, monthly, and yearly cycle, are caused by gravitational influences of the moon, sun, and planets. Tsunamis are not generated by this kind of gravitational forces and are unrelated to the tides, although the tidal level does influence a tsunami striking a coastal area.

The phenomenon we call a tsunami is a series of water waves of extremely long wavelength and long period, generated in an ocean by a geophysical disturbance that displaces the water within a short period of time. Waves are formed as the displaced water mass, which acts under the influence of gravity, attempts to regain its equilibrium. Tsunamis are primarily associated with submarine earthquakes in oceanic and coastal regions. However, landslides, volcanic eruptions, and even impacts of objects from outer space (such as meteorites, asteroids, and comets) can also trigger tsunamis.

Tsunamis are usually characterized as shallow-water waves or long waves, which are different from wind-generated waves, the waves many of us have observed on a beach. Wind waves of 5–20-s period (T = time interval between two successive wave crests or troughs) have wavelengths ($\lambda = T^2(g/2\pi)$ distance between two successive wave crests or troughs) of c. 40–620 m. On the other hand, a tsunami can have a wave period in the range of 10 min to 1 h and a wavelength in excess of 200 km in a deep ocean basin. A wave is characterized as a shallow-water wave when the water depth is less than 5% of the wavelength. The forward and backward water motion under the shallow-water wave is felted throughout the entire water column. The shallow water wave is also sensitive to the change of water depth. For instance, the speed (celerity) of a shallow-water wave is equal to the square root of the product

of the gravitational acceleration ($9.81\,\mathrm{m\,s^{-2}}$) and the water depth. Since the average water depth in the Pacific Ocean is 5 km, a tsunami can travel at a speed of about $800\,\mathrm{km\,h^{-1}}$ ($500\,\mathrm{mi\,h^{-1}}$), which is almost the same as the speed of a jet airplane. A tsunami can move from the West Coast of South America to the East Coast of Japan in less than 1 day.

The initial amplitude of a tsunami in the vicinity of a source region is usually quite small, typically only a meter or less, in comparison with the wavelength. In general, as the tsunami propagates into the open ocean, the amplitude of tsunami will decrease for the wave energy is spread over a much larger area. In the open ocean, it is very difficult to detect a tsunami from aboard a ship because the water level will rise only slightly over a period of 10 min to hours. Since the rate at which a wave loses its energy is inversely proportional to its wavelength, a tsunami will lose little energy as it propagates. Hence in the open ocean, a tsunami will travel at high speeds and over great transoceanic distances with little energy loss.

As a tsunami propagates into shallower waters near the coast, it undergoes a rapid transformation. Because the energy loss remains insignificant, the total energy flux of the tsunami, which is proportional to the product of the square of the wave amplitude and the speed of the tsunami, remains constant. Therefore, the speed of the tsunami decreases as it enters shallower water and the height of the tsunami grows. Because of this 'shoaling' effect, a tsunami that was imperceptible in the open ocean may grow to be several meters or more in height.

When a tsunami finally reaches the shore, it may appear as a rapid rising or falling water, a series of breaking waves, or even a bore. Reefs, bays, entrances to rivers, undersea features, including vegetations, and the slope of the beach all play a role modifying the tsunami as it approaches the shore. Tsunamis rarely become great, towering breaking waves. Sometimes the tsunami may break far offshore. Or it may form into a bore, which is a step-like wave with a steep breaking front, as the tsunami moves into a shallow bay or river. **Figure 1** shows the incoming 1946 tsunami at Hilo, Hawaii.

The water level on shore can rise by several meters. In extreme cases, water level can rise to more than 20 m for tsunamis of distant origin and over 30 m for tsunami close to the earthquake's epicenter. The first wave may not always be the largest in the series of waves. In some cases, the water level will fall significantly first, exposing the bottom of a bay

Figure 1 1946 tsunami at Hilo, Hawaii (Pacific Tsunami Museum). Wave height may be judged from the height of the trees.

or a beach, and then a large positive wave follows. The destructive pattern of a tsunami is also difficult to predict. One coastal area may see no damaging wave activity, while in a neighboring area destructive waves can be large and violent. The flooding of an area can extend inland by 500 m or more, covering large expanses of land with water and debris. Tsunamis may reach a maximum vertical height onshore above sea level, called a runup height, of 30 m.

Since scientists still cannot predict accurately when earthquakes, landslides, or volcano eruptions will occur, they cannot determine exactly when a tsunami will be generated. But, with the aid of historical records of tsunamis and numerical models, scientists can get an idea as to where they are most likely to be generated. Past tsunami height measurements and computer modeling can also help to forecast future tsunami impact and flooding limits at specific coastal areas.

Historical and Recent Tsunamis

Tsunamis have been observed and recorded since ancient times, especially in Japan and the Mediterranean areas. The earliest recorded tsunami occurred in 2000 BC off the coast of Syria. The oldest reference of tsunami record can be traced back to the sixteenth century in the United States.

During the last century, more than 100 tsunamis have been observed in the United States alone. Among them, the 1946 Alaskan tsunami, the 1960 Chilean tsunami, and the 1964 Alaskan tsunami were the three most destructive tsunamis in the US history. The 1946 Aleutian earthquake ($M_w = 7.3$) generated catastrophic tsunamis that attacked the Hawaiian Islands after traveling about 5 h and killed 159 people.

(The magnitude of an earthquake is defined by the seismic moment, M_0 (dyn cm), which is determined from the seismic data recorded worldwide. Converting the seismic moment into a logarithmic scale, we define $M_w = (1/1.5)\log_{10}M_0 - 10.7$.) The reported property damage reached $26 million. The 1960 Chilean tsunami waves struck the Hawaiian Islands after 14 h, traveling across the Pacific Ocean from the Chilean coast. They caused devastating damage not only along the Chilean coast (more than 1000 people were killed and the total property damage from the combined effects of the earthquake and tsunami was estimated as $417 million) but also at Hilo, Hawaii, where 61 deaths and $23.5 million in property damage occurred (see **Figure 2**). The 1964 Alaskan tsunami triggered by the Prince William Sound earthquake ($M_w = 8.4$), which was recorded as one of the largest earthquakes in the North American continent, caused the most destructive damage in Alaska's history. The tsunami killed 106 people and the total damage amounted to $84 million in Alaska.

Within less than a year between September 1992 and July 1993, three large undersea earthquakes strike the Pacific Ocean area, causing devastating tsunamis. On 2 September 1992, an earthquake of magnitude 7.0 occurred *c.* 100 km off the Nicaraguan coast. The maximum runup height was recorded as 10 m and 168 people died in this event. A few months later, another strong earthquake ($M_w = 7.5$) attacked the Flores Island and surrounding area in Indonesia on 12 December 1992. It was reported that more than 1000 people were killed in the town of Maumere alone and two-thirds of the population of Babi Island were swept away by the tsunami. The maximum runup was estimated as 26 m. The final toll of this Flores earthquake stood at 1712 deaths and more than 2000 injures. Exactly

Figure 2 The tsunami of 1960 killed 61 people in Hilo, destroyed 537 buildings, and damages totaled over $23 million.

7 months later, on 12 July 1993, the third strong earthquake ($M_w = 7.8$) occurred near the Hokkaido Island in Japan (Hokkaido Tsunami Survey Group 1993). Within 3–5 min, a large tsunami engulfed the Okushiri coastline and the central west of Hokkaido, impinging extensive property damages, especially on the southern tip of Okushiri Island in the town of Aonae. The runup heights on the Okushiri Island were thoroughly surveyed and they varied between 15 and 30 m over a 20-km stretch of the southern part of the island, with several 10-m spots on the northern part of the island. It was also reported that although the runup heights on the west coast of Hokkaido are not large (less than 10 m), damage was extensive in several towns. The epicenters of these three earthquakes were all located near residential coastal areas. Therefore, the damage caused by subsequent tsunamis was unusually large.

On 17 July 1998, an earthquake occurred in the Sandaun Province of northwestern Papua New Guinea, about 65 km northwest of the port city of Aitape. The earthquake magnitude was estimated as $M_w = 7.0$. About 20 min after the first shock, Warapo and Arop villages were completely destroyed by tsunamis. The death toll was at over 2000 and many of them drowned in the Sissano Lagoon behind the Arop villages. The surveyed maximum runup height was 15 m, which is much higher than the predicted value based on the seismic information. It has been suggested that the Papua New Guinea tsunami could be caused by a submarine landslide.

The most devastating tsunamis in recent history occurred in the Indian Ocean on 26 December 2004.

An earthquake of $M_w = 9.0$ occurred off the west coast of northern Sumatra. Large tsunamis were generated, severely damaging coastal communities in countries around the Indian Ocean, including Indonesia, Thailand, Sri Lanka, and India. The estimated tsunami death toll ranged from 156 000 to 178 000 across 11 nations, with additional 26 500–142 000 missing, most of them presumed dead.

Tsunami Generation Mechanisms

Tsunamigentic Earthquakes

Most tsunamis are the results of submarine earthquakes. The majority of earthquakes can be explained in terms of plate tectonics. The basic concept is that the outermost part the Earth consists of several large and fairly stable slabs of solid and relatively rigid rock, called plates (see **Figure 3**). These plates are constantly moving (very slowly), and rub against one another along the plate boundaries, which are also called faults. Consequently, stress and strain build up along these faults, and eventually they become too great to bear and the plates move abruptly so as to release the stress and strain, creating an earthquake. Most of tsunamigentic earthquakes occur in subduction zones around the Pacific Ocean rim, where the dense crust of the ocean floor dives beneath the edge of the lighter continental crust and sinks down into Earth's mantle. These subduction zones include the west coasts of North and South America, the coasts of East Asia (especially Japan), and many Pacific island chains (**Figure 3**). There are different types of faults along subduction margins. The interplate fault usually accommodates a large relative motion between two tectonic plates and the overlying plate is typically pushed upward. This upward push is impulsive; it occurs very quickly, in a few seconds. The ocean water surface responds immediately to the upward movement of the seafloor and the ocean surface profile usually mimics the seafloor displacement (see **Figure 4**). The interplate fault in a subduction zone has been responsible for

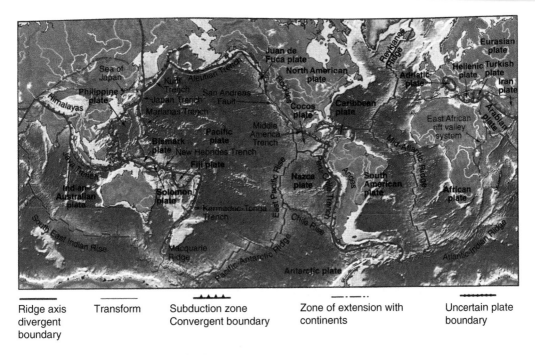

| Ridge axis divergent boundary | Transform | Subduction zone Convergent boundary | Zone of extension with continents | Uncertain plate boundary |

Figure 3 Major tectonic plates that make up the Earth's crust.

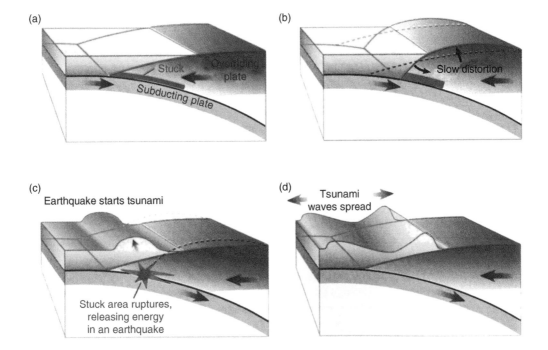

Figure 4 Sketches of the tsunami generation mechanism caused by a submarine earthquake. An oceanic plate subducts under an overriding plate (a). The overriding plate deforms due to the relative motion and the friction between two tectonic plates (b). The stuck area ruptures, releasing energy in an earthquake (c). Tsunami waves are generated due to the vertical seafloor displacement (d).

most of the largest tsunamis in the twentieth century. For example, the 1952 Kamchatka, 1957 Aleutian, 1960 Chile, 1964 Alaska, and 2004 Sumatra earthquakes all generated damaging tsunamis not only in the region near the earthquake epicenter, but also on faraway shores.

For most of the interplate fault ruptures, the resulting seafloor displacement can be estimated based on the dislocation theory. Using the linear elastic theory, analytical solutions can be derived from the mean dislocation field on the fault. Several parameters defining the geometry and strength of the fault rupture need to be specified. First of all, the mean fault slip, D, is calculated from the seismic moment M_0 as follows:

$$M_0 = \mu DS \qquad [1]$$

where S is the rupture area and μ is the rigidity of the Earth at the source, which has a range of $6-7 \times 10^{11}$ dyn cm^{-2} for interplate earthquakes. The seismic moment, M_0, is determined from the seismic data recorded worldwide and is usually reported as the Harvard Centroid-Moment-Tensor (CMT) solution within a few minutes of the first earthquake tremor. The rupture area is usually estimated from the aftershock data. However, for a rough

estimation, the fault plane can be approximated as a rectangle with length L and width W. The aspect ratio L/W could vary from 2 to 8. To find the static displacement of the seafloor, we need to assign the focal depth d, measuring the depth of the upper rim of the fault plane, the dip angle δ, and the slip angle λ of the dislocation on the fault plane measured from the horizontal axis (see **Figure 5**). For an oblique slip on a dipping fault, the slip vector can be decomposed into dip-slip and strike-slip components. In general, the magnitude of the vertical displacement is less for the strike-slip component than for the dip-slip component. The closed form expressions for vertical seafloor displacement caused by a slip along a rectangular fault are given by Mansinha and Smylie.

For more realistic fault models, nonuniform stress-strength fields (i.e., faults with various kinds of barriers, asperities, etc.) are expected, so that the actual seafloor displacement may be very complicated compared with the smooth seafloor displacement computed from the mean dislocation field on the fault. As an example, the vertical seafloor displacement caused by the 1964 Alaska earthquake is sketched in **Figure 6**. Although several numerical models have considered geometrically complex faults, complex slip distributions, and elastic layers of variable thickness, they are not yet disseminated in

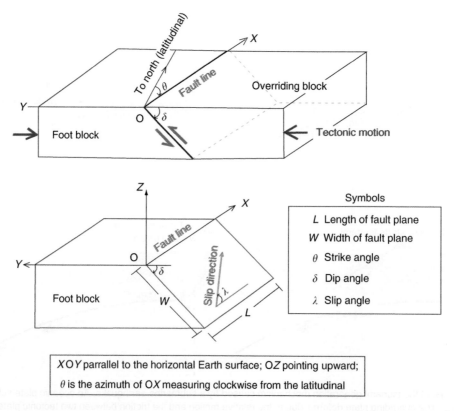

Figure 5 A sketch of fault plane parameters.

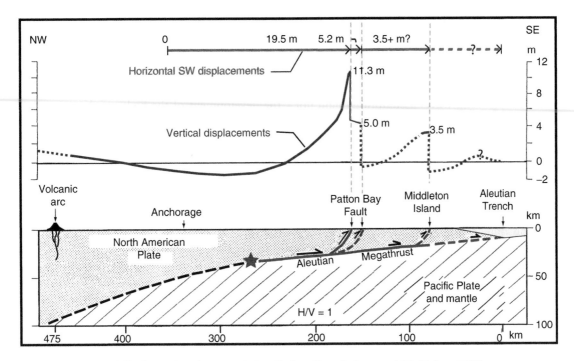

Figure 6 A sketch of 1964 Alaska earthquake generated vertical seafloor displacement (G. Plafker, 2006).

tsunami research. One of the reasons is that our knowledge in source parameters, inhomogeneity, and nonuniform slip distribution is too incomplete to justify using such a complex model.

Certain earthquakes referred to as tsunami earthquakes have slow faulting motion and very long rupture duration (more than several minutes). These earthquakes occur along the shallow part of the interplate thrust or décollement near the trench (the wedge portion of the thin crust above the interface of the continental crust and the ocean plate). The wedge portion consisting of thick deformable sediments with low rigidity, and the steepening of rupture surface in shallow depth all favor the large displacement of the crust and possibility of generating a large tsunami. Because of the extreme heterogeneity, accurate modeling is difficult, resulting in large uncertainty in estimated seafloor displacement.

Landslides and Other Generation Mechanisms

There are occasions when the secondary effects of earthquakes, such as landslide and submarine slump, may be responsible for the generation of tsunamis. These tsunamis are sometimes disastrous and have gained increasing attentions in recent years. Landslides are generated when slopes or sediment deposits become too steep and they fail to remain in equilibrium and motionless. Once the unstable conditions are present, slope failure can be triggered by storm, earthquakes, rains, or merely continued

deposition of materials on the slope. Alternative mechanisms of sediment instability range between soft sediment deformations in turbidities, to rotational slumps in cohesive sediments. Certain environments are particularly susceptible to the production of landslides. River delta and steep underwater slopes above submarine canyons are likely sites for landslide-generated tsunamis. At the time of the 1964 Alaska earthquake, numerous locally landslide-generated tsunamis with devastating effects were observed. On 29 November 1975, a landslide was triggered by a 7.2 magnitude earthquake along the southeast coast of Hawaii. A 60-km stretch of Kilauea's south coast subsided 3.5 m and moved seaward 8 m. This landslide generated a local tsunami with a maximum runup height of 16 m at Keauhou. Historically, there have been several tsunamis whose magnitudes were simply too large to be attributed to the coseismic seafloor movement and landslides have been suggested as an alternative cause. The 1946 Aleutian tsunami and the 1998 Papua New Guinea tsunami are two significant examples.

In terms of tsunami generation mechanisms, two significant differences exist between submarine landslide and coseismic seafloor deformation. First, the duration of a landslide is much longer and is in the order of magnitude of several minutes or longer. Hence the time history of the seafloor movement will affect the characteristics of the generated wave and needs to be included in the model. Second, the effective size of the landslide region is usually much smaller than the coseismic seafloor deformation zone. Consequently, the typical wavelength of the tsunamis generated by a submarine landslide is also shorter, that is, *c*. 1–10 km. Therefore, in some cases, the shallow-water (long-wave) assumption might not be valid for landslide-generated tsunamis.

Although they are rare, the violent geological activities associated with volcanic eruptions can also generate tsunamis. There are three types of tsunami-generation mechanism associated with a volcanic eruption. First, the pyroclastic flows, which are mixtures of gas, rocks, and lava, can move rapidly off an island and into an ocean, their impact displacing seawater and producing a tsunami. The second mechanism is the submarine volcanic explosion, which occurs when cool seawater encounters hot volcanic magma. The third mechanism is due to the collapse of a submarine volcanic caldera. The collapse may happen when the magma beneath a volcano is withdrawn back deeper into the Earth, and the sudden subsidence of the volcanic edifice displaces water and produces a tsunami. Furthermore, the large masses of rock that accumulate on the sides of volcanoes may suddenly slide down the slope into the sea, producing tsunamis. For example, in 1792, a large mass of the mountain slided into Ariake Bay in Shimabara on Kyushu Island, Japan, and generated tsunamis that reached a height of 10 m in some places, killing a large number of people.

In the following sections, our discussions will focus on submarine earthquake-generated tsunamis and their coastal effects.

Modeling of Tsunami Generation, Propagation, and Coastal Inundation

To mitigate tsunami hazards, the highest priority is to identify the high-tsunami-risk zone and to educate the citizen, living in and near the risk zone, about the proper behaviors in the event of an earthquake and tsunami attack. For a distant tsunami, a reliable warning system, which predicts the arrival time as well as the inundation area accurately, can save many lives. On the other hand, in the event of a near-field tsunami, the emergency evacuation plan must be activated as soon as the earth shaking is felt. This is only possible, if a predetermined evacuation/inundation map is available. These maps should be produced based on the historical tsunami events and the estimated 'worst scenarios' or the 'design tsunamis'. To produce realistic and reliable inundation maps, it is essential to use a numerical model that calculates accurately tsunami propagation from a source region to the coastal areas of concern and the subsequent tsunami runup and inundation.

Numerical simulations of tsunami have made great progress in the last 50 years. This progress is made possible by the advancement of seismology and by the development of the high-speed computer. Several tsunami models are being used in the National Tsunami Hazard Mitigation Program, sponsored by the National Oceanic and Atmospheric Administration (NOAA), in partnership with the US Geological Survey (USGS), the Federal Emergency Management Agency (FEMA), to produce tsunami inundation and evacuation maps for the states of Alaska, California, Hawaii, Oregon, and Washington.

Tsunami Generation and Propagation in an Open Ocean

The rupture speed of fault plane during earthquake is usually much faster than that of the tsunami. For instance, the fault line of the 2004 Sumatra earthquake was estimated as 1200-km long and the rupture process lasted for about 10 min. Therefore, the rupture speed was *c*. 2–3 km s^{-1}, which is considered as a relatively slow rupture speed and is still about 1 order of magnitude faster than the speed of tsunami (0.17 km s^{-1} in a typical water depth of 3 km). Since the compressibility of water is negligible, the initial free surface response to the seafloor deformation due to fault plane rupture is instantaneous. In other words, in terms of the tsunami propagation timescale, the initial free surface profile can be approximated as having the same shape as the seafloor deformation at the end of rupture, which can be obtained by the methods described in the previous section. As illustrated in **Figure 6**, the typical cross-sectional free surface profile, perpendicular to the fault line, has an N shape with a depression on the landward side and an elevation on the ocean side. If the fault plane is elongated, that is, $L \gg W$, the free surface profile is almost uniform in the longitudinal (fault line) direction and the generated tsunamis will propagate primarily in the direction perpendicular to the fault line. The wavelength is generally characterized by the width of the fault plane, W.

The measure of tsunami wave dispersion is represented by the depth-to-wavelength ratio, that is, $\mu^2 = h/\lambda$, while the nonlinearity is characterized by the amplitude-to-depth ratio, that is, $\varepsilon = A/h$. A tsunami generated in an open ocean or on a continental shelf could have an initial wavelength of several tens to hundreds of kilometers. The initial tsunami wave height may be on the order of magnitude of several meters. For example, the 2004 Indian Ocean tsunami had a typical wavelength of 200 km in the Indian Ocean basin with an amplitude of 1 m. The water depth varies from several hundreds of meters on the continental shelf to several kilometers in the open ocean. It is quite obvious that during the early stage of tsunami propagation both the nonlinear and frequency dispersion effects are small and can be ignored. This is particularly true for the 2004 Indian tsunami. The bottom frictional force and Coriolis force have even smaller effects and can be also neglected in the generation area. Therefore, the linear shallow water (LSW) equations are adequate equations describing the initial stage of tsunami generation and propagation.

As a tsunami propagates over an open ocean, wave energy is spread out into a larger area. In general, the tsunami wave height decreases and the nonlinearity remains weak. However, the importance of the frequency dispersion begins to accumulate as the tsunami travels a long distance. Theoretically, one can estimate that the frequency dispersion becomes important when a tsunami propagates for a long time:

$$t \gg t_d = \sqrt{\frac{h}{g}\left(\frac{\lambda}{h}\right)^3} \qquad [2]$$

or over a long distance:

$$x \gg x_d = t_d \sqrt{gh} = \frac{\lambda^3}{h^2} \qquad [3]$$

In the case of the 2004 Indian Ocean tsunami, $t_d \approx 700\,\mathrm{h}$ and $x_d \approx 5 \times 10^5\,\mathrm{km}$. In other words, the frequency dispersion effect will only become important when tsunamis have gone around the Earth several times. Obviously, for a tsunami with much shorter wavelength, for example, $\lambda \approx 20\,\mathrm{km}$, this distance becomes relatively short, that is, $x_d \approx 5 \times 10^2\,\mathrm{km}$, and can be reached quite easily. Therefore, in modeling transoceanic tsunami propagation, frequency dispersion might need to be considered if the initial wavelength is short. However, nonlinearity is seldom a factor in the deep ocean and only becomes significant when the tsunami enters coastal region.

The LSW equations can be written in terms of a spherical coordinate system as:

$$\frac{\partial \zeta}{\partial t} + \frac{1}{R\cos\varphi}\left[\frac{\partial P}{\partial \psi} + \frac{\partial}{\partial \varphi}(\cos\varphi Q)\right] = -\frac{\partial h}{\partial t} \qquad [4]$$

$$\frac{\partial P}{\partial t} + \frac{gh}{R\cos\varphi}\frac{\partial \zeta}{\partial \psi} = 0 \qquad [5]$$

$$\frac{\partial Q}{\partial t} + \frac{gh}{R}\frac{\partial \zeta}{\partial \varphi} = 0 \qquad [6]$$

where (ψ, φ) denote the longitude and latitude of the Earth, R is the Earth's radius, ζ is free surface elevation, P and Q the volume fluxes ($P = hu$ and $Q = hv$, with u and v being the depth-averaged velocities in longitude and latitude direction, respectively), and h the water depth. Equation [4] represents the depth-integrated continuity equation, and the time rate of change of water depth has been included. When the fault plane rupture is approximated as an instantaneous process and the initial free surface profile is prescribed, the water depth remains time-invariant during tsunami propagation and the right-hand side becomes zero in eqn [4].

The 2004 Indian Ocean tsunami provided an opportunity to verify the validity of LSW equations for modeling tsunami propagation in an open ocean. For the first time in history, satellite altimetry measurements of sea surface elevation captured the Indian Ocean tsunami. About 2 h after the earthquake occurred, two NASA/French Space Agency joint mission satellites, *Jason-1* and *TOPEX/Poseidon*, passed over the Indian Ocean from southwest to northeast (*Jason-1* passed the equator at 02:55:24UTC on 26 December 2004 and *TOPEX/Poseidon* passed the equator at 03:01:57UTC on 26 December 2004) (see **Figure 7**). These two altimetry satellites measured sea surface elevation with accuracy better than 4.2 cm.

Using the numerical model COMCOT (Cornell Multi-grid Coupled Tsunami Model), numerical simulations of tsunami propagation over the Indian Ocean with various fault plane models, including a transient seafloor movement model, have been carried out. The LSW equation model predicts accurately the arrival time of the leading wave and is insensitive of the fault plane models used. However, to predict the trailing waves, the spatial variation of seafloor deformation needs to be taken into consideration. In **Figure 8**, comparisons between LSW results with an optimized fault plane model and *Jason-1/TOPEX* measurements are shown. The excellent agreement between the numerical results and satellite data provides a direct evidence for the validity of the LSW modeling of tsunami propagation in deep ocean.

Coastal Effects – Inundation and Tsunami Forces

Nonlinearity and bottom friction become significant as a tsunami enters the coastal zone, especially during the runup phase. The nonlinear shallow water (NLSW) equations can be used to model certain aspects of coastal effects of a tsunami attack. Using the same notations as those in eqns [4]–[6], the NLSW

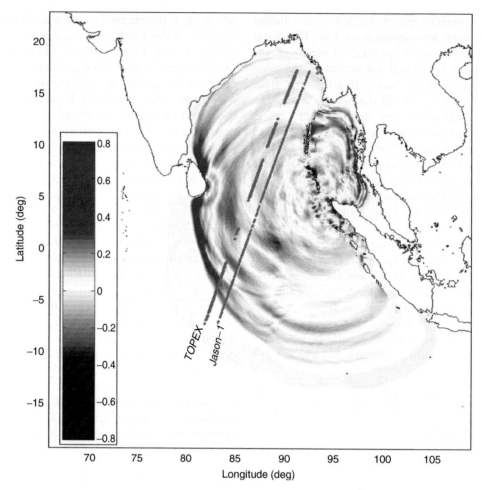

Figure 7 Satellite tracks for *TOPEX* and *Jason-1*. The colors indicate the numerically simulated free surface elevation in meter at 2 h after the earthquake struck.

equations in the Cartesian coordinates are

$$\frac{\partial \zeta}{\partial t} + \frac{\partial P}{\partial x} + \frac{\partial Q}{\partial y} = 0 \qquad [7]$$

$$\frac{\partial P}{\partial t} + \frac{\partial}{\partial x}\left(\frac{P^2}{H}\right) + \frac{\partial}{\partial y}\left(\frac{PQ}{H}\right) + gH\frac{\partial \zeta}{\partial x} + \tau_x H = 0 \qquad [8]$$

$$\frac{\partial Q}{\partial t} + \frac{\partial}{\partial x}\left(\frac{PQ}{H}\right) + \frac{\partial}{\partial y}\left(\frac{Q^2}{H}\right) + gH\frac{\partial \zeta}{\partial y} + \tau_y H = 0 \qquad [9]$$

The bottom frictional stresses are expressed as

$$\tau_x = \frac{gn^2}{H^{10/3}}P(P^2 + Q^2)^{1/2} \qquad [10]$$

$$\tau_y = \frac{gn^2}{H^{10/3}}Q(P^2 + Q^2)^{1/2} \qquad [11]$$

where n is the Manning's relative roughness coefficient. For flows over a sandy beach, the typical value for the Manning's n is 0.02.

Using a modified leapfrog finite difference scheme in a nested grid system, COMCOT is capable of solving both LSW and NLSW equations simultaneously in different regions. For the nested grid system, the inner (finer) grid adopts a smaller grid size and time step compared to its adjacent outer (larger) grid. At the beginning of a time step, along the interface of two different grids, the volume flux, P and Q, which is product of water depth and depth-averaged velocity, is interpolated from the outer (larger) grids into its inner (finer) grids. And at the end of this time step, the calculated water surface elevations, ζ, at the inner finer grids are averaged to update those values of the larger grids overlapping the finer grids, which are used to compute the volume fluxes at next time step in the outer grids. With this procedure, COMCOT can capture near-shore

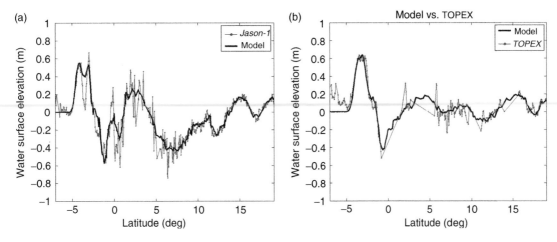

Figure 8 Comparisons between optimized fault model results and *Jason-1* measurements (a)/*TOPEX* measurements (b).

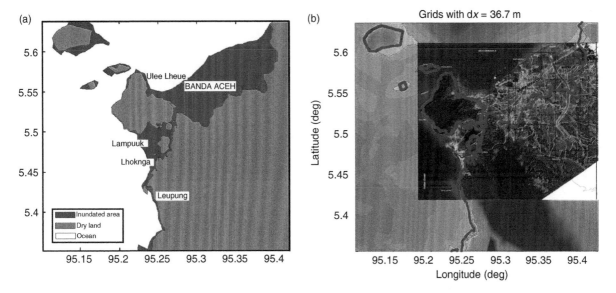

Figure 9 Calculated inundation areas (a) and overlaid with *QUICKBIRD* image (b) in Banda Aceh, Indonesia.

features of a tsunami with a higher spatial and temporal resolution and at the same time can still keep a high computational efficiency.

To estimate the inundation area caused by a tsunami, COMCOT adopts a simple moving boundary scheme. The shoreline is defined as the interface between a wet grid and its adjacent dry grids. Along the shoreline, the volume flux is assigned to be zero. Once the water surface elevation at the wet grid is higher than the land elevation in its adjacent dry grid, the shoreline is moved by one grid toward the dry grid and the volume flux is no longer zero and need to be calculated by the governing equations.

COMCOT, coupled up to three levels of grids, has been used to calculate the runup and inundation areas at Trincomalee Bay (Sri Lanka) and Banda Aceh (Indonesia). Some of the numerical results for Banda Aceh are shown here.

The calculated inundation area in Banda Aceh is shown in **Figure 9**. The flooded area is marked in blue, the dry land region is rendered in green, and the white area is ocean region. The calculated inundation area is also overlaid with a satellite image taken by *QUICKBIRD* in **Figure 9(b)**.

In the overlaid image, the thick red line indicates the inundation line based on the numerical simulation. In the satellite image, the dark green color (vegetation) indicates areas not affected by the tsunami and the area shaded by semitransparent red color shows flooded regions by this tsunami. Obviously, the calculated inundation area matches reasonably well with the satellite image in the neighborhood of Lhoknga and the western part of Banda Aceh. However, in the region of eastern Banda Aceh, the simulations significantly underestimate the inundation area. However, in general, the agreement between the numerical simulation and the satellite observation is surprisingly good.

In **Figure 10**, the tsunami wave heights in Banda Aceh are also compared with the field measurements by two Japan survey teams. On the coast between Lhoknga and Leupung, where the maximum height is measured more than 30 m, the numerical results match very well with the field measurements. However, beyond Lhoknga to the north, the numerical results, in general, are only half of the measurements, except in middle regions between Lhoknga and Lampuuk.

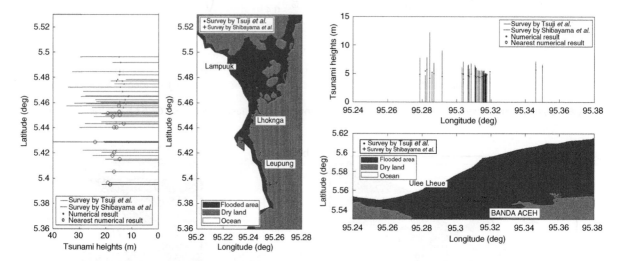

Figure 10 Tsunami heights on eastern and northern coast of Banda Aceh, Indonesia. The field survey measurements are from Tsuji *et al.* (2005) and Shibayama *et al.* (2005).

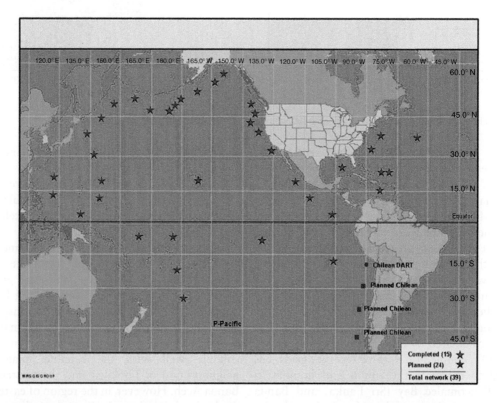

Figure 11 The locations of the existing and planned Deep-Ocean Assessment and Reporting of Tsunamis (DART) system in the Pacific Ocean (NOAA magazine, 17 Apr. 2006).

Tsunami Hazard Mitigation

The ultimate goal of the tsunami hazard mitigation effort is to minimize casualties and property damages. This goal can be met, only if an effective tsunami early warning system is established and a proper coastal management policy is practiced.

Tsunami Early Warning System

The great historical tsunamis, such as the 1960 Chilean tsunami and the 1964 tsunami generated near Prince William Sound in Alaska, prompted the US government to develop an early warning system in the Pacific Ocean. The Japanese government has also developed a tsunami early warning system for the entire coastal community around Japan. The essential information needed for an effective early warning system is the accurate prediction of arrival time and wave height of a forecasted tsunami at a specific location. Obviously, the accuracy of these predictions relies on the information of the initial water surface displacement near the source region, which is primarily determined by the seismic data. In many historical events, including the 2004 Indian Ocean tsunami, evidences have shown that accurate seismic data could not be verified until those events were over. To delineate the source region problem, in the United States, several federal agencies and states

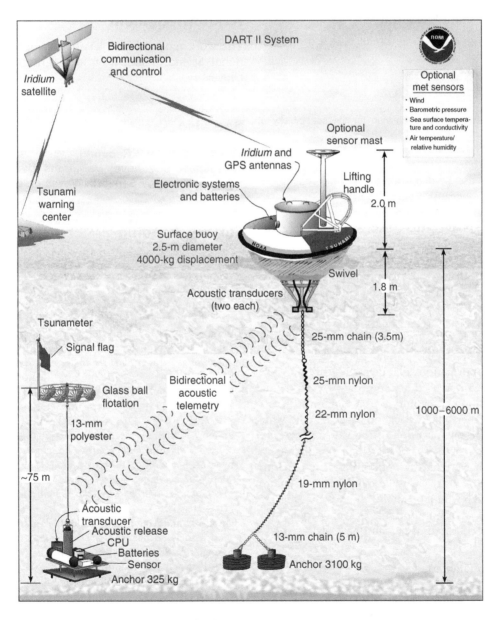

Figure 12 A sketch of the second-generation DART (II) system.

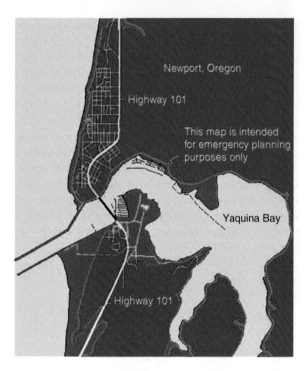

Figure 13 Tsunami inundation map for the coastal city of Newport, Oregon.

have joined together to create a warning system that involves the use of deep-ocean tsunami sensors to detect the presence of a tsunami.

These deep-ocean sensors have been deployed at different locations in the Pacific Ocean before the 2004 Indian Ocean tsunami. After the 2004 Indian Ocean tsunami, several additional sensors have been installed and many more are being planned (see **Figure 11**). The sensor system includes a pressure gauge that records and transmits the surface wave signals instantaneously to the surface buoy, which sends the information to a warning center via *Iridium* satellite (**Figure 12**). In the event of a tsunami, the information obtained by the pressure gauge array can be used as input data for modeling the propagation and evolution of a tsunami.

Although there have been no large Pacific-wide tsunamis since the inception of the warning system, warnings have been issued for smaller tsunamis, a few of which were hardly noticeable. This tends to give citizens a lazy attitude toward a tsunami warning, which would be fatal if the wave was large. Therefore, it is very important to keep people in a danger areas educated of tsunami hazards.

Coastal Inundation Map

Using numerical modeling, hazards in areas vulnerable to tsunamis can be assessed, without the area ever having experienced a devastating tsunami.

These models can simulate a 'design tsunami' approaching a coastline, and they can predict which areas are most at risk to being flooded. The tsunami inundation maps are an integral part of the overall strategy to reduce future loss of life and property. Emergency managers and local governments of the threatened communities use these and similar maps to guide evacuation planning.

As an example, the tsunami inundation map (**Figure 13**) for the coastal city of Newport (Oregon) was created using the results from a numerical simulation using a design tsunami. The areas shown in orange are locations that were flooded in the numerical simulation

Acknowledgment

The work reported here has been supported by National Science Foundation with grants to Cornell University.

See also

Glacial Crustal Rebound, Sea Levels and Shorelines. Sea Level Variations Over Geologic Time. Seismology Sensors. Waves on Beaches.

Further Reading

Geist EL (1998) Local tsunami and earthquake source parameters. *Advances in Geophysics* 39: 117–209.

Hokkaido Tsunami Survey Group (1993) Tsunami devastates Japanese coastal region. *EOS Transactions of the American Geophysical Union* 74: 417–432.

Kajiura K (1981) Tsunami energy in relation to parameters of the earthquake fault model. *Bulletin of the Earthquake Research Institute, University of Tokyo* 56: 415–440.

Kajiura K and Shuto N (1990) Tsunamis. In: Le Méhauté B and Hanes DM (eds.) *The Sea: Ocean Engineering Science*, pp. 395–420. New York: Wiley.

Kanamori H (1972) Mechanism of tsunami earthquakes. *Physics and Earth Planetary Interactions* 6: 346–359.

Kawata Y, Benson BC, Borrero J, *et al.* (1999) Tsunami in Papua New Guinea was as intense as first thought. *EOS Transactions of the American Geophysical Union* 80: 101, 104–105.

Keating BH and Mcguire WJ (2000) Island edifice failures and associated hazards. *Special Issue: Landslides and Tsunamis. Pure and Applied Geophysics* 157: 899–955.

Liu PL-F, Lynett P, Fernando H, *et al.* (2005) Observations by the International Tsunami Survey Team in Sri Lanka. *Science* 308: 1595.

Lynett PJ, Borrero J, Liu PL-F, and Synolakis CE (2003) Field survey and numerical simulations: A review of the

1998 Papua New Guinea tsunami. *Pure and Applied Geophysics* 160: 2119–2146.

Mansinha L and Smylie DE (1971) The displacement fields of inclined faults. *Bulletin of Seismological Society of America* 61: 1433–1440.

Satake K, Bourgeois J, Abe K, *et al.* (1993) Tsunami field survey of the 1992 Nicaragua earthquake. *EOS Transactions of the American Geophysical Union* 74: 156–157.

Shibayama T, Okayasu A, Sasaki J, *et al.* (2005) The December 26, 2004 Sumatra Earthquake Tsunami, Tsunami Field Survey in Banda Aceh of Indonesia. http://www.drs.dpri.kyoto-u.ac.jp/sumatra/indonesia-ynu/indonesia_survey_ynu_e.html (accessed Feb. 2008).

Synolakis CE, Bardet J-P, Borrero JC, *et al.* (2002) The slump origin of the 1998 Papua New Guinea tsunami. *Proceedings of Royal Society of London, Series A* 458: 763–789.

Tsuji Y, Matsutomi H, Tanioka Y, *et al.* (2005) Distribution of the Tsunami Heights of the 2004 Sumatra Tsunami in Banda Aceh measured by the Tsunami Survey Team. http://www.eri.u-tokyo.ac.jp/namegaya/sumatera/surveylog/eindex.htm (accessed Feb. 2008).

von Huene R, Bourgois J, Miller J, and Pautot G (1989) A large tsunamigetic landslide and debris flow along the Peru trench. *Journal of Geophysical Research* 94: 1703–1714.

Wang X and Liu PL-F (2006) An analysis of 2004 Sumatra earthquake fault plane mechanisms and Indian Ocean tsunami. *Journal of Hydraulics Research* 44(2): 147–154.

Yeh HH, Imamura F, Synolakis CE, Tsuji Y, Liu PL-F, and Shi S (1993) The Flores Island tsunamis. *EOS Transactions of the American Geophysical Union* 74: 369–373.

MAGNETICS

F. J. Vine, University of East Anglia, Norwich, UK

Introduction

Since World War II it has been possible to measure the variations in the intensity of the Earth's magnetic field over the oceans from aircraft or ships. In the 1950s the first detailed magnetic survey of an oceanic area, in the north-east Pacific, revealed a remarkable 'grain' of linear magnetic anomalies, quite unlike the anomaly pattern observed over the continents. In the 1960s it was realized that these linear anomalies result from a combination of seafloor spreading and reversals of the Earth's magnetic field. Hence they provide a detailed record of both the evolution of the ocean basins, and the timing of reversals of the Earth's magnetic field, during the past 160 million years. In addition, because of the dipolar nature of the field and the dominance of 'fossil' magnetization in the oceanic crust, the linear anomalies formed at midocean ridge crests, and the anomalies developed over isolated submarine volcanoes, can sometimes yield paleomagnetic information, such as the latitude at which these features were formed.

Units

In the SI system the unit of magnetic induction, or flux density (which geophysicists refer to as 'field intensity'), is the tesla (T). Because the magnitude of the Earth's magnetic field and magnetic anomalies is very small compared to 1 T, they are usually specified in nanoteslas (nT) $(1\,nT = 10^{-9}\,T)$. Fortunately, an old geophysical unit called the gamma is equivalent to 1 nT.

History of Measurement

William Gilbert, one-time physician to Elizabeth I of England, is thought to have been the first person to realize that the form of the Earth's magnetic field is essentially the same as that about a uniformly magnetized sphere. This is also equivalent to the field about a bar magnet (or magnetic dipole) placed at the center of the Earth, and aligned along the rotational axis (**Figure 1**). Certainly in terms of the written historical record he was the first person to propose this, in his Latin text *De Magnete*, published in 1600. Presumably, with the extension of European exploration to more southerly latitudes in the late fifteenth and the sixteenth centuries, mariners had problems with their compasses that Gilbert realized could be explained if the vertical component of the Earth's magnetic field varies with latitude. Accurate measurements of the direction of the Earth's magnetic field at London date from Gilbert's time. Measurement of the strength or intensity of the field, however, was not possible until the early part of the nineteenth century. The equipment used then, and for the following one hundred years or so, included a delicate suspended magnet system and required accurate orientation and leveling before a measurement could be made. Measurements on a moving platform

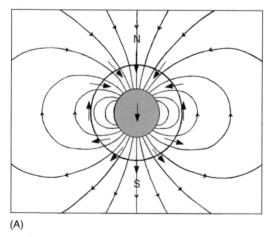
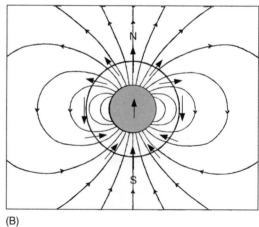

(A)   (B)

Figure 1 The (A) normal (present-day) and (B) reversed states of the dipolar magnetic field of the Earth. Shaded area shows the Earth's core; heavy arrows indicate directions of the field at the Earth's surface.

such as a ship were extremely difficult, therefore, and with the advent of iron and steel ships became impossible because of the magnetic fields associated with the ships themselves.

Meanwhile, measurements on land had revealed that although the Earth's magnetic field is, to a first approximation, equivalent to that about an axial dipole as envisaged by Gilbert, there is a considerable nondipole component, on average 20% of the dipole field. Moreover, the field is changing in time, albeit slightly and slowly, in both intensity and direction. As rather more than 70% of the Earth's surface is covered by water and there were few magnetic measurements in these areas, detailed mapping of the field at the surface worldwide was seriously hampered. In 1929 the Carnegie Institution of Washington went to the length of building a wooden research ship, the *Carnegie*, to map the Earth's magnetic field in oceanic areas. Similarly the then USSR commissioned a wooden research ship, the *Zarya*, in 1956. However, by this time new electronic instruments had been developed that were capable of making continuous measurements of the total intensity of the Earth's magnetic field from aircraft and ships.

Among the many projects instigated by the Allies in World War II, to counter the submarine menace, was the Magnetic Airborne Detector (MAD) project. The outcome was the development of the fluxgate magnetometer. To increase the sensitivity of the instrument, much of the Earth's magnetic field is 'backed off' by a solenoid producing a biasing field. Initially it was not possible to produce a constant biasing field and the instruments tended to 'drift,' which meant that they were not ideal for scientific purposes. After the war these instruments were redeployed for use in conducting aeromagnetic surveys over land areas, in connection with oil and mineral exploration, and then modified for use from ships. In both instances the detector was housed in a 'fish' that was towed, so as to remove it from the magnetic fields associated with the ship or aircraft.

In the 1950s the proton-precession magnetometer was developed, which had the advantage of achieving the same or somewhat better sensitivity (about 1 part in 50 000) without the problem of drift. Since 1970 even more sensitive magnetometers have been developed – the optical absorption magnetometers. Although in some ways superseded by these magnetometers based on proton and electron precession, fluxgates are still widely used because they have the advantage of measuring the component of the field directed along the axis of the detector rather than the total ambient field. This property makes them particularly useful in satellites, for example, where they are often used as components in orientation systems, at the same time yielding measurements of the Earth's magnetic field.

Measurements of the Earth's magnetic field in oceanic areas are, therefore, now relatively routine, whether they be from satellites, aircraft, ships, submersibles, or remotely operated or deeply towed vehicles.

Nature of the Earth's Magnetic Field

The differences between the measured magnetic field about the Earth and that predicted for a central and axially aligned dipole are considerable. Best-known is the difference between the magnetic poles – where the field is directed vertically – and the rotational, geographic, poles of the Earth. As a result, for most points on the Earth's surface there is an angular difference between the directions to true north and to magnetic north. Mariners refer to this as the magnetic variation; scientists refer to it as the magnetic declination. The centered dipole that best fits the observed field predicts a field strength of 30 000 nT around the equator and a maximum value of 60 000 nT at both poles. The actual field departs considerably from this, as can be seen in **Figure 2**. The intensity and direction of the field, or any of their components (such as magnetic variation) at any one point, vary with time, typically by tens of nanoteslas and a few minutes of arc per year. This is known as the secular variation of the field. The form of the Earth's magnetic field and its secular variation is thought to derive from the fact that it is generated in the outer, fluid core of the Earth, which is metallic (largely iron and nickel) and hence a good electrical conductor. Convective motions of this fluid conductor carrying electrical currents and interacting with magnetic fields produce a dynamo-like effect and an external magnetic field. The essential axial symmetry of this field is probably determined by the influence of the Coriolis force on the precise nature of the convective motions. Historical, archeomagnetic and paleomagnetic data suggest that although the field at any one time departs considerably from that predicted by a geocentric axial dipole, when averaged over several thousand years, the mean field is very close to that about such a dipole. On even longer, geological, timescales the field intermittently reverses its polarity completely (**Figure 1**), probably within a period of approximately 5000 years. The length of intervals of a particular polarity varies widely from a few tens of thousands of years to a few tens of millions of years.

The secular variation and changes in the polarity of the field result from dynamic processes deep in the

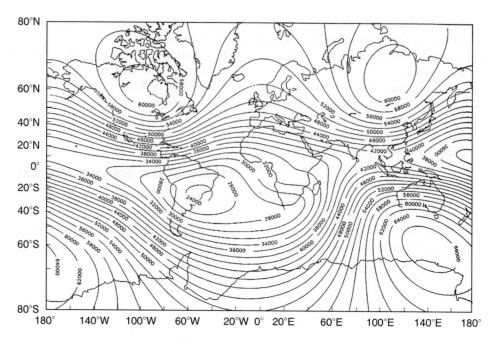

Figure 2 Total intensity of the Earth's magnetic field (in nT) at the Earth's surface (for epoch 1980). (Reproduced from Langel RA, in *Geomagnetism* Vol. 1, JA Jacobs (ed.) © 1987 Academic Press, by permission of the publisher.).

Earth's interior. There are also shorter-period variations of the field with time that are of external origin, essentially a result of the interaction of the solar wind with the Earth's magnetic field. The most relevant of these in the present context is the daily or diurnal variation of the field. This is a smooth variation that is greatest during daylight hours and typically has an amplitude of a few tens of nanoteslas. It can be greater, however, near the magnetic equator and poles. Increased solar activity can produce higher-amplitude and more irregular variations, and intense sunspot activity produces global magnetic storms; high-amplitude, short-period variations during which magnetic surveying has to be discontinued.

Reduction of Magnetic Data

Measurements of the Earth's magnetic field over oceanic areas are corrected for the present-day spatial and time variations of the field described above in order to obtain residual 'anomalies' in the field. These are caused by magnetization contrasts at or near the Earth's surface, i.e., in the upper lithosphere. Such anomalies should therefore yield information on the magnetization and structure of the oceanic crust.

For measurements made at or near the Earth's surface, i.e. from ships, aircraft, or submersibles, the secular and diurnal variation of the field can often be ignored because these effects are very small compared to the amplitudes of the anomalies being mapped. However, should a very accurate survey be required, secular variation has to be taken into account if surveys made at different times are being combined and the observations should be corrected for diurnal variation using records from nearby land stations or moored buoys. If there are sufficient 'cross-overs' during the survey (i.e. repeat measurements at the same point), it may also be possible, indeed preferable, to use these to correct for diurnal variation. It may also be necessary to correct for any magnetic effect of the moving platform itself. If present, this effect will vary according to the direction of travel.

For many purposes, however, the above corrections are so small that they can be ignored. The final correction, the removal of the main or 'regional' field of the Earth, that is, the field generated in the Earth's core, must always be applied. In theory this should be simple. The depth to the core–mantle boundary, 2900 km, means that the field originating in the core should have a smooth, long-wavelength variation at or above the Earth's surface. The magnetization contrasts within a few tens of kilometers of the Earth's surface will produce anomalies of much shorter wavelength. Between these two source regions any magnetic minerals in the mantle are at a temperature above their Curie temperature and are effectively nonmagnetic. In practice, because of its complexity and because it is changing with time, it has proved difficult to accurately define the main field of the Earth, i.e. that originating in the core, and the way in which it is changing with time. The first attempt to define a global 'International Geomagnetic Reference Field' was made in the 1960s and, although revised and greatly improved at five-year intervals since then, its level, if not its gradients, can still be seen to be slightly incorrect for certain oceanic areas. As a result, particularly in the past, the regional field for particular profiles or surveys has been obtained by fitting a smooth long-wavelength curve or surface to the observed data.

Once the long-wavelength 'regional field' has been removed from magnetic data, the resulting 'residual' or 'total-field' anomalies are assumed to result from magnetization contrasts within the upper, magnetic, part of the Earth's lithosphere (**Figure 3**).

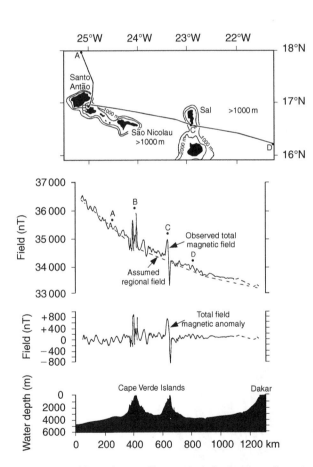

Figure 3 Magnetic profile recorded by a fluxgate magnetometer between the Cape Verde islands and Dakar, Senegal. The ship's track close to the islands is shown in the upper part of the diagram. The dashed line indicates the regional field used to calculate total field magnetic anomalies. The high-amplitude, short-wavelength anomalies close to the Cape Verde Islands reflect the presence of highly magnetic volcanic rocks at shallow depth. (Reproduced from Heezen BC, *et al.*, in *Deep-Sea Research*, Vol. 1 © 1953 Elsevier Science, by permission of the publisher.)

Magnetization of Ocean Floor Rocks

All minerals, and hence all rocks, exhibit magnetic properties. However, the magnetization that most rock types acquire in the relatively weak magnetic field of the Earth is insufficient to produce significant anomalies in the Earth's magnetic field, particularly when this is measured at some distance from the rocks, as is typically the case in oceanic areas. For rocks to be capable of producing appreciable anomalies they must contain more than a few percent by volume of 'ferromagnetic' minerals, that is, certain oxides and sulfides containing iron, notably magnetite (Fe_3O_4). Most sediments and 'acid' (silica-rich) igneous rocks, such as granite, do not meet this criterion. Basic (silica-deficient) igneous rocks such as basalts, and the coarser-grained but chemically equivalent gabbros, and ultrabasic rocks such as peridotite do contain a higher proportion of iron oxides and are capable of producing anomalies. Metamorphic rocks, formed when preexisting rocks are subjected to high temperatures and/or pressures, are typically weakly magnetized except for some formed from basic or ultrabasic igneous rocks.

Apart from its sedimentary veneer, the upper part of the oceanic lithosphere consists almost entirely of basic and ultrabasic rocks, i.e. basalts, gabbros, and peridotites. This is a consequence of the way in which it is formed by the process of seafloor spreading. At midocean ridge crests the ultrabasic peridotite of the Earth's mantle undergoes partial melting, producing basic magma that rises and collects as a magma chamber within oceanic crust. Solidification of such magma chambers ultimately forms the main crustal layer of gabbro, but not before some ultrabasic rocks have formed at the base of the chamber from the accumulation of first formed crystals, and magma has been extruded through near vertical fissures to form pillow basalts on the seafloor. Solidification of the magma in these fissures forms a layer consisting of dikes between the gabbro and the basalts.

Thus, because of the rock types present, the oceanic crust and upper mantle are relatively strongly magnetized and capable of producing large-amplitude anomalies in the Earth's magnetic field, even when measured at sea level. The thickness of the magnetic layer is determined by the depth to the Curie point isotherm. Because of the way in which the oceanic lithosphere is formed, by spreading about ridge crests, this varies from a few kilometers depth within the crust at ridge crests to a depth of approximately 40 km in oceanic lithosphere that is 100 million years, or more, in age. In places the thermal regime associated with seafloor spreading is modified by mantle 'hot spots.' As a result there is an enhanced degree of partial melting of the mantle and the larger volumes of magma produced extrude on to the seafloor to form seamounts, oceanic islands and, exceptionally, oceanic plateaux. These all involve an appreciable thickening of the oceanic crust, but the rock types involved are all essentially basic and potentially strongly magnetic (**Figure 3**).

Observed Anomalies

The marked contrast in the way in which continental and oceanic crust are formed, and hence in the predominant rock types in each setting, gives rise to a striking difference in the character of the total field anomalies developed over the two types of lithosphere. Within the continents the variety of rock types in mountain belts and their juxtaposition by folding and faulting produces magnetization contrasts and anomalies that delineate the general trend of the belt. Areas of igneous activity that include basic igneous rocks are characterized by very large-amplitude and typically short-wavelength anomalies, and sedimentary basins and extensive areas of granite are quiet magnetically. This pattern of

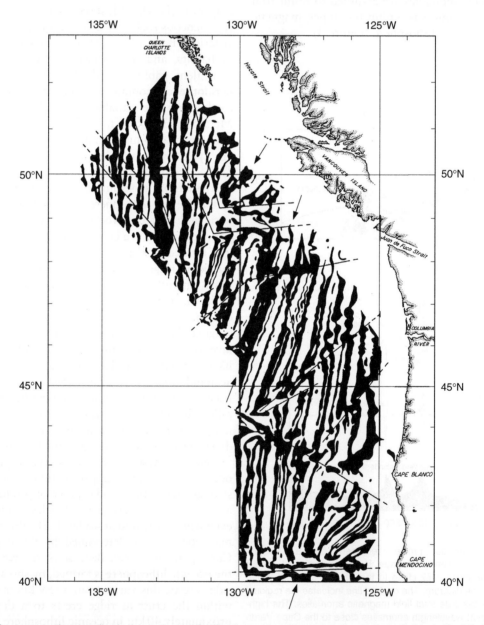

Figure 4   Linear magnetic anomalies in the north–east Pacific. Areas of positive anomaly are shown in black. Straight lines indicate faults offsetting the anomaly pattern; arrows, the axes of three short ridge lengths in the area – from north to south, the Explorer, Juan de Fuca and Gorda ridges. (Based on Figure 1 of Raff AD and Mason RG in *Bull. Geol. Soc. Amer.*, Vol 72. © 1961 *Geological Society of America*. Reproduced by permission of the publisher.)

anomalies is characteristic of the continental shelves out to the continental slope, the true geological boundary between the continents and the oceans, although, in that the shelves are typically underlain by a great thickness of sediments, they are often magnetically 'quiet.'

Deep sea areas, that is, those underlain by oceanic lithosphere, are characterized by remarkably linear and parallel anomalies that extend for hundreds of kilometers and are truncated and offset by fracture zones (**Figure 4**). The fracture zones are formed by the transform faults that offset the crest of the midocean ridge system. Thus the linear magnetic anomalies parallel the ridge crests. The anomalies are remarkable for their linearity, their high amplitude and the steep magnetic gradients that separate highs from lows. Any explanation of them in terms of linear structures and/or lateral variations in rock type within the oceanic crust is extremely improbable. It transpires that they result from a combination of sea floor spreading and reversals of the Earth's magnetic field. As new oceanic crust and upper mantle form at a ridge crest they acquire a permanent (remanent) magnetization which parallels the ambient direction of the field. If, as spreading occurs, the Earth's magnetic field reverses, then the ribbon of newly formed oceanic lithosphere along the whole length of the spreading ridge system acquires a remanent magnetisation in the opposite direction. It is these contrasts between normally and reversely magnetized material which produce the high amplitude linear anomalies and the steep gradients between them.

Rates of seafloor spreading vary greatly for different ridges and the interval between reversals of the Earth's magnetic field is also very variable throughout geological time. However, rates of spreading are typically a few tens of millimeters per year and the average polarity interval is about 0.5 million years. Thus typical linear anomalies are 10–20 km in width. Initially, spreading rates could only be reliably determined for the past 3.5 million years. For this period the reversal timescale had been independently determined from measurements of the age and polarity of remanent magnetization of both subaerial lava flows and deep-sea sediments, and it is clearly reproduced in the anomalies recorded across midocean ridge crests (**Figure 5**). With the dating of older oceanic crust by the international Deep Sea Drilling Program it became possible to deduce spreading rates at earlier times and to calibrate the timescale of reversals of the Earth's magnetic field implied by the older linear anomalies. In this way the geomagnetic reversal timescale for the past 160 million years has been deduced (**Figure 6**).

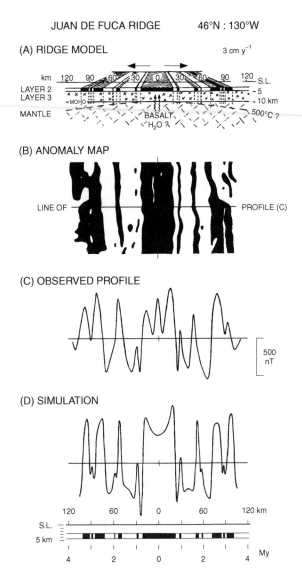

Figure 5 (A) Schematic crustal model for the Juan de Fuca Ridge, south-west of Vancouver Island. Shaded material in layer 2 is normally magnetized; unshaded material is reversely magnetized. SL = sea level. (B) Part of the summary map of magnetic anomalies recorded over the Juan de Fuca Ridge (**Figure 4**). (C) Total field magnetic anomaly profile along the line indicated in (B). (D) Computed profile assuming the model and reversal timescale for the past 3.5 million years. (Reproduced from Vine FJ in *The History of the Earth's Crust*. RA Phinney (ed.). © 1968 Princeton University Press, by permission of the publisher.)

In recording the times at which the Earth's magnetic field has reversed its polarity, the linear magnetic anomalies also serve as time or growth lines that reveal the evolution of the ocean basins in terms of seafloor spreading (**Figure 7**). Thus it is possible to accurately reconstruct the most recent phase of continental drift (during the past 180 million years) when a former supercontinent was split up to form the present-day continents and the Atlantic and Indian Oceans. The Pacific Ocean was formed during the

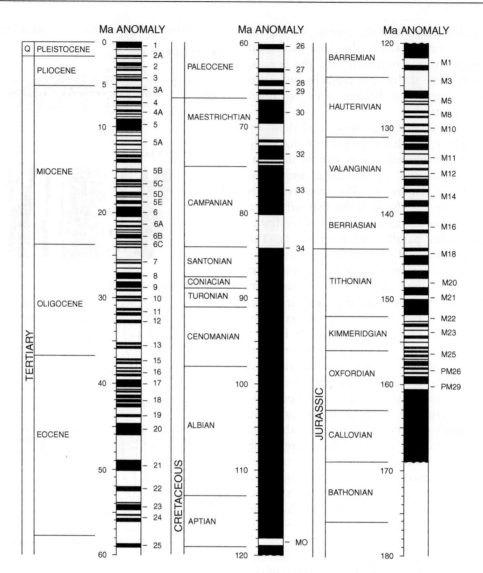

Figure 6 Geomagnetic polarity timescale for the past 160 million years. (Reproduced from Jones EJW, *Marine Geophysics.* © 1999 John Wiley and Sons, by permission of the publisher.)

same period, and in the process dispersed marginal fragments of the supercontinent around its rim where they are now recognized as 'suspect terranes.' In contrast to this very detailed record of spreading and drift for the past 180 million years there is no such record for the previous 96% of geological time, and earlier phases of drift and mountain building have to be deduced from the more complex and fragmentary geological record within the remaining 40% of the Earth's surface that is covered by continental crust.

Paleomagnetic Information Contained in Oceanic Magnetic Anomalies

As a result of the dipolar nature of the Earth's magnetic field, whereby its inclination to the horizontal varies systematically from the equator to the poles (**Figure 1**), anomalies in the total field over a relatively simple and symmetrical feature such as the central, normally magnetized ribbon of crust at a midocean ridge crest typically have an asymmetry (**Figure 8**). The exceptions occur at the poles and across ridges trending exactly north–south, where the anomaly is a symmetrical high, and over an east–west trending ridge at the Equator, where the anomaly is a symmetrical low. For all other latitudes, and all orientations other than north–south, the degree of asymmetry, or the 'phase-shift' of the anomaly, is a function of the latitude and orientation. As a result of spreading, all older linear anomalies, unless formed about a north–south trending, east–west spreading ridge, will now be at a different latitude from that at which they were formed, and the direction of their remanent magnetization will be

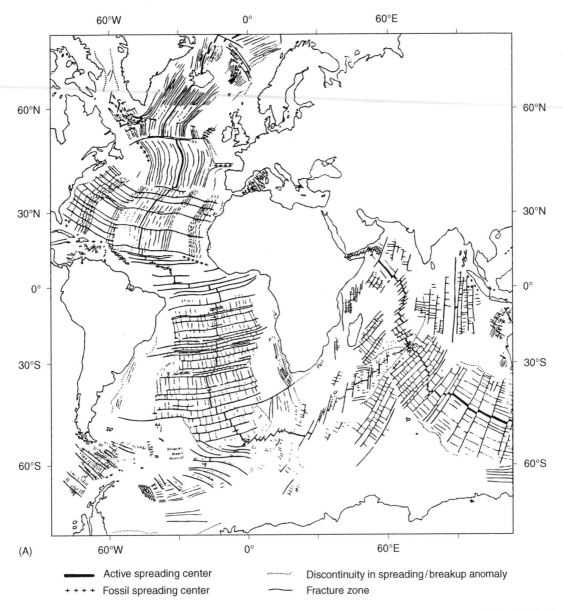

Active spreading center .... Discontinuity in spreading / breakup anomaly
+ + + + Fossil spreading center —— Fracture zone

Figure 7 A compilation of the trends of linear magnetic anomalies in (A) the Atlantic and Indian Oceans and (B) the Pacific Ocean. (Reproduced from Jones EJW, *Marine Geophysics*. © 1999 John Wiley and Sons, by permission of the publisher.)

different from the direction of the ambient magnetic field. As a consequence, the asymmetry of the anomaly is different from what one would predict for similarly directed remanence and ambient field. This difference between the observed and predicted phase shift reflects the latitudinal change. Although the interpretation of such data is somewhat ambiguous if the orientation of the anomaly is thought to have changed since the time of formation, it does provide paleolatitude, i.e. paleomagnetic, information for oceanic areas, which is otherwise rather sparse. As with all paleomagnetic data, they can be used to test independently derived models for the 'absolute' motion of plates and plate boundaries, such as ridge crests, across the face of the Earth.

In theory the ambiguity of paleolatitude determination mentioned above can be removed by carrying out the analysis on anomalies of the same age on either side of a particular ridge. An intriguing result of such studies, however, is that in some cases different latitudes of formation are deduced for the same anomaly on either side of the ridge. In that they were formed at the same time and at the same ridge crest, this cannot be so, and the result is giving us yet more information on the geometry or magnetization of the source region. Such a result could be produced by a decay in the intensity of the Earth's magnetic field or an increase in the number of magnetic excursions or very short-lived reversals as a particular polarity interval progresses, and/or a more complex

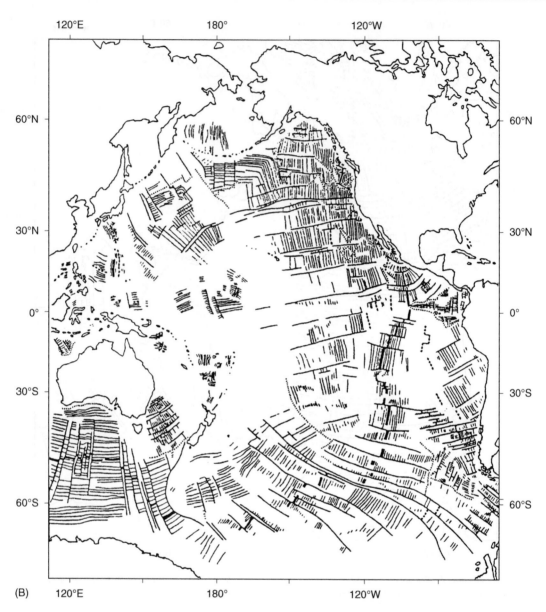

Figure 7 (Continued)

geometry for the boundaries between normally and reversely magnetized material in the oceanic lithosphere (**Figure 9**). One would predict inwardly sloping boundaries in the upper crust as a result of the lateral extrusion and vertical accretion of lava flows, and outwardly sloping boundaries in the lower crust reflecting the location of the Curie point isotherm at a ridge crest. Increasingly this latter interpretation seems to be the more likely.

The isolated magnetic anomalies that are associated with submarine volcanoes and oceanic islands can also, sometimes, yield useful paleomagnetic information. If it can be assumed that the edifice is uniformly magnetized throughout, then, knowing the topographic shape of the feature, the magnetic

anomaly over it can be analyzed to yield a direction and intensity of magnetization. Strictly, this is the sum of the induced and remanent magnetization of the feature, but the former is almost certainly small compared to the latter and the direction of magnetization deduced can be considered to be a good approximation to the direction of the remanent magnetization and hence treated as a paleomagnetic result. It seems probable that many of these features, particularly the large edifices, were formed during more than one polarity interval, in which case the method is inapplicable. However, many seamounts, particularly those formed during the long interval of normal polarity during mid-Cretaceous time, do yield satisfactory results. Examples are the New

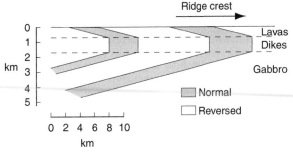

Figure 9 Postulated geometry of normal/reverse magnetization contrasts in the oceanic crust, which could explain the anomalous phase shift of certain linear anomalies.

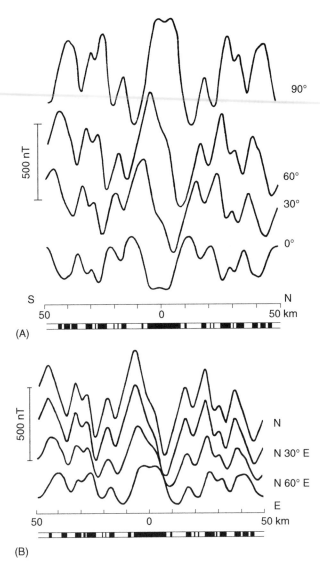

Figure 8 Variation of the magnetic anomaly pattern (A) with geomagnetic latitude (all profiles are N–S; angles refer to magnetic inclination; no vertical exaggeration) and (B) with direction of the profile at fixed latitude (magnetic inclination is 45° in all cases; no vertical exaggeration). (Reproduced from Kearey P and Vine FJ. *Global Tectonics*. © 1996 Blackwell Science, by permission of the publisher.)

England seamounts of the north–west Atlantic, and the Musician seamounts of the central Pacific. These yield paleomagnetic pole positions that are consistent with other results for the mid-Cretaceous for the North American and Pacific plates, respectively.

Conclusion

Thus, because of the dominance of remanent magnetization in the oceanic crust, and the relative simplicity of the way in which it is formed by seafloor spreading and volcanic activity, it records both the history of reversals of the Earth's magnetic field, and the lateral and latitudinal displacement of the crust during the past 160 million years. This record can be played back by measuring the anomalies in the intensity of the Earth's magnetic field over the oceans at the present day.

See also

Deep-Sea Drilling Results. Geomagnetic Polarity Timescale. Mid-Ocean Ridge Geochemistry and Petrology. Mid-Ocean Ridge Seismic Structure. Mid-Ocean Ridge Tectonics, Volcanism, and Geomorphology. Propagating Rifts and Microplates. Seamounts and Off-Ridge Volcanism. Seismic Structure.

Further Reading

Bullard EC and Mason RG (1963) The magnetic field over the oceans. In: Hill MN (ed.) *The Sea*, vol. 3, pp. 175–217. London: Wiley-Interscience.

Harrison CGA (1981) Magnetism of the oceanic crust. In: Emiliani C (ed.) *The Sea*, vol. 7, pp. 219–239. Wiley: New York.

Jones EJW (1999) *Marine Geophysics*. Chichester: Wiley.

Kearey P and Vine FJ (1996) *Global Tectonics*. Oxford: Blackwell Science.

Vacquier V (1972) *Geomagnetism in Marine Geology*. Amsterdam: Elsevier.

GEOMAGNETIC POLARITY TIMESCALE

C. G. Langereis and W. Krijgsman, University of Utrecht, Utrecht, The Netherlands

Introduction

Dating and time control are essential in all geoscientific disciplines, since they allows us to date, and hence correlate, rock sequences from widely different geographical localities and from different (marine and continental) realms. Moreover, accurate time control allows to understand rates of change and thus helps in determining the underlying processes and mechanisms that explain our observations. Biostratigraphy of different faunal and floral systems has been used since the 1840s as a powerful correlation tool giving the geological age of sedimentary rocks. Radiometric dating, originally applied mostly igneous rocks, has provided numerical ages; this method has become increasingly sophisticated and can now — in favorable environments — also be used on various isotopic decay systems in sediments. We are concerned with the application of magnetostratigraphy: the recording of the ancient geomagnetic field that reveals in lava piles and sedimentary sequences, intervals with different polarity. This polarity can either be normal, that is parallel to the present-day magnetic field (north-directed), or reversed (south-directed) (**Figure 1**). As a rule, it appears that these successive intervals of different polarity show an irregular thickness pattern, caused by the irregular duration of the successive periods of either normal or reversed polarity of the field. This produces a 'bar code' in the rock record that often is distinctive. Polarity intervals have a mean duration of some 300 000 y during the last 35 My, but large variations occur, from 20 000 y to several million years. If one can construct a calibrated 'standard' or a so-called 'geomagnetic polarity timescale' (GPTS), dated by radiometric methods and/or by orbital tuning, one can match the observed pattern with this standard and hence derive the age of the sediments. Magnetostratigraphy with correlation to the GPTS has become a standard tool in ocean sciences.

The Paleomagnetic Signal

The Earth's magnetic field is generated in the liquid outer core through a dynamo process that is maintained by convective fluid motion. At the surface of the Earth, the field can conveniently be described as a dipole field, which is equivalent to having a bar magnet at the center of the Earth. Such a dipole accounts for approximately 90% of the observed field; the remaining 10% derives from higher-order terms: the nondipole field. At any one time, the best-fitting

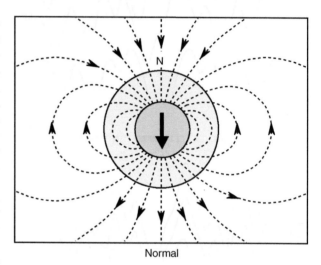

Normal

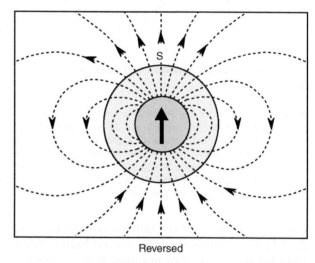

Reversed

Figure 1 Schematic representation of the geomagnetic field of a geocentric axial dipole ('bar magnet'). During normal polarity of the field the average magnetic north pole is at the geographic north pole, and a compass aligns along magnetic field lines. Historically, we refer to the north pole as the pole attracting the 'north-seeking' needle of a compass, but physically it is a south pole. During normal polarity, the inclination is positive (downward-directed) in the Northern Hemisphere and negative (upward-directed) in the Southern Hemisphere. Conversely, during reversed polarity, the compass needle points south, and the inclination is negative in the Northern and positive in the Southern Hemisphere.

geocentric dipole axis does not coincide exactly with the rotational axis of the Earth, but averaged over a few thousand years we may treat the dipole as both geocentric and axial.

The most distinctive property of the Earth's magnetic field is that it can reverse its polarity: the north and south poles interchange. Paleomagnetic studies of igneous rocks provided the first reliable information on reversals. In 1906, Brunhes observed lava flows magnetized in a direction approximately anti-parallel to the present geomagnetic field, and suggested that this was caused by a reversal of the field itself, rather than by a self-reversal mechanism of the rock. In 1929, Matuyama demonstrated that young Quaternary lavas were magnetized in the same direction as the present field (normal polarity), whereas older lavas were magnetized in the opposite direction. His study must be regarded as the first magnetostratigraphic investgation. Initially, it was believed that the field reversed periodically, but as more (K/Ar dating plus paleomagnetic) results of lava flows became available, it became clear that geomagnetic reversals occur randomly. It is this random character that fortuitously provides the distinctive 'fingerprints' and gives measured polarity sequences their correlative value.

A polarity reversal typically takes several thousands of years, which on geological timescales is short and can be taken as globally synchronous. The field itself is sign invariant: the same configuration of the geodynamo can produce either a normal or reversed polarity. What causes the field to reverse is still the subject of debate, but recent hypotheses suggests that lateral changes in heat flow at the core–mantle boundary play an important role. Although polarity reversals occur at irregular times, over geological time spans the reversal frequency can change considerably. For instance, the polarity reversal frequency has increased from approximately $1 \, \text{My}^{-1}$ some 80 My ago to $5 \, \text{My}^{-1}$ in more recent times. During part of the Creataceous, no reversals occurred at all from 110 to 80 Ma: the field showed a stable normal polarity during some 30 My. Such long periods of stable polarity are called Superchrons, and the one in the Cretaceous is also recognized as the Cretaceous Normal Quiet Zone in ocean-floor magnetic anomalies.

The ancient geomagnetic field can be reconstructed from its recording in rocks during the geological past. Almost every type of rock contains magnetic minerals, usually iron oxide/hydroxides or iron sulfides. During the formation of rocks, these magnetic minerals (or more accurately: their magnetic domains) statistically align with the then ambient field, and will subsequently be 'locked in,' preserving the direction of the field as natural

remanent magnetization (NRM): the paleomagnetic signal. The type of NRM depends on the mechanism of recording the geomagnetic signal, and we distinguish three basic types: TRM, CRM, and DRM.

TRM, or thermoremanent magnetization, is the magnetization acquired when a rock cools through the Curie temperature of its magnetic minerals. Curie temperatures of the most common magnetic minerals are typically in the range 350–700°C. Above this temperature, the magnetic domains align instantaneously with the ambient field. Upon cooling, they are locked: the magnetic minerals carry a remanence that usually is very stable over geological time spans. Any subsequent change of the direction of the ambient field cannot change this remanence. Typically, TRM is acquired in igneous rocks.

CRM, or chemical remanent magnetization, is the magnetization acquired when a magnetic mineral grows through a critical 'blocking diameter' or grain size. Below this critical grain size, the magnetic domains can still align with the ambient field; above it, the field will be locked and the acquired remanence may again be stable over billions of years. CRM may be acquired under widely different circumstances, e.g., during slow cooling of intrusive rocks, during metamorphosis or (hydrothermal) fluid migration, but also during late diagenetic processes such as weathering processes through formation of new magnetic minerals. A particularly important mechanism of CRM acquisition occurs in marine sediments during early diagenesis: depending on the redox conditions, iron-bearing minerals may dissolve, and iron may become mobile. If the mobilized iron subsequently encounters oxic conditions, it may precipitate again and form new magnetic minerals that then acquire a CRM.

DRM, or detrital remanent magnetization, is the magnetization acquired when magnetic grains of detrital origin already carrying TRM or CRM are deposited. The grains statistically align with the ambient field as long as they are in the water column or in the soft water-saturated topmost layer of the sediment (**Figure 2**). Upon compaction and dewatering, the grains are mechanically locked — somewhere in a 'lock-in depth zone' — and will preserve the direction of the ambient field.

The Magnetic Signal in Marine Sediments

In sediments, it is often assumed that the natural remanent magnetization is due to DRM. In the past, there has been some debate on the mechanisms of DRM acquisition, and the term pDRM or post-depositional DRM has often been used, because the lock-in zone has a certain depth. Sediment at this

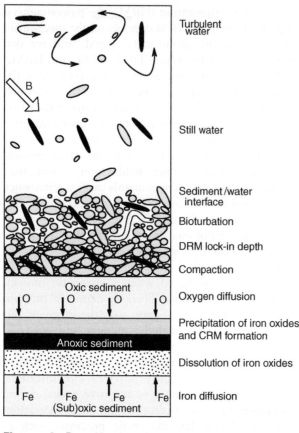

Figure 2 Depositional remanent magnetization (DRM) acquired in sediments involves a continuum of physical and chemical processes. Detrital magnetic minerals (black) will be statistically aligned along the ambient geomagnetic field (B) in still water and in the soft and bioturbated water-saturated sediment just below the sediment–water interface. Upon compaction and dewatering, the grains are mechanically locked, preserving the direction of the field. Early diagenetic processes such as sulfate reduction may dissolve iron-bearing minerals. Upon encountering a more oxic environment, iron may precipitate as iron oxides, which will acquire a chemical remanent magnetisation (CRM). The thus-acquired CRM in this layer may have a much later age than the depositional age of this layer.

organic matter, existing magnetic minerals may dissolve and iron may be mobilized and precipitate elsewhere. These geochemical processes may results in the (partial) removal of the original paleomagnetic signal and in the acquisition of CRM in a particular zone (much) later than the deposition of the sediment in this zone. Clearly, such processes may severely damage the fidelity of the paleomagnetic record, and may offset the position of a reversal boundary by a distance in sediment that can correspond to a time of up to tens of thousands of years. However, in 'suitable' sediments the damage is usually restricted, and the paleomagnetic signal of such sediments may be considered as reliable and near-depositional. Suitable environments are generally those with a sufficiently high sedimentation rate, a significant detrital input, and a predominantly oxic environment. Therefore, it is often necessary to check the origin of the NRM using rock magnetic and geochemical methods.

As a rule, the total NRM is composed of different components. Ideally, the primary NRM — that originating from near the time of deposition — has been conserved, but often this original signal is contaminated with 'viscous' remanence components, referred to as VRM. Such a VRM may result from partial realignment of 'soft' magnetic domains in the present-day field or from low-temperature oxidation of magnetic minerals. It is generally easily removed through magnetic 'cleaning,' which consists of a routine laboratory treatment called demagnetization; details can be found in any standard textbook on paleomagnetism. Despite these pitfalls in sedimentary paleomagnetic records, sediments — in contrast to igneous rocks — have in principle the distinctive advantage of providing a continuous record of the geomagnetic field, including the history of geomagnetic polarity reversals. Paleomagnetic studies of the sediments will then reveal the pattern of geomagnetic reversals during deposition.

depth is slightly older than that the sediment–water interface. Hence, the acquisition of NRM is always slightly delayed with respect to sediment age. For practical purposes (in magnetostratigraphy) this usually has no serious consequences. Nevertheless, the concept of a purely detrital remanent magnetization an oversimplification of the real world. We therefore prefer to use the term depositional remanent magnetization (DRM) to refer to a continuum of physical and chemical processes that occur during and shortly after deposition. The acquisition of a DRM thus depends both on detrital input of magnetic minerals and the new formation of such minerals in the sediment through early diagenetic processes (**Figure 2**). Early diagenesis is widespread and occurs in virtually every sedimentary environment. Depending on the role of

The Geomagnetic Polarity Timescale (GPTS)

Surveys over the ocean basins carried out from the 1950s onward found linear magnetic anomalies, parallel to mid-ocean ridges, using magnetometers towed behind research vessels. During the early 1960s, it was suggested, and soon confirmed, that these anomalies resulted from the remanent magnetization of the oceanic crust. This remanence is acquired during the process of seafloor spreading, when uprising magma beneath the axis of the mid-ocean ridges cools through its Curie temperature (**Figure 3**) in the ambient geomagnetic field, thus acquiring its direction and polarity. The continuous process of rising and cooling of magma at the ridge results in magnetized crust of alternating normal and reversed polarity that produces a slight increase or decrease of the measured field — the marine magnetic anomalies. It was also found that the magnetic anomaly pattern is generally symmetric on both sides of the ridge, and, most importantly, that it provides a wonderfully continuous 'tape recording' of the geomagnetic reversal sequence.

A major step in constructing a time series of polarity reversals was taken in 1968 by Heirztler and co-workers. They used a long profile from the southern Atlantic Ocean and, extrapolating from a known age for the lower boundary of the Gauss Chron (**Figure 4**), they constructed a geomagnetic polarity timescale under the assumption of constant spreading. Subsequent revisions mostly used the

anomaly profile of Heirtzler, often adding more detail from other ocean basins, and appending additional calibration points. These calibration points are derived from sections on land, which, first, must contain a clear fingerprint of magnetic reversals that can be correlated to the anomaly profile, and second, contain rocks that can be reliably dated by means of radiometric methods. The GPTS is then derived by linear interpolation of the anomaly pattern between these calibration points, again under the assumption of constant spreading rate between those points.

The development of the GPTS (**Figure 4**) shows increasing detail and gradually improved age control. Periods of a predominant (normal or reversed) polarity are called chrons, and the four youngest ones are named after individuals: Brunhes (normal), who suggested field reversal; Matuyama (reversed), who proved this; Gauss (normal), who

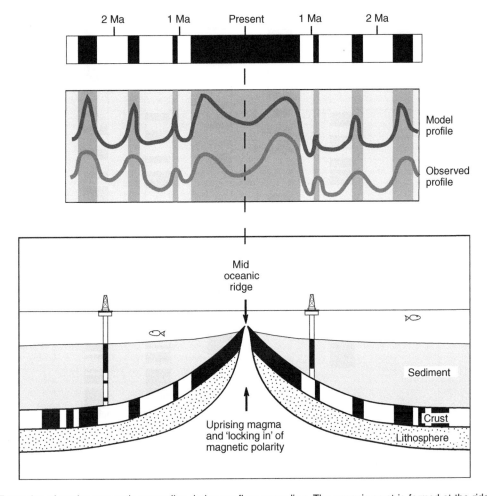

Figure 3 Formation of marine magnetic anomalies during seafloor spreading. The oceanic crust is formed at the ridge crest, and while spreading away from the ridge it is covered by an increasing thickness of oceanic sediments. The black (white) blocks of oceanic crust represent the original normal (reversed) polarity thermoremanent magnetization (TRM) acquired upon cooling at the ridge. The black and white blocks in the drill holes represent normal and reversed polarity DRM acquired during deposition of the marine sediments. The model profile (grey) represents computed magnetic anomalies produced by the block model of TRM polarity (top); the observed profile (dark) is the observed sea-level magnetic anomaly profile due to the magnetized oceanic crust.

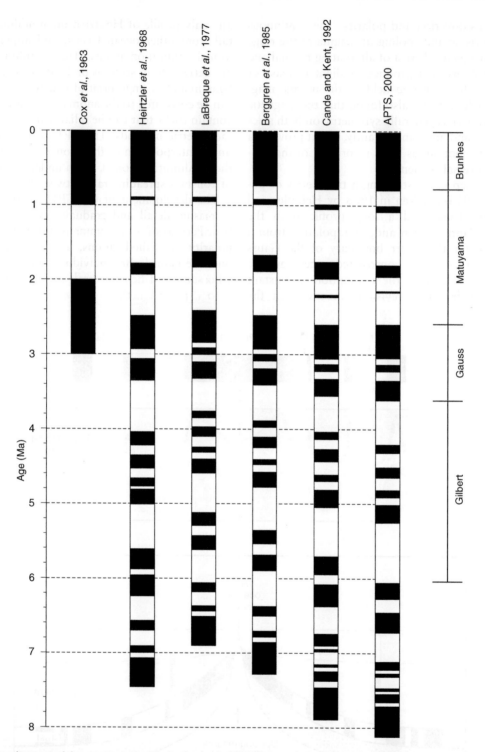

Figure 4 Development of the geomagnetic polarity timescale (GPTS) through time. The initial assumption of periodic behavior (in 1963) was soon abandoned as new data became available. The first modern GPTS based on marine magnetic anomaly patterns was established in 1968 by Heirtzler and co-workers. Subsequent revisions show improved age control and increased resolution. A major breakthrough came with the astronomical polarity timescale (APTS), in which each individual reversal is accurately dated.

mathematically described the field; and Gilbert (reversed), who discovered that the Earth itself is a huge magnet. Chrons may contain short intervals of opposite polarity called subchrons, which are named

after the locality where they were discovered; for example, the normal Olduvai subchron within the Matuyama reversed chron is named after Olduvai Gorge in Tanzania, and the Kaena reversed subchron within the Gauss normal chron after Kaena Point on Hawaii. Older chrons were not named but numbered, according to the anomaly numbers earlier given by Heirtzler, which has led to a confusing nomenclature of chrons and subchrons.

A major step forward was taken by Cande and Kent in 1992, who thoroughly revised the magnetic anomaly template over the last 110 My. They constructed a synthetic flow line in the South Atlantic, using a set of chosen anomalies that were taken as tie points. The intervals between t hese tie points were designated as category I intervals. On these intervals they projected (stacks of) the best-quality profiles surveyed in this ocean basin, providing category II intervals. Since the spreading of the Atlantic is slow, they subsequently filled in the category II intervals with high-resolution profiles from fast-spreading ridges (their category III). This enabled them to include much more detail on short polarity intervals (or subchrons); see, for instance, the increase of detail around 7 Ma in **Figure 4**. Very short and low-intensity anomalies still have an uncertain origin. They may represent very short subchrons of the field, as has been proven for some of them (e.g., the Cobb Mt. subchron at 1.21 Ma, or the Réunion subchron at 2.13–2.15 Ma), or may just represent intensity fluctuations of the geomagnetic field causing the oceanic crust to be less (or more) strongly magnetized. Because of their uncertain or unverified nature, these were called cryptochrons. In addition, Cande and Kent developed a consistent (sub)chron nomenclature that is now used as the standard. In total, they used nine calibration points, but they made a break with tradition by using, for the first time, an astronomically dated age tie point for the youngest one: the Gauss/Matuyama boundary. The correlation of the GPTS to global biostratigraphic zonations is covered extensively by Berggren *et al.* (1995).

The template of magnetic anomaly patterns from the ocean floor has remained central for constructing the GPTS from the late Cretaceous onward (110–0 Ma). Only recently, the younger part of the GPTS has been based on direct dating of each individual reversal through the use of orbitally tuned timescales. In their most recent version of the GPTS, Cande and Kent included the astronomical ages for all reversal boundaries for the past 5.3 My. The CK95 or Cande and Kent (1995) geomagnetic polarity timescale is at present the most widely used standard. Polarity timescales for the Mesozoic rely on some of the oldest magnetic anomaly profiles, down to the late Jurassic, and on dated magnetostratigraphies of sections on land.

The Astronomical Polarity Timescale (APTS)

The latest development in constructing a GPTS comes from orbital tuning of the sediment record. It differs essentially from the conventional GPTS in the sense that each reversal boundary — or any other geological boundary for that matter, e.g. biostratigraphic datum levels or stage and epoch boundaries — is dated individually. This has provided a breakthrough in dating of the geological record and has the inherent promise of increasing understanding of the climate system, since cyclostratigraphy and subsequent orbital tuning rely on decphering and understanding environmental changes driven by climate change, which in turn is orbitally forced.

The fact that the age of each reversal is directly determined, rather than interpolated between calibration points, has important consequences for (changes in) spreading rates of plat pairs. Rather than having to assume constant spreading rates between calibration points, one can now accurately determine these rates, and small changes therein. Indeed, Wilson found that the use of astronomical ages resulted in very small and physically realistic spreading rate variations. As a result, the discrepancy in between plate motion rates from the global plate tectonic model (NUVEL-1) and those derived from geodesy has become much smaller. Meanwhile NUVEL-1 has been updated (to NUVEL-1A) to incorporate the new astronomical ages.

Another application is the dating of Pleistocene, Pliocene, and Miocene, and older stage boundaries, many of which have been defined in the Mediterranean. The availability of a good astrochronology has effectively become a condition for the definition of a Global Boundary Stratotype Section and Point (GSSP). An example is shown in **Figure 5**, showing the Tortonian–Messinian GSSP that has recently been defined in the Atlantic margin basin of western Morocco.

Perhaps one of the most promising areas of the application of astrochronology is in the bed-to-bed correlation of the two different realms of oceans and continent. Climate forcing may be expected to have a different expression in the different realms because of the different nature of their sedimentary environments. A recently established and refined orbital timescale for the loess sequences of northern China relies upon the correlation of detailed monsoon records to the astronomical solutions and the oceanic oxygen isotope records. An important finding was that the straightforward use of magnetostratigraphy and correlation to the GPTS cannot provide a sufficiently accurate age model for comparison with the

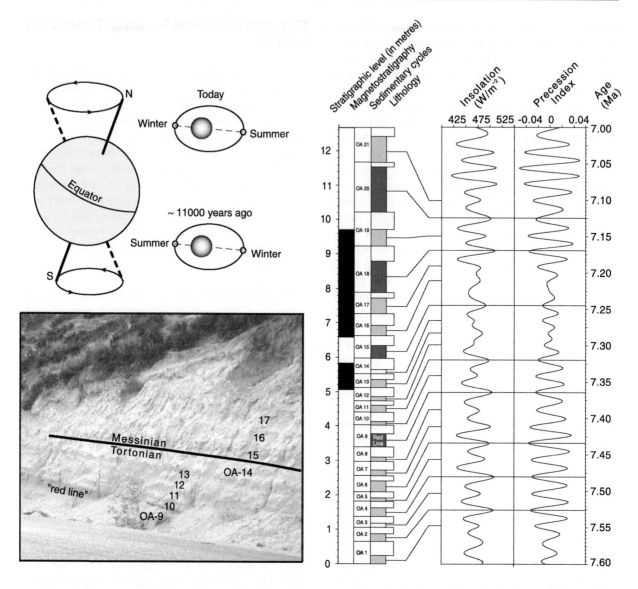

Figure 5 Magnetostratigraphy, cyclostratigraphy, and astrochronology of the marine Oued Akrech section from the Atlantic margin of Morocco, which defines the Tortonian–Messinian Global Stratotype Section and Point (GSSP). The sedimentary cycles represent a cyclically changing environment that correlates with variations in insolation. Insolation is strongly related to climatic precession (upper left panel), which induces cyclic changes in seasonal contrast, reflected one-on-one in the sedimentary cycles. (After Hilgen FJ *et al.* (2000) *Episodes* 23(3): 172–178.)

ocean record, since the analysis of the astrochronological framework demonstrates considerable downward displacement of reversal boundaries because of delayed lock-in of the NRM. With the new chronology and its direct correlation to the oceanic record, it is now possible to analyse terrestrial paleomonsoon behavior for the past 2.6 My and compare it to climate proxies from the marine realm. This may give important information, for instance, on leads and lags of various systems in response to climate change, on phase relations with insolation, or on the relation between global ice volume and monsoonal climate.

See also

Magnetics.

Further Reading

Berggren WA, Kent DV, Aubry MP, and Hardenbol J (1995) *Geochronology, Time Scales and Global Stratigraphic Correlations: A Unified Temporal Framework for an Historical Geology.* Society of Economic Paleontologists and Mineralogist, Special Volume 54.

Butler RF (1992) *Paleomagnetism, Magnetic Domains to Geologic Terranes.* Boston: Blackwell Scientific.

Cande SC and Kent DV (1992) A new geomagnetic polarity Time-Scale for the Late Cretaceous and Cenozoic. *Journal of Geophysical Research* 97: 13917–13951.

Cande SC and Kent DV (1995) Revised calibration of the Geomagnetic Polarity Time Scale for the Late Cretaceous and Cenozoic. *Journal of Geophysical Research* 100: 6093–6095.

Gradstein FM, Agterberg FP, Ogg JG, *et al.* (1995) A Triassic, Jurassic and Cretaceous time scale. In: Berggren WA, Kent DV, Aubry M and Hardenbol J (eds) *Geochronology, Time Scales and Stratigraphic Correlation.* Society of Economic Paleontologists and Mineralogist, Special Volume 54: 95–128.

Hilgen FJ, Krijgsman W, Langereis CG, and Lourens LJ (1997) Breakthrough made in dating of the geological record. *EOS, Transactions of the AGU* 78: 285–288.

Heirtzler JR, Dickson GO, Herron EM, Pitman WC III, and Le Pichon X (1968) Marine magnetic anomalies, geomagnetic field reversals, and motions of the ocean floor and continents. *Journal of Geophysical Research* 73: 2119–2136.

Heslop D, Langereis CG, and Dekkers MJ (2000) A new astronomical time scale for the loess deposits of Northern China. *Earth and Planetory Science Letters* 184: 125–139.

Maher BA and Thompson R (1999) *Quaternary Climates, Environments and Magnetism.* Cambridge: Cambridge University Press.

Opdyke ND and Channell JET (1996) *Magnetic Stratigraphy.* San Diego: Academic Press.

Shackleton NJ, Crowhurst SJ, Weedon GP, and Laskar J (1999) Astronomical calibration of Oligocene–Miocene time. *Philosophical Transactions of the Royal Society of London, A* 357: 1907–1929.

Wilson DS (1993) Confirmation of the astronomical calibration of the magnetic polarity time scale from seafloor spreading rates. *Nature* 364: 788–790.

MARINE DEPOSITS

ACOUSTIC MEASUREMENT OF NEAR-BED SEDIMENT TRANSPORT PROCESSES

P. D. Thorne and P. S. Bell, Proudman Oceanographic Laboratory, Liverpool, UK

Introduction: Sediments and Why Sound Is Used

Marine sediment systems are complex, frequently comprising mixtures of different particles with non-cohesive (sands) and cohesive (clays and muds) properties. The movement of sediments in coastal waters impacts on many marine processes. Through the actions of accretion, erosion, and transport, sediments define most of our coastline. Their deposition and resuspension by waves and tidal currents in estuarine and nearshore environments control seabed morphology. Fine sediments, which act as reservoirs for nutrients and contaminants and as regulators of light transmission through the water column, have significant impact on water chemistry and on primary production. Therefore, an improved understanding of sediment dynamics in coastal waters has relevance to a broad spectrum of marine science ranging from physical and chemical processes, to the complex biological and ecological structures supported by sedimentary environments. However, it is commonly acknowledged that our capability to describe the coupled system of the bed, the hydrodynamics, and the sediments themselves is still relatively primitive and often based on empiricism. Recent advances in observational technologies now allow sediment processes to be investigated with greater detail and precision than has previously been the case. The combined use of acoustics, laser and radar, both at large and small scales, is facilitating exciting measurement opportunities. The enhancement of computing capabilities also allows us to make use of more complex coupled sediment–hydrodynamic models, which, linked with the emerging observations, provide new openings for model development.

It is readily acknowledged by sedimentologists that the presently available commercial instrumentation does not satisfy a number of requirements for near-bed sediment transport processes studies. Here we focus on the development of acoustics to fulfill some of these needs. The question could be asked: Why use acoustics for such studies? Sediment transport can be thought of as dynamic interactions between: (1) the seabed morphology, (2) the sediment field, and (3) the hydrodynamics. These three components interrelate with each other in complex ways, being mutually interactive and interdependent as illustrated schematically as a triad in **Figure 1**. The objective therefore is to measure this interacting triad with sufficient resolution to study the dominate mechanisms. Acoustics uniquely offers the prospect of being able to nonintrusively provide profiles of the flow, the suspension field, and the bed topography. This exclusive combination of being able to measure all three components of the sediment triad, co-located and simultaneously, has been and is the driving force for applying acoustics to sediment transport processes.

The idea of using sound to study fundamental sediment processes in the marine environment is attractive, and, in concept, straightforward. A pulse of high-frequency sound, typically in the range 0.5–5 MHz in frequency, and centimetric in length, is transmitted downward from a directional sound source usually mounted a meter or two above the bed. As the pulse propagates down toward the bed, sediment in suspension backscatters a proportion of the sound and the bed generally returns a strong echo. The amplitude of the signal backscattered from the suspended sediments can be used to obtain vertical profiles of the suspended concentration and particle size. Utilizing the rate of change of phase of the backscattered signal provides profiles of the three orthogonal components of flow. The strong echo from the bed can be used to measure the bed forms.

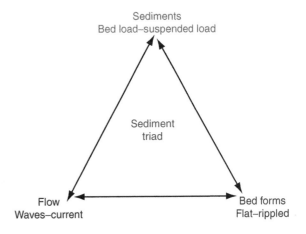

Figure 1 Illustration of the sediment processes triad and their interactions.

Acoustics therefore has the potentiality to provide profile measurements of near-bed sediment processes, with sufficient spatial and temporal resolution to allow turbulence and intrawave processes to be probed; this coupled with the bedform morphology observations provides sedimentologists and coastal engineers with an extremely powerful tool to advance understanding of sediment entrainment and transport. All of this is delivered with almost no influence on the processes being measured, because sound is the instrument of measurement.

Some Historical Background

For over two decades the vision of a number of people involved in studying small-scale sediment processes in the coastal zone has been to attempt to utilize the potential of acoustics to simultaneously and nonintrusively measure seabed morphology, suspended sediment particle size and concentration profiles, and profiles of the three components of flow, with the required resolution to observe the perceived dominant sediment processes. The capabilities to measure the three components have developed at different rates, and it is only in the past few years that the potentiality of an integrated acoustic approach for measuring the triad has become realizable. A schematic of the vision is shown in **Figure 2**.

The figure shows a visualization of the application of acoustics to sediment transport processes. 'A' is a multifrequency acoustic backscatter system (ABS), consisting, in this case, of three downward-looking narrow beam transceivers. The differential scattering characteristic of the suspended particles with frequency is used to obtain profiles of suspended sediment particle size and concentration profiles. 'B' is a three-axis coherent Doppler velocity profiler (CDVP) for measuring co-located profiles of the three orthogonal components of flow velocity; two horizontal and one vertical. It consists of an active narrow beam transceiver pointing vertically downward and two passive receivers having a wide beam width in the vertical and a narrow beam width in the horizontal. This system uses the rate of change of phase from the backscattered signal to obtain the three velocity components. 'C' is a pencil beam transceiver which rotates about a horizontal axis and functions as an acoustic ripple profiler (ARP). This is used to extract the bed echo and provide profiles of the bed morphology along a transect. These measurements are used to obtain, for example, ripple height and wavelength, and assess bed roughness. 'D' is a high-resolution acoustic bed (ripple) sector scanner (ARS) for imaging the local bed features. Although the ARS does not provide quantitative measurements of bed form height, it does provide the spatial distribution and this can be very useful when used in conjunction with the ARP. 'E' is a rapid backscatter ripple profiler system (BSARP) for measuring the instantaneous relationship between bed forms and the suspended sediments above it.

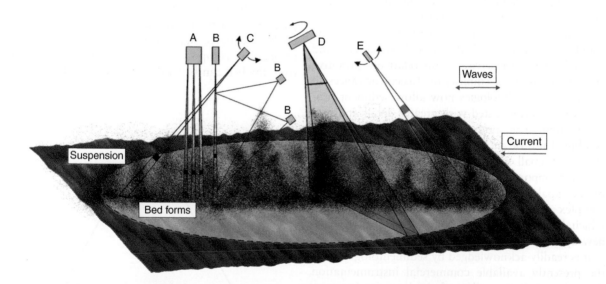

Figure 2 A vision of the application of acoustics to sediment transport processes. A, multifrequency acoustic backscatter for measuring suspended sediment particle size and concentration profiles. B, coherent Doppler velocity profiler for measuring the three orthogonal components of flow velocity. C, bed ripple profiler for measuring the bed morphology along a transect. D, high-resolution sector scanner for imaging the local bed features. E, backscatter scanning system for measuring the relationship between bed form morphology and suspended sediments.

What Can Be Measured?

The Bed

Whether the bed is rippled or flat has a profound influence on the mechanism of sediment entrainment into the water column. Steep ripples are associated with vortices lifting sediment well away from the bed, while for flat beds sediments primarily move in a confined thin layer within several grain diameters of the bed. Therefore knowing the form of the bed is a central component in understanding sediment transport processes. The development of the ARP and the ARS has had a significant impact on how we interpret sediment transport observations. These specifically designed systems typically either provide quantitative measurements of the evolution of a bed profile with time, the ARP, or generate an image of the local bed features over an area, the ARS.

Figure 3 shows data collected with a 2-MHz narrow pencil beam ARP in a marine setting. The figure shows the variability of a bed form profile, over nominally a 3-m transect, covering a 24-h period. Over this period the bed was subject to both tidal currents and waves and the figure shows the complex evolution of the bed with periods of ripples and less regular bed forms. The figure clearly shows the detailed quantitative measurements of the bed that can be obtained with the ARP.

Acoustic ripple scanners, ARSs, are based on sector-scanning technology, which has been specifically adapted for high-resolution images of bed form morphology. They typically have a frequency of around 1–2 MHz with beam widths of about 1° in the horizontal and 30° in the vertical. As the pulse is backscattered from the bed, the envelope of the signal is measured and usually displayed as image intensity.

An example of the data collected by an ARS is shown in **Figure 4**. As can be seen, this provides an aerial image of the bed, clearly showing the main bed features. The advantage of the ARS is the area coverage that is obtained, as opposed to a single line profile with the ARP; however, direct information on the height of features within the image cannot readily be extracted.

Ideally one would like to combine the two instruments and recently such systems have become available. This is essentially an ARP which also rotates horizontally through 180° and therefore allows a three-dimensional (3-D) measurements of the bed. The sonar gathers a single swath of data in the vertical plane and then rotates the transducer around the vertical axis and repeats the process until a circular area underneath the sonar has been scanned in a sequence of radial spokes. An example of data collected by a 3-D ARP operating at a frequency of 1.1 MHz is shown in **Figure 5**.

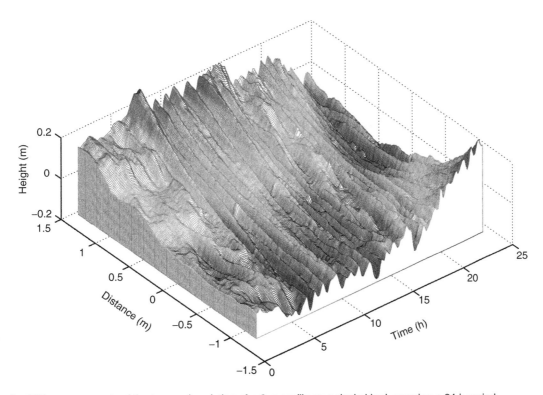

Figure 3 ARP measurements of the temporal evolution of a 3-m profile on a rippled bed, covering a 24-h period.

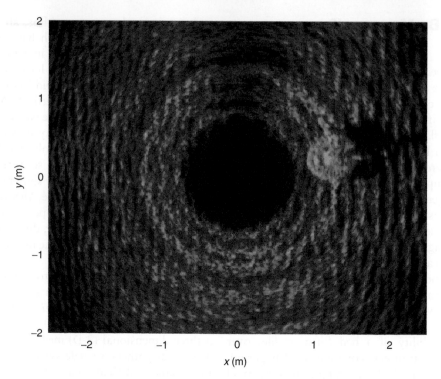

Figure 4 Image of ripples on the bed in a large-scale flume facility, the Delta Flume, collected using an acoustic ripple sector scanner, ARS. The 0.5-m-diameter circle at the center right of the image is one of the feet of the instrumented tripod used to collect the measurements, while the dark area at the center of the image is a blind spot directly beneath the sonar.

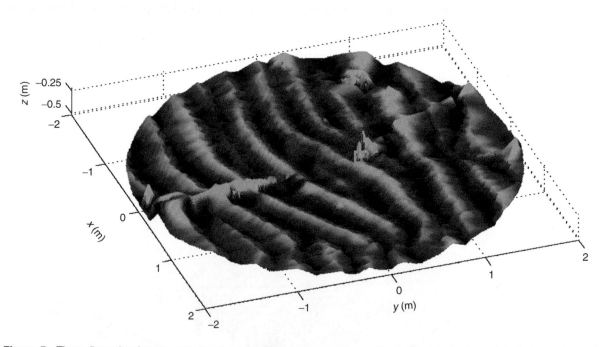

Figure 5 Three-dimensional measurements of a rippled bed collected in the marine environment using a 3-D ripple scanner, 3-D ARP. The artifacts in the image are associated with reflections from the instrumented tripod used to collect the measurements.

The data shown in the figure were collected during a recent marine deployment. The bed surface shown is based on a 3-D scan, with a 100 vertical swaths at 1.8° intervals, each of which comprised 200 acoustic samples at 0.9° intervals and spanning 180°. The plot clearly shows a rippled bed. There are one or two artifacts in the plot; these are associated with reflection from the rig on which the 3-D ARP was mounted. However, the figure plainly contains information on both the horizontal and vertical dimensions of the ripples and can therefore be used to precisely define quantitatively the features of the bed over an area. The 3-D ARP is a substantial advance on both the ARP and the ARS.

The Flow

The success of acoustic Doppler current profilers (ADCPs), which typically provide mean current profiles with decimeter spatial resolution, and more recently the acoustic Doppler velocimeter (ADV), which measures, subcentimetric, subsecond, three velocity components at a single height, has stimulated interest in using acoustics to measure near-bed velocity profiles. The objective is to use the same backscattered signal as used by the ABSs, but process the rate of change of phase of the signal (rather than the amplitude as used by ABSs) to obtain velocity profiles with comparable spatial and temporal resolution to ABSs. The phase technique is utilized in CDVPs and the phase approach has been the preferred method for obtaining high spatial and temporal resolution velocity profiles. An illustration of a three-axis CDVP is shown in **Figure 6**. A narrow-beam, downwardly pointing transceiver, Tz, transmits a pulse of sound. The scattered signal is picked up by Tz, and two passive receivers Rx and Ry which are orthogonal to each other and have a wide beam in the vertical and narrow in the horizontal.

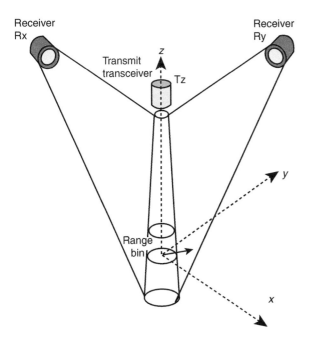

Figure 6 Schematic of the transducer arrangement for a three-axis CDVP.

To examine the capability of such a system, measurements from a three-axis CDVP have been compared with a commercially available ADV. The system had a spatial resolution of 0.04 m, operated over a range of 1.28 m, and provided 16-Hz velocity measurements of the vertical and two horizontal components of the flow. **Figure 7** shows a typical example of CDVP and ADV time series, power spectral density, and probability density function plots for **u**, the streamwise flow. The velocities presented in **Figure 7** show CDVP results that compare very favorably with the ADV measurements; having time series, spectra, and probability distributions in general agreement. There are differences in the spectrum; the CDVP spectra begins to depart from the ADV above about 4 Hz, with the CDVP measuring larger spectral components at the higher frequencies. This trend was common to all the records and is a limitation of the CDVP system used to obtain the data, rather than an intrinsic limitation to the technique. **Figure 8** illustrates the capability of the CDVP for flow visualization in the marine environment. The figure shows mean zeroed velocity vectors, **u–w**, **v–w**, and **u–v**, plotted over a 5-s time period, between 0.05 and 0.7 m above the bed. The length of the velocity vectors is indicated in the figure. A single-point measurement instrument such as an ADV can provide the time-varying velocity vectors at a single height above the bed; however, the spatial profiling which is achievable with the three-axis CDVP provides a capability to visualize structures in the flow. The structures seen in **Figure 8** are associated with combined turbulent and wave flows. This type of plot exemplifies the value of developing a three-axis CDVP with co-located measurement volumes, since it clearly illustrates the fine-scale temporal and spatial flow structures which can be measured in the near-bed flow regime. Linking such measurements with ABS profiles of particle size and concentration will provide a very powerful tool for studying near-bed fluxes and sediment transport processes.

The Suspended Sediments

Multifrequency acoustic backscattering, ABS, can be used to obtain profiles of mean particle size and concentration. The ABS is the only system available that profiles both parameters rapidly and simultaneously. Also the bed echo references the profile to the local bed position. This is important because all sediment transport formulas use the bed as the reference point and predict profiles of suspension parameters relative to the bed location. Examples of the results that can be obtained are shown

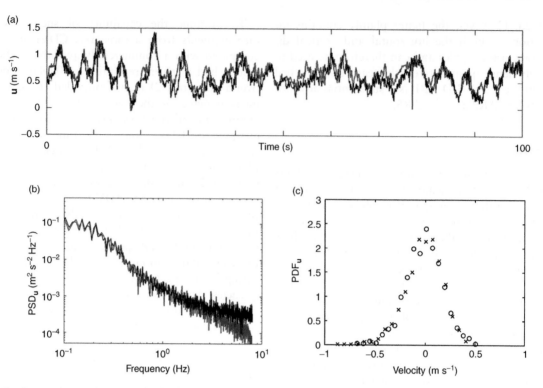

Figure 7 Comparison of the streamwise flow, **u**, measured by an ADV (red) and a CDVP (black). (a) The velocities measured at 16 Hz over 100 s. (b) The power spectra of the zero-mean velocities. (c) The probability density functions of the zero-mean velocities for the ADV (open circles) and the CDVP (crosses).

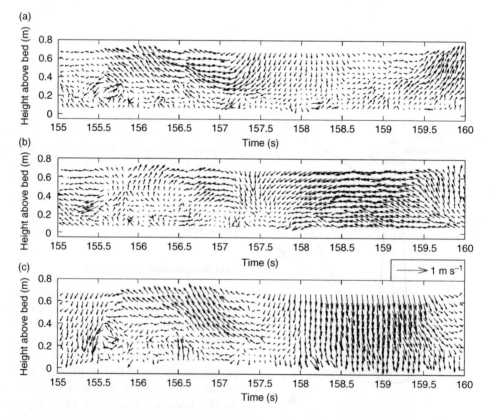

Figure 8 Demonstration of the capability of the triple-axis CDVP to provide visualizations of intrawave and turbulent flow. Plots in (a)–(c) show a time series over a 5 s period of the zero-mean velocities displayed as vectors **u**–**w**, **v**–**w**, and **u**–**v**, respectively.

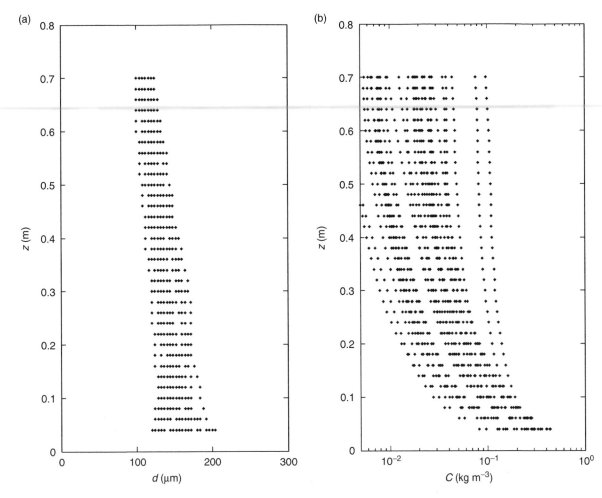

Figure 9 Profiles of suspended sediments: (a) particle size and (b) concentration.

in **Figures 9** and **10**. These observations were collected off Santa Cruz Pier, California, as a seasonal storm passed through the area over the period of a couple of days. **Figure 9** shows profiles of the suspended sediment particle size and concentration. The profiles of particle size are seen to be relatively consistent with a mean diameter of c. 180 µm near the bed and reducing to c. 120 µm at 0.7 m above the bed. The variability in size is relatively limited and due to changing bed and hydrodynamic conditions as the storm passed by.

In **Figure 9(b)** the suspended concentrations are comparable in their form, although they have absolute values that vary by greater than an order of magnitude. This variation in suspended concentration is associated with the changing conditions as the storm passed through the observational area. It is interesting to note that over the period even though there is a large variation in concentration, the particle size remains nominally consistent. **Figure 10** shows the temporal variation in particle size and

concentration with height above the bed. It can be clearly seen that some of the periods of increased particle size are associated with substantial suspended sediment events, as one might expect, however, there are one or two events where the correlation is not as clear. **Figures 9** and **10** clearly illustrate the capability of ABS to simultaneously measure profiles of concentration and particle size and the combination of both significantly adds to the assessment and development of sediment transport formulas.

A Case Study of Waves over a Rippled Bed

Here the use of acoustics is illustrated by application to a specific experimental study. Over large areas of the continental shelf outside the surf zone, sandy seabeds are covered with wave-formed ripples. If the ripples are steep, the entrainment of sediments into the water column, due to the waves, is considered

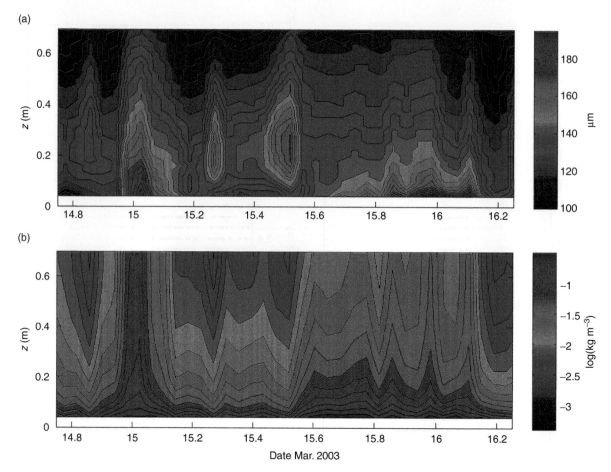

Figure 10 Measurements of the temporal variations with height above the bed, of: (a) particle size with the color bar scaled in microns and (b) logarithmic concentration with the color bar scaled relative to $1.0\,\mathrm{kg\,m^{-3}}$.

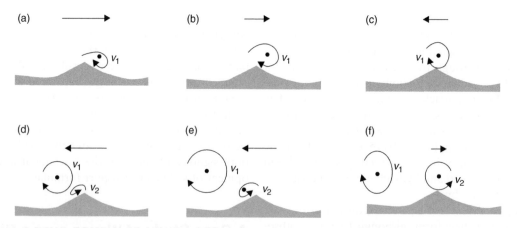

Figure 11 A schematic of vortex sediment entrainment by waves over a steeply rippled bed. The arrows show the direction and relative magnitude of the near-bed wave velocity. v_1 and v_2 are the lee slope-generated vortices.

to be primarily associated with the generation of vortices. This process is illustrated in **Figure 11**. As shown in **Figures 11(a)** and **11(b)**, a spinning parcel of sediment-laden water, v_1, is formed on the leeside of the ripple at the peak positive velocity in the wave cycle. This sediment-rich vortex is then thrown up into the water column at around flow reversal (**Figures 11(c)** and **11(d)**), carrying sediment well away from the bed and allowing it to be transported by the flow. At the same time, a sediment-rich vortex,

v_2, is being formed on the opposite side of the ripple due to the reversed flow. As shown in **Figures 11(d)–11(f)**, v_2 grows, entrains sediment, becomes detached, and moves over the crest at the next flow reversal carrying sediments into suspension. The main feature of the vortex mechanism is that sediment is carried up into the water column twice per wave cycle at flow reversal. This mechanism is completely different to the flat bed case where maximum near-bed concentration is at about the time of maximum flow velocity.

To study this fundamental process of sediment entrainment, experiments were conducted in one of the world's largest man-made channels, specifically constructed for such sediment transport studies, the Delta Flume; this is located at the De Voorst Laboratory of the Delft Hydraulics in the north of the Netherlands. The flume is shown in **Figure 12(a)**; it is 230 m in length, 5 m in width, and 7 m in depth, and it allows waves and sediment transport to be studied at full scale. A large wave generator at one end of the flume produced waves that propagated along the flume, over a sandy bed, and dissipated on a beach at the opposite end. The bed was comprised of coarse sand, with a mean grain diameter by weight of 330 µm, and this was located approximately halfway along the flume in a layer of thickness 0.5 m and length 30 m. In order to make the acoustic and other auxiliary measurements, an instrumented tripod platform was developed and is shown in **Figure 12(b)**. The tripod, STABLE II (Sediment Transport and Boundary Layer Equipment II), used an ABS to measure profiles of particle size and concentration, a pencil beam ARP to measure the bed forms, and, in this case, electromagnetic current meters (ECMs) to measure the horizontal and vertical flow components. **Figure 12(c)** shows a wave propagating along the flume with STABLE II submerged in water of depth 4.5 m, typical of coastal zone conditions.

To investigate and then model the vortex entrainment process it was necessary to establish at the outset whether or not the surface waves were generating ripples on the bed in the Delta Flume. Using an ARP a 3-m transect of the bed was measured over time. The results of the observations over a 90-min recording period are shown in **Figure 13**. Clearly ripples were formed on the bed and the ripples were mobile. To obtain flow separation and hence vortex formation requires a ripple steepness (ripple height/

(a)   (b)   (c)

Figure 12 (a) Photograph of the Delta Flume showing the sand bed at approximately midway along the flume and the wave generator at the far end of the flume. (b) The instrumented tripod platform, STABLE II, used to make the measurements. (c) A surface wave propagating along the flume.

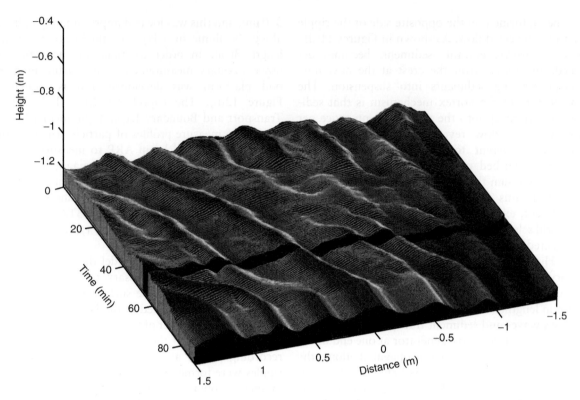

Figure 13 Acoustic ripple profiler measurements of a transect of the bed, over time, in the Delta Flume.

ripple wavelength) of the order of 0.1 or greater; an analysis of the observations showed that this was indeed the case.

Using the ABS, some of the most detailed full scale measurements of sediment transport over a rippled bed under waves were captured. These measurements from the Delta Flume are shown in **Figure 14**. The images shown were constructed over a 20-min period as a ripple passed beneath the ABS. The suspended concentrations over a ripple, at the same velocity instants during the wave cycle, were combined to generate a sequence of images of the concentration over the ripple with the phase of the wave. Four images from the sequence have been shown to illustrate the measured vortex entrainment. The length and direction of the arrows in the figure give the magnitude and direction of the wave velocity, respectively. Comparison of **Figure 14** with **Figure 11** shows substantial similarities. In **Figure 14(a)**, there can be observed the development of a high-concentration event at high flow velocity above the lee slope of the ripple, v_1. In **Figure 14(b)**, as the flow reduced in strength, the near-bed sediment-laden parcel of fluid travels up the leeside of the ripple toward the crest. As the flow reverses, this sediment-laden fluid parcel, v_1, travels over the crest and expands. As the reverse flow increases in strength (**Figure 14(d)**), the parcel v_1 begins to lift

away from the bed and a new sediment-laden lee vortex, v_2, is initiated on the lee slope of the ripple.

In order to capture the essential features of these data within a relatively simple, and hence practical, 1-DV (one-dimensional in the vertical) model, the data has first been horizontally averaged over one ripple wavelength at each phase instant during the wave cycle. The resulting pattern of sediment suspension contours is shown in the central panel of **Figure 15**, while the upper panel shows the oscillating velocity field measured at a height of 0.3 m above the bed. The concentration contours shown here are relative to the ripple crest level, the mean (undisturbed) bed level being at height $z = 0$. The measured concentration contours presented in **Figure 15** show two high concentration peaks near the bed that propagate rapidly upward through a layer of thickness corresponding to several ripple heights. The first, and the strongest, of these peaks occurs slightly ahead of flow reversal, while the second, weaker and more dispersed peak, is centered on flow reversal. The difference in the strengths of the two peaks reflects the greater positive velocity that can be seen to occur beneath the wave crest (time $= 0$ s) than beneath the wave trough (time $= 2.5$ s). Between the two concentration peaks the sediment settles rapidly to the bed. Maybe rather unexpectedly this settling effect occurs at the times of strong forward and backward velocity at measurement

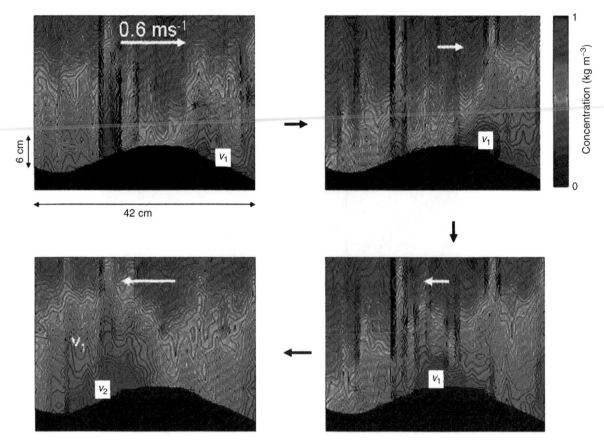

Figure 14 Acoustic imaging of suspended sand entrainment over a rippled bed due to waves, at four phases of the wave velocity. The length of the white arrow in each plot gives the magnitude and direction of the near-bed wave velocity.

levels well above the bed. The underlying mechanism of sediment entrainment by vortices shed at or near flow reversal is clearly evident in the spatially averaged measurements shown in **Figure 15**.

Any conventional 'flat rough bed' model that attempts to represent the above sequence of events in the suspension layer runs into immediate and severe difficulties, since such models predict maximum near-bed concentration at about the time of maximum flow velocity, and not at flow reversal. Here therefore, for the first time in a 1-DV model, it has been attempted to capture these effects realistically through the use of a strongly time-varying eddy viscosity that represents the timing and strength of the upward mixing events due to vortex shedding. The model initially predicts the size of the wave-induced ripples and the size of the grains found in suspension, and then goes on to solve numerically the equations governing the upward diffusion and downward settling of the suspended sediment. The resulting concentration contours in the present case are shown on the lower panel of **Figure 15**. The essential two-peak structure of the eddy shedding process can be seen to be represented rather well, with the initial

concentration peak being dominant. The decay rate of the concentration peaks as they go upward is also represented quite well, though a phase lag develops with height that is not seen to the same extent in the data. Essentially, the detailed acoustic observations of sediment entrainment under waves over ripples of moderate steepness have begun to establish a new type of 1-DV modeling, thereby allowing the model to go on to be used for practical prediction purposes in the rippled regime, which is the bed form regime of most importance over wide offshore areas in the coastal seas.

Discussion and Conclusions

The aim of this article has been to illustrate the application of acoustics in the study of near-bed sediment processes. It was not to detail the theoretical aspects of the work, which can be found elsewhere. To this end, measurements of bed forms, the hydrodynamics, and the movement of sediments have been used. These results show that acoustics is progressively approaching the stage where it

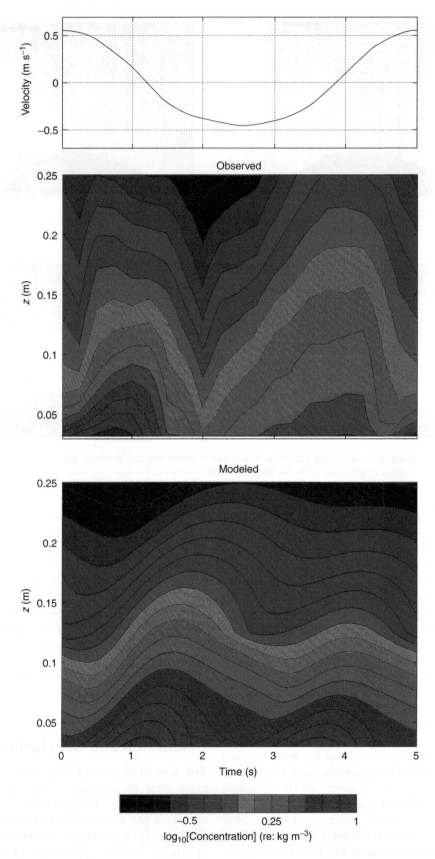

Figure 15 Measurement and modeling of suspended sediments with height *z* above a rippled bed under a 5-s period wave.

can measure nonintrusively, co-located and simultaneously, with high temporal and spatial resolution, all three components of the interacting sediment triad. **Figure 16** shows the instrumentation used in a further recent deployment in the Delta Flume. This shows the convergence of the instrumentation and also its use in conjunctions with instruments such as laser *in situ* scattering transmissometry (LISST).

Although substantial advances have been made in the past two decades, the application of acoustics to sediment transport processes is still in an ongoing developmental phase and there are limitations and shortcomings that need to be overcome, and further applications explored. Although there have been few reports to date on data collected using 3-D ARP, such systems are now becoming available. The 3-D ARP is a substantial development over the single-line ARP and the ARS, and should make a considerable contribution to the measurement and understanding of the formation and development of bed forms. There have been a number of reports on single-axis CDVP; however, it is the three-axis CDVP which is the way ahead. Again, these instruments are coming online, though they are still very much a research tool. However, the concept has now essentially been

proven and its use with a vengeance in sediment studies will begin to make an important impact in the next couple of years. It was with the concept of using acoustics to measure suspended sediment concentration that its application to the other components of the small-scale sedimentation processes triad followed. To date the use of sound to measure suspended sediment concentration and particle size has been successful when systems have been deployed over nominally homogeneous sandy beds. However, all who use acoustics recognize that the marine sedimentary environment is frequently much more complicated, and suspensions of cohesive sediments and combined cohesive and noncohesive sediments are common. To employ acoustics quantitatively in mixed and cohesive environments requires the development of a description of the scattering properties of suspensions of cohesive sediments and sediment mixtures. This would be interesting and very valuable work, and should significantly extend the deployment regime over which acoustic backscattering can be employed quantitatively.

In conclusion, the objective of this article has been to describe the role of acoustics, in near-bed sediment transport studies. It is clearly acknowledged

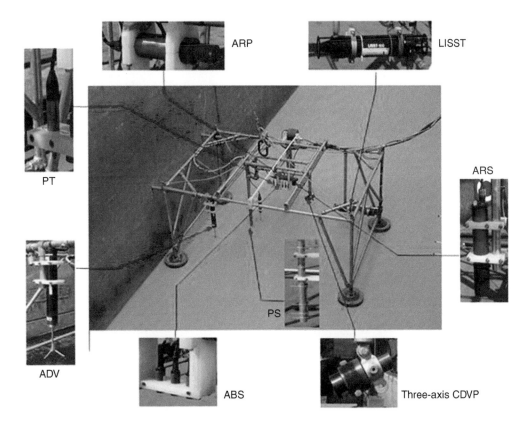

Figure 16 The instrumentation used in a recent Delta Flume experiment. LISST, laser *in situ* scattering transmissometry, PS, pumped samples, PT, pressure transducer.

that acoustics is one of a number of technologies advancing our capabilities to probe sediment processes. However, its nonintrusive profiling ability, coupled with its capability to measure all three components of the sediment dynamics triad, make it a unique and very powerful tool for studying the fundamental mechanisms of sediment transport.

See also

Acoustics in Marine Sediments. Offshore Sand and Gravel Mining.

Further Reading

Crawford AM and Hay AE (1993) Determining suspended sand size and concentration from multifrequency acoustic backscatter. *Journal of the Acoustical Society of America* 94(6): 3312–3324.

Davies AG and Thorne PD (2005) Modelling and measurement of sediment transport by waves in the vortex ripple regime. *Journal of Geophysical Research* 110: C05017 (doi:1029/2004JC002468).

Hay AE and Mudge T (2005) Principal bed states during SandyDuck97: Occurrence, spectral anisotropy, and the bed storm cycle. *Journal of Geophysical Research* 110: C03013 (doi:10.1029/2004JC002451).

Thorne PD and Hanes DM (2002) A review of acoustic measurements of small-scale sediment processes. *Continental Shelf Research* 22: 603–632.

Vincent CE and Hanes DM (2002) The accumulation and decay of near-bed suspended sand concentration due to waves and wave groups. *Continental Shelf Research* 22: 1987–2000.

Zedel L and Hay AE (1999) A coherent Doppler profiler for high resolution particle velocimetry in the ocean: Laboratory measurements of turbulence and particle flux. *Journal of Atmosphere and Ocean Technology* 16: 1102–1117.

Relevant Websites

http://www.aquatecgroup.com
– Aquatec Group Ltd.
http://www.marine-electronics.co.uk
– Marine Electronics Ltd.
http://www.pol.ac.uk
– POL Research, Proudman Oceanographic Laboratory.

ACOUSTICS IN MARINE SEDIMENTS

T. Akal, NATO SACLANT Undersea Research Centre, La Spezia, Italy

Introduction

Because of the ease with which sound can be transmitted in sea water, acoustic techniques have provided a very powerful means for accumulating knowledge of the environment below the ocean surface. Consequently, the fields of underwater acoustics and marine seismology have both used sound (seismo-acoustic) waves for research purposes.

The ocean and its boundaries form a composite medium, which support the propagation of acoustic energy. In the course of this propagation there is often interaction with ocean bottom. As the lower boundary, the ocean bottom is a multilayered structure composed of sediments, where acoustic energy can be reflected from the interface formed by the bottom and subbottom layers or transmitted and absorbed. At low grazing angles, wave guide phenomena become significant and the ocean bottom, covered with sediments of different physical characteristics, becomes effectively part of the wave guide. Depending on the frequency of the acoustic energy, there is a need to know the acoustically relevant physical properties of the sediments from a few centimeters to hundreds of meters below the water/sediment interface.

Underwater acousticians and civil engineers are continuously searching for practical and economical means of determining the physical parameters of the marine sediments for applications in environmental and geological research, engineering, and underwater acoustics. Over the past three decades much effort has been put into this field both theoretically and experimentally, to determine the physical properties of the marine sediments. Experimental and forward/inverse modeling techniques indicate that the acoustic wave field in the water column and seismo-acoustic wave field in the seafloor can be utilized for remote sensing of the physical characteristics of the marine sediments.

Sediment Structure as an Acoustic medium

Much of the floor of the oceans is covered with a mixture of particles of sediments range in size from boulder, gravels, coarse and fine sand to silt and clay, including materials deposited from chemical and biological products of the ocean, all being saturated with sea water. Marine sediments are generally a combination of several components, most of them coming from the particles eroded from the land and the biological and chemical processes taking place in sea water. Most of the mineral particles found in shallow and deep-water areas, have been transported by runoff, wind, and ice and subsequently distributed by waves and currents.

After these particles have been formed, transported, and transferred, they are deposited to form the marine sediments where the physical factors such as currents, dimensions and shapes of particles and deposition rate influence the spatial arrangements and especially sediment layering. Particles settle to the ocean floor and remain in place when physical forces are not sufficiently strong to move them. In areas with strong physical forces (tidal and ocean currents, surf zones etc.) large particles dominate, whereas in low motion energy areas (ocean basins, enclosed bays) small particles dominate.

During the sedimentation process these particles, based on the physical and chemical interparticle forces between them, form the sedimentary acoustic medium: larger particles (e.g., sands) by direct contact forces; small particles (e.g., clays and fine silts) by attractive electrochemical forces; and silts, remaining between sands and clays are formed by the combination of these two forces. The amount of the space between these particles is the result of different factors, mainly size, shape, mineral content and the packing of the particles determined by currents and the overburden pressure present on the ocean bottom.

Figure 1 is an example of a core taken very carefully by divers, to ensure an undisturbed internal structure of the sediment sample. Sediment structures have been quantified by using X-ray computed tomography to obtain values of density with a millimeter resolution for the full three-dimensional volume. The image shown in **Figure 1** is a false color 3D reconstruction of a core sample at a site where sediments consist of sandy silt (75% sand, 15% silt, and 10% clay) and shell pieces.

The results of the X-ray tomography of the same core, can also be shown on an X-ray cross-section slice along the center of the core (**Figure 2A**) and the corresponding two-dimensional spectral density levels for that cross-section (**Figure 2B**). The complex

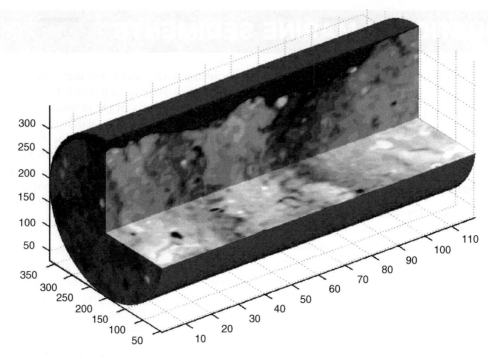

Figure 1 Three-dimensional reconstruction of a sediment core sample containing 75% sand, 15% silt and 10% clay. Scale in mm.

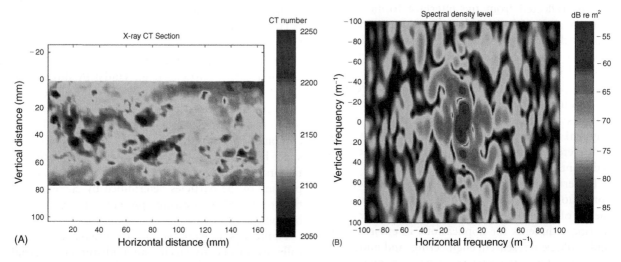

Figure 2 (A) X-ray cross-section slice along the center of the core shown in **Figure 1**. (B) 2-D spectral density levels for the cross-section shown in (A).

structure of the sediments can be with seen strong heterogeneity (local density fluctuation) of the medium that controls the interaction of the seismo-acoustic energy. In addition to the complex fine structure described above, the seafloor can also show complex layering (**Figure 3A**) or a simpler structure (**Figure 3B**). These structures result from the lowering of sea levels during the glaciation of the Pleistocene epoch during which sand was deposited over wide areas of the continental shelves. Unconsolidated

sediments subsequently covered the shelves as the sea level are in postglacial times.

Biot–Stoll Model

Various theories have been developed to describe the geoacoustic response of marine sediments. The most comprehensive theory is based on the Biot model as elaborated by Stoll. This model takes into account various loss mechanisms that affect the response of

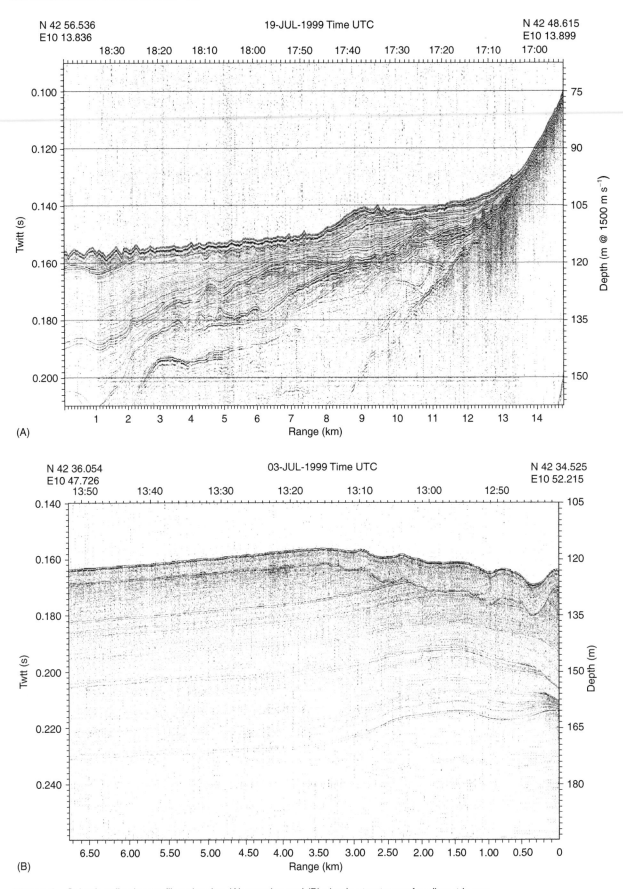

Figure 3 Seismic reflection profiles showing (A) complex and (B) simple structures of sediment layers.

porous sediments that are saturated with fluid. The Biot–Stoll theory shows that acoustic wave velocity and attenuation in porous, fluid-saturated sediments depend on a number of parameters including porosity, mean grain size, permeability, and the properties of the skeletal frame.

According to the Biot–Stoll theory, in an unbounded, fluid-saturated porous medium, there are three types of body wave. Two of these are dilatational (compressional) and one is rotational (shear). One of the compressional waves ('first kind') and the shear wave are similar to body waves in an elastic medium. In a compressional wave of the 'first kind', the skeletal frame and the sea water filling the pore space move nearly in phase so that the attenuation due to viscous losses becomes relatively small. In contrast, due to out-of-phase movement of the frame and the pore fluid, the compressional wave of the 'second kind' becomes highly attenuated. The Biot theory and its extensions by Stoll have been used by many researchers for detailed description of the acoustic wave–sediment interaction when basic input

Table 1 Basic input parameters to Biot–Stoll model

Frequency-independent Variables	
Porosity (%)	P
Mass density of grains	ρ_r
Mass density of pore fluid	ρ_f
Bulk modulus of sediment grains	K_r
Bulk modulus of pore fluid	K_f
Variables affecting global fluid motion	
Permeability	k
Viscosity of pore fluid	η
Pore-size parameter	a
Structure factor	α
Variables controlling frequency-dependent response of frame	
Shear modulus of skeletal frame	$\bar{\mu} = \mu_r(\omega) + i\mu_i(\omega)$
Bulk modulus of skeletal frame	$\bar{K}_b = K_{br}(\omega) + iK_{bi}(\omega)$

parameters such as those shown in **Table 1** are available.

Seismo-acoustic Waves in the Vicinity of the Water–Sediment Interface

As mentioned above, when acoustic energy interacts with the seafloor, the energy creates two basic types of deformation: translational (compressional) and rotational (shear). Solution of the equations of the wave motion shows that each of these types of deformation travels outward from the source with its own velocity. Wave type, velocity, and propagation direction vary in accordance with the physical properties and dimensions of the medium.

The ability of seafloor sediments to support the seismo-acoustic energy depends on the elastic properties of the sediment, mainly the bulk modulus (incompressibility, K) and the shear modulus (rigidity, G). These parameters are related to the compressional and shear velocities C_P and C_S respectively, by:

$$C_P = [(K + 4G/3)\rho]^{1/2}$$

$$C_S = (G/\rho)^{1/2}$$

where, ρ is the bulk density. **Table 2** shows basic seismic–acoustic wave types and their velocities related to elastic parameters

These two types of deformations (compressional and shear) belong to a group of waves (body waves) that propagate in an unbounded homogeneous medium. However, in nature the seafloor is bounded and stratified with layers of different physical properties. Under these conditions, propagating energy undergoes characteristic conversions every time it interacts with an interface: propagation velocity,

Table 2 Basic seismo-acoustic wave types and elastic parameters

Basic wave type		Wave velocity
Body wave	Compressional	$C_P = [(K + 4G/3)/\rho]^{1/2}$
	Shear	$C_S = (G/\rho)^{1/2}$
Ducted wave	Love	$C_L = (G/\rho)^{1/2}$
Surface wave	Scholte	$C_{SCH} = (G/\rho)^{1/2}$
Elastic parameters in terms of wave velocities (C) and bulk density (ρ)		
Bulk modulus (incompressibility)	$K = \rho(C_P^2 - 4C_S^2/3)$	
Compressibility	$\beta = 1/K$	
Young's modulus	$E = 2C_S^2\rho(1 + \sigma)$	
Poisson's ratio (transv./long. strain)	$\sigma = (3E - \rho C_P^2)/(3E + 2\rho C_P^2)$	
Shear modulus (rigidity)	$G = \rho C_S^2$	
Lame's constant	$\lambda = \rho(C_P^2 - 2C_S^2)$	

energy content, and spectral structure and propagation direction changes. In addition, other types of waves, i.e., ducted waves and surface waves, may be generated. These basic types of waves and their characteristics together with their arrival structure as synthetic seismograms for orthogonal directions are illustrated in **Figure 4**.

Body Waves

These waves propagate within the body of the material, as opposed to surface waves. External forces can distort solids in two different ways. The first involves the compression of the material without changing its shape; the second implies a change in shape without changing its volume (distortion). From earthquake seismology, these compressional and distortion waves are called primus (P) and secoundus (S), respectively, for their arrival sequence on earthquake records. However, the distortional waves are very often called shear.

Compressional waves   Compressional waves involve compression of the material in such a manner that the particles move in the direction of propagation.

Shear waves   Shear waves are those in which the particle motion is perpendicular to the direction of propagation. These waves can be generated at a layer interface by the incidence of compressional waves at other than normal incidence. Shear wave energy is polarized in the vertical or horizontal planes, resulting in vertically polarized shear waves (SV) and horizontally polarized shear waves (SH). However, if the interfaces are close (relative to a wavelength), one cannot distinguish between body and surface waves.

Ducted Waves

Love waves   Love waves are seismic surface waves associated with layering; they are characterized by horizontal motion perpendicular to the direction of propagation (SH wave). These waves can be considered as ducted shear waves traveling within the duct of the upper sedimentary layer where total reflection occurs at the boundaries; thus the waves represent energy traveling by multiple reflections.

Interface Waves

Seismic interface waves travel along or near an interface. The existence of these waves demands the combined action of compressional and shear waves. Thus at least one of the media must be solid, whereas

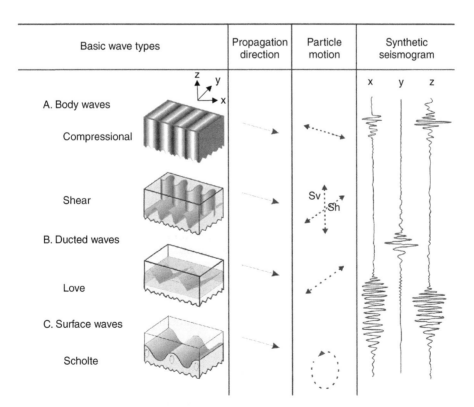

Figure 4 Basic seismo-acoustic waves in the vicinity of a water–sediment interface.

the other may be a solid (Stoneley wave), a liquid (Scholte wave), or a vacuum (Rayleigh wave). For two homogeneous half-spaces, interface waves are characterized by elliptical particle motion confined to the radial/vertical plane and by a constant velocity that is always smaller than the shear wave.

When different types of seismic waves propagate and interact with the layered sediments they are partly converted into each other and their coupling may create mixed wave types in the vicinity of the interface. **Figure 4** shows the basic seismo-acoustic waves in the vicinity of a water/sediment interface. Basic characteristics of these waves together with their particle motion and synthetic seismograms at three orthogonal directions (x, y and z) are also illustrated in the same Figure. Under realistic conditions, in which the seafloor cannot be considered to be homogeneous, isotropic, nor a half-space, some of these waves become highly attenuated or travel together, making identification very difficult. In fact, the interface waves shown in **Figure 4** are for a layered seafloor, where the dispersion of the signal is evident (homogeneous half-space would not give any dispersion).

Under realistic conditions, i.e., for an inhomogeneous, bounded and anisotropic seafloor, some of these waves convert from one to another. The different wave types may travel with different speeds or together, and they generally have different attenuation. As an example, **Figure 5** shows signals from an explosive source (0.5 kg trinitrotoluene (TNT))

received by a hydrophone and three orthogonal geophones placed on the seafloor, at a distance of 1.5 km from the source. The broadband signal of a few milliseconds duration generated by the explosive source is dispersed over nearly 18 s demonstrating the arrival structure of different types of waves. The characteristics of these waves are indicated in the figure for four different sensors. They can be identified in order of their arrival time as: (a) head wave; (b) water arrival (compressional wave); (c) interface wave; (d) Love waves.

Seafloor Roughness

The roughness of the water–sediment interface and layers below is another important parameter that needs to be considered in sediment acoustics. The seafloor contains a wide spectrum of topographic roughness, from features of the order of tens of kilometers, to those of the order of millimeters. The shape of the seafloor and its scattering effect on acoustic signals is be covered here.

Techniques to Measure Geoacoustic Parameters of Marine Sediments

The geoacoustic properties of the seafloor defined by the compressional and shear wave velocities, their attenuation, together with the knowledge of the material bulk density, and their variation as a function of depth, are the main parameters needed to solve the acoustic wave equation. To be able to determine these properties of the seabed, different techniques have been developed using samples taken from the seafloor, instruments and divers conducting measurements *in situ*, and remote techniques measuring seismo-acoustic waves and inverting this information with realistic models into sediment properties. Some of the current methods of obtaining geoacoustic parameters of the marine sediments are briefly described here.

Laboratory Measurements on Sediment Core Samples

Most of our knowledge of the physical properties of sediments is acquired through core sampling. A large number of measurements on marine sediments have been made in the past. In undisturbed sediment core samples, under laboratory conditions, density and compressional velocity can be measured with accuracy, and having measured values of density and compressional velocity, the bulk modulus can be selected as the third parameter, where it is can be calculated (**Table 1**).

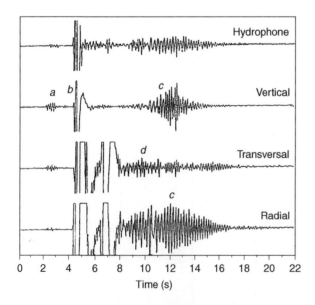

Figure 5 Signals from an explosive source of 0.5 kg trinitrotoluene received by a hydrophone and three orthogonal geophones at 1.5 km distance (a, Head wave; b, Compressional wave; c, Interface wave; d, Love wave).

There are several laboratory techniques available to measure some of the sediment properties. However, the reliability of such measurements can be degraded by sample disturbance and temperature and pressure changes. In particular, the acoustic properties are highly affected by the deterioration of the chemical and mechanical bindings caused by the differences in temperature and pressure between the sampling and the laboratory measurements. Controlling the relationships between various physical parameters can be used to check the accuracy of measured parameters of sediment properties. It has been shown that the density and porosity of sediments have a relationship with compressional velocity, and different empirical equations between them have been established.

Over the past three decades, at the NATO Undersea Research Centre (SACLANTCEN), a large number of laboratory and *in situ* measurements have been made of the physical properties of the seafloor sediments. These measurements have been conducted on the samples with the same techniques as when the same laboratory methods were applied. This data set with a great consistency of the hardware and measurement technique, has been used to demonstrate the physical characteristics of the sediment that affect acoustic waves.

The relationship between measured physical parameters From 300 available cores, 20 000 measured data samples for the density and porosity, and 10 000 samples for the compressional velocity were obtained. To be able to handle this large data set taken from different oceans at different water depths, all bulk density and compressional velocity data were converted into relative density (ρ) and relative velocity (C) with respect to the *in situ* water values:

$$\rho = \rho_S/\rho_W$$
$$C = C_S/C_W$$

where, ρ_S is sediment bulk density, ρ_W is water density, C_S is sediment compressional velocity and C_W is water compressional velocity. The data are not only from the water–sediment interface but cover sedimentary layers of up to 10 m deep.

Relative density and porosity The density and porosity of the marine sediment are least affected during coring and laboratory handling of the samples. Porosity is given by the percentage volume of the porous space and sediment bulk density by the weight of the sample per unit volume. The relationship between porosity and bulk density has been investigated by many authors with fewer data

than used here and shown to have a strong linear correlation. Theoretically this linearity only exists if the dry densities of the mineral particles are the same for all marine sediments. The density of the sediment would then be the same as the density of the solid material at zero porosity, and the same as the density of the water at 100% porosity. **Figure 6** shows this relationship.

Porosity and relative compressional wave velocity The relationship between porosity and compressional wave velocity has received much attention in the literature because porosity can be measured easily and accurately. Data from the SACLANTCEN sediment cores giving the relationship between porosity and compressional wave velocity are shown in **Figure 7**. As shown in **Figure 6** due to the linear relation between density and porosity, the relationship between density and compressional wave velocity is similar to the porosity compressional wave velocity relation.

In situ Techniques

There are several *in situ* techniques available that use instruments lowered on to the seafloor mainly by means of submersibles, remotely operated vehicles (ROVs), and autonomous underwater vehicles AUVs and divers. The first deep-water, *in situ* measurements of sediment properties were made from the bathyscaph *Trieste* in 1962. These measurements provided accurate results due to the minimum disturbance of temperature and pressure changes compared to bringing the sample to the surface and for

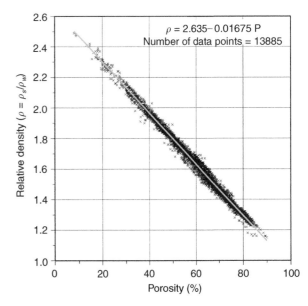

Figure 6 Relationship between relative density and porosity.

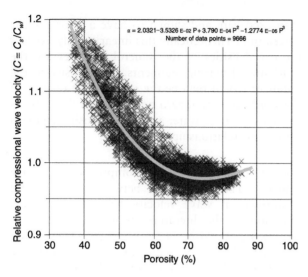

Figure 7 Relationship between relative compressional wave velocity and porosity.

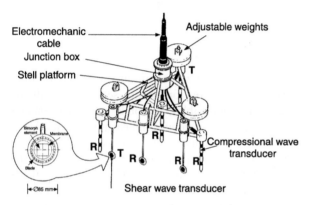

Figure 8 *In Situ* Sediment Acoustic Measurement System (ISSAMS) for near-surface *in situ* geoacoustic measurements.

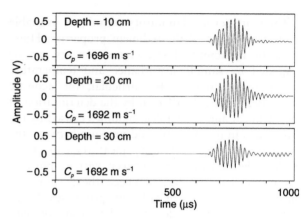

Figure 9 Examples of received compressional wave signals from fine sand sediment.

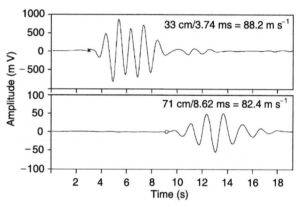

Figure 10 Examples of signals recorded from two shear wave receivers in hard-packed fine sand sediments.

analysis in the laboratory. The most reliable direct geoacoustic measurement techniques for marine sediments are *in situ* techniques. Some of the approaches used in recent years are discussed below.

Near-surface method A system has been developed to measure sediment geoacoustic parameters, including compressional and shear wave velocities and their attenuation at tens of centimeters below the sediment–water interface. **Figure 8** shows the main features of the *In situ* Sediment Acoustic Measurement System (ISSAMS). Shear and compressional wave probes are attached to a triangular frame that uses weights to force the probes into the sediment. In very shallow water, divers can be used to insert the probes into the sediment, whereas in deeper water, a sleeve system (not shown) allows the ISSAMS to penetrate into the seafloor.

Using the compressional and shear wave transducers measurements are made with a continuous wave (cw) pulse technique where the ratio of measured transducer separation and pulse arrival time yields the wave velocity. Samples of compressional and shear wave data are shown in **Figure 9** and **10** respectively. **Table 3** gives a comparison of laboratory and *in situ* values of compressional and shear wave velocities from two different types of Adriatic Sea sediment.

Cross-hole method Measurements as a function of depth in sediments can be made with boreholes, using either single or cross-hole techniques. Boreholes are made by divers using water–air jets to penetrate thin-walled plastic tubes for cross-hole measurements. **Figure 11** shows the experimental set up for the cross-hole measurements. The source is in the form of an electromagnetic mallet securely coupled to the inner wall of one of the plastic tubes with a hydraulic clamping device.

Table 3 Comparison of laboratory and *in situ* measurements

Sediment type	Porosity (%)	Mean grain size (ϕ)	Wave velocity (ms^{-1})			
			In situ (C_P)	Laboratory (C_P)	In situ (C_S)	Laboratory (C_S)
Sand	37	3.5	1557–1568	1580–1604	78–82	50
Mud	68	8.6	1467–1488	1468–1487	27–31	15

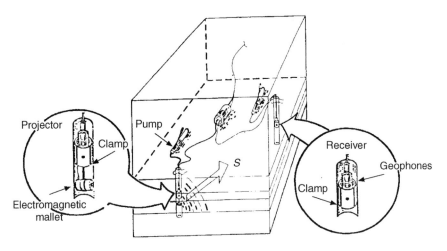

Figure 11 The experimental setup for cross-hole measurements.

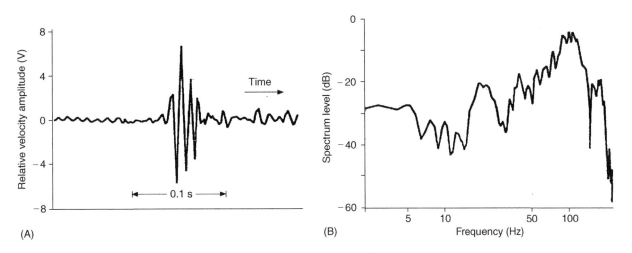

Figure 12 Cross-hole shear wave signal (A) and its spectrum (B) for a silty-clay bottom in the Ligurian Sea.

With a separation range of 2.9 m, moving-coil geophone receivers are also coupled to the inner wall of the second plastic tube with a hydraulic clamping device. The electromagnetic mallet generates a point source on the thin plastic tube wall, which results in a multipolarized transient signal to the sediment. Depending on the orientation of receiving sensors, vector components of the propagating compressional and shear waves are received and analyzed for velocity and attenuation parameters. Examples of a time series and a frequency spectrum for a shear wave signal received by a geophone with a natural frequency of 4.5 Hz are shown in **Figure 12**.

Figure 13 shows shear wave velocity as a function of depth obtained in the Ligurian Sea. It can be seen that the shear wave velocities are around $60\,m^{-1}$ s at the sediment interface and increase with depth.

Remote Sensing Techniques

Even though *in situ* techniques provide the most reliable data, they are usually more time consuming and expensive to make and they are limited to small areas. Remote sensing and inversion to obtain geoacoustic parameters can cover larger areas in less time and provide reliable information. These techniques are based on the use of a seismo-acoustic signal received by sensors on the seafloor and/or in the water column. **Figure 14** illustrates a characteristic shallow water signal from an explosion received by a hydrophone close to the sea bottom. Three different techniques to extract information relative to bottom parameters from these signals are described briefly below.

Reflected waves

The half-space seafloor. When the seafloor consists of soft unconsolidated sediments, due to its very low shear modulus it can be treated as fluid.

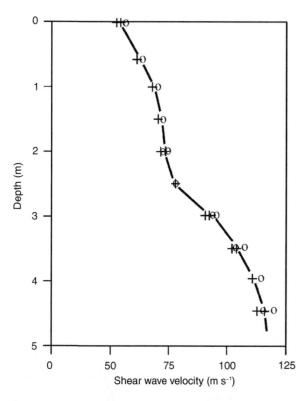

Figure 13 Shear wave velocity profile obtained from cross-hole measurements.

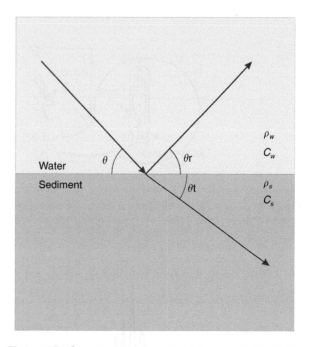

Figure 15 Geometry and notations for a simple half-space water–sediment interface.

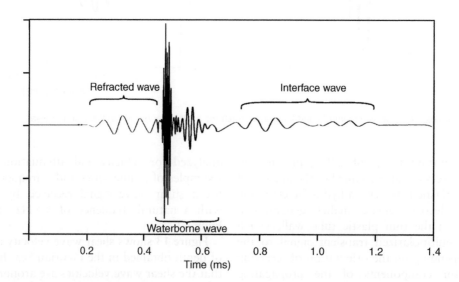

Figure 14 A characteristic shallow-water signal.

Reflection may occur whenever an acoustic wave is incident on an interface between two media. The amount of energy reflected, and its phase relative to the incident wave, depend on the ratio between the physical properties on the opposite sides of the interface. It is possible to calculate frequency independent reflection loss as the ratio between the amplitude of the shock pulse and the peak of the first reflection (after correcting for phase shift, absorption in the water column, and differences in spreading loss). If the relative density $\rho = \rho_S/\rho_W$, and the relative compressional wave velocity

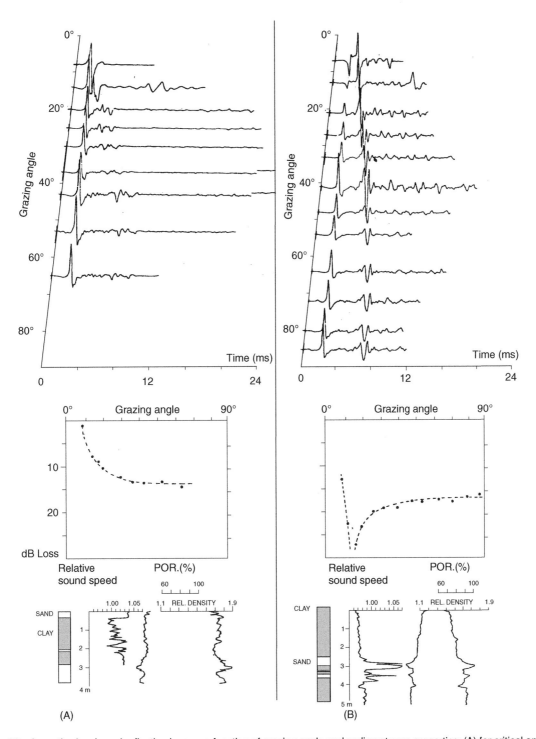

Figure 16 Acoustic signals and reflection loss as a function of grazing angle and sediment core properties: (A) for critical angle; (B) for angle of incidence case.

$C = C_S/C_W$, are used to present the contrast between the two media, reflection coefficient for such a simple environmental condition can be written as:

$$R = \frac{\rho\sin\Theta - \sqrt{(1/C^2 - \cos^2\Theta)}}{\rho\sin\Theta - \sqrt{(1/C^2 - \cos^2\Theta)}}$$

Where Θ is the grazing angle with respect to the interface as shown in **Figure 15**. For an incident path normal to a reflecting horizon, i.e., $\Theta = 90°$, the reflection coefficient is:

$$R = \rho C - 1/\rho C + 1$$

Critical angle case. When the velocity of compressional wave velocity is greater in the sediment layer ($C>1$), as the grazing angle is decreased, a unique value is reached at which the acoustic energy totally reflects back to the water column. This is known as the critical angle and is given by:

$$\Theta_{cr} = \arccos(1/C)$$

When the grazing angle is less than this critical angle, all the incident acoustic energy is reflected.

However, the phase of the reflected wave is then shifted relative to the phase of the incidence wave by an angle varying from 0° to 180° and is given as:

$$\Phi = -2\arctan\frac{\sqrt{(\cos^2\Theta - 1/C^2)}}{\rho\sin\Theta}$$

Figure 16 shows measured and calculated reflection losses ($20\log R$) for this simple condition together with the basic physical properties of the core sample taken in the same area for explosive signals at different grazing angles.

Angle of intromission case. Especially in deep-water sediments the sound velocity in the top layer of the bottom is generally less than in the water above ($C<1$). In such conditions there is an angle of incidence at which all of the incident energy is transmitted into the sedimentary layer and the reflection coefficient becomes zero:

$$\cos\Theta_i = \sqrt{\left(\frac{\rho^2 - 1/C^2}{\rho^2 - 1}\right)}$$

The phase shift is 0° when the ray angle is greater than the intromission angle and 180° when it is smaller. Thus, acoustical characteristics of the bottom, such as the critical angle, the angle of incidence, the phase shift,

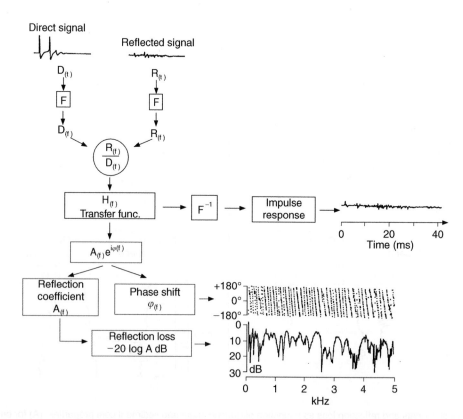

Figure 17 Acoustic reflection data and analysis technique for a layered seafloor.

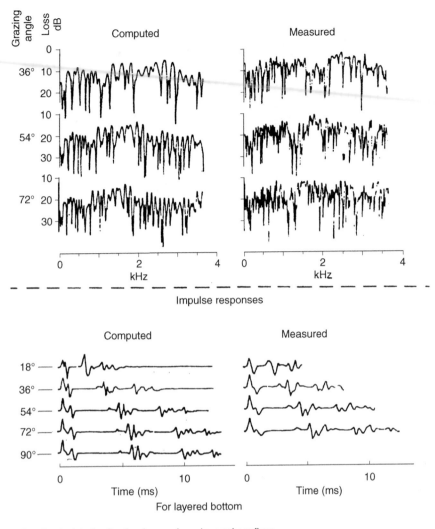

Figure 18 Measured and calculated reflection losses for a layered seafloor.

and the reflection coefficient, are primarily influenced by the relative density and relative sound velocity of the environment as a function of the ray angle.

The layered seafloor. Since the seafloor is generally layered, a simple peak amplitude approximation cannot be implemented because of the frequency dependence. In this case one calculates the transfer function (or reflection coefficient) from the convolution of the direct reference signal with the reflected signal. Examples of the phase shift and reflection loss as a function of frequency are shown in **Figure 17**. The reflectivity can also be described in the time domain by the impulse response, which is the inverse Fourier transform of the transfer function as shown in the same figure. This type of data can be utilized for inverse modeling to obtain the unknown parameters of the sediments. **Figure 18** is an example of measured and calculated reflection losses and impulse responses for a layered seafloor.

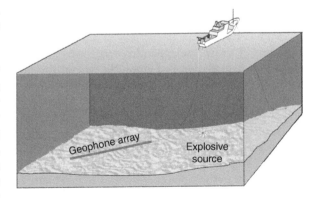

Figure 19 Experimental setup to measure seismo-acoustic waves on the seafloor.

Refracted waves Techniques developed for remote sensing of the uppermost sediments (25–50 m below the seafloor) utilize broad-band sources (small

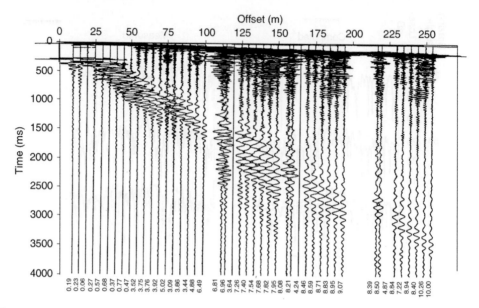

Figure 20 Signals received by geophone array.

explosives) and an array of geophones deployed on the seafloor. To obtain estimates of the bottom properties as a function of depth, both refracted compressional and shear waves as well as interface waves are analyzed. Inversion of the data is carried out using modified versions of techniques developed by earthquake seismologists to study dispersed Rayleigh-waves and refracted waves. **Figure 19** illustrates the basic experimental setup and **Figure 20** the signals received by an array of 24 geophones that permit studies of both interface and refracted waves.

These data are analyzed and inverted to obtain both compressional and shear wave velocities as a function of depth in the seafloor. Studies of attenuation and lateral variability are also possible using the same data set.

Figure 21 shows the expanded early portion of the geophone data shown in **Figure 20**, where, the first arriving energy out to a range of about 250 m has been fitted with a curve showing that compressional waves refracted through the sediments just beneath the seafloor travel faster as they penetrate more deeply into the sediment. At zero offset, the slope indicates a velocity of 1505 m s^{-1} whereas at a range offset at 250 m the slope corresponds to a velocity of 1573 m s^{-1}. At ranges over 250 m, a strong head wave becomes the first arrival and, the interpretation would be that there is an underlying rock layer with compressional wave velocity of about 4577 m s^{-1}.

A compressional wave velocity–depth curve for the upper part of the seafloor can be derived from the first arrivals shown in **Figure 21** using the classical Herglotz–Bateman–Wieckert integration method.

The slope of the travel-time curve (fitted parabola) gives the rate of change of the range with respect to time that is also the velocity of propagation of the diving compressional wave at the level of its deepest penetration (turning point) into the sediment. At each range Δ, the depth corresponding to the deepest penetration is then calculated using the following integral

$$z(V) = 1/\pi \int_0^\Delta \cosh^{-1}(V(dt/dx))dx$$

where

$$1/V = (dt/dx)_{x=\Delta}$$

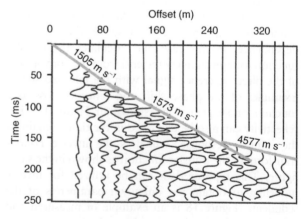

Figure 21 Expanded early portion of the data shown in **Figure 20**.

The result is the solid velocity–depth curve shown in **Figure 22**.

Interface waves In order to obtain a shear wave velocity–depth profile from the data, later arrivals corresponding to dispersed interface waves may be utilized (**Figure 20**). The portion of each individual signal corresponding to the interface-wave arrival can be processed using multiple filter analysis to create a group velocity dispersion diagram (Gabor diagram). The result of applying this technique to a dispersed signal is a filtered time signal whose envelope reaches a maximum at the group velocity arrival time for a selected frequency. The envelope is computed by taking the quadrature components of the inverse Fourier transform of the filtered signal. Filtering is carried out at many discrete frequencies over selected frequency bands. Once the arrival times are converted in to velocity, the envelopes are arranged in a matrix and contoured and dispersion curves are obtained by connecting the maximum values of the contour diagram (**Figure 23**).

Having obtained the dispersion characteristics of the interface waves, the geoacoustic model, made of a stack of homogeneous layers with different compressional and shear wave velocities for each layer that predict the measured dispersion curve is determined. **Figure 24** illustrates a number of examples from the Mediterranean sea covering data from soft clays to hard sands.

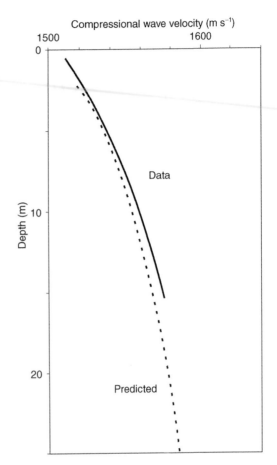

Figure 22 Compressional wave velocity versus depth curves derived from data and predicted (dashed line) by the model.

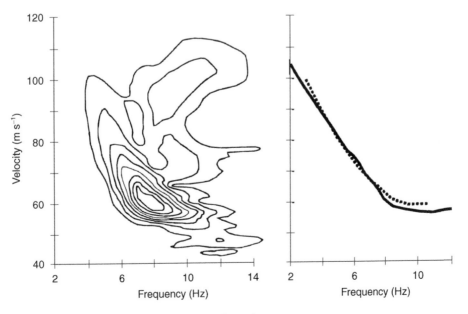

Figure 23 Dispersion diagram and measured and predicted dispersion curves.

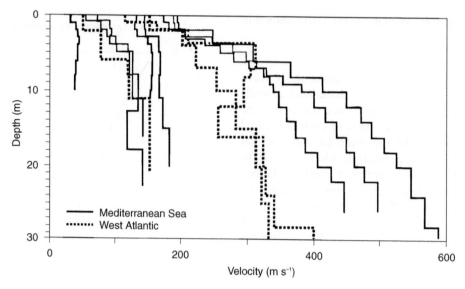

Figure 24   Summary of shear wave velocity profiles from the Mediterranean Sea.

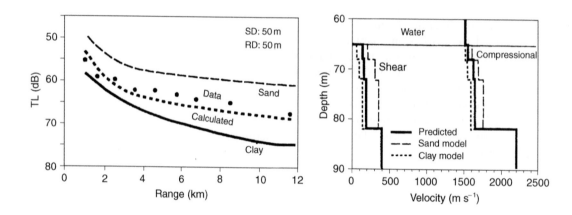

Figure 25   Comparison of transmission loss (TL) data and SAFARI prediction. Effects of changing bottom parameters from sand to clay is also shown together with the input profiles utilized for SAFARI predictions.

Transmission Loss Technique

The seafloor is known to be the controlling factor in low-frequency shallow water acoustic propagation. Forward modeling is performed with models giving exact solutions to the wave equation, i.e., SAFARI where, compressional and shear wave velocities, the attenuation factors associated with these waves, and the sediment density as a function of depth are the main input parameters. Acoustic energy propagating through a shallow water channel interacts with the seafloor causing partitioning of waterborne energy into different types of seismic and acoustic waves. The propagation and attenuation of these waves observed in such an environment are strongly dependent on the physical characteristics of the sea bottom. Transmission loss (TL), representing the

amount of energy lost along an acoustic propagation path, carries the information relative to the environment through which the wave is propagating. **Figure 25** shows a comparison of TL data and model predictions together with the input parameters used at 400 Hz. The effects of changes in bottom parameters are also shown in the figure. This technique becomes extremely useful when seafloor information is sparse.

Conclusions

Acoustic/seismic characteristics of the marine sediments have been of interest to a wide range of activities covering commercial operations involving trenching and cable lying, construction of offshore foundations, studies of slope stability, dredging and military applications like mine and submarine detection. Experimental and theoretical work over the years has shown that it is possible to determine the geoacoustic properties of sediments by different techniques. The characteristics of the marine sediments and techniques to obtain information about these characteristics have been briefly described. The studies so far conducted indicate that direct and indirect methods developed over the last four decades may give sufficient information to deduce some of the fundamental characteristics of the marine sediments. It is evident that still more research needs to be done to develop these techniques for fast and reliable results.

See also

Calcium Carbonates. Clay Mineralogy. Ocean Margin Sediments. Seismic Structure

Further Reading

Akal T and Berkson JM (eds.) (1986) *Ocean Seismo-Acoustics*. New York and London: Plenum press.

Biot MA (1962) Generalized theory of acoustic propagation in porous dissipative media. *Journal of Acoustical Society of America* 34: 1254–1264.

Brekhovskikh LM (1980) *Waves in Layered Media*. New York: Academic Press.

Cagniard L (1962) *Reflection and Refraction of Progressive Seismic Waves*. New York: McGraw-Hill.

Grant FS and West GF (1965) *Interpretation and Theory in Applied Geophysics*. New York: McGraw Hill.

Hampton L (ed.) (1974) *Physics of Sound in Marine Sediments*. New York and London: Plenum Press.

Hovem JM, Richardson MD, and Stoll RD (eds.) (1992) *Shear Waves in Marine Sediments*. Dordrecht: Kluwer Academic.

Jensen FB, Kuperman WA, Porter MB, and Schmidt H (1999) *Computational Ocean Acoustics*. New York: Springer-Verlag.

Kuperman WA and Jensen FB (eds.) (1980) *Bottom Interacting Ocean Acoustics*. New York and London: Plenum Press.

Lara-Saenz A, Ranz-Guerra C and Carbo-Fite C (eds) (1987) *Acoustics and Ocean Bottom*. II F.A.S.E. Specialized Conference. Inst. De Acustica, Madrid.

Pace NG (ed.) (1983) *Acoustics and the Sea-Bed*. Bath: Bath University Press.

Pouliquen E, Lyons AP, Pace NG *et al.* (2000) *Backscattering from unconsolidated sediments above 100 kHz*. In: Chevret P and Zakhario ME (ed.) Proceedings of the fifth European Conference on Under water Acoustics, ECUA 2000 Lyon, France.

Stoll RD (1989) *Lecture Notes in Earth Sciences*. New York: Springer-Verlag.

CLAY MINERALOGY

H. Chamley, Université de Lille 1, Villeneuve d'Ascq, France

Introduction

Clay constitutes the most abundant and ubiquitous component of the main types of marine sediments deposited from outer shelf to deep sea environments. The clay minerals are conventionally comprised of the $<2\,\mu m$ fraction, are sheet- or fiber-shaped, and adsorb various proportions of water. This determines a high buoyancy and the ability for clay to be widely dispersed by marine currents, despite its propensity for forming aggregates and flocs. Clay minerals in the marine environments are dominated by illite, smectite, and kaolinite, three families whose chemical composition and crystalline status are highly variable. The marine clay associations may include various amounts and types of other species, namely chlorite and random mixed layers, but also vermiculite, palygorskite, sepiolite, talc, pyrophyllite, etc. The clay mineralogy of marine sediments is therefore very diverse according to depositional environments, from both qualitative and quantitative points of view.

As clay minerals are considered to be dependent on chemically concentrated environments, and as they commonly form in surficial conditions on land especially through weathering and soil-forming processes, their detrital versus authigenic origin in marine sediments has been widely debated. The transition from continental fresh to marine saline water, marked by a rapid increase of dissolved chemical elements, was the central point of discussion and arose from both American and European examples. In fact the mineralogical changes recorded at the land-to-sea transition are either important or insignificant, are characterized in estuarine sediments by various, sometimes opposite trends impeding consistent geochemical explanations, and often vanish in open marine sediments. The changes observed at the fresh-to-saline water transition in the clay mineral composition essentially proceed from differential settling processes or from mixing between different sources, and not from chemical exchanges affecting the crystalline network. Such a historical debate underlines the interest in investigating the sensitive clay mineral associations for understanding and reconstructing environmental conditions. This article will consider the general distribution and significance of clay minerals in recent sediments, some depositional and genetic environments, and a few examples of the use of clay assemblages to reconstruct paleoclimatic and other paleoenvironmental changes.

General Distribution and Significance

As a result of extensive reviews made by both American and Russian research teams the general characters of the clay mineral distribution in deep sea sediments have been known since the late 1970s. The maps published by various authors demonstrate the dominant control of terrigenous sources, which comprise either soils and paleosoils or rocks. The impact of soils on the marine clay sedimentation is largely dependent on weathering intensity developing on land, and therefore on the climate. For instance, kaolinite mostly forms under intense warm, humid conditions characterizing the intertropical regions, and prevails in the clay fraction of corresponding marine sediments. By contrast chlorite and illite chiefly derive from physical weathering of crystalline and diagenetic sedimentary rocks outcropping widely in cold regions, and therefore occur abundantly in high latitude oceans. The kaolinite/chlorite ratio in marine sediments constitutes a reliable indicator of chemical hydrolysis versus physical processes in continental weathering profiles and therefore of climatic variations occurring on the land masses.

Other clay minerals are also able to bear a clear climatic message, as for instance the amount of random mixed layers and altered smectite in temperate regions, the crystalline status of illite in temperate to warm regions, and the abundance of soil-forming Al-Fe smectite in subarid regions. Detailed measurements on X-ray diffraction diagrams, electron microscope observations and geochemical analyses allow precise characterization of the different continental climatic environments from data obtained on detrital sedimentary clays.

Some terrigenous clay minerals in recent sediments reflect both climatic and non-climatic influences. For instance, the distribution of illite (**Figure 1**), a mineral that primarily derives from the erosion of mica-bearing rocks, shows increased percentages in high latitude oceans due to predominant physical

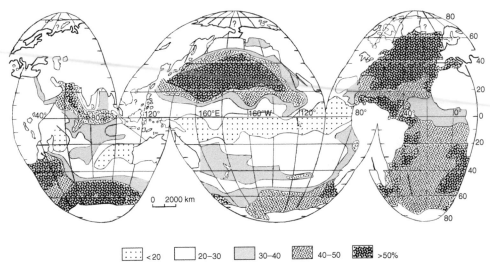

$\boxed{\vdots}$ <20 $\boxed{}$ 20–30 $\boxed{}$ 30–40 $\boxed{}$ 40–50 $\boxed{}$ >50%

Figure 1 Worldwide distribution of illite in the clay fraction of surface sediments in the ocean. (After Windom, 1976. Reproduced with permission from Chamley, 1989.)

weathering, but also in a few low latitude regions depending on active erosion of tectonically rejuvenated, high altitude domains (e.g., supply by Indus and Ganges river drainage systems of Himalayan material to the northern Indian Ocean). The abundance of illite in the Atlantic Ocean, especially in its high latitude and northern parts, is due to several converging causes: cold to temperate climate, extensive outcrops of crystalline and metamorphic rocks, active erosion and river input, relative narrowness of the ocean favoring the ubiquitous transportation of the mineral particles, etc. Abundant illite percentages centered on the 30° parallel of latitude in the North Pacific result from aeolian supply by high altitude jet streams blowing from eastern Asia, and subsequent rainfall above the ocean. The general distribution of illite in marine sediments therefore proceeds from direct and indirect climatic control, meteorological conditions, petrographic and tectonic characteristics, physiography, river influx, etc.

All clay minerals may potentially be reworked from continental outcrops and transported over long distances until they settle on the ocean bottom. This is the case for nearly all geochemical types of smectite minerals (except perhaps for some very unstable ferriferous varieties formed in dense saline brines of the Red Sea), and also of palygorskite and sepiolite, two fibrous species wrongly suspected to not undergo significant transport. For instance, palygorskite and sepiolite are widely transported by wind and or water and deposited as detrital aggregates around the Tertiary basins bordering Africa and Arabia, where they initially formed under arid and evaporative conditions.

The clay mineral family whose distribution is the most complex and dependent on various detrital and autochthonous processes is the smectite group. Moderately crystalline smectites of diverse chemical types form pedogenically by chemical weathering under temperate conditions (essentially by degradation of illite and chlorite), and are supplied by erosion to sediments of mid-latitude regions where they are associated with various types and amounts of random mixed layers. Climate is also the dominant factor in warm, subarid regions where Al-Fe smectite forms in vertisolic soils and is reworked towards the ocean. Fairly high percentages of Fe-smectite characterize the low latitude eastern Pacific basins, where clay minerals in the clay-sized fraction are accessory relative to Fe and Mn oxides, and result from *in situ* hydrogenous genesis. In addition, smectites of Fe, Mg, and even Al types may form by alteration of volcanic rocks, a process which is more intense in well drained, subaerial conditions (hydrolysis) than in submarine environments (halmyrolysis).

The diversity of the factors controlling the distribution of clay minerals in modern deep sea sediments is widely used to trace the influence of continental climate, geological and petrographic sources, tectonics, morphological barriers, etc., and also to identify the nature, direction and intensity of transportation agents. As an example, the distribution of smectite and illite in the western Indian Ocean depends on different source provinces as well as on land geology, climate, volcanism, aeolian and marine currents (**Figure 2**). The terrigenous sources and climatic conditions relieved by north-to-south or south-to-north surface to deep currents are responsible for

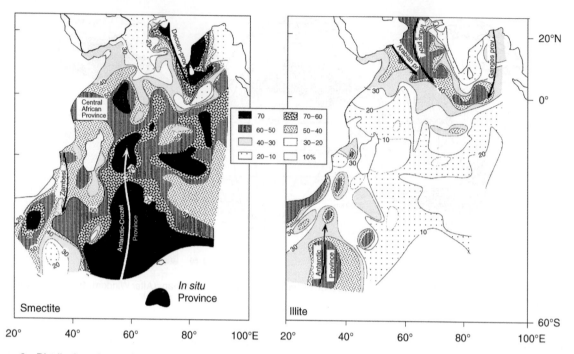

Figure 2 Distribution of smectite and illite in the western Indian Ocean, and related source provinces. (After Kolla *et al.*, 1976. Reproduced with permission from Chamley, 1989.)

long-distance transportation of Antarctic-derived smectite in the Crozet and Madagascar basins, of abundant volcanogenic smectite derived from Deccan traps erosion off the Indian coasts, of Himalayan illite in the Indus and Ganges deep sea fans, of illite associated with up to 30% palygorskite off Arabian and especially on submarine ridges (i.e., aeolian supply), and of illite associated with soil-derived kaolinite off Southeastern Africa. Both illite and smectite are dominantly inherited from various terrestrial rocks and soils, including Antarctic outcrops responsible for illite dominance to the west of the Indian Ocean (35°C) and for smectite dominance to the east (45–75°E). An *in situ* smectite-rich province located in the southern ocean around 55°S and 70°E is attributed to the submarine alteration of volcanic rocks. Volcanic contributions are also suspected in the Central Indian basin and in the vicinity of Indonesia. Of course such investigations constitute very useful guidelines for reconstructing past climatic, oceanographic, and physiographic conditions.

Marine Autochthonous Processes

From Volcanic to Hydrothermal and Hydrogenous Environments

Until the 1970s, the submarine weathering of **volcanic** material (basalt, glass, ash) was often considered to be responsible for important *in situ*

formation of clay minerals, especially of smectite, in deep sea sediments. Effectively basalt altered by surficial oxidation and hydration may give way to Mg-smectite, sometimes Fe-smectite, frequently associated with celadonite (a glauconite-like Fe-Al micaceous species), phillipsite (a Na-rich zeolite), calcium carbonates, Fe-Mn oxyhydroxides, etc. The more amorphous, the smaller sized and the more porous the volcanic material (e.g., pumiceous ashes), the more intense the submarine formation of clay. In fact the clay minerals resulting from halmyrolysis of volcanic material are quantitatively limited and essentially located at close vicinity to this material (e.g., altered volcaniclastites or basalts); they are unable to participate in a large way in the formation of the huge amounts of clay incorporated in deep-sea sediments. Additional arguments contradicting the importance of volcanic contribution to deep-sea clay consist of the frequent absence of correlation between the presence of volcanic remains and that of smectite, and in the non-volcanogenic chemistry of most marine smectites (e.g., aluminum content, rare earth elements, strontium isotopes). The shape of smectite particles observed by electron microscopy is typical of volcanic influence only in restricted regions marked by high volcanic activity, especially explosive activity. Notice that local overgrowths of lath systems oriented at 60° from each other may characterize marine clay particles and especially smectites, but they are neither specifically related to volcanic environments nor associated with noticeable increase of smectite proportion or specific change in the clay chemical or isotopic composition. The intrusion of basalt sills in soft marine sediments may determine some metamorphic effects and the very local formation of ordered mixed layers (corrensite), chlorite, and associated non-clay minerals.

The **hydrothermal** impact on deep-sea sedimentation is fundamentally characterized by *in situ* precipitation of Fe-Mn oxyhydroxides relatively depleted in accessory transition elements (Co, Cu, Ni), and locally by the deposition of massive sulfides near the vents where hot and chemically concentrated water merges. The autochthonous clay minerals in such environments are marked by various species depending on fluid temperature, oxidation-reduction processes, and fluid/rock ratio. For instance, drilling holes in Pacific hydrothermal systems show different mineral evolutions. In the hydrothermal mounds of the Galapagos spreading center, the fluids are rich in silicon and iron and of a low temperature (20°–30°C) throughout the 30 m-thick sedimentary column; this gives way in oxidized conditions to the precipitation of Fe-smectite as greenish layers interbedded in

biogenic oozes that at depth evolve into glauconite by addition of potassium (**Figure 3A**). By contrast the detrital to authigenic deposits of the Middle Valley of Juan de Fuca ridge show on a 40 m-thick series the *in situ* formation from high temperature Mg-rich fluids (200°C) of a downwards sequence characterized by saponite (a Mg-smectite), corrensite (a regular chlorite-smectite mixed layer), swelling chlorite, and chlorite (**Figure 3B**). At this site geochemical and isotope investigations reflect a noticeable downhole increase of temperature and strong changes in the fluid composition.

A more widespread process consists of the **hydrogenous** formation of clay at the sediment–seawater interface, in deep-sea environments characterized by water depths >4000 m, insignificant terrigenous supply, and very low sedimentation rate (<1 mm/1000 years). This is particularly the case for some Central and South Pacific basins. The sediments mostly consist of reddish-brown oozes rich in Fe and Mn oxides (i.e., 'deep sea red clay'). There iron-rich smectites of the nontronite group may form in significant proportions, probably due to long-term low temperature interactions between (1) metal

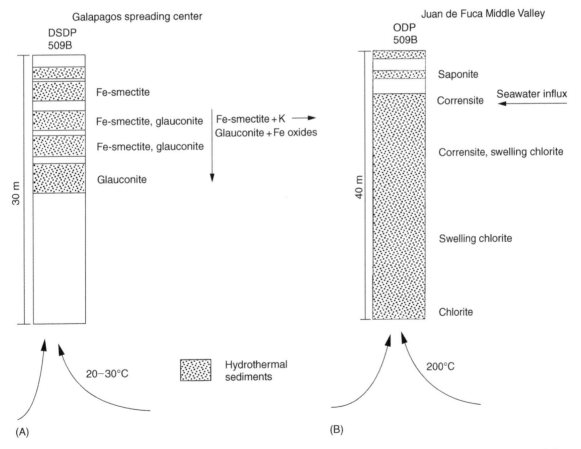

Figure 3 Schematic vertical distribution of typically hydrothermal clay minerals in the sedimentary systems of (A) the Galapagos spreading center, and (B) the Juan de Fuca Middle Valley. (Reproduced with permission from Buatier and Karpoff, 1995.)

Table 1   Examples of chemical composition of hydrothermal to hydrogenous smectites in Central and South Pacific sediments

Type of smectite	Tetrahedra			Octahedra			Interlayers		
	Si	Al	Fe	Al	Fe	Mg	Ca	NH$_4$	K
Pure hydrothermal (Galapagos mounds field)	3.94	0.06	–	0.03	1.59	0.38	0.03	–	0.36
Hydrothermal and hydrogenous (Bauer Deep)	3.97	0.03	–	0.44	1.07	0.54	0.05	–	0.06
Hydrogenous>hydrothermal (Galapagos spreading centre)	3.97	0.03	–	1.12	0.48	0.37	0.09	–	0.11
Pure hydrogenous (?) (North Marquises fracture zone)	3.37	0.63	–	0.39	1.12	0.46	0.46	–	0.17

(Reproduced with permission from Chamley, 1989.)

oxyhydroxides supplying the iron, (2) sea-water supplying the magnesium and other minor to trace elements, (3) biogenic silica supplying the silicon, and (4) allochthonous accessory particles (e.g., aeolian clay) supplying the other chemical elements (e.g., Al). Notice that the distinction between pure hydrothermal and pure hydrogenous clay minerals forming on the deep-sea floor necessitates detailed chemical analyses (**Table 1**) and often additional microprobe and isotope investigations.

To summarize, the distribution of clay minerals in deep sea deposits marked by active volcanic-hydrothermal activity and by very low sedimentation rates depends on various and complex *in situ* influences among which the hydrogenous processes quantitatively prevail. The distinction of these autochthonous influences is complicated both in the vicinity of land masses where terrigenous supply becomes active, and in shallower areas where biogenic influences may intervene more intensely (e.g., Nazca plate, southeast Pacific).

Ferriferous Clay Granules

Iron-rich clay granules are traditionally called glauconite, which is somewhat incorrect as glauconite is a specific clay mineral, whereas clay granules may include various iron-bearing clay species. Ferriferous clay granules form on continental margins at water depths not exceeding 1000 m, and comprise two major types characterized by specific colors, clay minerals, and habits. Glaucony, the most widespread type, constitutes dark green to brown clayey aggregates, and may comprise different varieties of iron-rich illite- and smectite-like minerals such as glauconite (Fe- and K-rich illitic clay), Fe-smectite, and Fe illite-smectite mixed layers. Glaucony may form at latitudes as high as 50° and in water depths as great as 1000 m, but usually occurs in 150–300 m water depths at the shelf-slope transition of temperate-warm to equatorial regions. Verdine, which is less ubiquitous and has been identified more recently, constitutes light green to light brown granules

characterized by phyllite V or odinite, a ferriferous clay mineral of the kaolinite family (described by G.S. Odin, who has developed outstanding investigations on clay granules). Verdine forms in rather shallow water sediments (maximum 50–80 m) of intertropical regions, and depends on the supply of abundant dissolved iron by low latitude rivers.

Ferriferous clay granules form at the sediment–water interface and evolve at burial depths rarely exceeding a few decimeters. They develop in semi-confined environments at the expense of various substrates submitted to 'greening': chiefly fecal pellets and microfossil chambers (e.g., foraminifera), calcareous or siliceous bioclasts, minerals (especially micas), and rock debris. The formation of glaucony (which somewhat leads to diffuse habits), occurs in successive stages marked by a rapid and strong enrichment of iron and then potassium, a volume increase causing external cracks, and the obliteration of the initial shape (**Figure 4**). The formation of verdine still has to be documented, but both clay granule types correspond to true authigenic formation rather than to transformation of pre-existing clay minerals. The chemical evolution of ferriferous clay granules vanishes either after a long exposure at the sediment–water interface (10^5–10^6 years for glaucony), or after significant burying.

Organic Environments

The influence of living organisms on clay-rich sediments is mainly marked by physical processes referred to as bioturbation, and concerns various marine environments, especially on continental shelves. Chemical modifications of clay associations are only occasionally reported and seem to affect the crystalline status of chlorite and associated random mixed layer clays locally through ingestion and digestion processes of shallow water crustaceans, annelids or copepods. The chemical interactions developing in digestive tracts between clay minerals and organic acids appear to have small quantitative

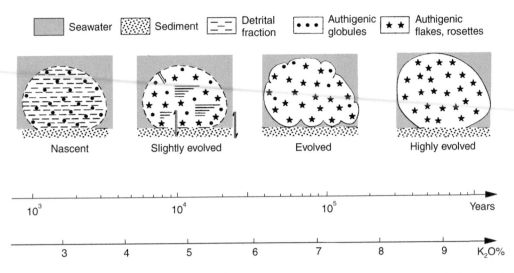

Figure 4 Successive stages of glaucony formation from a pre-existing substrate. (Reproduced with permission from Odin, 1998.)

effects, as the marine clay associations are roughly the same as the terrestrial associations.

The chemical impact on clay mineral stability of the organic matter incorporated in deep marine sediments is variable. Most sedimentary series containing significant amounts of dispersed organic matter (i.e., 1–3%) do not display any specific clay mineral composition. For example, this is the case for black shales deposited during the Cretaceous period in the Atlantic, where clay mineral associations may comprise vulnerable species such as smectite and palygorskite, the abundance and crystalline status of which vary independently of the content and distribution of the organic matter. In contrast, the sapropels developing in the eastern Mediterranean during the late Cenozoic era, especially in Quaternary high sea level stages, show some *in situ* degradation processes of the detrital clay minerals (**Figure 5**). Submarine alteration affects the mineral species in successive stages depending on

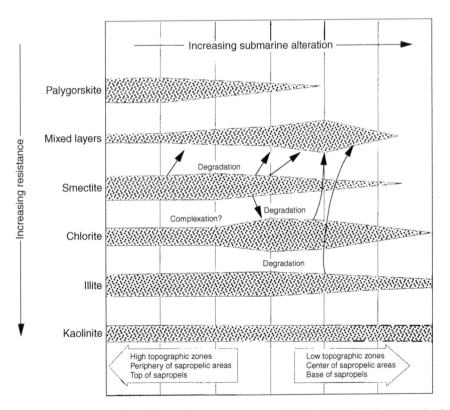

Figure 5 Characters of the clay mineral degradation in Quaternary sapropels of the eastern Mediterranean basins.

their ability to resist acid conditions. Palygorskite is the more vulnerable species and kaolinite the more resistant. The degradation of clay assemblages tends to increase toward the central and deepest parts of marine basins, in depressed morphological zones, and at the base of the decimeter-to-meter thick sapropels. The degradation of clay minerals under organic conditions has occurred close to the sediment–water interface and appears to depend on the chemical nature and evolution stage of the terrestrial and marine organic matter.

Paleoenvironmental Expression

Clay mineral assemblages of sediments successively deposited in marine basins express various environmental messages related to the geological history. A few examples from recent Quaternary to late Cenozoic series will be considered here. Similar messages may be preserved in much older series of Mesozoic and even Paleozoic ages, provided that the diagenetic imprint due to lithostatic overburden, geothermal gradient, and fluid circulation has remained moderate. Clay-rich, low permeability sedimentary formations 2–3 km thick and submitted to normal heat flow (c. 30°C/km) are usually prone to preserve such paleoenvironmental characteristics.

Climate

As clay minerals at the surface of the Earth are dominantly formed through pedogenic processes depending on climate and are particularly subjected to surficial erosion and reworking, their assemblages successively deposited in a given sedimentary basin are *a priori* able to reflect successive climatic conditions that prevailed on adjacent land masses. This implies that very little post-depositional, i.e., diagenetic changes have affected the clay assemblages after their storage in sediments. This is observed to be the case in many series drilled or cored in the oceans. The climatic message borne by clay has been documented by numerous investigations, and corroborated by the comparable range of variations recorded in the nature and proportions of clay minerals in both present-day soils outcropping at various latitudes and marine sedimentary columns. Marine clay mineral assemblages basically express the type and intensity of continental weathering, which depend predominantly on the ion leaching through the action of humidity and temperature, and secondarily on seasonal rainfall and drainage conditions.

Quaternary glacial–interglacial alternations caused terrestrial alternation of physical and chemical weathering processes, and this was reflected in the clay assemblages successively brought to marine sediments through soil erosion and river or wind transport. Sedimentary levels contemporary with cold periods are usually characterized by more abundant rock-derived minerals such as richly crystalline illite, chlorite, smectite and associated feldspars reworked from active physical weathering. Warm, humid periods generally correspond to increased supply of soil-derived kaolinite and metal oxides, poorly crystalline smectite and various random mixed layer clay minerals. For instance, the terrigenous fraction of hemipelagic sediments deposited from 500 000 to 100 000 years ago in the Northwestern Atlantic off New Jersey and dominantly derived from the erosion of Appalachian highlands shows increased proportions of chlorite in glacial isotopic stages, and of kaolinite in interglacial stages. This is clearly expressed by the kaolinite/chlorite ratio (**Figure 6**). Paleoclimatic reconstructions from clay mineral data are available for various geological periods, as for instance the passage since about 40 Ma from a non-glacial world dominated by chemical weathering (smectite, kaolinite) to a glacial world in which physical weathering was greater (chlorite, illite). The comparison of climatic curves provided by clay minerals and other indicators (oxygen isotopes, micro-faunas or -floras, magnetic susceptibility, etc.) allows a better understanding of the nature, intensity, and effect of the different factors characterizing the terrestrial and marine climate in given regions during given geological intervals.

High resolution studies show that clay assemblages may express terrestrial climatic variations at a centennial scale or even less, and that the influence of Earth's orbital parameters varies to different extents according to the latitude. For example, the clay minerals data of Quaternary North Atlantic deep sea sediments were submitted to cross-correlation spectral analyses on 5.5–14 m-long cores encompassing the last 300 000 years. The mineral composition displays a general 100 000-year cyclic signal (eccentricity) in the whole 45°–60°N range, a 41 000-year signal (obliquity) at highest latitudes related to dominant aeolian supply, and a 23 000-year signal (precession) at mid-latitudes related to dominant transport by marine currents (**Table 2**).

The paleoclimatic expression by clay mineral successions is direct or indirect, i.e., it either indicates the climate that actually prevailed at a given period, or reflects other events depending on climate: migration of lithospheric plates across successive climatic zones, varying extension of ice caps controlling the surficial erosion, variations in the marine circulation regime due to changing latitudinal and

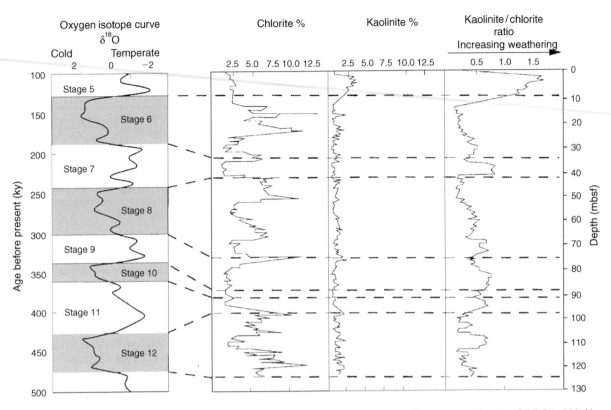

Figure 6 Comparison and climatic significance of clay mineral and oxygen isotope data from stages 12 to 5 at ODP Site 902, New Jersey continental margin. (Reproduced with permission from Vanderaveroet *et al.*, 1999.)

vertical heat transfers. The direct paleoclimatic reconstructions from clay mineral data are all the more reliable since the marine basins investigated are preserved from important erosion of paleosoils, changes in detrital sources, differential settling processes, longitudinal oceanic currents, and major geomorphological changes.

Marine Currents

The different marine water masses may carry the small and light clay mineral particles over long distances, and therefore leave an imprint within the sediments at the depth range they are moving. This has been demonstrated for late Quaternary sediments of the southwestern Atlantic, where the southward-flowing North Atlantic deep water is enriched in kaolinite supplied from rivers draining the intertropical South American continent, and the northward flowing Antarctic Atlantic bottom water supplies chlorite and smectite issuing from southernmost Argentina and Antarctica. Paleocurrent reconstructions from clay data exist mainly about Atlantic and Southern Oceans, which are marked by numerous and distinct terrigenous sources, vertical mixing and longitudinal heat transfers, and Tertiary to Quaternary changing

conditions of the superimposed water masses volume and celerity.

Tectonic Activity

The tectonic instability determines some changes in the composition of clay mineral assemblages which are usually much more important than those due to climate or circulation. First, the subpermanent rejuvenation by neotectonics of continental relief increases the erosion potential and therefore impedes the development of pedogenic blankets where clay minerals tend to be in equilibrium with current climatic conditions. Such a chronic tectonic activity explains the abundance of rock-derived illite and chlorite in equatorial Indian Ocean basins depending on Himalayan output. Second, a continental tectonic uplift determines changes in the nature of clay minerals eroded from rocky substrates, while submarine uplift may determine morphological barriers to the clay transfer. This was the case for the Hellenic Trench in the eastern Mediterranean during late Pliocene to early Pleistocene periods, when the combined uplift of Peloponnese and of Mediterranean ridge both increased the terrigenous input of European illite and chlorite and blocked the supply of African palygorskite. Due to their sensitivity to

Table 2 General relationships between the clay mineral distribution and the three main Earth's orbital frequency bands according to latitude, from cross-correlation spectral analysis of X-ray diffraction data on North Atlantic cores

| Core | SU 90-08 | | | SU 90-12 | | | SU 90-38 | | | SU 90-33 | | |
| Latitude | 44°N | | | 51°N | | | 54°N | | | 60°N | | |
Orbital parameters	**E**	**O**	**P**	**E**	**O**	**P**	**E**	**O**	**P**	**E**	**O**	**P**
Illite	H	–	V	H	–	V	V	H	–	H	V	–
Chlorite	V	–	V	V	–	V	V	V	–	V	V	–
Kaolinite	H	–	V	H	–	–	V	V	–	H	V	–
Illite-vermiculite random mixed layer	–	–	–	–	–	V	–	–	–	–	–	–

E, eccentricity band, 100000 year; O, obliquity band, 41000 year; P, precession band, 23000 year; H, high variance power; V, very high variance power. Maximum correlations in bold characters. (Reproduced with permission from Bout-Roumazeilles *et al.*, 1997.)

geomorphological changes and their aptitude for long distance transportation, clay minerals are able to express slight and progressive epeirogenic changes as well as very remote tectonic events.

See also

Deep-Sea Drilling Results. Hydrothermal Vent Deposits.

Further Reading

Bout-Roumazeilles V, Debrabant P, Labeyrie L, Chamley H, and Cortijo E (1997) Latitudinal control of astronomical forcing parameters on the high-resolution clay mineral distribution in the 45°–60° N range in the North Atlantic Ocean during the past 300,000 years. *Paleoceanography* 12: 671–686.

Buatier MD and Karpoff AM (1995) Authigenése et évolution d'argiles hydrothermales océaniques: exemples des monts des Galapagos et des sédiments de la ride de Juan de Fuca. *Bulletin de la Société Géologique de France* 166: 123–136.

Chamley H (1989) *Clay Sedimentology*. Berlin: Springer-Verlag.

Hoffert M (1980) Les 'argiles rouges des grands fonds' dans le Pacifique centre-est. *Sciences géologique*. Strasbourg, Mem 61: 257.

Millot G (1970) *Geology of Clays*. Berlin: Springer-Verlag.

Odin GS (ed.) (1988) *Green Marine Clays. Developments in Sedimentology*, 45, Amsterdam: Elsevier.

Robert C and Chamley H (1992) Late Eocene-early Oligocene evolution of climate and marine circulation: deep-sea clay mineral evidence. American Geophysical Union. *Antarctic Research Series* 56: 97–117.

Vanderaveroet P, Averbuch O, Deconinck JF, and Chamley H (1999) A record of glacial/interglacial alternations in Pleistocene sediments off New Jersey expressed by clay mineral, grain-size and magnetic susceptibility data. *Marine Geology* 159: 79–92.

Weaver CE (1999) *Clays, Muds, and Shales. Developments in Sedimentology*, 44. Amsterdam: Elsevier.

Windom HL (1976) Lithogenous material in marine sediments. *Chemical Oceanography* vol. 5, pp. 103–135. New York: Academic Press.

CALCIUM CARBONATES

L. C. Peterson, University of Miami, Miami, FL, USA

Introduction

The ocean receives a continual input of calcium from riverine and groundwater sources and from the hydrothermal alteration of oceanic crust at mid-ocean ridge spreading centers. Balancing this input is the biological precipitation of calcium carbonate ($CaCO_3$) by shell-and skeleton-building organisms in both shallow marine and open-ocean environments. In the deep sea, the primary contributors to the carbonate budget of open-ocean sediments are the skeletal remains of calcareous plankton that have settled down from the surface after death. Seafloor sediments consisting of more than 30% by weight calcium carbonate are traditionally referred to as calcareous or carbonate ooze; such oozes accumulate at the rate of 1–4 cm per 1000 years and cover roughly half of the ocean bottom. Carbonate oozes are the most widespread biogenous sediments in the ocean.

While the biological production of calcium carbonate in oversaturated surface waters determines the input of carbonate to the deep sea, it is the dissolution of carbonate in undersaturated deep waters that has the dominant control on calcium carbonate accumulation in the open ocean. Since carbonate production rates in the surface ocean today greatly exceed the rate of supply of calcium, this 'compensation' through dissolution must occur in order to keep the system in steady-state. Increased dissolution at depth is largely a function of the effect of increasing hydrostatic pressure on the solubility of carbonate. However, superimposed on this bathymetric effect are regional preservation patterns related to differences in carbonate input and the carbonate chemistry of deep water masses. Carbonate oozes in the deep sea serve as a major reservoir of calcium and carbon dioxide on the Earth's surface. Their spatial and temporal accumulation patterns in the marine stratigraphic record are thus a primary source of data about the carbonate chemistry and circulation of past oceans, as well as of the global geochemical cycle of CO_2.

Carbonate Producers

The most important carbonate producers in the open ocean are planktonic coccolithophorids and foraminifera, unicellular phytoplankton and zooplankton respectively, which inhabit the upper few hundred meters of the water column (**Figure 1**). Coccolithophorids are the dominant carbonate-precipitating organisms on Earth. During part of their life cycle, they produce a skeletal structure (the coccosphere) consisting of loosely interlocking plates, often button-like in appearance, known as coccoliths. Deep-sea carbonates generally contain only the individual coccoliths, as the intact coccospheres are rarely preserved. Foraminifera produce a calcareous shell, or 'test', a few hundred microns in size that sinks after death or reproduction to the sea floor. Both coccolithophorids and the planktonic foraminifera construct their skeletal elements out of the mineral calcite, the more stable polymorph of $CaCO_3$. Calcareous sediments dominated by one or the other component are termed coccolith oozes or foraminiferal oozes, although in reality most carbonate-rich sediments are a mixture of both.

Coccolithophorids made their first appearance in the geological record in the earliest Jurassic, while planktonic foraminifers evolved somewhat later in the middle Jurassic. The appearance of these two dominant pelagic carbonate producers, and their rapid diversification in the Cretaceous, would have had major effects upon the carbonate geochemistry of the oceans. Before this, most carbonate was deposited in shallow seas, accounting for the high proportion of limestones among older rocks on the continents. Since the Mesozoic, deep-ocean basins have become enormous sinks for carbonate deposition.

Smaller contributions to the deep-sea carbonate budget come from a variety of other sources. Pteropods, free-swimming pelagic gastropods, construct a relatively large (several millimeters) but delicate shell out of the metastable form of $CaCO_3$ known as aragonite. However, while pteropods can be unusually abundant in certain environments, the increased solubility of aragonite leads to very restricted preservation of the shells and pteropod oozes are relatively rare in the ocean. In the vicinity of shallow, tropical carbonate platforms such as the Bahamas or Seychelles Bank, shedding of aragonitic bank-top sediments derived from algal and coral production can lead to aragonite-rich 'periplatform oozes' in deep waters around the perimeters of the platform.

(A)

(B)

Figure 1 Carbonate oozes in the deep sea are dominated by the skeletal remains of (A) planktonic foraminifera (× 50 magnification) and (B) coccolithophorids (× 6000 magnification). Specimens shown here were isolated from a Caribbean sediment core.

In general, contributions from bottom-dwelling organisms (e.g. benthonic foraminifera, ostracods, micromollusks) are negligible in deep-sea sediments.

Carbonate Distribution and Dissolution

The distribution of carbonate sediments in the ocean basins is far from uniform (**Figure 2**). If it were possible to drain away all of the ocean's water, carbonate oozes would be found draped like snow over the topographic highs of the seafloor and to be largely absent in the deep basins. The lack of carbonate-rich sediments in the deepest parts of the world's oceans has been recognized since the earliest investigations. Although surface productivity and dilution by noncarbonate sediment sources can locally influence the concentration of carbonate in

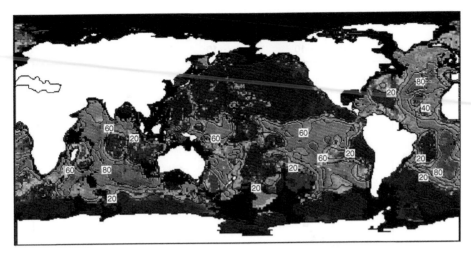

Figure 2 Global distribution of calcium carbonate (weight-% CaCO₃) in surface sediments of the ocean. Data compilation from Archer (1996); reproduced with permission from the American Geophysical Union.

deep-sea sediments, the clear-cut relationship between calcium carbonate content and water depth indicates that carbonate dissolution plays the major role in governing carbonate distribution patterns. To a first approximation, the dissolution of carbonate on the seafloor is a function of the corrosiveness or saturation state of the overlying bottom waters.

The amount of calcium carbonate that will dissolve in sea water if thermodynamic equilibrium is reached is governed by the following reaction:

$$CaCO_3(s) \leftrightarrow Ca^{2+}(aq) + CO_3^{2-}(aq)$$

At equilibrium, the rate of carbonate dissolution is equal to the rate of its precipitation and the sea water is said to be saturated with respect to the carbonate phase. In the deep sea, the degree of calcium carbonate saturation (D) can be expressed as:

$$D = \frac{[Ca^{2+}]_{seawater} \times [CO_3^{2-}]_{seawater}}{[Ca^{2+}]_{saturation} \times [CO_3^{2-}]_{saturation}}$$

where $[Ca^{2+}]_{seawater}$ and $[CO_3^{2-}]_{seawater}$ are the *in situ* concentrations in the water mass of interest and $[Ca^{2+}]_{saturation}$ and $[CO_3^{2-}]_{saturation}$ are the concentrations of these ions at equilibrium, or saturation, at the same conditions of pressure and temperature. Since shell formation and dissolution cause the concentration of $[Ca^{2+}]$ to vary by less than 1% in the ocean, the degree of calcium carbonate saturation (D) can be simplified and expressed in terms of the concentration of the carbonate ions only:

$$D = \frac{[CO_3^{2-}]_{seawater}}{[CO_3^{2-}]_{saturation}}$$

D is thus a measure of the degree to which a seawater sample is saturated with respect to calcite or aragonite, and so provides a measure of the strength of the driving force for dissolution. Values of $D > 1$ indicate oversaturation while values of $D < 1$ indicate undersaturation and a tendency for calcium carbonate to dissolve. Since the saturation carbonate ion concentration increases with increasing pressure and decreasing temperature, calcium carbonate is more soluble in the deep sea than at the surface. At the depth in the water column where $D = 1$, the transition from oversaturated to undersaturated conditions is reached. This depth is known as the saturation horizon (**Figure 3**). Aragonite is always more soluble than calcite, and its respective saturation horizon is shallower, because the saturation carbonate ion concentration for aragonite is always higher for the same conditions of pressure and temperature.

Observations from studies of surface sediments have allowed definition of regionally varying levels in the ocean at which pronounced changes in the presence or preservation of calcium carbonate result from the depth-dependent increase of dissolution on the seafloor. The first such level to be identified was simply the depth boundary in the ocean separating carbonate-rich sediments above from carbonate-free sediments below. This level is termed the calcite (or carbonate) compensation depth (CCD) and represents the depth at which the rate of carbonate dissolution on the seafloor exactly balances the rate of carbonate supply from the overlying surface waters. Because the supply and dissolution rates of carbonate differ from place to place in the ocean, the depth of the CCD is variable. In the Pacific, the CCD is typically found at depths between about 3500 and 4500 m. In the North Atlantic and parts of the South Atlantic, it is found

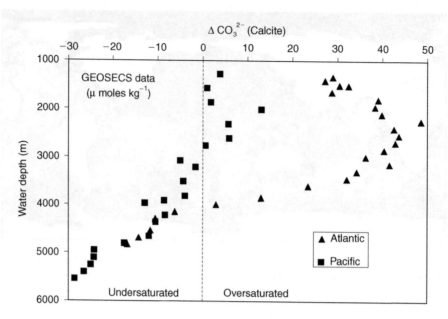

Figure 3 Bathymetric profiles of calcium carbonate (calcite) saturation for hydrographic stations in the Atlantic and Pacific Oceans (data from Takahashi *et al.* 1980). Carbonate saturation here is expressed as ΔCO_3^{2-}, defined as the difference between the *in situ* carbonate ion concentration and the saturation carbonate ion concentration at each depth $\Delta CO_3^{2-} = [CO_3^{2-}]_{seawater} - [CO_3^{2-}]_{saturation}$. The saturation horizon corresponds to the transition from waters oversaturated to waters undersaturated with respect to calcite ($\Delta CO_3^{2-} = 0$). This level is deeper in the Atlantic than in the Pacific because Pacific waters are CO_2-enriched and $[CO_3^{2-}]$-depleted as a result of thermohaline circulation patterns and their longer isolation from the surface. The Atlantic data are from GEOSECS Station 59 (30°12'S, 39°18'W); Pacific data come from GEOSECS Station 235 (16°45'N, 161°23'W).

closer to a depth of about 5000 m. Close to continental margins the CCD tends to shoal, although much of this apparent rise can be attributed to carbonate dilution by terrigenous input from the continents. Rarely does carbonate ooze accumulate on seafloor that is deeper than about 5 km.

In practice, the CCD is identified by the depth transition from carbonate ooze to red clay or siliceous ooze that effectively defines the upper limit of the zone of no net CaCO$_3$ accumulation on the seafloor. Given the practical difficulty (e.g. analytical precision, redeposition) of determining the depth level at which the carbonate content of sediment goes to zero, some investigators choose instead to recognize a carbonate critical depth (CCrD), defined as the depth level at which carbonate contents drop to <10% of the bulk sediment composition. The CCrD lies systematically and only slightly shallower than the CCD. A similar boundary to the CCD can be recognized marking the lower depth limit of aragonite-bearing sediment in the ocean, the aragonite compensation depth or ACD. Because of the greater solubility of aragonite as compared with calcite, the ACD is always much shallower than the CCD.

Above the CCD, the level at which significant dissolution of carbonate first becomes apparent is called the lysocline. As originally defined, the term

lysocline was used to describe the depth level where a pronounced decrease in the preservation of foraminiferal assemblages is observed. It thus marks a facies boundary separating well-preserved from poorly preserved assemblages on the seafloor. This level is now more specifically referred to as the foraminiferal lysocline to differentiate it from the coccolith lysocline and pteropod lysocline, which may differ in depth because of varying resistance to dissolution or differences in solubility (in the case of the aragonitic pteropods). In addition, it is customary to recognize a sedimentary or carbonate lysocline as the depth at which a noticeable decrease in the carbonate content of the sediment begins to occur.

In theory, the lysocline records the sedimentary expression of the saturation horizon, that is the depth-dependent transition from waters oversaturated to waters undersaturated with respect to carbonate solubility (**Figure 4**). The lysocline thus marks the top of a depth zone, bounded at the bottom by the CCD, over which the bulk of carbonate dissolution in the ocean is expected to occur in response to saturation state-driven chemistry. The thickness of this sublysocline zone, as indicated by the vertical separation between the lysocline and CCD, is variable and is governed by the rate of carbonate supply, the actual dissolution gradient, and

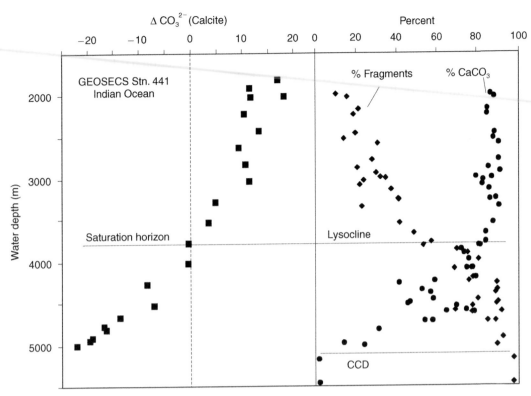

Figure 4 Comparison of carbonate saturation profile for the eastern equatorial Indian Ocean with measurements of foraminiferal fragmentation and carbonate content (weight-%) from depth-distributed modern sediment samples in this region. The saturation horizon with respect to calcite ($\Delta CO_3^{2-} = 0$) occurs locally in the water column at a depth of 3800 m. This level corresponds with both the foraminiferal lysocline and carbonate lysocline as recognized in the sediments. The carbonate compensation depth (CCD) in this region is found at a depth of approximately 5000 m. Increased foraminiferal fragmentation and decreases in sedimentary carbonate content are the result of dissolution and carbonate loss below the lysocline. Carbonate saturation data are from GEOSECS Station 441 (5°2'S, 91°47'E; Takahashi *et al.* 1980); modern sediment data are from Peterson and Prell (1985).

potentially by noncarbonate dilution in certain regions of the ocean.

While the term lysocline was originally used to define a preservational boundary, it has also been used in a fundamentally different sense to denote the depth at which dissolution rates of carbonate on the seafloor greatly accelerate. Whether these levels may or may not coincide, and the nature of their relationship to the saturation horizon or 'chemical lysocline', has been the subject of much discussion and debate. One of the reasons for uncertainty in this regard is the fact that both the carbonate content (%) of a sediment sample and the preservation of the calcareous microfossil assemblages there in can be surprisingly poor indicators of the extent to which dissolution has occurred. For example, the loss of carbonate (L) from sediment, expressed as a weight percentage of the total sediment, is given by:

$$L = 100(1 - R_o/R)$$

where R_o and R are the initial and final values of the noncarbonate (or residual) material. Thus, for a

sample initially containing 95% carbonate and a R_o value of 5%, 50% of the carbonate in the sample must be dissolved in order to double the noncarbonate fraction and reduce the carbonate content to 90%. Since the carbonate fraction of the pelagic rain in the open ocean often approaches 95%, this inherent insensitivity means that significant loss of carbonate can occur before detectable changes in the carbonate content are observed. As a consequence, the carbonate lysocline, traditionally defined as the level where the carbonate content of sediments begins to sharply decrease with water depth, may lie deeper than the depth at which significant loss of carbonate to dissolution actually begins to occur.

Dissolution leads to an increase in surface area during the etching of carbonate skeletal material. Etching produces roughness and widens pores, leading to weakening and ultimately to breakage. Because of their larger size, planktonic foraminifera have usually been the subject of dissolution studies that focus on the preservation state of the microfossils themselves. Planktonic foraminifera have a wide range of morphologic characteristics that enhance their abilities to remain suspended in the upper water column while alive. These same characteristics largely dictate their resistance to dissolution after death. Taxa living in warm, tropical surface waters, where density is generally low, tend to be open-structured with thin shells and porous walls. Taxa that live deeper in cooler, denser subsurface waters, or in colder surface waters at high latitudes, tend to be more heavily calcified with thicker shells and small or closed up pores. On the seafloor, the thin-shelled, more fragile species tend to dissolve more readily than the robust taxa. In effect, this means that individual species each have their own 'lysocline', which can be offset shallower or deeper from the foraminiferal lysocline determined from the total assemblage. There are additional consequences of this selective preservation of taxa that must be considered in paleo-oceanographic or paleoclimatic studies. For example, the selective preservation of more heavily calcified taxa tends to impart a generally 'cooler' appearance to the overall microfossil population and can bias attempts to derive paleotemperature information from seafloor assemblages, as well as other population properties such as diversity.

For carbonate particles produced in the upper ocean, settling rates play an important role in their distribution and preservation. Smaller planktonic foraminifers settle at about $150-250 \, \text{m d}^{-1}$, while larger ($>250 \, \mu\text{m}$) foraminifers may settle as much as $2000 \, \text{m d}^{-1}$. These rates are rapid enough that little dissolution is thought to occur in the water column. Solitary coccoliths, on the other hand, sink at rates of 0.3 to $\sim10 \, \text{m d}^{-1}$, slow enough that dissolution within the water column should theoretically prevent their ever reaching the ocean bottom. However, sediment trap studies have shown that transport by fecal pellets is the dominant process by which small phytoplankton skeletons are transferred to the seafloor. Protection offered by the organic fecal pellet covering may also protect the coccoliths after deposition and account for the fact that the coccolith lysocline is generally observed to lie somewhat deeper than the foraminiferal lysocline.

While the seafloor depths of the lysocline and CCD can be readily identified from sedimentary criteria, this information is of limited use without realistic knowledge of the rates at which calcium carbonate is lost from the sediments to dissolution. In practice, it is much easier to determine carbonate accumulation in the deep sea than it is to estimate carbonate loss. Yet the latter information is clearly needed in order to close sediment budgets and to reconstruct changes in the carbonate system.

Carbonate-rich sediments deposited above the saturation horizon should experience little in the way of saturation-driven dissolution because they lie in contact with waters oversaturated with respect to calcite. Nevertheless, evidence for significant supralysoclinal dissolution has been found in a number of studies. Much of this dissolution at shallower water depths is thought to be driven by chemical reactions associated with the degradation of organic carbon in the sediments. Organic carbon arriving at the seafloor is generally respired as CO_2 or remineralized to other organic compounds by benthic organisms. The metabolic CO_2 generated by organisms that live within the sediment can contribute to the dissolution of calcite even above the lysocline by increasing the chemical corrosivity of the pore waters. Studies of organic matter diagenesis in deep-sea sediments suggest that rates of supralysoclinal dissolution vary greatly with location, ranging from minimal loss to $>40\%$ calcite loss by weight. Temporal and spatial changes in the rain rate of organic carbon relative to carbonate can affect this process.

Whether above or below the lysocline, carbonate dissolution is mostly confined to the bioturbated surface sediment layer (typically $\leq10 \, \text{cm}$ in the deep sea). As carbonate is depleted from this bioturbated layer, older 'relict' carbonate is entrained from the sediments below. This results in 'chemical erosion' and can produce substantial hiatuses or gaps in the record. Dissolution, and hence erosion, eventually stops when nonreactive materials fill up the mixed layer and isolate the underlying sediment from the overlying water. Many clay layers interbedded within carbonate-rich sequences are likely produced by this mechanism; the resulting lithologic contrasts often show up as subsurface seismic horizons which can be traced for long distances and tell a story of changing dissolution gradients and carbonate chemistry in the past.

Basin-to-Basin Fractionation in the Modern Ocean

Superimposed on the general depth-dependent decrease of carbonate accumulation observed everywhere in the deep sea are preservation patterns that differ between the major ocean basins. Today, carbonate-rich sediments tend to accumulate in the Atlantic Ocean, while more carbonate-poor sediments are generally found at comparable water depths in the Indian and Pacific Oceans. This modern pattern is largely the product of the ocean's thermohaline circulation and has been termed 'basin-to-basin fractionation'. In the Atlantic, deep and bottom waters tend to be produced at high latitudes because cold temperatures and high sea surface salinities lead to the formation of dense water masses that sink and spread at depth. These young, relatively well oxygenated and $[CO_3^{2-}]$-enriched waters tend to depress the depth of the saturation horizon and allow carbonate to accumulate over much of the Atlantic basin, as manifested by a deep lysocline and CCD. In contrast, neither the Indian nor Pacific Oceans today experience surface conditions that allow deep or bottom waters to form; water masses at depth in these basins largely originate in the Atlantic sector as part of what is sometimes described as the ocean's conveyor belt circulation, with a general upwelling of waters from depth balancing the formation and sinking of deep waters in the Atlantic source areas. Since deep and bottom waters in the Indian and Pacific Oceans are further removed from their modern source areas in the Atlantic, they tend to be CO_2-enriched and $[CO_3^{2-}]$-depleted because of their greater age and the cumulative effects of organic matter remineralization along their flow path. In particular, the *in situ* decrease in $[CO_3^{2-}]$ concentration leads to an increase in undersaturation of the water masses and a progressive shoaling of the saturation horizon (**Figure 3**). Thus, Indian and Pacific deep waters are generally more corrosive to the biogenic carbonate phases than Atlantic waters at comparable depth, the lysocline and CCD are shallower, and a smaller area of the seafloor experiences conditions suitable for carbonate preservation and accumulation. This pronounced modern pattern of basin-to-basin fractionation is illustrated by the fact that roughly 65% of the present Atlantic seafloor is covered by carbonate ooze, while only 54% of the Indian Ocean floor and 36% of the Pacific Ocean floor share that distinction. Naturally, if thermohaline circulation patterns have changed in the past, then carbonate preservation and accumulation patterns will change accordingly. The mapping and reconstruction of such trends has emerged as a powerful paleoceanographic tool.

Temporal Changes in Carbonate Accumulation and Preservation

The patterns of carbonate accumulation and preservation in the deep sea contain important information about the chemistry and fertility of ancient oceans. Numerous studies have now shown that variations in the carbonate system have occurred on a variety of timescales, both within and between ocean basins. On a local or even regional scale, such variations can often be used as a correlation tool. This has come to be known as 'preservation stratigraphy'.

A number of criteria have commonly been used as indicators of the intensity of carbonate dissolution in deep-sea sediments. Variations in the measured carbonate content of sediments are commonly used to correlate between cores in a region, but are difficult to interpret strictly in terms of dissolution and changing deep-water chemistry. This is because the weight percent carbonate content of a sample can also be affected by changing carbonate input (i.e. surface production) and by dilution from non-carbonate sources. More useful are indices based on some direct measure of preservation state, such as the percentage of foraminiferal fragments in a sample relative to whole shells (**Figure 5**). However, while clearly recording dissolution, preservation-based indices can also be affected by other factors, including ecologic changes that may introduce variable proportions of solution-susceptible species into a region over time.

Because carbonate dissolution is a depth-dependent process, it is best studied where existing seafloor topography allows for sampling of sediments over a broad depth range. Given this sampling strategy, one way to circumvent the problems of using measured carbonate content and other relative dissolution indices (e.g. fragmentation) is to calculate carbonate accumulation histories for the individual sampling locations and examine depth-dependent differences in accumulation rates and patterns. To do so requires an accurate knowledge of sedimentation rates (e.g. cm per thousand years) and measurements of sediment bulk density (in $g\,cm^{-3}$), in addition to the data on carbonate content. The product of these three measures yields a mass accumulation rate for the carbonate component expressed in g per cm^2 per thousand years. Differences in accumulation between depth-distributed sites can provide insights into dissolution gradients and carbonate loss.

As the relative importance of calcium supply from weathering and carbonate production vary through time, the depth of the CCD must adjust to control dissolution and to keep calcium levels in balance. Studies of CCD behavior during the Cenozoic (**Figure 6**) have generally shown that CCD fluctuations were similar in the various ocean basins and were likely to have been driven by a global mechanism, such as a change in sea level and/or hypsometry of the ocean basins or a change in supply of calcium to the oceans. There are, however, clear ocean-to-ocean differences in this general pattern that are likely to have been the result of changes in regional productivity and the interbasinal exchange of deep and surface waters. By examining such differences, estimates of past circulation and of the relative differences in carbonate productivity in different regions can be determined from regional offsets in the depth of the CCD.

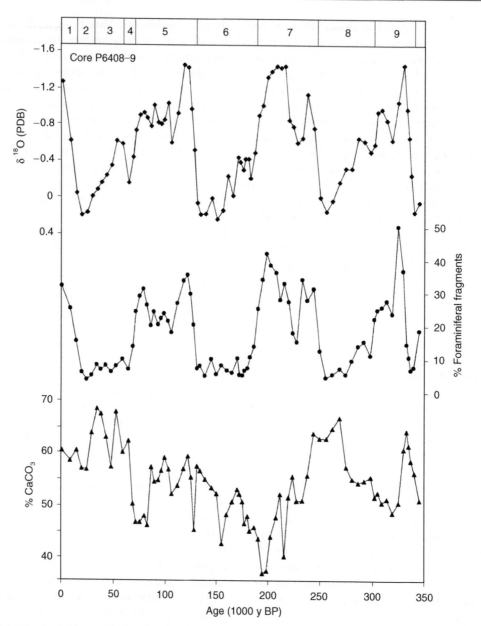

Figure 5 Measurements of foraminiferal fragmentation and calcium carbonate content (weight-%) spanning the last 350 000 years in Caribbean sediment core P6408-9. Stratigraphy and age control come from the oxygen isotope ($\delta^{18}O$) record shown at the top of the figure; odd-numbered stages are warm, interglacial intervals and even-numbered stages indicate cold, glacial climates with greatly expanded Northern Hemisphere ice cover. Variations in the ratio of foraminiferal fragments to whole shells can be directly related to the intensity of carbonate dissolution on the seafloor. Greatly increased preservation (i.e. decreased numbers of fragments)during cold, glacial stages indicates reductions in the chemical corrosivity of deep Caribbean waters in response to climate and ocean circulation changes. Note that variations in carbonate content at this location are not as clearly linked to climate-induced changes in deep-water chemistry as the fragmentation record. This is because the carbonate content of the sediments can also be affected by carbonate productivity at the surface and by dilution on the seafloor by noncarbonate sediment types. (Unpublished data from L. Peterson.)

Seafloor Diagenesis

With time and burial, carbonate oozes undergo a progressive sequence of diagenesis and are transformed first to chalk and then to limestone through a combination of gravitational compaction, dissolution, reprecipitation, and recrystallization. Porosity is reduced from about 70% in typical unconsolidated carbonate oozes to roughly 10% in cemented limestones, while overall volume decreases by about one-third. Drilling results have shown that the transformation from ooze to chalk typically

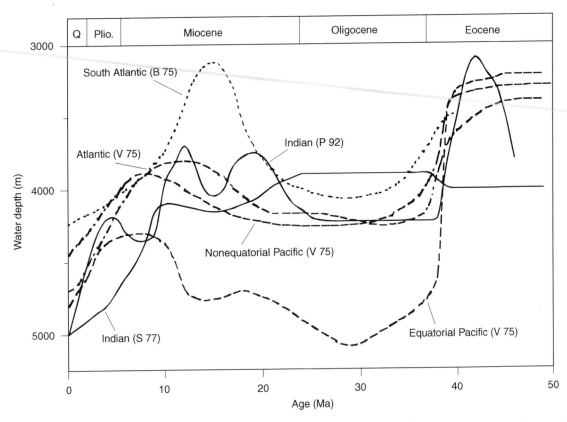

Figure 6 Compilation of reconstructed variations in the depth of the CCD from selected studies covering the last 50 million years for different oceanic regions. The overall similarity of the CCD behavior between regions suggests a common forcing mechanism, such as global sea level or a long-term change in the supply of calcium to the ocean. Variations between the oceans are probably the result of differences in regional surface productivity and deep circulation patterns. Cited CCD studies include: V75, van Andel (1975); B75, Berger and Roth (1975); S77, Sclater *et al.* (1977); P92, Peterson *et al.* (1992).

occurs within a few hundred meters of burial, while limestones are produced by further cementation under about 1 km of burial. Although the transformation of ooze to chalk to limestone is the expected diagenetic sequence, smaller scale reversals in lithification are often observed. Such reversals in pattern have led to the concept of diagenetic potential, which simply states that different sediments will take different lengths of time to reach equal stages of lithification depending upon the original character of the deposited sediment. Such factors as the original proportions of coccoliths to foraminifera (affecting grain size), the amount of dissolution experienced before burial, sedimentation rates, and numerous other subtle factors can influence the diagenetic potential of a carbonate sediment. To the extent that these factors reflect original oceanographic conditions, the sub-bottom acoustic reflectors that result from changing lithification state and diagenetic potential preserve a history of paleoceanographic events that can often be traced across large regions within ocean basins.

See also

Acoustics in Marine Sediments.

Further Reading

Archer DE (1996) An atlas of the distribution of calcium carbonate in sediments of the deep sea. *Global Biogeochemical Cycles* 10: 159–174.

Arrhenius G (1988) Rate of production, dissolution and accumulation of biogenic solids in the ocean. *Palaeogeography, Palaeoclimatology and Palaeoecology* 67: 1119–1146.

Berger WH (1976) Biogenous deep sea sediments: production, preservation and interpretation. In: Riley JP and Chester R (eds.) *Chemical Oceanography*, vol. 5, pp. 266–388. London: Academic Press.

Berger WH and Roth PH (1975) Oceanic micropaleontology: progress and prospect. *Reviews of Geophysics and Space Physics* 13: 561–585.

Broecker WS and Peng T-H (1982) *Tracers in the Sea.* Palisades, NY: Lamont-Doherty Geological Observatory Press.

Emerson S and Archer DE (1992) Glacial carbonate dissolution cycles and atmospheric pCO_2: a view from the ocean bottom. *Paleoceanography* 7: 319–331.

Jahnke RA, Craven DB, and Gaillard J-F (1994) The influence of organic matter diagenesis on $CaCO_3$ dissolution at the deep-sea floor. *Geochimica Cosmochimica Acta* 58: 2799–2809.

Milliman JD (1993) Production and accumulation of calcium carbonate in the ocean: budget of a nonsteady state. *Global Biogeochemical Cycles* 7: 927–957.

Peterson LC and Prell WL (1985) Carbonate dissolution in recent sediments of the eastern equatorial Indian Ocean: Preservation patterns and carbonate loss above the lysocline. *Marine Geology* 64: 259–290.

Peterson LC, Murray DW, Ehrmann WU, and Hempel P (1992) Cenozoic carbonate accumulation and compensation depth changes in the Indian Ocean. In: Duncan RA, Rea DK, Kidd RB, von Rad U, and Weissel JK (eds.) *Synthesis of Results from Scientific Drilling in the Indian Ocean*, Geophysical Monograph 70, pp. 311–333. Washington, DC: American Geophysical Union.

Schlanger SO and Douglas RG (1974) The pelagic ooze-chalk-limestone transition and its implications for marine stratigraphy. In: Hsü KJ and Jenkyns HC (eds.) *Pelagic Sediments on Land and Under the Sea*, Special Publication of the International Association of Sedimentologists, 1, pp. 117–148. Oxford: Blackwell.

Sclater JG, Abbott D, and Thiede J (1977) Paleobathymetry and sediments of the Indian Ocean. In: Heirtzler JR, Bolli HM, Davies TA, Saunders JB, and Sclater JG (eds.) *Indian Ocean Geology and Biostratigraphy*, pp. 25–60. Washington, DC: American Geophysical Union.

Takahashi T, Broecker WS, Bainbridge AE, and Weiss RF (1980) *Carbonate Chemistry of the Atlantic, Pacific and Indian Oceans: The Results of the GEOSECS Expeditions, 1972–1978*, Lamont-Doherty Geological Observatory Technical Report 1, CU-1-80.

van Andel TH (1975) Mesozoic-Cenozoic calcite compensation depth and the global distribution of calcareous sediments. *Earth and Planetary Science Letters* 26: 187–194.

BENTHIC FORAMINIFERA

A. J. Gooday, Southampton Oceanography Centre, Southampton, UK

Introduction

Foraminifera are enormously successful organisms and a dominant deep-sea life form. These amoeboid protists are characterized by a netlike (granuloreticulate) system of pseudopodia and a life cycle that is often complex but typically involves an alternation of sexual and asexual generations. The most obvious characteristic of foraminifera is the presence of a shell or 'test' that largely encloses the cytoplasmic body and is composed of one or more chambers. In some groups, the test is constructed from foreign particles (e.g., mineral grains, sponge spicules, shells of other foraminifera) stuck together ('agglutinated') by an organic or calcareous/organic cement. In others, it is composed of calcium carbonate (usually calcite, occasionally aragonite) or organic material secreted by the organism itself.

Although the test forms the basis of foraminiferal classification, and is the only structure to survive fossilization, the cell body is equally remarkable and important. It gives rise to the complex, highly mobile, and pervasive network of granuloreticulose pseudopodia. These versatile organelles perform a variety of functions (locomotion, food gathering, test construction, and respiration) that are probably fundamental to the ecological success of foraminifera in marine environments.

As well as being an important component of modern deep-sea communities, foraminifera have an outstandingly good fossil record and are studied intensively by geologists. Much of their research uses knowledge of modern faunas to interpret fossil assemblages. The study of deep-sea benthic foraminifera, therefore, lies at the interface between biology and geology. This articles addresses both these facets.

History of Study

Benthic foraminifera attracted the attention of some pioneer deep-sea biologists in the late 1860s. The monograph of H.B. Brady, published in 1884 and based on material collected in the *Challenger* round-the-world expedition of 1872–76, still underpins our knowledge of the group. Later biological expeditions added to this knowledge. For much of the 1900s, however, the study of deep-sea foraminifera was conducted largely by geologists, notably J.A. Cushman, F.B. Phleger, and their students, who amassed an extensive literature dealing with the taxonomy and distribution of calcareous and other hard-shelled taxa. In recent decades, the emphasis has shifted toward the use of benthic species in paleoceanographic reconstructions. Interest in deep-sea foraminifera has also increased among biologists since the 1970s, stimulated in part by the description of the Komokiacea, a superfamily of delicate, soft-shelled foraminifera, by O.S. Tendal and R.R. Hessler. This exclusively deep-sea taxon is a dominant component of the macrofauna in some abyssal regions.

Morphological and Taxonomic Diversity

Foraminifera are relatively large protists. Their tests range from simple agglutinated spheres a few tens of micrometers in diameter to those of giant tubular species that reach lengths of 10 cm or more. However, most are a few hundred micrometers in size. They exhibit an extraordinary range of morphologies (**Figures 1** and **2**), including spheres, flasks, various types of branched or unbranched tubes, and chambers arranged in linear, biserial, triserial, or coiled (spiral) patterns. In most species, the test has an aperture that assumes a variety of forms and is sometimes associated with a toothlike structure. The komokiaceans display morphologies not traditionally associated with the foraminifera. The test forms a treelike, bushlike, spherical, or lumpish body that consists of a complex system of fine, branching tubules (**Figure 2A–C**).

The foraminifera (variously regarded as a subphylum, class, or order) are highly diverse with around 900 living genera and an estimated 10 000 described living species, in addition to large numbers of fossil taxa. Foraminiferal taxonomy is based very largely on test characteristics. Organic, agglutinated, and different kinds of calcareous wall structure serve to distinguish the main groupings (orders or suborders). At lower taxonomic levels, the nature and position of the aperture and the number, shape, and arrangement of the chambers are important.

Figure 1 Scanning electron micrographs of selected deep-sea foraminifera (maximum dimensions are given in parentheses). (A) *Epistominella exigua*; 4850 m water depth, Porcupine Abyssal Plain, NE Atlantic (190 μm). (B) *Nonionella iridea*; 1345 m depth, Porcupine Seabight, NE Atlantic (110 μm). (C) *Nonionella stella*; 550 m depth, Santa Barbara Basin, California Borderland (220 μm). (D) *Brizalina tumida*; 550 m depth, Santa Barbara Basin, California Borderland (680 μm). (E) *Melonis barleaanum*; 1345 m depth, Porcupine Seabight, NE Atlantic (450 μm). (F) *Hormosina* sp., 4495 m depth, Porcupine Abyssal Plain (1.5 mm). (G) *Pyrgoella* sp.; 4550 m depth, foothills of Mid-Atlantic Ridge (620 μm). (H) *Technitella legumen*; 997–1037 m depth, NW African margin (8 mm). (A)–(E) and (G) have calcareous tests, (F) and (H) have agglutinated tests. (C) and (D), photographs courtesy of Joan Bernhard.

Methodology

Qualitative deep-sea samples for foraminiferal studies are collected using nets (e.g., trawls) that are dragged across the seafloor. Much of the *Challenger* material studied by Brady was collected in this way. Modern quantitative studies, however, require the use of coring devices. The two most popular corers used in the deep sea are the box corer, which obtains a large (e.g., $0.25 \, \text{m}^2$) sample, and the multiple corer, which collects simultaneously a battery of up to 12 smaller cores. The main advantage of the multiple corer is that it obtains the sediment–water interface in a virtually undisturbed condition.

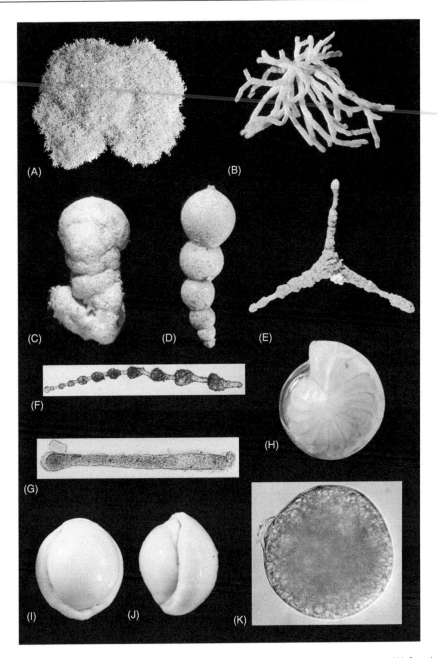

Figure 2 Light micrographs of deep-sea foraminifera (maximum dimensions are given in parentheses). (A) Species of *Lana* in which pad-like test consists of tightly meshed system of fine tubules; 5432 m water depth, Great Meteor East region, NE Atlantic (7.4 mm). (B) *Septuma* sp.; same locality (2 mm). (C) *Edgertonia* mudball; same locality (3.8 mm). (D) *Hormosina globulifera*; 4004 m depth, NW African margin (6.4 mm). (E) *Rhabdammina parabyssorum*; 3392 m depth, Oman margin, NW Arabian Sea (18 mm). (F) *Leptohalysis* sp.; 3400 m depth, Oman margin, NW Arabian Sea (520 μm). (G) Minute species of *Hyperammina*; 3400 m depth, Oman margin, NW Arabian Sea (400 μm). (H) *Lenticularia* sp.; 997–1037 m depth, NW African margin (2.5 mm). (I, J) *Biloculinella* sp.; 4004 m depth, NW African margin (3 mm). (K) Spherical allogromiid; 3400 m depth, Oman margin, NW Arabian Sea (105 μm). Specimens illustrated in (A)–(G) have agglutinated tests, in (H)–(J) calcareous tests and in (K) an organic test. (A)–(C) belong to the superfamily Komokiacea.

Foraminifera are extracted from sieved sediment residues. Studies are often based on dried residues and concern 'total' assemblages (i.e. including both live and dead individuals). To distinguish individuals that were living at the time of collection from dead tests, it is necessary to preserve sediment samples in either alcohol or formalin and then stain them with rose Bengal solution. This colors the cytoplasm red and is most obvious when residues are examined in water. Stained assemblages provide a snapshot of the foraminifera that were living when the samples were collected. Since the live assemblage varies in both time and space, it is also instructive to examine the dead assemblage that provides an averaged view of the foraminiferal fauna. Deep-sea foraminiferal assemblages are typically very diverse and therefore faunal data are often condensed mathematically by using multivariate approaches such as principal components or factor analysis.

The mesh size of the sieve strongly influences the species composition of the foraminiferal assemblage retained. Most deep-sea studies have been based on $> 63\,\mu m$, $125\,\mu m$, $150\,\mu m$, $250\,\mu m$, or even $500\,\mu m$ meshes. In recent years, the use of a fine $63\,\mu m$ mesh has become more prevalent with the realization that some small but important species are not adequately retained by coarser sieves. However, the additional information gained by examining fine fractions must be weighed against the considerable time and effort required to sort foraminifera from them.

Ecology

Abundance and Diversity

Foraminifera typically make up $> 50\%$ of the soft-bottom, deep-sea meiofauna (**Table 1**). They are also often a major component of the macrofauna.

In the central North Pacific, for example, foraminifera (mainly komokiaceans) outnumber all metazoans combined by at least an order of magnitude. A few species are large enough to be easily visible to the unaided eye and constitute part of the megafauna. These include the tubular species *Bathysiphon filiformis*, which is sometimes abundant on continental slopes (**Figure 3**). Some xenophyophores, agglutinated protists that are probably closely related to the foraminifera, are even larger (up to $24\,cm$ maximum dimension!). These giant protists may dominate the megafauna in regions of sloped topography (e.g., seamounts) or high surface productivity. In well-oxygenated areas of the deep-seafloor, foraminiferal assemblages are very species rich, with well over 100 species occurring in relatively small volumes of surface sediment (**Figure 4**). Many are undescribed delicate, soft-shelled forms. There is an urgent need to describe at least some of these species as a step toward estimating global levels of deep-sea species diversity. The common species are often widely distributed, particularly at abyssal depths, although endemic species undoubtedly also occur.

Table 1 The percentage contribution of foraminifera to the deep-sea meiofauna at sites where bottom water is well oxygenated

Area	Depth (m)	Percentage of foraminifera	Number of samples
NW Atlantic			
Off North Carolina	500–2500	11.0–90.4	14
Off North Carolina	400–4000	7.6–85.9	28
Off Martha's Vineyard	146–567	3.4–10.6	4
NE Atlantic			
Porcupine Seabight	1345	47.0–59.2	8
Porcupine Abyssal Plain[a]	4850	61.8–76.3	3
Madeira Abyssal Plain[a]	4950	61.4–76.1	3
Cape Verde Abyssal Plain[a]	4550	70.2	1
Off Mauretania	250–4250	4–27	26
46°N, 16–17°W	4000–4800	0.5–8.3	9
Indian Ocean			
NW Arabian Sea[b]	3350	54.4	1
Pacific			
Western Pacific	2000–6000	36.0–69.3	11
Central North Pacific	5821–5874	49.5	2
Arctic	1000–2600	14.5–84.1	74
Southern Ocean	1661–1680	2.2–23.7	2

[a] Data from Gooday AJ (1996) Epifaunal and shallow infaunal foraminiferal communities at three abyssal NE Atlantic sites subject to differing phytodetritus input regimes. *Deep-Sea Research I* 43: 1395–1421.
[b] Data from Gooday AJ, Bernhard JM, Levin LA and Suhr SB (2000) Foraminifera in the Arabian Sea oxygen minimum zone and other oxygen-deficient settings: taxonomic composition, diversity, and relation to metazoan faunas. *Deep-Sea Research II* 47: 25–54.
Based on Gooday AJ (1986) Meiofaunal foraminiferans from the bathyal Porcupine Seabight (northeast Atlantic): size structure, standing stock, taxonomic composition, species diversity and vertical distribution in the sediment. *Deep-Sea Research* 35: 1345–1373; with permission from Elsevier Science.

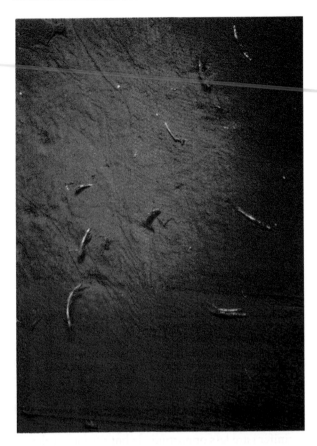

Figure 3 *Bathysiphon filiformis*, a large tubular agglutinated foraminifer, photographed from the Johnson Sealink submersible on the North Carolina continental slope (850 m water depth). The tubes reach a maximum length of about 10 cm. (Photograph courtesy of Lisa Levin.)

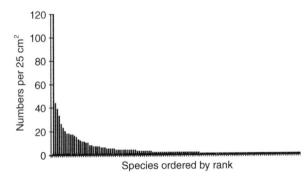

Figure 4 Deep-sea foraminiferal diversity: all species from a single multiple corer sample collected at the Porcupine Abyssal Plain, NE Atlantic (4850 m water depth), ranked by abundance. Each bar represents one 'live' (rose Bengal-stained) species. The sample was 25.5 cm² surface area, 0–1 cm depth, and sieved on a 63 μm mesh sieve. It contained 705 'live' specimens and 130 species.

Foraminifera are also a dominant constituent of deep-sea hard-substrate communities. Dense populations encrust the surfaces of manganese nodules as well as experimental settlement plates deployed on the sea floor for periods of months. They include various undescribed matlike taxa and branched tubular forms, as well as a variety of small coiled agglutinated species (many in the superfamily Trochamminacea), and calcareous forms.

Role in Benthic Communities

The abundance of foraminifera suggests that they play an important ecological role in deep-sea communities, although many aspects of this role remain poorly understood. One of the defining features of these protists, their highly mobile and pervasive pseudopodial net, enables them to gather food particles very efficiently. As a group, foraminifera exhibit a wide variety of trophic mechanisms (e.g., suspension feeding, deposit feeding, parasitism, symbiosis) and diets (herbivory, carnivory, detritus feeding, use of dissolved organic matter). Many deep-sea species appear to feed at a low trophic level on organic detritus, sediment particles, and bacteria. Foraminifera are prey, in turn, for specialist deep-sea predators (scaphopod mollusks and certain asellote isopods), and also ingested (probably incidentally) in large numbers by surface deposit feeders such as holothurians. They may therefore provide a link between lower and higher levels of deep-sea food webs.

Some deep-sea foraminifera exhibit opportunistic characteristics – rapid reproduction and population growth responses to episodic food inputs. Well-known examples are *Epistominella exigua*, *Alabaminella weddellensis* and *Eponides pusillus*. These small (generally <200 μm), calcareous species feed on fresh algal detritus ('phytodetritus') that sinks through the water column to the deep-ocean floor after the spring bloom (a seasonal burst of phytoplankton primary production that occurs most strongly in temperate latitudes). Utilizing energy from this labile food source, they reproduce rapidly to build up large populations that then decline when their ephemeral food source has been consumed. Moreover, certain large foraminifera can reduce their metabolism or consume cytoplasmic reserves when food is scarce, and then rapidly increase their metabolic rate when food again becomes available. These characteristics, together with the sheer abundance of foraminifera, suggest that their role in the cycling of organic carbon on the deep-seafloor is very significant.

The tests of large foraminifera are an important source of environmental heterogeneity in the deep sea, providing habitats and attachment substrates for other foraminifera and metazoans. Mobile infaunal species bioturbate the sediment as they move through it. Conversely, the pseudopodial systems of foraminifera may help to bind together and stabilize deep-sea sediments, although this has not yet been clearly demonstrated.

Microhabitats and Temporal Variability

Like many smaller organisms, foraminifera reside above, on and within deep-sea sediments. Various factors influence their overall distribution pattern within the sediment profile, but food availability and geochemical (redox) gradients are probably the most important. In oligotrophic regions, the flux of organic matter (food) to the seafloor is low and most foraminifera live on or near the sediment surface where food is concentrated. At the other extreme, in eutrophic regions, the high organic-matter flux causes pore water oxygen concentrations to decrease rapidly with depth into the sediment, restricting access to the deeper layers to those species that can tolerate low oxygen levels. Foraminifera penetrate most deeply into the sediment where organic inputs are of intermediate intensity and the availability of food and oxygen within the sediment is well balanced.

Underlying these patterns are the distributions of individual species. Foraminifera occupy more or less distinct zones or microenvironments ('microhabitats'). For descriptive purposes, it is useful to recognize a number of different microhabitats: epifaunal and shallow infaunal for species living close to the sediment surface (upper 2 cm); intermediate infaunal for species living between about 1 cm and 4 cm (**Figure 5**); and deep infaunal for species that occur at depths

down to 10 cm or more (**Figure 6**). A few deep-water foraminifera, including the well-known calcareous species *Cibicidoides wuellerstorfi*, occur on hard substrates (e.g., stones) that are raised above the sediment–water interface (elevated epifaunal microhabitat). There is a general relation between test morphotypes and microhabitat preferences. Epifaunal and shallow infaunal species are often trochospiral with large pores opening on the spiral side of the test; infaunal species tend to be planispiral, spherical, or ovate with small, evenly distributed pores. It is important to appreciate that foraminiferal microhabitats are by no means fixed. They may vary between sites and over time and are modified by the burrowing activities of macrofauna. Foraminiferal microhabitats should therefore be regarded as dynamic rather than static. This tendency is most pronounced in shallow-water settings where environmental conditions are more changeable and macrofaunal activity is more intense than in the deep sea.

The microhabitats occupied by species reflect the same factors that constrain the overall distribution patterns of foraminifera within the sediment. Epifaunal and shallow infaunal species cannot tolerate low oxygen concentrations and also require a diet of relatively fresh organic matter. Deep infaunal foraminifera are less opportunistic but are more tolerant of oxygen depletion than are species living close to the sediment–water interface (**Figure 6**). It has been suggested that species of genera such as *Globobulimina* may consume either sulfate-reducing bacteria or labile organic matter released by the metabolic

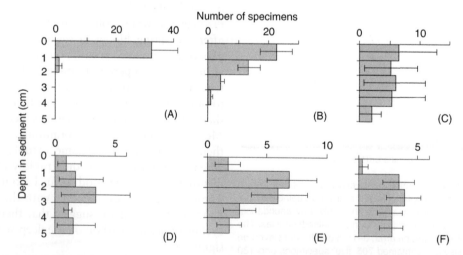

Figure 5 Vertical distribution patterns within the top 5 cm of sediment of common foraminiferal species ('live', rose Bengal-stained specimens) in the Porcupine Seabight, NW Atlantic (51°36′N, 13°00′W; 1345 m water depth). Based on >63 μm sieve fraction. (A) *Ovammina* sp. (mean of 20 samples). (B) *Nonionella iridea* (20 samples). (C) *Leptohalysis* aff. *catenata* (7 samples). (D) *Melonis barleeanum* (9 samples). (E) *Haplophragmoides bradyi* (19 samples). (F) '*Turritellella*' *laevigata* (21 samples). (Amended and reprinted from Gooday AJ (1986) Meiofaunal foraminiferans from the bathyal Porcupine Seabight (northeast Atlantic): size structure, standing stock, taxonomic composition, species diversity and vertical distribution in the sediment. *Deep-Sea Research* 35: 1345–1373; permission from Elsevier Science.)

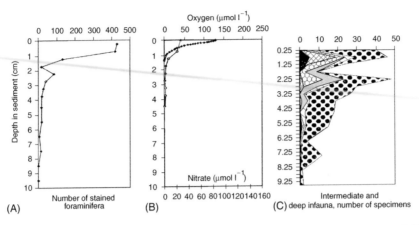

Figure 6 Vertical distribution of (A) total 'live' (rose Bengal-stained) foraminifera), (B) pore water oxygen and nitrate concentrations, and (C) intermediate and deep infaunal foraminiferal species within the top 10 cm of sediment on the north-west African margin (21°28.8′N, 17°57.2′W, 1195 m). All foraminiferal counts based on > 150 μm sieve fraction, standardized to a 34 cm³ volume. Species are indicated as follows: *Pullenia salisburyi* (black), *Melonis barleeanum* (crossed pattern), *Chilostomella oolina* (honeycomb pattern), *Fursenkoina mexicana* (grey), *Globobulimina pyrula* (diagonal lines), *Bulimina marginata* (large dotted pattern). (Adopted and reprinted from Jorissen FJ, Wittling I, Peypouquet JP, Rabouille C and Relexans JC (1998) Live benthic foraminiferal faunas off Cape Blanc, northwest Africa: community structure and microhabitats. *Deep-Sea Research I* 45: 2157–2158; with permission from Elsevier Science.)

activities of these bacteria. These species move closer to the sediment surface as redox zones shift upward in the sediment under conditions of extreme oxygen depletion. Although deep-infaunal foraminifera must endure a harsh microenvironment, they are exposed to less pressure from predators and competitors than those occupying the more densely populated surface sediments.

Deep-sea foraminifera may undergo temporal fluctuations that reflect cycles of food and oxygen availability. Changes over seasonal timescales in the abundance of species and entire assemblages have been described in continental slope settings (**Figure 7**). These changes are related to fluctuations in pore water oxygen concentrations resulting from episodic (seasonal) organic matter inputs to the seafloor. In some cases, the foraminifera migrate up and down in the sediment, tracking critical oxygen levels or redox fronts. Population fluctuations also occur in abyssal settings where food is a limiting ecological factor. In these cases, foraminiferal population dynamics reflect the seasonal availability of phytodetritus ('food'). As a result of these temporal processes, living foraminifera sampled during one season often provide an incomplete view of the live fauna as a whole.

Environmental Controls on Foraminiferal Distributions

Our understanding of the factors that control the distribution of foraminifera on the deep-ocean floor is very incomplete, yet lack of knowledge has not

prevented the development of ideas. It is likely that foraminiferal distribution patterns reflect a combination of influences. The most important first-order factor is calcium carbonate dissolution. Above the carbonate compensation depth (CCD), faunas include calcareous, agglutinated, and allogromiid taxa. Below the CCD, calcareous species are almost entirely absent. At oceanwide or basinwide scales, the organic carbon flux to the seafloor (and its seasonality) and bottom-water hydrography appear to be particularly important, both above and below the CCD.

Studies conducted in the 1950s and 1960s emphasized bathymetry (water depth) as an important controlling factor. However, it soon became apparent that the bathymetric distribution of foraminiferal species beyond the shelf break is not consistent geographically. Analyses of modern assemblages in the North Atlantic, carried out in the 1970s, revealed a much closer correlation between the distribution of foraminiferal species and bottom-water masses. For example, *Cibicidoides wuellerstorfi* was linked to North Atlantic Deep Water (NADW) and *Nuttallides umbonifera* to Antarctic Bottom Water (AABW). At this time, it was difficult to explain how slight physical and chemical differences between water masses could influence foraminiferal distributions. However, recent work in the south-east Atlantic, where hydrographic contrasts are strongly developed, suggests that the distributions of certain foraminiferal species are controlled in part by the lateral advection of water masses. In the case of *N. umboniferus* there is good evidence that the main

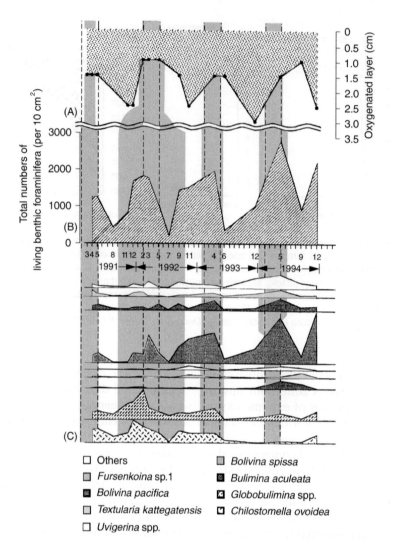

Figure 7 Seasonal changes over a 4-year period (March 1991 to December 1994) in (A) the thickness of the oxygenated layer, (B) the total population density of live benthic foraminifera, and (C) the abundances of the most common species at a 1450 m deep site in Sagami Bay, Japan. (Reprinted from Ohga T and Kitazato H (1997) Seasonal changes in bathyal foraminiferal populations in response to the flux of organic matter (Sagami Bay, Japan). *Terra Nova* 9: 33–37; with permission from Blackwell Science Ltd.)

factor is the degree of undersaturation of the bottom water in calcium carbonate. This abyssal species is found typically in the carbonate-corrosive (and highly oligotrophic) environment between the calcite lysocline and the CCD, a zone that may coincide approximately with AABW. Where water masses are more poorly delineated, as in the Indian and Pacific Oceans, links with faunal distributions are less clear.

During the past 15 years, attention has focused on the impact on foraminiferal ecology of organic matter fluxes to the seafloor. The abundance of dead foraminiferal shells >150 μm in size correlates well with flux values. There is also compelling evidence that the distributions of species and species associations are linked to flux intensity. Infaunal species, such as *Melonis barleeanum*, *Uvigerina peregrina*,

Chilostomella ovoidea and *Globobulimina affinis*, predominate in organically enriched areas, e.g. beneath upwelling zones. Epifaunal species such as *Cibicidoides wuellerstorfi* and *Nuttallides umbonifera* are common in oligotrophic areas, e.g. the central oceanic abyss. In addition to flux intensity, the degree of seasonality of the food supply (i.e., whether it is pulsed or continuous) is a significant factor. *Epistominella exigua*, one of the opportunists that exploit phytodetritus, occurs in relatively oligotrophic areas where phytodetritus is deposited seasonally.

Recent analysis of a large dataset relating the relative abundance of 'live' (stained) foraminiferal assemblages in the north-east Atlantic and Arctic Oceans to flux rates to the seafloor has provided a quantitative framework for these observations. Although species are associated with a wide flux range, this range diminishes as a species become relatively more abundant and conditions become increasingly optimum for it. When dominant occurrences (i.e., where species represent a high percentages of the fauna) are plotted against flux and water depth, species fall into fields bounded by particular flux and depth values (**Figure 8**). Despite a good deal of overlap, it is possible to distinguish a series of dominant species that succeed each other bathymetrically on relatively eutrophic continental slopes and other species that dominate on the more oligotrophic abyssal plains.

Other environmental attributes undoubtedly modify the species composition of foraminiferal assemblages in the deep sea. Agglutinated species with tubular or spherical tests are found in areas where the seafloor is periodically disturbed by strong currents capable of eroding sediments. Forms projecting into the water column may be abundant where steady flow rates convey a continuous supply of suspended food particles. Other species associations may be linked to sedimentary characteristics.

Low-Oxygen Environments

Oxygen availability is a particularly important ecological parameter. Since oxygen is consumed during the degradation of organic matter, concentrations of oxygen in bottom water and sediment pore water are inversely related to the organic flux derived from surface production. In the deep sea, persistent oxygen depletion ($O_2 < 1\,ml\,l^{-1}$) occurs at bathyal depths ($<1000\,m$) in basins (e.g., on the California Borderland) where circulation is restricted by a sill and in areas where high primary productivity resulting from the upwelling of nutrient-rich water leads to the development of an oxygen minimum zone (OMZ; e.g., north-west Arabian Sea and the Peru margin). Subsurface sediments also represent an oxygen-limited setting, although oxygen penetration is generally greater in oligotrophic deep-sea sediments than in fine-grained sediments on continental shelves.

On the whole, foraminifera exhibit greater tolerance of oxygen deficiency than most metazoan taxa, although the degree of tolerance varies among species. Oxygen probably only becomes an important limiting factor for foraminifera at concentrations well below $1\,ml\,l^{-1}$. Some species are abundant at levels of $0.1\,ml\,l^{-1}$ or less. A few apparently live in permanently anoxic sediments, although anoxia sooner or later results in death when accompanied by high concentrations of hydrogen sulfide. Oxygen-deficient areas are characterized by high foraminiferal densities but low, sometimes very low (<10), species numbers. This assemblage structure (high dominance, low species richness) arises because (i) low oxygen

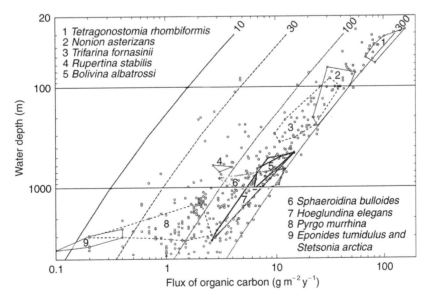

Figure 8 Dominant 'live' (rose Bengal-stained) occurrences of foraminiferal species in relation to water depth and flux or organic carbon to seafloor in the North Atlantic from the Guinea Basin to the Arctic Ocean. Each open circle corresponds to a data point. The polygonal areas indicate the combination of water depth and flux conditions under which nine different species are a dominant faunal component. The diagonal lines indicate levels of primary production (10, 30, 100, 300 g m^{-2} y^{-1}) that result in observed flux rates. Based on $>250\,\mu m$ sieve fraction plus 63–250 μm fraction from Guinea Basin and Arctic Ocean. (Reprinted from Altenbach AV, Pflaumann U, Schiebel R et al. (1999) Scaling percentages and distribution patterns of benthic foraminifera with flux rates of organic carbon. *Journal of Foraminiferal Research* 29: 173–185; with permission from The Cushman Foundation.)

concentration acts as a filter that excludes non-tolerant species and (ii) the tolerant species that do survive are able to flourish because food is abundant and predation is reduced. Utrastructural studies of some species have revealed features, e.g., bacterial symbionts and unusually high abundances of peroxisomes, that may be adaptations to extreme oxygen depletion. In addition, mitachondria-laden pseudopodia have the potential to extend into overlying sediment layers where some oxygen may be present.

Many low-oxygen-tolerant foraminifera belong to the Orders Rotaliida and Buliminida. They often have thin-walled, calcareous tests with either flattened, elongate biserial or triserial morphologies (e.g., *Bolivina, Bulimina, Globobulimina, Fursenkoina, Loxotomum, Uvigerina*) or planispiral/lenticular morphologies (e.g., *Cassidulina, Chilostomella, Epistominella, Loxotomum, Nonion, Nonionella*). Some agglutinated foraminifera, e.g., *Textularia,*

Trochammina (both multilocular), *Bathysiphon*, and *Psammosphaera* (both unilocular), are also abundant. However, miliolids, allogromiids, and other soft-shelled foraminifera are generally rare in low-oxygen environments. It is important to note that no foraminiferal taxon is currently known to be confined entirely to oxygen-depleted environments.

Deep-Sea Foraminifera in Paleo-Oceanography

Geologists require proxy indicators of important environmental variables in order to reconstruct ancient oceans. Benthic foraminifera provide good proxies for seafloor parameters because they are widely distributed, highly sensitive to environmental conditions, and abundant in Cenozoic and Cretaceous deep-sea sediments (note that deep-sea

Table 2 Benthic foraminiferal proxies or indicators (both faunal and chemical) useful in paleo-oceanographic reconstruction

Environmental parameter/property	Proxy or indicator	Remarks
Water depth	Bathymetric ranges of abundant species in modern oceans	Depth zonation largely local although broad distinction between shelf, slope and abyssal depth zones possible
Distribution of bottom water masses	Characteristic associations of epifaunal species	Relations between species and water masses may reflect lateral advection
Carbonate corrosiveness of bottom water	Abundance of *Nuttallides umbonifera*	Corrosive bottom water often broadly corresponds to Antarctic Bottom Water
Deep-ocean thermohaline circulation	Cd/Ca ratios and δ^{13}C values for calcareous tests	Proxies reflect 'age' of bottom watermasses; i.e., period of time elapsed since formation at ocean surface
Oxygen-deficient bottom-water and pore water	Characteristic species associations; high-dominance, low-diversity assemblages	Species not consistently associated with particular range of oxygen concentrations and also found in high-productivity areas
Primary productivity	Abundance of foraminiferal tests > 150 μm	Transfer function links productivity to test abundance (corrected for differences in sedimentation rates between sites) in oxygenated sediments
Organic matter flux to seafloor	(i) Assemblages of high productivity taxa (e.g. *Globobulimina, Melonis barleeanum*) (ii) Ratio between infaunal and epifaunal morphotypes (iii) Ratio between planktonic and benthic tests	Assemblages indicate high organic matter flux to seafloor, with or without corresponding decrease in oxygen concentrations
Seasonality in organic matter flux	Relative abundance of 'phytodetritus species'	Reflects seasonally pulsed inputs of labile organic matter to seafloor
Methane release	Large decrease (2–3‰) in δ^{13}C values of benthic and planktonic tests	Inferred sudden release of ^{12}C enriched methane from clathrate deposits following temperature rise

sediments older than the middle Jurassic age have been destroyed by subduction, except where preserved in ophiolite complexes).

Foraminiferal faunas, and the chemical tracers preserved in the tests of calcitic species, can be used to reconstruct a variety of paleoenvironmental parameters and attributes. The main emphasis has been on organic matter fluxes and bottom-water/pore water oxygen concentrations (inversely related parameters), the distribution of bottom-water masses, and the development of thermohaline circulation (**Table 2**). Modern deep-sea faunas became established during the Middle Miocene (10–15 million years ago), and these assemblages can often be interpreted in terms of modern analogues. This approach is difficult or impossible to apply to sediments from the Cretaceous and earlier Cenozoic, which contain many foraminiferal species that are now extinct. In these cases, it can be useful to work with test morphotypes (e.g., trochospiral, cylindrical, biserial/triserial) rather than species. The relative abundance of infaunal morphotypes, for example, has been used as an index of bottom-water oxygenation or relative intensities of organic matter inputs. The trace element (e.g., cadmium) content and stable isotope (δ^{13}C; i.e., the deviation from a standard ^{12}C : ^{12}C ratio) chemistry of the calcium carbonate shells of benthic foraminifera provide powerful tools for making paleo-oceanographic reconstructions, particularly during the climatically unstable Quaternary period.

The cadmium/calcium ratio is a proxy for the nutrient (phosphate) content of sea water that reflects abyssal circulation patterns. Carbon isotope ratios also reflect deep-ocean circulation and the strength of organic matter fluxes to the seafloor.

It is important to appreciate that the accuracy with which fossil foraminifera can be used to reconstruct ancient deep-sea environments is often limited. These limitations reflect the complexities of deep-sea foraminiferal biology, many aspects of which remain poorly understood. Moreover, simple relationships between the composition of foraminiferal assemblages and environmental variables are elusive, and it is often difficult to identify faunal characteristics that can be used as precise proxies for paleo-oceanographic parameters. For example, geologists often wish to establish paleobathymetry. However, the bathymetric distributions of foraminiferal species are inconsistent and depend largely on the organic flux to the seafloor, which decreases with increasing depth (**Figure 8**) and is strongly influenced by surface productivity. Thus, foraminifera can be used only to discriminate in a general way between shelf, slope, and abyssal faunas, but not to estimate precise paleodepths. Oxygen concentrations and organic matter inputs are particularly problematic. Certain species and morphotypes dominate in low-oxygen habitats that also are usually characterized by high organic loadings. However, the same foraminifera may occur in organically enriched settings where oxygen levels are

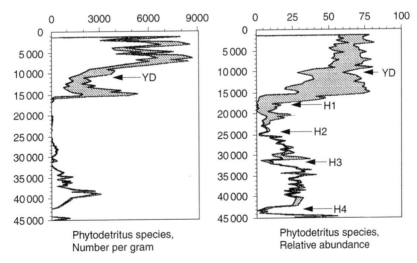

Phytodetritus species,
Number per gram

Phytodetritus species,
Relative abundance

Figure 9 (A) Absolute (specimens per gram of dry sediment) and (B) relative (percentage) abundances of *Alabaminella weddellensis* and *Epistominella exigua* (>63 μm fraction) in a long-sediment core from the North Atlantic (50°41.3′N, 21°51.9′W, 3547 m water depth). In modern oceans, these two species respond to pulsed inputs of organic matter ('phytodetritus') derived from surface primary production. Note that they increased in abundance around 15 000 years ago, corresponding to the main Northern Hemisphere deglaciation and the retreat of the Polar Front. Short period climatic fluctuations (YD = Younger Dryas; H1–4 = Heinrich events, periods of very high meltwater production) are also evident in the record of these two species. (Reprinted from Thomas E, Booth L, Maslin M and Shackleton NJ (1995). Northeast Atlantic benthic foraminifera during the last 45 000 years: change in productivity seen from the bottom up. *Paleoceanography.* 10: 545–562; with permission from the American Geophysical Union.)

not severely depressed, making it difficult for paleo-oceanographers to disentangle the influence of these two variables. Finally, biological factors such as microhabitat preferences and the exploitation of phytodetrital aggregates ('floc') influence the stable isotope chemistry of foraminiferal tests.

There are many examples of the use of benthic foraminiferal faunas to interpret the geological history of the oceans. Only one is given here. Cores collected at 50°41′N, 21°52′W (3547 m water depth) and 58°37′N, 19°26′W (1756 m water depth) were used by E. Thomas and colleagues to study changes in the North Atlantic over the past 45 000 years. The cores yielded fossil specimens of two foraminiferal species, *Epistominella exigua* and *Alabaminella weddellensis*, both of which are associated with seasonal inputs of organic matter (phytodetritus) in modern oceans. In the core from 51°N, these 'phytodetritus species' were uncommon during the last glacial maximum but increased sharply in absolute and relative abundance during the period of deglaciation 15 000–16 000 years ago (**Figure 9**). At the same time there was a decrease in the abundance of *Neogloboquadrina pachyderma*, a planktonic foraminifer found in polar regions, and an increase in the abundance of *Globigerina bulloides*, a planktonic species characteristic of warmer water. These changes were interpreted as follows. Surface primary productivity was low at high latitudes in the glacial North Atlantic, but was much higher to the south of the Polar Front. At the end of the glacial period, the ice sheet shrank and the Polar Front retreated northwards. The 51°N site was now overlain by more productive surface water characterized by a strong spring bloom and a seasonal flux of phytodetritus to the seafloor. This episodic food source favored opportunistic species, particularly *E. exigua* and *A. weddellensis*, which became much more abundant both in absolute terms and as a proportion of the entire foraminiferal assemblage.

Conclusions

Benthic foraminifera are a major component of deep-sea communities, play an important role in ecosystem functioning and biogeochemical cycling, and are enormously diverse in terms of species numbers and test morphology. These testate (shell-bearing) protists are also the most abundant benthic organisms preserved in the deep-sea fossil record and provide powerful tools for making paleo-oceanographic reconstructions. Our understanding of their biology has advanced considerably during the last two decades, although much remains to be learnt.

Further Reading

Fischer G and Wefer G (1999) *Use of Proxies in Paleoceanography: Examples from the South Atlantic.* Berlin: Springer-Verlag.

Gooday AJ, Levin LA, Linke P, and Heeger T (1992) The role of benthic foraminifera in deep-sea food webs and carbon cycling. In: Rowe GT and Pariente V (eds.) *Deep-Sea Food Chains and the Global Carbon Cycle*, pp. 63–91. Dordrecht: Kluwer Academic.

Jones RW (1994) *The Challenger Foraminifera.* Oxford: Oxford University Press.

Loeblich AR and Tappan H (1987) *Foraminiferal Genera and their Classification*, vols 1, 2. New York: Van Nostrand Reinhold.

Murray JW (1991) *Ecology and Palaeoecology of Benthic Foraminifera.* New York: Wiley; Harlow: Longman Scientific and Technical.

SenGupta BK (ed.) (1999) *Modern Foraminifera.* Dordrecht: Kluwer Academic.

Tendal OS and Hessler RR (1977) An introduction to the biology and systematics of Komokiacea. *Galathea Report* 14: 165–194, plates 9–26.

Van der Zwan GJ, Duijnstee IAP, den Dulk M, *et al.* (1999) Benthic foraminifers:: proxies or problems? A review of paleoecological concepts. *Earth Sciences Reviews* 46: 213–236.

COCCOLITHOPHORES

T. Tyrrell, National Oceanography Centre, Southampton, UK
J. R. Young, The Natural History Museum, London, UK

Introduction

Coccolithophores (**Figure 1**) are a group of marine phytoplankton belonging to the division Haptophyta. Like the other free-floating marine plants (phytoplankton), the coccolithophores are microscopic (they range in size between about 0.003 and 0.040 mm diameter) single-celled organisms which obtain their energy from sunlight. They are typically spherical in shape. They are distinguished from other phytoplankton by their construction of calcium carbonate ($CaCO_3$) plates (called coccoliths) with which they surround their cells. While not quite the only phytoplankton to use $CaCO_3$ (there are also some calcareous dinoflagellates), they are by far the most numerous; indeed they are one of the most abundant of phytoplankton groups, comprising in the order of 10% of total global phytoplankton biomass.

The first recorded observations of coccoliths were made in 1836 by Christian Gottfried Ehrenberg, a founding figure in micropaleontology. The name 'coccoliths' (Greek for 'seed-stones') was coined by Thomas Henry Huxley (famous as 'Darwin's bulldog') in 1857 as he studied marine sediment samples. Both Ehrenberg and Huxley attributed coccoliths to an inorganic origin. This was soon challenged by Henry Clifton Sorby and George Charles Wallich who inferred from the complexity of coccoliths that they must be of biological origin, and supported this with observations of groups of coccoliths aggregated into empty spheres.

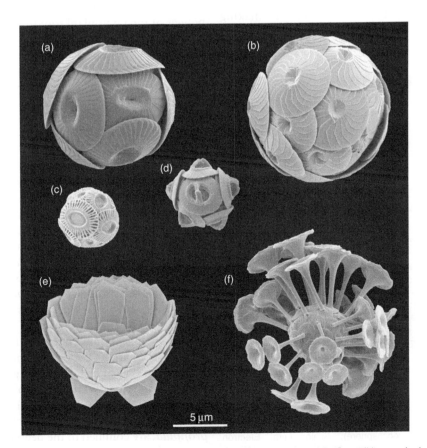

Figure 1 Electron microscope images of some major coccolithophore species: (a) *Coccolithus pelagicus*, (b) *Calcidiscus quadriperforatus*, (c) *Emiliania huxleyi*, (d) *Gephyrocapsa oceanica*, (e) *Florisphaera profunda*, (f) *Discosphaera tubifera*.

Species and Distribution

Approximately 200 species of coccolithophore have been formally described, separated into 65 genera. However, the true number of authentic modern coccolithophore species is rather unclear, for a couple of reasons. First, it is now realized that pairs of species, previously thought to be distinct and rather unrelated, are actually different life-cycle stages of the same species; coccolithophores typically have life cycles in which the haploid (single set of chromosomes, as in sex cells) and diploid (double set of chromosomes) phases can form different coccolith types: 'heterococcoliths' during the diploid life stage, and 'holococcoliths' during the haploid life stage. This type of life cycle has long been known from classic studies of laboratory cultures. It has only recently been appreciated that it is a very widespread pattern, as a result of observations of combination coccospheres representing the transition between the two life-cycle phases, that is to say possessing half a covering of heterococcoliths and half a covering of holococcoliths (**Figure 2**). Fifty of the 200 described coccolithophores are taxa known only from their holococcolith-producing phase and so may prove to be part of the life cycle of a heterococcolith-producing species. The second factor making diversity estimates difficult is that recent research combining studies of fine-scale morphology, biogeography, and molecular genetics has suggested that many described species are actually clusters of a few closely related, but genetically distinct, species. Indeed, as a result of such studies, numerous additional morphotypes have been recognized and await formal description.

Coccolithophores occur widely throughout the world's oceans, with the exception of the very-high-latitude polar oceans. Most individual species have more restricted biogeographical ranges than the range of the group as a whole, but still typically have interoceanic distributions. Unlike diatoms (the other major group of phytoplankton that make hard mineral shields), they are absent from almost all freshwater rivers and lakes. They occur in the brackish (more saline than freshwater but less saline than seawater) Black Sea, but not in the brackish Baltic Sea. In contrast once more to diatoms, coccolithophores are almost exclusively planktonic. There are very few bottom-dwelling species, even at shallow depths experiencing adequate light levels. Most species today live in warm, nutrient-poor conditions of the subtropical oceanic gyres, where they form a prominent component of the phytoplankton; there are fewer species that inhabit coastal and temperate or subpolar waters.

Emiliania huxleyi (**Figure 1(c)**) is the best-known species, primarily because it forms intense blooms which are clearly visible in satellite images, appearing as pale turquoise swirls in the ocean (**Figure 3**). While *E. huxleyi* frequently dominates phytoplankton counts in seawater samples, at least in terms of numbers of cells, their cells (and therefore also the coccoliths that surround them) are rather small, with the cells about 5 μm across and the coccoliths about 3 μm long. No other coccolithophore species regularly forms blooms, although occasional blooms of other species, for instance *Gephyrocapsa oceanica* and *Coccolithus pelagicus*, have been recorded. Many other species, for example *Calcidiscus quadriperforatus* and *Umbilicosphaera sibogae*, are most successful in low-productivity waters but do not bloom there. Although these species are almost always much less numerous than *E. huxleyi* in water samples, they are on the other hand also significantly larger, with typical cell diameters greater than 10 μm and correspondingly larger coccoliths. Most species of coccolithophores are adapted to life in the surface mixed layer, but some species, such as *Florisphaera profunda*, are confined to the deep photic zone where they make an important contribution to the 'shade flora'.

Figure 2 A combination coccosphere, upper half (inner layer) heterococcoliths, lower half (outer layer) holococcoliths, of the species *Calcidiscus quadriperforatus* (the two stages were previously regarded as two separate species – *Calcidiscus leptoporus* and *Syracolithus quadriperforatus* – prior to discovery of this combination coccosphere). Scale = 1 μm. Electron microscope image provided by Markus Geisen (Alfred-Wegener-Institute, Germany).

Figure 3 *SeaWiFS* satellite image from 15 June, 1998 of *E. huxleyi* blooms (the turquoise patches) along the west coast of Norway and to the southwest of Iceland. The perspective is from a point over the Arctic Ocean, looking southward down the North Atlantic. The Greenland ice sheet is visible in the center foreground. Imagery provided with permission by GeoEye and NASA SeaWiFS Project.

In oligotrophic surface waters, typical concentrations of coccolithophores are in the range 5000–50 000 cells per liter. To put this in context, a teaspoonful (5 ml) of typical surface open ocean seawater will contain between 25 and 250 coccolithophore cells. Blooms of *E. huxleyi* have been defined as concentrations exceeding 1 million cells per liter; the densest bloom ever recorded, in a Norwegian fiord, had a concentration of 115 000 000 cells per liter. Blooms of *E. huxleyi* can cover large areas; the largest ever recorded bloom occurred in June 1998 (see **Figure 3**) in the North Atlantic south of Iceland and covered about 1 million km^2, 4 times the area of the United Kingdom.

Coccoliths

Coccolithophores, and the coccolith shields with which they surround themselves, are incredibly small. And yet, despite their small size, coccoliths are elegant and ornate structures, which, if the water chemistry is suitable, are produced reliably with few malformations. This efficient manufacture occurs at a miniature scale: the diameter of an *E. huxleyi* coccolith 'spoke' (**Figure 1(c)**) is of the order 50 nm, considerably smaller than the wavelength range of visible light (400–700 nm). Calcite is mostly transparent to visible light (unsurprisingly, given that coccolithophores are photosynthetic) and the small coccoliths are often at the limit of discrimination, even under high magnification. However, under cross-polarized light, coccoliths produce distinctive patterns which are closely related to their structure. As a result most coccoliths can be accurately identified by light microscopy. However, the details and beauty of coccoliths can only be properly appreciated using electron microscopy (**Figure 1**).

Coccolithophores synthesize different types of coccoliths during different life-cycle stages. Here we concentrate on the heterococcoliths associated with the diploid life stage. These heterococcoliths are formed from crystal units with complex shapes, in contrast to holococcoliths which are constructed out of smaller and simpler crystal constituents. Coccoliths are typically synthesized intracellularly (within a vesicle), probably one at a time, and subsequently extruded to the cell surface. The time taken to form a single coccolith can be less than 1 h for *E. huxleyi*. Coccoliths continue to be produced until a complete coccosphere covering (made up of maybe 20 coccoliths, depending on species) is produced.

Most coccolithophores construct only as many coccoliths as are required in order to provide a complete single layer around their cell. *Emiliania huxleyi* is unusual in that, under certain conditions, it overproduces coccoliths; many more coccoliths are built than are needed to cover the cell. In these conditions, multiple layers of coccoliths accumulate around the *E. huxleyi* cell until the excessive covering eventually becomes unstable and some of the coccoliths slough off to drift free in the water. The large number of unattached coccoliths accompanying an *E. huxleyi* bloom contributes to a great extent to the turbidity of the water and to the perturbations to optics that make the blooms so apparent from space.

Curiously, the functions of coccoliths are still uncertain. It is probable that a major function is to provide some protection from grazing by zooplankton, but many alternative hypotheses have also been advanced. For instance, the coccoliths may increase the rate of sinking of the cells through the water (and therefore also enhance the rate at which nutrient-containing water flows past the cell surface) or they may provide protection against the entry of viruses or bacteria to the cell. At one time it was thought that coccoliths might provide protection against very high light intensities, which could explain the resistance to photoinhibition apparent in *E. huxleyi*, but various experimental results make this explanation unlikely. One species, *F. profunda*, a member of the deeper 'shade flora', orients its coccoliths in such a way that they conceivably act as a light-focusing apparatus maximizing photon capture in the darker waters it inhabits (**Figure 1(e)**). Some species produce trumpet-like protrusions from each coccolith (**Figure 1(f)**), again for an unknown purpose. Currently there is a paucity of hard data with which to discriminate between the various hypotheses for coccolith function, and the diversity of coccolith morphology makes it likely that they have been adapted to perform a range of functions.

Life Cycle

Many details are still obscure, and data are only available from a limited number of species, but it appears that most coccolithophores alternate between fully armored (heterococcolith-covered) diploid life stages and less-well-armoured (either holococcolith-covered or else naked) haploid phases. Both phases are capable of indefinite asexual reproduction, which is rather unusual among protists. That sexual reproduction also occurs fairly frequently is evidenced by the observation of significant genetic diversity within coccolithophore blooms. Bloom populations do not consist of just one clone (just one genetic variant of the organism). Coccolithophore gametes (haploid stages) are radically different from those of larger (multicellular) organisms in the sense of being equipped for an independent existence: they can move about, acquire energy (photosynthesize), and divide asexually by binary fission. Naked diploid phases can be induced in cultures, but these may be mutations which are not viable in the wild. There are no confirmed identifications of resting spore or cyst stages in coccolithophores.

Coccolithophores, in common with other phytoplankton, experience only an ephemeral existence. Typical life spans of phytoplankton in nature are measured in days. Comparison of the rate at which $CaCO_3$ is being produced in open ocean waters (as measured by the rate of uptake of isotopically labeled carbon), to the amounts present (the 'standing-stock'), has led to the calculation that the average turnover (replacement) time for $CaCO_3$ averages about 3 days, ranging between a minimum of <1 and a maximum of 7 days at different locations in the Atlantic Ocean. This implies that if a surface-dwelling coccolithophore synthesizes coccoliths on a Monday, the coccoliths are fairly unlikely to still be there on the Friday, either because they have re-dissolved or else because they have sunk down to deeper waters.

The genome of one species, E. huxleyi, has recently been sequenced, but at the time of writing its analysis is at an early stage.

Calcification

Calcification is the synthesis of solid calcium carbonate from dissolved substances, whether passively by spontaneous formation of crystals in a supersaturated solution (inorganic calcification) or actively through the intervention of organisms (biocalcification). The building of coccoliths by coccolithophores is a major fraction of the total biocalcification taking place in seawater. Inorganic calcification is not commonplace or quantitatively significant in the global budget, with the exception of 'whitings' that occur in just a few unusual locations in the world's oceans, such as the Persian Gulf and the Bahamas Banks. The chemical equation for calcification is

$$Ca^{2+} + 2HCO_3^- \Rightarrow CaCO_3 + H_2O + CO_2$$

Heterococcoliths are constructed out of calcite (a form of calcium carbonate; corals by contrast synthesize aragonite, which has the same chemical composition but a different lattice structure). Heterococcolith calcite typically has a very low magnesium content, making coccoliths relatively dissolution-resistant (susceptibility to dissolution increases with increasing magnesium content).

Dissolved inorganic carbon in seawater is comprised of three different components: bicarbonate ions (HCO_3^-), carbonate ions (CO_3^{2-}), and dissolved CO_2 gas ($CO_2(aq)$), of which it appears that bicarbonate or carbonate ions are taken up to provide the carbon source for $CaCO_3$ (coccoliths have a $\delta^{13}C$ isotopic composition that is very different from dissolved CO_2 gas). The exact physiological mechanisms of calcium and carbon assimilation remain to be established. Calcification (coccolith genesis) is stimulated by light but inhibited in most cases by plentiful nutrients. Separate experiments have found that increased rates of calcification in cultures can be induced by starving the cultures of phosphorus, nitrogen, and zinc. Low levels of magnesium also enhance calcification, and high levels inhibit it, but in this case probably because Mg atoms can substitute for Ca atoms in the crystalline lattice and thereby 'poison' the lattice. Calcification shows the opposite response to levels of calcium, unsurprisingly. Progressive depletion of calcium in the growth medium induces progressively less normal (smaller and more malformed) coccoliths. The calcification to photosynthesis (C:P) ratio in nutrient-replete, Ca-replete cultures is often in the vicinity of 1:1 (i.e., more or less equivalent rates of carbon uptake into the two processes). Low levels of iron appear to depress calcification and photosynthesis equally.

Measurements at sea suggest that the total amount of carbon taken up by the whole phytoplankton community to form new $CaCO_3$ is rather small compared to the total amount of carbon taken up to form new organic matter. Both calcification carbon demand and photosynthetic carbon demand have recently been measured on a long transect in the Atlantic Ocean and the ratio of the two was found to average 0.05; or, in other words, for every 20 atoms of carbon taken up by phytoplankton, only one on average was taken up into solid $CaCO_3$.

Ecological Niche

In addition to our lack of knowledge about the exact benefit of a coccosphere, we also have rather little definite knowledge as to the ecological conditions that favor coccolithophore success. There is certainly variation between species, with some being adapted to relatively eutrophic conditions (although diatoms invariably dominate the main spring blooms in temperate waters, as well as the first blooms in nutrient-rich, recently upwelled waters) and some to oligotrophic conditions. Most species are best adapted to living near to the surface, but some others to the darker conditions prevailing in the thermocline. Most species today live in warm, nutrient-poor, open ocean conditions; the highest diversity occurs in subtropical oceanic gyres, whereas lower diversity occurs in coastal and temperate waters.

Much of our knowledge of coccolithophore physiology and ecology comes from studies of *E. huxleyi*, which has attracted more scientific interest than the other coccolithophore species because of its ease of culturing and the visibility of its blooms from space. The ability to map bloom distributions from space provides unique information on the ecology of this species. Blooms of the species *E. huxleyi* occur preferentially in strongly stratified waters experiencing high light levels. Coccolithophore success may be indirectly promoted by exhaustion of silicate, due to exclusion of the more competitive diatoms. By analogy with diatoms, whose success is contingent on silicate availability for their shell building, coccolithophores might be expected to be more successful at high $CaCO_3$ saturation state $\Omega \; (= [CO_3^{2-}][Ca^{2+}]/K_{sp})$, because the value of Ω controls inorganic calcification and dissolution. Such a dependency would render coccolithophores vulnerable to ocean acidification, as discussed further below. It was formerly thought that *E. huxleyi* was particularly successful in phosphate-deficient waters, but a reassessment has suggested that this is not a critical factor. Many coccolithophores are restricted to the warmer parts of the oceans, although this may be coincidental rather than due to a direct temperature effect. *Emiliania huxleyi* is found to grow well at low iron concentrations, in culture experiments.

Biogeochemical Impacts

Coccolithophores assimilate carbon during photosynthesis, leading to similar biogeochemical impacts to other phytoplankton that do not possess mineral shells. They also, however, assimilate carbon into biomass.

Following death, some of the coccolith $CaCO_3$ dissolves in the surface waters inhabited by coccolithophores, with the rest of the coccolith $CaCO_3$ sinking out of the surface waters within zooplankton fecal pellets or marine snow aggregates. The exact means by which some coccoliths are dissolved in near-surface waters are unclear (dissolution within zooplankton guts may be important), but regardless of mechanisms several lines of evidence suggest that near-surface dissolution does occur. The size of coccoliths precludes the likelihood of single coccoliths sinking at all rapidly under gravity, because of the considerable viscosity of water with respect to such small particles (Stokes' law). Stokes' law can be overcome if coccoliths become part of larger aggregates, either marine snow or zooplankton fecal pellets. Another possible fate for coccoliths is to become incorporated into the shells of tintinnid microzooplankton, which when grazing on coccolithophores make use of the coccoliths in their own shells (**Figure 4**). Regardless of their immediate fate, the coccoliths must eventually either dissolve or else sink toward the seafloor.

The construction of $CaCO_3$ coccoliths (calcification) leads to additional impacts, over and above those associated with the photosynthesis carried out by all species. The first and perhaps the most important of these is that $CaCO_3$ contains carbon and the vertical downward flux of coccoliths thereby removes carbon from the surface oceans. It might be expected that this would lead to additional removal of CO_2 from the atmosphere to the oceans, to replace that taken up into coccoliths, but in fact, because of the complex effect of calcification ($CaCO_3$ synthesis) on seawater chemistry, the production of coccoliths actually increases the partial pressure of CO_2 in surface seawater and promotes outgassing rather than ingassing. Determining the exact nature and magnitude of the overall net effect is complicated by a possible additional role of coccoliths as 'ballast' (coccoliths are denser than water and hence when

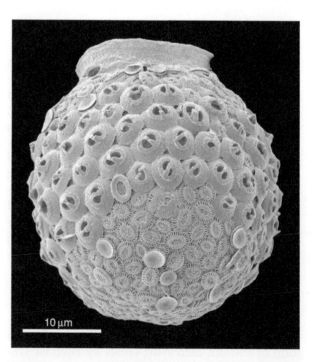

Figure 4 Tintinnid lorica (casing) with embedded coccoliths.

incorporated into aggregates of particulate fecal material may drag down extra organic carbon into the ocean interior).

Microscopic examination of seafloor sediments (if shallow enough that the $CaCO_3$ does not dissolve) and of material caught in sediment traps has revealed that much of the calcium carbonate in the samples consists of coccoliths. The flux of coccoliths probably accounts for *c.* 50% of the total vertical $CaCO_3$ flux in open ocean waters (in other words, about 50% of the inorganic carbon pump), with foraminifera shells responsible for most of the rest. It is usually not the most numerous species (*E. huxleyi*) but rather larger species (e.g., *Calcidiscus quadriperforatus* and *Coccolithus pelagicus*) that make the greatest contributions to the total coccolith flux.

Coccolithophores also impact on climate in other ways, ones that are unconnected with carbon. Coccolithophores are intense producers of a chemical called dimethylsulfoniopropionate (DMSP). The production of DMSP leads eventually (via several chemical transformations) to additional cloud condensation nuclei in the atmosphere and thereby to increased cloud cover.

Coccoliths also scatter light, polarizing it in the process. They do not reflect or block light (this would clearly be disadvantageous for the photosynthetic cell underneath), but the difference between the refractive indices of water and of calcium carbonate means that the trajectories of photons are deflected by encounters with coccoliths. A small proportion of the scattering (deflection) events are through angles greater than 90°, leading to photons being deflected into upward directions and eventually passing back out through the sea surface. Because of this light-scattering property of coccoliths, their bulk effect is to make the global oceans slightly brighter than they would otherwise be. It has been calculated that the Earth would become slightly dimmer (the albedo of the Earth would decrease by about 0.1% from its average global value of about 30%) were coccolithophores to disappear from the oceans. The effect of coccoliths in enhancing water brightness is seen in its most extreme form during coccolithophore blooms (**Figure 3**).

The Past

Coccolithophores are currently the dominant type of calcifying phytoplankton, but further back in time there were other abundant calcifying phytoplankton, for instance the nannoconids, which may or may not have been coccolithophores. The fossil calcifying phytoplankton are referred to collectively as calcareous nannoplankton.

The first calcareous nannoplankton are seen in the fossil record *c.* 225 Ma, in the late Triassic period. Abundance and biodiversity increased slowly over time, although they were at first restricted to shallow seas. During the early Cretaceous (145–100 Ma), calcareous nannoplankton also colonized the open ocean. They reached their peak, both in terms of abundance and number of different species (different morphotypes) in the late Cretaceous (100–65 Mya). 'The Chalk' was formed at this time, consisting of thick beds of calcium carbonate, predominantly coccoliths. Thick deposits of chalk are most noticeable in various striking sea cliffs, including the white cliffs of Dover in the United Kingdom, and the Isle of Rugen in the Baltic Sea. The chalk deposits were laid down in the shallow seas that were widespread and extensive at that time, because of a high sea level.

Calcareous nannoplankton, along with other biological groups, underwent long intervals of slowly but gradually increasing species richness interspersed with occasional extinction events. Their heyday in the late Cretaceous was brought to an abrupt end by the largest extinction event of all at the K/T boundary (65 Ma), at which point ~93% of all species (~85% of genera) suddenly went extinct. Although biodiversity recovered rapidly in the early Cenozoic, calcareous nannoplankton have probably never since re-attained their late Cretaceous levels.

Because the chemical and isotopic composition of coccoliths is influenced by the chemistry of the seawater that they are synthesized from, coccoliths from ancient sediments have the potential to record details of past environments. Coccoliths are therefore a widely used tool by paleooceanographers attempting to reconstruct the nature of ancient oceans. Some of the various ways in which coccoliths are put to use in interpreting past conditions are as follows: (1) elemental ratios such as Sr/Ca and Mg/Ca are used to infer past seawater chemistry, ocean productivity, and temperatures; (2) the isotopic composition ($\delta^{13}C$, $\delta^{18}O$) of the calcium carbonate is used to infer past carbon cycling, temperatures, and ice volumes; (3) the species assemblage of coccoliths (some species assemblages are characteristic of eutrophic conditions, some of oligotrophic conditions) is used to infer trophic status and productivity. Some of the organic constituents of coccolithophores are also used for paleoenvironmental reconstructions. In particular, there is a distinctive group of ketones, termed long-chain alkenones, which are specific to one family of coccolithophores and closely related haptophytes. These alkenones tend to survive degradation in sediments, and the ratio of one type of alkenones to another (the U^{37}_K index) can be used to estimate past ocean temperatures. Calcareous nannofossils are also extremely useful in determining the age of different layers in cores of ocean sediments (biostratigraphy).

The Future

The pH of the oceans is falling (they are becoming increasingly acidic), because of the invasion of fossil fuel-derived CO_2 into the oceans. Surface ocean pH has already dropped by 0.1 units and may eventually drop by as much as 0.7 units, compared to pre-industrial times, depending on future CO_2 emissions. The distribution of dissolved inorganic carbon (DIC) between bicarbonate, carbonate, and dissolved CO_2 gas changes with pH in such a way that carbonate ion concentration (and therefore saturation state, Ω) is decreasing even as DIC is increasing due to the invading anthropogenic CO_2. It is predicted that, by the end of this century, carbonate ion concentration and Ω may have fallen to as little as 50% of preindustrial values. If emissions continue for decades and centuries without regulation then the surface oceans will eventually become undersaturated with respect to calcium carbonate, first with respect to the more soluble aragonite used by corals, and some time later also with respect to the calcite formed by coccolithophores.

There has been an increasing appreciation over the last few years that declining saturation states may well have significant impacts on marine life, and, in particular, on marine organisms that synthesize $CaCO_3$. Experiments on different classes of marine calcifiers ($CaCO_3$ synthesizers) have demonstrated a reduction in calcification rate in high CO_2 seawater. One such experiment showed a strong decline in coccolithophore calcification rate (and a notable increase in the numbers of malformed coccoliths) at high CO_2 (low saturation state), although some other experiments have obtained different results. If coccolithophore biocalcification is controlled by Ω then the explanation could be linked to the importance of Ω in controlling inorganic calcification, although coccolithophores calcify intracellularly and so such a link is not guaranteed. At the time of writing, further research is being undertaken to determine whether, as the oceans become more acidic, coccolithophores will continue to be able to synthesize coccoliths and subsequently maintain them against dissolution.

Our ability to predict the consequences of ocean acidification on coccolithophores is hampered by our poor understanding of the function of coccoliths (what they are for, and therefore how the cells will be affected by their absence), and also by our poor understanding of the possibilities for evolutionary adaptation to a low-pH ocean. These constraints can be overcome to an extent by examining the geological history of coccolithophores, and their (in)ability to survive previous acid ocean events in Earth history. Although coccoliths (and other calcareous nannofossils) have been widely studied by geologists, it is only recently that there has been a concerted effort to study their species turnover through events in Earth history when the oceans were more acidic than now.

Although many authors have taken the success of coccolithophores during the high-CO_2 late Cretaceous as reassuring with respect to their future prospects, the reasoning is fallacious. Levels of calcium are thought to have been higher than now during the Cretaceous, and the CCD (the depth at which $CaCO_3$ disappears from sediments due to dissolution, which is a function of deep-water Ω) was only slightly shallower than today, indicating that Cretaceous seawater conditions were not analogous to those to be expected in a future high-CO_2 world.

It turns out that coccolithophores survived the Paleocene–Eocene Thermal Maximum event (thought to more closely resemble the predicted future) fairly well, with a modest increase in extinction rates matched by a similar increase in speciation rates. On the other hand, the environmental changes at the Cretaceous–Tertiary boundary (the K/T impact event), which also appears to have induced acidification, led to a mass extinction of 93% of all coccolithophore species, as well as to extinction of many other calcifying marine organisms including ammonites. It is necessary to more accurately characterize the environmental changes that took place across such events, in order to better determine how well they correspond to the ongoing and future ocean acidification.

See also

Calcium Carbonates. Benthic Foraminifera.

Further Reading

Gibbs SJ, Bown PR, Sessa JA, Bralower TJ, and Wilson PA (2007) Nannoplankton extinction and origination across the Paleocene–Eocene Thermal Maximum. *Science* 314: 1770–1773 (doi: 10.1126/science.1133902).

Holligan PM, Fernandez E, Aiken J, *et al.* (1993) A biogeochemical study of the coccolithophore *Emiliania huxleyi* in the North Atlantic. *Global Biogeochemical Cycles* 7: 879–900.

Paasche E (2002) A review of the coccolithophorid *Emiliania huxleyi* (Prymnesiophyceae), with particular reference to growth, coccolith formation, and calcification–photosynthesis interactions. *Phycologia* 40: 503–529.

Poulton AJ, Sanders R, Holligan PM, *et al.* (2006) Phytoplankton mineralization in the tropical and subtropical Atlantic Ocean. *Global Biogeochemical Cycles* 20: GB4002 (doi: 10.1029/2006GB002712).

Riebesell U, Zonderva I, Rost B, Tortell PD, Zeebe RE, and Morel FMM (2000) Reduced calcification in marine plankton in response to increased atmospheric CO_2. *Nature* 407: 634–637.

Thierstein HR and Young JR (eds.) (2004) *Coccolithophores: From Molecular Processes to Global Impact*. Berlin: Springer.

Tyrrell T, Holligan PM, and Mobley CD (1999) Optical impacts of oceanic coccolithophore blooms. *Journal of Geophysical Research, Oceans* 104: 3223–3241.

Winter A and Siesser WG (eds.) (1994) *Coccolithophores*. Cambridge, UK: Cambridge University Press.

Young JR, Geisen M, Cros L, *et al.* (2003) *Special Issue: A Guide to Extant Coccolithophore Taxonomy. Journal of Nannoplankton Research* 1–125.

Relevant Websites

http://cics.umd.edu
 – Blooms of the Coccolithophorid *Emiliania huxleyi* in Global and US Coastal Waters, CICS.
http://www.ucl.ac.uk
 – Calcareous Nannofossils, MIRACLE, UCL.
http://www.nanotax.org
 – Calcareous Nannofossil Taxonomy.
http://www.emidas.org
 – Electronic Microfossil Image Database System.
http://www.noc.soton.ac.uk
 – *Emiliania huxleyi* Home Page, National Oceanography Centre, Southampton.
http://www.nhm.ac.uk
 – International Nannoplankton Association page, hosted at Natural History Museum website.

DEEP-SEA DRILLING METHODOLOGY

K. Moran, University of Rhode Island, Narragansett, RI, USA

Introduction

The technology developed and used in past scientific drilling programs, the Deep Sea Drilling Project and the Ocean Drilling Program, has now been expanded in the current Integrated Ocean Drilling Program (IODP). These technologies include innovative drilling methods, sampling tools and procedures, *in situ* measurement tools, and seafloor observatories. This new IODP technology will be used to drill deeper into the seafloor than was possible in the previous scientific programs. The first drilling target of the IODP using the new technology is the seismogenic zone offshore Japan, a location deep in the Earth (7–14 km) where earthquakes are generated.

Drilling Technology

The Deep Sea Drilling Project and the Ocean Drilling Program used the same basic drilling technology, the open hole method. Today, the IODP has extended this capability to include closed hole methods, known as riser drilling.

Open Hole or Nonriser Drilling

Drilling is the process of establishing a borehole. The open hole method uses a single drill pipe that hangs from the drill ship's derrick, a tall framework positioned over the drill hole used to support the drill pipe. The drill pipe is rotated using drilling systems, specifically a hydraulically powered top drive located above the drill floor of the ship. Surface sea water is flushed through the center of the pipe to lubricate the rotating bit that cuts the rock and then flushes sediment and rock cuttings away to the seafloor (**Figure 1**). Open hole refers to the resulting borehole which remains open to the ocean during drilling. This method is also called a riserless drilling system. Important parts of the deep-water drilling system are a drilling derrick that is large and strong enough to hang a long length of drill pipe reaching deep ocean and subseafloor depths (up to 8 km); a system that rotates the drill pipe; a motion compensator that isolates the

ship's motion from the drill pipe; and a pump that flushes sea water through the drill pipe.

Open hole methods are successfully used in all of the Earth's oceans (**Figure 2**). Scientific ocean-drilling achievements include drilling in very deep water (6 km) and to >2 km below the seafloor (**Table 1**). Although there have been many achievements using these methods, there are also limitations. Although the exact depth limit of open-hole drilling method is not yet known, it is likely limited to 2–4 km below the seafloor. This limitation exists because the drill fluid must be modified to a lower density so that the deep cuttings can be lifted from the bit and flushed out of the hole when drilling deep into the seafloor. Another

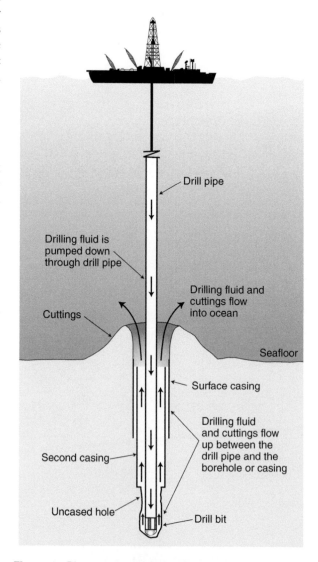

Drill pipe

Drilling fluid is pumped down through drill pipe

Drilling fluid and cuttings flow into ocean

Cuttings

Seafloor

Surface casing

Drilling fluid and cuttings flow up between the drill pipe and the borehole or casing

Second casing

Uncased hole

Drill bit

Figure 1 Diagram of a nonriser drilling system.

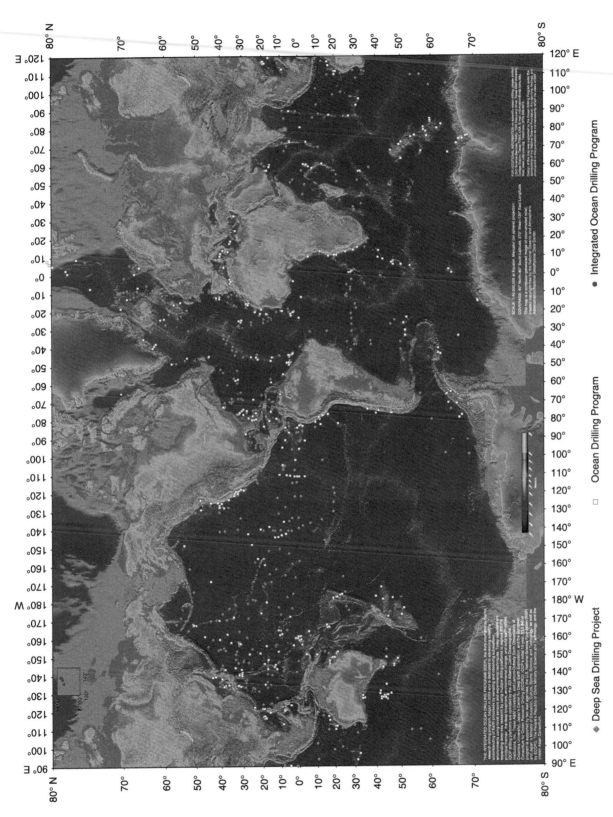

Figure 2 Map of all sites drilled by the Deep Sea Drilling Project (DSDP), the Ocean Drilling Program (ODP), and the IODP.

Table 1 Fact sheet about the *JOIDES Resolution,* the research vessel used by the Ocean Drilling Program

Total number of days in port		445 days
Total number of days at sea		4751 days
Total distance traveled		507 420 km
Total number of holes drilled		1445 holes
Deepest water level drilled	Leg 129	5980 m
Deepest hole drilled	Leg 148, Hole 504B (South-eastern Pacific Ocean, off coast of Ecuador)	2111 m
Total amount of core		180 880 m
Most core recovered on single	Leg 175 – Benguela (15 Aug.–10 Oct. 1997)	8003 m
Northern-most site drilled	Leg 113, site 911	Latitude 80.4744° N Longitude 8.2273° E
Southern-most site drilled	Leg 151, site 693	Latitude 70.8315° S Longitude 14.5735° W
Year and place of constitution	1978	Halifax, Nova Scotia, Canada
Laboratories and other scientific equipment installed	1984	Pascagoula, Mississippi
Gross tonnage		9719 tons
Net tonnage		2915 tons
Engines/generators		Seven 16 cyl index Diesel 5@2100 kW (2815 hp) 2@15500 kW (2010 hp)
Length		143 m
Beam		21 m
Derrick		62 m
Speed		11 knots
Crusing range		120 days
Scientific and technical party		50 people
Ship's Crew		65 people
Laboratory space		1115 m^2
Drill string		8838 m

limitation of the open hole method is that drilling must be restricted to locations where hydrocarbons are unlikely to be encountered. In an open hole, there is no way to control the drilling fluid pressure. In locations where oil and gas may exist, the formations are frequently overpressured (similar to a champagne bottle). If these formations would be punctured with an open hole system, the drill pipe would act like a straw that connects this overpressured zone in the rock to the ocean and the ship. This type of puncture is called a 'blow-out' and is a serious drilling hazard. The explosion as gases are vented through the straw to the ship's drill floor could cause serious damage, or worse yet, the change in the density of the seawater as the gas bubbles are released into the overlying ocean could cause the ship to sink. Without a system to control the pressure in the borehole, there is no way to prevent a blow-out.

Closed Hole or Riser Drilling

The new deep-sea drilling technology used in IODP is a closed system, also known as riser drilling. This technology has been used, in shallow to intermediate water depths, by the offshore oil industry to explore for, and produce oil and gas.

Riser drilling uses two pipes: a drill pipe similar to that used for open hole drilling and a wider diameter riser pipe that surrounds the drill pipe and is cemented into the seafloor (**Figure 3**). The system is closed because drill fluid (seawater and additives) is pumped down the drillpipe (to lubricate the bit and flush rock cuttings away from the bit) and then returned to the ship via the riser. With riser drilling, the drill fluid density can be varied and borehole pressure can be monitored and controlled, thus overcoming the two limitations of open hole drilling. The single limitation of riser drilling is water depth. The current water depth limit of riser technology is approximately 3 km. This water depth limit occurs because the riser, filled with drill fluid and cuttings, puts a large amount of pressure on the rock. This pressure is greater than the strength of the rock; thus, the rock breaks apart under the riser pressure, causing the drilling system to fail.

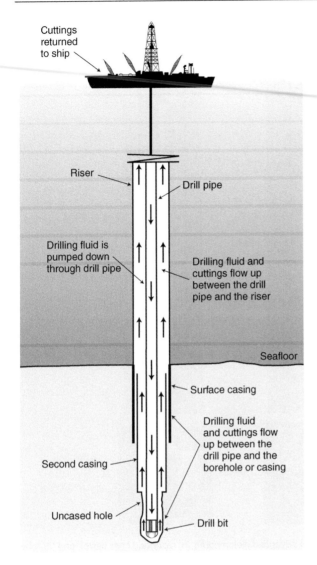

Cuttings returned to ship

Riser

Drill pipe

Drilling fluid is pumped down through drill pipe

Drilling fluid and cuttings flow up between the drill pipe and the riser

Seafloor

Surface casing

Drilling fluid and cuttings flow up between the drill pipe and the borehole or casing

Second casing

Uncased hole

Drill bit

Figure 3 Diagram of a riser drilling system.

Deep-Sea Sampling

Scientific ocean drilling not only requires a borehole, but – more importantly – the recovery of high-quality core samples taken as continuously as possible. Recovering sediment and rock samples from below the seafloor in deep water requires the use of wireline tools. Wireline tools are pumped down the center of the drill pipe to the bottom (called the bottom hole assembly) where they are mechanically latched into place near the drill bit in preparation for sampling. Different types of tools are advanced into the geological formation and take a core sample in different ways, depending on the type of sediment or rock. After the tool samples the rock formation, it is unlatched from the bottom hole assembly with a mechanical device, called an overshot, that is sent down the pipe on a wire. The overshot is used to

unlatch the wireline tool, return it to the ship, and recover the core sample.

In scientific ocean drilling, three standard wireline core sampling tools are used: the advanced piston corer, the extended core barrel, and the rotary core barrel.

The piston corer is advanced into the sediment ahead of the drill bit using pressure applied by shipboard pumps through the drill pipe (**Figure 4(a)**). The drill fluid pressure in the pipe is increased until the corer shoots into the sediment. After the corer is shot 10 m ahead of the bit, it is recovered using the wireline overshot. The drill pipe and bit are then advanced by rotary drilling another 10 m, in preparation for taking another core sample. The piston corer is designed to recover undisturbed core samples of soft ooze and sediments up to 250 m below the seafloor. In soft to hard sediments, the piston corer can achieve 100% recovery. However, because the cores are taken sequentially, sediment between consecutive cores may not be recovered. To ensure that a continuous sedimentary section is recovered, particularly for paleoclimate studies, the IODP drills a minimum of three boreholes at one site. The positions of the breaks between consecutive core samples are staggered in each borehole so that if sediment is not sampled in one borehole at a core break depth, it will be recovered in the second or third borehole.

The extended core barrel is a modification of the oil industry's rotary corer and is designed to recover core samples of sedimentary rock formation (**Figure 4(b)**). Typically, the extended core barrel is deployed at depths below the seafloor at which the sediment is too hard for sampling by the piston corer. The extended corer uses the rotation of the drill string to deepen the borehole and cut the core sample. The cutting action is done with a small bit attached to the core barrel. An innovation of this tool is an internal spring that allows the core barrel's smaller bit to extend ahead of the drill bit in softer formations. In hard formations, where greater cutting action is needed, the spring is compressed and the small core bit rotates with the main drill bit.

The rotary core barrel is a direct descendant of the rotary coring system used in the oil industry and is similar to the extended core barrel in its retracted mode. The rotary corer is designed to recover core samples from medium to very hard formations, including igneous rock. The corer uses the rotation of the drill string and the main drill bit to deepen the hole and cut the core sample (**Figure 4(c)**).

In all drilling operations, there are times when the drill pipe must be recovered to the ship, for example, to change a worn bit or to install different bottom hole equipment. Before pulling the pipe out of the

(a)   (b)   (c)

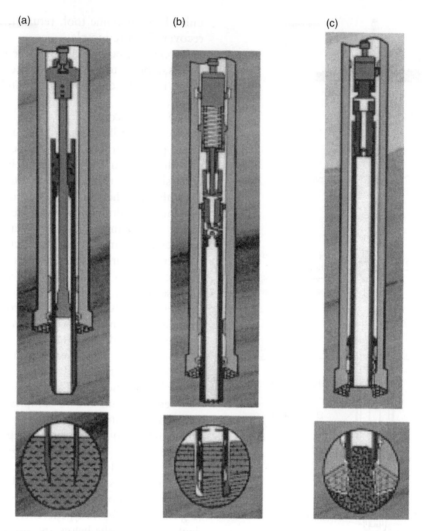

Figure 4 Diagrams of the Ocean Drilling Program's coring tools: (a) advanced piston corer; (b) extended core barrel; and (c) rotary core barrel.

borehole, a re-entry cone (**Figure 5**) is dropped down the outside of the drill pipe where it free-falls to the seafloor. The cone is used as a guide to re-enter the borehole with a new bit or equipment. The cone is a very small target in deep water (1–6 km) and ancillary tools are needed to locate it for re-entry into the borehole. The cone is first located acoustically, using seafloor transponders. To precisely pinpoint the cone, a video camera is lowered with the drill pipe to visually pinpoint the location and drop the drill pipe back into the borehole.

Special corers are used to sample unusual and difficult formations. For example, gas hydrates, ice-like material that is stable under high pressure and low temperature, commonly occur in deep water below the seafloor and require special samplers. Gas hydrates have generated much public interest since they can contain methane gas trapped within their structure, which is thought to be a potential future energy source. However, when gas hydrate is sampled, it must be kept at *in situ* pressure conditions to maintain the integrity of the core. Thus, a pressure core sampler is used to sample hydrates. The sampler is similar to the extended corer in that it has its own bit, but it has an internal valve that closes before the sampler is removed from the formation. The closed valve maintains the sample at *in situ* pressure conditions.

Drilling Measurements

Once sampling is completed, logging tools are lowered into the borehole to measure the *in situ* geophysical and chemical properties of the formation. In IODP, logging tools, developed for use in the oil industry, are most commonly leased from Schlumberger. The origins of logging go back to 1911 when the science of geophysics was new and was just beginning to be used

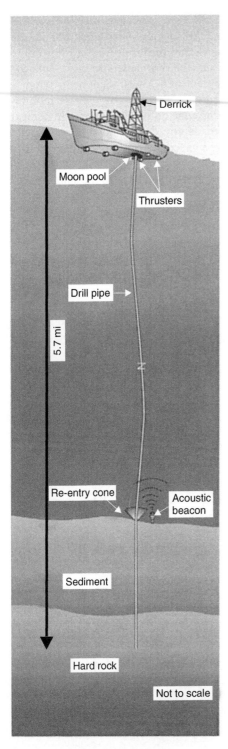

Figure 5 Diagram of the ship, drill pipe, and re-entry cone used in the drilling process.

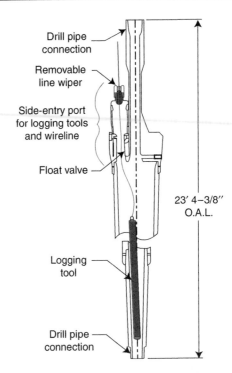

Figure 6 Diagram of the side-entry sub technology.

to explore the internal structure of the Earth. Conrad and Marcel Schlumberger, the founders of Schlumberger, conceived the idea that electrical measurements could be used to detect ore (precious minerals). Working at first alone and then with a number of associates, they extended the electrical prospecting technique from the surface to the oil well. Now, the use of electric prospecting, called logging, is widely accepted as a standard method in oil exploration.

Logging tools are lowered into the borehole using a cable that also transmits the data, in real time, to the ship. The term logging refers to the type of data collected. For example, a borehole log is a record or ledger of the sediment and rock encountered while drilling. Logs are geophysical and chemical records of the borehole. The logging tools typically comprise transmitters and sensors or a sensor alone encased in a robust stainless steel tube. Examples of tool measurements include electrical resistivity, gamma ray attenuation, natural gamma, acoustic velocity, and magnetic susceptibility. Log data provide an almost continuous record of the sediment and rock formation along length of the borehole.

Log data are of high quality when collected in the open hole, outside of the drill pipe. The most common method for deploying logging tools is to deploy a device that releases the drill bit from the drill pipe once all drilling and sampling operations are completed. The drill bit falls to the bottom of the borehole and is left there. The drill pipe is retracted and only 75–100 m of pipe is left at the top of the borehole to keep the upper, loose part of the borehole stable. Logging tools are lowered through the drill pipe to the bottom of the borehole. Log data are acquired by slowly raising the tool up the borehole at a constant speed. When boreholes are unstable and the walls are collapsing into the open hole, another method is used, unique to scientific ocean drilling. In unstable conditions, the drill pipe cannot be retracted to within 75 m of the seafloor without borehole walls collapsing and blocking or bridging the hole. In these situations, after the bit is released, the logging tools are lowered inside the drill pipe to the bottom of the hole. The drill pipe is retracted only enough to expose the logging tools to the open hole, while protecting the remainder of the borehole walls. Then, drill pipe is retracted at the same speed as logging tools are pulled up through the borehole. The technology developed that allows for this unique operation is called the side-entry sub (**Figure 6**). When inserted as a part of the drill string, the cable, to which the tools are attached, exits the drill pipe at the side-entry sub, positioned well below the ship. In the way, the logging cable does not interfere with removal of drill pipe.

Data in the borehole are also collected using wireline-deployed tools. Sediment temperature is measured with a temperature sensor mounted inside the cutting edge of the piston corer. In addition, other tools are deployed that do not recover a sample. They are pushed into the sediment ahead of the drill bit, left in place for 10–15 min to record temperature, and then pulled back to the ship, where the data are downloaded and analyzed. Special wireline tools have also been used to measure pressure and fluid flow properties of sediment and rock.

IODP also uses an oil industry-developed method for logging sediment and igneous rocks – logging-while-drilling or LWD. In this method, some of the

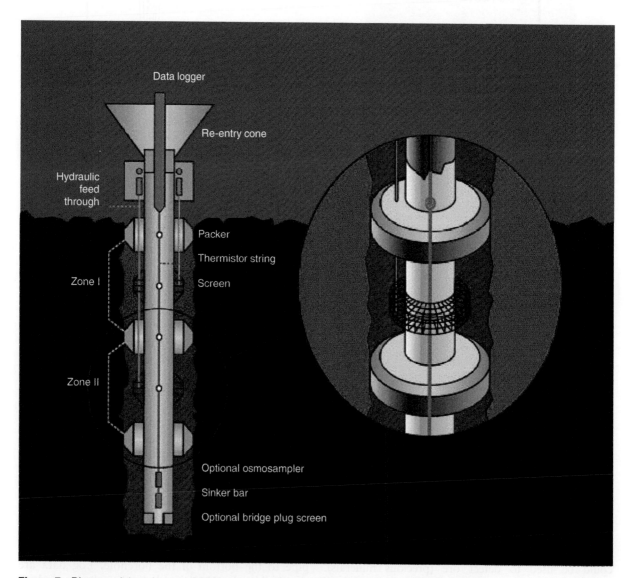

Figure 7 Diagram of the advanced CORK system used to seal instruments in the borehole.

same wireline logging sensors are repackaged and attached to the drill string, immediately behind the drill bit. The sensors log the geophysical properties of the sediment and rock as drilling proceeds, thus eliminating problems associated with borehole collapse.

Deep Seafloor Observatories

The circulation obviation retrofit kit (CORK) is a seafloor observatory that measures pressure, temperature, and fluid composition – important parameters for the study of the dynamics of deep-sea hydrologic systems. CORKs are installed by the IODP for measurements over long periods of time (months to years). Since 1991, observatories have been installed on the deep seafloor in different settings, for example, at mid-ocean ridge hydrothermal systems and at active margins.

The CORKs are installed by the drill ship. After a borehole is drilled, a CORK is installed to seal instruments in the borehole away from the overlying ocean (**Figure 7**). The CORK has two major parts: the CORK body that provides the seal and an instrument cable, for measuring fluid pressures, and temperatures that hangs from the CORK into the borehole. A data recorder is included with the instrument cable. The data recorders have sufficient battery power and memory for up to 5 years of operation. Data are recovered from CORKs using manned submersibles or remotely operated vehicles. The instruments in the CORK measure pressure and temperature spaced along a cable that extends into the sealed borehole. The CORK also includes a valve above the seal where borehole fluids can be sampled. Advanced CORKs are also used to isolate specific and measure the properties of different sediment or rock zones.

The IODP installs another type of long-term seafloor observatory for earthquake studies. Seismic monitoring instruments are installed in deep boreholes located in seismically active regions, for example, off the coast of Japan. These data are used to help established predictive measures to prevent loss of life and damage to cities during large earthquakes.

Deep-sea seismic observatories contain a strainmeter, two seismometers, a tiltmeter, and a temperature sensor. The observatories have replaceable data-recording devices and batteries like CORKs, and are serviced by remotely operated vehicles. Eventually, real-time power supply and data retrieval will be possible when some of the observatories are connected to nearby deep-sea fiber-optic cables.

Summary

Deep-sea scientific drilling applies innovative sampling, instrument, and observatory technologies to the study of Earth system science. These range from the study of Earth's past ocean and climate conditions using high-quality sediment cores, to the study of earthquakes and tectonic processes using logging tools and seafloor observatories, to exploring gas hydrates (a potential future energy source) using specialized sampling tools.

In 2004, the IODP succeeded two earlier scientific programs, Deep Sea Drilling and Ocean Drilling. The IODP operates two ships: the D/V JOIDES Resolution and the D/V Chikyu and leases ships for special operations in shallow water and ice-covered seas. The IODP is supported by the US, Japan, and Europe.

See also

Deep Submergence, Science of. Deep-Sea Drilling Results. Manned Submersibles, Deep Water.

Further Reading

DSDP (1969–86) Initial Reports of the Deep Sea Drilling Project. Washington, DC: US Government Printing Office. http://www.deepseadrilling.org/i_reports.htm (accessed Mar. 2008).

Integrated Ocean Drilling Program (2001) Earth, Oceans and Life: Scientific Investigation of the Earth System Using Multiple Drilling Platforms and New Technologies, Integrated Ocean Drilling Program, Initial Science Plan, 2003–2013.

Joint Oceanographic Institutions (1996) *Understanding Our Dynamic Earth: Ocean Drilling Program Long Range Plan*. Washington, DC: Joint Oceanographic Institutions.

Oceanography Society (2006) *Special Issue: The Impact of the Ocean Drilling Program. Oceanography* 19(4).

ODP (1985–present) Proceedings of the Ocean Drilling Program, Initial Reports, vols. 101–210. College Station, TX: ODP. http://www-odp.tamu.edu/publications (accessed Mar. 2008).

ODP (1985–present) Proceedings of the Ocean Drilling Program, Scientific Results, vols. 101–210. College Station, TX: ODP. http://www-odp.tamu.edu/publications (accessed Mar. 2008).

ODP (2004–present) Proceedings of the Integrated Ocean Drilling Program, vols. 300–. College Station, TX: Integrated Ocean Drilling Program.

WHOI (1993–94) *Oceanus: 25 Years of Ocean Drilling*, vol. 36, no. 4. Woods Hole, MA: Woods Hole Oceanographic Institution.

DEEP-SEA DRILLING RESULTS

J. G. Baldauf, Texas A&M University, College Station, TX, USA

Introduction

Modern scientific ocean drilling commenced over forty years ago with the inception of Project Mohole. This project was named for its goal of coring a 5–6 km borehole through thin oceanic crust, continuing through the Mohorovicic Discontinuity and into the earth's mantle. Project Mohole was active from 1957 through 1966. Although it did not achieve its objective, the Project demonstrated the means for coring in the oceans for scientific purposes and in doing so planted the seed for future decades of scientific ocean drilling.

The current era of scientific coring commenced with the creation of the Joint Oceanographic Institutions for Deep Earth Sampling (JOIDES) in 1964. In 1965, JOIDES completed its first scientific drilling program on the Blake Plateau using the drilling vessel *Caldrill*. This initial experiment led to the development of the Deep Sea Drilling Project (DSDP) managed by Scripps Institution of Oceanography, California. The scientific objective of DSDP was expanded, compared to that of Project Mohole, and included the recovery of sediments and rocks from throughout the World Ocean to improve the understanding of the natural processes active on this planet. Central to DSDP was the scientific research vessel *Glomar Challenger* (**Figure 1**) operated by Global Marine Inc.

During the 15 years (1968–1983) of operations, DSDP advanced the scientific frontier by completing 96 scientific expeditions throughout the World Ocean (**Figure 2**). These expeditions contributed to

Figure 1 Scientific Research Vessel *Glomar Challenger* operated by Global Marine Inc. for the Deep Sea Drilling Project from 1968 to 1983.

advancement of the earth sciences. For example, DSDP confirmed Alfred Wegner's theory of continental drift, and provided fundamental evidence in support of plate tectonics, established the timing of northern and southern hemisphere glaciations, documented the desiccation of the Mediterranean Sea, documented the northward migration of India and its collision with Asia, determined the history of the major oceanic gateways and their impact on ocean circulation, and improved the understanding of oceanic crust formation and subduction processes.

The overwhelming success of DSDP and the numerous scientific questions still remaining provided the framework for the current scientific drilling program, the Ocean Drilling Program (ODP). This new program, established in 1984, continues to explore the history of the earth as recorded in the rocks and sediments beneath the World Ocean. The centerpiece of ODP is the scientific research vessel

JOIDES Resolution (**Figure 3**) operated by Transocean SedcoForex for Texas AM University. To date 92 scientific expeditions have been completed by ODP (**Figure 4**).

Science Initiatives

The earth system is complex and dynamic with numerous variables and a multitude of forcing and response mechanisms. The sediments preserved beneath the World Ocean record the tempo and variation of the climate system at annual to millennial scales. Likewise, oceanic crust records the environment at the time of its formation. The understanding of these variables and the naturally occurring process active in the Earth's environment and in the Earth's interior continue to evolve based on results from scientific ocean drilling. The

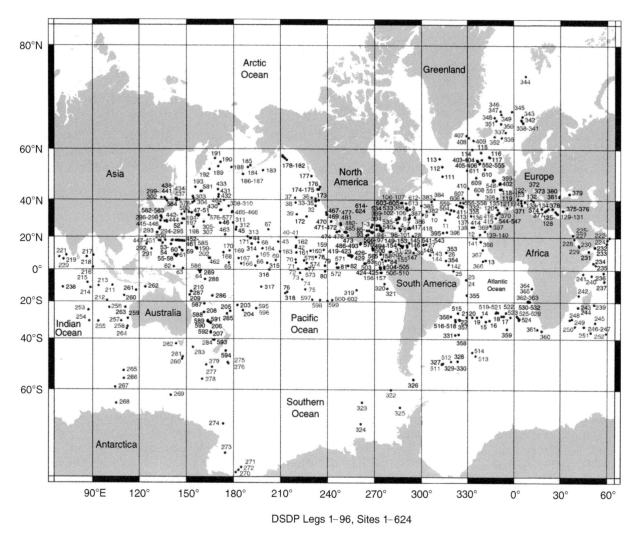

DSDP Legs 1–96, Sites 1–624

Figure 2 Geographic location of the sites occupied during scientific expeditions of the Deep Sea Drilling Project from 1968 to 1983.

Figure 3   Scientific Research Vessel *JOIDES Resolution* operated by Transocean SedcoForex for the Ocean Drilling Program 1984 to present.

partitioning of the Earth's processes into these two themes is in part arbitrary as the external environment is closely linked to that of the Earth's interior.

Dynamics of the Earth's Environment

The Earth's climate system consists of processes active in and between five regimes, space, atmosphere, ocean, cryosphere, and crust (**Figure 5**). For example, variation in incoming solar radiation influences the atmosphere-ocean coupling through changes in precipitation, evaporation and heat exchange. In addition, ice-sheets (and in turn sea level), sea ice, albedo, and terrestrial and oceanic biomass also respond to solar radiation changes.

Similarly, variations in the dimensions and shape of the ocean basins, through crust formation on destruction or the opening or closing of oceanic gateways, influence oceanic circulation, as well as the salinity and temperature of specific water masses. These changes subsequently impact the distribution of nutrients, regional climates, and the Earth's biomass.

The complexity of the climate system is only now beginning to be realized, in part through the ground truthing of climatic models and the historical perspective at various (annual to millennial) scales provide through ocean drilling. ODP's contributions to understanding the Earth's environment are extensive. For example, coring off northern Florida provided evidence to support the Cretaceous/Tertiary Boundary meteorite impact theory and its causal impact on widespread extinction of the Earth's biota. Numerous other cruises have provided insight as to the complexity of the climate system, investigating scientific themes such as the desertification of Africa, implications on climate of the uplift of Tibet plateau, refinement of the glacial history of both hemispheres, including the climate periodicity in icehouse and

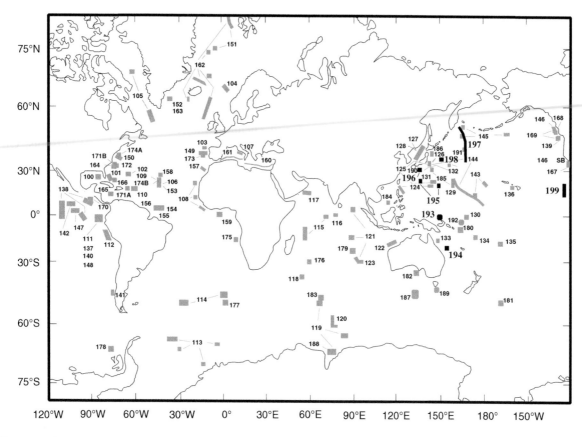

Figure 4 Geographic locations of the 91 scientific expeditions currently completed by ODP from 1984 to the present. Legs indicated in black were completed in 1990 and 1991.

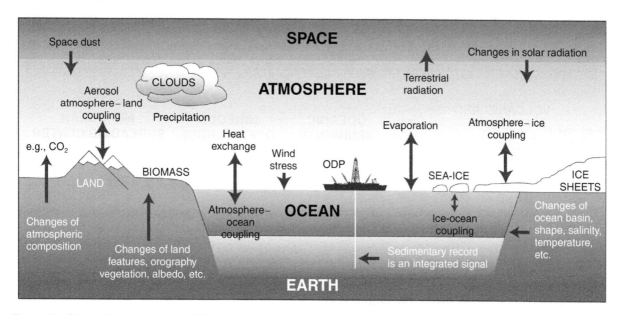

Figure 5 Schematic representation of the major components of the earth's climatic system. (Adapted from Crowley and North (1992) and the ODP long-range plan (1996).)

greenhouse worlds, and understanding of the history of sea level.

In addition, ODP continues to investigate the extent of the biosphere based on evidence of organisms deep within Earth's crust. The discovery of organisms in volcanic crust is significant as it extends the depth of the biosphere and indicates that life is possible in extreme environments.

Three scientific themes associated with the dynamics of the Earth's environment continue to be explored by ODP. These themes include understanding the Earth's changing climate, the causes and effects of sea level change, and fluids and bacteria as agents of change.

Dynamics of the Earth's Interior

The global cycling of mass and energy in the Earth's interior and the extrusion of this material into the Earth's exterior environment impacts global geochemical budgets (**Figure 6**). For example, the formation of oceanic crust at the mid-ocean spreading centers plays an important role in mantle dynamics, geochemical fluxes, and heat exchange within and between the Earth's internal and external environments. Similarly, the subduction of a lithospheric plate at a convergent margin results in the melting of the plate, which in turn contributes to the composition and circulation dynamics of the mantle. Associated with subduction zones are volcanic systems through which magma and gases are extruded to Earth's exterior, contributing to the chemical fluxes of the oceans and atmosphere.

Understanding of the processes associated with crustal formation and plate subduction has been enhanced through the recovery of crustal rocks from beneath the oceans. For example, ODP has improved the understanding of the chemical flux by coring within subduction zones to understand the processes associated with subduction, including chemical and mass balances and fluid flow associated with the interface between the two plates, referred to the Decollement zone.

Coring of large igneous provinces (LIPs) such as the Kerguelen Plateau and the Ontong Java Plateau has also provided insight into the mantle dynamics and chemical fluxes. These LIPs formed from the injection of magma through volcanic hotspots. Obtaining crustal samples from these regions enhances the understanding of chemical composition and fluxes, mantle dynamics, and the impact of the formation of these features on oceanic and atmospheric chemistry.

ODP has also recovered over 500 m of gabbro from a single hole on the South-west Indian Ridge. This sequence represents the most continuous stratigraphic sequence of lower crust from oceanic basement, allowing insight into the chemical composition and structure of oceanic crust.

Two other areas of interest for ODP have been the processes associated with the formation of ore bodies and the formation of gas hydrates – frozen methane crystallized with water. Of particular interest is the origin and history of such deposits and their potential influence on global chemical fluxes.

Two scientific themes remain central to ODP investigations: exploring the transfer of heat and material to and from earth's interior and investigating deformation of the lithosphere and earthquake processes.

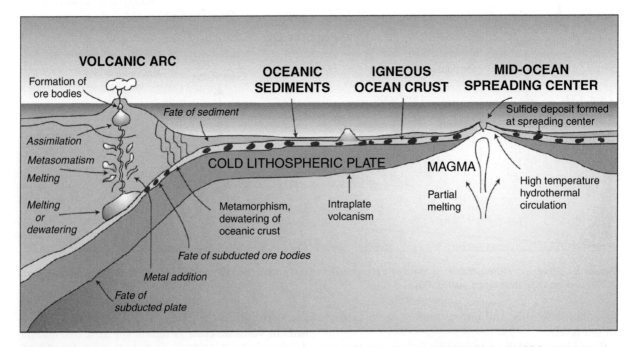

Figure 6 Schematic representation of the major components of the earth's tectonic cycle. (Adapted from the ODP long-range plan 1996.)

Technological Advances

Scientific advances made through ocean drilling are closely linked to advances in technology. New tools such as the development of seafloor observatories, enhanced coring and logging prototypes, seafloor equipment, and state-of-the-art laboratory equipment have provided new opportunities to advance the scientific frontier.

Seafloor Observatories

A reentry cone seal, also referred to as CORK, has been developed and used to seal an ODP hole, thus preventing flow into or out of the borehole. The CORK provides a means for monitoring borehole temperature and pressure as well as recovery of borehole fluid samples. A typical CORK configuration (**Figure 7A** and **B**) is characterized by the thermistor string used for the collection of pressure and temperature data, a borehole fluid sampler for collection of borehole fluids, and a data logger for the recording and storage of data from the thermistor string until it can be downloaded via a submersible. The next-generation borehole seals, known as the ACORK, are currently being developed. The ACORK will allow subdivision of the borehole into isolated segments, allowing monitoring of fluid flow for given horizons or intervals.

Coring and Logging Tools

Advanced piston corer (APC) This is a hydraulically actuated piston corer designed to recover undistributed core samples from soft sediments with enhanced core quality and core recovery. Initially developed during the latter phases of DSDP and enhanced during ODP, the APC has become the mainstay for the recovery of high-resolution sedimentary records for paleoceanographic and climate studies.

Pressure core sampler (PCS) The PCS was developed for the retrieval of core samples from the ocean floor while maintaining near *in situ* pressures up to 10 000 psi (69 MPa). This tool continues to be critical for investigating gas-bearing sediments, such as gas hydrates, for the analysis of biogeochemical cycling.

Formation microscanner (FMS) Adapted from industry, this logging tool provides an oriented, two-dimensional, high-resolution image of the variations in microresistivity around the borehole

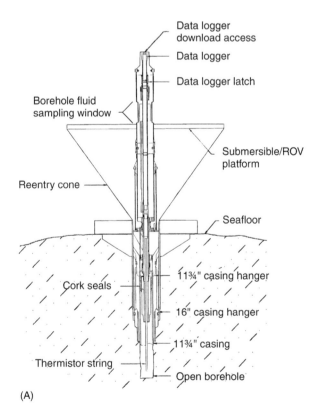

(A)   (B)

Figure 7 (A) Schematic of a reentry cone seal, also referred to as a CORK. The tool provides a means for monitoring borehole temperature and pressure as well as recovery of borehole fluid samples. (B) Underwater view of the CORK landed in a reentry cone.

wall. Collected data allow the correlation of coring and logging depth, orientation of cores and location of the cored sections when recovery is less than 100%, mapping of sedimentary structures, and interpretation of depositional environments.

Seafloor Equipment

Seafloor equipment such as a reentry cone and the hard rock guide base have been developed and implemented to achieve scientific objectives at specific sites. The reentry cone is a permanent seafloor installation, which serves as a conduit for reentry of the borehole and a platform for supporting various casing strings. Often the drill pipe is pulled from the hole to replace coring bits or to change coring tools. A temporary reentry cone referred to as a 'free fall funnel' is also used to allow bit changes.

The hard rock guide base (HRGB) was developed to focus the direction of the drill bit into hard, irregular seafloor surfaces that are otherwise undrillable. The difficulty in drilling is both the inability to spud or start a hole as insufficient weight could be applied to the bit and the tendency of the bit to 'walk' downhill when trying to start a hole on a sloping surface. The HRGB, when placed on the seafloor, provides support for the drill string to start a hole. Recent technological development of the Hard Rock Reentry System (HRRS) allows a cased reentry hole to be established in bare hard rock without the use of the HRGB.

Laboratory Equipment

Mutisensor track (MST)

The MST allows measurements of cores at centimeter scales for examination of changes in physical parameters resulting from changes in lithology, composition, porosity, density, and magnetic susceptibility of the sediment collected. Changes in these properties result from changing oceanographic conditions at the time of deposition or from post-deposition processes. The high-resolution records obtained with the MST also allow correlation of data sets from offset holes to ensure completeness of the geological record. In addition, the high-resolution data collected (100–1000 y) can be correlated to data collected through the logging of the borehole to allow direct correlation of laboratory measurements with those from the borehole.

Facilities

The scientific research vessel *JOIDES Resolution* is a dynamically positioned drilling vessel capable of maintaining position over specific locations while coring in water depths down to 8200 m. The vessel was built in Halifax, Nova Scotia, Canada in 1978 and has a length of 143 m, a breadth of 21 m, a gross tonnage of 7539, and a derrick that towers 61.5 m above the water line. A computer-controlled automated dynamic positioning system regulates 12 thrusters in addition to the main propulsion system to maintain the position of the vessel directly over the drill site. *JOIDES Resolution* was originally operated as an oil exploration vessel. In 1984, ODP converted the vessel into a scientific research vessel by removing the shipboard riser system and adding scientific laboratories. Operating statistics of the *JOIDES Resolution* are shown in **Table 1**.

Unique to this vessel is the 7-story laboratory structure housing state-of-the-art scientific equipment for use in the studies of geochemistry, microbiology, paleomagnetism, paleontology, petrology, physical properties, sedimentology, downhole measurements, and marine geophysics. This structure also includes support facilities for electronic repair, computers, photography, database management, communications, and conference facilities.

Cores and data collected during each scientific cruise are stored at shore-based facilities for future research by members of the international scientific community. ODP maintains four core repositories,

Table 1 Significant operational highlights of the *JOIDES Resolution* from 1985 to 1999 (Leg 100–183)

Deepest hole penetrated	2111 m below the seafloor – Holes 504B completed during Leg 148, eastern Pacific.
Shallowest operational water depth	37.5 m – Leg 143, north-west Pacific.
Deepest operational water depth	5980 m – Leg 129, western Pacific.
Most core recovered during a single expedition	8003 m – Leg 175, south-east Atlantic.
Total number of sites visited	535 – ODP Legs 100–183.
Total number of holes cored	1417 – ODP Legs 100–183.
Total core cored	251 017 m – ODP Leg 100 through Leg 183.
Total core recovered	170 770 m – ODP Leg 100 through Leg 183.

three in North America and one in Europe. The North American repositories include the East Coast Repository (ECR) at Lamont Doherty Earth Observatory (LDEO), which stores cores from the Atlantic Ocean through Leg 150; the Gulf Coast Repository (GCR) at Texas A&M University (TAMU), Texas, which houses cores from the Pacific and Indian Oceans, and the West Coast Repository (WCR) at Scripps Institution of Oceanography, California, which houses cores from the Pacific and Indian Ocean collected during the Deep Sea Drilling Project. The Bremen Core Repository at the University Bremen, Germany, houses cores obtained from the Atlantic since Leg 151. Data collected during each cruise are stored in one of two locations. Downhole measurement (logging) and site survey data are housed at LDEO. All other data collected during a cruise are housed at TAMU.

Scientific Expeditions

Using the *JOIDES Resolution*, ODP recovers sediments or rocks from beneath the World Ocean. Typically two types of coring tools, the Rotary Core Barrel (RCB) and the Advanced Piston Corer (APC), are used to cut the oceanic sediments and rocks. The RCB, typical of the petroleum industry, is used for penetration of hard rocks or sediment, while the APC is used in the less-indurated, softer sediment. The APC tool is typically used to recover the upper several hundred meters of the sediment sequence, with the subsequent sequence cored using the RCB. Once collected, the cores are returned to the ship via a wireline. Once on board the 9.5 m cores are sectioned into 1.5 m lengths for ease of handling within the shipboard laboratories.

Following coring operations, the borehole is often logged using standard industrial tools. Logging provides data on the variation in the physical and chemical properties of the sediments and rocks directly from the walls of the borehole. Seafloor laboratories may also be established by instrumenting a borehole with thermistors, water samplers, or seismometers for long term, multiyear monitoring.

Each cruise addresses a specific scientific theme based on a rigorous review of proposals submitted by members of the international science community. The duration of a leg depends on the specific objective. Generally each leg is two months in duration.

The ship crew consists of about 106 individuals of whom 51 are members of the scientific party. The remaining contingent supports ship and coring operations. The scientific party consists of international scientists from participating member countries and a technical support staff from TAMU.

International Partnership

Eight international members representing 22 countries currently provide funding for the Ocean Drilling Program. The budget for the program is about US$46 million annually with the US National Science Foundation contributing about 66% of the required funds. The remaining funds are provided by five additional full members, each contributing about US$3 million annually, and two associate members each contributing between US$0.5 and 2 million annually.

Full members of the program are the Australia/Canada/Korea/Chinese Taipei Consortium for Ocean Drilling, the European Science Foundation Consortium for Ocean Drilling (Belgium, Denmark, Finland, Iceland, Ireland, Italy, Norway, Portugal, Spain, Sweden, Switzerland, and The Netherlands), Germany, Japan, the United Kingdom, and the United States. Associate members include France and the People's Republic of China.

Management Structure

The ODP management team consists of the Prime Contractor, Joint Oceanographic Institution (JOI), the Science Operator at Texas A&M University, and the Wireline Operator at Lamont Doherty Earth Observatory (**Figure 8**). As the prime contractor from the National Science Foundation, JOI is responsible for overall management of the program, including scientific planning, operations, and

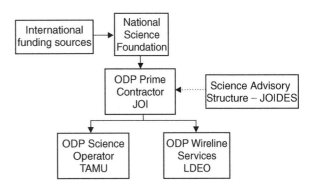

Figure 8 ODP management structure including the National Science Foundation, Joint Oceanographic Institutes (JOI), Texas A&M University (TAMU) and Lamont Doherty Earth Observatory (LDEO). Science advice is provided to JOI through the Joint Oceanographic Institutes for Deep Earth Sampling (JOIDES) panels.

responding to recommendations from the science advisory structure.

Texas A&M University is subcontracted through JOI to implement the scientific program developed by the science advisory structure. As Science Operator, TAMU is responsible for shipboard operations, including cruise staffing, maintenance and support of shipboard laboratories, data acquisition, engineering development, publication, core curation, and shipboard logistic (clearance, safety review, etc.)

Lamont Doherty Earth Observatory is subcontracted through JOI to implement the wireline-logging program for each scientific cruise. As Wireline Operator, LDEO is responsible for standard logging, specialized logging, and log analysis support services and database. LDEO is also responsible for the ODP Site Survey Data Bank at LDEO, which archives and distributes site survey data for ODP.

Advisory Structure

The ODP advisory structure (**Figure 9**) provides advice on the scientific program and on logistical activity and program facilities. Scientific guidance is provided by two JOIDES advisory panels, one focused on the Earth's environment and the second focused on Earth's interior. The remaining advisory panels provide guidance for shipboard operations and logistics, such as shipboard laboratories, pollution prevention and safety, and site locations.

Scientific Guidance

The success of ODP is in the bottom-up approach when it comes to determining the scientific programs

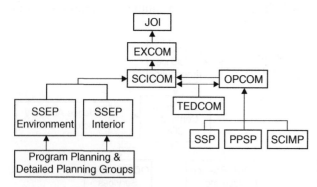

Figure 9 JOIDES Advisory structure consisting of two science panels (SSEP) and four additional panels technology TEDCOM, Site Survey (SSP), pollution prevention and safety (PPSP) and shipboard measurements (SCIMP) providing guidance and advice to the SCICOM committee (SCICOM) and the Operations subcommittee (OPCOM). Recommendations of SCICOM are forwarded to the Executive committee (EXCOM) for endorsement and to JOI for implementation.

most likely to advance the scientific frontier. Individuals or groups of scientists in the international community submit a proposal to answer specific scientific questions. Each proposal is reviewed by the Science Advisory Panels (SSPS) and recommendations are made to the Science Committee (SCICOM), which is responsible for ranking each proposal based on the overall contribution the proposed program would make to understanding Earth history. SCICOM forwards to the Operations Subcommittee (OPCOM), the highest-ranked proposals for scheduling consideration, taking into account available budget, weather constraints, and efficiency of scheduling when comparing days on site versus days in transit between sites. OPCOM sends back to SCICOM a proposed schedule, which is then ratified by SCICOM and forward to the Executive Committee (EXCOM) for approval. Upon approval, EXCOM forwards the schedule to JOI and subsequently to the TAMU and LDEO for implementation.

Logistics/Infrastructure

In addition to science planning, the JOIDES Advisory structure provides recommendations on site locations, pollution prevention and safety, shipboard measurements and technological developments. The Site Survey Panel (SSP) is responsible for reviewing all site location data and forwarding recommendations to OPCOM concerning the state of readiness of proposals under consideration for implementation. Similarly, the Pollution Prevention and Safety panel reviews each proposal for safety concerns. This panel is focused on reducing risks associated with the potential occurrence of hydrocarbons to an acceptable level. Both SSP and PPSP make recommendations to OPCOM for consideration when considering proposals for scheduling.

The Scientific Measurements Panel (SCIMP) provides advice and guidance on the shipboard laboratories, primarily pertaining to data acquisition, database storage, and data retrieval. In addition, this panel provides recommendations on publications and databases.

The Technology Committee (TEDCOM) works closely with the Engineering development teams at TAMU and LDEO to provide guidance on technological developments.

Future Directions

Like the Deep Sea Drilling Project before it, the present ODP has a defined duration, with the present program ending 30 September 2003. Although the program as currently configured will end, the need for continued scientific research through ocean drilling remains great. Recognizing this continued demand, the international partners are actively planning for a new program of scientific ocean drilling.

The future program is envisioned to utilize multiple platforms, with both Japan and United States providing platforms for the Integrated Ocean Drilling Program (IODP). Japan will provide a new raiser vessel capable of completing deep (> 3 km) stratigraphic holes. The United States will provide a nonriser vessel to continue work similar to that currently completed by the *JOIDES Resolution*. In addition to these two platforms, additional alternate platforms will be included for operating in regions not accessible by the other two vessels, such as the Arctic Ocean, shallow water continental margins, and coral reefs, among others. It is envisioned that this new era of scientific ocean drilling will commence in late 2003.

See also

Deep Submergence, Science of. Deep-Sea Drilling Methodology. Manned Submersibles, Deep Water.

Further Reading

Bleil U and Thiede J (1988) *Geological History of the Polar Oceans: Arctic versus Antarctic. NATO Series C, Mathematical and Physical Sciences*, vol. 308. Dordrecht: Kluwer Academic.

Coffin MF and Eldholm O (1994) Large igneous provinces: crustal structure, dimensions, and external consequences. *Reviews of Geophysics* 32: 1–36.

Cullen V (1993/94) 25 years of Ocean Drilling. *Oceanus* 36(4).

Dickens GR, Paull CK, Wallace P and the Leg 164 Science party (1997) Direct measurement of *in situ* methane quantities in a large gas hydrate reservoir. *Nature* 385: 426–428

Lomask M (1976) *Project Mohole and JOIDES, A Minor Miracle, An Informal History of the National Science Foundation*, National Science Foundation. Washington, DC: pp. 167–197.

Hsu KJ (1992) *Challenger at Sea, a Ship that Revolutionized Earth Science*. Princeton, NJ: Princeton University Press.

Kastner M, Elderfield H, and Martin JB (1991) Fluids in convergent margins: what do we know about their composition, origin, role in diagenesis and importance of oceanic chemical fluxes? *Philosophical Transactions of the Royal Society(London)* 335: 275–288.

Kennett JP and Ingram BL (1995) A 20,000 year record of ocean circulation and climate change from Santa Barbara basin. *Nature* 377: 510–513.

Parkes RJ, Cragg SJ, Bale JM, *et al.* (1994) Deep bacterial biosphere in Pacific Ocean sediments. *Nature* 371: 410–413.

Summerhayes CP, Prell WL, and Emeis KC (1992) *Upwelling Systems: Evolution since the Early Miocene. Geological Society Special Publication*, 64. London: Geological Society of London.

Summerhayes CP and Shackleton NJ (1986) *North Atlantic Paleoceanography*. Geological Society Special Publication 21. London: Blackwell Scientific.

Warme JE, Douglad RG, and Winterer EL (1981) *The Deep Sea Drilling Project: a Decade of Progress*, Special Publication 32. Tulsa, OK: Society of Economic Paleontologists and Mineralogists.

SEDIMENT CHRONOLOGIES

J. K. Cochran, State University of New York, Stony Brook, NY, USA

Introduction

Although the stratigraphic record preserved in deep-sea sediments can span up to 200 Ma, techniques of isotopic dating commonly used to extract sediment accumulation time scales are useful for only a fraction of this range. In addition, the temporal record is blurred by the mixing activities of the benthic fauna living in the upper centimeters of the sediment column. Radionuclide distributions in the sediments provide the most straightforward way of resolving mixing and accumulation rates in deep-sea sediment over the past ~ 5–7 Ma. The basis for these techniques is the supply of radionuclides to the oceanic water column, followed by their scavenging onto sinking particles and transport to the sediment–water interface. Decay of the radionuclides following burial provides chronometers with which mixing and accumulation rates can be determined.

Radionuclide Supply to the Sediment–Water Interface

Table 1 lists the most frequently used radionuclides for determining chronologies of deep-sea sediments. Many of these are members of the naturally occurring ^{238}U and ^{235}U decay series. Both ^{238}U and ^{235}U, as well as ^{234}U, are supplied to the oceans by rivers and are stably dissolved in sea water as the uranyl tricarbonate species $[UO_2(CO_3)_3]^{-4}$. In sea water these three U isotopes decay to ^{234}Th, ^{231}Pa, and ^{230}Th, respectively. The extent of removal of these radionuclides from the oceanic water column is a function of the rate of scavenging relative to the rate of decay. ^{234}Th has a relatively short half-life and can be effectively scavenged in near-surface and near-bottom waters of the open ocean and in the near-shore. ^{230}Th and ^{231}Pa, on the other hand, both have long half-lives and are efficiently scavenged. (While removal of ^{230}Th is nearly quantitative in the open ocean water column, ^{231}Pa shows some spatial variations in the extent of scavenging, with more effective removal at ocean margins.)

^{210}Pb is another ^{238}U decay series radionuclide that has been applied to deep-sea sediment chronologies. ^{210}Pb is produced from dissolved ^{226}Ra in sea water but is also added to the surface ocean from the atmosphere, where it is produced from decay of ^{222}Rn. Like thorium and protactinium, ^{210}Pb is scavenged from sea water and carried to the sediments in association with sinking particles. Owing to its short half-life, ^{210}Pb has been used principally to determine the rate at which the surface sediments are mixed by organisms.

Two other radionuclides that are supplied to the oceans from the atmosphere are ^{14}C and ^{10}Be. Both are produced in the atmosphere from the interaction of cosmic rays with atmospheric gases. (^{14}C also has been produced from atmospheric testing of atomic weapons.) ^{14}C is transferred from the dissolved inorganic carbon pool to calcium carbonate tests and to organic matter and is carried to the sea floor with sinking biogenic particles. ^{10}Be is scavenged onto particle surfaces, much like thorium and protactinium.

Radionuclides produced in association with atmospheric testing of atomic weapons provide pulse-input tracers to the oceans. Both ^{137}Cs and $^{239,240}Pu$ have been introduced to the oceans in this fashion, and their input peaked in 1963–64 as a consequence of the imposition of the ban on atmospheric weapons testing. Fractions of the oceanic inventories of both cesium and plutonium have been transferred to deep-sea sediments via scavenging onto sinking particles. In deep-sea sediments, the distributions of plutonium and ^{137}Cs are useful for constraining rates of particle mixing.

Radionuclides are commonly measured by detection of the α, β or γ emissions given off when they decay. This approach takes advantage of the fact that the radioactivity (defined as λN, the product of the decay constant λ and the number of atoms N) is often more readily measurable than the number of atoms (i.e., the concentration). As radiation interacts with matter, ions are produced and radiation detection involves measuring the electric currents that result. Both gas-filled and solid-state detectors are used. Measurement of radioactivity often involves chemical separation and purification of the element of interest, followed by preparation of an appropriate source for counting. Recent advances in mass spectrometry permit direct determination of atom concentrations for uranium, plutonium, and long-lived thorium isotopes by thermal ionization mass spectrometry (TIMS), as well as radiocarbon and ^{10}Be using tandem accelerators as mass spectrometers.

Table 1 Radionuclides useful in determining chronologies of deep-sea sediments

Radionuclide	Half-life	Source	Use	Useful time range
^{234}Th	24 days	Dissolved ^{238}U	Particle mixing	100 days
^{210}Pb	22 years	Dissolved ^{226}Ra, atmospheric deposition	Particle mixing	100 years
^{14}C^a	5730 years	Cosmogenic production	Sediment accumulation	35 000 years
^{231}Pa	32 000 years	Dissolved ^{235}U	Sediment accumulation	150 000 years
^{230}Th	75 000 years	Dissolved ^{234}U	Sediment accumulation	400 000 years
^{10}Be	1.5×10^6 years	Cosmogenic production		7×10^6 years
^{239,240}Pu	6600, 24 000 years	Anthropogenic: atomic weapons testing	Particle mixing	Since input (1954)
^{137}Cs	30 years	Anthropogenic: atomic weapons testing	Particle mixing	Since input (1954)

^{a} ^{14}C also has an anthropogenic source from atmospheric testing of atomic weapons.

Principles of Determination of Chronologies

Once deposited at the sediment–water interface, particle–reactive radionuclides are subject to decay as well as downward transport by burial and particle mixing. These processes are represented by the general diagenetic equation applied to radionuclides:

$$\frac{\partial A}{\partial t} = D_B \frac{\partial^2 A}{\partial x^2} - S\frac{\partial A}{\partial x} - \lambda A \qquad [1]$$

where A is the nuclide radioactivity (dpm/cm^3 sediment), D_B is the particle mixing coefficient (cm^2/x), S is the accumulation rate (cm/y), λ is the decay constant (y), x is depth in the sediment column (with $x = 0$ taken to be the sediment–water interface), and t is time. Certain underlying assumptions are made in the formulation of eqn [1]. These include no chemical mobilization of the radionuclide in the sediment column and constant sediment porosity.

Particle mixing of deep-sea sediments by benthic organisms is often parametrized as an eddy diffusion-like process, although nonlocal models invoking mixing at discrete depths also have been applied. Except in sediments deposited in anoxic basins, mixing of deep-sea sediments by organisms is commonly active in the upper 2–10 cm of the sediment column, possibly because this near-interface zone contains the most recently deposited organic material. Evidence from multiple profiles of long-lived radionuclides in deep-sea sediments suggests that particle mixing by organisms generally does not extend below the surficial mixed zone. This pattern is in contrast to that observed in estuarine and coastal sediments, which can be mixed to depths in excess of 1 m by organisms. Such deep mixing perturbs radionuclide profiles and makes extraction of sediment chronologies difficult in coastal sediments.

For the uranium and thorium decay series radionuclides, the assumption is usually made that the depth profiles are in steady state (i.e., invariant with time) because production and supply from the overlying water column are continuous. The solution to eqn [1] can be written as

$$A(x) = C \exp(\alpha x) + F \exp(\beta x) \qquad [2]$$

If sediments are mixed to a depth L (cm) the constants in eqn [2] can be evaluated with the boundary conditions $A = A_0$ at $x = 0$ and $D_B (\delta A/\delta x) = 0$ at $x = L$.

$$F = \frac{-A_0\, \alpha \exp(\alpha L)}{\beta \exp(\beta L) - \alpha \exp(\alpha L)} \qquad [3]$$

$$C = A_0 - F \qquad [4]$$

$$\alpha = \frac{S + \sqrt{S^2 + 4\lambda D_B}}{2D_B} \qquad [5]$$

$$\beta = \frac{S - \sqrt{S^2 + 4\lambda D_B}}{2D_B} \qquad [6]$$

If the depth of the mixed layer is greater than the penetration depth of the tracers, an approximation to the solution of eqn [1] is given by

$$A(x) = A_0 \exp\left[\left(\frac{S - \sqrt{S^2 + 4\lambda D_B}}{2D_B}\right)(x)\right] \qquad [7]$$

If particle mixing is negligible ($D_B = 0$) below this mixed zone, eqn [1] reduces to

$$\frac{\partial A}{\partial t} = S\frac{\partial A}{\partial x} - \lambda A \qquad [8]$$

For the condition, $A = A_0$ at $x = L$, eqn [8] is solved as

$$A(x) = A_0 \exp\left(-\frac{\lambda(x - L)}{S}\right) \qquad [9]$$

Values of the sediment accumulation rate are determined from eqn [9] by plotting $\ln A$ versus depth (x), such that

$$\ln A = \ln A_L - \lambda(x - L)/S \qquad [10]$$

where A_L is the activity at the base of the mixed layer and $(x - L)$ is depth below the mixed layer. The sediment accumulation rate S is thus determined from the slope of the $\ln A$–x plot.

Deep-sea sediments often contain detrital minerals supplied to the oceans by riverine and atmospheric transport. For the radionuclides in the uranium decay series, these minerals contain small amounts of the parent radionuclides ^{238}U, ^{234}U, ^{235}U, and ^{226}Ra. Activities of ^{234}Th, ^{230}Th, ^{231}Pa, and ^{210}Pb will decrease to equilibrium with the parent activity. For chronometric purposes, the parent activity is subtracted from the measured daughter activity to obtain a quantity that can be used in eqns [1] and [8]. This quantity, termed the 'excess' activity, corresponds to that scavenged from sea water, and given sufficient time (~ 5 half-lives) approaches zero with depth in the sediment column. Indeed, the useful time range of the uranium series radionuclides, as well as the cosmogenic chronometers, is approximately 5 half-lives, after which the activity is only $\sim 3\%$ of the initial value. **Table 1** gives the useful time ranges of the radionuclides commonly measured in deep-sea sediments.

Application of eqn [1] to anthropogenic radionuclides such as ^{137}Cs or $^{239,240}Pu$ does not permit the assumption of steady state because these nuclides were supplied at varying rates with time. Non-steady-state solutions have been formulated, but require the assumption of an input function for the radionuclides to the sediment–water interface. Models including both a constant input since the peak introduction of ^{137}Cs and plutonium to the ocean or a pulse input (maximum in 1963–64) have been used. The validity of these input scenarios is a significant limitation on the use of anthropogenic radionuclides in sediment mixing studies.

Examples of Radionuclide Profiles in Deep-sea Sediments

Figure 1 shows a profile of excess ^{210}Pb in a sediment core taken in a sediment pond in the Mid-Atlantic Ridge. ^{210}Pb in this core is mixed to a depth of 8 cm.

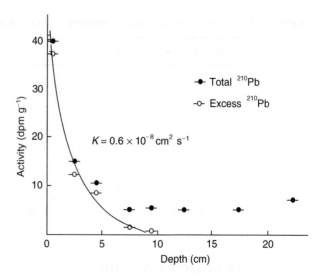

Figure 1 Excess ^{210}Pb activity versus depth in a sediment core from the Mid-Atlantic Ridge. The activity is mixed to ~ 8 cm by the benthic fauna. The rate of mixing (D_B in eqn [7]) is 0.6×10^{-8} cm^2 s^{-1} (~ 0.2 cm^2 a^{-1}). (Reprinted from Nozaki Y, Cochran JK, Turekian KK and Keller G. Radiocarbon and ^{210}Pb distribution in submersible-taken deep-sea cores from Project FAMOUS. *Earth and Planetary Science Letters*, vol. 34, pp. 167–173, copyright 1977, with permission from Elsevier Science.)

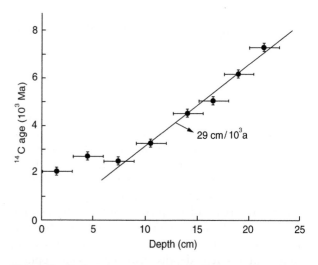

Figure 2 Radiocarbon age versus depth in a sediment core from the Mid-Atlantic Ridge. The age is homogenized in the upper 8 cm owing to mixing by the benthic fauna (see **Figure 1**). Below 8 cm, a sediment accumulation rate of 2.9 cm ka^{-1} is calculated (eqn [9]). (Reprinted from Nozaki Y, Cochran JK, Turekian KK and Keller G. Radiocarbon and ^{210}Pb distribution in submersible-taken deep-sea cores from Project FAMOUS. *Earth and Planetary Science Letters*, vol. 34, pp. 167–173, copyright 1977, with permission from Elsevier Science.)

Such a depth of mixing is quite typical of deep-sea sediments, and below this depth the sediment is undisturbed by mixing by the benthic fauna. **Figure 2** shows the radiocarbon profile in the same core. The rate of sediment accumulation may be calculated from the gradient in radiocarbon ages with depth. Indeed, radiocarbon is unique among the chronometers considered here in providing absolute ages for a given depth in the sediment column. This is possible because radiocarbon can be related to the activity in pre-industrial, pre-bomb carbon to provide an absolute age for the carbon fraction being analyzed. All the other chronometers discussed herein provide relative ages by relating the activity at depth to that at the sediment–water interface or the base of the mixed zone (eqn [9]).

For a long-lived radionuclide such as ^{230}Th, mixing will tend to homogenize the activity in the mixed zone. Below that depth, ^{230}Th will decrease consistently with its decay constant and the sediment accumulation rate (eqn [9]). **Figure 3** shows excess ^{230}Th profiles in three deep-sea cores from the Pacific Ocean. The mixing of the surficial layers is quite clear from the profile. The gradient in activity with depth below the mixed zone yields sediment accumulation rates of 0.14 to 0.30 cm per 1000 years. Sediment accumulation rates of deep-sea sediments determined by the excess ^{230}Th and ^{231}Pa methods typically range from millimeters to centimeters per 1000 years.

Profiles of the anthropogenic radionuclides ^{137}Cs and ^{239,240}Pu in a sediment core of the deep Pacific Ocean are shown in **Figure 4**. A non-steady-state solution to eqn [1] must be applied to these profiles because the radionuclides have been added to the oceans only since 1945 and the profiles are evolving in time as the radionuclides are removed from the overlying water column and added to the sediment–water interface. Particle mixing rates determined from these profiles are 0.36 or 1.4 cm^2 a^{-1} depending on the input function chosen. A pulse input of the radionuclides at the time of maximum fallout to the earth's surface (1963) provides mixing rates that are most similar to that obtained from ^{210}Pb in the same core. Mixing rates of deep-sea sediments determined from short-lived and recently input radionuclides are generally <1 cm^2 y^{-1}. (In shallow water sediments, mixing rates can be two orders of magnitude greater than observed in the deep sea.) The rate and depth of mixing of sediments determines the extent to which changes in paleoceanographic indicators (e.g., oxygen isotopes) can be resolved.

Long-lived radionuclides such as ^{10}Be offer the opportunity to extend radionuclide chronologies of deep-sea sediments to several million years. Recent advances in the measurement of ^{10}Be by accelerator mass spectrometry (AMS) permit analysis of small samples and high-quality chronologies to be determined using this radionuclide. Longer chronologies are especially useful in interpreting the record of parameters such as oxygen or carbon isotopes that

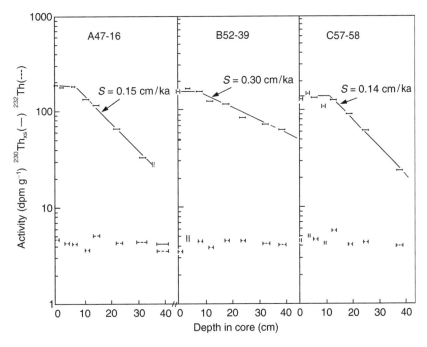

Figure 3 Excess ^{230}Th activities versus depth in three cores from the north equatorial Pacific. The activities are homogenized in the upper ~10 cm as a consequence of particle mixing by the benthic fauna. Accumulation rates calculated from the decreasing portions of the profiles are 0.14–0.3 cm ka^{-1}. (Reprinted from Cochran JK and Krishnaswami S. Radium, thorium, uranium and ^{210}Pb in deep-sea sediment and sediment pore water from the North Equator Pacific. *American Journal of Science*, vol. 280, pp. 847–889, copyright 1980, with permission from American Journal of Science.)

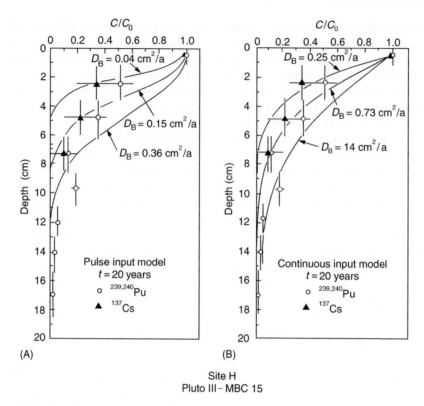

Site H
Pluto III – MBC 15

Figure 4 Activities of ^{239,240}Pu and ^{137}Cs versus depth in a sediment core from the equatorial Pacific. The activities are normalized to the value in the surficial depth interval and are modeled using pulse and continuous inputs of these anthropogenic radionuclides to the sediment–water interface. The profiles are the result of mixing by the benthic fauna. (Reprinted from Cochran JK Particle mixing rates in sediments of the eastern Equatorial Pacific: Evidence from ^{210}Pb, ^{239,240}Pu and ^{137}Cs distributions at MANOP sites. *Geochimica et Casmochimica Acta*, vol. 49, pp. 1195–1210, copyright 1985, with permission from Elsevier Science.)

are linked to paleoceanographic changes. Indeed it has become common to use the now well-established stratigraphy of oxygen isotopes to 'date' depth horizons of deep-sea sediments, yet it is important to recognize that the oxygen isotope stratigraphy was first established through the use of uranium series radionuclides (principally excess ^{230}Th).

Final mention must be made of the dating of horizons preserved in deep-sea sediments via the potassium–argon method. The method is based on the decay of ^{40}K (half-life $= 1.2 \times 10^9$ y) to stable ^{40}Ar, a noble gas. The method is useful only for materials whose initial argon was lost when the rock was formed. Subsequent production of ^{40}Ar in the rock is from ^{40}K decay and the ^{40}Ar/^{40}K ratio serves as an indicator of the rock's age. The method can be used to date volcanic materials that are deposited at the sediment–water interface, for example, as volcanic dust or ash associated with a volcanic eruption. Because of the long half-life of ^{40}K, this method has potential for dating sediments on long timescales, but because of the particular requirements (volcanic material deposited at the sediment–water interface), it is not often possible to use it.

See also

Ocean Margin Sediments.

Further Reading

Berner RA (1980) *Early Diagonesis: A Theortical Approach*. Princeton: Princeton University Press: 241pp.

Boudreau BP (1997) *Diagenetic Models and Their Implementation*, 414 pp. Heidelberg; Germany: Springer-Verlag

Cochran JK (1992) The oceanic chemistry of the uranium- and thorium-series nuclides, In: Ivanovich M and Harmon RS (eds.) *Uranium Series Disequilibrium: Applications to Earth, Marine, and Environmental Sciences* 2nd edn. pp. 334–395. Oxford: Oxford University Press.

Huh C-A and Kadko DC (1992) Marine sediments and sedimentation processes, In: Ivanovich M and Harmon RS (eds.) *Uranium Series Disequilibrium: Applications to Earth, Marine, and Environmental Sciences* 2nd edn. pp. 460–486. Oxford: Oxford University Press.

Libes SM (1992) *Marine Biogeochemistry*, pp. 517–556. New York: Wiley.

Turekian KK (1996) *Global Environmental Change.* Englewood Cliffs, NJ: Prentice-Hall.

Turekian KK and Cochran JK (1978) Determination of marine chronologies using natural radionuclides. In: Riley JP and Chester R (eds.) *Chemical Oceano-graphy*, Vol. 7, pp. 313–360. New York: Academic Press.

Turekian KK, Cochran JK, and DeMaster DJ (1978) Bioturbation in deep-sea deposits: rates and consequences. *Oceanus* 21: 34–41.

DEEP-SEA SEDIMENT DRIFTS

D. A. V. Stow, University of Southampton, Southampton, UK

Introduction

The recognition that sediment flux in the deep ocean basins might be influenced by bottom currents driven by thermohaline circulation was first proposed by the German physical oceanographer George Wust in 1936. His, however, was a lone voice, decried by other physical oceanographers and unheard by most geologists. It was not until the 1960s, following pioneering work by the American team of Bruce Heezen and Charlie Hollister, that the concept once more came before a critical scientific community, but this time with combined geological and oceanographic evidence that was irrefutable.

A seminal paper of 1966 demonstrated the very significant effects of contour-following bottom currents (also known as contour currents) in shaping sedimentation on the deep continental rise off eastern North America. The deposits of these currents soon became known as contourites, and the very large, elongate sediment bodies made up largely of contourites were termed sediment drifts. Both were the result of semipermanent alongslope processes rather than downslope event processes. The ensuing decade saw a profusion of research on contourites and bottom currents in and beneath the present-day oceans, coupled with their inaccurate identification in ancient rocks exposed on land.

By the late 1970s and early 1980s, the present author had helped establish the standard facies models for contourites, and demonstrated the direct link between bottom current strength and nature of the contourite facies, especially grain size. Discrimination was made between contourites and other deep-sea facies, such as turbidites deposited by catastrophic downslope flows and hemipelagites that result from continuous vertical settling in the open ocean. Since then, much progress has been made on the types and distribution of sediment drifts, the nature and variability of bottom currents, and the correct identification of fossil contourites.

Of particular importance has been the work at Cambridge University in decoding the often very subtle signatures captured in contourites in terms of variation in deep-sea paleocirculation. As this is closely linked to climate, the drift successions of ocean basins hold one of the best records of past climate change. This clear environmental significance, together with the recognition that sandy contourites are potential reservoirs for deep-sea oil and gas, has spurred much current research in the field.

Bottom Currents

At the present day, deep-ocean bottom water is formed by the cooling and sinking of surface water at high latitudes and the deep slow thermohaline circulation of these polar water masses throughout the world's ocean (**Figures 1** and **2**). Antarctic Bottom Water (AABW), the coldest, densest, and hence deepest water in the oceans, forms close to and beneath floating ice shelves around Antarctica, with localized areas of major generation such as the Weddell Sea. Once formed at the surface, partly by cooling and partly as freezing sea water leaves behind water of greater salinity, AABW rapidly descends the continental slope, circulates eastwards around the continent and then flows northwards through deep-ocean gateways into the Pacific, Atlantic and Indian Oceans.

Arctic Bottom Water (ABW) forms in the vicinity of the subpolar surface water gyre in the Norwegian and Greenland Seas and then overflows intermittently to the south through narrow gateways across the Scotland–Iceland–Greenland topographic barrier. It mixes with cold deep Labrador Sea water as it flows south along the Greenland–North American continental margin. Above these bottom waters, the ocean basins are compartmentalized into water masses with different temperature, salinity, and density characteristics.

Bottom waters generally move very slowly (1–2 cm s^{-1}) throughout the ocean basins, but are significantly affected by the Coriolis Force, which results from the Earth's spin, and by topography. The Coriolis effect is to constrain water masses against the continental slopes on the western margins of basins, where they become restricted and intensified forming distinct Western Boundary Undercurrents that commonly attain velocities of 10–20 cm s^{-1} and exceed 100 cm s^{-1} where the flow is particularly restricted. Topographic flow constriction is greater on steeper slopes as well as through narrow passages or gateways on the deep seafloor.

Bottom currents are a semipermanent part of the thermohaline circulation pattern, and sufficiently

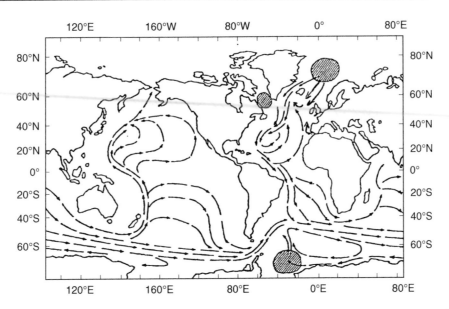

Figure 1 Global pattern of abyssal circulation. Shaded areas are regions of production of bottom waters. (After Stow *et al.*, 1996).

competent in parts to erode, transport and deposit sediment. They are also highly variable in velocity, direction, and location. Mean flow velocity generally decreases from the core to the margins of the current, where large eddies peel off and move at high angles or in a reverse direction to the main flow. Tidal, seasonal, and less regular periodicities have been recorded during long-term measurements, and complete flow reversals are common. Variation in kinetic energy at the seafloor results in the alternation of short (days to weeks) episodes of high velocity known as benthic storms, and longer periods (weeks to months) of lower velocity. Benthic storms lead to sediment erosion and the resuspension of large volumes of sediment into the bottom nepheloid layer. They appear to correspond to episodes of high surface kinetic energy due to local storms.

Deep and intermediate depth water is also formed from relatively warm surface waters that are subject to excessive evaporation at low latitudes, and hence to an increase in relative density. This process is generally most effective in semi-enclosed marginal seas and basins. The Mediterranean Sea is currently the principal source of warm, highly saline, intermediate water, that flows out through the Strait of Gibraltar and then northwards along the Iberian and north European margin. At different periods of Earth history warm saline bottom waters will have been equally or more important than cold water masses.

Contourite Drifts

Contourite accumulations can be grouped into five main classes on the basis of their overall morphology: (I) contourite sheet drifts; (II) elongate mounded drifts; (III) channel-related drifts; (IV) confined drifts; and (V) modified drift–turbidite systems (**Table 1, Figure 3**). It is important to note, however, that these distinctive morphologies are simply type members within a continuous spectrum, so that all hybrid types may also occur. They are also found at all depths within the oceans, including all deep-water (>2000 m) and mid-water (300–2000 m) settings. Those current-controlled sediment bodies that occur in shallower water (50–300 m) on the outer shelf or uppermost slope are not considered contourite drifts *sensu stricto*. The occurrence and geometry of these different types is controlled principally by five interrelated factors: the morphological context or bathymetric framework; the current velocity and variability; the amount and type of sediment available; the length of time over which the bottom current processes have operated; and modification by interaction with downslope processes and their deposits.

Contourite Sheet Drifts

These form extensive very low-relief accumulations, either as part of the fill of basin plains or plastered against the continental margin. They comprise a layer of more or less constant thickness (up to a few hundred meters) that covers a large area, but that demonstrates a very slight decrease in thickness towards its margins, i.e., having a very broad low-mounded geometry. The internal seismofacies is typically one of low amplitude, discontinuous reflectors or, in some parts, is more or less transparent. They may be covered by large fields of sediment waves, as in the case of the South Brazilian and

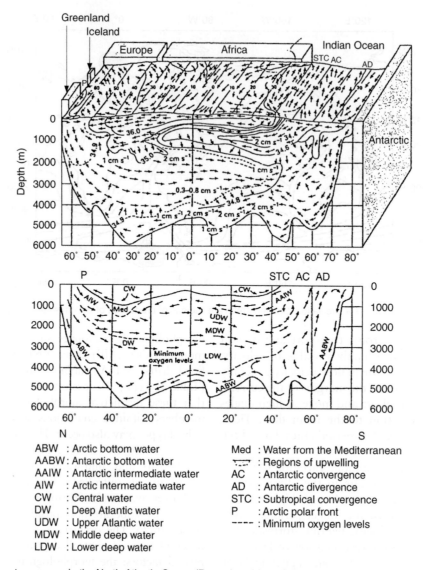

Figure 2 Bottom water masses in the North Atlantic Ocean (Reproduced from Stow *et al.*, 1996).

Argentinian basins where they are also capped in the central region by giant elongate bifurcated drifts.

The different hydrological and morphological contexts define either abyssal sheets or slope sheets (also known as plastered drifts). The former carpet the floors of abyssal plains and other deep-water basins including those of the South Atlantic and the central Rockall trough in the north-east Atlantic. The basin margin relief partially traps the bottom currents and determines a very complex gyratory circulation. Slope sheets occur near the foot of slopes where outwelling or downwelling bottom currents exist, such as in the Gulf of Cadiz as a result of the deep Mediterranean Sea Water outwelling at an intermediate water level into the Atlantic, or around the Antarctic margins as a result of the formation and downwelling of cold AABW. They are also found plastered against the slope at any level, particularly

where gentle relief and smooth topography favors a broad nonfocused bottom current, such as along the Hebrides margin and Scotian margin.

Abyssal sheet drifts typically comprise fine-grained contourite facies, including silts and muds, biogenic-rich pelagic material, or manganiferous red clay, interbedded with other basin plain facies. Accumulation rates are generally low – around $2-4\,\mathrm{cm\,ky^{-1}}$. Slope sheets are more varied in grain size, composition and rates of accumulation. Thick sandy contourites have been recovered from base-of-slope sheets in the Gulf of Cadiz, and rates of over $20\,\mathrm{cm}$ $\mathrm{ky^{-1}}$ (1000 years) are found in sandy–muddy contourite sheets on the Hebridean slope.

Elongate Mounded Drifts

This type of contourite accumulation is distinctly mounded and elongate in shape with variable

Table 1 Drift morphology, classification and dimensions

Drift type	Subdivisions	Size (km²)	Examples
Contourite sheet drift	Abyssal sheet	10^5–10^6	Argentine basin; Gloria Drift
	Slope (plastered sheet)	10^3–10^4	Gulf of Cadiz; Campos margin
	Slope (patch) sheets	10^3	
Elongated mounded drift	Detached drift	10^3–10^5	Eirek drift; Blake drift
	Separated drift	10^3–10^4	Feni drift; Faro drift
Channel-related drift	Patch-drift	10–10^3	North-east Rockall trough
	Contourite-fan	10^3–10^5	Vema Channel exit
Confined drift		10^3–10^5	Sumba drift; East Chatham rise
Modified drift–turbidite systems	Extended turbidite bodies	10^3–10^4	Columbia levee South Brazil Basin; Hikurangi fandrift
	Sculptured turbidite bodies	10^3–10^4	South-east Weddell Sea
	Intercalated turbidite–contourite bodies	Can be very extensive	Hatteras rise

dimensions: lengths from a few tens of kilometers to over 1000 km, length to width ratios of 2 : 1 to 10 : 1, and thicknesses up to 2 km. They may occur anywhere from the outer shelf/upper slope, such as those east of New Zealand to the abyssal plains, depending on the depth at which the bottom current flows. They are very common throughout the North Atlantic, but also occur in all the other ocean basins and some marginal seas. One or both lateral margins are generally flanked by distinct moats along which the flow axis occurs and which experience intermittent erosion and nondeposition. Elongate drifts associated with channels or confined basins are classified separately.

Both the elongation trend and direction of progradation are dependent on an interaction between the local topography, the current system and intensity, and the Coriolis Force. Elongation is generally parallel or subparallel to the margin, with both detached and separated types recognized, but progradation can lead to parts of the drift being elongated almost perpendicular to the margin. Internal seismic character reflects the individual style of progradation, typically with lenticular, convex-upward depositional units overlying a major erosional discontinuity. Fields of migrating sediment waves are common.

Sedimentation rates depend very much on the amount and supply of material to the bottom currents. On average, rates are greater than for sheet drifts, being between 2 and 10 cm ky^{-1}, but may range from <2 cm ky^{-1} for open ocean pelagic biogenic-rich drifts, to >60 cm ky^{-1}, for some marginal drifts (e.g., along the Hebridean margin). The sediment type also varies according to input, including biogenic, volcaniclastic, and terrigenous types. Grain size varies from muddy to sandy as a result of long-term fluctuations in bottom current strength.

Channel-related Drifts

This type of contourite deposit is related to deep channels, passageways or gateways through which the bottom circulation is constrained so that flow velocities are markedly increased (e.g., Vema Channel, Kane Gap, Samoan Passage, Almirante Passage, Faroe-Shetland Channel etc.). Gateways are very important narrow conduits that cut across the sills between ocean basins and thereby allow the exchange of deep and intermediate water masses. In addition to significant erosion and scouring of the passage floor, irregular discontinuous sediment bodies are deposited on the floor and flanks of the channel, as axial and lateral patch drifts, and at the downcurrent exit of the channel, as a contourite fan.

Patch drifts are typically small (a few tens of square kilometers in area, 10–150 m thick) and either irregular in shape or elongate in the direction of flow. They can be reflector-free or with a more chaotic seismic facies, and may have either a sheet or mounded geometry. Contourite fans are much larger cone-shaped deposits, up to 100 km or more in width and radius and 300 m in thickness (e.g., the Vema contourite fan).

Channel floor deposits include patches of coarse-grained (sand and gravel) lag contourites, mud–clast contourites and associated hiatuses that result from substrate erosion, as well as patch drifts of finer-grained muddy and silty contourites where current velocities are locally reduced. Manganiferous mud contourites and nodules are also typical in places. Accumulation rates range from very low, due to nondeposition and erosion, to as much as 10 cm ky^{-1} in some patch drifts and contourite fans.

Confined Drifts

Relatively few examples are currently known of drifts confined within small basins. These typically

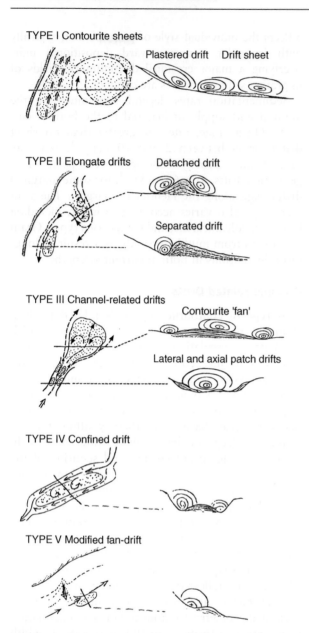

TYPE I Contourite sheets

Plastered drift   Drift sheet

TYPE II Elongate drifts   Detached drift

Separated drift

TYPE III Channel-related drifts

Contourite 'fan'

Lateral and axial patch drifts

TYPE IV Confined drift

TYPE V Modified fan-drift

Figure 3 Contourite drift models. (Modified from Faugeres *et al.*, 1999.)

occur in tectonically active areas, such as the Sumba drift in the Sumba forearc basin of the Indonesian arc system, the Meiji drift in the Aleutian trench and an unnamed drift in the Falkland Trough. Apart from their topographic confinement, the gross seismic character appears similar to mounded elongate drifts with distinct moats along both margins. Sediment type and grain size depend very much on the nature of input to the bottom current system.

Modified Drift-turbidite Systems: Process Interaction

The interaction of downslope and alongslope processes and deposits at all scales is the normal condition on the margins as well as within the central parts of present ocean basins. Interaction with slow pelagic and hemipelagic accumulation is also the norm, but these deposits do not substantially affect the drift type or morphology. Over a relatively long timescale, there has been an alternation of periods during which either downslope or alongslope processes have dominated as a result of variations in climate, sealevel and bottom circulation coupled with basin morphology and margin topography. This has been particularly true since the late Eocene onset of the current period of intense thermohaline circulation, and with the marked alternation of depositional style reflecting glacial–interglacial episodes during the past 2 My (million years).

At the scale of the drift deposit, this interaction can have different expressions as exemplified in the following examples.

1. Scotian Margin: regular interbedding of thin muddy contourite sheets deposited during interglacial periods and fine-grained turbidites dominant during glacials; marked asymmetry of channel levees on the Laurentian Fan, with the larger levees and extended tail in the direction of the dominant bottom current flow.

2. Cape Hatteras Margin: complex imbrication of downslope and alongslope deposits on the lower continental rise, that has been referred to as a companion drift-fan.

3. The Chatham–Kermadec Margin: the deep western boundary current in this region scours and erodes the Bounty Fan south of the Chatham Rise and directly incorporates fine-grained material from turbidity currents that have traveled down the Hikurangi Channel. This material, together with hemipelagic material, is swept north from the downstream end of the turbidity current channel to form a fan-drift deposit.

4. West Antarctic Peninsula Margin: eight large sediment mounds, elongated perpendicular to the margin and separated by turbidity current channels, have an asymmetry that indicates construction by entrainment of the suspended load of down-channel turbidity currents within the ambient south-westerly directed bottom currents and their deposition downcurrent.

5. Hebridean Margin: complex pattern of intercalation of downslope (slides, debrites, and turbidites), alongslope contourites and glaciomarine hemipelagites in both time and space; the alongslope distribution of these mixed facies types by the northward-directed slope current has led to the term composite slope-front fan for the Barra Fan.

Erosional Discontinuities

The architecture of deposits within a drift is complex, stressing variations of the processes and accumulation rates linked to changes in current activity. In many cases, the history of contourite drift construction is marked by an alternation of periods of sedimentation and erosion or nondeposition, the latter corresponding to a greater instability of and/or a drastic change in current regime. The result is the superposition of depositional units whose general geometry is lenticular and whose limits correspond to major discontinuities, that are more or less strongly erosive. These discontinuities can be traced at the scale of the accumulation as a whole and are marked by a strong-amplitude continuous reflector, commonly marking a change in seismofacies linked to variation in current strength. Such extensive and synchronous discontinuities are typical of most drifts. The principal characteristics of drifts evident in seismic records are shown in **Figure 4**.

Contourite Sediment Facies

Several different contourite facies can be recognized on the basis of variations in grain size and composition. These are listed and briefly described below and illustrated in **Figures 5** and **6**.

- Siliciclastic contourites (muddy, silty, sandy and gravel-rich variation)
- Shale-clast/shale-chip contourites (all compositions possible)
- Volcaniclastic contourites (muddy–silty–sandy variations)
- Calcareous biogenic contourites (calcilutite, -siltite, -arenite variations)
- Siliceous biogenic contourites (mainly sand grade)
- Manganiferous muddy contourites (+ manganiferous nodules/pavements)

Muddy contourites

These are homogeneous, poorly bedded and highly bioturbated, with rare primary lamination (partly destroyed by bioturbation), and irregular winnowed concentrations of coarser material. They have a silty-clay grain size, poor sorting, and a mixed terrigenous (or volcaniclastic)–biogenic composition. The components are in part local, including a pelagic contribution, and in part far-traveled.

Silty contourites

These, which are also referred to as mottled silty contourites commonly show bioturbational mottling to indistinct discontinuous lamination, and are gradationally interbedded with both muddy and sandy contourite facies. Sharp to irregular tops and bases of silty layers are common, together with thin lenses of coarser material. They have a poorly sorted clayey-sandy silt size and a mixed composition.

Sandy contourites

These occur as both thin irregular layers and as much thicker units within the finer-grained facies and are generally thoroughly bioturbated throughout. In some cases, rare primary horizontal and cross-lamination is preserved (though partially destroyed by bioturbation), together with irregular erosional contacts and coarser concentrations or lags. The mean grain size is normally no greater than fine sand, and sorting is mostly poor due to bioturbational mixing, but more rarely clean and well-sorted sands occur. Both positive and negative grading may be present. A mixed terrigenous–biogenic composition is typical, with evidence of abrasion, fragmented bioclasts and iron oxide staining.

Gravel-rich contourites

These are common in drifts at high latitudes as a result of input from ice-rafted material. Under relatively low-velocity currents, the gravel and coarse sandy material remains as a passive input into the contourite sequence and is not subsequently reworked to any great extent by bottom currents. Gravel lags indicative of more extensive winnowing have been noted from both glacigenic contourites and from shallow straits, narrow moats, and passageways, where gravel pavements are

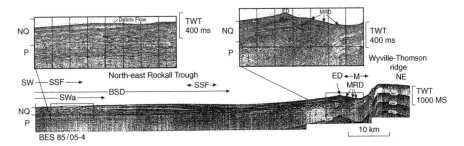

Figure 4 Seismic profiles of actual drift systems.

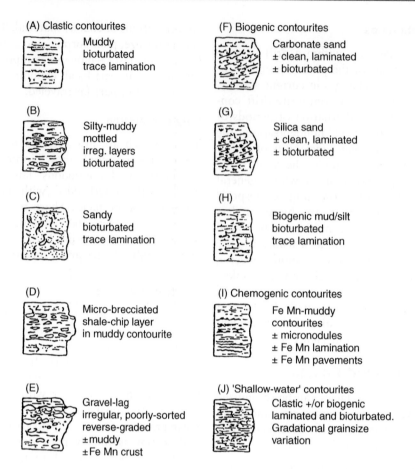

Figure 5 Contourite facies models for clastic, biogenic, chemogenic, and 'shallow-water' contourites. (Reproduced from Stow *et al.*, 1996.)

locally developed in response to high-velocity bottom current activity

Shale-clast or shale-chip layers

These have been recognized in both muddy and sandy contourites from relatively few locations. They result from substrate erosion under relatively strong bottom currents, where erosion has led to a firmer substrate and in some cases burrowing on the omission surface has helped to break up the semi-firm muds.

Calcareous and siliceous biogenic contourites

These occur in regions of dominant pelagic biogenic input, including open ocean sites and beneath areas of upwelling. In most cases bedding is indistinct, but may be enhanced by cyclic variations in composition, and primary sedimentary structures are poorly developed or absent, in part due to thorough bioturbation as in siliciclastic contourites. In rare cases, the primary lamination appears to have been well preserved. The mean grain size is most commonly silty clay, clayey silt or muddy-sandy, poorly sorted and with a distinct sand-size fraction representing the coarser biogenic particles that have not been too fragmented during transport. The composition is typically pelagic to hemipelagic, with nannofossils and foraminifera as dominant elements in the calcareous contourites and radiolaria or diatoms dominant in the siliceous facies. Many of the biogenic particles are fragmented and stained with either iron oxides or manganese dioxde. There is a variable admixture of terrigenous or volcaniclastic material.

Manganiferous contourites

These manganiferous or ferromanganiferous-rich horizons are common. This metal enrichment may occur as very fine dispersed particles, as a coating on individual particles of the background sediment, as fine encrusted horizons or laminae, or as micronodules. It has been observed in both muddy and biogenic contourites from several drifts.

Bottom-current influence

It is important to recognize that bottom currents will influence, to a greater or lesser extent, other deep-water sediments, particularly pelagic, hemipelagic,

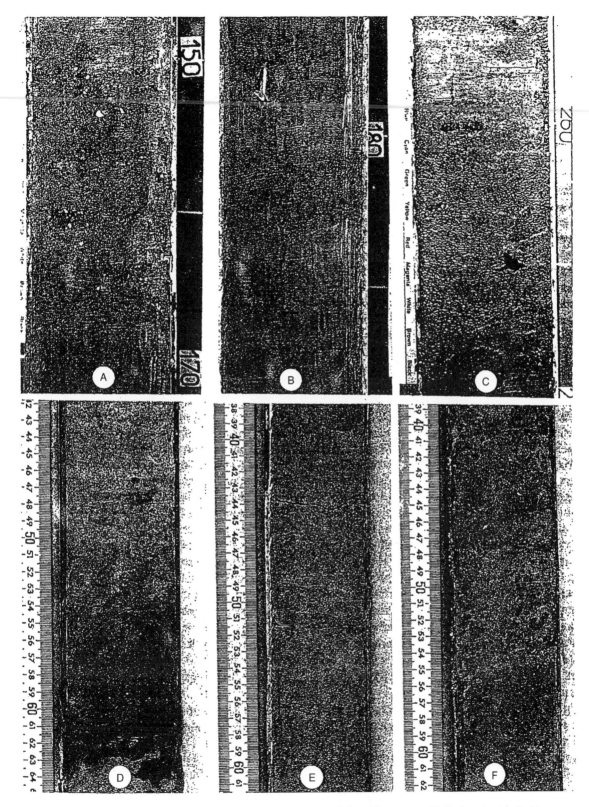

Figure 6 Photographs of contourite facies from cores drilled through existing drift systems. Vertical scales labelled in cm.

turbiditic, and glacigenic, both during and after deposition. Where the influence is marked and deposition occurs in a drift, then the sediment is termed contourite. Where the influence is less severe, so that features of the original deposit type remain dominant, then the sediment is said to have been influenced by bottom currents, as in bottom-current reworked turbidites. Some more-laminated facies, as well as the thin, clean, cross-laminated sands originally described from the north-east American margin, are most likely of this type.

Contourite Sequences and Current Velocity

Muddy, silty, and sandy contourites, of siliciclastic, volcaniclastic, or mixed composition, commonly occur in composite sequences or partial sequences a few decimeters in thickness. The ideal or complete sequence shows overall negative grading from muddy through silty to sandy contourites and then positive grading back through silty to muddy contourite facies (**Figure 7**). Such sequences of grain size and facies variation are now widely recognized, although not always fully developed, and are most probably related to long-term fluctuations in the mean current velocity. Not enough data exist to be certain of the timescale of these cycles, though some evidence points towards 5000–20 000 cycles for certain marginal drifts.

The occurrence of widespread hiatuses in the deep-ocean sediment record is best related to episodes of particularly intense bottom currents. More locally,

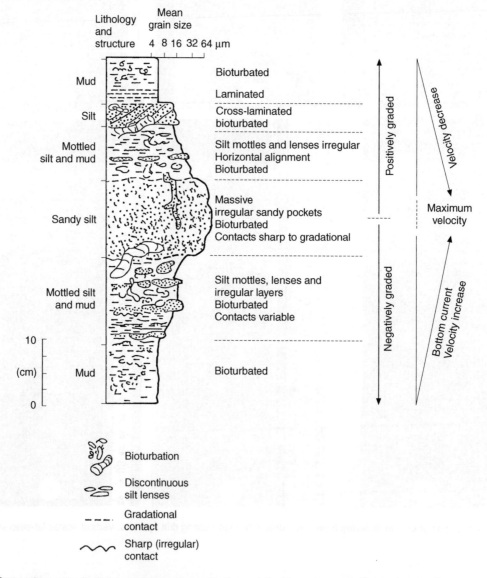

Figure 7 Composite contourite facies model showing grain size variation through a mud–silt–sand contourite sequence. (Modified from Stow *et al.*, 1996.)

such strong currents result in significant sediment winnowing and the accumulation of sand, gravel, and shale-clast contourites. Thick units of sandy contourites together with sandy turbidites reworked by the bottom current are potentially important as hydrocarbon reservoirs where suitably buried in association with source rocks.

Biogenic contourites typically occur in similar sequences of a decimetric scale that show distinct variation in biogenic/terrigenous ratio, generally linked to the grain size variation. This cyclic facies pattern has a longer timescale, in the few examples from which there is good dating, and is closely analogous to the Milankovitch cyclicity recognized in many pelagic and hemipelagic successions. It is, therefore, believed to be driven by the same mechanism of orbital forcing superimposed on changes in bottom-current velocity.

The link between contourite sequences and changes in paleoclimate and paleocirculation is an extremely important one. Where such sequences can be correctly decoded then a more accurate understanding of the paleo-ocean and its environment can be built up.

See also

Ocean Margin Sediments.

Further Reading

Faugeres JC, Stow DAV, Imbert P, Viana A, and Wynn RB (1999) Seismic features diagnostic of contourite drifts. *Marine Geology*, 162: 1–38.

Heezen BC, Hollister CD, and Ruddiman WF (1966) Shaping the continental rise by deep geostrophic contour currents. *Science* 152: 502–508.

McCave IN, Manighetti B, and Robinson SG (1995) Sortable silt and fine sediment size/composition slicing: parameters for paleocurrent speed and paleoceanography. *Paleoceanography* 10: 593–610.

Nowell ARM and Hollister CD (eds.) (1985) Deep ocean sediment transport – preliminary results of the high energy benthic boundary layer experiment. *Marine Geology* 66.

Pickering KT, Hiscott RN, and Hein FJ (1989) *Deep-Marine Environments: Clastic Sedimentation and Tectonics*. London: Unwin Hyman.

Stow DAV and Faugeres JC (eds.) (1993) Contourites and Bottom Currents, *Sedimentary Geology*, Special Volume 82, 1–310.

Stow DAV, Reading HG, Collinson J (1996) Deep seas. In: Reading HG (ed.) *Sedimentary Environments and Facies*. 3rd edn, pp. 380–442. Blackwell Science Publishers.

Stow DAV and Faugeres JC (eds.) (1998) Contourites, turbidites and process interaction. *Sedimentary Geology Special Issue* 115.

Stow DAV and Mayall M (eds.) (2000) Deep-water sedimentary systems: new models for the 21st century. *Marine and Petroleum Geology*, Special Volume 17.

BIOTURBATION

D. H. Shull, Western Washington University, Bellingham, WA, USA

© 2009 Elsevier Ltd. All rights reserved.

Introduction

Activities of organisms inhabiting seafloor sediments (termed benthic infauna) are concealed from visual observation but their effects on sediment chemical and physical properties are nevertheless apparent. Sediment ingestion, the construction of pits, mounds, fecal pellets, and burrows, and the ventilation of subsurface burrows with overlying water significantly alter rates of chemical reactions and sediment–water exchange, destroy signals of stratigraphic tracers, bury reactive organic matter, exhume buried chemical contaminants, and change sediment physical properties such as grain size, porosity, and permeability. Biogenic sediment reworking resulting in a detectable change in sediment physical and chemical properties is termed bioturbation. It is critical to account for bioturbation when calculating chemical fluxes at the sediment–water interface and when interpreting chemical profiles in sediments. In the narrowest sense, bioturbation refers to the biogenic transport of particles that destroys stratigraphic signals. In the broader sense it can refer to biogenic transport of pore water and changes in sediment physical properties due to organism activities as well. Bioturbation and its effects on sediment chemistry and stratigraphy is a natural consequence of adaptation by organisms to living and foraging in sediments.

Particle Bioturbation

Deposit feeding, the ingestion of particles comprising sedimentary deposits, is the dominant feeding strategy in muddy sediments. In fact, since the vast majority of the ocean is underlain by muddy sediments, deposit feeding is the dominant feeding strategy on the majority of the Earth's surface. Because digestible organic matter typically comprises less than 1% of sediments by mass, to meet their metabolic needs deposit feeders exhibit rapid sediment ingestion rates, averaging roughly three body weights per day. Deposit feeders adapted to living in sediments with relatively low organic matter concentrations tend to exhibit elevated ingestion rates; there is no free lunch even for deposit feeders. Rates of deposit feeding of individual organisms increase with increasing body size so that bioturbation rates in some sedimentary deposits may be controlled by a handful of larger species. Deposit feeders employ a wide variety of strategies to collect particles for food, but reworking modes due to deposit feeding can be broken down into the following categories: conveyor-belt feeding where particles are collected at depth and deposited at the sediment surface; subductive feeding, where particles are collected at or near the sediment surface and deposited at depth; and interior feeding where particles are collected and deposited within the sediment column. These feeding modes transport particles the length of the organism's body or the length of its burrow. Some species of deposit feeders also ingest and egest sediment near the sediment surface, resulting in horizontal movement of particles but limited vertical displacement. Due to rapid particle ingestion rates and relatively large horizontal and vertical transport distances, deposit feeding is considered to be the primary agent of bioturbation.

Benthic organisms also rework sediments through burrow formation. Muddy sediments behave more like elastic solids than granular material. A benthic burrower in muddy sediments uses its burrowing apparatus (bivalve foot, polychaete proboscis, amphipod carapace, or other burrowing tool) like a wedge to create and propagate cracks in sediments. After an organism passes through a crack, sediments tend to rebound viscoelastically resulting in relatively little net movement of particles compared to deposit feeding. An exception to this generality is burrowing by large epibenthic predators including skates, rays, and benthic-feeding marine mammals such as gray whales and walrus, which can cause extensive reworking in sediment patches where they are feeding.

From a particle's perspective, bioturbation consists of relatively short-lived intervals of particle movement due to deposit feeding or burrowing interspersed by relatively long periods during which the particles remain at rest. When particles pass through animal guts, in addition to changing location, the particles may be aggregated into fecal pellets (particles surrounded by or embedded in mucopolymers). When constructing burrows, some infauna produce mucopolymer glue to form sturdy burrow walls, locking particles into place for an extended period of time. Transport of subsurface particles to the sediment surface by conveyor-belt feeding results in downward gravitative movement of particles within the sediment column as subsurface feeding voids are filled with sediment from above. Within a particular sedimentary habitat many particle reworking mechanisms occur simultaneously.

There are many ways to quantify mathematically the ensemble of particle motions that results in bioturbation. Traditionally, bioturbation has been modeled as though it were analogous to diffusion. This means that the collection of particle motions resembles a large number of small random displacements. Under these assumptions, bioturbation is included in the general diagenetic equation as a biodiffusion coefficient, D_B. Ignoring vertical gradients in porosity and sediment compaction, the rate of change of a chemical tracer, C, in the vertical spatial dimension, x, can be represented as follows:

$$\frac{\partial C}{\partial t} = D_B \frac{\partial^2 C}{\partial x^2} - u\frac{\partial C}{\partial x} + \sum R, \quad x < L \quad [1]$$

where D_B is the biodiffusion coefficient ($\text{cm}^2\,\text{yr}^{-1}$), u is the sediment accumulation rate ($\text{cm}\,\text{yr}^{-1}$), and $\sum R$ represents the sum of chemical reactions. In the absence of specific information on bioturbation mechanisms, it is often assumed that D_B is constant throughout the reworked layer to the depth L. Below the depth L, D_B is zero. The advantage of this formulation is that all the various particle reworking processes are quantified by one parameter, D_B. The nondimensional Peclet number, $Pe = uLD_B^{-1}$, quantifies the relative importance of bioturbation and sediment accumulation in determining the profile of C within the reworked layer. Values of Pe less than 1 indicate a strong influence of bioturbation. Table 1 summarizes the general pattern of variation in D_B, u, L, and Pe among benthic provinces at different water depths. The depth of the reworked layer, L, shows little systematic variation among habitats, averaging 10 cm. Although we would expect considerable variation in Pe, the low values in each province indicate that bioturbation is generally important throughout the ocean. An easy-to-

remember rule of thumb regarding bioturbation rates is that D_B varies from $c.\ 0.01$ to $100\,\text{cm}^2\,\text{yr}^{-1}$ from deep-sea to shallow-water depths. This variation is correlated with increased abundance of infauna, greater rates of food supply, and (with the exception of polar regions) elevated average bottom-water temperatures with decreasing water depth.

Because bioturbation mechanisms can transport particles relatively large distances, roughly the length of the reworked zone, L, and because particle trajectories are often nonrandom, the biodiffusion coefficient is not appropriate for modeling the effects of bioturbation on transport of some tracers. A more general model of particle mixing that includes longer-distance, nonrandom particle trajectories is the nonlocal bioturbation model. Again neglecting variation in porosity:

$$\frac{\partial C}{\partial t} = \int\limits_0^L K(x', x; t)C(x')\mathrm{d}x$$

$$- C(x)\int\limits_0^L K(x, x'; t)\mathrm{d}x' - u\frac{\partial C}{\partial x} + \sum R \quad [2]$$

where K is the exchange function (dimensions: time^{-1}) that quantifies the rate of particle movement from one depth, x, to any other depth, x'. The first term on the right-hand side gives the concentration change at depth x due to the delivery of a particle tracer from other depths, x'. The second term gives the concentration change at depth x due to transport of a tracer from depth x to other depths x'. The other terms are defined as in eqn [1]. The exchange function, K, can potentially quantify a complex ensemble of bioturbation mechanisms. Analogs of eqn [2] that rely on discrete mathematics exist. In one dimension, nonlocal transport can be modeled as a transition matrix in which the rows of the transition matrix correspond to depths in the sediment and the matrix elements quantify the probability of transport of a tracer among depths. Multiple-dimensional automaton models can simulate complex modes of particle transport in both vertical and horizontal dimensions. These more complex models can better capture the complexities of bioturbation but sacrifice the one-parameter simplicity of eqn [1].

There are two common approaches for determining the values of the bioturbation parameters in these models. Mixing parameters can be estimated from direct measurements of deposit-feeder ingestion rates and organism burrowing rates. More often, these parameters are estimated by measuring sediment-bound tracers with known inputs to the sediment and

Table 1 Variation in the biodiffusion coefficient, D_B, sedimentation rate, u, and the Peclet number, Pe, characteristic of different benthic environments. A Peclet number greater than 1 indicates sediment accumulation is more important than bioturbation

	D_B		u		L	Pe	
Shallow water	10	100	0.1	1	10	0.01	1
Cont. Shelf	0.1	10	0.01	0.5	10	0.01	50
Slope	0.05	1	0.001	0.05	10	0.01	10
Deep sea	0.01	0.5	0.0001	0.01	10	0.002	10

A Peclet number less than 1 indicates that bioturbation is more important for transport relative to sedimentation.

known reaction rates ($\sum R$). Mixing parameters are then calculated by fitting measured tracer profiles to profiles calculated by use of the appropriate mathematical model. Useful bioturbation tracers include excess activities of naturally occurring particle-reactive radionuclides such as ^{234}Th, ^{210}Pb, or ^{7}Be. These radionuclides have a relatively continuous source either from the atmosphere or from the overlying water column, are rapidly scavenged onto particles, and sink to the seafloor (*see* Sediment Chronologies). In addition, chlorophyll *a*, artificial tracers added to the sediment surface as an impulse such as glass beads or fluorescent luminophores, or other exotic identifiable material with a known rate of input are used as tracers of bioturbation. The profile of excess ^{210}Pb in **Figure 1** illustrates several effects of bioturbation on a tracer profile. The rate of bioturbation in the top 6 cm is rapid enough to completely mix excess ^{210}Pb within this layer. The subsurface peak at 15–16 cm indicates subsurface deposition of surficial material. Below 16 cm, the slope of the profile is set by the rate of sediment accumulation and radioactive decay of ^{210}Pb.

Bioturbation has important consequences for sediment stratigraphy, chemistry, and biology. Bioturbation can homogenize tracers within the reworked layer (**Figure 1**). Bioturbation acts as a low-pass filter, destroying information deposited on short timescales, but preserving longer-term trends.

Bioturbation makes it generally difficult or impossible to resolve timescales of less than 10^3 years stratigraphically in deep-sea sediment cores. If the bioturbation mechanism is not known, it is difficult to separate changes in the input signal from changes due to mixing (**Figure 2**). If mixing is not complete, and the bioturbation mechanism is known, it may be possible to deconvolve the input signal to the stratigraphic record, although detailed information will be lost. If bioturbation in the surface reworked zone completely homogenizes a tracer, then knowing the mixing mechanism is unimportant. Once pancake batter is thoroughly mixed, for example, it no longer contains information on how the mixing was performed.

Bioturbation has important consequences for sediment geochemistry. Bioturbation buries reactive organic matter. Subductive deposit feeders selectively feed on organic-rich particles near the sediment surface and deposit them at depth, perhaps as food caches. In the presence of horizontal transport of organic matter, or suspension-feeding benthos that locally enhance organic matter deposition through biodeposit formation, bioturbation can greatly enhance the organic matter inventory in sediments. Burial of organic matter exposes it to different oxidants, changing the degradation pathway. In particular, reworking of Mn and Fe oxides cycles them between oxidative and reducing environments,

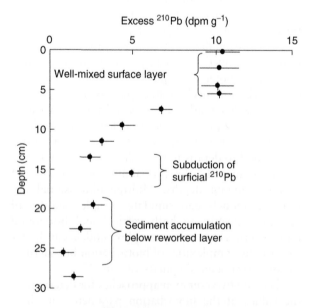

Figure 1 Excess ^{210}Pb activity vs. depth in a sediment core from Narragansett Bay, Rhode Island. Data with permission from Shull DH (2001) Transition-matrix model of bioturbation and radionuclide diagenesis. *Limnology and Oceanography* 46(4): 905–916. Copyright (2001) by the American Society of Limnology and Oceanography, Inc.

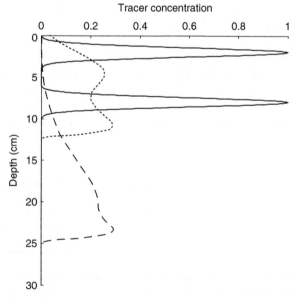

Figure 2 Changes in the profile of a hypothetical conservative tracer present initially as two narrow subsurface peaks, as predicted from eqn [1]. $D_B = 0.1\,\text{cm}^2\,\text{yr}^{-1}$, $u = 0.1\,\text{cm}\,\text{yr}^{-1}$, $L = 10\,\text{cm}$. Solid line: tracer profile at $t = 0$. Dotted line: $t = 25$ years. Dashed line: $t = 150$ years.

resulting in enhanced anaerobic degradation of organic matter.

Bioturbation changes sediment properties as well. Pelletization changes the sediment grain size distribution. Furthermore, bioturbation rates are particle-size-dependent. Size-selective feeding by deposit feeders results in biogenic graded bedding with lag layers of large sediment particles either at the sediment surface or at depth, depending upon the bioturbation mechanism. Formation of pellets and burrows increases sediment porosity, counteracting the effects of sediment compaction. Sediment surface manifestations of bioturbation such as pits, mounds, and tubes alter seafloor roughness and flow characteristics of the benthic boundary layer, roughly doubling the drag compared to a hydrodynamically smooth bed.

Pore-Water Bioirrigation

Most benthic infauna maintain a burrow that connects to the sediment–water interface to facilitate respiration, feeding, defecation, and other metabolic processes. These burrows exist in a range of geometries including vertical cylinders, U- or J-shaped tubes, or branching networks. In deep-sea sediments, dissolved oxygen can penetrate 30 m into the sediment. Near the shore, however, oxygen penetration is quite variable and in muddy sediments it often penetrates no farther than a few millimeters. To meet their metabolic requirements for oxygen, infauna ventilate their burrows by thrashing their bodies, flapping their appendages, by peristalsis, by beating cilia, or by oscillating like pistons in their tubes. These ventilation mechanisms result in intermittent burrow flushing, which exchanges a portion of the fluid inside the burrow with overlying water. In this way, organisms in the sediment can flush out metabolic wastes and toxins such as hydrogen sulfide that have accumulated in their burrows and they can restock the burrow water with dissolved oxygen. Observations of organisms in artificial tubes maintained in the laboratory indicate that burrow ventilation is episodic, with ventilation frequencies ranging from once per hour to 10 or more ventilation events per hour. Deposit-feeding infauna generally ventilate less frequently than suspension-feeding infauna, which pump water through their burrows for both respiration and food capture.

The sediments into which these burrows are built are mixtures of particles and interconnected pore water. Surficial sediments may possess porosities (defined as the volume of interconnected pore water per unit volume of sediment) in excess of 90%. Thus, surface sediments generally contain more pore water than particles. The rate of molecular diffusion of solutes through pore water is reduced relative to diffusion in a free solution because the solutes must follow a winding path through the particles, called sediment tortuosity. Particle bioturbation mechanisms redistribute this pore water along with the particles, but since rates of pore-water transport, inferred from dissolved tracer distributions, are typically an order of magnitude higher than rates of particle bioturbation, particle reworking is a relatively unimportant mechanism for transporting pore water. Rather, burrow ventilation seems to be the most important biogenic mechanism of pore-water transport. The consequences of burrow ventilation on pore-water transport in the surrounding sediments (termed bioirrigation) depend upon sediment permeability.

Sandy sediment typically possesses high enough permeability to allow advective flow of pore water through the interconnected pore space surrounding sediment particles. Under these conditions, the pressure head within a burrow created by burrow ventilation activities can drive pore-water flow from the burrow into surrounding sediments. The velocity of this flow can be calculated using Darcy's law:

$$u_d = -\frac{k}{\mu}(\nabla P - \rho g \nabla x) \qquad [3]$$

where u_d is the Darcy velocity, k is sediment permeability, μ is the dynamic viscosity of pore water, P is pressure, ρ is the pore-water density, g is gravity, and ∇ is a gradient operator (e.g., $\partial/\partial x$, $\partial/\partial y$). The velocity of pore water, u, is related to the Darcy velocity, $u_d = u\varphi^{-1}$, where φ = porosity. Substituting eqn [3] into the general diagenetic equation gives the expected change in concentration of a pore-water tracer subjected to an advection velocity driven by burrow ventilation, molecular diffusion, and chemical reactions:

$$\frac{\partial C}{\partial t} = D'_M \frac{\partial^2 C}{\partial x^2} - u\frac{\partial C}{\partial x} + \sum R, \quad x < L \qquad [4]$$

where D'_M is the molecular diffusion coefficient, corrected for tortuosity.

In contrast to sandy sediments, permeability of mud is generally too low to permit significant pore-water advection so that $u = 0$. Thus, pore-water transport in muddy sediments is dominated by molecular diffusion. Burrow ventilation in muddy sediments enhances pore-water transport by changing the diffusive geometry. **Figure 3** shows the geometry of idealized equally spaced vertical

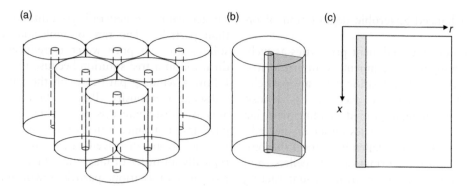

Figure 3 Idealized burrow geometry underlying eqn [5]. (a) Burrows as close-packed cylinders. (b) Rectangular plane intersecting an average burrow microenvironment. (c) The $x - r$ domain of eqn [5]. The shaded rectangle represents the burrow, while the unshaded rectangle represents the surrounding sediment. Reproduced from Aller RC (1980) Quantifying solute distributions in the bioturbated zone of marine sediments by defining an average microenvironment. *Geochimica et Cosmochimica Acta* 44(12): 1955–1965, with permission from Elsevier.

burrows embedded into sediment. If these burrows were rapidly flushed so that the solute concentration within the burrows were equal to the solute concentration in the overlying water, then the corresponding diagenetic equation governing pore-water transport in the vertical dimension, x, and in the radial dimension, r, would be given by

$$\frac{\partial C}{\partial t} = D'_M\left(\frac{\partial^2 C}{\partial x^2} + \frac{1}{r}\frac{\partial}{\partial r}\left(r\frac{\partial C}{\partial r}\right)\right) + \sum R \qquad [5]$$

The diffusion operator within the parentheses is similar to the diffusion operator in eqn [4], but quantifies molecular diffusion in both the x and r dimensions. A one-dimensional diagenetic model that incorporates the effects of bioirrigation on pore-water transport can be derived from eqn [5]:

$$\frac{\partial C}{\partial t} = D'_M\frac{\partial^2 C}{\partial x^2} - \alpha(C - C_0) + \sum R \qquad [6]$$

where $\alpha(day^{-1})$ is the coefficient of nonlocal bioirrigation, and C_0 is the concentration of the solute tracer in overlying water. The nonlocal bioirrigation coefficient, α, in eqn [6] treats bioirrigation as both a source and a sink for solutes at depth.

The value of the bioirrigation exchange rate, α, is usually determined by measuring dissolved pore-water tracers with known inputs and reaction kinetics. The most commonly used radionuclide tracer of bioirrigation is ^{222}Rn. Produced within sediments from the decay of its parent ^{226}Ra, ^{222}Rn is a soluble noble gas. Pore-water exchange with overlying water results in lower ^{222}Rn activity in sediment pore waters than would be expected, compared to the activity of its parent ^{226}Ra. The shape of the ^{222}Rn profile and the magnitude of the ^{222}Rn deficit relative to ^{226}Ra are used to determine rates of bioirrigation

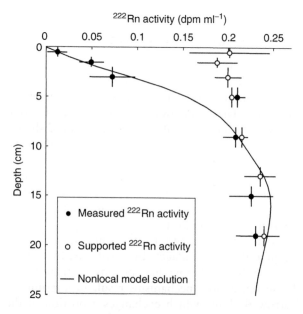

Figure 4 Measured ^{222}Rn activity and supported ^{222}Rn activity (produced from the decay of ^{226}Ra within sediment particles) vs. depth in a sediment core from Boston Harbor, Massachusetts. Horizontal error bars represent standard error from three replicate cores. The bioirrigation rate, α (day^{-1}), was modeled as the exponential function. $\alpha = 3.8e^{-x}$, and the modeled profile was calculated from eqn [6] and from Shull DH, previously unpublished.

(**Figure 4**). Other tracers of pore-water exchange include inert solutes such as bromide or dissolved nutrients such as ammonium, nitrate, or silicate, if reaction kinetics can be estimated.

Bioirrigation has important implications for sediment geochemistry. Bioirrigation accelerates sediment–water fluxes, changes rates of elemental cycling, catalyzes oxidation reactions in the sediment, changes vertical and horizontal gradients of pore-water solutes, elevates levels of dissolved oxygen, and reduces exposure of organisms to metabolic wastes. By changing the redox geometry of sediments, bioirrigation can significantly alter rates of redox-sensitive reactions that occur in sediments such as nitrification, denitrification, sulfate reduction, and mercury methylation.

Bioturbation and the Ecology and Evolution of Benthic Communities

Bioturbation has numerous effects on benthic community structure. In muddy sediments, bioturbation by deposit feeders appears to reduce densities of suspension feeders. Conveyor-belt bioturbation can displace surface-dwelling benthos. Bioturbation changes the depth distribution of organic matter and can increase the inventory and quality of food for deposit feeders in sediments. It can increase nutrient fluxes leading to elevated rates of benthic primary production and increased microbial productivity as well. Furthermore, elevated nutrient recycling between sediments and overlying water helps to maintain water-column productivity in estuaries and other shallow-water marine environments.

Marine benthic habitats of the late Neoproterozoic and early Phanerozoic (600–500 Ma) were very different from benthic habitats that existed later. Seafloors at this time were characterized by well-developed microbial mats, as suggested by studies of sedimentary fabric preserved in the geologic record. These extensive microbial mats and associated sea-floor fauna, such as immobile suspension-feeding helicopacoid echinoderms, became scarce or extinct in the Cambrian. The substantial change that occurred in seafloor communities around this time, termed the 'Cambrian substrate revolution', is thought to be caused by the development of bioturbation. It is hypothesized that the emergence of both bioturbation and predation around this time led to the extinction of nonburrowing taxa and influenced subsequent development of animal body plans during the Cambrian. Bioturbation also made a new food resource, buried organic matter, accessible to deposit feeders while radically changing the geochemistry of the seafloor.

See also

Ocean Margin Sediments. Sediment Chronologies.

Further Reading

Aller RC (1980) Quantifying solute distributions in the bioturbated zone of marine sediments by defining an average microenvironment. *Geochimica et Cosmochimica Acta* 44(12): 1955–1965.

Aller RC (1982) The effects of macrobenthos on chemical properties of marine sediments and overlying waters. In: McCall PL and Tevesz MJS (eds.) *Animal–Sediment Relations*, pp. 53–102. New York: Plenum.

Boudreau BP and Jorgensen BB (2001) *The Benthic Boundary Layer: Transport Processes and Biogeochemistry*. Oxford, UK: Oxford University Press.

Dorgan KM, Jumars PA, Johnson BD, Boudreau BP, and Landis E (2005) Burrowing by crack propagation: Efficient locomotion through muddy sediments. *Nature* 433: 475.

Lohrer AM, Thrush SF, and Gibbs MM (2004) Bioturbators enhance ecosystem function through complex biogeochemical interactions. *Nature* 431: 1092–1095.

Meysman F, Boudreau BP, and Middelburg JJ (2003) Relations between local, non-local, discrete and continuous models of bioturbation. *Journal of Marine Research* 61: 391–410.

Shull DH (2001) Transition-matrix model of bioturbation and radionuclide diagenesis. *Limnology and Oceanography* 46(4): 905–916.

METHANE HYDRATE AND SUBMARINE SLIDES

J. Mienert, University of Tromsø, Tromsø, Norway

Introduction

Methane hydrate is an ice-like substance composed of water molecules forming a rigid lattice of cages that enclose single molecules of gas, mainly methane. They occur in vast quantities beneath the ocean floor, and they are stable under high-pressure and low-temperature conditions existing along many continental margins (**Figure 1**). In siliciclastic margins, sediment sliding and slumping can generate simple or complex slide scars and seabed morphologies (**Figure 2**). The gravity-driven movements of sediments masses vary in volume from a few cubic kilometers to several thousands of cubic kilometers. Giant slides reach run-out distances close to 1000 km, such as the Storegga slide on the mid-Norwegian margin. Average slope angles are low and vary typically between 2° and 4°. A slide is distinguished from a slump based on the Skampton ratio h/l, where h is the thickness of the sliding block or depth and l is the length of the slide. Most observed submarine mass movements appear to be translational with Skampton ratios <0.15. Slumps are rotational with Skampton ratios >0.33 and may coexist with translational slides creating mingled slope-failure generations.

Seabed bathymetry and reflection seismic data from continental slopes and rises of the Atlantic, Pacific, and Indian Oceans and the Black Sea reveal major slide scars and slumps overlying gas hydrate deposits. Slides and slumps shape the morphology of world ocean margins and represent a major mechanism for transferring enormous amounts of sediment material over a relatively short time, between hours and years, into the deep sea. In most of the past slope-failure events, the trigger mechanisms of the slides and slumps are unknown and lead to speculations about various mechanisms. Many of the large slides are associated with surprisingly low slope angles (<4°), much less than are required to destabilize the seafloor under static conditions. It appears that failure mechanisms have the highest possibility of triggering slumps during sea-level lowstands. However, this is not true for some of the largest slides known on continental margins; for example, the Storegga and Traenadjupet slides on the mid-Norwegian margin occurred during the last 8000 years within a period of rapid sea-level rise.

The role of methane hydrate in the development of slides and slumps is an intriguing question in ocean

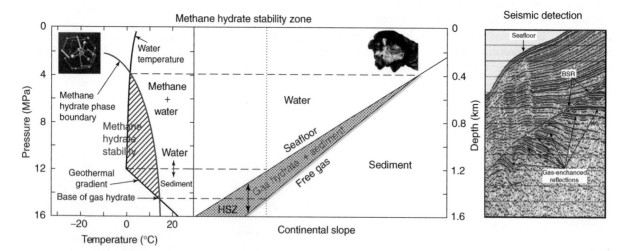

Figure 1 Methane hydrate stability zone as a function of water temperature, geothermal and pressure gradient. A methane hydrate I structure is shown on the upper left and a methane hydrate sample on the upper right. The hydrate stability zone (HSZ) increases with water depth (pressure) while the base of the hydrate stability zone (BHSZ) is evident from the seismic detection by a bottom simulating reflector (BSR). The hydrate occurrence zone (HOZ) where hydrates occupy the pore space of sediments lies between the seafloor and the BSR. Adapted from Kayen RE and Lee HJ (1991) Pleistocene slope instability of gas-hydrate laden sediment on the Beaufort Sea margin. *Marine Technology* 10: 32–40.

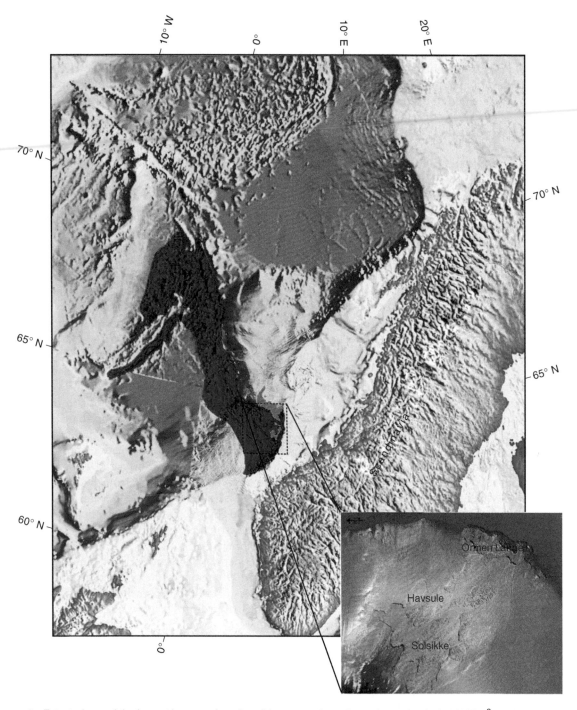

Figure 2 Extent of one of the largest known submarine slides on continental margins and today's 4000 km² gas hydrate province inferred from BSR distributions (yellow). The slide affected an area of approximately 95 000 km² and removed 2500–3500 km³ of sediments. The main slide scar is shown on the inlet figure with the main headwall at the continental shelf and several headwalls and individual glide planes downslope.

and climate research. The interest stems mainly from the hypothesis that large releases of methane from marine gas hydrate reservoirs generate continental margin instability. If methane hydrate dissociates, gas and fresh water are released. Based on the amount of dissociated hydrate, this volume increase can generate pore-fluid overpressure and zones of reduced shear strength; thus glide planes develop in continental margin sediments. These zones develop critical preconditions for a slope to fail. Continental margin instability leading to slides or slumps and abrupt release of methane, a greenhouse gas, may be sufficiently large enough to trigger major climatic changes. Substantial uncertainties exit in the amount of gas hydrate in space and time, and in the dynamics of dissociation processes controlled mainly by ocean pressure (sea level) and temperature (hydro- and geothermal), and material properties of the sub-seafloor. They lead to many developments of regional and global scenarios and a basis for a continuing debate.

One of the main debates centers on the explanation of observations of massive injections of isotopically light carbon into the oceans and atmosphere in Cenozoic and Mesozoic times, and whether the observations in the geological record are sufficient to propose a major release of methane from gas hydrate. Accumulating seabed observations from various areas along continental margins indicate massive slumping during late Cenozoic and Mesozoic times. Many have wondered which processes could account for giant slumps ($>1000 \, km^2$), aside from the classical preconditioning factors and trigger mechanisms such as oversteepening of slopes, rapid sediment loading, and, most importantly, earthquakes. The idea of linking methane hydrate and slumps became widespread after 1981 following a discovery of a slump/gas hydrate area on the eastern US continental margin. Further observations where slump locations coincide with present gas hydrate reservoirs suggested linkages, which in turn are considered as possible agents for rapid releases of methane. But the picture is more complex because climatically induced increases in sedimentation rate and changes in sediment type can create zones of weakness. The geological evidence to discriminate the mechanisms is generally missing after an offshore slide or slump has occurred. From the existing knowledge, it is difficult to extract any overarching role of methane hydrate in slope stability and slumping.

One solution may arise from investigations of slump timing and frequency. Since the temperature–pressure regime on continental margins controls the stability of methane hydrates on a large scale, either of these can alter the stability of hydrates. This knowledge implies that methane hydrate reservoirs in continental margins should react vigorously and almost simultaneously with global change in the ocean environments, that is, sea level and temperature. Yet, current knowledge of all geophysically inferred gas hydrate occurrences and all drilled gas hydrate reservoirs underlines that we have no reason to consider hydrates to be either continuously or homogenously distributed in continental margins. Geophysical and geochemical data demonstrate that methane hydrate accumulations are a result of fluid migration and their pathways. Attempts to verify

coupled processes between hydrate dissociation and slumps from geological records are ongoing and are notably important for understanding climate-induced geohazards. It requires a knowledge of where, how much, and how far accumulations of gas hydrate extend in continental margins. Equally important is an adequate knowledge of changes in slope stability during hydrate formation and dissociation processes in a variety of geological environments from high to low latitudes.

Methane Hydrate Stability and Seismic Detection

Methane hydrates belong to compounds named 'clathrates' after the Latin *clathratus*, meaning encaged. Gases that form hydrates occur mainly in two distinct structures: structure I hydrates form with natural gases containing molecules smaller than propane. They are found *in situ* in continental margin sediments with biogenic gases that consist mostly of methane but rarely contain carbon dioxide and hydrogen sulfide. Structure II hydrates form from thermogenic gases that, in addition to methane, incorporate molecules larger than ethane but smaller than pentane. The Gulf of Mexico is an area of massive but local development of structure II hydrates, where salt tectonics and related structures control focused fluid flow from hydrocarbon reservoirs. It is therefore of less relevance for extended slides and slumps. For oceanic methane hydrates and slumps, structure I hydrates are considered to be important.

The methane hydrate stability depends mainly on pressure (P), temperature (T), and the presence of gas in excess of solubility as a function of P and T. The stability of hydrates is more susceptible to changes in temperature than in pressure. Methane hydrate equilibrium conditions are complex, depending also on the texture of the host sediments, pore water salinity, and type of gas entrapped. Laboratory results show that sand is more conducive to hydrate formation than silt and clay. Hydrate formation in sands starts at a lower pressure (shallower water depth) compared to silt and clay. Beneath the hydrate stability zone (HSZ), the solubility of gas controls the amount of free gas. Lateral and vertical variations in texture and mineralogical properties of sediment contribute to a heterogeneous distribution of hydrate in the HSZ and free gas beneath it.

It is generally accepted that the depth of the HSZ can often be found by seismic detection of the base of the hydrate stability zone (BHSZ). The base is generally inferred from the presence of a bottom-simulating seismic reflector (BSR), though the absence of a BSR does not rule out the presence of hydrates. A phased-reversed seismic reflector originates at this hydrate–gas phase boundary due to a distinct acoustic impedance contrast. The impedance contrast is caused by the low p-wave velocity of gas-bearing sediments beneath the higher p-wave velocities of hydrate-bearing sediments. The bottom-simulating behavior of the hydrate–gas phase boundary suggests a direct dependence on geothermal gradients, that is, temperature. Geological controls on hydrate formation may explain the often-observed discrepancies between the calculated BHSZ and the calculated and observed BSR depth. Observations of BSR's beneath slump scars became an argument for inferring a link between methane hydrates and slumps, but this is clearly not sufficient as a sole indicator.

Natural Trigger Mechanisms of Slides from Geological Records

Most submarine landslides occur unobserved on the seabed along continental margins. We, therefore, deduct the processes involved on the basis of seabed bathymetry, sub-seabed paleomorphology, and sediment cores a long time after an event takes place. Only a very few submarine landslide events are directly documented. They caused submarine cable breaks or started as an underwater slope failure and retrogressively approached the coastline. Much less information exists about their trigger mechanisms, and the most widely used example for a continental slope failure is the 1929 Grand Banks earthquake. It appeared to trigger a slump that subsequently developed into a turbidity current.

Another condition that is most commonly associated with submarine slides is the rapid accumulation of thick sedimentary deposits over under-consolidated sediments, which by generating excess pore pressure reduces the effective stress that holds the sediment grains together. The magnitude of a combined instantaneous loading and a progressive excess pore pressure buildup appear to generate weakened sediment layers. The resulting reduction in shear strength allows sediments to move down very gentle slopes (<1°). Good examples are the well-studied areas of the Mississippi River delta. On formerly glaciated continental margins such as the passive Norwegian margin, thick layers of siliceous ooze of Miocene age exist in the sub-seabed. Their low density provides conditions for high compressibility and thus pore pressure builds up when subjected to rapid sediment loading during ice sheet advances of Pleistocene times. Cyclic sedimentation rates with high inputs of dense material during glacial times and low sediment input of more permeable material during interglacial times may explain the long-term instability of this margin. Megaslides seem to occur at a frequency of approximately 100 ky after each of the major ice ages since the onset of continental shelf glaciations at 500 ka. It is speculated that earthquake loading is the final trigger for the initiation of the slides at the end of each glaciation.

Steeper slopes or internally derived seepage forces might aid to fracturing the seabed leading to slope failures. Examples come from salt tectonics and flow of salt and mud in the sub-seabed of the Gulf of Mexico and the western Mediterranean of the Nile River delta. Up to now, the analysis of the forces of gravity and earthquake loading have shown that they are often not great enough to be the sole cause of failure along the slopes of continental margins. The controls on most identified slides are still obscure though earthquakes have been documented for a few events. Gas hydrate dissociation is another factor that is becoming more frequently used to decipher what controls slope failure.

Methane Hydrate Decomposition as a Natural Trigger Mechanism

In fully saturated structure I hydrates there is one molecule of methane present for every six molecules of water. In theory, when methane hydrate dissociates this results in an increase of volume and pressure buildup, because $1 m^3$ of methane hydrate dissociation develops into $164 m^3$ of methane gas at standard temperature and pressure conditions. The effect of dissociation pore pressure depends on water depth; if we decompose $1 m^3$ of gas hydrate at 900 m water depth (90 atm) we will get $1.82 m^3$ of free gas while at 1500-m water depth (150 atm) only $1.09 m^3$ of free gas develops. Shallow-water methane hydrate reservoirs are expected to develop a larger volume of free methane and are therefore able to generate higher excess pore pressure per unit volume than deep-water methane hydrate reservoirs depending on the rate of dissociation and the permeability of the sediment. As a result, upper continental margin slopes are more vulnerable to gas hydrate dissociation than lower slopes. Dissociaition of methane hydrates into liquid and free gas can build up excess pore pressure that reduces the shear strength of the sediments, thereby causing a natural trigger mechanism for a slope to fail.

A change in gas solubility generated by variations in temperature and/or pressure is generally ignored as a factor in methane hydrate and slumps. A geological model for a retrogressive (bottom-up) failure surface generated by hydrate dissolution due to gas solubility, and a sketch for a progressive (top-down) failure surface by the hydrate dissociation due to a temperature increase illustrates the resulting differences in slope instabilities (**Figure 3**). There are numerous fundamental problems remaining with our observations and theories. We know that the degree of hydrate saturation in the pore space of sediments will determine the total volume and pore-pressure increase during decomposition. Observations indicate a heterogeneous distribution of hydrates within the hydrate stability zone (HSZ) in which, except for massive hydrates, the space occupied by hydrates is usually low (<5%) with local peaks (up to <15%). Thus, there appears to be no continuous or homogenous methane hydrate occurrence zone (HOZ) leading to a glide plane. The thickness of the HSZ increases with water depth, and theoretical gas–hydrate-bearing sediments could occur in a zone of ~400 m below the seafloor (mbsf) in deep water of ~2000 m.

Quantifying *in situ* distributions of methane hydrates in three dimensions is difficult because of the effect of the heterogeneous distribution sediment type and texture. Laboratory experiments and the physics involved in modeling hydrate formation and dissociation are complex, and simplified assumptions have been made which take into consideration the time dependence of the processes. The hydrate formation rates beneath the seabed depend on the fluid flow from beneath the HSZ. Here, it is the availability of water and gas which in turn depends on diffusion rates and seepage conditions. On the other hand, hydrate dissociation requires only temperature increase and/or pressure reduction, but these occur at different rates in nature. To add to this complexity, the dissipation rates of excess gas and water and thus excess pore pressure depend on the diffusion coefficient and permeability of the sub-seabed.

Let us consider a simple hypothetical scenario. A drop in sea level during glacial times reduces the hydrostatic pressure, leading to a thinning of the HSZ and a potential thickening of the free gas zone beneath. As a consequence, the potential for

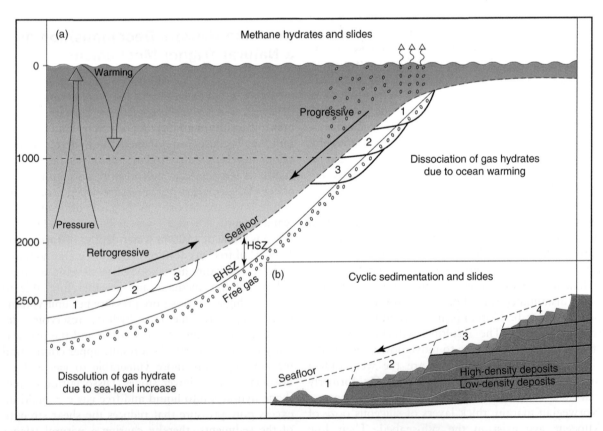

Figure 3 (a) Methane hydrates affected by ocean warming on the upper continental margin leading to progressive slide mechanisms while methane hydrates affected by sea level changes have the potential for retrogressive slide mechanisms. (b) Cyclic sedimentation with high density and less permeable sediments over less density and more permeable sediments causes overpressurized layers and characteristic glide planes indicating retrogressive sliding (see also **Figure 2**).

pore-pressure buildup and slumps increases. The emission of methane as part of the slump process would become greater with the frequency of slumps. Methane releases would enhance the greenhouse gases and provide a negative feedback to growing glaciations, thus becoming a potential terminator to glaciations during Quaternary times. The opposite can be drawn from methane releases during interglacial times. Ocean warming is superior to sea level rise affecting gas hydrate stability, methane hydrate melting, and the potential for slump processes and methane release increases. Such positive feedback may create an amplifier for global warming. Yet, like many other negative–positive feedback loop models, such classical scenarios remain to be proven and are mostly conjectural.

Methane Hydrates in Regions of Slope Instability: Searching for a Link

The most needed information is how excess pore pressure induced by the dissociation of methane hydrate is linked to the spatial and temporal changes in methane hydrate concentrations in continental margins. Such key information is far from being obtained. There is hardly any direct evidence that the observed slides and slumps are triggered by methane hydrate dissociation. Up to now, observations and possible connections between methane hydrate and slumps exist in selected areas of nearly all ocean margins. Evidence is provided from slumps of the eastern US Atlantic continental slope; slides and slumps of the continental rise and slope of SW Africa; giant slides and slumps of the continental slope of Norway; slides and slumps of the continental slope

on the Alaska Beaufort Sea; slides on the northern California continental margin; slides on the southern Caspian Sea margins; slumps of the western continental margin of India; slope failures along the Chile–Peru margins; slope failures in the northwestern Sea of Okhotsk; slope failures on the Amazon fan of the Brazil margin; slope failures on the northern Black Sea margin; slope failures on the Gulf of Cadiz margin (**Figure 4**).

Long-term instability is evident from the New Jersey margin of the East Coast of the United States. Within the Paleogene, four periods – near the Cretaceous–Tertiary boundary, Palaeocene–Eocene boundary, top of Lower Eocene, and Middle Eocene – show slope failures. They are identified in reflection seismic records. Since all of the events occur close to major sea level lowstands, methane hydrate dissociation accompanied by excess pore pressure is regarded as a potential factor for the initiation of the slides and slumps.

Linking methane hydrates with slope failures also remains speculative during Neogene times. Dating the event of many of the slope failures is still a problem and therefore large uncertainties exist in relating slides and slumps to a specific ocean's sea level and/or temperature condition. It has been suggested that for the past 45 ky, glacial-period slope failures occur mainly in low latitudes associated with sea level lowstand. It is during periods of change that hydrate is most likely to dissociate or to experience dissolution, depending on which direction of sea level and/or water temperature leads. This case is used to propose that reduced hydrostatic pressure triggered dissociation of methane hydrate and slope failure. In contrast, ocean warming during rising sea level conditions in the northern high latitudes is used

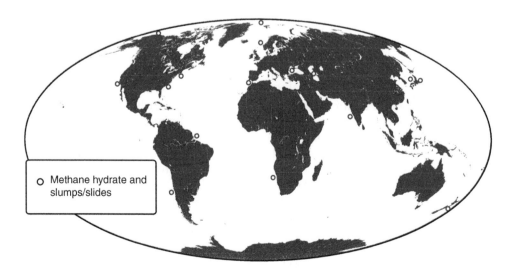

Figure 4   Global distribution of major submarine landslides overlying areas of today's known gas hydrate provinces.

to argue for dissociation of methane hydrate contributing to slope failure. Another approach for dating slides is based on turbidite sequences deposited in ocean basins that are a consequence of slides and slumps. Dating of turbidites resulted in an equally complex picture. At the Norwegian margin, the inferred slides occurred during (1) periods of sea level lowstand or an advance or presence of the Fennoscandian Ice Sheet at the shelf break, or (2) at periods of rapid sea level rise accompanied by ocean warming. Seabed side-scan sonar and bathymetry data from the northern rim of the giant Storegga slide on the mid-Norwegian margin show cracks in the seabed upslope. The BSR, which crosses bedding planes and is projected toward the shelf edge pinches out at the upper slope where the HSZ intersects the seabed in the area of the cracks. Other empirical support of BSR occurrences in areas of slumps exists particularly from the west coast off India, Nankai trough, and the US Atlantic continental slope.

A Link between Slide Headwall and Excess Pore Pressure

Ocean conditions and hence the stability of methane hydrate along most continental margins have changed significantly since the Last Glacial Maximum (LGM). In terms of methane hydrate stability, both the pressure related to eustatic sea level changes and bottom water temperature (BWT) changes in intermediate water masses have to be addressed.

The hypothetical scenario mentioned before is replaced by an actual scenario for the Norwegian margin, where detailed information about ocean and seabed conditions exists. The combined effect of eustatic sea level rise of ~120 m and BWT increases of ~5 °C, documented since the last LGM, change hydrate stability conditions along the entire continental slope. In water depth of >800 m, BHSZ deepens slightly due to the sea level rise. No temperature increase is indicated for deeper seabed. In contrast, the effect of sea level rise on the HSZ is, for localities above 800-m water depth, more than compensated by a rise in BWT. The combination of two such opposing or counteracting effects in terms of gas hydrate stability conditions (sea level rise is favorable for HSZ while BWT rise is not) predicts a distinct reduction in the extent and thickness of the HSZ in shallower water between 300 and 700 m since the LGM. Such results illustrate that shallower water depth areas are more sensitive to changes in sea level and water temperature than deeper localities. From this we may conclude that deeper water areas are less sensitive and methane hydrate dissociation may

trigger slumps favorably along the upper continental margins. Combining the dissociation model and the observation of a BSR along reflection seismic profiles crossing the Storegga headwall supports the hypothesis of excess pore pressure buildup at slide headwalls and thus slide progradation (from upper to lower slope). However, it contradicts many geomorphologic observations along continental margins including the well-studied Norwegian margin, which indicates retrogressive sliding (from the lower to the upper slope). Discussions about the seemingly opposing results are ongoing and lead to suggestions that excess pore pressure can also result from methane hydrate dissolution. This argument is based on thermodynamic calculations. Models show that due to a temperature and pressure increase, hydrate dissolution may start at the top of the hydrate occurrence zone. Simulation also documents that this process, due to a change in gas solubility, can occur at the origin of a retrogressive failure initiated at the lower slope. No conclusions exist as yet but the effect of dissolution, that is, the gas entering the pore water, and dissociation on deep and shallow water hydrates appear to be of importance for understanding all aspects of triggering retrogressive versus progressive slides.

Methane Blasts in the Past?

Current debates concern the explanation of massive injections of isotopically light carbon to the oceans and atmosphere in Cenozoic and Mesozoic times but also beyond. Distinct negative $\delta^{13}C$ excursions measured in samples from various locations reflect a rapid injection of massive quantities of ^{12}C-rich carbon to the ocean and atmosphere. Support for a hypothesis of methane blasts from gas hydrate dissociation has grown since agreements among scientists that other reservoirs cannot supply a sufficient mass of carbon over the relatively short geological timescales represented by the $\delta^{13}C$ spikes. Abrupt negative $\delta^{13}C$ excursions are reported from the Neoproterozoic; the early Paleozoic; the Permian/Triassic; the Triassic/Jurassic; the late and early Jurassic; the early Cretaceous (Aptian); the Paleocene–Eocene boundary; and the Quaternary. The Quaternary has the advantage to be documented by both high-resolution climate ice core and ocean sediment core records. The Quaternary methane within the trapped air in the Greenland and Antarctica ice cores, and the $\delta^{13}C$ excursions from both planktonic and benthic foraminifera created a heated and ongoing debate about the sources of methane. What caused the rapid and large atmospheric methane increase in the atmosphere during interstadials, and the negative $\delta^{13}C$ excursions in the Santa Barbara Basin offshore California? What caused the rapid shifts to positive excursions during stadials? Which sources can react fast enough and at the same time are so large that shifts reach several hundred ppbv (parts per billion by volume) in atmospheric methane and several per mil (−1‰ to −6‰) in ocean carbon isotopes?

Though indirect methane measurements of atmospheric methane concentration from ice cores go back more than a hundred thousand years, direct atmospheric measurements of methane only go back to 1970. Coming from the last glaciation into the deglaciation, the atmospheric methane as recorded in ice cores increased from $c.$ 400 to 700 ppbv, but from 1800 to 1990 it reached 1700 ppbv. It is still increasing and eventually most methane (CH_4) will end up as CO_2 and water. Evidently, CH_4 and CO_2 changes do track most of the rapid climate shifts. Up to now, terrestrial measurements of the global methane budget provide clear hints that the main drivers of methane appear to be the terrestrial sources (natural, anthropogenic) and sinks (soil microbes). A general consensus exists that continental wetlands are a major methane source. At the same time there is a major unknown, which is the role of the deep and shallow ocean environment where microbes are thriving on methane, and where methane hydrate dissociation and slumps may release significant amounts of methane. These processes are now becoming recognized as part of a complex feedback system. The suggested methane blasts of the past were certainly helpful in increasing our awareness for a holistic approach.

Discussion

Massive and rapid releases of methane from hydrates via slumping may have the potential for entering the atmosphere. The natural formation of methane hydrate beneath the oceans, however, may buffer the volume of methane entering the ocean and atmosphere. Our present observations from continental margin methane hydrate, BSRs, and slumps are still insufficient to quantify any of these major processes on a global or even regional scale. A breakthrough may come from a less obvious field, the methane records from ice cores. They may open new pathways for one of the two competing explanations for abrupt CH_4 increases. One working hypothesis uses the sudden release of methane from marine hydrates while the other explanation makes the terrestrial biosphere responsible for rapid methane releases. Developments of today's arguments and theories for

linkages of methane hydrate, slumps, and climate greatly depend on an understanding of the Quaternary atmospheric CH_4 record coming from ice cores. Equally important are ocean records such as negative $\delta^{13}C$ spikes in fossil records of planktonic and benthic foraminifera, and the age and frequency of slumps associated with gas hydrate reservoirs. Quaternary ice core records may soon allow confirmation or rejection of the proposed scenarios for methane release from ocean margins due to the distinct deuterium/hydrogen record of marine hydrates. There is hope that δD_{CH4} provides a means of discriminating between methane sources from land or marine methane hydrate reservoirs. In Cenozoic and earlier times, the geological records of slumps and negative $\delta^{13}C$ excursions are less clear. An underlying understanding about the role of both methane hydrate and slumps for methane releases from past ocean hydrate reservoirs will remain a challenge for the years to come. An answer as to whether or not the dissociation/dissolution of massive methane hydrates contributed to slumping and rapid increases of atmospheric greenhouse gases would give us an important perspective on how likely it is that we shall encounter such geohazards in the future. It is not unreasonable to expect that the amount of methane hydrates and related pore pressure buildup during dissociation and dissolution processes in slide-dominated ocean margins is soon to be quantified. Today's evaluations show that an ocean warming leading to a temperature increase at the BHSZ can lead to dissociation of gas hydrates and generation of high excess pore pressure. Sea level fall can generate similar effects. More research will establish whether this is a viable phenomenon leading to assessments of today's methane hydrate and slope stability. Improved techniques for deciphering the origin of methane from ice cores, quantifications of methane records in oceans from isotopically light carbon spikes, and determinations of the timing of submarine slides and slumps in Cenozoic times will provide new information for the development of theories (models, hypotheses, etc.) for this potentially dynamic part of ocean margins.

See also

Slides, Slumps, Debris Flow and Turbidity Currents.

Further Reading

Carpenter G (1981) Coincident sediment slump/clathrate complexes on the US Atlantic continental slope. *Geo-Marine Letters* 1: 29–32.

Dickens GR (2001) The potential volume of oceanic methane hydrates with variable external conditions. *Organic Geochemistry* 32: 1179–1193.

Hampton MA, Lee HJ, and Locat J (1996) Submarine landslides. *Reviews of Geophysics* 34(1): 33–60.

Henriet JP and Mienert J (eds.) (1998) *Geological Society, London, Special Publications No. 137: Gas Hydrates – Relevance to World Margin Stability and Climate Change*. London: Geological Society.

Higgins JA and Schrag DP (2006) Beyond methane: Towards a theory for the Paleocene–Eocene thermal maximum. *Earth and Planetary Science Letters* 245: 523–537.

Katz ME, Cramer BS, Mountain GS, Katz S, and Miller KG (2001) Uncorking the bottle: What triggered the Paleocene/Eocene thermal maximum methane release? *Paleoceanography* 16: 549–562.

Kayen RE and Lee HJ (1991) Pleistocene slope instability of gas-hydrate laden sediment on the Beaufort Sea margin. *Marine Technology* 10: 32–40.

Kennett JP, Cannariato KG, Hendy IL, and Behl RJ (2003) *American Geophysical Union, Special Publication, Vol. 54: Methane Hydrates in Quaternary Climate Change – The Clathrate Gun Hypothesis*, 216pp. Washington, DC: American Geophysical Union.

Maslin M, Owen M, Day S, and Long D (2004) Linking continental-slope failures and climate change: Testing the clathrate gun hypothesis. *Geology* 32: 53–56.

McIver RD (1982) Role of naturally occurring gas hydrates in sediment transport. *AAPG Bulletin* 66: 789–792.

Mienert J, Vanneste M, Bünz S, Andreassen K, Haflidason H, and Sejrup HP (2005) Ocean warming and gas hydrate stability on the mid-Norwegian margin at the Storegga slide. *Marine and Petroleum Geology* 22: 233–244.

Mulder TH and Cochonat P (1996) Classification of offshore mass movements. *Journal of Sedimentary Geology* 66: 43–57.

Paull CK, Buelow WJ, Ussler W, III, and Borowski WS (1996) Increased continental margin slumping frequency during sea-level lowstands above gas hydrate-bearing sediments. *Geology* 24: 143–146.

Sloan ED and Koh C (2008) *Clathrate Hydrates of Natural Gases*, 3rd edn., Chemical Industries Series/119. Boca Raton, FL: CRC Press, Taylor and Francis Group. New York: Dekker.

Sower T (2006) Late quaternary atmospheric CH_4 isotope record suggests marine clathrates are stable. *Science* 311(5762): 838–840.

Sultan N, Cochonat P, Foucher JP, and Mienert J (2004) Effect of gas hydrates melting on seafloor slope instability. *Marine Geology* 213: 379–401.

Xu W and Ruppel C (1999) Predicting the occurrence, distribution, and evolution of methane gas hydrate in porous marine sediments. *Journal of Geophysical Research* 104(B3): 5081–5096.

TURBULENCE IN THE BENTHIC BOUNDARY LAYER

R. Lueck and L. St. Laurrent, University of Victoria, Victoria, BC, Canada
J. N. Moum, Oregon State University, Corvallis, OR, USA

Introduction

Fluids do not slip at solid boundaries. The fluid velocity changes from 0 to a speed that matches the 'far field' in a transition, or boundary layer, where friction and shear (the rate of change of velocity with distance from the boundary) are strong. The thickness of the ocean bottom (benthic) boundary layer is determined by the bottom stress and the rate of rotation of the Earth. The benthic boundary layer is usually thin ($O(10\,\text{m})$) compared to typical ocean depths of $\sim 4000\,\text{m}$. However, in coastal regions which are shallow, and where currents and thus friction are relatively strong compared to the deep ocean, the benthic boundary layer may span most of the water column.

The boundary layer can be separated into several layers within which some forces are much stronger than others. Neglect of the weaker forces leads to scaling and parametrization of the flow within each layer. The benthic boundary layer is usually considered to consist of (1) an outer or Ekman layer in which rotation and turbulent friction (Reynolds stress) are important; (2) a very thin ($O(10^{-3}\,\text{m})$) viscous layer right next to the boundary where molecular friction is important; and (3) a transitional layer between these, usually called the logarithmic layer, in which turbulent friction is important (**Figure 1**). The pressure gradient is an important force in all the three layers. Because the velocity profile within the logarithmic layer must match smoothly with both the Ekman layer above and the viscous layer below, it will be considered last.

This discussion is framed in the context of a neutrally-stratified ocean remote from the free surface. Additional constraints due to stratification and proximity to the free surface are noted later.

The Ekman Layer

Most of the open ocean is essentially frictionless and in geostrophic balance, being well described by a balance between the Coriolis force which pushes the flow to the right (in the Northern Hemisphere) and the pressure gradient which keeps it from veering (**Figure 2(a)**). This picture changes as we approach the bottom. Friction acts against the flow and decreases the velocity U. However, the pressure gradient remains and is not completely balanced by the Coriolis force fU. The current backs leftward so that friction, which is directed against the current, establishes a balance of forces in the horizontal plane (**Figure 2(b)**). Progressively closer to the bottom, the increasing friction slows the flow and brings it to a complete halt right at the bottom while also further backing the flow direction. A vertical profile of the two components of the horizontal velocity might look like those depicted in **Figure 3**.

The equations of motion and their boundary conditions are

$$-fV = -\frac{1}{\rho}\frac{\partial P}{\partial x} + \frac{1}{\rho}\frac{\partial \tau_x}{\partial z}, \quad fU = -\frac{1}{\rho}\frac{\partial P}{\partial y} + \frac{1}{\rho}\frac{\partial \tau_y}{\partial z}$$
$$U = U_g, V = V_g, \tau_x = \tau_y = 0 \text{ as } z \to \infty \quad [1]$$
$$U = V = 0 \text{ at } z = 0$$

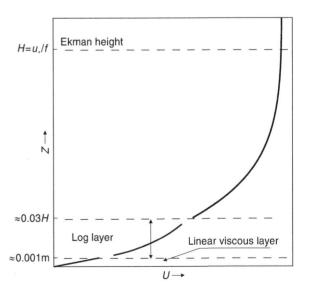

Figure 1 A conceptual sketch of the three sublayers forming the bottom boundary layer. The pressure-gradient, friction, and Coriolis forces are in balance in the Ekman layer while only friction and pressure-gradient forces are significant in the logarithmic and viscous layers. In the logarithmic layer, friction stems predominantly from the Reynolds stress of turbulence, whereas in the viscous layer it comes mainly from molecular effects.

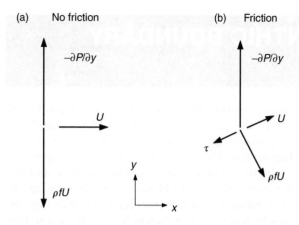

Figure 2 Plan view of the balance of forces in the geostrophic flow far above the bottom (a) and in the Ekman layer (b). The current, U, is directed to the right in the positive x-direction. Far above the bottom, the pressure gradient in the y-direction is balanced by Coriolis force in the opposite direction and this force is always directed to the right of the current (in the Northern Hemisphere). Within the Ekman layer, friction, τ, acts against the current. A balance of forces in 'both' the x- and y-directions is only possible if the current backs anticlockwise when viewed from above.

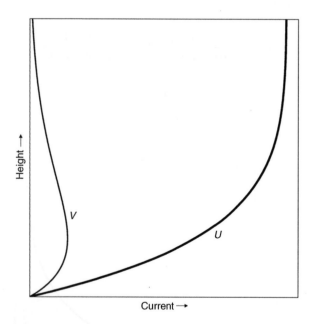

Figure 3 A conceptual velocity profile that may result from the effect of friction as depicted in **Figure 2**. A positive current component, V, is directed to the left of the geostrophic current.

where we have assumed that the vertical velocity, W, is zero (flat bottom), taken the bottom at $z = 0$, assumed that both components of the stress (τ_x, τ_y) vanish far above the bottom, and assigned the x- and y-components of the geostrophic velocity to U_g and V_g, respectively. The flow is geostrophic far above the bottom, that is

$$fU_g = -\frac{1}{\rho}\frac{\partial P}{\partial y}, \quad -fV_g = -\frac{1}{\rho}\frac{\partial P}{\partial x} \qquad [2]$$

and if the density is homogeneous within the boundary layer, the pressure gradient is independent of height within this layer. Substituting [2] into [1] gives the so-called Ekman equation for the boundary layer, namely

$$-f(V - V_g) = \frac{1}{\rho}\frac{\partial \tau_x}{\partial z}, \quad f(U - U_g) = \frac{1}{\rho}\frac{\partial \tau_y}{\partial z} \qquad [3]$$

It is convenient to assume that the bottom stress has no y-component so that the bottom stress $\tau_0 = \tau_x(0)$ is directed entirely in the x-direction, that is, $\tau_y(0) = 0$. Solving [3] for the velocity profile requires the relationship between stress and velocity, which is a major focus of boundary layer research. Fortunately, the height above the bottom over which friction is important can be determined using only dimensional analysis. For example, the x-component of velocity must be some function, F, of the parameters and variables in [3] and its boundary condition, $\tau_x(0) = \tau_0$, so that

$$U = F(\rho, \tau_0, f, z) \qquad [4]$$

All the four variables in [4] cannot be independent. For example, ρ and τ_0 must always appear as a ratio because they are the only ones with the dimension of mass. The root of this ratio,

$$u_* \equiv \sqrt{\frac{\tau_0}{\rho}} \qquad [5]$$

has dimensions of velocity and is called the 'friction velocity'. It represents a scale for the turbulent velocity fluctuations in the boundary layer. The only other independent variable is

$$H \equiv \frac{1}{f}\sqrt{\frac{\tau_0}{\rho}} = \frac{u_*}{f} \qquad [6]$$

and this is the only dimensional group that can be used to nondimensionalize z, the height above the bottom. Thus, the velocity profile must be

$$\begin{aligned} U - U_g/u_* &= F_u(z/H) \\ (V - V_g)/u_* &= F_v(z/H) \end{aligned} \qquad [7]$$

where F_u and F_v are universal functions. Equation [7] is usually called the 'velocity defect law'. The order of magnitude of the height H of the boundary layer, that is, the scale over which friction is important, is usually called the Ekman height. The actual height to which friction is important is within a factor of order unity of H. The Ekman height can also be considered the transition height; for $z \ll H$, friction dominates over the Coriolis force while above this level, the reverse holds. An important effect of rotation is that the thickness of the BBL does not grow in the downstream direction (for uniform bottom conditions) whereas the boundary layer over a nonrotating and flat surface grows downstream.

Numerical values for the Ekman height can be derived from a traditional formulation of the bottom stress in terms of a drag coefficient, such as

$$\tau_0 = \rho C_D U_g^2 \qquad [8]$$

where the drag coefficient, C_D, must depend on the bottom characteristics, such as roughness. Typical values are $C_D \approx 0.002$. Using a geostrophic flow of $U_g \approx 0.1 \, \mathrm{m \, s^{-1}}$ commonly found in the open ocean and $f = 1 \times 10^{-4} \, \mathrm{s^{-1}}$ gives a friction velocity of $u_* = 4.5 \times 10^{-3} \, \mathrm{m \, s^{-1}}$ and an Ekman height of $H = 45 \, \mathrm{m}$ which is 100 times smaller than the average ocean depth. The friction layer is thus thin compared to the ocean depth, as assumed.

Ekman solved [3] almost a century ago for the special case of a stress proportional to the shear. That is,

$$\tau_x = -\rho K_V \frac{\partial U}{\partial z}, \quad \tau_y = -\rho K_V \frac{\partial V}{\partial z} \qquad [9]$$

where K_V is the eddy viscosity. The mathematically elegant spiral predicted by [3] and [9] is presented in standard textbooks on fluid mechanics. However, the predicted profile is not directly observed due to a number of complicating factors, such as complex boundary geometries, temporal variability in stress acting in the boundary layer, nonconstant eddy viscosity, and nonlinear dynamics. Despite these difficulties, very nice Ekman spirals have been documented when data of sufficient quantity and quality have been carefully analyzed. Trowbridge and Lentz provide an excellent contemporary example of bottom boundary layer observations and analysis (see Further Reading). They show that Ekman balance dynamics are recovered with adequate time averaging. They also show how the traditional Ekman equations presented above must be modified to include important buoyancy effects occurring in a stratified boundary layer. An additional review of interest, focused more on the Ekman spiral extending from the ocean surface boundary layer, is given by Rudnick.

Viscous Sublayer

Very near to a smooth bottom, $z \ll H$, a layer forms in which momentum is transferred only by molecular diffusion. In general, the stress

$$\tau_x = \rho \nu \frac{dU}{dz} - \rho \overline{u'w'} \qquad [10]$$

is the sum of the shear stress due to molecular friction (first term on the right-hand side of [10]) and the Reynolds stress $-\rho \overline{u'w'}$, where $\nu = \mu/\rho$ is the kinematic molecular viscosity ($\approx 1 \times 10^{-6} \, \mathrm{m^2 \, s^{-1}}$). The covariance $\overline{u'w'}$ of horizontal, u', and vertical, w', velocity fluctuations leads to a transfer of momentum from the fluid toward the wall. Very near the wall, vertical velocity fluctuations are strongly suppressed (no normal-flow boundary condition) and the Reynolds stress is negligible compared to molecular friction.

The Ekman height, H, is not an appropriate parameter for nondimensionalizing the height above the bottom in the thin viscous layer. Rather, the viscous scale is used:

$$\delta = \nu/u_* \qquad [11]$$

Using [3] the nondimensionalized momentum balance is

$$-\frac{\delta}{Hu_*}(V - V_g) = \frac{\partial(\tau_x/u_*^2)}{\partial(z/\delta)}$$
$$\frac{\delta}{Hu_*}(U - U_g) = \frac{\partial(\tau_y/u_*^2)}{\partial(z/\delta)} \qquad [12]$$

To estimate the magnitude of the terms on the right-hand side of [12], we note the following. From [8], the ratio of the geostrophic speed to friction velocity is related to the drag coefficient by $U_g/u_* = C_D^{-1/2}$, and this equals approximately 25. The velocity is at most comparable to the geostrophic velocity, so the factor $(U - U_g)/u_*$ is no more than about 25. Even for very weak flows, the terms in [12] are smaller than $O(10^{-3})$. Thus, the vertical divergence of the stress is zero and the stress itself is constant.

When the stress stems entirely from molecular friction, the only possible velocity profile is a linear one that has a shear which is commensurate with the bottom stress, that is,

$$\frac{U}{u_*} = \frac{z u_*}{\nu} = z_+ \qquad [13]$$

Laboratory observations of flow over smooth surfaces show that [13] holds to about $z_+ \approx 5$ and this innermost region is called the 'viscous sublayer'. A typical dimensional thickness for the viscous sublayer is $5\nu/u_* \approx 0.001$ m. Thus, this layer never extends more than a few millimeters above the bottom. Most of the ocean bottom is not 'smooth' compared to this scale.

The Wall Layer

Further above the bottom but still well within the extent of the Ekman layer, for $\nu/u_* \ll z \ll H = u_*/f$, neither the Ekman height, H, nor the molecular viscosity, v, can be relevant parameters controlling the velocity profile. The only parameter that can nondimensionalize the vertical height is either the thickness of the viscous sublayer or the characteristic height of bottom roughness features, z_0. Equation [12] is still the appropriate nondimensional momentum balance if we substitute z_0 for δ. The left-hand side of [12] is no longer as small as for the viscous sublayer but it is still small compared to unity, and the stress can be taken as constant. Thus, the wall layer and the viscous sublayer are usually called the constant-stress layer. The stress [10], however, is now entirely due to the Reynolds stress. Because the bottom stress has no component in the y-direction, there is also no bottom velocity in this direction.

The only parameters that control the velocity profile are the bottom stress and the roughness height. On purely dimensional grounds, we have near the wall:

$$V/u_* = 0$$
$$U/u_* = g_2(z/z_0) \qquad [14]$$

where g_2 is a yet to be determined universal function. Equation [14] is the 'law of the wall' for rough bottoms. The law of the wall must be matched to the velocity-defect law [7] and this is usually done by matching the shear rather than the velocity itself. The result is that

$$\frac{V_g}{u_*} = -F_v(0) = -A$$
$$\frac{U}{u_*} = \frac{1}{\kappa}\ln\left(\frac{z}{z_0}\right) \qquad [15]$$
$$\frac{U_g}{u_*} = \frac{1}{\kappa}\ln\left(\frac{H}{z_0}\right) - C$$

where $\kappa = 0.4$ is von Karman's constant and atmospheric observations indicate that $A \approx 12$ and $C \approx 4$. These equations are valid for $z/z_0 \gg 1$ and $z/H \ll 1$ 'simultaneously'. Thus, the velocity increases logarithmically with increasing height and this profile

ultimately turns into an 'Ekman'-like spiral that matches the geostrophic flow at $z = O(H)$. A thin viscous sublayer may underlie this profile if the bottom is very smooth, in which case z_0 is chosen to match the profile given by [13] for the same bottom stress.

It is frequently convenient to express the stress in terms of an eddy viscosity and the shear such as in [9]. However, a constant stress and a logarithmic velocity profile make the eddy viscosity proportional to height, namely

$$K = u_*\kappa z \qquad [16]$$

Thus, a constant eddy diffusivity is not a good model for the wall layer and may well be inappropriate in much of the Ekman layer.

The Reynolds stress in the presence of a shear leads to the production of turbulent kinetic energy (TKE) within the wall layer. It is thought that almost all of the TKE is dissipated locally and that the rate of dissipation is given by

$$\varepsilon = -\overline{u'w'}\frac{\partial U}{\partial z} = \frac{u_*^3}{\kappa z} \qquad [17]$$

Thus, profiles of the rate of dissipation of kinetic energy provide an alternate measure of the bottom stress to that which can be derived from the velocity profile.

Observations

Values of the bottom stress are required for two major purposes: as a boundary condition for flows above the bottom and for the prediction of sediment motions. The near-bottom velocity profile [15] provides a convenient method for estimating the bottom stress through a fitting of U against the logarithm of z. This profile method is the one most frequently used to estimate the bottom stress. Point current meters have been placed within a few meters of the bottom and, under the assumption that they are within the logarithmic region, the bottom stress was estimated from as few as a pair of current meters. Some bottom velocity 'profile' measurements were accompanied by concurrent measurements of the turbulent fluctuations of along-flow and vertical velocity components. The covariance of these fluctuations, $-\rho\overline{u'w'}$, is an unambiguous measure of the Reynolds stress and, when this stress is extrapolated to the bottom, it usually agrees closely with the stress (ρu_*^2) inferred from the slope of the logarithmic velocity profile. (Readers are referred to the article by Trowbridge and Lentz in 'Further Reading', an excellent source of citations to past observational studies.)

Taking profiles of velocity within the BBL is very difficult. Consequently, there is very little observational evidence on the form of the velocity profile. One of the best deep-ocean velocity profiles was taken in the North Atlantic Western Boundary Current over the Blake Outer Ridge and reached to within 5 m of the bottom (**Figure 4**). The potential density was homogeneous within 250 m of the bottom and so the pressure gradient was independent of height as assumed in [3]. The current in the upper parts of the homogeneous layer was $0.22 \, \mathrm{m \, s^{-1}}$ and directed along the isobaths (approximately southward). The along-slope current had a very slight maximum at 40 m, decreased sharply below 15 m, and dropped to $0.18 \, \mathrm{m \, s^{-1}}$ at 5 m. The full decay to zero current at the bottom was not resolved for instrumental reasons. The cross-slope current was negligible further than 50 m above the bottom. It increased to $0.025 \, \mathrm{m \, s^{-1}}$ at 5 m and was consistently directed to the left of the along-slope current (approximately eastward). The veering of the velocity vector with height above the bottom was like that depicted in **Figures 2** and **3** and reached a maximum of $8°$ at the lowest observation located at 5 m. Simultaneous measurements of the rate of dissipation of TKE indicate that the turbulence was negligible for heights greater than 50 m above the bottom. The dissipation rate decreased monotonically with increasing height up to 50 m. Above this height, it was small and fairly uniform. Thus, the frictional layer was 50-m thick and 5 times thinner than the homogeneous layer. It is common to find different heights for the homogeneous ('mixed') and the turbulent ('mixing') layers. The height of the Ekman layer, H, predicted by [6] was 120 m and the actual height to which friction was important was close to the expected value of $\kappa H = 50 \, \mathrm{m}$, where $\kappa = 0.41$ is the von Karman constant.

The height of the logarithmic layer [15] has not been extensively surveyed and based on the scaling arguments it must be small compared to the Ekman height. Measurements in a tidal channel indicate that profiles depart from a logarithmic form at about 3–4% of the Ekman height. The height of the constant stress layer cannot be greater than the logarithmic profile height.

For horizontally homogeneous bottom roughness, such as flat sand and fine gravel, the roughness height, z_0, is c. 30 times smaller than the actual roughness. The notion is that the velocity profile reaches zero somewhere below the highest bottom features. Thus, there must be considerable local variations of the velocity profile for heights less than $z \approx 30 z_0$ and [15] represents a horizontally averaged velocity profile. The constancy of z_0 is not well established for any particular site nor does it increase consistently with increasing bottom roughness. Cheng *et al.* found a systematic decrease in z_0 with increasing speed above $0.2 \, \mathrm{m \, s^{-1}}$ and attributed this to the onset of sediment motions and its smoothing effect upon the bottom.

The bottom roughness is seldom horizontally homogeneous and the major contribution to roughness comes from bedforms (e.g., ripples and sand waves) and other features with horizontal scales far greater than the largest pieces of bottom material. Thus, bottom profiles well above $z \approx 30 z_0$ should show horizontal variations (**Figure 5**). For example, a

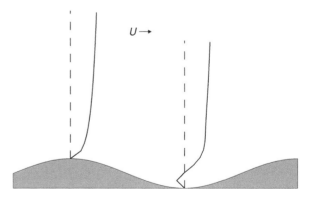

Figure 4 A sketch of the along-, U, and across-isobath, V, flow over the Blake Outer Ridge in the North Atlantic Western Boundary Current as reported by Stahr and Sanford (1999). Dashed lines within 5 m of the bottom are hypothetical extensions.

Figure 5 Conceptual sketch of spatial variations in the vertical profile of velocity over bedforms with long horizontal scales, such as sand waves. The vertical and dashed lines give a zero-velocity reference. The flow accelerates and stream-lines compress on the 'upwind' side of crests and the flow decelerates and its streamlines dilate on the lee side. This causes a pressure drop in the flow direction. If slopes are steep, flow separation and back flow may occur in the troughs and over the leesides as depicted for the right profile.

wavy bottom may appear locally to have a roughness scale commensurate with the bottom material (such as sand) but at a height comparable to the amplitude of sand waves, the bottom turns 'rough' as the turbulent eddies respond to the larger horizontal-scale structures on the bottom and not just the local features. Additional drag will be exerted on the flow by the pressure differences across sand waves (or other obstacles) due to stream-line asymmetry and outright flow separation when the slope on the lee side of objects is very steep. This is usually labeled 'form drag' due to its similarity to the drag on bluff bodies. This feature was first observed by Chriss and Caldwell in 1982 in profiles taken over the continental shelf off Oregon. They found two logarithmic layers with differing slopes (**Figure 6a**). The lower layer extended to 0.1 m, and its logarithmic slope implies a friction velocity of $u_* = 0.004$ m s^{-1}. This layer appears to be associated with skin friction over a fairly smooth surface. The upper layer reached to at least 0.2 m and its much greater slope is indicative of stress due to form drag. Form drag can also result when boundary layer turbulence is produced by wave-like variations in the seabed. Sanford and Lien

report on measurements from a tidal channel, where sand ripples of 0.3-m amplitude and 16-m wavelength were present and oriented span-wise to the flow direction. They find a double logarithmic velocity profile (**Figure 6b**) similar to that observed by Chriss and Caldwell in 1982. The slope of the velocity–log z relation increased by a factor of 2 near $z = 4$ m, even though the seabed amplitude variations were much smaller than this height. The effect of long horizontal-scale features on the flow over the bottom is still being investigated.

An alternate method of estimating the bottom stress is provided by the dissipation profile technique. Profiles of the rate of dissipation have verified the inverse height dependence predicted by [17] for heights of up to 10 m. However, when the estimates of bottom stress derived from dissipation profiles are compared to the stress estimated from a fit of the velocity profile to a logarithmic form, the dissipation-based estimates are typically 3 times smaller. Momentum budgets for bottom streams such as the Mediterranean outflow are consistent with the drag determined from the velocity profile but not with the drag inferred from dissipation profiles. There is still

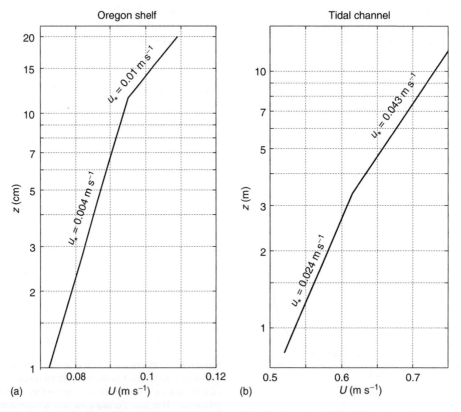

Figure 6 A sketch of velocity profiles plotted against the logarithm of height above the bottom based on data reported by Chriss and Caldwell (1982) and Sanford and Lien (1999). Approximately 10 data points were available for each regression (a) whereas data from about 100 different depths were used for (b). Both profiles imply a jump by a factor of 2 in friction velocity and an increase by a factor of 4 in stress for the upper logarithmic layer compared to the lower layer. The grey-shaded line is the law-of-the-wall scaling modified by proximity to the free surface.

no satisfactory explanation for such discrepancies. Recent observations on the Oregon shelf include those of Nash and Moum, who document the hydraulic production of bottom boundary layer turbulence at a topographic bump in the presence of a coastal jet. Their measurements appear to be dominated by a form drag stress-layer, with $u_* \simeq 0.005 - 0.01$ m s^{-1} estimated using the dissipation rate method. They find that the resulting form drag can be of sufficient magnitude to break the geostrophic balance of flow near the bump. Nash and Moum also document that the boundary layer properties can change significantly due variations in forcing and stratification over both short ($O(1$ day)) and long ($O(1$ year)) timescales.

Modifications due to Stratification and Proximity of the Free Surface

In the absence of stratification and a close upper boundary, turbulence length scales increase linearly from $z = 0$. However, turbulence scales are attenuated by stratification and by boundaries. Most commonly, unstratified near-bottom layers are capped by stratified layers. In these cases, the length scales of the turbulence cannot increase unbounded and are attenuated throughout the boundary layer (Perlin *et al.*, 2005). In shallow tidal channels (such as is shown in **Figure 6b**), a similar effect ensues. This offers an alternate explanation to the two logarithmic layer model, in which a single velocity scale (u_*) describe the full velocity profile (shown in grey in **Figure 6b**).

Sub-sea Permafrost.

Further Reading

Cheng RT, Ling C-H, and Gartner JW (1999) Estimates of bottom roughness length and bottom shear stress in South San Francisco Bay, California. *Journal of Geophysical Research* 104: 7715–7728.

Chriss TM and Caldwell DR (1982) Evidence for the influence of form drag on bottom boundary layer flow. *Journal of Geophysical Research* 87: 4148–4154.

Dewey RK and Crawford WR (1988) Bottom stress estimates from vertical dissipation rate profiles on the continental shelf. *Journal of Physical Oceanography* 18: 1167–1177.

Johnson GC, Lueck RG, and Sanford TB (1995) Stress on the Mediterranean outflow plume. Part 2: Turbulent dissipation and shear measurements. *Journal of Physical Oceanography* 24: 2072–2083.

Lueck RG and Huang D (1999) Dissipation measurement with a moored instrument in a swift tidal channel. *Journal of Atmospheric and Oceanic Technology* 16: 1499–1505.

Nash JD and Moum JN (2001) Internal hydraulic flows on the continental shelf: High drag states over a small bank. *Journal of Geophysical Research* 106: 4593–4612.

Perlin A, Moum JN, Klymak JM, Levine MD, Boyd T, and Kosro PM (2005) A modified law-of-the-wall applied to oceanic bottom boundary layers. *J. Geophysics Research* 110: doi:10.1029/2004JC002310.

Rudnick D (2003) Observations of momentum transfer in the upper ocean: Did Ekman get it right? In: Muller P and Garrett C (eds.) *Proceedings of the 'Aha Huliko'a Hawaiian Winter Workshop*, pp. 163–170. Honolulu, HI: University of Hawaii.

Sanford TB and Lien R-C (1999) Turbulent properties in a homogeneous tidal bottom boundary layer. *Journal of Geophysical Research* 104: 1245–1257.

Stahr FR and Sanford TB (1999) Transport and bottom boundary layer observations of the North Atlantic deep western boundary current at the Blake outer ridge. *Deep-Sea Research* 46: 205–243.

Trowbridge JH and Lentz SJ (1998) Dynamics of the bottom boundary layer on the North California Shelf. *Journal of Physical Oceanography* 28: 2075–2093.

SUB-SEA PERMAFROST

T. E. Osterkamp, University of Alaska, Alaska, AK, USA

Introduction

Sub-sea permafrost, alternatively known as submarine permafrost and offshore permafrost, is defined as permafrost occurring beneath the seabed. It exists in continental shelves in the polar regions (**Figure 1**). When sea levels are low, permafrost aggrades in the exposed shelves under cold subaerial conditions. When sea levels are high, permafrost degrades in the submerged shelves under relatively warm and salty boundary conditions. Sub-sea permafrost differs from other permafrost in that it is relic, warm, and generally degrading. Methods used to investigate it include probing, drilling, sampling, drill hole log analyses, temperature and salt measurements, geological and geophysical methods (primarily seismic and electrical), and geological and geophysical models. Field studies are conducted from boats or, when the ocean surface is frozen, from the ice cover. The focus of this article is to review our understanding of sub-sea permafrost, of processes ocurring within it, and of its occurrence, distribution, and characteristics.

Sub-sea permafrost derives its economic importance from current interests in the development of offshore petroleum and other natural resources in the continental shelves of polar regions. The presence and characteristics of sub-sea permafrost must be considered in the design, construction, and operation of coastal facilities, structures founded on the seabed, artificial islands, sub-sea pipelines, and wells drilled for exploration and production.

Scientific problems related to sub-sea permafrost include the need to understand the factors that control its occurrence and distribution, properties of warm permafrost containing salt, and movement of heat and salt in degrading permafrost. Gas hydrates that can occur within and under the permafrost are a potential abundant source of energy. As the sub-sea permafrost warms and thaws, the hydrates destabilize, producing gases that may be a significant source of global carbon.

Nomenclature

'Permafrost' is ground that remains below 0°C for at least two years. It may or may not contain ice. 'Ice-bearing' describes permafrost or seasonally frozen soil that contains ice. 'Ice-bonded' describes ice-bearing material in which the soil particles are mechanically cemented by ice. Ice-bearing and ice-bonded material may contain unfrozen pore fluid in addition to the ice. 'Frozen' implies ice-bearing or ice-bonded or both, and 'thawed' implies non-ice-bearing. The 'active layer' is the surface layer of sediments subject to annual freezing and thawing in areas underlain by permafrost. Where seabed temperatures are negative, a thawed layer ('talik') exists near the seabed. This talik is permafrost but does not contain ice because soil particle effects, pressure, and the presence of salts in the pore fluid can depress the freezing point 2°C or more. The boundary between a thawed region and ice-bearing permafrost is a phase boundary. 'Ice-rich' permafrost contains ice in excess

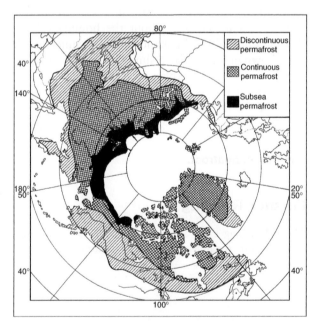

Figure 1 Map showing the approximate distribution of sub-sea permafrost in the continental shelves of the Arctic Ocean. The scarcity of direct data (probing, drilling, sampling, temperature measurements) makes the map highly speculative, with most of the distribution inferred from indirect measurements, primarily water temperature, salinity, and depth (100 m depth contour). Sub-sea permafrost also exists near the eroding coasts of arctic islands, mainlands, and where seabed temperatures remain negative. (Adapted from Pewe TL (1983). *Arctic and Alpine Research* 15(2):145–156 with the permission of the Regents of the University of Coloroado.)

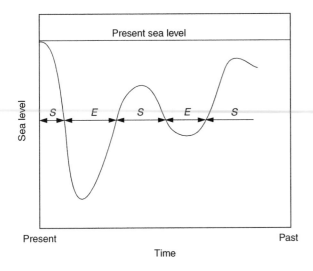

Figure 2 Schematic sea level curve during the last glaciation. The history of emergence (E) and submergence (S) can be combined with paleotemperature data on sub-aerial and sub-sea conditions to construct an approximate thermal boundary condition for sub-sea permafrost at any water depth.

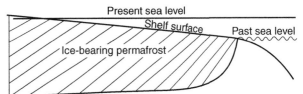

Figure 3 A schematic illustration of ice-bearing sub-sea permafrost in a continental shelf near the time of minimum sea level. Typical thicknesses at the position of the present shoreline would have been about 400–1000 m with shelf widths that are now typically 100–600 km.

of the soil pore spaces and is subject to settling on thawing.

Formation and Thawing

Repeated glaciations over the last million years or so have caused sea level changes of 100 m or more (Figure 2). When sea levels were low, the shallow continental shelves in polar regions that were not covered by ice sheets were exposed to low mean annual air temperatures (typically −10 to −25°C). Permafrost aggraded in these shelves from the exposed ground surface downwards. A simple conduction model yields the approximate depth (X) to the bottom of ice-bonded permafrost at time t, (eqn [1]).

$$X(t) = \sqrt{\frac{2K(T_e - T_g)t}{h}} \qquad [1]$$

K is the thermal conductivity of the ice-bonded permafrost, T_e is the phase boundary temperature at the bottom of the ice-bonded permafrost, T_g is the long-term mean ground surface temperature during emergence, and h is the volumetric latent heat of the sediments, which depends on the ice content. In eqn [1], K, h, and T_e depend on sediment properties. A rough estimate of T_g can be obtained from information on paleoclimate and an approximate value for t can be obtained from the sea level history (Figure 2). Eqn [1] overestimates X because it neglects geothermal heat flow except when a layer of ice-bearing permafrost from the previous transgression remains at depth. Timescales for permafrost growth are such that

hundreds of meters of permafrost could have aggraded in the shelves while they were emergent (Figure 3).

Cold onshore permafrost, upon submergence during a transgression, absorbs heat from the seabed above and from the geothermal heat flux rising from below. It gradually warms (Figures 4 and 5), becoming nearly isothermal over timescales up to a few millennia (Figure 4, time t_3). Substantially longer times are required when unfrozen pore fluids are present in equilibrium with ice because some ice must thaw throughout the permafrost thickness for it to warm.

A thawed layer develops below the seabed and thawing can proceed from the seabed downward, even in the presence of negative mean seabed temperatures, by the influx of salt and heat associated with the new boundary conditions. Ignoring seabed erosion and sedimentation processes, the thawing rate at the top of ice-bonded permafrost during submergence is given by eqn [2].

$$\dot{X}_{top} = \frac{J_t}{h} - \frac{J_f}{h} \qquad [2]$$

J_t is the heat flux into the phase boundary from above and J_t is the heat flux from the phase boundary into the ice-bonded permafrost below. J_t depends on the difference between the long-term mean temperature at the seabed, T_s, and phase boundary temperature, T_p, at the top of the ice-bonded permafrost. For $J_t = J_t$, the phase boundary is stable. For $J_t < J_t$, refreezing of the thawed layer can occur from the phase boundary upward. For thawing to occur, T_s must be sufficiently warmer than T_p to make $J_t > J_t$. T_s is determined by oceanographic conditions (currents, ice cover, water salinity, bathymetry, and presence of nearby rivers). T_p is determined by hydrostatic pressure, soil particle effects, and salt concentration at the phase boundary (the combined effect of *in situ* pore fluid salinity, salt transport from the seabed through the thawed layer, and changes in concentration as a result of freezing or thawing).

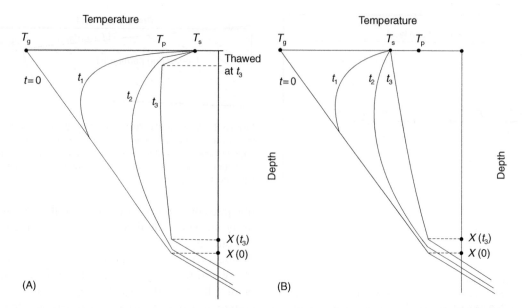

Figure 4 Schematic sub-sea permafrost temperature profiles showing the thermal evolution at successive times (t_1, t_2, t_3) after submergence when thawing occurs at the seabed (A) and when it does not (B). T_g and T_s are the long-term mean surface temperatures of the ground during emergence and of the seabed after submergence. T_p is the phase boundary temperature at the top of the ice-bonded permafrost. $X(0)$ and $X(t_3)$ are the depths to the bottom of ice-bounded permafrost at times $t = 0$ and t_3. (Adapted with permission from Lachenbruch AH and Marshall BV (1977) Open File Report 77-395. US Geological Survey, Menlo Park, CA.)

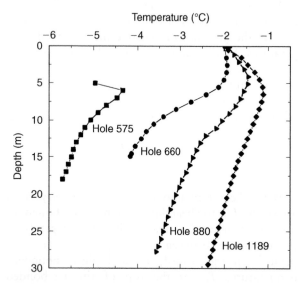

Figure 5 Temperature profiles obtained during the month of May in sub-sea permafrost near Barrow, Alaska, showing the thermal evolution with distance (equivalently time) offshore. Hole designation is the distance (m) offshore and the shoreline erosion rate is about 2.4 m y^{-1}. Sea ice freezes to the seabed within 600 m of shore. (Adapted from Osterkamp TE and Harrison WD (1985) Report UAGR-301. Fairbanks, AK: Geophysical Institute, University of Alaska.)

Nearshore at Prudhoe Bay, $\dot{X}_{\text{top}}$ varies typically from centimeters to tens of centimeters per year while farther offshore it appears to be on the order of millimeters per year. The thickness of the thawed layer at the seabed is typically 10 m to 100 m, although values of less than a meter have been observed. At some sites, the thawed layer is thicker in shallow water and thinner in deeper water.

Sub-sea permafrost also thaws from the bottom by geothermal heat flow once the thermal disturbance of the transgression penetrates there. The approximate thawing rate at the bottom of the ice-bonded permafrost is given by eqn [3].

$$\dot{X}_{\text{bot}} \cong \frac{J_g}{b} - \frac{J_f'}{b} \qquad [3]$$

J_g is the geothermal heat flow entering the phase boundary from below and J_f' is the heat flow from the phase boundary into the ice-bonded permafrost above. J_f' becomes small within a few millennia except when the permafrost contains unfrozen pore fluids. $\dot{X}_{\text{bot}}$ is typically on the order of centimeters per year. Timescales for thawing at the permafrost table and base are such that several tens of thousands of years may be required to completely thaw a few hundred meters of sub-sea permafrost.

Modeling results and field data indicate that impermeable sediments near the seabed, low T_s, high ice contents, and low J_g favor the survival of ice-bearing sub-sea permafrost during a transgression. Where conditions are favorable, substantial thicknesses of ice-bearing sub-sea permafrost may have survived previous transgressions.

Characteristics

The chemical composition of sediment pore fluids is similar to that of sea water, although there are detectable differences. Salt concentration profiles in thawed coarse-grained sediments at Prudhoe Bay (**Figure 6**) appear to be controlled by processes occurring during the initial phases of submergence. There is evidence for highly saline layers within ice-bonded permafrost near the base of gravels overlying a fine-grained sequence both onshore and offshore. In the Mackenzie Delta region, fluvial sand units deposited during regressions have low salt concentrations (**Figure 7**) except when thawed or when lying under saline sub-sea mud. Fine-grained mud sequences from transgressions have higher salt concentrations. Salts increase the amount of unfrozen pore fluids and decrease the phase equilibrium temperature, ice content, and ice bonding. Thus, the sediment layering observed in the Mackenzie Delta region can lead to unbonded material (clay) between layers of bonded material (fluvial sand).

Thawed sub-sea permafrost is often separated from ice-bonded permafrost by a transition layer of ice-bearing permafrost. The thickness of the ice-bearing layer can be small, leading to a relatively sharp (centimeters scale) phase boundary, or large, leading to a diffuse boundary (meters scale). In general, it appears that coarse-grained soils and low salinities produce a sharper phase boundary and fine-grained soils and higher salinities produce a more diffuse phase boundary.

Sub-sea permafrost consists of a mixture of sediments, ice, and unfrozen pore fluids. Its physical and mechanical properties are determined by the individual properties and relative proportions. Since ice and unfrozen pore fluid are strongly temperature dependent, so also are most of the physical and mechanical properties.

Ice-rich sub-sea permafrost has been found in the Alaskan and Canadian portions of the Beaufort Sea and in the Russian shelf. Thawing of this permafrost can result in differential settlement of the seafloor that poses serious problems for development.

Processes

Submergence

Onshore permafrost becomes sub-sea permafrost upon submergence, and details of this process play a major role in determining its future evolution. The rate at which the sea transgresses over land is determined by rising sea levels, shelf topography, tectonic setting, and the processes of shoreline erosion, thaw settlement, thaw strain of the permafrost, seabed erosion, and sedimentation. Sea levels on the polar continental shelves have increased more than 100 m in the last 20 000 years or so. With shelf widths of 100–600 km, the average shoreline retreat rates would have been about 5 to 30 m y^{-1}, although maximum rates could have been much larger. These average rates are comparable to areas with very rapid shoreline retreat rates observed today on the Siberian and North American shelves. Typical values are 1–6 m y^{-1}.

It is convenient to think of the transition from sub-aerial to sub-sea conditions as occurring in five regions (**Figure 8**) with each region representing different thermal and chemical surface boundary conditions. These boundary conditions are successively applied to the underlying sub-sea permafrost during a transgression or regression. Region 1 is the onshore permafrost that forms the initial condition for sub-sea permafrost. Permafrost surface temperatures range down to about $-15°C$ under current sub-aerial conditions and may have been 8–10°C colder during glacial times. Ground water is generally fresh, although salty lithological units may exist within the permafrost as noted above.

Region 2 is the beach, where waves, high tides, and resulting vertical and lateral infiltration of sea water produce significant salt concentrations in the active layer and near-surface permafrost. The active layer and temperature regime on the beach differ from those on land. Coastal banks and bluffs are a trap for wind-blown snow that often accumulates in insulating drifts over the beach and adjacent ice cover.

Region 3 is the area where ice freezes to the seabed seasonally, generally where the water depth is less than about 1.5–2 m. This setting creates unique thermal boundary conditions because, when the ice freezes into the seabed, the seabed becomes conductively coupled to the atmosphere and thus very cold. During summer, the seabed is covered with shallow, relatively warm sea water. Salt concentrations at the seabed are high during winter because of salt rejection from the growing sea ice and restricted circulation under the ice, which eventually freezes into the seabed. These conditions create highly saline brines that infiltrate the sediments at the seabed.

Region 4 includes the areas where restricted under-ice circulation causes higher-than-normal sea water salinities and lower temperatures over the sediments. The ice does not freeze to the seabed or only freezes to it sporadically. The existence of this region depends on the ice thickness, on water depth, and on flushing processes under the ice. Strong currents or steep bottom slopes may reduce its extent or eliminate it.

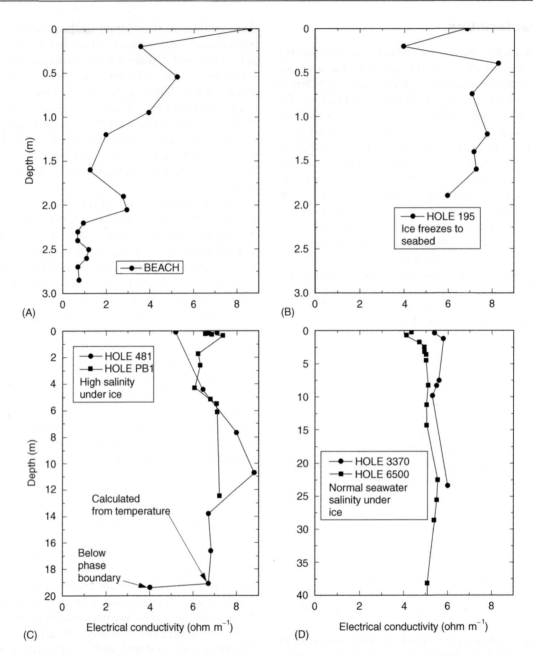

Figure 6 Electrical conductivity profiles in thawed, relatively uniform coarse-grained sediments. The holes lie along a line extending offshore near the West Dock at Prudhoe Bay, Alaska except for PB1, which is in the central portion of Prudhoe Bay. Hole designation is the distance (m) offshore. On the beach (A), concentrations decrease by a factor of 5 at a few meters depth. There are large variations with depth and concentrations may be double that of normal seawater where ice freezes to the seabed (B) and where there is restricted circulation under the ice (C). Farther offshore (D), the profiles tend to be relatively constant with depth with concentrations about the same or slightly greater than the overlying seawater. (Adapted from Iskandar IK *et al.* (1978) *Proceedings, Third International Conference on Permafrost*, Edmonton, Alberta, Canada, pp. 92–98. Ottawa, Ontario: National Research Council.)

The setting for region 5 consists of normal sea water over the seabed throughout the year. This results in relatively constant chemical and thermal boundary conditions.

There is a seasonal active layer at the seabed that freezes and thaws annually in both regions 3 and 4. The active layer begins to freeze simultaneously with the formation of sea ice in shallow water. Brine drainage from the growing sea ice increases the water salinity and decreases the temperature of the water at the seabed because of the requirement for phase equilibrium. This causes partial freezing of the less saline pore fluids in the sediments. Thus, it is not necessary for the ice to contact the seabed for the seabed to freeze. Seasonal changes in the pore fluid salinity show that the partially frozen active layer

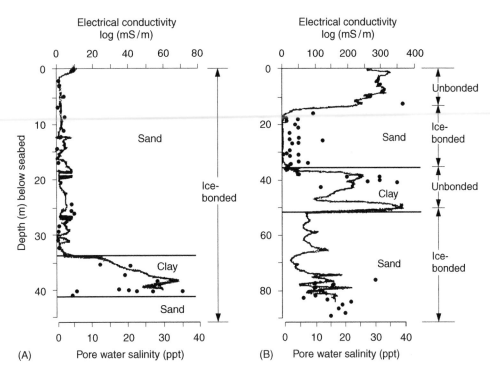

Figure 7 Onshore (A) and offshore (B) (water depth 10 m) electrical conductivity log (lines) and salinity (dots) profiles in the Mackenzie Delta region. The onshore sand–clay–sand sequence can be traced to the offshore site. Sand units appear to have been deposited under sub-aerial conditions during regressions and the clay unit under marine conditions during a transgression. At the offshore site, the upper sand unit is thawed to the 11 m depth. (Adapted with permission from Dallimore SR and Taylor AE (1994). *Proceedings, Sixth International Conference on Permafrost*, Beijing, vol. 1, pp.125–130. Wushan, Guangzhou, China: South China University Press.)

redistributes salts during freezing and thawing, is infiltrated by the concentrated brines, and influences the timing of brine drainage to lower depths in the sediments. These brines, derived from the growth of sea ice, provide a portion or all of the salts required for thawing the underlying sub-sea permafrost in the presence of negative sediment temperatures.

Depth to the ice-bonded permafrost increases slowly with distance offshore in region 3 to a few meters where the active layer no longer freezes to it (**Figure 8**) and the ice-bonded permafrost no longer couples conductively to the atmosphere. This allows the permafrost to thaw continuously throughout the year and depth to the ice-bonded permafrost increases rapidly with distance offshore (**Figure 8**).

The time an offshore site remains in regions 3 and 4 determines the number of years the seabed is subjected to freezing and thawing events. It is also the time required to make the transition from sub-aerial to relatively constant sub-sea boundary conditions. This time appears to be about 30 years near Lonely, Alaska (about 135 km southeast of Barrow) and about 500–1000 years near Prudhoe Bay, Alaska. **Figure 9** shows variations in the mean seabed temperatures with distance offshore near Prudhoe Bay where the shoreline retreat rate is about $1 \, \mathrm{m \, y^{-1}}$.

The above discussion of the physical setting does not incorporate the effects of geology, hydrology, tidal range, erosion and sedimentation processes, thaw settlement, and thaw strain. Regions 3 and 4 are extremely important in the evolution of sub-sea permafrost because the major portion of salt infiltration into the sediments occurs in these regions. The salt plays a strong role in determining T_p and, thus, whether or not thawing will occur.

Heat and Salt Transport

The transport of heat in sub-sea permafrost is thought to be primarily conductive because the observed temperature profiles are nearly linear below the depth of seasonal variations. However, even when heat transport is conductive, there is a coupling with salt transport processes because salt concentration controls T_p. Our lack of understanding of salt transport processes hampers the application of thermal models.

Thawing in the presence of negative seabed temperatures requires that T_p be significantly lower than T_s, so that generally salt must be present for thawing to occur. This salt must exist in the permafrost on submergence and/or be transported from the seabed

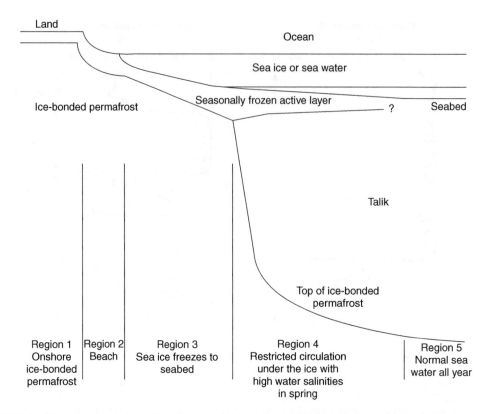

Figure 8 Schematic illustration of the transition of permafrost from sub-aerial to sub-sea conditions. There are five potential regions with differing thermal and chemical seabed boundary conditions. Hole 575 in **Figure 5** is in region 3 and the rest of the holes are in region 4. (Adapted from Osterkamp TE (1975) Report UAGR-234. Fairbanks, AK: Geophysical Institute, University of Alaska.)

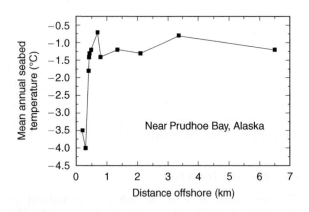

Figure 9 Variation of mean annual seabed temperatures with distance offshore. Along this same transect, about 6 km offshore from Reindeer Island in 17 m of water, the mean seabed temperature was near −1.7°C. Data on mean annual seabed salinities in regions 3 and 4 do not appear to exist. (Adapted from Osterkamp TE and Harrison WD (1985) Report UAGR-301. Fairbanks, AK: Geophysical Institute, University of Alaska.)

to the phase boundary. The efficiency of salt transport through the thawed layer at the seabed appears to be sensitive to soil type. In clays, the salt transport process is thought to be diffusion, a slow process; and in coarse-grained sands and gravels, pore fluid convection, which (involving motion of fluid) can be rapid.

Diffusive transport of salt has been reported in dense overconsolidated clays north of Reindeer Island offshore from Prudhoe Bay. Evidence for convective transport of salt exists in the thawed coarse-grained sediments near Prudhoe Bay and in the layered sands in the Mackenzie Delta region. This includes rapid vertical mixing as indicated by large seasonal variations in salinity in the upper 2 m of sediments in regions 3 and 4 and by salt concentration profiles that are nearly constant with depth and decrease in value with distance offshore in region 5 (**Figures 5–7**). Measured pore fluid pressure profiles (**Figure 10**) indicate downward fluid motion. Laboratory measurements of downward brine drainage velocities in coarse-grained sediments indicate that these velocities may be on the order of $100 \, \mathrm{m \, y^{-1}}$.

The most likely salt transport mechanism in coarse-grained sediments appears to be gravity-driven convection as a result of highly saline and dense brines at the seabed in regions 3 and 4. These brines infiltrate the seabed, even when it is partially frozen, and move rapidly downward. The release of relatively fresh and buoyant water by thawing ice at the phase boundary may also contribute to pore fluid motion.

Occurrence and Distribution

The occurrence, characteristics, and distribution of sub-sea permafrost are strongly influenced by regional and local conditions and processes including the following.

1. Geological (heat flow, shelf topography, sediment or rock types, tectonic setting)
2. Meteorological (sub-aerial ground surface temperatures as determined by air temperatures, snow cover and vegetation)
3. Oceanographic (seabed temperatures and salinities as influenced by currents, ice conditions, water depths, rivers and polynyas; coastal erosion and sedimentation; tidal range)
4. Hydrological (presence of lakes, rivers and salinity of the ground water)
5. Cryological (thickness, temperature, ice content, physical and mechanical properties of the onshore permafrost; presence of sub-sea permafrost that has survived previous transgressions; presence of ice sheets on the shelves)

Lack of information on these conditions and processes over the long timescales required for permafrost to aggrade and degrade, and inadequacies in the theoretical models, make it difficult to formulate reliable predictions regarding sub-sea permafrost. Field studies are required, but field data are sparse and investigations are still producing surprising results indicating that our understanding of sub-sea permafrost is incomplete.

Pechora and Kara Seas

Ice-bonded sub-sea permafrost has been found in boreholes with the top typically up to tens of meters below the seabed. In one case, pure freshwater ice was found 0.3 m below the seabed, extending to at least 25 m. These discoveries have led to difficult design conditions for an undersea pipeline that will cross Baydaratskaya Bay transporting gas from the Yamal Peninsula fields to European markets.

Laptev Sea

Sea water bottom temperatures typically range from $-0.5°C$ to $-1.8°C$, with some values colder than $-2°C$. A 300–850 m thick seismic sequence has been found that does not correlate well with regional tectonic structure and is interpreted to be ice-bonded permafrost. The extent of ice-bonded permafrost appears to be continuous to the 70 m isobath and widespread discontinuous to the 100 m isobath. Depth to the ice-bonded permafrost ranges from 2 to 10 m in water depths from 45 m to the shelf edge. Deep taliks may exist inshore of the 20 m isobath. A shallow sediment core with ice-bonded material at its base was recovered from a water depth of 120 m. Bodies of ice-rich permafrost occur under shallow water at the locations of recently eroded islands and along retreating coastlines.

Bering Sea

Sub-sea permafrost is not present in the northern portion except possibly in near-shore areas or where shoreline retreat is rapid.

Chukchi Sea

Seabed temperatures are generally slightly negative and thermal gradients are negative, indicating ice-bearing permafrost at depth within 1 km of shore near Barrow, Rabbit Creek and Kotzebue.

Alaskan Beaufort Sea

To the east of Point Barrow, bottom waters are typically $-0.5°C$ to $-1.7°C$ away from shore, shoreline erosion rates are rapid (1–10 m y^{-1}) and sediments are thick. Sub-sea permafrost appears to be thicker in the Prudhoe Bay region and thinner west of Harrison Bay to Point Barrow. Ice layers up

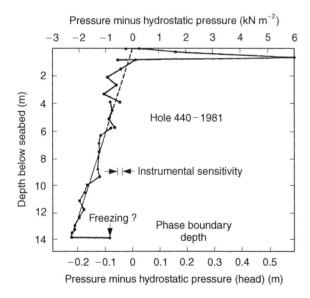

Figure 10 Measured *in situ* pore fluid pressure minus calculated hydrostatic pressure through the thawed layer in coarse-grained sediments at a hole 440 m offshore in May 1981 near Prudhoe Bay, Alaska. High pressure at the 0.54 m depth is probably related to seasonal freezing and the solid line is a least-squares fit to the data below 5.72 m. The negative pressure head gradient (-0.016) indicates a downward component of pore fluid velocity. (Adapted from Swift DW *et al.* (1983) *Proceedings, Fourth International Conference on Permafrost*, Fairbanks, Alaska, vol. 1, pp. 1221–1226. Washington DC: National Academy Press.)

to 0.6 m in thickness have been found off the Sagavanirktok River Delta.

Surface geophysical studies (seismic and electrical) have indicated the presence of layered ice-bonded sub-sea permafrost. A profile of sub-sea permafrost near Prudhoe Bay (**Figure 11**) shows substantial differences in depth to ice-bonded permafrost between coarse-grained sediments inshore of Reindeer Island and fine-grained sediments farther offshore. Offshore from Lonely, where surface sediments are fine-grained, ice-bearing permafrost exists within 6–8 m of the seabed out to 8 km offshore (water depth 8 m). Ice-bonded material is deeper (~15 m). In Elson Lagoon (near Barrow) where the sediments are fine-grained, a thawed layer at the seabed of generally increasing thickness can be traced offshore.

Mackenzie River Delta Region

The layered sediments found in this region are typically fluvial sand and sub-sea mud corresponding to regressive/transgressive cycles (see **Figure 12**). Mean seabed temperatures in the shallow coastal areas are generally positive as a result of warm river water, and negative farther offshore. The thickness of ice-bearing permafrost varies substantially as a result of a complex history of transgressions and regressions, discharge from the Mackenzie River, and possible effects of a late glacial ice cover. Ice-bearing permafrost in the eastern and central Beaufort Shelf exceeds 600 m. It is thin or absent beneath Mackenzie Bay and may be only a few hundred meters thick toward the Alaskan coast. The upper surface of

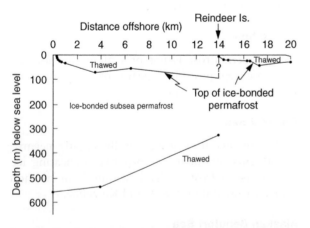

Figure 11 Sub-sea permafrost profile near Prudhoe Bay, Alaska, determined from drilling and well log data. The sediments are coarse-grained with deep thawing inshore of Reindeer Island and fine-grained with shallow thawing farther offshore. Maximum water depths are about 8 m inshore of Reindeer Island and 17 m about 6 km north. (Adapted from Osterkamp TE *et al.* (1985) *Cold Regions Science and Technology* 11: 99–105, 1985, with permission from Elsevier Science.)

ice-bearing permafrost is typically 5–100 m below the seabed and appears to be under control of seabed temperatures and stratigraphy.

The eastern Arctic, Arctic Archipelago and Hudson Bay regions were largely covered during the last glaciation by ice that would have inhibited permafrost growth. These regions are experiencing isostatic uplift with permafrost aggrading in emerging shorelines.

Antarctica

Negative sediment temperatures and positive temperature gradients to a depth of 56 m below the seabed exist in McMurdo Sound where water depth is 122 m and the mean seabed temperature is −1.9°C. This sub-sea permafrost did not appear to contain any ice.

Models

Modeling the occurrence, distribution and characteristics of sub-sea permafrost is a difficult task. Statistical, geological, analytical, and numerical models are available. Statistical models attempt to combine geological, oceanographic, and other information into algorithms that make predictions about sub-sea permafrost. These statistical models have not been very successful, although new GIS methods could potentially improve them.

Geological models consider how geological processes influence the formation and development of sub-sea permafrost. It is useful to consider these models since some sub-sea permafrost could potentially be a million years old. A geological model for the Mackenzie Delta region (**Figure 12**) has been developed that provides insight into the nature and complex layering of the sediments that comprise the sub-sea permafrost there.

Analytical models for investigating the thermal regime of sub-sea permafrost include both one- and two-dimensional models. All of the available analytical models have simplifying assumptions that limit their usefulness. These include assumptions of one-dimensional heat flow, stable shorelines or shorelines that undergo sudden and permanent shifts in position, constant air and seabed temperatures that neglect spatial and temporal variations over geological timescales, and constant thermal properties in layered sub-sea permafrost that is likely to contain unfrozen pore fluids. Neglect of topographical differences between the land and seabed, geothermal heat flow, phase change at the top and bottom of the sub-sea permafrost, and salt effects also limits their application. An analytical model exists that addresses the coupling between heat and

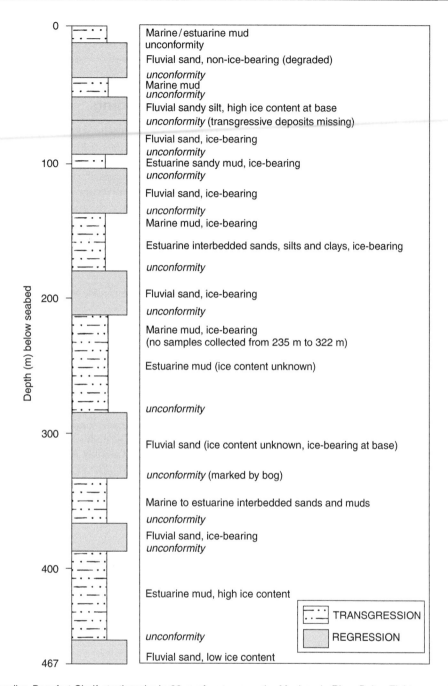

Figure 12 Canadian Beaufort Shelf stratigraphy in 32 m of water near the Mackenzie River Delta. Eight regressive/transgressive fluvial sand/marine mud cycles are shown. It is thought that, except for thawing near the seabed, the ice-bearing sequence has been preserved through time to the present. (Adapted with permission from Blasco S (1995) GSC Open File Report 3058. Ottawa, Canada: Geological Survey of Canada.)

salt transport but only for the case of diffusive transport with simplifying assumptions. Nevertheless, these analytical models appear to be applicable in certain special situations and have shaped much of the current thinking about sub-sea permafrost.

Two-dimensional numerical thermal models have addressed most of the concerns related to the assumptions in analytical models except for salt transport. Models have been developed that address salt transport via the buoyancy of fresh water generated by thawing ice at the phase boundary. Models for the infiltration of dense sea water brines derived from the growth of sea ice into the sediments do not appear to exist.

Successful application of all models is limited because of the lack of information over geological timescales on initial conditions, boundary conditions, material properties, salt transport and the coupling of heat and salt transport processes. There is also a lack of areas with sufficient information and measurements to fully test model predictions.

Nomenclature

h	Volumetric latent heat of the sediments (1 to $2 \times 10^8 \ \mathrm{J\ m^{-3}}$)
J_g	Geothermal heat flow entering the bottom phase boundary from below
J_f	Heat flow from the bottom phase boundary into the ice-bonded permafrost above
J_t	Heat flux into the top phase boundary from above
J_f	Heat flux from the top phase boundary into the ice-bonded permafrost below
K	Thermal conductivity of the ice-bonded permafrost (1 to 5 $\mathrm{W\,m^{-1}\,K^{-1}}$)
t	Time
T_s	Long-term mean temperature at the seabed
T_p	Phase boundary temperature at the top of the ice-bonded permafrost (0 to -2°C)
T_e	Phase boundary temperature at the bottom of the ice-bonded permafrost (0 to -2°C)
T_g	Long-term mean ground surface temperature during emergence
$X(t)$	Depth to the bottom of ice-bonded permafrost at time, t
$\dot{X}_{top}$	Thawing rate at top of ice-bonded permafrost during submergence
$\dot{X}_{bot}$	Thawing rate at the bottom of ice-bonded permafrost during submergence

See also

Glacial Crustal Rebound, Sea Levels and Shorelines. Methane Hydrate and Submarine Slides. Mid-Ocean Ridge Tectonics, Volcanism, and Geomorphology. Sea Level Variations Over Geologic Time.

Further Reading

Dallimore SR and Taylor AE (1994) Permafrost conditions along an onshore–offshore transect of the Canadian Beaufort Shelf. *Proceedings, Sixth International Conference on Permafrost, Beijing, vol. 1, pp 125–130.* South China University Press: Wushan, Guangzhou, China.

Hunter JA, Judge AS, MacAuley HA, *et al.* (1976) *The Occurrence of Permafrost and Frozen Sub-sea Bottom Materials in the Southern Beaufort Sea. Beaufort Sea Project, Technical Report 22.* Ottawa: Geological Survey Canada.

Lachenbruch AH, Sass JH, Marshall BV, and Moses TH Jr (1982) Permafrost, heat flow, and the geothermal regime at Prudhoe Bay, Alaska. *Journal of Geophysical Research* 87(B11): 9301–9316.

Mackay JR (1972) Offshore permafrost and ground ice, Southern Beaufort Sea, Canada. *Canadian Journal of Earth Science* 9(11): 1550–1561.

Osterkamp TE, Baker GC, Harrison WD, and Matava T (1989) Characteristics of the active layer and shallow sub-sea permafrost. *Journal of Geophysical Research* 94(C11): 16227–16236.

Romanovsky NN, Gavrilov AV, Kholodov AL, *et al.* (1998) *Map of predicted offshore permafrost distribution on the Laptev Sea Shelf. Proceedings, Seventh International Conference on Permafrost, Yellowknife, Canada, pp. 967–972.* University of Laval, Quebec: Center for Northern Studies.

Sellmann PV and Hopkins DM (1983) *Sub-sea permafrost distribution on the Alaskan Shelf. Final Proceedings, Fourth International Conference on Permafrost, Fairbanks, Alaska, pp. 75–82.* Washington, DC: National Academy Press.

COASTAL ENVIRONMENT

LAND–SEA GLOBAL TRANSFERS

F. T. Mackenzie and L. M. Ver, University of Hawaii, Honolulu, HI, USA

Introduction

The interface between the land and the sea is an important boundary connecting processes operating on land with those in the ocean. It is a site of rapid population growth, industrial and agricultural practices, and urban development. Large river drainage basins connect the vast interiors of continents with the coastal zone through river and groundwater discharges. The atmosphere is a medium of transport of substances from the land to the sea surface and from that surface back to the land. During the past several centuries, the activities of humankind have significantly modified the exchange of materials between the land and sea on a global scale – humans have become, along with natural processes, agents of environmental change. For example, because of the combustion of the fossil fuels coal, oil, and gas and changes in land-use practices including deforestation, shifting cultivation, and urbanization, the direction of net atmospheric transport of carbon (C) and sulfur (S) gases between the land and sea has been reversed. The global ocean prior to these human activities was a source of the gas carbon dioxide (CO_2) to the atmosphere and hence to the continents. The ocean now absorbs more CO_2 than it releases. In pre-industrial time, the flux of reduced sulfur gases to the atmosphere and their subsequent oxidation and deposition on the continental surface exceeded the transport of oxidized sulfur to the ocean via the atmosphere. The situation is now reversed because of the emissions of sulfur to the atmosphere from the burning of fossil fuels and biomass on land. In addition, river and groundwater fluxes of the bioessential elements carbon, nitrogen (N), and phosphorus (P), and certain trace elements have increased because of human activities on land. For example, the increased global riverine (and atmospheric) transport of lead (Pb) corresponds to its increased industrial use. Also, recent changes in the concentration of lead in coastal sediments appear to be directly related to changes in the use of leaded gasoline in internal combustion engines. Synthetic substances manufactured by modern society, such as pesticides and pharmaceutical products, are now appearing in river

and groundwater flows, thus moving toward the sea in a greater degree than before.

The Changing Picture of Land–Sea Global Transfers

Although the exchange through the atmosphere of certain trace metals and gases between the land and the sea surface is important, rivers are the main purveyors of materials to the ocean. The total water discharge of the major rivers of the world to the ocean is $36\,000\,km^3\,yr^{-1}$. At any time, the world's rivers contain about 0.000 1% of the total water volume of $1459 \times 10^6\,km^3$ near the surface of the Earth and have a total dissolved and suspended solid concentration of $c.$ 110 and $540\,ppm\,l^{-1}$, respectively. The residence time of water in the world's rivers calculated with respect to total net precipitation on land is only 18 days. Thus the water in the world's rivers is replaced every 18 days by precipitation. The global annual direct discharge to the ocean of groundwater is about 10% of the surface flow, with a recent estimate of $2400\,km^3\,yr^{-1}$. The dissolved constituent content of groundwater is poorly known, but one recent estimate of the dissolved salt groundwater flux is $1300 \times 10^6\,t\,yr^{-1}$.

The chemical composition of average river water is shown in **Table 1**. Note that the major anion in river water is bicarbonate, HCO_3^-; the major cation is calcium, Ca^{2+}, and that even the major constituent concentrations of river water on a global scale are influenced by human activities. The dissolved load of the major constituents of the world's rivers is derived from the following sources: about 7% from beds of salt and disseminated salt in sedimentary rocks, 10% from gypsum beds and sulfate salts disseminated in rocks, 38% from carbonates, and 45% from the weathering of silicate minerals. Two-thirds of the HCO_3^- in river waters are derived from atmospheric CO_2 via the respiration and decomposition of organic matter and subsequent conversion to HCO_3^- through the chemical weathering of silicate (~30% of total) or carbonate (~70% of total) minerals. The other third of the river HCO_3^- comes directly from carbonate minerals undergoing weathering.

It is estimated that only about 20% of the world's drainage basins have pristine water quality. The organic productivity of coastal aquatic environments has been heavily impacted by changes in the

Table 1 Chemical composition of average river water

By continent	River water concentration[a] ($mg\,l^{-1}$)									TOC^{b}	TDN^{b}	TDP^{b}	Water discharge ($10^{3}\,km^{3}\,yr^{-1}$)	Runoff ratio[c]
	Ca^{2+}	Mg^{2+}	Na^{+}	K^{+}	Cl^{-}	SO_4^{2-}	HCO_3^-	SO_2	TDS^{b}					
Africa														
Actual	5.7	2.2	4.4	1.4	4.1	4.2	36.9	12.0	60.5				3.41	0.28
Natural	5.3	2.2	3.8	1.4	3.4	3.2	26.7	12.0	57.8					
Asia														
Actual	17.8	4.6	8.7	1.7	10.0	13.3	67.1	11.0	134.6				12.47	0.54
Natural	16.6	4.3	6.6	1.6	7.6	9.7	66.2	11.0	123.5					
S. America														
Actual	6.3	1.4	3.3	1.0	4.1	3.8	24.4	10.3	54.6				11.04	0.41
Natural	6.3	1.4	3.3	1.0	4.1	3.5	24.4	10.3	54.3					
N. America														
Actual	21.2	4.9	8.4	1.5	9.2	18.0	72.3	7.2	142.6				5.53	0.38
Natural	20.1	4.9	6.5	1.5	7.0	14.9	71.4	7.2	133.5					
Europe														
Actual	31.7	6.7	16.5	1.8	20.0	35.5	86.0	6.8	212.8				2.56	0.42
Natural	24.2	5.2	3.2	1.1	4.7	15.1	80.1	6.8	140.3					
Oceania														
Actual	15.2	3.8	7.6	1.1	6.8	7.7	65.6	16.3	125.3				2.40	
Natural	15.0	3.8	7.0	1.1	5.9	6.5	65.1	16.3	120.3					
World average														
Actual	14.7	3.7	7.2	1.4	8.3	11.5	53.0	10.4	110.1	12.57	0.574	0.053	37.40	0.46
Natural (unpolluted)	13.4	3.4	5.2	1.3	5.8	5.3	52.0	10.4	99.6	9.89	0.386	0.027	37.40	0.46
Pollution	1.3	0.3	2.0	0.1	2.5	6.2	1.0	0.0	10.5	2.67	0.187	0.027		
World % pollutive	9%	8%	28%	7%	30%	54%	2%	0%	10%	21%	33%	50%		

[a] Actual concentrations include pollution. Natural concentrations are corrected for pollution.
[b] TDS, total dissolved solids; TOC, total organic carbon; TDN, total dissolved nitrogen; TDP, total dissolved phosphorus.
[c] Runoff ratio = average runoff per unit area/average rainfall.

Revised after Meybeck M (1979) Concentrations des eaux fluviales en elements majeurs et apports en solution aux oceans. *Revue de Geologie Dynamique et de Geographie Physique* 21: 215–246; Meybeck M (1982) Carbon, nitrogen, and phosphorus transport by world rivers. *American Journal of Science* 282: 401–450; Meybeck M (1983) C, N, P and S in rivers: From sources to global inputs. In: Wollast R, Mackenzie FT, and Chou L (eds.) *Interactions of C, N, P and S Biogeochemical Cycles and Global Change*, pp. 163–193. Berlin: Springer.

dissolved and particulate river fluxes of three of the major bioessential elements found in organic matter, C, N, and P (the other three are S, hydrogen (H), and oxygen (O)). Although these elements are considered minor constituents of river water, their fluxes may have doubled over their pristine values on a global scale because of human activities. Excessive river-borne nutrients and the cultural eutrophication of freshwater and coastal marine ecosystems go hand in hand. In turn, these fluxes have become sensitive indicators of the broader global change issues of population growth and land-use change (including water resources engineering works) in the coastal zone and upland drainage basins, climatic change, and sea level rise.

In contrast to the situation for the major elements, delivery of some trace elements from land to the oceans via the atmosphere can rival riverine inputs. The strength of the atmospheric sources strongly depends on geography and meteorology. Hence the North Atlantic, western North Pacific, and Indian Oceans, and their inland seas, are subjected to large atmospheric inputs because of their proximity to both deserts and industrial sources. Crustal dust is the primary terrestrial source of these atmospheric inputs to the ocean. Because of the low solubility of dust in both atmospheric precipitation and seawater and the overwhelming inputs from river sources, dissolved sources of the elements are generally less important. However, because the oceans contain only trace amounts of iron (Fe), aluminum (Al), and manganese (Mn) (concentrations are in the ppb level), even the small amount of dissolution in seawater (~10% of the element in the solid phase) results in eolian dust being the primary source for the dissolved transport of these elements to remote areas of the ocean. Atmospheric transport of the major nutrients N, silicon (Si), and Fe to the ocean has been hypothesized to affect and perhaps limit primary productivity in certain regions of the ocean at certain times. Modern processes of fossil fuel combustion and biomass burning have significantly modified the atmospheric transport from land to the ocean of trace metals like Pb, copper (Cu), and zinc (Zn), C in elemental and organic forms, and nutrient N.

As an example of global land–sea transfers involving gases and the effect of human activities on the exchange, consider the behavior of CO_2 gas. Prior to human influence on the system, there was a net flux of CO_2 out of the ocean owing to organic metabolism (net heterotrophy). This flux was mainly supported by the decay of organic matter produced by phytoplankton in the oceans and part of that transported by rivers to the oceans. An example overall reaction is:

$$C_{106}H_{263}O_{110}N_{16}S_2P + 141O_2 \Rightarrow 106CO_2 \\ + 6HNO_3 + 2H_2SO_4 + H_3PO_4 + 120H_2O \quad [1]$$

Carbon dioxide was also released to the atmosphere due to the precipitation of carbonate minerals in the oceans. The reaction is:

$$Ca^{2+} + 2HCO_3^- \Rightarrow CaCO_3 + CO_2 + H_2O \quad [2]$$

The CO_2 in both reactions initially entered the dissolved inorganic carbon (DIC) pool of seawater and was subsequently released to the atmosphere at an annual rate of about 0.2×10^9 t of carbon as CO_2 gas. It should be recognized that this is a small number compared with the 200×10^9 t of carbon that exchanges between the ocean and atmosphere each year because of primary production of organic matter and its subsequent respiration.

Despite the maintenance of the net heterotrophic status of the ocean and the continued release of CO_2 to the ocean–atmosphere owing to the formation of calcium carbonate in the ocean, the modern ocean and the atmosphere have become net sinks of anthropogenic CO_2 from the burning of fossil fuels and the practice of deforestation. Over the past 200 years, as CO_2 has accumulated in the atmosphere, the gradient of CO_2 concentration across the atmosphere–ocean interface has changed, favoring uptake of anthropogenic CO_2 into the ocean. The average oceanic carbon uptake for the decade of the 1990s was $c.$ 2×10^9 t annually. The waters of the ocean have accumulated about 130×10^9 t of anthropogenic CO_2 over the past 300 years.

The Coastal Zone and Land–Sea Exchange Fluxes

The global coastal zone environment is an important depositional and recycling site for terrigenous and oceanic biogenic materials. The past three centuries have been the time of well-documented human activities that have become an important geological factor affecting the continental and oceanic surface environment. In particular, historical increases in the global population in the areas of the major river drainage basins and close to oceanic coastlines have been responsible for increasing changes in land-use practices and discharges of various substances into oceanic coastal waters. As a consequence, the global C cycle and the cycles of N and P that closely interact with the carbon cycle have been greatly affected. Several major perturbations of the past three centuries of the industrial age have affected the processes of transport from land, deposition of terrigenous materials, and *in situ* production of organic matter in coastal zone environments. In addition, potential future changes in oceanic circulation may have significant effects on the biogeochemistry and CO_2 exchange in the coastal zone.

The coastal zone is that environment of continental shelves, including bays, lagoons, estuaries, and near-shore banks that occupy 7% of the surface area of the ocean $(36 \times 10^{12} \, m^2)$ or *c.* 9% of the volume of the surface mixed layer of the ocean $(3 \times 10^{15} \, m^3)$. The continental shelves average 75 km in width, with a bottom slope of $1.7 \, m \, km^{-1}$. They are generally viewed as divisible into the interior or proximal shelf, and the exterior or distal shelf. The mean depth of the global continental shelf is usually taken as the depth of the break between the continental shelf and slope at *c.* 200 m, although this depth varies considerably throughout the world's oceans. In the Atlantic, the median depth of the shelf-slope break is at 120 m, with a range from 80 to 180 m. The depths of the continental shelf are near 200 m in the European section of the Atlantic, but they are close to 100 m on the African and North American coasts. Coastal zone environments that have high sedimentation rates, as great as $30–60 \, cm \, ky^{-1}$ in active depositional areas, act as traps and filters of natural and human-generated materials transported from continents to the oceans via river and groundwater flows and through the atmosphere. At present a large fraction ($\sim 80\%$) of the land-derived organic and inorganic materials that are transported to the oceans is trapped on the proximal continental shelves. The coastal zone also accounts for 30–50% of total carbonate and 80% of organic carbon accumulation in the ocean. Coastal zone environments are also regions of higher biological production relative to that of average oceanic surface waters, making them an important factor in the global carbon cycle. The higher primary production is variably attributable to the nutrient inflows from land as well as from coastal upwelling of deeper ocean waters.

Fluvial and atmospheric transport links the coastal zone to the land; gas exchange and deposition are its links with the atmosphere; net advective transport of water, dissolved solids and particles, and coastal upwelling connect it with the open ocean. In addition, coastal marine sediments are repositories for much of the material delivered to the coastal zone. In the last several centuries, human activities on land have become a geologically important agent affecting the land–sea exchange of materials. In particular, river and groundwater flows and atmospheric transport of materials to the coastal zone have been substantially altered.

Bioessential Elements

Continuous increase in the global population and its industrial and agricultural activities have created four major perturbations on the coupled system of the biogeochemical cycles of the bioessential elements C, N, P, and S. These changes have led to major alterations in the exchanges of these elements between the land and sea. The perturbations are: (1) emissions of C, N, and S to the atmosphere from fossil-fuel burning and subsequent partitioning of anthropogenic C and deposition of N and S; (2) changes in land-use practices that affect the recycling of C, N, P, and S on land, their uptake by plants and release from decaying organic matter, and the rates of land surface denudation; (3) additions of N and P in chemical fertilizers to cultivated land area; and (4) releases of organic wastes containing highly reactive C, N, and P that ultimately enter the coastal zone. A fifth major perturbation is a climatic one: (5) the rise in mean global temperature of the lower atmosphere of about 1 °C in the past 300 years, with a projected increase of about 1.4–5.8 °C relative to 1990 by the year 2100. **Figure 1** shows how the fluxes associated with these activities have changed during the past three centuries with projections to the year 2040.

Partially as a result of these activities on land, the fluxes of materials to the coastal zone have changed historically. **Figure 2** shows the historical and projected future changes in the river fluxes of dissolved inorganic and organic carbon (DIC, DOC), nitrogen (DIN, DON), and phosphorus (DIP, DOP), and fluxes associated with the atmospheric deposition and denitrification of N, and accumulation of C in organic matter in coastal marine sediments. It can be seen in **Figure 2** that the riverine fluxes of C, N, and P all increase in the dissolved inorganic and organic phases from about 1850 projected to 2040. For example, for carbon, the total flux (organic + inorganic) increases by about 35% during this period. These increased fluxes are mainly due to changes in land-use practices, including deforestation, conversion of forest to grassland, pastureland, and urban centers, and regrowth of forests, and application of fertilizers to croplands and the subsequent leaching of N and P into aquatic systems.

Inputs of nutrient N and P to the coastal zone which support new primary production are from the land by riverine and groundwater flows, from the open ocean by coastal upwelling and onwelling, and to a lesser extent by atmospheric deposition of nitrogen. New primary production depends on the

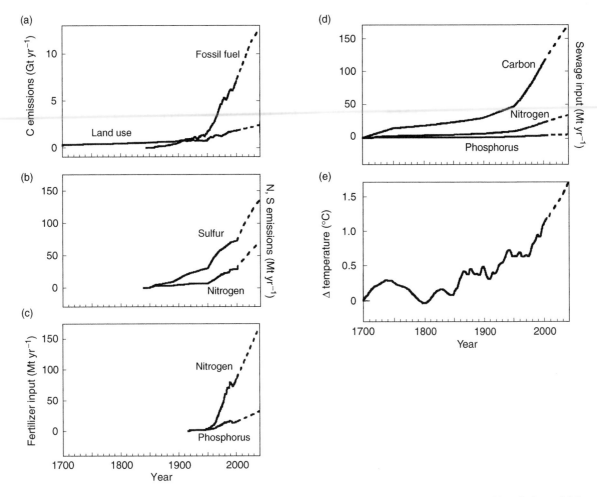

Figure 1 Major perturbations on the Earth system over the past 300 years and projections for the future: (a) emissions of CO_2 and (b) gaseous N and S from fossil-fuel burning and land-use activities; (c) application of inorganic N and P in chemical fertilizers to cultivated land; (d) loading of highly reactive C, N, and P into rivers and the coastal ocean from municipal sewage and wastewater disposal; and (e) rise in mean global temperature of the lower atmosphere relative to 1700. Revised after Ver LM, Mackenzie FT, and Lerman A (1999) Biogeochemical responses of the carbon cycle to natural and human perturbations: Past, present, and future. *American Journal of Science* 299: 762–801.

availability of nutrients from these external inputs, without consideration of internal recycling of nutrients. Thus any changes in the supply of nutrients to the coastal zone owing to changes in the magnitude of these source fluxes are likely to affect the cycling pathways and balances of the nutrient elements. In particular, input of nutrients from the open ocean by coastal upwelling is quantitatively greater than the combined inputs from land and the atmosphere. This makes it likely that there could be significant effects on coastal primary production because of changes in ocean circulation. For example, because of global warming, the oceans could become more strongly stratified owing to freshening of polar oceanic waters and warming of the ocean in the tropical zone. This could lead to a reduction in the intensity of the oceanic thermohaline circulation (oceanic circulation owing to differences in density of water masses, also popularly known as the 'conveyor belt') and hence the rate at which nutrient-rich waters upwell into coastal environments.

Another potential consequence of the reduction in the rate of nutrient inputs to the coastal zone by upwelling is the change in the CO_2 balance of coastal waters: reduction in the input of DIC to the coastal zone from the deeper ocean means less dissolved CO_2, HCO_3^-, and CO_3^{2-} coming from that source. With increasing accumulation of anthropogenic CO_2 in the atmosphere, the increased dissolution of atmospheric CO_2 in coastal water is favored. The combined result of a decrease in the upwelling flux of DIC and an enhancement in the transfer of atmospheric CO_2 across the air–sea interface of coastal waters is a lower saturation state for coastal waters with respect to the carbonate minerals calcite, aragonite, and a variety of magnesian calcites. The

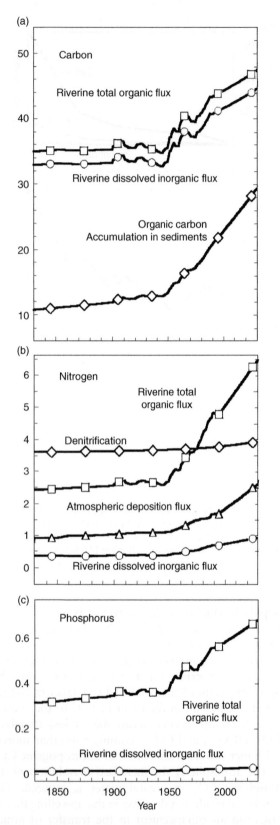

Figure 2 Past, present, and predicted fluxes of carbon, nitrogen, and phosphorus into or out of the global coastal margin, in 10^{12} mol yr^{-1}.

lower saturation state in turn leads to the likelihood of lower rates of inorganic and biological precipitation of carbonate and hence deposition and accumulation of sedimentary carbonate.

In addition, the present-day burial rate of organic carbon in the ocean may be about double that of the late Holocene flux, supported by increased fluxes of organic carbon to the ocean via rivers and groundwater flows and increased *in situ* new primary production supported by increased inputs of inorganic N and P from land and of N deposited from the atmosphere. The organic carbon flux into sediments may constitute a sink of anthropogenic CO_2 and a minor negative feedback on accumulation of CO_2 in the atmosphere.

The increased flux of land-derived organic carbon delivered to the ocean by rivers may accumulate there or be respired, with subsequent emission of CO_2 back to the atmosphere. This release flux of CO_2 may be great enough to offset the increased burial flux of organic carbon to the seafloor due to enhanced fertilization of the ocean by nutrients derived from human activities. The magnitude of the CO_2 exchange is a poorly constrained flux today. One area for which there is a substantial lack of knowledge is the Asian Pacific region. This is an area of several large seas, a region of important river inputs to the ocean of N, P, organic carbon, and sediments from land, and a region of important CO_2 exchange between the ocean and the atmosphere.

Anticipated Response to Global Warming

From 1850 to modern times, the direction of the net flux of CO_2 between coastal zone waters and the atmosphere due to organic metabolism and calcium carbonate accumulation in coastal marine sediments was from the coastal surface ocean to the atmosphere (negative flux, **Figure 3**). This flux in 1850 was on the order of $-0.2 \times 10^9\,\mathrm{t\,yr}^{-1}$. In a condition not disturbed by changes in the stratification and thermohaline circulation of the ocean brought about by a global warming of the Earth, the direction of this flux is projected to remain negative (flux out of the coastal ocean to the atmosphere) until early in the twenty-first century. The increasing partial pressure of CO_2 in the atmosphere because of emissions from anthropogenic sources leads to a reversal in the gradient of CO_2 across the air–sea interface of coastal zone waters and, hence, invasion of CO_2 into the coastal waters. From that time on the coastal ocean will begin to operate as a net sink (positive flux) of atmospheric CO_2 (**Figure 3**). The role of the open ocean as a sink for anthropogenic CO_2 is slightly reduced while that of the coastal oceans

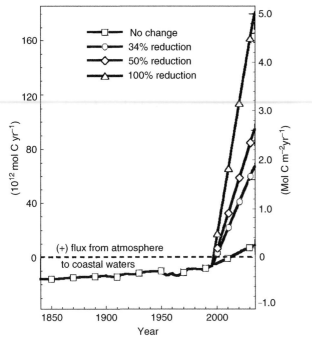

Figure 3 The net flux of CO_2 between coastal zone waters and the atmosphere due to organic metabolism and calcium carbonate accumulation in coastal marine sediments, under three scenarios of changing thermohaline circulation rate compared to a business-as-usual scenario, in units of $10^{12}\,mol\,C\,yr^{-1}$.

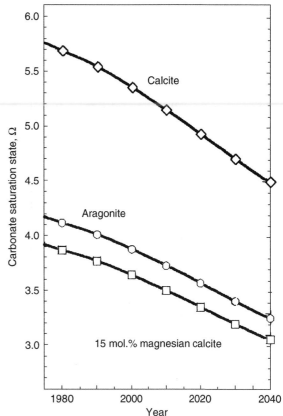

Figure 4 Changes in saturation state with respect to carbonate minerals of surface waters of the coastal ocean projected from 1999 to 2040. Calculations are for a temperature of 25 °C.

increases. The net result is the maintenance of the role of the global oceans as a net sink for anthropogenic CO_2.

The saturation index (Ω) for calcite or aragonite (both $CaCO_3$) is the ratio of the ion activity product IAP in coastal waters to the respective equilibrium constant K at the *in situ* temperature. For aqueous species, $IAP = aCa^{2+} \times aCO_3^{2-}$ (where a is the activity; note that for 15 mol.% magnesian calcite, the IAP also includes the activity of the magnesium cation, aMg^{2+}). Most coastal waters and open-ocean surface waters currently are supersaturated with respect to aragonite, calcite, and magnesian calcite containing 15 mol.% Mg, that is, $\Omega_{calcite}$, $\Omega_{aragonite}$, and $\Omega_{15\%magnesian\ calcite}$ are >1. Because of global warming and the increasing land-to-atmosphere-to-seawater transport of CO_2 (due to the continuing combustion of fossil fuels and burning of biomass), the concentration of the aqueous CO_2 species in seawater increases and the pH of the water decreases slightly. This results in a decrease in the concentration of the carbonate ion, CO_3^{2-}, resulting in a decrease in the degree of supersaturation of coastal zone waters. **Figure 4** shows how the degree of saturation might change into the next century because of rising atmospheric CO_2 concentrations. The overall reduction in the saturation state of coastal

zone waters with respect to aragonite from 1997 projected to 2040 is about 16%, from 3.89 to 3.26.

Modern carbonate sediments deposited in shoal-water ('shallow-water') marine environments (including shelves, banks, lagoons, and coral reef tracts) are predominantly biogenic in origin derived from the skeletons and tests of benthic and pelagic organisms, such as corals, foraminifera, echinoids, mollusks, algae, and sponges. One exception to this statement is some aragonitic muds that may, at least in part, result from the abiotic precipitation of aragonite from seawater in the form of whitings. Another exception is the sand-sized, carbonate oöids composed of either aragonite with a laminated internal structure or magnesian calcite with a radial internal structure. In addition, early diagenetic carbonate cements found in shoal-water marine sediments and in reefs are principally aragonite or magnesian calcite. Thus carbonate production and accumulation in shoal-water environments are dominated by a range of metastable carbonate minerals associated with skeletogenesis and abiotic processes, including calcite, aragonite, and a variety of magnesian calcite compositions.

With little doubt, as has been documented in a number of observational and experimental studies, a reduction in the saturation state of ocean waters will lead to a reduction in the rate of precipitation of both inorganic and skeletal calcium carbonate. Conversely, increases in the degree of supersaturation and temperature will increase the precipitation rates of calcite and aragonite from seawater. During global warming, rising sea surface temperatures and declining carbonate saturation states due to the absorption of anthropogenic CO_2 by surface ocean waters result in opposing effects. However, experimental evidence suggests that within the range of temperature change predicted for the next century due to global warming, the effect of changes in saturation state will be the predominant factor affecting precipitation rate. Thus decreases in precipitation rates should lead to a decrease in the production and accumulation of shallow-water carbonate sediments and perhaps changes in the types and distribution of calcifying biotic species found in shallow-water environments.

Anticipated Response to Heightened Human Perturbation: The Asian Scenario

In the preceding sections it was shown that the fluxes of C, N, and P from land to ocean have increased because of human activities (refer to **Table 1** for data comparing the actual and natural concentrations of C, N, P, and other elements in average river water). During the industrial era, these fluxes mainly had their origin in the present industrialized and developed countries. This is changing as the industrializing and developing countries move into the twenty-first century. A case in point is the countries of Asia.

Asia is a continent of potentially increasing contributions to the loading of the environment owing to a combination of such factors as its increasing population, increasing industrialization dependent on fossil fuels, concentration of its population along the major river drainage basins and in coastal urban centers, and expansion of land-use practices. It is anticipated that Asia will experience similar, possibly even greater, loss of storage of C and nutrient N and P on land and increased storage in coastal marine sediments per unit area than was shown by the developed countries during their period of industrialization. The relatively rapid growth of Asia's population along the oceanic coastal zone indicates that higher inputs of both dissolved and particulate organic nutrients may be expected to enter coastal waters.

A similar trend of increasing population concentration in agricultural areas inland, within the drainage basins of the main rivers flowing into the ocean, is also expected to result in increased dissolved and particulate organic nutrient loads that may eventually reach the ocean. Inputs from inland regions to the ocean would be relatively more important if no entrapment or depositional storage occurred en route, such as in the dammed sections of rivers or in alluvial plains. In the case of many of China's rivers, the decline in sediment discharge from large rivers such as the Yangtze and the Yellow Rivers is expected to continue due to the increased construction of dams. The average decadal sediment discharge from the Yellow River, for example, has decreased by 50% from the 1950s to the 1980s. If the evidence proposed for the continental United States applies to Asia, the damming of major rivers would not effectively reduce the suspended material flow to the ocean because of the changes in the erosional patterns on land that accompany river damming and more intensive land-use practices. These flows on land and into coastal ocean waters are contributing factors to the relative importance of autotrophic and heterotrophic processes, competition between the two, and the consequences for carbon exchange between the atmosphere and land, and the atmosphere and ocean water. The change from the practices of land fertilization by manure to the more recent usage of chemical fertilizers in Asia suggests a shift away from solid organic nutrients and therefore a reduced flow of materials that might promote heterotrophy in coastal environments.

Sulfur is an excellent example of how parts of Asia can play an important role in changing land–sea transfers of materials. Prior to extensive human interference in the global cycle of sulfur, biogenically produced sulfur was emitted from the sea surface mainly in the form of the reduced gas dimethyl sulfide (DMS). DMS was the major global natural source of sulfur for the atmosphere, excluding sulfur in sea salt and soil dust. Some of this gas traveled far from its source of origin. During transport the reduced gas was oxidized to micrometer-size sulfate aerosol particles and rained out of the atmosphere onto the sea and continental surface. The global sulfur cycle has been dramatically perturbed by the industrial and biomass burning activities of human society. The flux of gaseous sulfur dioxide to the atmosphere from the combustion of fossil fuels in some regions of the world and its conversion to sulfate aerosol greatly exceeds natural fluxes of sulfur gases from the land surface. It is estimated that this flux for the year 1990 was equivalent to 73×10^6 $t \, yr^{-1}$, nearly 4 times the natural DMS flux from the ocean. This has led to a net transport of sulfur from the land to the ocean via the atmosphere, completely reversing the flow direction in preindustrial times. In addition, the sulfate aerosol content of the atmosphere derived from human activities has increased. Sulfate aerosols affect global climate directly as particles that scatter incoming solar radiation and indirectly as cloud condensation nuclei (CCNs), which lead to an increased number of cloud droplets and an increase in the solar reflectance of clouds. Both effects cause the cooling of the planetary surface. As can be seen in **Figure 5** the eastern Asian region is an important regional source of sulfate aerosol because of the combustion of fossil fuels,

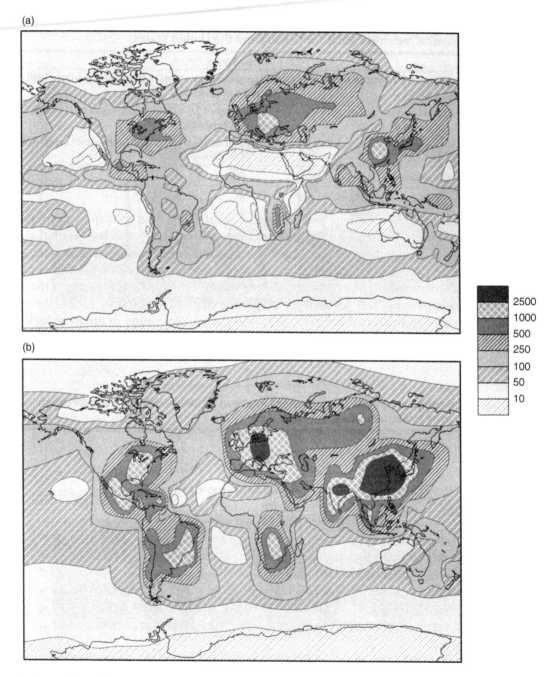

Figure 5 Comparison of the magnitude of atmospheric sulfur deposition for the years 1990 (a) and 2050 (b). Note the large increases in both spatial extent and intensity of sulfur deposition in both hemispheres and the increase in importance of Asia, Africa, and South America as sites of sulfur deposition between 1990 and 2050. The values on the diagrams are in units of kg S m^{-2} yr^{-1}. Revised after Mackenzie FT (1998) *Our Changing Planet: An Introduction to Earth System Science and Global Environmental Change*. Upper Saddle River, NJ: Prentice Hall; Rodhe H, Langner J, Gallardo L, and Kjellström E (1995) Global transport of acidifying pollutants. *Water, Air and Soil Pollution* 85: 37–50.

particularly coal. This source is predicted to grow in strength during the early- to mid-twenty-first century (**Figure 5**).

Conclusion

Land–sea exchange processes and fluxes of the bioessential elements are critical to life. In several cases documented above, these exchanges have been substantially modified by human activities. These modifications have led to a number of environmental issues including global warming, acid deposition, excess atmospheric nitrogen deposition, and production of photochemical smog. All these issues have consequences for the biosphere – some well known, others not so well known. It is likely that the developing world, with increasing population pressure and industrial development and with no major changes in agricultural technology and energy consumption rates, will become a more important source of airborne gases and aerosols and materials for river and groundwater systems in the future. This will lead to further modification of land–sea global transfers. The region of southern and eastern Asia is particularly well poised to influence significantly these global transfers.

See also

Coastal Topography, Human Impact on.

Further Reading

Berner EA and Berner RA (1996) *Global Environment: Water, Air and Geochemical Cycles*. Upper Saddle River, NJ: Prentice Hall.

Galloway JN and Melillo JM (eds.) (1998) *Asian Change in the Context of Global Change*. Cambridge, MA: Cambridge University Press.

Mackenzie FT (1998) *Our Changing Planet: An Introduction to Earth System Science and Global Environmental Change*. Upper Saddle River, NJ: Prentice Hall.

Mackenzie FT and Lerman A (2006) *Carbon in the Geobiosphere – Earth's Outer Shell*. Dordrecht: Springer.

Meybeck M (1979) Concentrations des eaux fluviales en elements majeurs et apports en solution aux oceans. *Revue de Geologie Dynamique et de Geographie Physique* 21: 215–246.

Meybeck M (1982) Carbon, nitrogen, and phosphorus transport by world rivers. *American Journal of Science* 282: 401–450.

Meybeck M (1983) C, N, P and S in rivers: From sources to global inputs. In: Wollast R, Mackenzie FT, and Chou L (eds.) *Interactions of C, N, P and S Biogeochemical Cycles and Global Change*, pp. 163–193. Berlin: Springer.

Rodhe H, Langner J, Gallardo L, and Kjellström E (1995) Global transport of acidifying pollutants. *Water, Air and Soil Pollution* 85: 37–50.

Schlesinger WH (1997) *Biogeochemistry: An Analysis of Global Change*. San Diego, CA: Academic Press.

Smith SV and Mackenzie FT (1987) The ocean as a net heterotrophic system: Implications from the carbon biogeochemical cycle. *Global Biogeochemical Cycles* 1: 187–198.

Ver LM, Mackenzie FT, and Lerman A (1999) Biogeochemical responses of the carbon cycle to natural and human perturbations: Past, present, and future. *American Journal of Science* 299: 762–801.

Vitousek PM, Aber JD, and Howarth RW (1997) Human alteration of the global nitrogen cycle: Sources and consequences. *Ecological Applications* 7(3): 737–750.

Wollast R and Mackenzie FT (1989) Global biogeochemical cycles and climate. In: Berger A, Schneider S, and Duplessy JC (eds.) *Climate and Geo-Sciences*, pp. 453–473. Dordrecht: Kluwer Academic Publishers.

BEACHES, PHYSICAL PROCESSES AFFECTING

A. D. Short, University of Sydney, Sydney, Australia

Introduction

Beach systems occur on shorelines throughout the world and are a product of three basic factors – wave, sediment, and substrate. They are, however, also influenced by tides and impacted by a range of geological, biotic, chemical, and temperature factors. Beach systems therefore occupy a considerable range of environments, from the picturesque tropical beaches, to those exposed to the full forces of North Atlantic storms, to the usually frozen shores of the Arctic Ocean. In addition formative waves can range from low to extreme, and sediment from fine sand to boulders. This article looks at how the two main processes – waves and tides – interact with sediment to form three types of beach systems: wave-dominated, tide-modified, and tide-dominated, the three representing all the world's beach types. It then considers some of the major factors that modify beach systems on a global and regional scale, including geology, temperature, biota, and chemical processes.

Beaches

Beaches are a wave-deposited accumulation of sediment lying between wave base and the limit of wave uprush or swash. The wave base is the seaward limit or depth from which average waves can entrain and transport sediment shoreward. The essential requirements for a beach to form are therefore an underlying substrate, sediment, and waves. In two-dimensions as waves shoal across the nearshore zone, between wave base and the breaker zone or shoreline, they generate a concave upward profile. As the waves transform at breaking to swash, the swash builds a steep beach face or swash zone, which caps the beach system.

In reality beaches are usually more complex. Waves can range in size from a few centimeters to several meters, and period from 1 to 20 s, and additional processes, particularly tides influence both the waves and the resulting beach morphology. Beaches can also be composed of sediments ranging from fine sand to boulders. In addition, the underlying substrate, known as the geological inheritance provides both the two-dimensional foundations, as well as potential two- and three-dimensional irregularities such as reefs and headlands, all of which influence wave transformation, sediment transport, and two- and three-dimensional beach morphology. Finally when waves break across a surf zone, they generate three-dimensional flows that produce a more complex three-dimensional surf zone and beach morphology, that lies between the nearshore and swash zones. Beaches, therefore, all contain a nearshore zone dominated by wave shoaling processes; they may contain a surf zone dominated by wave breaking and surf zone currents, and finally a swash zone dominated by wave uprush and backwash (**Figure 1**).

Beaches also range from relatively simple low-energy, two-dimensional, tideless systems with waves surging against a uniform shoreline, to more complex higher energy systems, incorporating surf zones with three-dimensional circulation, such as rip currents, associated with highly rhythmic bars and shorelines. In addition geological inheritance can introduce a wide spectrum of bedrock and sedimentology inputs that cause additional wave transformation through variable wave refraction and attenuation, as well as influence sediment type and mobility.

This article will cover the full range of physical beach systems from the low to high wave energy wave-dominated beaches. It will then address the influence of increasing tide range in the tide-modified and tide-dominated beaches. Finally it will briefly review the role of geology (e.g., headlands and reefs) and other physical, chemical, and biotic factors in modifying beach systems.

Wave-dominated Beaches

Wave-dominated beaches occur in environments where waves are high relative to the tide range. This can be defined quantitatively by the relative tide range (RTR)

$$RTR = TR/Hb \qquad [1]$$

where TR is the spring tide range and Hb the average breaker wave height. When RTR < 3 beaches are wave dominated, when 3 < RTR < 15 they are tide-modified and when the RTR > 15 they become tide-dominated.

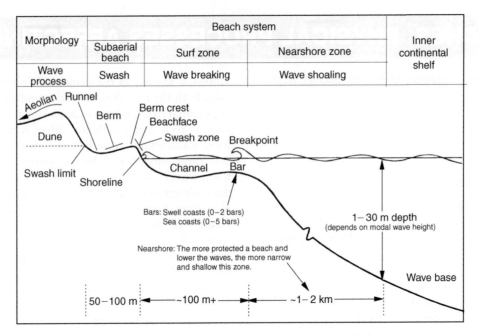

Figure 1 Definition sketch of a high-energy wave-dominated beach system. The beach contains a subaerial beach dominated by swash, a surf zone dominated by breaking waves and the nearshore zone across which waves shoal. (Reproduced from Short, 1999.)

Within any beach environment, wave height and period and sediment size can all vary substantially. In order to accommodate the potentially wide range of wave height–period–sediment combinations **Figure 2** presents a sensitivity plot which defines the three wave-dominated beach types, reflective, intermediate and dissipative, using the dimensionless fall velocity

$$\Omega = Hb/TWs \qquad [2]$$

where T is wave period (s) and W_s is sediment fall velocity (ms^{-1}).

Reflective Beaches

Reflective beaches occur when $\Omega < 1$, which requires a combination of lower waves, longer periods and particularly coarser sands. They occur on sandy open swell coasts when waves average less than 0.5 m, and on all coasts when beach sediments are coarse sand or coarser, including all gravel through boulder beaches. On gravel–boulder reflective beaches waves may exceed a few meters. They are, however, all characterized by a nearshore zone of wave shoaling that extends to the shoreline. Waves then break by plunging and/or surging across the base of the beach face. The ensuing strong swash rushes up the beach, combining with the coarse sediments to build a steep beach face, commonly capped by well-developed beach cusps and/or a berm (**Figures 3** lower and **4**). When the sediment consists of a range of grain sizes, the coarser grains accumulate as a coarser steep step

below the zone of wave breaking, at the base of the beach face.

Reflective beaches therefore consist of a concave upward nearshore zone composed of wave ripples and relatively fine sand across which waves shoal. As the waves break at the base of the beach there may be a coarser step up to a few decimeters high, then a relatively steep (4–10°) beach face, possibly capped with cusps. The cusps are a product of cellular circulation on the high tide beach resulting from subharmonic edge waves produced from the interaction of the incoming and reflected backwash. The high degree of incident wave reflection off the beach face is responsible for naming of this beach type, i.e., reflective. Apart from the cosmetic beach cusps and swash circulation, if present, these are essentially two-dimensional beaches with no alongshore variation in either processes or morphology. On sand beaches they also represent the lower energy end of the beach spectrum and as such are relatively stable systems, only experiencing beach change during periods of higher waves which tend to erode the berm and deposit it as an attached low tide bar or terrace (**Figure 3** lower). During following lower waves the terrace quickly migrates back up onto the beach as a new berm or cusps.

Intermediate Beaches

Intermediate beaches are called such as they represent a suite of beach types between the lower energy reflective and higher energy dissipative. They are the

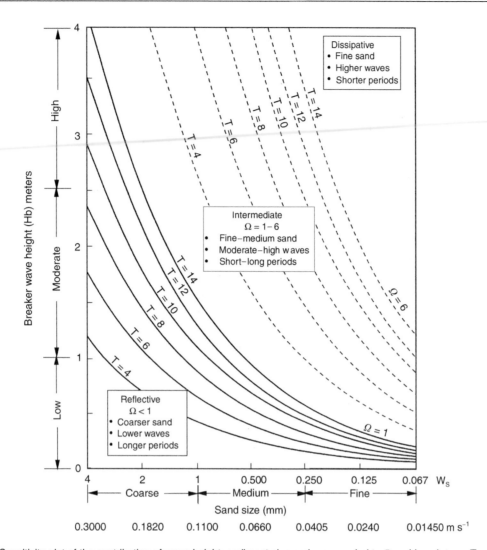

Figure 2 Sensitivity plot of the contribution of wave height, sediment size and wave period to Ω and beach type. To use the chart determine the breaker height, period (T) and grain size/fall velocity (phi or m s^{-1}). Read off the wave height and grain size, then use the period to determine where the boundary of reflective/intermediate, or intermediate/dissipative beaches lie. $\Omega = 1$ along solid T lines, and $\Omega = 6$ along dashed T lines. Below the solid lines $\Omega < 1$ the beach is reflective, above the dashed lines $\Omega > 6$ the beach is dissipative, between the solid and dashed lines Ω is between 1 and 6 and the beach is intermediate. (Reproduced from Short, 1999.)

beaches that form under moderate (Hb > 0.5 m) to high waves, on swell and seacoasts, in fine to medium sand (**Figure 2**). The two most distinguishing characteristics of intermediate beaches is first a surf zone, and second cellular rip circulation, usually associated with rhythmic beach topography. Since intermediate beaches can occur across a wide range of wave conditions, they consist of four beach states ranging from the lower energy low tide terrace to the high energy longshore bar and trough (**Figure 3**).

Intermediate beaches are controlled by processes related to wave dissipation across the surf zone which transfers energy from incident waves with periods of 2–20 s, to longer infragravity waves with periods > 30 s. As a consequence, incoming long waves associated with wave groupiness, increase in energy and amplitude across the surf zone and are manifest at the shoreline as wave set up (crest) and set down (trough). They then reflect off the beach leading to an interaction between the incoming and outgoing waves to produce a standing wave across the surf zone. It is believed that standing edge waves trapped in the surf zone are responsible for the cellular circulation that develops into rip current circulation, that in turn is responsible for the high degree of spatial and temporal variability in intermediate beach morphodynamics.

Low tide terrace   Low tide terrace (or ridge and runnel) beaches are characterized by a continuous attached bar or terrace located at low tide (**Figure 5**). Usually smaller waves (0.5–1 m) break

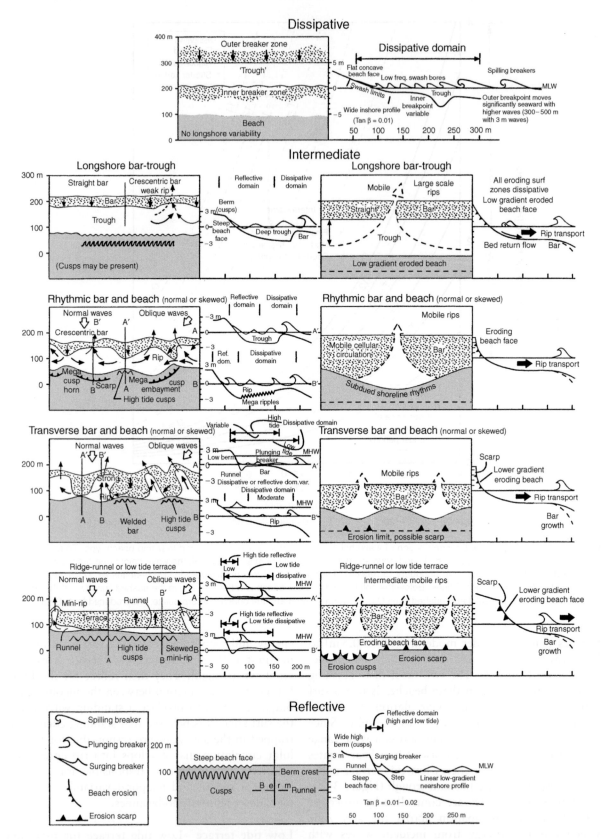

Figure 3 Three-dimensional sequence of wave-dominated beach changes for accretionary (left) and erosional (right) wave conditions. The sequence ranges from dissipative (top), through intermediate, to reflective (lower). (Reproduced from Short, 1999.)

Figure 4 Pearl Beach near Sydney, Australia, is a modally reflective beach composed of medium to coarse sand and receiving usually lower swell waves. Note the very narrow zone of wave breaking, well-developed beach cusps and cusp-controlled segregated swash circulation (A.D. Short).

Figure 6 Well-developed transverse bar and rip beach with high rhythmic shoreline (megacusp horns and embayments), attached bars and well-developed rip channels, Johana Beach, Victoria (A.D. Short).

Figure 5 A wide, well-developed low tide terrace at Crowdy Head, New South Wales. Note the transition from transverse bar and rips (foreground) to mini-rips, then continuous low tide terrace (A.D. Short).

across the bar at low tide, whereas at high tide they may remain unbroken and surge up the reflective high tide beach face. At mid to low tide cellular circulation in the surf zone usually produces weak rip current flows, which may be transitory across a planar bar, or may reside in shallow rip channels (mini-rips) (**Figure 3**).

Transverse bar and rip Transverse bar and rip beaches form under moderate waves (1–1.5 m) on swell coasts. They consist of well-developed rip channels, which are separated by shallow bars, which are attached and perpendicular or transverse to the beach. Variable wave breaking and refraction across the shallow bars and deeper rip channels leads to a longshore variation in swash height and approach,

which reworks the beach to form prominent megacusp horns, in lee of bars, and embayments, in lee of channels (**Figures 3** and **6**). Water tends to flow shoreward over the bars, then into the rip feeder channels. The flow moves close to the shoreline and converges laterally in the rip embayment. It then moves seaward in the rip channel as a relatively narrow (a few meters), strong flow (0.5–$1\,\mathrm{m\ s^{-1}}$), called a rip current. The current usually moves through the surf zone and beyond the breakers as a rip head. This beach state has extreme spatial–longshore variation in wave breaking, surf zone and swash circulation, and beach and surf zone topography, leading to a high unstable and variable beach system.

Rhythmic bar and beach The rhythmic bar and beach state forms during periods of moderate to high waves (1.5–2 m) on swell coasts. The high waves lead to greater surf zone discharges that require deeper and wider rip feeder channels and rip channels to accommodate the flows. Rips flow in well-developed rip channels, separated by transverse bars, but the bars are detached from the beach by the wider feeder channels (**Figure 7**). The bar is highly rhythmic moving seaward where rips exit the surf, and shoreward over the bars, whereas the shoreline is also highly rhythmic for the reason stated above. It is not uncommon for the more energetic swash in the rip embayments to scarp the beach face, while the adjoining horns are accreting (**Figure 3**).

Longshore bar and trough The longshore bar and trough systems are a product of periods of higher waves which excavate a continuous longshore trough between the bar and the beach. Waves break heavily on the outer bar, reform in the trough and

Figure 7 Rhythmic bar and beach at Egmond, The Netherlands. Note the shoreline rhythms and detached crescentic bar (A.D. Short).

Figure 8 Aracaju Beach in north-east Brazil is a well-developed double bar dissipative beach produced by persistent trade wind waves breaking across a fine sand beach. Note the straight beach scarp, a characteristic of dissipative beach erosion. Photograph by AD Short.

then break again at the shoreline, often producing a steep reflective beach face (coarser sand, **Figure 3**) or a low tide terrace (finer sand). On swell coasts waves must exceed 2 m to produce this intermediate beach state. The high level of surf zone discharge excavates the deep trough to accommodate the flow. Surf zone circulation is both cellular rip flows, as well as increasingly shore normal bed return flows (see below).

Dissipative Beaches

Dissipative beaches represent the high-energy end of the beach spectrum. They occur in areas of high waves, prefer short wave periods, and must have fine sand. They are relatively common in exposed sea environments where occasional periods of high, short storm waves produce multibarred dissipative beach systems, as in the North and Baltic seas. They also occur on high-energy midlatitude swell and storm wave coasts as in northwest USA, southern Africa, Australia, and New Zealand. On swell coasts waves must exceed 2–3 m for long periods to generate fully dissipative beaches. They are characterized by a wide long gradient beach face and surf zone, with two and more shore parallel bars forming across the surf zone (**Figures 3** upper and **8**). The low gradient is a product of both the fine sand, and the dominance of lower frequency infragravity swash and surf zone circulation, which act to plane down the beach. Their name comes from the fact that waves dissipate their energy across the many bars and wide surf zone. The dissipation of the incident wave energy leads to a growth in the longer infragravity energy, which becomes manifest as a strong set up and set down at the shoreline. The standing wave generated by the interaction of the incoming and outgoing waves may have two or more nodes across the surf zone. It is believed that the bar crests from under the standing wave nodes and troughs under the antinodes. Surf zone circulation is vertically segregated. Wave bores move water shoreward toward the surface of the water column. This water builds up against the shoreline as wave set up. As it sets down the return flows tends to concentrate toward the bed, which propels a current across the bed of the surf zone (below the wave bores) called bed return flow.

Like the reflective end of the beach spectrum dissipative beaches are remarkably stable systems. They are designed to accommodate high waves, and can accommodate still higher waves by simply widening their surf zone and increasing the amplitude of the standing waves, while periods of lower waves are often too short to permit substantial onshore sediment migration.

Bar Number

All dissipative beaches have at least two bars and in sea environments commonly have three or more, with up to 10 recorded in the Baltic Sea. In addition, whereas the reflective and intermediate beaches described above assume one bar, all may occur in the lee of an outer bar or bars. In energetic waves the number of bars can be empirically predicted by the bar parameter

$$B^* = \chi_s/g \tan \beta T^2 \qquad [3]$$

where χ_s is the distance to where nearshore gradient approaches zero, g is the gravitational constant and β the beach gradient. Bar number increases as the gradient and/or period decrease, with no bar occurring when $B^* < 20$, one bar between 20–50, and two or more bars > 50.

In two or more bar environments sequential wave breaking lowers the height across the surf zone. As a consequence, Ω (eqn [2]) decreases shoreward and the bar type adjusts to both the lower wave height and Ω. A common multibar sequence is a dissipative outer (storm) bar, with more rhythmic (intermediate) inner bar/s, and a lower energy beach (reflective/low tide terrace).

Modes of Beach Change and Erosion

Figure 3 illustrates an idealized sequence of wave-dominated beaches from dissipative to reflective. It also suggests that beaches can change from one type or state to another, either though beach accretion (movement downward, on left side) or beach erosion (upward, on right side). Beach change, whether erosion or accretion is simply a manifestation of a beach adapting to changing wave conditions by attempting to become more reflective and move sediment shoreward (lower waves) or more dissipative and move sediment seaward (higher waves). Shoreward sediment movement normally requires onshore bar migration, leading in **Figure 3**, from a longshore bar and trough, to a rhythmic bar and beach, the bars then attaching to the beach in the transverse bar and rip, the rips infilling at the low tide terrace, and finally migrating up onto the reflective beach. The rip channels tend to remain stationary during beach accretion, as they are topographically fixed by the bars. Conversely, higher waves and greater surf zone discharges require a wider and deeper surf zone to accommodate the flows. Sediment is eroded from the beach faces and inner surf zone (see scarps, **Figure 2**) and moved seaward to build a wider bar/s. Large erosional transitory rips and increasingly bed return flow transport the sediment seaward.

In nature, movement through this entire sequence is rare, requiring several weeks of low waves for accretion, and many days of high waves for the full

erosion. In reality waves rarely stay so low or so high long enough, and most beaches shift between one of two states as they adjust to ever-changing wave conditions, particularly in swell environments. In sea environments, calm periods between storms lead to a stagnation of often high-energy beach topography, which will await the next storm to be reactivated.

Tide-modified Beaches

Most of the world's beaches are affected by tide. On most open coasts where tides are low (<2 m) waves dominate and tidal impacts are minimized, as in the wave-dominated systems. However, as tide range increases and/or wave height decreases tidal influences become increasingly important. To accommodate these influences beaches, still by definition wave-formed, can be divided into three tide-modified and three tide-dominated types, as defined by eqn [1].

The major impact of increasing tide range is to shift the location of the shoreline between high and low tide, depending on the shoreline gradient, by tens to hundreds of meters. This shift not only moves the shoreline and accompanying swash zone, but also the surf zone, if present, and the nearshore zone. Whereas wave-dominated beaches have a 'fixed' swash–surf–nearshore zone, on tide-modified beaches they are more mobile. The net result is a smearing of the three dominant wave processes of shoaling (nearshore zone), breaking (surf zone) and swash (swash zone), as a section of intertidal beach can be exposed to all three processes at different states of the tide. Because all three zones are mobile, except for a brief period at high and low tide, there is a reduction in the time any one process can fully imprint its dominance on a particular part of the beach. As a consequence there is a tendency for swash processes to dominate only the spring high tide beach, for surf zone processes to dominate only the beach morphology around low tide, during the turn of the low tide, while the shoaling wave processes become increasingly dominant overall, producing a lower gradient more concave beach cross-section.

If we take the three wave-dominated beach types and increase tide range to an RTR of between 3 and 15 the following three beach types result.

Reflective Plus Low Tide Terrace

When $\Omega < 2$ and RTR > 3 the high tide beach remains reflective, much like its wave-dominated counterpart. However, the higher tide range leads to an exposure of the inner nearshore zone, which when swept by surf and swash processes becomes an

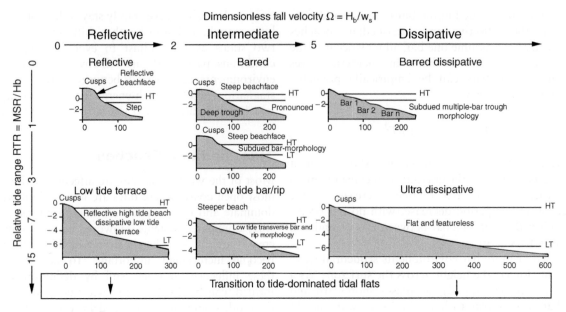

Figure 9 A simplified classification of beaches based on the dimensionless fall velocity and relative tide range. Wave-dominated beaches (top) have an RTR < 3, whereas tide-modified beach types have RTR 3–15 (lower). (Reproduced from Short, 1999.)

Figure 10 Reflective high tide beach and wide low tide terrace, Embleton Bay, Northumberland (A.D. Short).

attached low tide terrace (**Figure 9**). It has a sharp break in slope between the steep beach face and flatter terrace at the point where the water table exits from the beach (**Figure 10**). The low tide drainage from the beach is another feature of tide-modified beaches. At low tide waves dissipate across the flat terrace, whereas at high tide they surge across the steep reflective high tide beach.

Reflective Plus Low Tide Bar and Rips

In areas of moderate waves ($\Omega = 2$–5) and tide range (RTR $= 3$–7) the tide-modified beaches consist of a high tide reflective beach, a usually wider intertidal

zone, which may contain transitory swash ridges, and a low tide bar dominated by surf zone morphology, which may include transverse and rhythmic bars and rips. The surf zone processes, including the rips currents, are however only active at low tide. In this system the higher waves maintain a surf right across the systems as the tide rises and falls, but only during the low tide turn-around is there sufficient time for the cellular surf zone processes to imprint themselves on the morphology, producing surf zone topography essentially identical to the wave-dominated intermediate counterparts. However, there is no associated shoreline rhythmic forms, as the tide-modified surf zone is detached from the shoreline by as much as a few hundred meters (**Figure 11**). At mid to high tide the low tide morphology is dominated by shoaling waves as the surf zone moves rapidly shoreward.

Ultradissipative

Higher energy tide-modified beaches composed of fine sand are characterized by a wide, low gradient concave upward, flat, and featureless, beach and intertidal system (**Figures 9 and 12**). These beaches are swept by both high incident waves and their associated long waves (set up and set down), as well as shoaling waves at high tide. Even at low tide the waves cannot imprint their dissipative bar morphology on the topography. The only surface features usually present on the low tide beach are wave and swash ripples.

Figure 11 Nine Mile Beach, Queensland, showing a dye-highlighted low tide rip current, the 100 m wide intertidal zone with a low swash ridge, and the linear high tide reflective beach (right) (A.D. Short).

Figure 12 Rhossili Beach in Wales is an exposed high-energy tide-modified beach shown here at low tide. Note the wide, flat, featureless intertidal zone, and wide dissipative surf zone, which moves across the beach every six hours (A.D. Short).

Tide-dominated Beach

When the RTR exceeds 15, the tide range is more than 15 times the wave height. As the maximum global tide range is about 12 m, and usually much less, this means that most tide-dominated beaches also receive low (<1 m) to very low waves, and are commonly dominated by locally generated wind waves. Tide-dominated beaches in fact represent a transition from beach to tidal flats, though they must still maintain a high tide beach. Tide-dominated beaches are characterized by a usually steep high tide beach, and a wide, low gradient (<1°) sandy intertidal zone, which in temperate to tropical locations is usually bordered by subtidal seagrass meadows.

Beach and Sand Ridges

The beach and sand ridge type consists of a steep, reflective low-energy high tide beach, which is only active during spring high tides, and in seagrass locations is usually littered with seagrass debris. It is fronted by a low gradient sand flats which contain multiple, sinuous, shore-parallel, equally spaced, low amplitude sand ridges. On the Queensland coast the number of ridges range from one to 19 and averages seven. What causes the ridges is still not understood, and they are very different in size and spacing from the standing wave-generated multiple dissipative bars. They do, however, occur on the more exposed sand flat locations, compared to the next beach type.

Beach and Sand Flats

A beach and sand flats is just that; a very low energy periodically active high tide reflective beach, fronted by a wide flat, featureless intertidal sand flat. The only features on the sand flats are usually associated ground-water drainage from the high tide beach, and bioturbation particularly crab holes and balls.

Tidal Sand Flat

The final 'beach' type is the tidal sand flat, which is a beach in so far as it consists of sand and is exposed to sufficient waves to winnow out any finer sediments. It is flat and featureless; it may have irregular, discontinuous, shelly high tide beach deposits, interspersed occasionally with mangroves in tropical locations, or salt marsh in high latitudes. It is transitional between the beaches and the often muddy, tidal flats with associated inter- and supra-tidal flora.

Beach Modification

The above beach systems assume beaches are solely a product of the wave-tides and sediment regimes. Most beaches are, however, influenced to some degree by geological inheritance, and, depending on latitude, by temperature-controlled processes.

Geological Inheritance

The dominant physical factors modifying beaches are related to the role of what is generally termed geological inheritance, that is, pre-existing bedrock (or artificial structures) and sedimentological controls. Bedrock influences beaches as headlands, which form boundaries and determine beach length. In addition headlands and associated subaqueous substrate (reefs, rocky seabed) induce wave refraction and attenuation, which in turn influences beach shape and breaker wave height. Wave refraction around

Figure 13 A small embayed, headland-controlled beach, with a topographically controlled rip exiting against the headland, Point Peter, South Australia (A.D. Short).

Figure 14 Frozen tundra (dark patches left), beach covered by snow and ice (center) and shore-parallel ice ridges (right); Point Lay, north Alaska (A.D. Short).

headlands and reefs will tend to produce swash aligned beaches in their lee, as well as lower energy beaches. Wave attenuation over rock substrates will reduce wave energy and through breaker height influence beach type. Eroding basement rocks and cliffs can also supply sediment to the beach budget.

In energetic embayed (headland-controlled) beaches, beach systems can be classified as normal, where the headland exerts no influences, as in areas with low waves or on long beaches. On shorter beaches and during moderate waves the beaches become transitional, with normal surf zone circulation in the center, and headland-induced rips to either end. Finally in high energy situations (Hb > 3 m) and on shorter beaches (< few kilometers) the entire surf zone circulation becomes cellular, meaning it is controlled by the bedrock topography, leading to large-scale mega-rips draining the surf zone, and often aligned against a headland (**Figure 13**).

Temperature

Frozen beaches   Ocean temperature is responsible for two dramatic physical impacts on beach systems. In the high latitudes when air temperature falls below zero the beach surface freezes and when ocean temperature falls below − 4°C the water surface freezes. In places like the Arctic Ocean beach processes cease for several months of the year, as a frozen ocean surface and snow-covered beach negate any physical activity (**Figure 14**) other than tides. When the ice and snow finally melts, near normal beach processes return and control the long-term beach morphology.

Biotic influences   In the tropics, warm water (> 26°C) permits the prolific growth of coral and algal reefs in suitable shallow water environments.

Figure 15 Three bands of beachrock at Shoal Cape, Western Australia. Note the wave breaking over the successive reefs, leading to a low energy inner beach, on an exposed high-energy coast (A.D. Short).

Where these reefs lie seaward of beach systems, they lower the wave height through wave breaking and therefore lead to solely reflective beaches in their lee. The reefs also supply carbonate detritus to the beaches leading to coarse carbonate rich beach systems.

In temperate semiarid to arid shelf regions such as across southern Australia and South Africa, the prolific growth of shelf carbonate supplies the bulk of the beach sediments, leading to extensive mid-latitude carbonate coasts and beach systems.

Chemical influences   Water chemistry modifies tropical beaches through the formation of beachrock, the cementation of the intertidal sand grains by carbonate cement. This commonly occurs in all tropical beach locations and during beach erosion leads to exposure of the beachrock parallel to the shoreline.

On arid to temperate carbonate-rich coasts sub-aerial pedogenesis led to the formation of calcarenite in beach and dune systems, particularly during Pleistocene low sea levels. When these systems are subsequently drowned by rising sea level, they remain as dune or beach calcarenite, and are manifest as reefs, islands, and rocky bluffs and cliffs, all of which exert a geological influence on the shoreline (**Figure 15**).

See also

Geomorphology. Waves on Beaches.

Further Reading

Carter RWG (1988) *Coastal Environments. An Introduction to the Physical, Ecological and Cultural Systems of Coastlines.* London: Academic Press.

Hardisty J (1990) *Beaches: Form and Process.* London: Unwin and Hyman.

Horikawa K (ed.) (1988) *Nearshore Dynamics and Coastal Processes.* Tokyo: University of Tokyo Press.

King CAM (1972) *Beaches and Coasts.* London: Edward Arnold.

Komar PD (1998) *Beach Processes and Sedimentation.* 2nd edn. Englewood Cliffs, NJ: Prentice-Hall.

Pethick J (1984) *An Introduction to Coastal Geomorphology.* London: Edward Arnold.

Schwartz ML and Bird ECF (eds) (1990) Artificial Beaches. *Journal of Coastal Research Special Issue 6.*

Short AD (ed.) (1993) Beach and surf zone morphodynamics. *Journal of Coastal Research Special Issue 15.*

Short AD (ed.) (1999) *Handbook of Beach and Shoreface Morphodynamics.* Chichester: John Wiley.

COASTAL TOPOGRAPHY, HUMAN IMPACT ON

D. M. Bush, State University of West Georgia, Carrollton, GA, USA

O. H. Pilkey, Duke University, Durham, NC, USA

W. J. Neal, Grand Valley State University, Allendale, MI, USA

Introduction

The trademark of humans throughout time is the modification of the natural landscape. Topography has been modified from the earliest farming to the modern modifications of nature for transportation and commerce (e.g., roads, utilities, mining), and often for recreation, pleasure, and esthetics. While human modifications of the environment have affected vast areas of the continents, and small portions of the ocean floor, nowhere have human intentions met headlong with nature's forces as in the coastal zone.

A most significant change in human behavior since the 1950s has been the dramatic, rapid increase in population and nonessential development in the coastal zone (**Figure 1**). The associated density of development is in an area that is far more vulnerable and likely to be impacted by natural processes (e.g., wind, waves, storm-surge flooding, and coastal erosion) than most inland areas. Not only are more

Figure 1 The coastal population explosion has resulted in too many people and buildings crowded too close to the shoreline. As sea level rises, the shoreline naturally moves back and encounters the immovable structures of human development. In this example from San Juan, Puerto Rico, erosion in front of buildings has necessitated engineering of the shoreline.

people and development in harm's way, but the human modifications of the coastal zone (e.g., dune removal) have increased the frequency and severity of the hazards. Finally, coastal engineering as a means to combat coastal erosion and management of waterways, ports, and harbors has had profound and often deleterious effects on coastal environments. The endproduct is a total interruption of sediment interchange between land and sea, and a heavily modified topography. Natural hazard mitigation is now moving with a more positive, albeit small, approach by restoring natural features, such as beaches and dunes, and their associated interchangeable sediment supply.

The Scope of Human Impact on the Coast

The natural coastal zone is highly dynamic, with geomorphic changes occurring over several time scales. Equally significant changes are made by humans. On Ocean Isle, NC, USA, an interior dune ridge, the only one on the island, was removed to make way for development. The lowered elevation put the entire development in a higher hazard zone, with a corresponding greater risk for property damage from flooding and other storm processes.

Another example of change, impacting on property damage risk, can be seen in Kitty Hawk, North Carolina. A large shorefront dune once extended in front of the entire community. The dune was constructed in the 1930s by the Civilian Conservation Corps to halt shoreline erosion, and provide a 'protected' area along which to build a road. The modification was done before barrier island migration was understood. Erosion was assumed to be permanent land loss. The artificial dune actually increased erosion here by acting as a seawall in a long-term sense, blocking overwash sand which would have raised island elevation and brought sand to the backside of the island, although the dune did afford some protection for development. As a consequence, buildings by the hundreds were built in the lee of the dune. Fifty years on, however, the price is being paid. During the 1980s, the dune began to deteriorate due to storm penetration, and the 1991 Halloween northeaster finished the job by creating large gaps in the dune, resulting in flooding of portions of the community. The dune cannot be rebuilt in place because the old dune location is now occupied by the beach, backed up against the frontal road. Between the time of dune construction and 1991, the community had only experienced major flooding once, in the great 1962 Ash Wednesday storm. Between 1991 and 2000, the community was flooded four times.

The effect of shoreline engineering on a whole-island system is starkly portrayed by the contrast between Ocean City, MD and the next island to the south, Assateague Island, MD. It has taken several decades to be fully realized, but the impact of the jetties is now apparent. Assateague Island has moved back one entire island width due to sand trapping by an updrift jetty. Similar stories abound along the coast. The Charleston lighthouse, once on the

backside of Morris Island, SC, now stands some 650 m at sea; a sentinel that watched Morris Island rapidly migrate away after the Charleston Harbor jetties, built in 1898, halted the supply of sand to the island (**Figure 2**).

Human alterations of the natural environment have direct and indirect effects. Some types of human modifications to the coastal environment include: (1) construction site modification, (2) building and infrastructure construction, (3) hard shoreline stabilization, (4) soft shoreline stabilization, and (5) major coastal engineering construction projects for waterway, port, and harbor management and inlet channel alteration. Each of the modification types impacts the coastal environment in a variety of ways and also

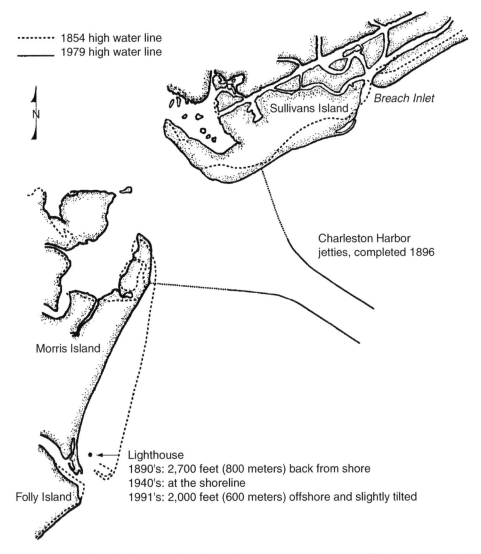

Figure 2 The jetties that stabilize Charleston Harbor, South Carolina, were completed in 1896. The jetties block the southward transport sediment along the beach. As a result, the islands to the north of the jetties have grown seaward slightly, but the islands to the south of the inlet have eroded back more than 1400 m. The Morris Island Lighthouse, once on the back side of the island, is now 650 m offshore.

has several direct and indirect effects. Some of the effects are obvious and intuitive, but many are surprising in that there can be a domino effect as one simple modification creates potential for damage and destruction by increasing the frequency and intensity of natural hazards at individual sites.

Construction Site Modification

Building sites are often flattened and vegetation is removed for ease of construction. Activities such as grading of the natural coastal topography include dune and forest removal. Furthermore, paving of large areas is common, as roads, parking lots, and driveways are constructed. Direct effects of building site modification, in addition to changes in the natural landform configuration, include demobilization of sediment in some places by paving and building footpaths, but also sediment mobilization by removal of vegetation. In either case, rates of onshore–offshore sediment transport and storm-recovery capabilities are changed, which can increase or decrease erosion rates as sediment supply changes.

Other common site modifications include excavating through dunes (dune notching) to improve beach access or sea views. This is particularly common at the ends of streets running toward the beach. After Hurricane Hugo in South Carolina in 1989, shore-perpendicular streets where dunes were notched at their ocean termini were seen to have acted as storm-surge ebb conduits, funneling water back to the sea and increasing scour and property damage. The same effect was noted after Hurricane Gilbert along the northern coast of the Yucatán Peninsula of Mexico in 1988.

Building and Infrastructure Construction

A variety of buildings are constructed in the coastal zone, ranging from single-family homes to high-rise hotels and commercial structures. Some of the common direct effects of building construction are alteration of wind patterns as the buildings themselves interact with natural wind flow, obstruction of sediment movement, marking the landward limit of the beach or dune, channelizing storm surge and storm-surge ebb flow, and reflection of wave energy. Indirect effects result from the simple fact that once there is construction in an area, people tend to want to add more construction, and to increase and improve infrastructure and services. As buildings become threatened by shoreline erosion, coastal engineering endeavors begin.

Roads, streets, water lines, and other utilities are often laid out in the standard grid pattern used inland, cutting through interior and frontal dunes instead of over and around coastal topography (**Figure 3**). Buildings block natural sediment flow (e.g., overwash) while the ends of streets and gaps between rigid buildings funnel and concentrate flow, accentuating the erosive power of flood waters. As noted above, during Hurricane Hugo, water, sand, and debris were carried inland along shore-perpendicular roads in several South Carolina communities. Storm-surge ebb along the shore-perpendicular roads caused scour channels, which undermined roadways and damaged adjacent houses and property. Even something as seemingly harmless as buried utilities may cause a problem as the excavation disrupts the substrate, resulting in a less stable topography after post-construction restorations.

Plugging dune gaps can be a part of nourishment and sand conservation projects. Because dunes are critical coastal geomorphic features with respect to property damage mitigation, they are now often protected, right down to vegetation types that are critical to dune growth. Prior to strict coastal-zone management regulations, however, frontal dunes were often excavated for ocean views or building sites, or notched at road termini for beach access. These artificially created dune gaps are exploited by waves and storm-surge, and by storm surge ebb flows. Wherever dune removal for development has occurred, the probability is increased for the likelihood of complete overwash and possible inlet formation.

Hard Shoreline Stabilization

Hard shoreline stabilization includes various fixed, immovable structures designed to hold an eroding shoreline in place. Hard stabilization is one of the most common modifiers of topography in the coastal zone and is discussed in more detail below. Seawalls, jetties, groins, and offshore breakwaters interrupt sediment exchange and reduce shoreline flexibility to respond to wave and tidal actions. Armoring the shoreline changes the location and intensity of erosion and deposition. Indirectly, hard shoreline stabilization gives a false sense of security and encourages increased development landward of the walls, placing more and more people and property at risk from coastal hazards including waves, storm surge, and wind. Eventual loss of the recreational beach as shoreline erosion continues and catches up with the static line of stabilization is almost a certainty. In addition, structures beget more structures as small walls or groins are replaced by larger and larger walls and groins.

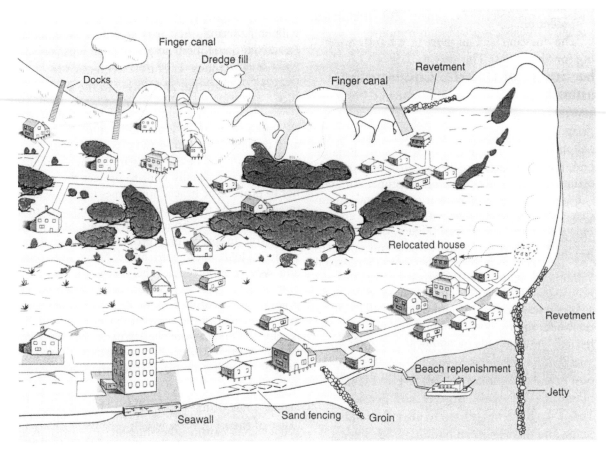

Figure 3 A compilation of many of the impacts humans have on the coastal topography. In this fictional barrier island, roads have been cut through excavated dunes, maritime forest removed for building sites, finger canals dredged, structures built too close to the water, and several types of coastal engineering projects undertaken.

Soft Shoreline Stabilization

The most common forms of soft shoreline stabilization are beach nourishment, dune building, sand fencing, beach bulldozing (beach scraping), and planting of vegetation to grow or stabilize dunes. Direct effects of such manipulations are changes in sedimentation rates and severity of erosion, and interruption of the onshore–offshore sediment transfer, similar in that respect to hard shoreline stabilization. Indirectly, soft shoreline stabilization may make it more difficult to recognize the severity of an erosion problem, i.e., 'masking' the erosion problem. Moreover, as with hard shoreline stabilization, development is actually encouraged in the high-hazard zone behind the beach.

Coastal Engineering Construction Projects

The construction of harbors, port facilities, waterways (e.g., shipping channels, canals) and inlet channel alterations significantly change the coastal outline as well as eliminating land topographic features or erecting artificial shorelines and dredge spoil banks. The Intracoastal Waterway of the Atlantic and Gulf Coasts is one of the longest artificial coastal modifications in the world. Large harbors in many places around the world represent significant alteration of the landscape. Many examples of coastal fill or artificial shorelines exist, but one of the best examples of such a managed shoreline is Chicago's 18 miles of continuous public waterfront. Major canals such as the Suez, Panama, Cape Cod, or Great Dismal Swamp Canal also represent major modifications in the coastal zone. The Houston Ship Channel made the city of Houston, Texas, a major port some 40 miles from the Gulf of Mexico.

Tidal inlets, either on the mainland or between barrier islands, can be altered by dredging, relocation, or artificial closure. Direct effects of dredging tidal inlets are changes in current patterns, which may change the location and degree of erosion and deposition events, and prevention of sand transfer across inlets. In either case, additional shoreline hardening is a common response.

The Scope of Coastal Engineering Impacts

Between 80 and 90% of the American open-ocean shoreline is retreating in a landward direction because of sea-level rise and coastal erosion. Because more static buildings are being sited next to this moving and constantly changing coastline, our society faces major problems. Various coastal engineering approaches to dealing with the coastal erosion problem have been developed (**Figure 3**). More than a century of experience with seawalls and other engineering structures in New Jersey and other coastal developments shows that the process of holding the shoreline in place leads to the loss of the beach, dunes, and other coastal landforms. The real societal issue is how to save both buildings and beaches. The action taken often leads to modifications to the coast that limit the natural flexibility of the coastal zone to respond to storms, that inhibit the natural onshore–offshore exchange of sand, and that interrupt the natural alongshore flow of sand.

Seawalls

Seawalls include a family of coastal engineering structures built either on land at the back of the beach or on the beach, parallel to the shoreline. Strictly defined, seawalls are free-standing structures near the surf-zone edge. The best examples are the giant walls of the northern New Jersey coast, the end result of more than a century of armoring the shoreline (**Figure 4**). If such walls are filled in behind with soil or sand, they are referred to as bulkheads. Revetments, commonly made of piled loose rock, are walls built up against the lower dune-face or land at the back of the beach. For the purpose of considering their alteration of topography both at their construction site and laterally, the distinction between the types of walls is gradational and unimportant, and the general term seawall is used here for all structures on the beach that parallel the shoreline.

Seawalls are usually built to protect the property, not to protect the beach. Sometimes low seawalls are intended only to prevent shoreline retreat, rather than to block wave attack on buildings. Seawalls are successful in preventing property damage if built strongly, high enough to avoid being overtopped, and kept in good repair. The problem is that a very high societal price is paid for such protection. That price is the eventual loss of the recreational beach and steepening of the shoreface or outer beach. This is why several states in the USA (e.g. Maine, Rhode Island, North Carolina, South Carolina, Texas, and Oregon) prohibit or place strict limits on shoreline armoring.

Three mechanisms account for beach degradation by seawalls. Passive loss is the most important. Whatever is causing the shoreline to retreat is unaffected by the wall, and the beach eventually retreats up against the wall. Placement loss refers to the emplacement of walls on the beach seaward of the high-tide line, thus removing part or all of the beach when the wall is constructed (**Figure 5**). Seawall placement was responsible for much of the beach loss in Miami Beach, Florida, necessitating a major beach nourishment project, completed in 1981. Active loss is the least understood of the beach

Figure 4   Cape May, New Jersey has been a popular seaside resort since 1800. Several generations of larger and larger seawalls have been built as coastal erosion caught up with the older structures. Today in many places there is no beach left in front of the seawall.

Figure 5   An example of placement loss in Virginia Beach, Virginia. The seawall was built out on the recreational beach, instantaneously narrowing the beach in front of the wall.

degradation mechanisms. Seawalls are assumed to interact with the surf during storms, which enhances the rate of beach loss. This interaction can occur in a number of ways including seaward reflection of waves, refraction of waves toward the end of the wall, and intensification of surf-zone currents.

By the year 2000, 50% of the developed shoreline on Florida's western (Gulf of Mexico) coast was armored, the same as the New Jersey coast. Similarly 45% of developed shoreline on Florida's eastern (Atlantic Ocean) coast was armored, in contrast to 27% for South Carolina, and only 6% of the developed North Carolina open-ocean shoreline. These figures represent the armored percentage of developed shorelines and do not include protected areas such as parks and National Seashores.

Shoreline stabilization is a difficult political issue because seawalls take as long as five or six decades to destroy beaches, although the usual time range for the beach to be entirely eroded at mid-to-high tide may be only one to three decades. Thus it takes a politician of some foresight to vote for prohibition of armoring. Another issue of political difficulty is that there is no room for compromise. Once a seawall is in place, it is rarely removed. The economic reasoning is that the wall must be maintained and even itself protected, so most walls grow higher and longer.

Groins and Jetties

Groins and jetties are walls or barriers built perpendicular to the shoreline. A jetty, often very long (thousands of feet), is intended to keep sand from flowing into a ship channel within an inlet and to reduce the cost of channel maintenance by dredging. Groins are much shorter structures built on straight stretches of beach away from inlets. Groins are intended to trap sand moving in longshore currents. They can be made of wood, stone, concrete, steel, or fabric bags filled with sand. Some designs are referred to as T-groins because the end of the structure terminates in a short shore-parallel segment.

Both groins and jetties are very successful sand traps. If a groin is working correctly, more sand should be piled up on one side of the groin than on the other. The problem with groins is that they trap sand that is flowing to a neighboring beach. Thus, if a groin is growing the topographic beach updrift, it must be causing downdrift beach loss. Per Bruun, past director of the Coastal Engineering program at the University of Florida, has observed that, on a worldwide basis, groins may be a losing proposition, i.e. more beach may be lost than gained by the use of groins. After one groin is built, the increased rate of

Figure 6 A groin field along Pawleys Island, South Carolina. Trapping of sand on the updrift side of a groin, and erosion of the beach on the downdrift side usually results in a sawtooth pattern to the beach. Note that in this example the beach is the same width on both sides of each groin, indicating little or no longshore transport of sand.

erosion effect on adjacent beaches has to be addressed. So other groins are constructed, in self-defense. The result is a series of groins sometimes extending for miles (**Figure 6**). The resulting groin field is a saw-toothed beach in plan view.

Groins fail when continued erosion at their landward end causes the groin to become detached, allowing water and sand to pass behind the groin. When detachment occurs, beach retreat is renewed and additional alteration of the topography occurs.

Jetties, because of their length, can cause major topographic changes. After jetty emplacement, massive tidal deltas at most barrier island inlents will be dispersed by wave activity. In addition, major buildout of the updrift and retreat of the downdrift shorelines may occur. In the case of the Charleston, SC, jetties noted earlier, beach accretion occurred on the updrift Sullivans Island and Isle of Palms.

Offshore Breakwaters

Offshore breakwaters are walls built parallel to the shoreline but at some distance offshore, typically a few tens of meters seaward of the normal surf zone. These structures dampen the wave energy on the 'protected' shoreline behind the breakwater, interrupting the longshore current and causing sand to be deposited and a beach to form. Sometimes these deposits will accumulate out to the breakwater, creating a feature like a natural tombolo. As in the case of groins, the sand trapped behind breakwaters causes a shortage of sediment downdrift in the directions of dominant longshore transport, leading to additional shoreline retreat (e.g. beach and dune loss, scarping of the fastland, accelerated mass wasting).

Beach Nourishment

Beach nourishment consists of pumping or trucking sand onto the beach. The goal of most communities is to improve their recreational beach, to halt shoreline erosion, and to afford storm protection for beachfront buildings. Many famous beaches in developed areas, in fact, are now artificial!

The beach or zone of active sand movement actually extends out to a water depth of 9–12 m below the low-tide line. This surface is referred to as the shoreface. With nourishment, only the upper beach is covered with new sand so that a steeper beach is created, i.e. the topographic profile is modified on land and offshore. This new steepened profile often increases the rate of erosion; in general, replenished beaches almost always disappear at a faster rate than their natural predecessors.

Beach scraping (bulldozing) should not be confused with beach nourishment. Beach sand is moved from the low-tide beach to the upper back beach (independent of building artificial dunes) as an erosion-mitigation technique. In effect this is beach erosion! A relatively thin layer of sand (≤ 30 cm) is removed from over the entire lower beach using a variety of heavy machinery (drag, grader, bulldozer, front-end loader) and spread over the upper beach. The objectives are to build a wider, higher, high-tide dry beach; to fill in any trough-like lows that drain across the beach; and to encourage additional sand to accrete to the lower beach.

The newly accreted sand in turn, can be scraped, leading to a net gain of sand on the manicured beach. An enhanced recreational beach may be achieved for the short term, but no new sand has been added to the system. Ideally, scraping is intended to encourage onshore transport of sand, but most of the sand 'trapped' on the lower beach is brought in by the longshore transport. Removal of this lower beach sand deprives downdrift beaches of their natural nourishment, steepens the beach topographic profile, and destroys beach organisms.

Dune building is often an important part of beach nourishment design, or it may be carried out independently of beach nourishment. Coastal dunes are a common landform at the back of the beach and part of the dynamic equilibrium of barrier beach systems. Although extensive literature exists about dunes, their protective role often is unknown or misunderstood. Frontal dunes are the last line of defense against ocean storm wave attack and flooding from overwash, but interior dunes may provide high ground and protection against penetration of overwash, and against the damaging effects of storm-surge ebb scour.

Human Impact on Sand Supply

In most of the preceding discussion the impact of humans on beaches and shoreline shape and position was emphasized. The beach plays a major role in supplying sand to barrier islands and, in fact, is important in supplying sand and gravel to any kind of upland, mainland, or island. In this sense, any topographic modification, however small, that affects the sand supply of the beach will affect the topography. In beach communities, sand is routinely removed from the streets and driveways after storms or when sand deposited by wind has accumulated to an uncomfortable level for the community. This sand would have been part of the island or coastal evolution process. Often, dunes are replaced by flat, well-manicured lawns. Sand-trapping dune vegetation is often removed altogether.

The previously mentioned Civilian Conservation Corps construction of the large dune line along almost the entire length of the Outer Banks of North Carolina is an example of a major topographic modification that had unexpected ramifications, namely the increased rate of erosion on the beach as well as on the backside of the islands. Prior to dune construction, the surf zone, especially during storms, expended its energy across a wide band of island surface which was overwashed several times a year. After construction of the frontal dune, wave energy was expended in a much narrower zone, leading to increased rates of shoreline retreat, and overwash no longer nourished the backside of the island. Now that the frontal dune is deteriorating, North Carolina Highway-12 is buried by overwash sand in a least a dozen places 1–4 times each year. Overwash sand is an important part of the island migration process, because these deposits raise the elevation of islands, and when sediment extends entirely across an island, widening occurs. If not for human activities, much of the Outer Banks would be migrating at this point in time, but because preservation of the highway is deemed essential to connect the eight villages of the southern Outer Banks, the NC Department of Transportation removes sand and places it back on the beach. As a result, the island fails to gain elevation.

Inlet formation also is an important part of barrier island evolution. Each barrier island system is different, but inlets form, evolve, and close in a manner to allow the most efficient means of moving water in and out of estuaries and lagoons. Humans interfere by preventing inlets from forming, by closing them after they open naturally (usually during storms), or by preventing their natural migration by construction of jetties. The net result is clogging of navigation channels by construction of huge tidal deltas and reduced water circulation and exchange between the sea and estuaries.

Globally shoreline change is being affected by human activity that causes subsidence and loss of sand supply. The Mississippi River delta is a classic example. The sediment discharge from the Mississippi River has been substantially reduced by upstream dam construction on the river and its tributaries. Large flood-control levees constructed along the lower Mississippi River prevent sediment from reaching the marshes and barrier islands along the rim of the delta in the Gulf of Mexico. Natural land subsidence caused by compaction of muds has added to the problem by creating a rapid (1–2 m per century) relative sea-level rise. Finally, maintaining the river channel south of New Orleans has extended the river mouth to the edge of the continental shelf, causing most remaining sediment to be deposited in the deep sea rather than on the delta. The end result is an extraordinary loss rate of salt marshes and very rapid island migration. The face of the Mississippi River delta is changing with remarkable rapidity.

Other deltas around the world have similar problems that accelerate changes in the shape of associated marshes and barrier islands. The Niger and Nile deltas have lost a significant part of their sediment supply because of trapping sand behind dams. Land loss on the Nile delta is permanent and not just migration of the outermost barrier islands. On the Niger delta the lost sediment supply is compounded by the subsidence caused by oil, gas, and water extraction. The barrier islands there are rapidly thinning.

Sand mining is a worldwide phenomenon whose quantitative importance is difficult to guage. Mining dunes, beaches, and river mouths for sand has reduced the sand supply to the shoreface, beaches, and barrier islands. In developing countries beach-sand mining is ubiquitous, while in developed countries beach and dune mining often is illegal and certainly less extensive, although still a problem. For example, sand mining has adversely affected the beaches of many West Indies nations going through the growing pains of development. Dune mining has been going on for so long that many current residents cannot remember sand dunes ever being present on the beaches, although they must have been there at one time, given the sand supply and the strong winds. For example, on the dual-island nation of Antigua and Barbuda, beach ridges – evidence of accumulating sand – can be observed on Barbuda, but are missing on Antigua. The beach ridges of Barbuda have survived to date only because it is much less heavily developed and populated. Sadly, Barbuda's beach ridges are being actively mined.

Puerto Rico is a heavily developed Caribbean island, much larger than Antigua or Barbuda, and with a more diversified economy. Many of Puerto Rico's dunes have been trucked away (**Figure 7**). East of the capital city of San Juan, large sand dunes were mined to construct the International Airport at Isla Verde by filling in coastal wetlands. As a result of removing the dunes, the highway was regularly overwashed and flooded during even moderate winter storms. First an attempt was made to rebuild the dune, then a major seawall was built to protect the lone coastal road.

Dredging or pumping sand from offshore seems like a quick and simple solution to replace lost beach sand; however, such operations must be considered with great care. The offshore dredge hole may allow larger waves to attack the adjacent beach. Offshore sand may be finer in grain size, or it may be

Figure 7 The dunes here near Camuy, Puerto Rico, used to be over 20 m high. After mining for construction purposes, all that remains is a thin veneer of sand over a rock outcrop.

composed of calcium carbonate, which breaks up quickly under wave abrasion. In all of these cases, the new beach will erode faster than the original beach. Dredging also may create turbidity that can kill bottom organisms. Offshore, protective reefs may be damaged by increased turbidity. Loss of reefs will mean faster beach erosion, as well as the obvious loss to the fishery habitat.

Sand can also be brought in from land sources by dump truck, but this may prove to be more expensive. Sand is a scarce resource, and beaches/dunes have been regarded as a source for mining rather than areas that need artificial replenishment. Past beach and dune mining may well be a principal cause of present beach erosion. In some cases, gravel may be better for nourishment than sand, but the recreational value of beaches declines when gravel is substituted.

Sand mining of beaches and dunes accounts for many of Puerto Rico's problem erosion areas. Such sand removal is now illegal, but permits are given to remove sand for highway construction and emergency repair purposes. However, the extraction limits of such permits are often exceeded – and illegal removal of sand for construction aggregate continues. In all cases, the sand removal eliminates natural shore protection in the area of mining, and robs from the sand budget of downdrift beaches, accelerating erosion. Even a small removal operation can set off a sequence of major shoreline changes.

The Caribe Playa Seabeach Resort along the southeastern coast illustrates just such a chain reaction. Located west of Cabo Mala Pascua, the resort has lost nearly 15 m of beach in recent years according to the owner. The problem dates to the days before permits and regulation, when an updrift property owner sold beach sand for 50 cents per dump truck load; a bargain by anyone's standards but a swindle to downdrift property owners. Where the sand was removed the beach eroded, resulting in shoreline retreat and tree kills. In an effort to restabilize the shore, and ultimately protect the highway, a rip-rap groin and seawall were constructed. Today, only a narrow gravel beach remains. Undoubtedly much of the aggregate in the concrete making up the buildings lining the shore, and now endangered by beach erosion, was beach sand. What extreme irony: taking sand from the beach to build structures that were subsequently endangered by the loss of beach sand.

Conclusion

The majority of the world's population lives in the coastal zone, and the percentage is growing. As this trend continues, the coastal zone will see increased impact of humans as more loss of habitats, more inlet dredging and jetties, continued sand removal, topography modification for building, sand starvation from groins and jetties, and the increased tourism and industrial use of coasts and estuaries. Our society's history illustrates the impact of humans as geomorphic agents, and nowhere is that fact borne out as it is in the coastal zone. The ultimate irony is that many of the human modifications on coastal topography actually decrease the esthetics of the area or increase the potential hazards.

See also

Beaches, Physical Processes Affecting.

Further Reading

Bush DM and Pilkey OH (1994) Mitigation of hurricane property damage on barrier islands: a geological view. *Journal of Coastal Research* Special issue no. 12: 311–326.

Bush DM, Pilkey OH, and Neal WJ (1996) *Living by the Rules of the Sea*. Durham, NC: Duke University Press.

Bush DM, Neal WJ, Young RS, and Pilkey OH (1999) Utilization of geoindicators for rapid assessment of coastal-hazard risk and mitigation. *Ocean and Coastal Management* 42: 647–670.

Carter RWG and Woodroffe CD (eds.) (1994) *Coastal Evolution: Late Quaternary Shoreline Morphodynamics*. Cambridge: Cambridge University Press.

Carter RWG (1988) *Coastal Environments: An Introduction to the Physical, Ecological, and Cultural Systems of Coastlines*. London: Academic Press.

Davis RA Jr (1997) *The Evolving Coast*. New York: Scientific American Library.

French PW (1997) *Coastal and Estuarine Management, Routledge Environmental Management Series*. London: Routledge Press.

Kaufmann W and Pilkey OH Jr (1983) *The Beaches are Moving: The Drowning of America's Shoreline*. Durham, NC: Duke University Press.

Klee GA (1999) *The Coastal Environment: Toward Integrated Coastal and Marine Sanctuary Management*. Upper Saddle River, NJ: Prentice Hall.

Nordstrom KF (1987) Shoreline changes on developed coastal barriers. In: Platt RH, Pelczarski SG, and Burbank BKR (eds.) *Cities on the Beach: Management Issues of Developed Coastal Barriers*, pp. 65–79. University of Chicago, Department of Geography, Research Paper no. 224.

Nordstrom KF (1994) Developed coasts. In: Carter RWG and Woodroffe CD (eds.) *Coastal Evolution: Late Quaternary Shoreline Morphodynamics*, pp. 477–509. Cambridge: Cambridge University Press.

Nordstrom KF (2000) *Beaches and Dunes of Developed Coasts*. Cambridge: Cambridge University Press.

Pilkey OH and Dixon KL (1996) *The Corps and the Shore*. Washington, DC: Island Press.

Platt RH, Pelczarski SG and Burbank BKR (eds.) (1987) *Cities on the Beach: Management Issues of Developed Coastal Barriers*. University of Chicago, Department of Geography, Research Paper no. 224.

Viles H and Spencer T (1995) *Coastal Problems: Geomorphology, Ecology, and Society at the Coast*. New York: Oxford University Press.

SEA LEVEL VARIATIONS OVER GEOLOGICAL TIME

M. A. Kominz, Western Michigan University,
Kalamazoo, MI, USA

Introduction

Sea level changes have occurred throughout Earth history. The magnitudes and timing of sea level changes are extremely variable. They provide considerable insight into the tectonic and climatic history of the Earth, but remain difficult to determine with accuracy.

Sea level, where the world oceans intersect the continents, is hardly fixed, as anyone who has stood on the shore for 6 hours or more can attest. But the ever-changing tidal flows are small compared with longer-term fluctuations that have occurred in Earth history. How much has sea level changed? How long did it take? How do we know? What does it tell us about the history of the Earth?

In order to answer these questions, we need to consider a basic question: what causes sea level to change? Locally, sea level may change if tectonic forces cause the land to move up or down. However, this article will focus on global changes in sea level. Thus, the variations in sea level must be due to one of two possibilities: (1) changes in the volume of water in the oceans or (2) changes in the volume of the ocean basins.

Sea Level Change due to Volume of Water in the Ocean Basin

The two main reservoirs of water on Earth are the oceans (currently about 97% of all water) and glaciers (currently about 2.7%). Not surprisingly, for at least the last three billion years, the main variable controlling the volume of water filling the ocean basins has been the amount of water present in glaciers on the continents. For example, about 20 000 years ago, great ice sheets covered northern North America and Europe. The volume of ice in these glaciers removed enough water from the oceans to expose most continental shelves. Since then there has been a sea level rise (actually from about 20 000 to about 11 000 years ago) of about 120 m (**Figure 1A**).

A number of methods have been used to establish the magnitude and timing of this sea level change. Dredging on the continental shelves reveals human

activity near the present shelf-slope boundary. These data suggest that sea level was much lower a relatively short time ago. Study of ancient corals shows that coral species which today live only in very shallow water are now over 100 m deep. The carbonate skeletons of the coral, which once thrived in the shallow waters of the tropics, yield a detailed picture of the timing of sea level rise, and, thus, the melting of the glaciers. Carbon-14, a radioactive isotope formed by carbon-12 interacting with high-energy solar radiation in Earth's atmosphere allows us to determine the age of Earth materials, which are about 30 thousand years old.

This is just the most recent of many, large changes in sea level caused by glaciers, (**Figure 1B**). These variations in climate and subsequent sea level changes have been tied to quasi-periodic variations in the Earth's orbit and the tilt of the Earth's spin axis. The record of sea level change can be estimated by observing the stable isotope, oxygen-18 in the tests (shells) of dead organisms. When marine microorganisms build their tests from the calcium, carbon, and oxygen present in sea water they incorporate both the abundant oxygen-16 and the rare oxygen-18 isotopes. Water in the atmosphere generally has a lower oxygen-18 to oxygen-16 ratio because evaporation of the lighter isotope requires less energy. As a result, the snow that eventually forms the glaciers is depleted in oxygen-18, leaving the ocean proportionately oxygen-18-enriched. When the microorganisms die, their tests sink to the seafloor to become part of the deep marine sedimentary record. The oxygen-18 to oxygen-16 ratio present in the fossil tests has been calibrated to the sea level change, which occurred from 20 000 to 11 000 years ago, allowing the magnitude of sea level change from older times to be estimated. This technique does have uncertainties. Unfortunately, the amount of oxygen-18 which organisms incorporate in their tests is affected not only by the amount of oxygen-18 present but also by the temperature and salinity of the water. For example, the organisms take up less oxygen-18 in warmer waters. Thus, during glacial times, the tests are even more enriched in oxygen-18, and any oxygen isotope record reveals a combined record of changing local temperature and salinity in addition to the record of global glaciation.

Moving back in time through the Cenozoic (zero to 65 Ma), paleoceanographic data remain excellent due to relatively continuous sedimentation on the ocean floor (as compared with shallow marine and

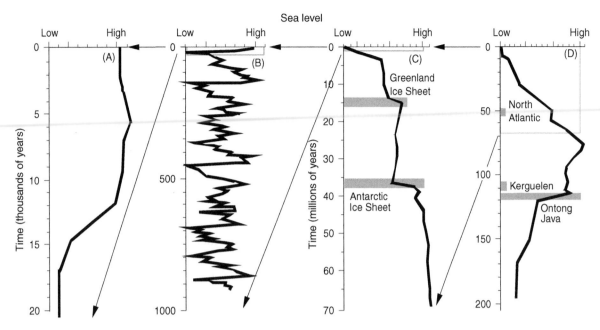

Figure 1   (A) Estimates of sea level change over the last 20 000 years. Amplitude is about 120 m. (B) Northern Hemisphere glaciers over the last million years or so generated major sea level fluctuations, with amplitudes as high as 125 m. (C) The long-term oxygen isotope record reveals rapid growth of the Antarctic and Greenland ice sheets (indicated by gray bars) as Earth's climate cooled. (D) Long-term sea level change as indicated from variations in deep-ocean volume. Dominant effects include spreading rates and lengths of mid-ocean ridges, emplacement of large igneous provinces (the largest, marine LIPs are indicated by gray bars), breakup of supercontinents, and subduction of the Indian continent. The Berggren *et al.* (1995) chronostratigraphic timescale was used in (C) and (D).

terrestrial sedimentation). Oxygen-18 in the fossil shells suggests a general cooling for about the last 50 million years. Two rapid increases in the oxygen-18 to oxygen-16 ratio about 12.5 Ma and about 28 Ma are observed (**Figure 1C**). The formation of the Greenland Ice Sheet and the Antarctic Ice Sheet are assumed to be the cause of these rapid isotope shifts. Where oxygen-18 data have been collected with a resolution finer than 20 000 years, high-frequency variations are seen which are presumed to correspond to a combination of temperature change and glacial growth and decay. We hypothesize that the magnitudes of these high-frequency sea level changes were considerably less in the earlier part of the Cenozoic than those observed over the last million years. This is because considerably less ice was involved.

Although large continental glaciers are not common in Earth history they are known to have been present during a number of extended periods ('ice house' climate, in contrast to 'greenhouse' or warm climate conditions). Ample evidence of glaciation is found in the continental sedimentary record. In particular, there is evidence of glaciation from about 2.7 to 2.1 billion years ago. Additionally, a long period of glaciation occurred shortly before the first fossils of multicellular organisms, from about 1

billion to 540 million years ago. Some scientists now believe that during this glaciation, the entire Earth froze over, generating a 'snowball earth'. Such conditions would have caused a large sea level fall. Evidence of large continental glaciers are also seen in Ordovician to Silurian rocks ($\sim$420 to 450 Ma), in Devonian rocks ($\sim$380 to 390 Ma), and in Carboniferous to Permian rocks ($\sim$350 to 270 Ma).

If these glaciations were caused by similar mechanisms to those envisioned for the Plio-Pleistocene (**Figure 1B**), then predictable, high-frequency, periodic growth and retreat of the glaciers should be observed in strata which form the geologic record. This is certainly the case for the Carboniferous through Permian glaciation. In the central United States, UK, and Europe, the sedimentary rocks have a distinctly cyclic character. They cycle in repetitive vertical successions of marine deposits, near-shore deposits, often including coals, into fluvial sedimentary rocks. The deposition of marine rocks over large areas, which had only recently been nonmarine, suggests very large-scale sea level changes. When the duration of the entire record is taken into account, periodicities of about 100 and 400 thousand years are suggested for these large sea level changes. This is consistent with an origin due to a response to changes in the eccentricity of the Earth's orbit. Higher-frequency cyclicity associated with the tilt of the spin axis and precession of the equinox is more difficult to prove, but has been suggested by detailed observations.

It is fair to say that large-scale (10 to > 100 m), relatively high-frequency (20 000–400 000 years; often termed 'Milankovitch scale') variations in sea level occurred during intervals of time when continental glaciers were present on Earth (ice house climate). This indicates that the variations of Earth's orbit and the tilt of its spin axis played a major role in controlling the climate. During the rest of Earth history, when glaciation was not a dominant climatic force (greenhouse climate), sea level changes corresponding to Earth's orbit did occur. In this case, the mechanism for changing the volume of water in the ocean basins is much less clear.

There is no geological record of continental ice sheets in many portions of Earth history. These time periods are generally called 'greenhouse' climates. However, there is ample evidence of Milankovitch scale variations during these periods. In shallow marine sediments, evidence of orbitally driven sea level changes has been observed in Cambrian and Cretaceous age sediments. The magnitudes of sea level change required (perhaps 5–20 m) are far less than have been observed during glacial climates. A possible source for these variations could be variations in average ocean-water temperature. Water expands as it is heated. If ocean bottom-water sources were equatorial rather than polar, as they are today, bottom-water temperatures of about 2°C today might have been about 16°C in the past. This would generate a sea level change of about 11 m. Other causes of sea level change during greenhouse periods have been postulated to be a result of variations in the magnitude of water trapped in inland lakes and seas, and variations in volumes of alpine glaciers. Deep marine sediments of Cretaceous age also show fluctuations between oxygenated and anoxic conditions. It is possible that these variations were generated when global sea level change restricted flow from the rest of the world's ocean to a young ocean basin. In a more recent example, tectonics caused a restriction at the Straits of Gibraltar. In that case, evaporation generated extreme sea level changes and restricted their entrance into the Mediterranean region.

Sea Level Change due to Changing Volume of the Ocean Basin

Tectonics is thought to be the main driving force of long-term (≥ 50 million years) sea level change. Plate tectonics changes the shape and/or the areal extent of the ocean basins.

Plate tectonics is constantly reshaping surface features of the Earth while the amount of water present has been stable for about the last four billion years. The reshaping changes the total area taken up by oceans over time. When a supercontinent forms, subduction of one continent beneath another decreases Earth's ratio of continental to oceanic area, generating a sea level fall. In a current example, the continental plate including India is diving under Asia to generate the Tibetan Plateau and the Himalayan Mountains. This has probably generated a sea level fall of about 70 m over the last 50 million years. The process of continental breakup has the opposite effect. The continents are stretched, generating passive margins and increasing the ratio of continental to oceanic area on a global scale (**Figure 2A**). This results in a sea level rise. Increments of sea level rise resulting from continental breakup over the last 200 million years amount to about 100 m of sea level rise.

Some bathymetric features within the oceans are large enough to generate significant changes in sea level as they change size and shape. The largest physiographic feature on Earth is the mid-ocean ridge system, with a length of about 60 000 km and a width of 500–2000 km. New ocean crust and lithosphere are generated along rifts in the center of these ridges. The ocean crust is increasingly old, cold, and dense away from the rift. It is the heat of ocean lithosphere formation that actually generates this feature. Thus, rifting of continents forms new ridges, increasing the proportionate area of young, shallow, ocean floor to older, deeper ocean floor (**Figure 2B**). Additionally, the width of the ridge is a function of the rates at which the plates are moving apart. Fast spreading ridges (e.g. the East Pacific Rise) are very broad while slow spreading ridges (e.g. the North Atlantic Ridge) are quite narrow. If the average spreading rates for all ridges decreases, the average volume taken up by ocean ridges would decrease. In this case, the volume of the ocean basin available for water would increase and a sea level fall would occur. Finally, entire ridges may be removed in the process of subduction, generating fairly rapid sea level fall.

Scientists have made quantitative estimates of sea level change due to changing ocean ridge volumes. Since ridge volume is dependent on the age of the ocean floor, where the age of the ocean floor is known, ridge volumes can be estimated. Seafloor magnetic anomalies are used to estimate the age of the ocean floor, and thus, spreading histories of the oceans 256. The oldest ocean crust is about 200 million years old. Older oceanic crust has been subducted. Thus, it is not surprising that quantitative estimates of sea level change due to ridge volumes are increasingly uncertain and cannot be calculated

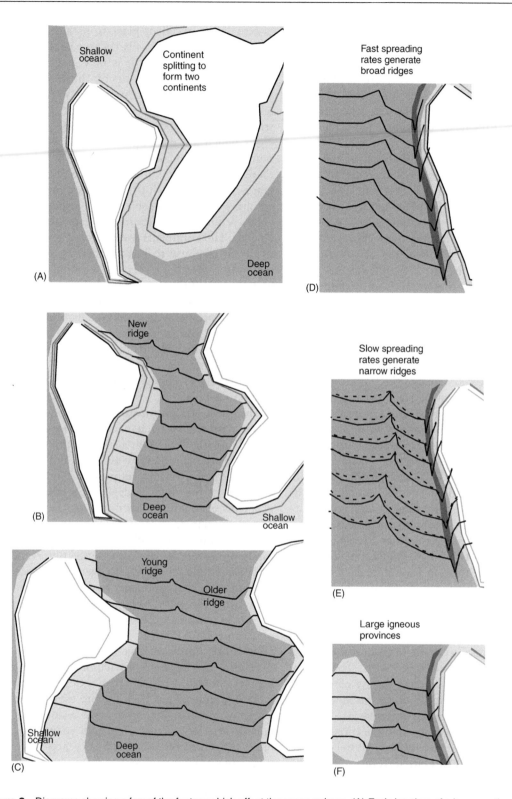

Figure 2 Diagrams showing a few of the factors which affect the ocean volume. (A) Early breakup of a large continent increases the area of continental crust by generating passive margins, causing sea level to rise. (B) Shortly after breakup a new ocean is formed with very young ocean crust. This young crust must be replacing relatively old crust via subduction, generating additional sea level rise. (C) The average age of the ocean between the continents becomes older so that young, shallow ocean crust is replaced with older, deeper crust so that sea level falls. (D) Fast spreading rates are associated with relatively high sea level. (E) Relatively slow spreading ridges (solid lines in ocean) take up less volume in the oceans than high spreading rate ridges (dashed lines in ocean), resulting in relatively low sea level. (F) Emplacement of large igneous provinces generates oceanic plateaus, displaces ocean water, and causes a sea level rise.

before about 90 million years. Sea level is estimated to have fallen about 230 m ($\pm$120 m) due to ridge volume changes in the last 80 million years.

Large igneous provinces (LIPs) are occasionally intruded into the oceans, forming large oceanic plateaus (*see* Igneous Provinces). The volcanism associated with LIPs tends to occur over a relatively short period of time, causing a rapid sea level rise. However, these features subside slowly as the lithosphere cools, generating a slow increase in ocean volume, and a long-term sea level fall. The largest marine LIP, the Ontong Java Plateau, was emplaced in the Pacific Ocean between about 120 and 115 Ma (**Figure 1D**). Over that interval it may have generated a sea level rise of around 50 m.

In summary, over the last 200 million years, long-term sea level change (**Figure 1D**) can be largely attributed to tectonics. Continental crust expanded by extension as the supercontinents Gondwana and Laurasia split to form the continents we see today. This process began about 200 Ma when North America separated from Africa and continues with the East African Rift system and the formation of the Red Sea. The generation of large oceans occurred early in this period and there was an overall rise in sea level from about 200 to about 90 million years. New continental crust, new mid-ocean ridges, and very fast spreading rates were responsible for the long-term rise (**Figure 1D**). Subsequently, a significant decrease in spreading rates, a reduction in the total length of mid-ocean ridges, and continent–continent collision coupled with an increase in glacial ice (**Figure 1C**) have resulted in a large-scale sea level fall (**Figure 1D**). Late Cretaceous volcanism associated with the Ontong Java Plateau, a large igneous province (*see* Igneous Provinces), generated a significant sea level rise, while subsequent cooling has enhanced the 90 million year sea level fall. Estimates of sea level change from changing ocean shape remain quite uncertain. Magnitudes and timing of stretching associated with continental breakup, estimates of shortening during continental assembly, volumes of large igneous provinces, and volumes of mid-ocean ridges improve as data are gathered. However, the exact configuration of past continents and oceans can only be a mystery due to the recycling character of plate tectonics.

Sea Level Change Estimated from Observations on the Continents

Long-term Sea Level Change

Estimates of sea level change are also made from sedimentary strata deposited on the continents. This is actually an excellent place to obtain observations of sea level change not only because past sea level has been much higher than it is now, but also because in many places the continents have subsequently uplifted. That is, in the past they were below sea level, but now they are well above it. For example, studies of 500–400 million year old sedimentary rocks which are now uplifted in the Rocky Mountains and the Appalachian Mountains indicate that there was a rise and fall of sea level with an estimated magnitude of 200–400 m. This example also exemplifies the main problem with using the continental sedimentary record to estimate sea level change. The continents are not fixed and move vertically in response to tectonic driving forces. Thus, any indicator of sea level change on the continents is an indicator of relative sea level change. Obtaining a global signature from these observations remains extremely problematic. Additionally, the continental sedimentary record contains long periods of non-deposition, which results in a spotty record of Earth history. Nonetheless a great deal of information about sea level change has been obtained and is summarized here.

The most straightforward source of information about past sea level change is the location of the strand line (the beach) on a stable continental craton (a part of the continent, which was not involved in local tectonics). Ideally, its present height is that of sea level at the time of deposition. There are two problems encountered with this approach. Unfortunately, the nature of land–ocean interaction at their point of contact is such that those sediments are rarely preserved. Where they can be observed, there is considerable controversy over which elements have moved, the continents or sea level. However, data from the past 100 million years tend to be consistent with calculations derived from estimates of ocean volume change. This is not saying a lot since uncertainties are very large (see above).

Continental hypsography (cumulative area versus height) coupled with the areal extent of preserved marine sediments has been used to estimate past sea level. In this case only an average result can be obtained, because marine sediments spanning a time interval (generally 5–10 million years) have been used. Again, uncertainties are large, but results are consistent with calculations derived from estimates of ocean volume change.

Backstripping is an analytical tool, which has been used to estimate sea level change. In this technique, the vertical succession of sedimentary layers is progressively decompacted and unloaded (**Figure 3A**). The resulting hole is a combination of the subsidence generated by tectonics and by sea level change

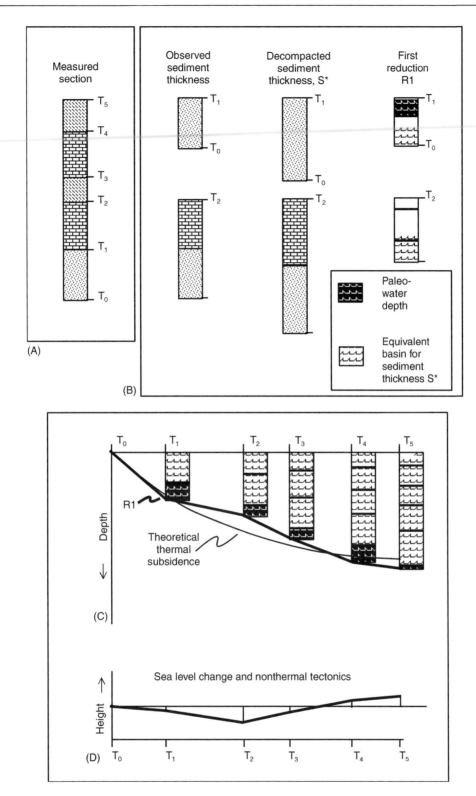

Figure 3 Diagrams depicting the backstripping method for obtaining sea level estimates in a thermally subsiding basin. (A) A stratigraphic section is measured either from exposed sedimentary rocks or from drilling. These data include lithologies, ages, and porosity. Note that the oldest strata are always at the base of the section (T0). (B) Porosity data are used to estimate the thickness that each sediment section would have had at the time of deposition (S*). They are also used to obtain sediment density so that the sediments can be unloaded to determine how deep the basin would have been in the absence of the sediment load (R1). This calculation also requires an estimate of the paleo-water depth (the water depth at the time of deposition). (C) A plot of R1 versus time is compared (by least-squares fit) to theoretical tectonic subsidence in a thermal setting. (D) The difference between R1 and thermal subsidence yields a quantitative estimate of sea level change if other, nonthermal tectonics, did not occur at this location.

(**Figure 3B**). If the tectonic portion can be established then an estimate of sea level change can be determined (**Figure 3C**). This method is generally used in basins generated by the cooling of a thermal anomaly (e.g. passive margins). In these basins, the tectonic signature is predictable (exponential decay) and can be calibrated to the well-known subsidence of the mid-ocean ridge.

The backstripping method has been applied to sedimentary strata drilled from passive continental margins of both the east coast of North America and the west coast of Africa. Again, estimates of sea level suggest a rise of about 100–300 m from about 200–110 Ma followed by a fall to the present level (**Figure 1D**). Young interior basins, such as the Paris Basin, yield similar results. Older, thermally driven basins have also been analyzed. This was the method used to determine the (approximately 200 m) sea level rise and fall associated with the breakup of a Pre-Cambrian supercontinent in earliest Phanerozoic time.

Million Year Scale Sea Level Change

In addition to long-term changes in sea level there is evidence of fluctuations that are considerably shorter than the 50–100 million year variations discussed above, but longer than those caused by orbital variations ($\leq$0.4 million years). These variations appear to be dominated by durations which last either tens

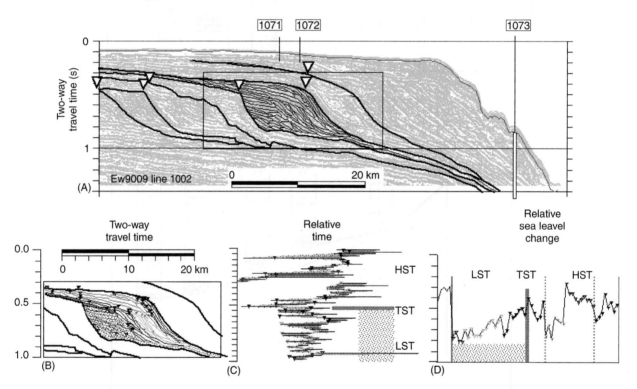

Figure 4 Example of the sequence stratigraphic approach to estimates of sea level change. (A) Multichannel seismic data (gray) from the Baltimore Canyon Trough, offshore New Jersey, USA (Miller *et al.*, 1998). Black lines are interpretations traced on the seismic data. Thick dark lines indicate third-order Miocene-aged (5–23 Ma) sequence boundaries. They are identified by truncation of the finer black lines. Upside-down deltas indicate a significant break in slope associated with each identified sequence boundary. Labeled vertical lines (1071–1073) show the locations of Ocean Drilling Project wells, used to help date the sequences. The rectangle in the center is analyzed in greater detail. (B) Detailed interpretation of a single third-order sequence from (A). Upside-down deltas indicate a significant break in slope associated with each of the detailed sediment packages. Stippled fill indicates the low stand systems tract (LST) associated with this sequence. The gray packages are the transgressive systems tract (TST), and the overlying sediments are the high stand systems tract (HST). (C) Relationships between detailed sediment packages (in B) are used to establish a chronostratigraphy (time framework). Youngest sediment is at the top. Each observed seismic reflection is interpreted as a time horizon, and each is assigned equal duration. Horizontal distance is the same as in (A) and (B). A change in sediment type is indicated at the break in slope from coarser near-shore sediments (stippled pattern) to finer, offshore sediments (parallel, sloping lines). Sedimentation may be present offshore but at very low rates. LST, TST, and HST as in (B). (D) Relative sea level change is obtained by assuming a consistent depth relation at the change in slope indicated in (B). Age control is from the chronostratigraphy indicated in (C). Time gets younger to the right. The vertical scale is in two-way travel time, and would require conversion to depth for a final estimate of the magnitude of sea level change. LST, TST, and HST as in (B). Note that in (B), (C), and (D), higher frequency cycles (probably fourth-order) are present within this (third-order) sequence. Tracing and interpretations are from the author's graduate level quantitative stratigraphy class project (1998, Western Michigan University).

of millions of years or a half to three million years. These sea level variations are sometimes termed second- and third-order sea level change, respectively. There is considerable debate concerning the source of these sea level fluctuations. They have been attributed to tectonics and changing ocean basin volumes, to the growth and decay of glaciers, or to continental uplift and subsidence, which is independent of global sea level change. As noted above, the tectonic record of subsidence and uplift is intertwined with the stratigraphic record of global sea level change on the continents. Synchronicity of observations of sea level change on a global scale would lead most geoscientists to suggest that these signals were caused by global sea level change. However, at present, it is nearly impossible to globally determine the age equivalency of events which occur during intervals as short as a half to two million years. These data limitations are the main reason for the heated controversy over third-order sea level.

Quantitative estimates of second-order sea level variations are equally difficult to obtain. Although the debate is not as heated, these somewhat longer-term variations are not much larger than the third-order variations so that the interference of the two signals makes definition of the beginning, ending and/or magnitude of second-order sea level change

equally problematic. Recognizing that our understanding of second- and third-order (million year scale) sea level fluctuations is limited, a brief review of that limited knowledge follows.

Sequence stratigraphy is an analytical method of interpreting sedimentary strata that has been used to investigate second- and third-order relative sea level changes. This paradigm requires a vertical succession of sedimentary strata which is analyzed in at least a two-dimensional, basinal setting. Packages of sedimentary strata, separated by unconformities, are observed and interpreted mainly in terms of their internal geometries (e.g. **Figure 4**). The unconformities are assumed to have been generated by relative sea level fall, and thus, reflect either global sea level or local or regional tectonics. This method of stratigraphic analysis has been instrumental in hydrocarbon exploration since its introduction in the late 1970s. One of the bulwarks of this approach is the 'global sea level curve' most recently published by Haq *et al.* (1987). This curve is a compilation of relative sea level curves generated from sequence stratigraphic analysis in basins around the world. While sequence stratigraphy is capable of estimating relative heights of relative sea level, it does not estimate absolute magnitudes. Absolute dating requires isotope data or correlation via fossil data into the

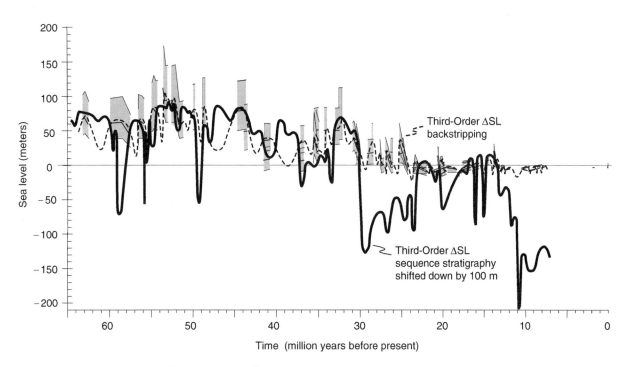

Figure 5 Million year scale sea level fluctuations. Estimates from sequence stratigraphy (Haq *et al.*, 1987; solid curve) have been shifted down by 100 m to allow comparison with estimates of sea level from backstripping (Kominz *et al.*, 1998; dashed curve). Where sediments are present, the backstripping results, with uncertainty ranges, are indicated by gray fill. Between backstrip observations, lack of preserved sediment is presumed to have been a result of sea level fall. The Berggren *et al.* (1995) biostratigraphic timescale was used.

chronostratigraphic timescale. However, the two-dimensional nature of the data allows for good to excellent relative age control.

Backstripping has been used, on a considerably more limited basis, in an attempt to determine million year scale sea level change. This approach is rarely applied because it requires very detailed, quantitative, estimates of sediment ages, paleo-environments and compaction in a thermal tectonic setting. A promising area of research is the application of this method to coastal plain boreholes from the mid-Atlantic coast of North America. Here an intensive Ocean Drilling Project survey is underway which is providing sufficiently detailed data for this type of analysis. Initial results suggest that magnitudes of million year scale sea level change are roughly one-half to one-third that reported by Haq *et al*. However, in glacial times, the timing of the cycles was quite consistent with those of this 'global sea level curve' derived by application of sequence stratigraphy (**Figure 5**). Thus, it seems reasonable to conclude that, at least during glacial times, global, third-order sea level changes did occur.

Summary

Sea level changes are either a response to changing ocean volume or to changes in the volume of water contained in the ocean. The timing of sea level change ranges from tens of thousands of years to over 100 million years. Magnitudes also vary significantly but may have been as great as 200 m or more. Estimates of sea level change currently suffer from significant ranges of uncertainty, both in magnitude and in timing. However, scientists are converging on consistent estimates of sea level changes by using very different data and analytical approaches.

Further Reading

Allen PA and Allen JR (1990) *Basin Analysis: Principless, Applications*. Oxford: Blackwell Scientific Publications.

Berggren WA, Kent DV, Swisher CC, and Aubry MP (1995) A revised Cenozoic geochronology and chronostratigraphy. In: Berggren WA, Kent DV, and Hardenbol J (eds.) *Geochronology, Time Scales and Global Stratigraphic Correlations: A Unified Temporal Framework for an Historical Geology*, SEPM Special Publication No. 54, 131–212.

Bond GC (1979) Evidence of some uplifts of large magnitude in continental platforms. *Tectonophysics* 61: 285–305.

Coffin MF and Eldholm O (1994) Large igneous provinces: crustal structure, dimensions, and external consequences. *Reviews of Geophysics* 32: 1–36.

Crowley TJ and North GR (1991) *Paleoclimatology*. Oxford Monographs on Geology and Geophysics, no. 18

Fairbanks RG (1989) A 17,000-year glacio-eustatic sea level record: influence of glacial melting rates on the Younger Dryas event and deep-ocean circulation. *Nature* 6250: 637–642.

Hallam A (1992) *Phanerozoic Sea-Level Changes*. NY: Columbia University Press.

Haq BU, Hardenbol J, and Vail PR (1987) Chronology of fluctuating sea levels since the Triassic (250 million years ago to present). *Science* 235: 1156–1167.

Harrison CGA (1990) Long term eustasy and epeirogeny in continents. In: Sea-Level Change, pp. 141–158. Washington, DC: US National Research Council Geophysics Study Committee.

Hauffman PF and Schrag DP (2000) Snowball Earth. *Scientific American* 282: 68–75.

Kominz MA, Miller KG, and Browning JV (1998) Long-term and short term global Cenozoic sea-level estimates. *Geology* 26: 311–314.

Miall AD (1997) *The Geology of Stratigraphic Sequences*. Berlin: Springer-Verlag.

Miller KG, Fairbanks RG, and Mountain GS (1987) Tertiary oxygen isotope synthesis, sea level history, and continental margin erosion. *Paleoceanography* 2: 1–19.

Miller KG, Mountain GS, Browning J, *et al.* (1998) Cenozoic global sea level, sequences, and the New Jersey transect; results from coastal plain and continental slope drilling. *Reviews of Geophysics* 36: 569–601.

Sahagian DL (1988) Ocean temperature-induced change in lithospheric thermal structure: a mechanism for long-term eustatic sea level change. *Journal of Geology* 96: 254–261.

Wilgus CK, Hastings BS, Kendall CG St C *et al.* (1988) *Sea Level Changes: An Integrated Approach*. Special Publication no. 42. Society of Economic Paleontologists and Mineralogists.

ROCKY SHORES

G. M. Branch, University of Cape Town, Cape Town, Republic of South Africa

Introduction

Intertidal rocky shores have been described as 'superb natural laboratories' and a 'cauldron of scientific ferment' because a rich array of concepts has arisen from their study. Because intertidal shores form a narrow band fringing the coast, the gradient between marine and terrestrial conditions is sharp, with abrupt changes in physical conditions. This intensifies patterns of distribution and abundance, making them readily observable. Most of the organisms are easily visible, occur at high densities, and are relatively small and sessile or sedentary. Because of these features, experimental tests of concepts have become a feature of rocky-shore studies, and the critical approach encouraged by scientists such as Tony Underwood has fostered rigor in marine research as a whole.

Rocky shores are a strong contrast with sandy beaches. On sandy shores, the substrate is shifting and unstable. Organisms can burrow to escape physical stresses and predation, but experience continual turnover of the substrate by waves. Most of the fauna relies on imported food because macroalgae cannot attach in the shifting sands, and primary production is low. Physical conditions are relatively uniform because waves shape the substrate. On rocky shores, by contrast, the physical substrate is by definition hard and stable. Escape by burrowing may not be impossible, but is limited to a small suite of creatures capable of drilling into rock. Macroalgae are prominent and *in situ* primary production is high. Rocks alter the impacts of wave action, leading to small-scale variability in physical conditions.

Research on rocky shores began with a phase describing patterns of distribution and abundance. Later work attempted to explain these patterns – initially focusing on physical factors before shifting to biological interactions. Integration of these focuses is relatively recent, and has concentrated on three issues: the relative roles of larval recruitment versus adult survival; the impact of productivity; and the effects of stress or disturbance on the structure and function of rocky shores.

Zonation

The most obvious pattern on rocky shores is an upshore change in plant and animal life. This often creates distinctive bands of organisms. The species making up these bands vary, but the high-shore zone is frequently dominated by littorinid gastropods, the upper midshore prevalently occupied by barnacles, and the lower section by a mix of limpets, barnacles, and seaweeds. The low-shore zone commonly supports mussel beds or mats of algae. Such patterns of zonation were of central interest to Jack Lewis in Britain, and to Stephenson, who pioneered descriptive research on zonation, first in South Africa and then worldwide.

In general, physical stresses ameliorate progressively down the shore. In parallel, biomass and species richness increase downshore. Three factors powerfully influence zonation: the initial settlement of larvae and spores; the effects of physical factors on the survival or movement of subsequent stages; and biotic interactions between species.

Settlement of Larvae or Spores

Many rocky-shore species have adults that are sessile, including barnacles, zoanthids, tubicolous polychaete worms, ascidians, and macroalgae. Many others, such as starfish, anemones, mussels, and territorial limpets, are extremely sedentary, moving less than a few meters as adults. For such species, settlement of the dispersive stages in their life cycles sets initial limits to their zonation (often further restricted by later physical stresses or biological interactions).

Some larvae selectively settle where adults are already present. Barnacles are a classic example. This gregarious behavior, which concentrates individuals in particular zones, has several possible advantages. The presence of adults must indicate a habitat suitable for survival. Furthermore, individuals of sessile species that practice internal fertilization (e.g., barnacles) are obliged to live in close proximity. Even species that broadcast their eggs and sperm will enhance fertilization if they are closely spaced, because sperm becomes diluted away from the point of release. Finally, adults may themselves shelter new recruits. As examples, larvae of the sabellariid reef-worm *Gunnarea capensis* that settle on adult colonies suffer less desiccation, and sporelings of kelps that settle among the holdfasts of adults experience less intense grazing than those that are isolated.

Cues used by larvae to select settlement sites are diverse. Barnacle larvae differ in their preferences for light intensity, water movement, substrate texture, and water depth. All of these responses influence the type of habitat or zone in which the larvae will settle. Most barnacle larvae are attracted to species-specific chemicals in the exoskeletons of their own adults, which persists on the substratum even after adults are eliminated. (Incidentally, this behavior is not just of academic interest: gregarious settlement of barnacles on the hulls of ships costs billions of dollars each year due to increased fuel costs caused by the additional drag of 'fouling' organisms.)

Cues influencing larval settlement can also be negative. Rick Grosberg elegantly demonstrated that the larvae of a wide range of sessile species that are vulnerable to overgrowth avoid settling in the presence of *Botryllus*, a compound ascidian known to be an aggressive competitor for space.

A different aspect of 'supply-side ecology' is the rate at which dispersive larvae or spores settle. The relative importance of recruit supply versus subsequent survival is a topic of intense research. In situations of low recruitment, rate of supply critically influences population and community dynamics. At high levels of recruitment, however, supply rates become less important than subsequent biological interactions such as competition.

Control of Zonation by Physical Factors

Early research on the causes of zonation focused strongly on physical factors such as desiccation, temperature, and salinity, all of which increase in severity upshore. Measurements showed a correlation between the zonation of species and their tolerance of extremes of these factors. In some cases – particularly for sessile species living at the top of the shore – physical conditions become so severe that they kill sections of the population, thus imposing an upper zonation limit by mortality.

For those species blessed with mobility, zonation is more often set by behavior than death, and most individuals live within the 'zone of comfort' that they can tolerate. One example illustrates the point. The trochid gastropod *Oxystele variegata* increases in size from the low- to the high-shore. This gradient is maintained by active migration, and animals transplanted to the 'wrong' zone re-establish themselves in their original zones within 24 hours. The underlying causes of this size gradient seem to be twofold: desiccation is too high in the upper shore for small individuals to survive there; and predation on adults is greatest in the lower shore.

Adaptations to minimize the effects of physical stresses are varied. Physiological adaptation and tolerance are one avenue of escape. Avoidance by concealment in microhabitats is another. Morphological adaptations are a third route. For instance, desiccation and heat stress can be reduced by large size (reducing the ratio of surface area to size), and by differences in shape, color, and texture (**Figure 1A**).

Physical factors can clearly limit the upper zonation of species. It is often difficult, however, to imagine them setting lower limits. For this, we turn to biological interactions.

Biological Interactions

Interactions between species – particularly competition and predation – only began to influence the thinking of intertidal rocky-shore ecologists in the mid 1950s. Extremely influential was Connell's work, exploring whether the zonation of barnacles in Scotland is influenced by competition. He noted that a high-shore species, *Chthamalus stellatus*, seldom penetrates down into the midshore, where another species, *Semibalanus balanoides*, prevails. Was competition from *Semibalanus* excluding *Chthamalus* from the midshore? In field experiments in which *Semibalanus* was eliminated, *Chthamalus* not only occupied the midshore, but survived and grew there better than in its normal high-shore zone. *Chthamalus* is more tolerant of physical stresses than *Semibalanus*, and can therefore survive in the high-shore, where it has a 'spatial refuge' beyond the limits of *Semibalanus*. In the midshore, however, *Semibalanus* thrives and competitively excludes *Chthamalus* by undercutting or overgrowing it.

Other forms of competition have since been discovered. For instance, territorial limpets defend patches of algae by aggressively pushing against other grazers. In areas where they occur densely, they profoundly influence the zonation of other species and the nature of their associated communities.

The role of predators came to the fore following the work of Bob Paine who showed that experimental removal of the starfish *Pisaster ochraceus* from open-coast shores in Washington State led to encroachment of the low-shore by mussels, which are usually restricted to the midshore. Thus, predation sets lower limits to the zonation of the

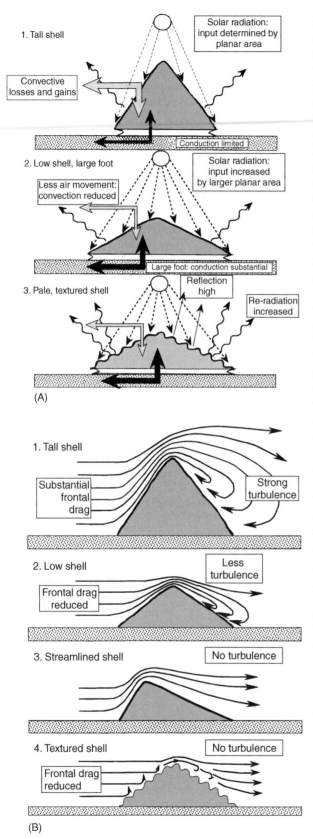

1. Tall shell

Solar radiation: input determined by planar area

Convective losses and gains

Conduction limited

2. Low shell, large foot

Solar radiation: input increased by larger planar area

Less air movement: convection reduced

Large foot: conduction substantial

3. Pale, textured shell

Reflection high

Re-radiation increased

(A)

1. Tall shell

Substantial frontal drag

Strong turbulence

2. Low shell

Frontal drag reduced

Less turbulence

3. Streamlined shell

No turbulence

4. Textured shell

No turbulence

Frontal drag reduced

(B)

Figure 1 (A) Effects of shell shape, color, and texture on heat uptake and loss. (B) Influence of shell proportions and texture on hydrodynamic drag.

mussels. More importantly, it was shown that the downshore advance of mussel beds reduced the number of space-occupying species there. In other words, predation normally prevents the competitively superior mussels from ousting subordinate species, thus maintaining a higher diversity of species. This concept – the 'predation hypothesis' – has since been broadened to include all forms of biological or physical disturbance. If disturbance is too great, few species survive and diversity is low. On the other hand, if it is absent or has little effect, a few species may competitively monopolize the system, reducing diversity. At intermediate levels of disturbance, diversity is highest – the 'intermediate disturbance hypothesis'. (Incidentally, the idea is not new. Darwin gives an accurate description of this effect in sheep-grazed meadows.)

Paine's work led to the idea of 'keystone predators': those that have a strong effect on community structure and function. Unfortunately the term has been debased by general application to any species that an author feels is somehow 'important'. Consequently, not all scientists are enamoured of the concept. More recently, it has been redefined to mean those species whose effects are disproportionately large relative to their abundance. This more usefully allows recognition of species that can be regarded as 'strong interactors', and which powerfully influence community dynamics.

Many researchers have shown that grazers (particularly limpets) profoundly influence algal zonation, excluding them from much of the shore by eliminating sporelings before they develop. Hawkins and Hartnoll suggested that interactions between grazers and algae depend on the upshore gradient in physical stress, and argued that low on the shore, algae are likely to be sufficiently productive to escape grazing and proliferate, forming large, adult growths that are relatively immune to grazing. In the mid- to high-shore, grazers become dominant and algae seldom develop beyond the sporeling stage. A more subtle reverse effect, that of algae-limiting grazers, has also been described. Low on the shore where productivity is high, algae form dense mats. This not only deprives grazers of a firm substrate for attachment, but also of their primary food source, namely microalgae. Grazers experimentally transplanted into low-shore algal beds starve in the midst of apparent plentitude.

The influences of grazers and predators extend beyond their direct impacts on prey. A bird consuming limpets has positive effects on algae that are grazed by the limpets. Such indirect effects occupy ecologists because their consequences are often difficult to predict. For example, experimental removal of a large, grazing chiton, *Katharina tunicata*, from the shores of Washington State might logically have been expected to improve the lot of intertidal limpets, on the grounds that elimination of a competitor must be good for the remaining grazers. In fact, elimination of the chiton led to starvation of the limpets because macroalgae proliferated, excluding microalgae on which the limpets depend. Indirectly, *Katharina* facilitates microalgae, thus benefiting limpets.

The mix of physical and biological controls affecting zonation was investigated by Bruce Menge in 1976 by using cages to exclude predators from plots in the high-, mid-, and low-shore. In the high-shore, only barnacles became established, irrespective of whether predators were present or absent. In the midshore, mussels became dominant and outcompeted barnacles, again independently of the presence or absence of predators. Competition ruled. In the low-shore, however, mussels only dominated where cages excluded the predators. Elsewhere, predators eliminated mussels, thus allowing barnacles to persist. These results elucidated the interplay between physical stress and biological control and were instrumental in formalizing 'environmental stress models'. These suggest that predation will only be important (exerting a 'top-down' control on community structure) when physical conditions are mild. As stress rises, first predation, and then competition, diminish in importance.

Wave Action

It has been shown that wave action is probably the most important factor affecting distribution patterns along the shore. In a negative sense, waves physically remove organisms, damage them by throwing up logs and boulders, reduce their foraging excursions, and increase the amount of energy devoted to clinging on. One manifestation is a reduction of grazer biomass (**Figure 2F**). Adaptations can, however, counter these adverse effects. Tenacity can be increased by cementing the shell to the rock face (e.g., oysters), developing temporary attachments (e.g., the byssus threads of mussels), or employing

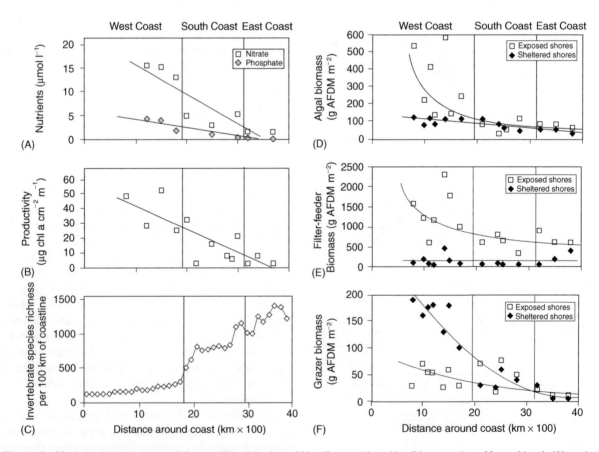

Figure 2 Moving west to east around the coastline of southern Africa (from northern Namibia to southern Mozambique), (A) nutrient input and (B) intertidal primary productivity decrease, (C) invertebrate species richness rises, and there are declines in the biomasses (ash-free dry mass) of (D) algae, (E) filter-feeders, and (F) grazers. Note that biomass is strongly influenced by wave action, either positively (algae and filter-feeders) or negatively (grazers). AFDM.

adhesion (e.g., the feet of limpets and chitons). Shape can be modified to reduce drag, turbulence, and lift (**Figure 1B**). Each organism is, however, a compromise between conflicting stresses. For instance, desiccation is reduced if a limpet has a small aperture; but this implies a smaller foot and thus less ability to cling to the rocks.

Wave action also brings benefits. It enhances nutrient supply, reduces predation and grazing, and increases food supply for filter-feeders. The biomass on South African rocky shores has been shown to rise steeply as wave action increases, mostly due to increases in filter-feeders (**Figure 2E**). Possible explanations include enhanced larval supply and reduced predation, but measurements of food abundance and turnover showed that wave action vitally enhances particulate food.

Wave action varies over short distances. Headlands and bays influence the magnitude of waves at a scale of kilometers, but even a single large boulder will alter wave impacts at a scale of centimeters to meters. As a result, community structure can be extremely patchy on rocky shores.

Productivity

'Nutrient/productivity models' (NPMs) attempt to explain community structure in terms of nutrient input and, thus, productivity. This 'bottom-up' approach concerns how the influences of primary production filter up to higher levels. Nutrient/productivity models were developed with terrestrial systems in mind, but their application to rocky shores has been enlightening. Terrestrial systems depend largely on the productivity of plants, which is usually limited by availability of nutrients and water. On rocky shores, neither constraint is necessarily relevant. The shore is washed by the rise and fall of the tide, which also imports particulate material that fuels filter-feeders independently of the productivity on the shore itself. Even so, local productivity does influence rocky-shore community structure. In one theory, the lower the productivity, the fewer the steps that can be supported in the trophic web. If production is very low, plant life cannot sustain grazers. As productivity rises, grazers may be supported, and begin to control the standing stocks of plants. Further increases may lead to three trophic levels, with predators controlling grazers, which thus lose their capacity to regulate plants.

Nowhere else in the world is the coastline better configured to test ideas about NPMs than in South Africa. The cold, nutrient-rich, upwelled Benguela Current bathes the west coast. The east coast receives fast-moving, warm, nutrient-poor waters from the southward-flowing Agulhas Current. Between the two, the Agulhas swings away from the south coast, creating conditions that are intermediate. From west to east, the coast has a strong gradient of nutrients and productivity (**Figure 2A, B**). As productivity drops, so do the average biomasses of algae, filter-feeders, and grazers (**Figure 2D–F**), and the total biomass. On the other hand, species richness rises (**Figure 2C**). At a more local scale, guano input on islands achieves the same effects (**Figure 3**).

Productivity also has more subtle effects on the functioning of rocky shores. For instance, the frequency of territoriality in limpets is inversely correlated with productivity. It seems that the need to defend patches of food diminishes as the ratio of productivity : consumption rises. Indirectly, this has profound effects on community dynamics, because these territorial algal species reduce species richness and biomass but greatly increase local productivity.

Increased productivity is, however, not an unmixed blessing. The upwelling that fuels coastal productivity also results in a net offshore movement of water. This may export the recruits of species with dispersive stages. The scarcity of barnacles on the west coast of South Africa may be one consequence. In California it has been shown that barnacle settlement is inversely related to an index of upwelling. Furthermore, nutrient input can lead to heavy blooms of phytoplankton (often called red or brown tides), that subsequently decay, causing anoxia or even development of hydrogen sulfide. Either eventuality is lethal, and mass mortalities ensue. Records of thousands of tons of rock lobsters spectacularly stranding themselves on the shore in a futile attempt to escape anoxic waters testify to the ecological and economic consequences.

Energy Flow

Flows of energy (or of any material such as carbon or nitrogen) through an ecosystem can be used to quantify rates of turnover and passages between elements of the food web. Developing a complete energy-flow model for a rocky shore is a formidable task. For at least the major species, energy uptake must be measured directly, or estimated by summing the requirements for growth and reproduction and the losses associated with respiration, excretion, and secretion. Critics of ecosystem energy-flow models emphasize the huge investment of research time needed to complete the task, and argue that rocky shores differ so much from place to place that a model describing one shore will often be inapplicable elsewhere.

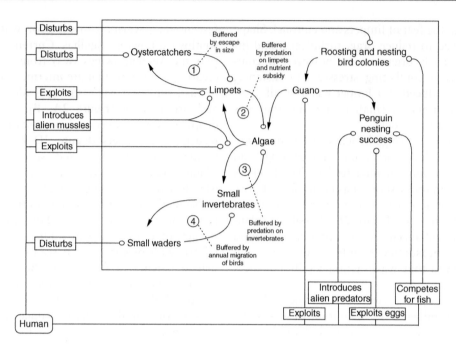

Figure 3 Interactions between organisms on nearshore islands on the west coast of South Africa. Natural interactions, and the processes buffering them (numbered 1–4), are shown inside the box; human impacts influencing them lie outside the box. Lines terminating in arrows and circles indicate positive and negative effects, respectively.

Nevertheless, energy-flow models of rocky shores have revealed features difficult to grasp by other approaches, as the following examples reveal. First, measurements show that most macroalgal productivity is not consumed by grazers, but adds to a detrital pool that fuels filter-feeders. On sheltered shores, production exceeds demands, and they are net exporters of material. On wave-beaten shores, however, the needs of filter-feeders far exceed intertidal production, and these shores depend on a subsidy of materials from the subtidal zone or from offshore. Tides, waves and currents play a vital role in turning over this supply of particulate materials. Intertidal standing stocks of particulate matter are in the order of $0.25\,g\,C\,m^{-3}$, far below the annual requirements of filter-feeders (about $500\,g\,C\,m^{-2}\,y^{-1}$). But if a hypothetical flow of $20\,m^3$ passes over each square meter of shore per day, it will supply $1825\,g\,C\,m^{-2}\,y^{-1}$. This stresses the need for small-scale hydrographic research to predict flows that are meaningful at the level of individual filter-feeders.

Second, some elements of the ecosystem have consistently been overlooked because of their small contribution to biomass. An obvious case is the almost invisible 'skin' of microbiota (sporelings, diatoms, bacteria, and fungi) coating rocks. On most shores, their share of the biomass is an apparently insignificant 0.1%. However, in terms of productivity, they contribute 12%, and most grazers depend on this food source.

Finally, there has always been a tacit assumption that sedentary intertidal grazers must depend on *in situ* algal productivity. However, on the west coast of South Africa, grazers reach extraordinarily high biomasses (up to $1000\,g$ wet flesh m^{-2}). Modeling shows that their needs greatly exceed *in situ* primary production. Instead, they survive by trapping drift and attached kelp. This subtidal material effectively subsidizes the intertidal system. Thus sustained, dense beds of limpets dominate sections of the shore, eliminating virtually all macroalgae and most other grazers. Both 'bottom-up' and 'top-down' processes are at play.

Energy-flow models are seldom absolute measures of how a system operates. Changes in time and space preclude this. Rather, their power lies in identifying bottlenecks, limitations, and overlooked processes, which can then be investigated by complementary approaches.

Integration of Approaches

Ecology as a whole, and that addressing rocky shores in particular, has suffered from polarized viewpoints. Classic examples are arguments over competition versus predation, or the merits of 'top-down' versus 'bottom-up' approaches. In reality, all of these are valid. What is needed is an integration that identifies the circumstances under which one or other model has greatest predictive power. A single example demonstrates the multiplicity of factors operating on intertidal rocky shores (**Figure 3**).

Islands off the west coast of South Africa support dense colonies of seabirds. These have two important effects on the dynamics of rocky shores. First, their guano fuels primary production. Second, oyster-catchers that aggregate on the islands prey on lim-pets. Indirect effects complicate this picture. Reduction of limpets adds to the capacity of sea-weeds to escape grazing, leading to luxurious algal mats. These sustain small invertebrates, in turn a source of food for waders. The limpets benefit in-directly from the guano because their growth rates and maximum sizes are increased by the high algal productivity. This allows some to reach a size where they are immune to predation by oystercatchers, and also increases their individual reproductive output. But this 'interaction web' embraces less obvious connections. Dense bird colonies only exist on the islands for two reasons: absence of predators, and food in the form of abundant fish, sustained ultim-ately by upwelling. Comparison with adjacent mainland rocky shores reveals a contrast: roosting and nesting birds are scarce or absent, oystercatchers are much less numerous, limpets are abundant but small, and algal beds are absent.

This case emphasizes the complexity of driving forces and the difficulty of making predictions about their consequences. Top-down and bottom-up ef-fects, direct and indirect impacts, productivity, grazing, predation, competition, and physical stres-ses all play their role.

Human Impacts

In one sense intertidal rocky-shore communities are vulnerable to human effects because they are ac-cessible and many of the species have no refuge. In another sense, they are relatively resilient to change. One reason is that humans seldom change the structure and physical factors influencing rocky shores. Tidal excursions and wave action, the two most important determinants of rocky-shore com-munity structure, are seldom altered. The physical rock itself is also rarely modified by human actions. These circumstances are an important contrast with systems such as mangroves, estuaries, and coral reefs, where the structure is determined by the biota. Re-move or damage mangrove forests, salt marshes, or corals, and the fundamental nature of these systems is changed and slow to recover. Estuaries are espe-cially vulnerable because their two most important physical attributes – input of riverine water and tidal

exchange – can be revolutionized by human actions. Even after massive abuse such as oil pollution, rocky shores recover relatively quickly once the agent of change is brought under control; and there is good evidence that the kinds of communities appearing after recovery resemble those originally present.

Humans impact rocky-shore communities in many ways, including trampling, harvesting, pollution, introduction of alien species and by altering global climate. Harvesting is of specific interest because it has taught much about the functioning of rocky shores.

Almost without exception, harvesting reduces mean size, density, and total reproductive output of the target species, although compensatory increases of growth rate and reproductive capacity of surviving individuals are not unusual. Of greater interest, however, are the consequences for community structure and dynamics. Clear demonstrations of this come from Chile, where intense artisanal harvesting occurs on rocky shores. In particular, a lucrative trade has developed for giant keyhole limpets, *Fis-surella* spp., and for a predatory muricid gastropod, *Concholepas concholepas*, colloquially known as 'loco.' Decimated along most of the coast, the nat-ural roles of these species are only evident inside marine protected areas. There, locos consume a small mussel, *Perumytilus purpuratus*, which other-wise outcompetes barnacles. Macroalgae and bar-nacles compete for space, but only on a seasonal basis. Keyhole limpets control macroalgae and out-compete smaller acmaeid limpets. *Perumytilus* acts as a settlement site for conspecifics and for recruits of keyhole limpets and locos. With a combination of predation, grazing, competition and facilitation, and both direct and indirect effects, the consequences of harvesting locos or keyhole limpets would have been impossible to resolve without the existence of marine protected areas. Even then, careful manipulative experiments inside and outside these areas were re-quired to disentangle these interacting effects.

A second issue of general interest is whether human impacts are qualitatively different from those of other species. The short answer is 'yes', and is best illustrated by a return to an earlier example – inter-actions between species on rocky shores of islands on the west coast of South Africa (**Figure 3**). In its un-disturbed state, each of the key interactions in this ecosystem is buffered in some way. Limpets are consumed by oystercatchers, but some escape by growing too large to be eaten, aided by the high primary production. Limpets and other invertebrates graze on algae, but their effects are muted by pre-dation on them and by the enhancement of algal growth by guano. Waders eat small seaweed-associ-ated invertebrates, but emigrate in winter.

For several reasons, human impacts are not constrained in these subtle ways. First, human populations do not depend on rocky shores in any manner limiting their own numbers. They can harvest these resources to extinction with impunity. Second, modern human effects are too recent a phenomenon for the impacted species to have evolved defenses. Thirdly, humans are supreme generalists. Simultaneously, they can act as predators, competitors, amensal disturbers of the environment, and 'commensal' introducers of alien species. Fourthly, money, not returns of energy, determines profitability. Fifthly, long-range transport means that local needs no longer limit supply and demand. Sixthly, technology denies resources any refuge.

Thus, humans supersede the ecological and evolutionary rules under which natural systems operate; and only human-imposed rules and constraints can replace them in meeting our self-proclaimed goals of sustainable use and maintenance of biodiversity.

See also

Beaches, Physical Processes Affecting. Waves on Beaches.

Further Reading

Branch GM and Griffiths CL (1988) The Benguela ecosystem Part V. The coastal zone. *Oceanography and Marine Biology Annual Review* 26: 395–486.

Castilla JC (1999) Coastal marine communities: trends and perspectives from human-exclusion experiments. *Trends in Ecology and Evolution* 14: 280–283.

Connell JH (1975) Some mechanisms producing structure in natural communities: a model and evidence from field experiments. In: Cody ML and Diamond JM (eds.) *Ecology and Evolution of Communities*, pp. 460–490. Cambridge, MA: Belknap Press.

Denny MW (1988) *Biology and the Mechanics of the Wave-swept Environment*. Princeton, NJ: Princeton University Press.

Hawkins SJ and Hartnoll RG (1983) Grazing of intertidal algae by marine invertebrates. *Oceanography and Marine Biology Annual Review* 21: 195–282.

Menge BA and Branch GM (2001) Rocky intertidal communities. In: Bertness MD, Gaines SL, and Hay ME (eds.) *Marine Community Ecology*. Sunderland: Sinauer Associates.

Moore PG and Seed R (1985) *The Ecology of Rocky Coasts*. London: Hodder and Stoughton.

Newell RC (1979) *Biology of Intertidal Animals*. Faversham, Kent: Marine Ecological Surveys.

Paine RT (1994) *Marine Rocky Shores and Community Ecology: An Experimentalist's Perspective*. Oldendorf: Ecology Institute.

Siegfried WR (ed.) (1994) *Rocky Shores: Exploitation in Chile and South Africa*. Berlin: Springer-Verlag.

Underwood AJ (1997) *Experiments in Ecology: their Logical Design and Interpretation Using Analysis of Variance*. Cambridge: Cambridge University Press.

SALT MARSHES AND MUD FLATS

J. M. Teal, Woods Hole Oceanographic Institution,
Rochester, MA, USA

Structure

Salt marshes are vegetated mud flats. They are above
mean sea level in the intertidal area where higher
plants (angiosperms) grow. Sea grasses are an exception to the generalization about higher plants
because they live below low tide levels. Mud flats are
vegetated by algae.

Geomorphology

Salt marshes and mud flats are made of soft sediments deposited along the coast in areas protected
from ocean surf or strong currents. These are long-
term depositional areas intermittently subject to
erosion and export of particles. Salt marsh sediments
are held in place by plant roots and rhizomes
(underground stems). Consequently, marshes are resistant to erosion by all but the strongest storms.
Algal mats and animal burrows bind mud flat sediments, although, even when protected along tidal
creeks within a salt marsh, mud flats are more easily
eroded than the adjacent salt marsh plain.

Salinity in a marsh or mud flat, reported in parts
per thousand (ppt), can range from about 40 ppt
down to 5 ppt. The interaction of the tides and
weather, the salinity of the coastal ocean, and the
elevation of the marsh plain control salinity on a
marsh or mud flat. Parts of the marsh with strong,
regular tides (1 m or more) are flooded twice a day,
and salinity is close to that of the coastal ocean.
Heavy rain at low tide can temporarily make the
surface of the sediment almost fresh. Salinity may
vary seasonally if a marsh is located in an estuary
where the river volume changes over the year. Salinity varies within a marsh with subtle changes in
surface elevation. Higher marshes at sites with
regular tides have variation between spring and neap
tides that result in some areas being flooded every
day while other, higher, areas are flooded less frequently. At higher elevations flooding may occur on
only a few days each spring tide, while at the highest
elevations flooding may occur only a few times a
year.

Some marshes, on coasts with little elevation
change, have their highest parts flooded only seasonally by the equinoctial tides. Other marshes occur
in areas with small lunar tides where flooding is
predominantly wind-driven, such as the marshes in
the lagoons along the Texas coast of the United
States. They are flooded irregularly and, between
flooding, the salinity is greatly raised by evaporation
in the hot, dry climate. The salinity in some of the
higher areas becomes so high that no rooted plants
survive. These are salt flats, high enough in the tidal
regime for higher plants to grow, but so salty that
only salt-resistant algae can grow there. The weather
further affects salinity within marshes and mud flats.
Weather that changes the temperature of coastal
waters or varying atmospheric pressure can change
sea level by 10 cm over periods of weeks to months,
and therefore affect the areas of the marsh that are
subjected to tidal inundation.

Sea level changes gradually. It has been rising since
the retreat of the continental glaciers. The rate of rise
may be increasing with global warming. For the last
10 000 years or so, marshes have been able to keep
up with sea level rise by accumulating sediment, both
through deposition of mud and sand and through
accumulation of peat. The peat comes from the
underground parts of marsh plants that decay slowly
in the anoxic marsh sediments. The result of these
processes is illustrated in **Figure 1**, in which the
basement sediment is overlain by the accumulated
marsh sediment. Keeping up with sea level rise creates a marsh plain that is relatively flat; the elevation
determined by water level rather than by the geological processes that determined the original, basement sediment surface on which the marsh
developed. Tidal creeks, which carry the tidal waters
on and off the marsh, dissect the flat marsh plain.

Organisms

The duration of flooding and the salinities of the
sediments and tidal waters control the mix of higher
vegetation. Competitive interactions between plants
and interactions between plants and animals further
determine plant distributions. Duration of flooding
duration controls how saturated the sediments will
be, which in turn controls how oxygenated or reduced the sediments are. The roots of higher plants
must have oxygen to survive, although many can
survive short periods of anoxia. Air penetrates into
the creekbank sediments as they drain at low tide.

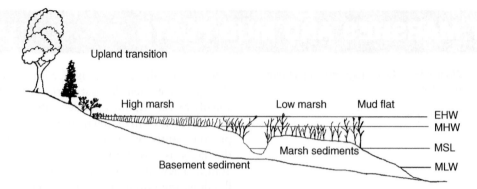

Figure 1 Cartoon of a typical salt marsh of eastern North America. The plants shown are mostly grasses and may differ in other parts of the world. MLW, mean low water; MSL, mean sea level; MHW, mean high water; EHW, extreme high water. The mud flat is shown as a part of the marsh but mud flats also exist independently of marshes.

Evapotranspiration from plants at low tide also removes water from the sediments and facilitates entry of air. Most salt marsh higher plants have aerenchyma (internal air passages) through which oxygen reaches the roots and rhizomes by diffusion or active transport from the above-ground parts. However, they also benefit from availability of oxygen outside the roots.

The species of higher plants that dominate salt marshes vary with latitude, salinity, region of the world, and tidal amplitude. They are composed of relatively few species of plants that have invested in the ability to supply oxygen to roots and rhizomes in reduced sediments and to deal with various levels of salt. Grasses are important, with *Spartina alterniflora* the dominant species from mid-tide to high-tide levels in temperate Eastern North America. *Puccinellia* is a dominant grass in boreal and arctic marshes. The less regularly flooded marshes of East Anglia (UK) support a more diverse vegetation community in which grasses are not dominant. The salty marshes of the Texas coast are covered by salt-tolerant *Salicornia* species. Adjacent to the upper, landward edge of the marsh lie areas flooded only at times when storms drive ocean waters to unusual heights. Some land plants can survive occasional salt baths, but most cannot. An extreme high-water even usually results in the death of plants at the marsh border.

Algae on the marsh and mud flat are less specialized. Depending upon the turbulence of the tidal water, macroalgae (seaweeds) may be present, but a diverse microalgal community is common. Algae live on or near the surface of the sediments and obtain oxygen directly from the air or water and from the oxygen produced by photosynthesis. Their presence on surface sediment is controlled by light. In highly turbid waters they are almost entirely limited to the intertidal flats. In clearer waters, they can grow below low-tide levels. Algae growing on the vegetated marsh plain and on the stems of marsh plants get less light as the plants mature. Production of these algae is greatest in early spring, before the developing vegetation intercepts the light.

Photosynthetic bacteria also contribute to marsh and mud flat production. Blue-green bacteria can be abundant enough to forms mats. Photosynthetic sulfur bacteria occupy a thin stratum in the sediment where they get light from above and sulfide from deeper reduced levels for their hydrogen source but are below the level of oxygen penetration that would kill them. These strange organisms are relicts from the primitive earth before the atmosphere contained oxygen.

Salt marsh animals are from terrestrial and marine sources; mud flat inhabitants are limited to marine sources. Insects, spiders, and mites live in marsh sediments and on marsh plants. Crabs, amphipods, isopods and shrimps, polychaete and oligochaete annelids, snails, and bivalves live in and on the sediments. Most of these marine animals have planktonic larval stages that facilitate movement between marshes and mud flats. Although burrowing animals, such as crabs that live at the water edge of the marsh, may be fairly large (2–15 cm), in general burrowers in marshes are smaller than those in mud flats, presumably because the root mats of the higher plants interfere with burrowing.

Fish are important faunal elements in regularly flooded salt marshes and mud flats. They can be characterized as permanent marsh residents; seasonal residents (species that come into the marsh at the beginning of summer as new post-larvae and live in the marsh until cold weather sets in); species that are primarily residents of coastal waters but enter the marshes at high tide; and predatory fish that come into marshes on the ebb tide to feed on the smaller fishes forced off the marsh plain and out of the smaller creeks by falling water levels.

A few mammals live in the marshes, including those that flee only the highest tides by retreating to land, such as voles, or those that make temporary refuges in tall marsh plants, such as raccoons. The North American muskrat builds permanent houses on the marsh from the marsh plants, although muskrats are typically found only in the less-saline marshes. Grazing mammals feed on marsh plants at low tide. In Brittany, lambs raised on salt marshes are specially valued for the flavor of their meat.

Many species of birds use salt marshes and mud flats. Shore bird species live in the marshes and/or use associated mud flats for feeding during migration. Northern harriers nest on higher portions of salt marshes and feed on their resident voles. Several species of rails dwell in marshes as do bitterns, ducks, and some wrens and sparrows. The nesting species must keep their eggs and young from drowning, which they achieve by building their nests in high vegetation, by building floating nests, and by nesting and raising their young between periods of highest tides.

Functions

Marshes, and to some extent mud flats, produce animals and plants, provide nursery areas for marine fishes, modify nutrient cycles, degrade organic chemicals, immobilize elements within their sediments, and modify wave action on adjacent uplands.

Production and Nursery

Plant production from salt marshes is as high as or higher than that of most other systems because of the ability of muddy sediments to serve as nutrient reservoirs, because of their exposure to full sun, and because of nutrients supplied by sea water. Although the plant production is food for insects, mites and voles, large mammalian herbivores that venture onto the marsh, a few crustacea, and other marine animals, most of the higher plant production is not eaten directly but enters the food web as detritus. As the plants die, they are attacked by fungi and bacteria that reduce them to small particles on the surface of the marsh. Since the labile organic matter in the plants is quickly used as food by the bacteria and fungi, most of the nutrient value of the detritus reaches the next link in the food chain through these microorganisms. These are digested from the plant particles by detritivores, but the cellulose and lignin from the original plants passes through them and is deposited as feces that are recolonized by bacteria and fungi. Besides serving as a food source in the marsh itself, a portion of the detritus–algae mixture is exported by tides to serve as a food source in the marsh creeks and associated estuary.

The primary plant production supports production of animals. Fish production in marshes is high. Resident fishes such as North American *Fundulus heteroclitus* live on the marsh plain during their first summer, survive low tide in tiny pools or in wet mud, feed on the tiny animals living on the detritus–algae–microorganism mix, and grow to migrate into small marsh creeks. At high tide they continue to feed on the marsh plain, where they are joined by the young-of-the-year of those species that use the marsh principally as a nursery area. The warm, shallow waters promote rapid growth and are refuge areas where they are protected from predatory fishes, but not from fish-eating herons and egrets.

The fishes are the most valuable export from the marshes to estuaries and coastal oceans. Some of the fishes are exported in the bodies of predatory fishes that enter the marsh on the ebb tide to feed. Many young fishes, raised in marshes, migrate offshore in the autumn after having spent the summer growing in the marsh.

Nutrient and Element Cycling

Nitrogen is the critical nutrient controlling plant productivity in marshes. Phosphorus is readily available in muddy salt marsh sediments and potassium is sufficiently abundant in sea water. Micronutrients, such as silica or iron, that may be limiting for primary production in deeper waters are abundant in marsh sediments. Thus nitrogen is the nutrient of interest for marsh production and nutrient cycling.

In marshes where nitrogen is in short supply, blue-green bacteria serve as nitrogen fixers, building nitrogen gas from the air into their organic matter. Nitrogen-fixing bacteria associated with the roots of higher plants serve the same function. Nitrogen fixation is an energy-demanding process that is absent where the supply is sufficient to support plant growth.

Two other stages of the nitrogen cycle occur in marshes. Organic nitrogen released by decomposition is in the form of ammonium ion. This can be oxidized to nitrate by certain bacteria that derive energy from the process if oxygen is present. Both nitrate and ammonium can satisfy the nitrogen needs of plants, but nitrate can also serve in place of oxygen for the respiration of another group of bacteria that release nitrogen gas as a by-product in the process called denitrification. Denitrification in salt marshes and mud flats is significant in reducing eutrophication in estuarine and coastal waters.

Phosphorus, present as phosphate, is the other plant nutrient that can be limiting in marshes, especially in regions where nitrogen is in abundant supply. It can also contribute to eutrophication of coastal waters, but phosphate is readily bound to sediments and so tends to be retained in marshes and mud flats rather than released to the estuary.

Sulfur cycling in salt marshes, while of minor importance as a nutrient, contributes to completing the production cycle. Sulfate is the second most abundant anion in sea water. In anoxic sediments, a specialized group of decomposing organisms living on the dead, underground portions of marsh plants can use sulfate as an electron acceptor – an oxygen substitute – in their respiration. The by-product is hydrogen sulfide rather than water. Sulfate reduction yields much less energy than respiration with oxygen or nitrate reduction, so these latter processes occur within the sediment surface, leaving sulfate reduction as the remaining process in deeper parts of the sediments. The sulfide carries much of the free energy not captured by the bacteria in sulfate reduction. As it diffuses to surface layers, most of the sulfide is oxidized by bacteria that grow using it as an energy source. A small amount is used by the photosynthetic sulfur bacteria mentioned above.

Pollution

Marshes, like the estuaries with which they tend to be associated, are depositional areas. They tend to accumulate whatever pollutants are dumped into coastal waters, especially those bound to particles. Much of the pollution load enters the coast transported by rivers and may originate far from the affected marshes. For example, much of the nitrogen and pesticide loading of marshes and coastal waters of the Mississippi Delta region of the United States comes from farming regions hundreds or thousands of kilometers upstream.

Many pollutants, both organic and inorganic, bind to sediments and are retained by salt marshes and mud flats. Organic compounds are often degraded in these biologically active systems, especially since many of them are only metabolized when the microorganisms responsible are actively growing on other, more easily degraded compounds. There are, unfortunately, some organics, the structures of which are protected by constituents such as chlorine, that are highly resistant to microbial attack. Some polychlorinated biphenyls (PCBs) have such structures, with the result that a PCB mixture will gradually lose the degradable compounds while the resistant components will become relatively more concentrated.

Metals are also bound to sediments and so may be removed and retained by marshes and mud flats. Mercury is sequestered in the sediment, while cadmium forms soluble complexes with chloride in sea water and is, at most, temporarily retained.

Since marshes and mud flats tend, in the long term, to be depositional systems, they remove pollutants and bury them as long as the sediments are not remobilized by erosion. Since mud flats are more easily eroded than marsh sediments held in place by plant roots and rhizomes, they are less secure long-term storage sites.

Storm Damage Prevention

While marshes and mud flats exist only in relatively protected situations, they are still subject to storm damage as are the uplands behind them. During storms, the shallow waters and the vegetation on the marshes offer resistance to water flow, making them places where wave forces are dissipated, reducing the water and wave damage to the adjoining upland.

Human Modifications

Direct Effects

Many marshes and mud flats in urban areas have been highly altered or destroyed by filling or by dredging for harbor, channel or marina development. Less intrusive actions can have large impacts. Since salt marshes and mud flats typically lie in indentations along the coast, the openings where tides enter and leave them are often sites of human modification for roads and railroads. Both culverts and bridges restrict flow if they are not large enough. Flow is especially restricted at high water unless the bridge spans the entire marsh opening, a rare situation because it is expensive. The result of restriction is a reduction in the amount of water that floods the marsh. The plants are submerged for a shorter period and to a lesser depth, and the floodwaters do not extend as far onto the marsh surface. The ebb flow is also restricted and the marsh may not drain as efficiently as in the unimpeded case. Poor drainage could freshen the marsh after a heavy rain and run-off. Less commonly, it could increase salinity after an exceptionally prolonged storm-driven high tide.

The result of the disturbance will be a change in the oxygen and salinity relations between roots and sediments. Plants may become oxygen-stressed and drown. Tidal restrictions in moist temperate regions usually result in a freshening of the sediment salinity. This favors species that have not invested in salt control mechanisms of the typical salt marsh plants. A widespread result in North America has been the spread of the common reed, *Phragmites australis*, a brackish-water and freshwater species. Common reed is a tall (3 m) and vigorous plant that can spread horizontally by rhizomes at 10 meters per year. Its robust stem decomposes more slowly than that of the salt marsh cord grass, *Spartina alterniflora* and as a result, it takes over a marsh freshened by tidal restrictions. Since its stems accumulate above ground and rhizomes below ground, it tends to raise the marsh level, fill in the small drainage channels, and reduce the value of the marsh for fish and wildlife. Although *Phragmites australis* is a valuable plant for many purposes (it is the preferred plant for thatching roofs in Europe), its takeover of salt marshes is considered undesirable.

The ultimate modification of tidal flow is restriction by diking. Some temperate marshes have been diked to allow the harvest of salt hay, valued as mulch because it lacks weed seeds. Since some diked marshes are periodically flooded in an attempt to maintain the desired vegetation, they are not completely changed and can be restored. Other marshes, such as those in Holland, have been diked and removed from tidal flow so that the land may be used for upland agriculture. Many marshes and mud flats have been modified to create salt pans for production of sea salt and for aquaculture. The latter is a greater problem in the tropics, where the impact is on mangroves rather than on the salt marshes of more temperate regions.

Indirect Effects

Upland diking The upper borders of coastal marshes were often diked to prevent upland flooding. People built close to the marshes to take advantage of the view. With experience they found that storms could raise the sea level enough to flood upland. The natural response was to construct a barrier to prevent flooding. Roads and railroads along the landward edges of marshes are also barriers that restrict upland flooding. They are built high enough to protect the roadbed from most flooding and usually have only enough drainage to allow rain runoff to pass to the sea. In both cases, the result is a barrier to landward migration of the marsh. As the relative sea level rises, sandy barriers that protect coastal marshes are flooded and, during storms, the sand is washed onto the marsh. As long as the marsh can also move back by occupying the adjacent upland it may be able to persist without loss in area, but if a barrier prevents landward transgression the marsh will be squeezed between the barrier and the rising sea and will eventually disappear. During this process, the drainage

structures under the barrier gradually become flow restrictors. The sea will flood the land behind the barrier through the culverts, but these are inevitably too small to permit unrestricted marsh development. When flooding begins, the culverts are typically fitted with tide gates to prevent whatever flooding and marsh development could be accommodated by the capacity of the culverts.

Changes in sediment loading Increases in sediment supplies can allow the marshes to spread as the shallow waters bordering them are filled in. The plant stems further impede water movement and enhance spread of the marsh. This assumes that storms do not carry the additional sediment onto the marsh plain and raise it above normal tidal level, which would damage or destroy rather than extend the marsh.

Reduced sediment supply can destroy a marsh. In a river delta where sediments gradually de-water and consolidate, sinking continually, a continuous supply of new sediment combined with vegetation remains, accumulating as peat, and maintains the marsh level. When sediment supplies are cut off, the peat accumulation may be insufficient to maintain the marsh at sea level. Dams, such as the Aswan Dam on the Nile, can trap sediments. Sediments can be channeled by levées so that they flow into deep water at the mouth of the river rather than spreading over the delta marshes, as is happening in the Mississippi River delta. In the latter case, the coastline of Louisiana is retreating by kilometers a year as a result of the loss of delta marshlands.

Introduction of foreign species Dramatic changes in the marshes and flats of England and Europe occurred after *Spartina alterniflora* was introduced from the east coast of North America and probably hybridized with the native *S. maritima* to produce *S. anglica*. The new species was more tolerant of submergence than the native forms and turned many mud flats into salt marshes. This change reduced populations of mud flat animals, many with commercial value, and reduced the foraging area for shore birds that feed on mud flats. A similar situation has developed in the last decades on the north-west coast of the United States, where introduced *Spartina alterniflora* is invading mud flats and reducing the available area for shellfish.

Marsh restoration Salt marshes and mud flats may be the most readily restored of all wetlands. The source and level of water is known. The vascular plants that will thrive are known and can be planted if a local seed source is not available. Many of the marsh animals have planktonic larvae that can invade the restored marsh on their own. Although many of the properties of a mature salt marsh take time to develop, such as the nutrient-retaining capacity of the sediments, these will develop if the marsh is allowed to survive.

Further Reading

Adam P (1990) *Salt Marsh Ecology*. Cambridge: Cambridge University Press.

Mitsch WJ and Gosselink JG (1993) *Wetlands*. New York: Wiley.

Peterson CH and Peterson NM (1972) *The Ecology of Intertidal Flats of North Carolina: A Community Profile*. Washington, DC: US Fish and Wildlife Service, Office of Biological Services, FWS/OBS-79/39.

Streever W (1999) *An International Perspective on Wetland Rehabilitation*. Dordrecht: Kluwer.

Teal JM and Teal M (1969) *Life and Death of the Salt Marsh*. New York: Ballentine.

Weinstein MP and Kreeger DA (2001) *Concepts and Controversies in Tidal Marsh Ecology*. Dordrecht: Kluwer.

Whitlatch RB (1982) *The Ecology of New England Tidal Flats: A Community Profile*. Washington, DC: US Fish and Wildlife Service, Biological Services Program, FES/OBS-81/01.

GLACIAL CRUSTAL REBOUND, SEA LEVELS AND SHORELINES

K. Lambeck, Australian National University, Canberra, ACT, Australia

Introduction

Geological, geomorphological, and instrumental records point to a complex and changing relation between land and sea surfaces. Elevated coral reefs or wave-cut rock platforms indicate that in some localities sea levels have been higher in the past, while observations elsewhere of submerged forests or flooded sites of human occupation attest to levels having been lower. Such observations are indicators of the relative change in the land and sea levels: raised shorelines are indicative of land having been uplifted or of the ocean volume having decreased, while submerged shorelines are a consequence of land subsidence or of an increase in ocean volume. A major scientific goal of sea-level studies is to separate out these two effects.

A number of factors contribute to the instability of the land surfaces, including the tectonic upheavals of the crust emanating from the Earth's interior and the planet's inability to support large surface loads of ice or sediments without undergoing deformation. Factors contributing to the ocean volume changes include the removal or addition of water to the oceans as ice sheets wax and wane, as well as addition of water into the oceans from the Earth's interior through volcanic activity. These various processes operate over a range of timescales and sea level fluctuations can be expected to fluctuate over geological time and are recorded as doing so.

The study of such fluctuations is more than a scientific curiosity because its outcome impacts on a number of areas of research. Modern sea level change, for example, must be seen against this background of geologically–climatologically driven change before contributions arising from the actions of man can be securely evaluated. In geophysics, one outcome of the sea level analyses is an estimate of the viscosity of the Earth, a physical property that is essential in any quantification of internal convection and thermal processes. Glaciological modeling of the behavior of large ice sheets during the last cold period is critically dependent on independent constraints on ice volume, and this can be extracted from the sea level information. Finally, as sea level rises and falls, so the shorelines advance and retreat. As major sea level changes have occurred during critical periods of human development, reconstructions of coastal regions are an important part in assessing the impact of changing sea levels on human movements and settlement.

Tectonics and Sea Level Change

Major causes of land movements are the tectonic processes that have shaped the planet's surface over geological time. Convection within the high-temperature viscous interior of the Earth results in stresses being generated in the upper, cold, and relatively rigid zone known as the lithosphere, a layer some 50–100 km thick that includes the crust. This convection drives plate tectonics — the movement of large parts of the lithosphere over the Earth's surface — mountain building, volcanism, and earthquakes, all with concomitant vertical displacements of the crust and hence relative sea level changes. The geological record indicates that these processes have been occurring throughout much of the planet's history. In the Andes, for example, Charles Darwin identified fossil seashells and petrified pine trees trapped in marine sediments at 4000 m elevation. In Papua New Guinea, 120 000-year-old coral reefs occur at elevations of up to 400 m above present sea level (**Figure 1**).

One of the consequences of the global tectonic events is that the ocean basins are being continually reshaped as mid-ocean ridges form or as ocean floor collides with continents. The associated sea level changes are global but their timescale is long, of the order 10^7 to 10^8 years, and the rates are small, less than 0.01 mm per year. **Figure 2** illustrates the global sea level curve inferred for the past 600 million years from sediment records on continental margins. The long-term trends of rising and falling sea levels on timescales of 50–100 million years are attributed to these major changes in the ocean basin configurations. Superimposed on this are smaller-amplitude and shorter-period oscillations that reflect more regional processes such as large-scale volcanism in an ocean environment or the collision of continents. More locally, land is pushed up or down episodically

Figure 1 Raised coral reefs from the Huon Peninsula, Papua New Guinea. In this section the highest reef indicated (point 1) is about 340 m above sea level and is dated at about 125 000 years old. Elsewhere this reef attains more than 400 m elevation. The top of the present sea cliffs (point 2) is about 7000 years old and lies at about 20 m above sea level. The intermediate reef tops formed at times when the rate of tectonic uplift was about equal to the rate of sea level rise, so that prolonged periods of reef growth were possible. Photograph by Y. Ota.

in response to the deeper processes. The associated vertical crustal displacements are rapid, resulting in sea level rises or falls that may attain a few meters in amplitude, but which are followed by much longer periods of inactivity or even a relaxation of the original displacements. The raised reefs illustrated in **Figure 1**, for example, are the result of a large number of episodic uplift events each of typically a meter amplitude. Such displacements are mostly local phenomena, the Papua New Guinea example extending only for some 100–150 km of coastline.

The episodic but local tectonic causes of the changing position between land and sea can usually be identified as such because of the associated seismic activity and other tell-tale geological signatures. The development of geophysical models to describe these local vertical movements is still in a state of infancy

and, in any discussion of global changes in sea level, information from such localities is best set aside in favor of observations from more stable environments.

Glacial Cycles and Sea Level Change

More important for understanding sea level change on human timescales than the tectonic contributions – important in terms of rates of change and in terms of their globality – is the change in ocean volume driven by cyclic global changes in climate from glacial to interglacial conditions. In Quaternary time, about the last two million years, glacial and interglacial conditions have followed each other on timescales of the order of 10^4–10^5 years. During interglacials, climate conditions were similar to those of today and sea levels were within a few meters of their present day position. During the major glacials, such as 20 000 years ago, large ice sheets formed in the northern hemisphere and the Antarctic ice sheet expanded, taking enough water out of the oceans to lower sea levels by between 100 and 150 m. **Figure 3** illustrates the changes in global sea level over the last 130 000 years, from the last interglacial, the last time that conditions were similar to those of today, through the Last Glacial Maximum and to the present. At the end of the last interglacial, at 120 000 years ago, climate began to fluctuate; increasingly colder conditions were reached, ice sheets over North America and Europe became more or less permanent features of the landscape, sea levels reached progressively lower values, and large parts of today's coastal shelves were exposed. Soon after the culmination of maximum glaciation, the ice sheets again disappeared, within a period of about

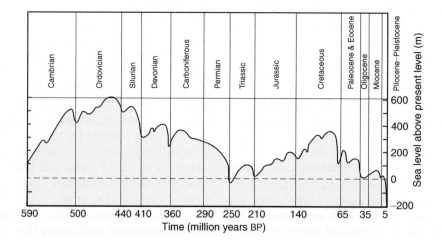

Figure 2 Global sea level variations through the last 600 million years estimated from seismic stratigraphic studies of sediments deposited at continental margins. Redrawn with permission from Hallam A (1984) Pre-Quaternary sea-level changes. *Annual Review of Earth and Planetary Science* 12: 205–243.

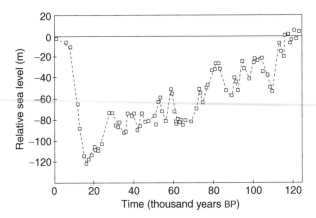

Figure 3 Global sea level variations (relative to present) since the time of the last interglacial 120 000–130 000 years ago when climate and environmental conditions were last similar to those of the last few thousand years. Redrawn with permission from Chapell J *et al.* (1996) Reconciliation of Late Quaternary sea level changes derived from coral terraces at Huon Peninsula with deep sea oxygen isotope records. *Earth and Planetary Science Letters* 141b: 227–236.

10 000 years, and climate returned to interglacial conditions.

The global changes illustrated in **Figure 3** are only one part of the sea level signal because the actual ice–water mass exchange does not give rise to a spatially uniform response. Under a growing ice sheet, the Earth is stressed; the load stresses are transmitted through the lithosphere to the viscous underlying mantle, which begins to flow away from the stressed area and the crust subsides beneath the ice load. When the ice melts, the crust rebounds. Also, the meltwater added to the oceans loads the seafloor and the additional stresses are transmitted to the mantle, where they tend to dissipate by driving flow to unstressed regions below the continents. Hence the seafloor subsides while the interiors of the continents rise, causing a tilting of the continental margins. The combined adjustments to the changing ice and water loads are called the glacio- and hydro-isostatic effects and together they result in a complex pattern of spatial sea level change each time ice sheets wax and wane.

Observations of Sea Level Change Since the Last Glacial Maximum

Evidence for the positions of past shorelines occurs in many forms. Submerged freshwater peats and tree stumps, tidal-dwelling mollusks, and archaeological sites would all point to a rise in relative sea level since the time of growth, deposition, or construction. Raised coral reefs, such as in **Figure 1**, whale bones cemented in beach deposits, wave-cut rock platforms and notches, or peats formed from saline-loving plants would all be indicative of a falling sea level since the time of formation. To obtain useful sea level measurements from these data requires several steps: an understanding of the relationship between the feature's elevation and mean sea level at the time of growth or deposition, a measurement of the height or depth with respect to present sea level, and a measurement of the age. All aspects of the observation present their own peculiar problems but over recent years a substantial body of observational evidence has been built up for sea level change since the last glaciation that ended at about 20 000 years ago. Some of this evidence is illustrated in **Figure 4**, which also indicates the very significant spatial variability that may occur even when, as is the case here, the evidence is from sites that are believed to be tectonically stable.

In areas of former major glaciation, raised shorelines occur with ages that are progressively greater with increasing elevation and with the oldest shorelines corresponding to the time when the region first became ice-free. Two examples, from northern Sweden and Hudson Bay in Canada, respectively, are illustrated in **Figure 4A**. In both cases sea level has been falling from the time the area became ice-free and open to the sea. This occurred at about 9000 years ago in the former case and about 7000 years later in the second case. Sea level curves from localities just within the margins of the former ice sheet are illustrated in **Figure 4B**, from western Norway and Scotland, respectively. Here, immediately after the area became ice-free, the sea level fell but a time was reached when it again rose. A local maximum was reached at about 6000 years ago, after which the fall continued up until the present. Farther away from the former ice margins the observed sea level pattern changes dramatically, as is illustrated in **Figure 4C**. Here the sea level initially rose rapidly but the rate decreased for the last 7000 years up to the present. The two examples illustrated, from southern England and the Atlantic coast of the United States, are representative of tectonically stable localities that lie within a few thousand kilometers from the former centers of glaciation. Much farther away again from the former ice sheets the sea level signal undergoes a further small but significant change in that the present level was first reached at about 6000 years ago and then exceeded by a small amount before returning to its present value. The two examples illustrated in **Figure 4D** are from nearby localities in northern Australia. At both sites present sea level was reached about 6000 years ago after a prolonged period of rapid rise with resulting highstands that are small in amplitude but geographically variable.

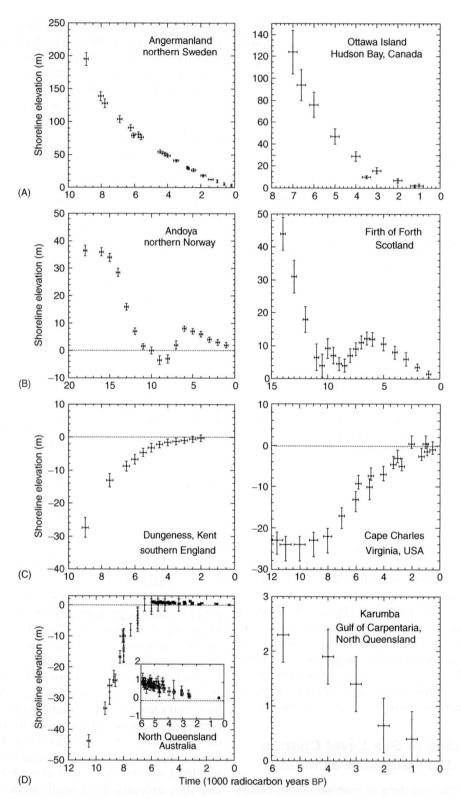

Figure 4 Characteristic sea level curves observed in different localities (note the different scales used). (A) From Ångermanland, Gulf of Bothnia, Sweden, and from Hudson Bay, Canada. Both sites lie close to centers of former ice sheet, over northern Europe and North America respectively. (B) From the Atlantic coast of Norway and the west coast of Scotland. These sites are close to former ice-sheet margins at the time of the last glaciation. (C) From southern England and Virginia, USA. These sites lie outside the areas of former glaciation and at a distance from the margins where the crust is subsiding in response to the melting of the nearby icesheet. (D) Two sites in northern Australia, one for the Coral Sea coast of Northern Queensland and the second from the Gulf of Carpentaria some 200 km away but on the other side of the Cape York Peninsula. At both localities sea level rose until about 6000 years ago before peaking above present level.

The examples illustrated in **Figure 4** indicate the rich spectrum of variation in sea level that occurred when the last large ice sheets melted. Earlier glacial cycles resulted in a similar spatial variability, but much of the record before the Last Glacial Maximum has been overwritten by the effects of the last deglaciation.

Glacio-Hydro-Isostatic Models

If, during the decay of the ice sheets, the meltwater volume was distributed uniformly over the oceans, then the sea level change at time t would be

$$\Delta\zeta_e(t) = (\text{change in ice volume/ocean surface area})$$
$$\times \rho_i/\rho_w \qquad [1a]$$

where ρ_i and ρ_w are the densities of ice and water, respectively. (See end of article – symbols used.) This term is the ice-volume equivalent sea level and it provides a measure of the change in ice volume through time. Because the ocean area changes with time a more precise definition is

$$\Delta\zeta_e(t) = \frac{dV_i}{dt}dt\frac{\rho_i}{\rho_0}\int_t\frac{\rho_i}{A_0(t)} \qquad [1b]$$

where A_0 is the area of the ocean surface, excluding areas covered by grounded ice. The sea level curve illustrated in **Figure 3** is essentially this function. However, it represents only a zero-order approximation of the actual sea level change because of the changing gravitational field and deformation of the Earth.

In the absence of winds or ocean currents, the ocean surface is of constant gravitational potential. A planet of a defined mass distribution has a family of such surfaces outside it, one of which – the geoid – corresponds to mean sea level. If the mass distribution on the surface (the ice and water) or in the interior (the load-forced mass redistribution in the mantle) changes, so will the gravity field, the geoid, and the sea level change. The ice sheet, for example, represents a large mass that exerts a gravitational pull on the ocean and, in the absence of other factors, sea level will rise in the vicinity of the ice sheet and fall farther away. At the same time, the Earth deforms under the changing load, with two consequences: the land surface with respect to which the level of the sea is measured is time-dependent, as is the gravitational attraction of the solid Earth and hence the geoid.

The calculation of the change of sea level resulting from the growth or decay of ice sheets therefore involves three steps: the calculation of the amount of water entering into the ocean and the distribution of this meltwater over the globe; the calculation of deformation of the Earth's surface; and the calculation of the change in the shape of the gravitational equipotential surfaces. In the absence of vertical tectonic motions of the Earth's surface, the relative sea level change $\Delta\zeta(\varphi,t)$ at a site φ and time t can be written schematically as

$$\Delta\zeta(\varphi,t) = \Delta\zeta_e(t) + \Delta\zeta_I(\varphi,t) \qquad [2]$$

where $\Delta\zeta_I(\varphi, t)$ is the combined perturbation from the uniform sea level rise term [1]. This is referred to as the isostatic contribution to relative sea level change.

In a first approximation, the Earth's response to a global force is that of an elastic outer spherical layer (the lithosphere) overlying a viscous or viscoelastic mantle that itself contains a fluid core. When subjected to an external force (e.g., gravity) or a surface load (e.g., an ice cap), the planet experiences an instantaneous elastic deformation followed by a time-dependent or viscous response with a characteristic timescale(s) that is a function of the viscosity. Such behavior of the Earth is well documented by other geophysical observations: the gravitational attraction of the Sun and Moon raises tides in the solid Earth; ocean tides load the seafloor with a time-dependent water load to which the Earth's surface responds by further deformation; atmospheric pressure fluctuations over the continents induce deformations in the solid Earth. The displacements, measured with precision scientific instruments, have both an elastic and a viscous component, with the latter becoming increasingly important as the duration of the load or force increases. These loads are much smaller than the ice and water loads associated with the major deglaciation, the half-daily ocean tide amplitudes being only 1% of the glacial sea level change, and they indicate that the Earth will respond to even small changes in the ice sheets and to small additions of meltwater into the oceans.

The theory underpinning the formulation of planetary deformation by external forces or surface loads is well developed and has been tested against a range of different geophysical and geological observations. Essentially, the theory is one of formulating the response of the planet to a point load and then integrating this point-load solution over the load through time, calculating at each epoch the surface deformation and the shape of the equipotential surfaces, making sure that the meltwater is appropriately distributed into the oceans and that the total ice–water mass is preserved. Physical inputs into the formulation are the time–space history of the ice sheets, a description of the ocean basins from which the water is extracted or into which it is added, and a description of the rheology, or response parameters, of the Earth. For the last requirement, the elastic properties of the Earth, as well as the density distribution with depth, have been determined from seismological and geodetic studies. Less well determined are the viscous properties of the mantle, and the usual procedure is to adopt a simple parametrization of the viscosity structure and to estimate the relevant parameters from analyses of the sea level change itself.

The formulation is conveniently separated into two parts for schematic reasons: the glacio-isostatic effect representing the crustal and geoid displacements due to the ice load, and the hydro-isostatic effect due to the water load. Thus

$$\Delta\zeta(\varphi,t) = \Delta\zeta_e(t) + \Delta\zeta_i(\varphi,t) + \Delta\zeta_w(\varphi,t) \qquad [3]$$

where $\Delta\zeta_i(\varphi,t)$ is the glacio-isostatic and $\Delta\zeta_w(\varphi,t)$ the hydro-isostatic contribution to sea level change. (In reality the two are coupled and this is included in the formulation). If $\Delta\zeta_i(\varphi,t)$ is evaluated everywhere, the past water depths and land elevations $H(\varphi,t)$ measured with respect to coeval sea level are given by

$$H(\varphi,t)H_0(\varphi) - \Delta\zeta(\varphi,t) \qquad [4]$$

where $H(\phi)$ is the present water depth or land elevation at location φ.

A frequently encountered concept is eustatic sea level, which is the globally averaged sea level at any time t. Because of the deformation of the seafloor during and after the deglaciation, the isostatic term $\Delta\zeta(\varphi,t)$ is not zero when averaged over the ocean at any time t, so that the eustatic sea level change is

$$\Delta\zeta_{eus}(t) = \Delta\zeta_e(t) + \langle\Delta\zeta_I\{\varphi,t\}\rangle_0$$

where the second term on the right-hand side denotes the spatially averaged isostatic term. Note that $\Delta\zeta_e(t)$ relates directly to the ice volume, and not $\Delta\zeta_{eus}(t)$.

The Anatomy of the Sea Level Function

The relative importance of the two isostatic terms in eqn. [3] determines the spatial variability in the sea level signal. Consider an ice sheet of radius that is much larger than the lithospheric thickness: the limiting crustal deflection beneath the center of the load is $I\rho_i/I\rho_m$ where I is the maximum ice thickness and $I\rho_m$ is the upper mantle density. This is the local isostatic approximation and it provides a reasonable approximation of the crustal deflection if the loading time is long compared with the relaxation time of the mantle. Thus for a 3 km thick ice sheet the maximum

deflection of the crust can reach 1 km, compared with a typical ice volume for a large ice sheet that raises sea level globally by 50–100 m. Near the centers of the formerly glaciated regions it is the crustal rebound that dominates and sea level falls with respect to the land.

This is indeed observed, as illustrated in **Figure 4A**, and the sea level curve here consists of essentially the sum of two terms, the major glacio-isostatic term and the minor ice-volume equivalent sea level term $\Delta\zeta_e(t)$ (**Figure 5A**). Of note is that the rebound continues long after all ice has disappeared, and this is evidence that the mantle response includes a memory of the earlier load. It is the decay time of this part of the curve that determines the mantle viscosity. As the ice margin is approached, the local ice thickness becomes less and the crustal rebound is reduced and at some stage is equal, but of opposite sign, to $\Delta\zeta_e(t)$. Hence sea level is constant for a period (**Figure 5B**) before rising again when the rebound becomes the minor term. After global melting has ceased, the dominant signal is the late stage of the crustal rebound and levels fall up to the present. Thus the oscillating sea level curves observed in areas such as Norway and Scotland (**Figure 4B**) are essentially the sum of two effects of similar magnitude but opposite sign. The early part of the observation contains information on earth rheology as well as on the local ice thickness and the globally averaged rate of addition of meltwater into the oceans. Furthermore, the secondary maximum is indicative of the timing of the end of global glaciation and the latter part of the record is indicative mainly of the mantle response.

At the sites beyond the ice margins it is the meltwater term $\Delta\zeta_e(t)$ that is important, but it is not the sole factor. When the ice sheet builds up, mantle material flows away from the stressed mantle region and, because flow is confined, the crust around the periphery of the ice sheet is uplifted. When the ice sheet melts, subsidence of the crust occurs in a broad zone peripheral to the original ice sheet and at these locations the isostatic effect is one of an apparent subsidence of the crust. This is illustrated in **Figure 5C**. Thus, when the ocean volumes have stabilized, sea level continues to rise, further indicating that the planet responds viscously to the changing surface loads. The early part of the observational record (e.g., **Figure 4C**) is mostly indicative of the rate at which meltwater is added into the ocean, whereas the latter part is more indicative of mantle viscosity.

In all of the examples considered so far it is the glacial-rebound that dominates the total isostatic adjustment, the hydro-isostatic term being present but comparatively small. Consider an addition of water that raises sea level by an amount D. The local

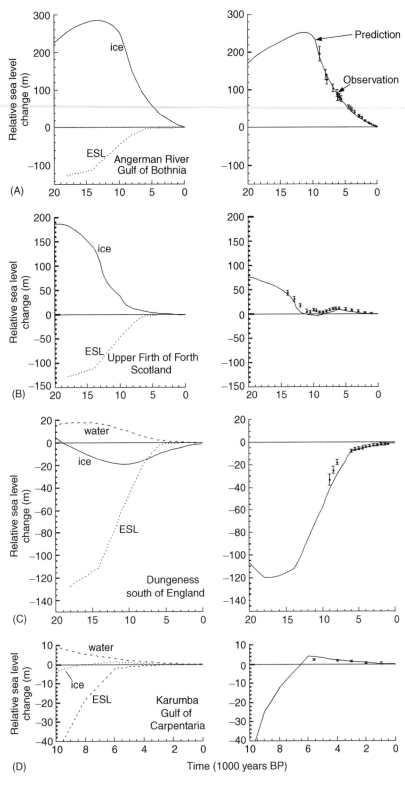

Figure 5 Schematic representation of the sea level curves in terms of the ice-volume equivalent function $\Delta\zeta_e(t)$ (denoted by ESL) and the glacio- and hydro-isostatic contributions $\Delta\zeta_i(\phi, t)$, $\Delta\zeta_w(\phi, t)$ (denoted by ice and water, respectively). The panels on the left indicate the predicted individual components and the panels on the right indicate the total predicted change compared with the observed values (data points). (A) For the sites at the ice center where $\Delta\zeta_i(\phi, t) \gg \Delta\zeta_e(t) \gg \Delta\zeta_w(\phi, t)$. (B) For sites near but within the ice margin where $|\Delta\zeta_i(\phi, t)| \sim |\Delta\zeta_e(t)| \gg |\Delta\zeta_w(\phi, t)|$, but the first two terms are of opposite sign. (C) For sites beyond the ice margin where $|\Delta\zeta_e(t)| > |\Delta\zeta_i(\phi, t)| \gg |\Delta\zeta_w(\phi, t)|$. (D) For sites at continental margins far from the former ice sheets where $|\Delta\zeta_w(\phi, t)| > |\Delta\zeta_i(t)|$. Adapted with permission from Lambeck K and Johnston P (1988). The viscosity of the mantle: evidence from analyses of glacial rebound phenomena. In: Jackson ISN (ed.) *The Earth's Mantle – Composition, Structure and Evolution*, pp. 461–502. Cambridge: Cambridge University Press.

isostatic response to this load is $D\rho_w/\rho_m$, where ρ_w is the density of ocean water. This gives an upper limit to the amount of subsidence of the sea floor of about 30 m for a 100 m sea level rise. This is for the middle of large ocean basins and at the margins the response is about half as great. Thus the hydro-isostatic effect is significant and is the dominant perturbing term at margins far from the former ice sheets. This occurs at the Australian margin, for example, where the sea level signal is essentially determined by $\Delta\zeta_e(t)$ and the water-load response (**Figure 4D**). Up to the end of melting, sea level is dominated by $\Delta\zeta_e(t)$ but thereafter it is determined largely by the water-load term $\Delta\zeta_w(\varphi, t)$, such that small highstands develop at 6000 years. The amplitudes of these highstands turn out to be strongly dependent on the geometry of the water-load distribution around the site: for narrow gulfs, for example, their amplitude increases with distance from the coast and from the water load at rates that are particularly sensitive to the mantle viscosity.

While the examples in **Figure 5** explain the general characteristics of the global spatial variability of the sea level signal, they also indicate how observations of such variability are used to estimate the physical quantities that enter into the schematic model (3). Thus, observations near the center of the ice sheet partially constrain the mantle viscosity and central ice thickness. Observations from the ice sheet margin partially constrain both the viscosity and local ice thickness and establish the time of termination of global melting. Observations far from the ice margins determine the total volumes of ice that melted into the oceans as well as providing further constraints on the mantle response. By selecting data from different localities and time intervals, it is possible to estimate the various parameters that underpin the sea level eqn. [3] and to use these models to predict sea level and shoreline change for unobserved areas. Analyses of sea-level change from different regions of the world lead to estimates for the lithospheric thickness of between 60–100 km, average upper mantle viscosity (from the base of the lithosphere to a depth of 670 km) of about $(1–5)10^{20}$ Pa s and an average lower mantle viscosity of about $(1–5)10^{22}$ Pa s. Some evidence exists that these parameters vary spatially; lithosphere thickness and upper mantle viscosity being lower beneath oceans than beneath continents.

Sea Level Change and Shoreline Migration: Some Examples

Figure 6 illustrates the ice-volume equivalent sea level function $\Delta\zeta_e(t)$ since the time of the last maximum glaciation. This curve is based on sea level indicators from a number of localities, all far from the former ice sheets, with corrections applied for the isostatic effects. The right-hand axis indicates the corresponding change in ice volume (from the relation [1b]). Much of this ice came from the ice sheets over North America and northern Europe but a not insubstantial part also originated from Antarctica. Much of the melting of these ice sheets occurred within 10 000 years, at times the rise in sea level exceeding 30 mm per year, and by 6000 years ago most of the deglaciation was completed. With this sea level function, individual ice sheet models, the formulation for the isostatic factors, and a knowledge of the topography and bathymetry of the world, it becomes possible to reconstruct the paleo shorelines using the relation [4].

Scandinavia is a well-studied area for glacial rebound and sea level change since the time the ice retreat began about 18 000 years ago. The observational evidence is quite plentiful and a good record of ice margin retreat exists. **Figure 7** illustrates examples for two epochs. The first (**Figure 7A**), at 16 000 years ago, corresponds to a time after onset of deglaciation. A large ice sheet existed over

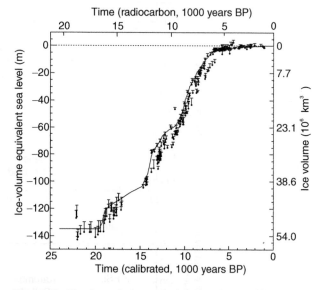

Figure 6 The ice-volume equivalent function $\Delta\zeta_e(t)$ and ice volumes since the time of the last glacial maximum inferred from corals from Barbados, from sediment facies from north-western Australia, and from other sources for the last 7000 years. The actual sea level function lies at the upper limit defined by these observations (continuous line). The upper time scale corresponds to the radiocarbon timescale and the lower one is calibrated to calendar years. Adapted with permission from Fleming K *et al.* (1998) Refining the eustatic sea-level curve since the LGM using the far- and intermediate-field sites. *Earth and Planetary Science Letters* 163: 327–342; and Yokoyama Y *et al.* (2000) Timing of the Last Glacial Maximum from observed sea-level minimum. *Nature* 406: 713–716.

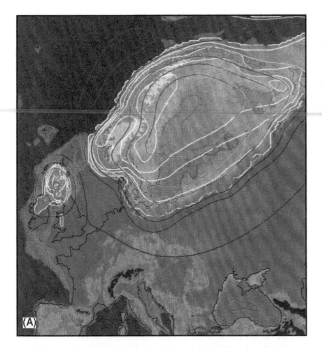

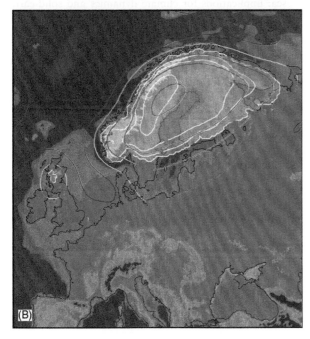

Figure 7   Shoreline reconstructions for Europe (A) at 16 000 and (B) at 10 500 years ago. The contours are at 400 m intervals for the ice thickness (white) and 25 m for the water depths less than 100 m. The orange, yellow, and red contours are the predicted lines of equal sea level change from the specified epoch to the present and indicate where shorelines of these epochs could be expected if conditions permitted their formation and preservation. The zero contour is in yellow; orange contours, at intervals of 100 m, are above present, and red contours, at 50 m intervals, are below present. At 10 500 years ago the Baltic is isolated from the Atlantic and its level lies about 25 m above that of the latter.

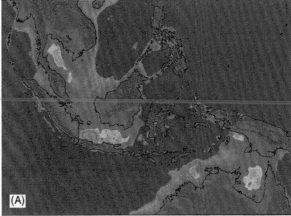

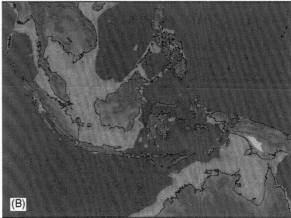

Figure 8   Shoreline reconstructions for South East Asia and northern Australia at the time of the last glacial maximum, (A) at about 18 000 years ago and (B) at 12 000 years ago. The water depth contours are the same as in **Figure 7**. The water depths in the inland lakes at 18 000 years ago are contoured at 25 m intervals relative to their maximum levels that can be attained without overflowing.

Scandinavia with a smaller one over the British Isles. Globally sea level was about 110 m lower than now and large parts of the present shallow seas were exposed, for example, the North Sea and the English Channel but also the coastal shelf farther south such as the northern Adriatic Sea. The red and orange contours indicate the sea level change between this period and the present. Beneath the ice these rebound contours are positive, indicating that if shorelines could form here they would be above sea level today. Immediately beyond the ice margin, a broad but shallow bulge develops in the topography, which will subside as the ice sheet retreats. At the second epoch selected (**Figure 7B**) 10 500 years ago, the ice has retreated and reached a temporary halt as the climate briefly returned to colder conditions; the Younger Dryas time of Europe. Much of the Baltic was then ice-free and a freshwater lake developed at some

25–30 m above coeval sea level. The flooding of the North Sea had begun in earnest. By 9000 years ago, most of the ice was gone and shorelines began to approach their present configuration.

The sea level change around the Australian margin has also been examined in some detail and it has been possible to make detailed reconstructions of the shoreline evolution there. **Figure 8** illustrates the reconstructions for northern Australia, the Indonesian islands, and the Malay–IndoChina peninsula. At the time of the Last Glacial Maximum much of the shallow shelves were exposed and deeper depressions within them, such as in the Gulf of Carpentaria or the Gulf of Thailand, would have been isolated from the open sea. Sediments in these depressions will sometimes retain signatures of these pre-marine conditions and such data provide important constraints on the models of sea level change. Part of the information illustrated in **Figure 6**, for example, comes from the shallow depression on the Northwest Shelf of Australia. By 12 000 years ago the sea has begun its encroachment of the shelves and the inland lakes were replaced by shallow seas.

Symbols used

$\Delta\zeta_e(t)$ Ice-volume equivalent sea level. Uniform change in sea level produced by an ice volume that is distributed uniformly over the ocean surface.

$\Delta\zeta_I(t)$ Perturbation in sea level due to glacioisostatic $\Delta\zeta_i(t)$ and hydro-isostatic $\Delta\zeta_w(t)$ effects.

$\Delta\zeta_{eus}(t)$ Eustatic sea level change. The globally averaged sea level at time t.

See also

Beaches, Physical Processes Affecting. Coral Reefs. Geomorphology. Rocky Shores. Salt Marshes and Mud Flats.

Further Reading

Lambeck K (1988) *Geophysical Geodesy: The Slow Deformations of the Earth*. Oxford: Oxford University Press.

Lambeck K and Johnston P (1999) The viscosity of the mantle: evidence from analysis of glacial-rebound phenomena. In: Jackson ISN (ed.) *The Earth's Mantle – Composition, Structure and Evolution*, pp. 461–502. Cambridge: Cambridge University Press.

Lambeck K, Smither C, and Johnston P (1998) Sea-level change, glacial rebound and mantle viscosity for northern Europe. *Geophysical Journal International* 134: 102–144.

Peltier WR (1998) Postglacial variations in the level of the sea: implications for climate dynamics and solid-earth geophysics. *Reviews in Geophysics* 36: 603–689.

Pirazzoli PA (1991) *World Atlas of Holocene Sea-Level Changes*. Amsterdam: Elsevier.

van de Plassche O (ed.) (1986) *Sea-Level Research: A Manual for the Collection and Evaluation of Data*. Norwich: Geo Books.

Sabadini R, Lambeck K, and Boschi E (1991) *Glacial Isostasy, Sea Level and Mantle Rheology*. Dordrecht: Kluwer.

WAVES ON BEACHES

R. A. Holman, Oregon State University, Corvallis, OR, USA

Introduction

Wave motions are one of the most familiar of oceanographic phenomena. The waves that we see on beaches were originally generated by ocean winds and storms, sometimes at long distances from their final destination. In fact, groups of waves, generated by large storms, have been tracked from the Southern Ocean near Australia all the way to Alaska.

Open ocean waves can be thought of as simple sinusoids that are superimposed to yield a realistic sea. Waves entering the nearshore, called incident waves, can have wave periods (T, the time between consecutive passages of wave crests) ranging from 2 to 20 s, with 10 s a typical value. Wave heights (H, the vertical distance from the trough to peak of a wave) can exceed 10 m, but are typically 1 m, representing an energy density, of $1250 \, \mathrm{J \, m^{-2}}$ (ρ is the density of sea water, g is the acceleration of gravity) and a flux of power impinging on the coast of about 10 kW per meter of coastline. Although this is a substantial amount of power, it is not enough to make broad commercial exploitation of wave power economical at the time of writing.

Of interest in this section are the dynamics of waves once they progress into the shallow beach environment such that the ocean bottom begins to restrict the water motions. Most people are familiar with refraction (the turning of waves toward the beach), wave breaking, and swash (the back and forth motion of the water's edge), but are less familiar with the other types of fluid motion that are generated near the beach.

Figure 1 illustrates schematically the evolution of ocean wave energy as it moves from deep water (top of the figure) through progressively shallower water toward the beach (bottom of the figure). Offshore, most energy lies in waves of roughly 10 s period (middle of the figure). However, the processes of shoaling distribute that energy to both higher frequencies (right half of the figure) and lower frequencies (left half of the figure) including mean flows. In general, these processes can be distinguished as those that occur offshore of the surf zone (where waves become

overly steep and break) and those that occur within the surf zone.

The axes of **Figure 1**, cross-shore position and frequency, are two of several variables that can be used to structure a discussion of near-shore fluid dynamics. Other important distinctions that will be made include whether the incident waves are monochromatic (single frequency) versus random (including a range of frequencies), depth-averaged versus depth-dependent, longshore uniform (requiring consideration of only one horizontal dimension, 1HD) versus long shore variable (2HD), and linear versus nonlinear.

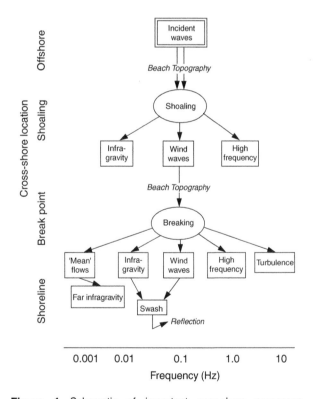

Figure 1 Schematic of important near-shore processes showing how the incident wave energy that drives the system evolves as the waves progress from offshore to the shoreline (top to bottom of figure). Wave evolution is grouped into processes occurring seaward of the breakpoint (labeled 'shoaling') and those within the surf zone (denoted 'breaking'). In both cases, energy is spread to lower (left) and higher (right) frequencies. The beach topography provides the bottom boundary condition for flow, so is important to wave processes. In turn, the waves move sediment, slowly changing the topography. Wind and tides may be important in some settings, but are not shown here.

Table 1 Kinematic relationships of linear waves[a]

Parameter	Shallow water[b]	General expression	Deep water[c]
Surface elevation	$\eta = \dfrac{H}{2}\cos(kX - \sigma t)$	$\eta = \dfrac{H}{2}\cos(kX - \sigma t)$	$\eta = \dfrac{H}{2}\cos(kX - \sigma t)$
Energy density	$E = \dfrac{1}{8}\rho g H^2$	$E = \dfrac{1}{8}\rho g H^2$	$E = \dfrac{1}{8}\rho g H^2$
Wave length	$L = T\sqrt{(gh)}$	$L = \dfrac{gT^2}{2\pi}\tanh(kh)$	$L_0 = \dfrac{gT^2}{2\pi}$
Phase velocity	$c = \sqrt{(gh)}$	$c = \dfrac{gT}{2\pi}\tanh(kh)$	$c_0 = \dfrac{gT}{2\pi}$
Group velocity	$c_g = c = \sqrt{(gh)}$	$c_g = cn = c\left[\dfrac{1}{2} + \dfrac{kh}{\sinh(2kh)}\right]$	$c_g = \dfrac{c}{2} = \dfrac{gT}{4\pi}$
Wave power, $P = Ec_g$	$P = \dfrac{1}{8}\rho g H^2\sqrt{(gh)}$	$\dfrac{\rho g H^2}{8}\dfrac{gT\tanh(kh)}{2\pi}\left[\dfrac{1}{2} + \dfrac{kh}{\sinh(2kh)}\right]$	$P = \dfrac{1}{8}\rho g H^2\dfrac{gT}{4\pi}$
Wave momentum flux	$S_{xx} = \dfrac{3}{2}E$	$S_{xx} = E\left[\dfrac{2kh}{\sinh(2kh)} + \dfrac{1}{2}\right]$	$S_{xx} = \dfrac{1}{2}E$
Horizontal velocity magnitude	$u = \dfrac{H}{2}\sqrt{\left(\dfrac{g}{h}\right)}$	$u = \dfrac{\pi H}{T}\dfrac{\cosh(k(z + h))}{\sinh(kh)}$	$u = \dfrac{\pi H}{T}e^{kz}$

[a]H is the peak to trough wave height and is twice the amplitude, a; k is the wave number $= 2\pi/L$; X is distance in the direction of propagation; σ is the frequency $= 2\pi/T$ where T is the wave period; ρ is the density of water; g is the acceleration of gravity; z is the depth below the surface within the water column of total depth h. Units of each quantity depend on units used in each equation.
[b]$h < L_0/20$. h is the water depth; L_0 is the deep water wavelength.
[c]$h < L_0/2$.

The Dynamics of Incident Waves

Much of the early progress in understanding nearshore waves was based on the examination of a monochromatic wave train, propagating onto a long shore uniform beach (1HD). Many observable properties can be explained in terms of a few principles including conservation of wave crests, of momentum, and of energy. Most dynamics are depth-averaged.

Table 1 lists a number of properties of monochromatic ocean waves, in the linear limit of infinitesimal wave amplitude (known as linear, or Airy wave theory). The general expressions (center column) contain complicated forms that can be substantially simplified for both shallow (depths less than 1/20 of the deep water wavelength, L_0) and deep (depths greater than 1/2 L_0) water limits.

The speed of wave propagation is known as the celerity or phase speed, c, to distinguish it from the velocity of the actual water particles. In deep water, c depends only on the wave period (independent of depth). However, as the wave enters shallower water, the wavelength decreases and the phase speed becomes slower (contrary to common belief, this is not a result of bottom friction). An interesting consequence of the slowing is wave refraction, the turning of waves toward the coast. For a wave approaching the coast at any angle, the end in shallower water will always progress more slowly than the deeper end. By propagating faster, the deeper end will begin to catch up to the shallow end, effectively turning the wave toward the beach (refraction). In shallow water, the general expression for celerity, $c = (gh)^{1/2}$, depends only on depth so that waves of all periods propagate at the same speed.

The energy density of Airy waves (energy per unit area) is the sum of kinetic and potential energy components and depends only on the square of the wave height (**Table 1**). Perhaps of more interest is the rate at which this energy is propagated by the wave train, known as the wave power, P, or wave energy flux. In deep water, wave energy progresses at half the speed of wave phase (individual wave crests will out-run the energy packet), whereas in shallow water energy travels at the same speed as wave phase and the flux depends only on $H^2 h^{1/2}$. Offshore of the surf zone, wave energy is conserved since there is no breaking dissipation and energy loss through bottom friction has been shown to be negligible except over very wide flat seas. Thus, as the depth, h, decreases, the wave height, H, must increase to conserve $H^2 h^{1/2}$. This is a phenomenon familiar to beachgoers as the looming up of a wave just before breaking.

The combined result of shoaling is reduced wavelength and increased wave height, hence waves that become increasingly steep and may become unstable and break. One criterion for breaking is that the increasing water particle velocities (u in **Table 1**) exceed the decreasing wave phase speed such that the water leaps ahead of the wave in a curling or plunging breaker. From the relationships in **Table 1**, it can be found that this occurs. when γ, the wave height to depth ratio ($\gamma = H/h$) exceeds a value of 2. Of course, for waves that have steepened to the point of breaking, the approximations of infinitesimal waves, inherent in Airy wave theory, are badly violated. However, observations show that γ does reach a limiting value of approximately 1 for monochromatic waves and 0.4 for a random wave field. As waves continue to break across the surf zone, the wave height decreases in a way that γ is approximately maintained, and the wave field is said to be saturated (cannot get any larger).

The above-saturation condition implies that wave heights will be zero at the shoreline and there will be no swash, in contradiction to common observation. Instead, it can be shown that very small amplitude waves, incident on a sloping beach, will not break unless their shoreline amplitude, a_s, exceeds a value determined by

$$\frac{\sigma^2 a_s}{g\beta^2} \leq \kappa \qquad [1]$$

where κ is an O(1) constant. For larger amplitudes, the wave amplitude at any cross-shore position is the sum of a standing wave contribution of this maximum value plus a dissipative residual that obeys the saturation relationship.

The ratio of terms on the left-hand side of eqn.[1] is important to a wide range of nearshore phenomena and is often re-written as the Iribarren number,

$$\xi_0 = \frac{\beta}{(H_s/L_0)^{1/2}} \qquad [2]$$

where the measure of wave amplitude is replaced by the offshore significant wave height,[1] H_s. This form clarifies the importance of beach steepness, made dynamically important by comparing it to the wave steepness, H_s/L_0. For very large values of ξ_0, the beach acts as a wall and is reflective to incident

waves (the non-breaking case from eqn. [1]). For smaller values, the presence of the sloping beach takes on increasing importance as the waves begin to break as the plunging breakers that surfers like, where water is thrown ahead of the wave and the advancing crest resembles a tube. Still smaller beach steepnesses (and ξ_0) are associated with spilling breakers in which a volume of frothy turbulence is pushed along with the advancing wave front.

Radiation Stress: the Forcing of Mean Flows and Set-up

The above discussion is based on the assumption of linearity, strictly true only for waves of infinitesimal amplitude. Because the dynamics are linear, energy in a wave of some particular period, say 10 s, will always be at that same period. In fact, once wave amplitude is no longer negligible, there are a number of nonlinear interactions that may transfer energy to other frequencies, for example to drive currents (zero frequency).

Nonlinear terms describe the action of a wave motion on itself and arise in the momentum equation from the advective terms, $u \cdot \nabla u$, and from the integrated effect of the pressure term. For waves, the time-averaged effect of these terms can easily be calculated and expressed in terms of the radiation stress, S, defined as the excess momentum flux due to the presence of waves. Since a rate of change of momentum is the equivalent of a force by Newton's second law, radiation stress allows us to understand the time-averaged force exerted by waves on the water column through which they propagate. A spatial gradient in radiation stress, for instance a larger flux of momentum entering a particular location than exiting, would then force a current.

Radiation stress is a tensor such that S_{ij} is the flux of i-directed momentum in the j-direction. For waves in shallow water, approaching the coast at an angle θ, the components of the radiation stress tensor are

$$S = \begin{bmatrix} S_{xx} & S_{xy} \\ S_{yx} & S_{yy} \end{bmatrix} = E \begin{bmatrix} (\cos^2\theta + 1/2) & \cos\theta\sin\theta \\ \cos\theta\sin\theta & (\sin^2\theta + 1/2) \end{bmatrix} \qquad [3]$$

where x is the cross-shore distance measured positive to seaward from the shoreline and y is the long-shore distance measured in a right-hand coordinate system with z positive upward from the still water level.

For a wave propagating straight toward the beach ($\theta = 0$) $S_{xx} = 3/2E$ is the shoreward flux of shoreward-directed momentum. The increase of wave height (hence energy, E) associated with shoaling outside the surf zone must be accompanied by an

---

[1] Although the peak to trough vertical distance for monochromatic waves is a unique and hence sensible measure of wave height, for random waves this scale is a statistical quantity, representing a distribution. A single measure, often chosen to represent the random wave Reld, is the significant wave height, Hs, defined as the average height of the largest one-third of the waves. This statistic was chosen historically as best representing the value that would be visually estimated by a semitrained observer. It is usually calculated as four times the standard deviation of the sea surface times series.

increasing shoreward flux of momentum (radiation stress). This gradient, in turn, provides an offshore thrust, pushing water away from the break point and yielding a lowering of mean sea level called set-down. As the waves start to break and decrease in height through the surf zone, the decreasing radiation stress pushes water against the shore until an opposing pressure gradient balances the radiation stress gradient. The resulting set-up at the shoreline, $\bar{\eta}$, a contributor to coastal erosion and flooding, is found to depend on the offshore significant wave height, H_s, as

$$\bar{\eta}_{max} = KH_s\xi_0 \qquad [4]$$

where K is found empirically to be 0.45.

If waves approach the beach at an angle, they also carry with them a shoreward flux of longshore-directed momentum, S_{yx}. Cross-shore gradients in this quantity, due to the breaking of waves in the surf zone, provide a net long shore force that accelerates a long shore current, $\bar{V}$ along the beach until the forcing just balances bottom friction. If the cross-shore structure of $\bar{V}$ is solved for, a discontinuity is evident at the seaward limit of the surf zone, where the radiation stress forcing jumps from zero (seaward of the break point) to a large value (where the wave just begin to break).

This discontinuity is an artifact of the fact that every wave breaks at exactly the same location for an assumed monochromatic wave forcing, and must be artificially smoothed by an assumed horizontal mixing for this case. However, a natural random wave field consists of an ensemble of waves with (for linear waves) a Rayleigh distribution of heights. Depth-limited breaking of such a wave field will be spread over a region from offshore, where a few largest waves break, to onshore where the smallest waves finally begin to dissipate. The spatially distributed nature of these contributions to the average radiation stress provides a natural smoothing, often obviating the need for additional horizontal smoothing.

Nonlinear Incident Waves

The above discussion dwelt on the nonlinear transfer of energy from incident waves to mean flows. Non-linearities will also transfer energy to higher frequencies, yielding a transformation of incident wave shape from sinusoidal to peaky and skewed forms. The Ursell number, $(H/L)(L/h)^3$, measures the strength of the nonlinearity. For monochromatic incident waves, this evolution was often modeled in terms of an ordered Stokes expansion of the wave form to produce a series of harmonics (multiples of the incident

wave frequency) that are locked to the incident wave. For waves with Ursell number of O(1), propagating in depths that are not large compared to the wave height, higher order theories must be used to model the finite amplitude dynamics. For a random sea under such theories, the total evolution of the spectrum must be found by summing the spectral evolution equations for all possible Fourier pairs (in other words, all frequencies in the sea can and will interact with all other frequencies). Such approaches are very successful in predicting the evolving shape and nonlinear statistics (important for driving sediment transport) for natural random wave fields outside the surf zone.

Vertically Dependent Processes

Depth-independent models are successful in reproducing many nearshore fluid processes but cannot explain several important phenomena, for example undertow, offshore-directed currents that exist in the lower part of the water column under breaking waves. The primary cause of depth dependence arises from wave-breaking processes. When waves break, the organized orbital motions break down, either through the plunge of a curling jet of water thrown ahead of the advancing wave or as a turbulent foamy mass (called a roller) carried on the advancing crest. Both processes originate at the surface but drive turbulence and bubbles into the upper part of the water column.

The transfer of momentum from wave motions to mean currents described by radiation stress gradients above does not account for the existence of an intermediate repository, the active turbulence of the roller, that decays slowly as it is carried with the progressing wave. This time delay causes a shift of the forcing of longshore currents, such that a current jet will occur landward of the location expected from study of the breaking locations of incident waves.

The other consequence of the vertical dependence of the momentum transfer is that the shoreward thrust provided by wave breaking is concentrated near the surface. Set-up, the upward slope of sea level against the shore, will balance the depth averaged wave forcing. However, due to the vertical structure of the forcing, shoreward flows are driven near the top of the water column and a balancing return flow, the undertow, occurs in the lower water column. Undertow strengths can reach $1\,m\,s^{-1}$.

2HD Flows – Circulation

All of the previous discussion was based on the assumption that all processes were long-shore uniform

(1 HD) so that no long-shore gradients existed. It is rare in nature to have perfect long shore uniformity. Most commonly, some variability (often strong) exists in the underlying bathymetry. This can lead to refractive focusing (the concentration of wave energy by refraction of waves onto a shallower area) and the forcing of long-shore gradients in wave height, hence of setup. Since setup is simply a pressure head, long-shore gradients will drive long-shore currents toward low points where the converging water will turn seaward in a jet called a rip current.

It is possible to develop long-shore gradients in wave height (hence rips) in the absence of long-shore variations in bathymetry. Interactions between two elements of the wave field (either two incident wave trains from different directions or an incident wave and an edge wave, defined below) can force rip currents if the interacting trains always occur with a fixed relative phase.

Infragravity Waves and Edge Waves

There is a further, very important consequence of the fact that natural wave fields are not monochromatic, but instead are random. For random waves, wave height is no longer constant but varies from wave to wave. Usually these variations are in the form of groups of five to eight waves, with heights gradually increasing then decreasing again. This observation is known as surf beat and is particularly familiar to surfers. A consequence of these slow variations is that the radiation stress of the waves is no longer constant, but also fluctuates with wave group time-scales and forces flows (and waves) in the near-shore with corresponding wave periods. These waves have periods of 30–300 s and are called infragravity waves, in analogy to infrared light being lower frequency than its visible light counterpart.

The direct forcing of infragravity motions described above can be thought of as a time-varying setup, with the largest waves in a group forcing shoreward flows that pile up in setup, followed by seaward flow as the setup gradients dominate over the weaker forcing of the small waves. If the modulations of the incident wave group are long shore uniform, this result of this setup disturbance will simply be a free (but low frequency) wave motion that propagates out to sea. However, in the normal case of wave groups with longshore (as well as time) variability, we can think of the response by tracing rays as the setup disturbance tries to propagate away. Rays that travel offshore at an angle to the beach will refract away from the beach normal (essentially the opposite of incident wave refraction, discussed

earlier). For rays starting at a sufficiently steep angle to the normal, refraction can completely turn the rays such that they re-approach and reflect from the shore in a repeating way and the energy is trapped within the shallow region of the beach. These trapped motions are called edge waves because the wave motions are trapped in the near-shore wave guide. (Any region wherein wave celerity is a minimum can similarly trap energy by refraction and is known as a wave guide. The deep ocean sound channel is a well-known example and allows propagation of trapped acoustic energy across entire ocean basins.) Motions that do not completely refract and thus are lost to the wave guide are called leaky modes.

The requirement that wave rays start at a sufficiently steep angle to be trapped by refraction can be expressed in terms of the long-shore component of wavenumber, k_y. For large k_y (waves with a large angle to the normal), rays will be trapped in edge waves whereas small k_y motions will be leaky modes. The cutoff between these is σ^2/g.

In the same sense that waves that slosh in a bathtub occur as a discrete set of modes that exactly fit into the tub, edge waves occur in a set of modes that exactly fit between reflection at the shoreline and an exponentially decaying tail offshore. The detailed form of the waves depends on the details of the bathymetry causing the refraction. However, for the example of a plane beach of slope $\beta (h = x\tan\beta)$, the cross-shore forms of the lowest four modes (mode numbers, $n = 0, 1, 2, 3$), are shown in **Figure 2** and are given in **Table 2**.

The existence of edge waves as a resonant mode of wave energy transmission in the near-shore has several impacts. First, the dispersion relation provides a selection for particular scales. For example, **Figure 3** shows a spectrum of infragravity wave energy collected at Duck, North Carolina, USA. The concentration of energy into very clear, preferred scales is striking and has led to suggestions that edge waves may be responsible for the generation of sand bars with corresponding scales. Second, because edge-wave energy is trapped in the near-shore, it can build to substantial levels even in the presence of weak, incremental forcing. Moreover, because edge-wave energy is large at the shoreline where the incident waves have decayed to their minimum due to breaking, edge waves may feasibly be the dominant fluid-forcing pattern on near-shore sediments in these regions.

The magnitudes of infragravity energy (including edge waves and leaky modes) have been found to depend on the relative beach steepness as expressed by the Iribarren number (eqn. [2]). For steep beaches (high ξ_0), very little infragravity energy can be

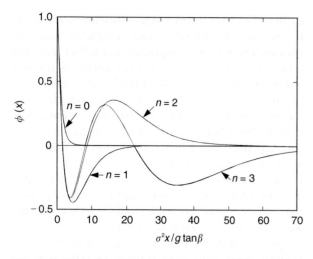

Table 3 Magnitude of infragravity waves on different beach types

Location	ξ_0	m^a
New South Wales, Australia	0.40	0.185 ± 0.157
Duck, NC, USA	0.43	0.237 ± 0.057
Martinique Beach, Canada	0.57	0.323 ± 0.073
Torrey Pines, CA, USA	0.60	0.554 ± 0.100
Santa Barbara, CA, USA	0.83	0.459 ± 0.089

[a]Least squares regression slope between the significant swash height (computed from the infragravity band energy only) and the offshore significant wave height.
Reproduced from Howd PA, Oltman-Shay J, and Holman RA (1991) Wave variance partitioning in the trough of a barred beach. Journal of Geophysical Research 96 (C7), 12781} 12795.)

Figure 2 Cross-shore structure of edge waves. Only the lowest four mode numbers of the larger set are shown. The mode number, n, describes the number of zero crossings of the modes. So, for example, a mode 1 edge wave always has a low offshore, opposite a shoreline high, and visa versa. Edge waves propagate along the beach.

Table 2 The average abundance of the refractory elements in the Earth's crust, and their degree of enrichment, relative to aluminum, in the oceans

$k_y < \dfrac{\sigma^2}{g}$, leaky modes	
$k_y > \dfrac{\sigma^2}{g}$, edge waves	

Velocity potential	$\Phi = \dfrac{ag}{\sigma}\phi(x)\cos(ky - \sigma t)$
Sea surface elevation	$\eta = -a\phi(x)\sin(ky - \sigma t)$
Cross-shore velocity	$u = \dfrac{ag}{\sigma}\dfrac{\partial\phi(x)}{\partial x}\cos(ky - \sigma t)$
Long-shore velocity	$v = -\dfrac{a\sigma}{\sin[(2n+1)\beta]}\phi(x)\sin(ky - \sigma t)$
Dispersion relationship	$L = \dfrac{gT^2}{2\pi}\sin[(2n+1)\beta]$

Cross-shore shape functions	n	$\phi(x)$
	0	$1 \cdot e^{-kx}$
	1	$(1 - 2kx) \cdot e^{-kx}$
	2	$(1 - 4kx + 2k^2x^2) \cdot e^{-kx}$
	3	$(1 - 6kx + 6k^2x^2 - 4/3k^3x^3) \cdot e^{-kx}$

generated and the beaches are termed reflective due to the high reflection coefficient for the incident waves. However, for low-sloping beaches (small ξ_0), infragravity energy can be dominant, especially compared to the highly dissipated incident waves. On the Oregon Coast of the USA, for example, swash spectra have been analyzed in which 99% of the variation is at infragravity timescales (making beachcombing an energetic activity). **Table 3** lists five

representative beach locations, with mean values of ξ_0 and of m, the linear regression slope between the measured significant swash magnitude,[2]R_s, in the infragravity band, and the offshore significant wave height, H_s.

Shear Waves

Up until the mid-1980s long shore currents were viewed as mean flows whose dynamics were readily described as in the above sections. However, field data from the Field Research Facility in Duck, North Carolina, provided surprising evidence that as long-shore currents accelerated on a beach with a well-developed sand bar, the resulting current was not steady but instead developed slow fluctuations in strength and a meandering pattern in space. Typical wave periods of these wave are hundreds of seconds and long-shore wavelengths are just hundreds of meters (**Figure 3**). These very low frequencies are called far infragravity waves, in analogy to the relationship of far infrared to infrared optical frequencies. However, the wavelengths are several orders of magnitude shorter than that which would be expected for gravity waves (e.g., edge waves or leaky modes) of similar periods.

These meanders have been named shear waves and arise due to an instability of strong currents, similar to the instability of a rising column of smoke. The name comes from the dependence of the dynamics (described briefly below) on the shear of the long-shore current (the cross-shore gradient of the long-shore current). A jet-like current with large shear, such as might develop on a barred beach where the wave forcing is concentrated over and near the bar,

---

[2]Swash oscillations are commonly expressed in terms of their vertical component.

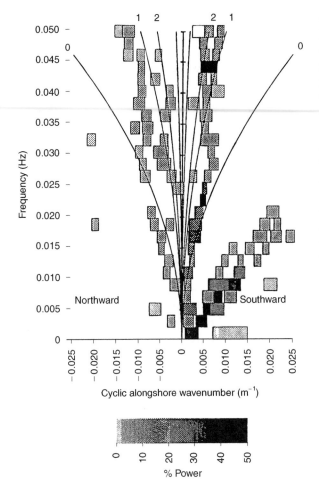

Figure 3 The spectrum of low frequency wave motions, as measured by current meters sampling the long shore component of velocity, from Duck, North Carolina, USA. The vertical axis corresponds to frequency and the horizontal axis to along shore component of wavenumber. Positive wavenumbers describe wave motions propagating along the beach to the south. Surprisingly, wave energy is not spread broadly in frequency–wavenumber space, but concentrates on specific ridges. Black lines, indicating the theoretical dispersion lines for edge waves (mode numbers marked at figure top), provide a good match to much of the data at low frequencies with some offset at higher frequency associated with Doppler shifting by the long-shore current. The concentration of low-frequency energy angling up to the right corresponds to shear waves. (After Oltman-Shay J and Guza RT (1987) Infragravity edge wave observations on two California beaches. *Journal of Physical Oceanography* 17(5): 644–663.)

can develop strong shear waves. In contrast, for a broad, featureless planar beach, the shear of any generated long-shore current will be weak so that shear wave energy may be undetectable. This explains why shear waves were not discovered until field experiments were carried out on barred beaches.

The instability by which shear waves are generated has a number of other analogs in nature. Large-scale coastal currents, flowing along a continental shelf, have been shown to develop similar instabilities although with very larges scales. Similarly, under wave motions, the bottom boundary layer, matching the moving wave oscillations of the water column interior with zero velocity at the fixed boundary, is also unstable.

It can be shown that a necessary condition for such an instability is the presence of an inflection point in the velocity profile (the spatial curvature of the current field changes sign), a requirement satisfied by long shore currents on a beach. In that case, cross-shore perturbations will extract energy from the mean long-shore current at a rate that depends on the strength of the current shear. Thus, these perturbations will grow to become first wave-like meanders, then if friction is not strong, to become a field of turbulent eddies. The extent of this evolution (wave-like versus eddy-like) is not yet known for natural beach environments.

Shear waves can clearly form an important component of the near-shore current field.) Root mean square (RMS) velocity fluctuations can reach 35 cm s^{-1} in both cross-shore and long-shore components of flow. This corresponds to an RMS swing of the current of 70 cm s^{-1}, with many oscillations much larger.

Conclusions

As ocean waves propagate into the shoaling waters of the nearshore, they undergo a wide range of changes. Most people are familiar with the refraction, shoaling and eventual breaking of waves in a near-shore surf zone. However, this same energy can drive strong secondary flows. Wave breaking pushes water shoreward, yielding a super-elevation at the shoreline that can accentuate flooding and erosion. Waves arriving at an angle to the beach will drive strong currents along the beach that can transport large amounts of sediment. Often these currents form circulation cells, with strong rip currents spaced along the beach.

Natural waves occur in groups, with heights that vary. The breaking of these fluctuating groups drives waves and currents at the same modulation timescale, called infragravity waves. These can be trapped in the nearshore by refraction as edge waves. Even long shore currents can develop instabilities called shear waves that drive meter-per-second fluctuations in the current strength with timescales of several minutes.

The apparent physics that dominates different beaches around the world often appears to vary. For example, on low-sloping energetic beaches, infra-gravity energy often dominates the surf zone, whereas shear waves can be very important on barred beaches. In fact, the physics is unchanging in these environments, with only the observable manifestations of that physics changing. The unification of these diverse observations through parameters such as the Iribarren number is an important goal for future research.

Nomenclature

E	wave energy density
H	wave height
H_s	significant wave height
L	wavelength
L_0	deep water wave length
P	wave power or energy flux
R_s	significant swash height
RMS	root mean square statistic
S	radiation stress (wave momentum flux)
T	wave period
$\bar{V}$	mean longshore current
X	distance coordinate in the direction of wave propagation
a	wave amplitude
a_s	wave amplitude at the shoreline
c	wave celerity, or phase velocity
c_g	wave group velocity
g	acceleration of gravity
h	water depth
k	wavenumber (inverse of wavelength)
n	ratio of group velocity to celerity
m	ratio of infragravity swash height to offshore wave height
n	edge wave mode number
u	water particle velocity under waves
v	long-shore component of wave particle velocity
x	cross-shore position coordinate
y	long-shore position coordinate
z	vertical coordinate
β	beach slope
γ	ratio of wave height to local depth for breaking waves
η	sea surface elevation
$theta$	angle of incidence of waves relative to normal
ξ_0	Iribarren number
ρ	density of water
σ	*radial frequency $2\pi/T$*
φ	cross-shore structure function for edge waves
$\bar{\eta}_{max}$	mean set-up at the shoreline
∇	gradient operator

See also

Beaches, Physical Processes Affecting.

SLIDES, SLUMPS, DEBRIS FLOW, AND TURBIDITY CURRENTS

G. Shanmugam, The University of Texas at Arlington, Arlington, TX, USA

Introduction

Since the birth of modern deep-sea exploration by the voyage of HMS *Challenger* (21 December 1872–24 May 1876), organized by the Royal Society of London and the Royal Navy, oceanographers have made considerable progress in understanding the world's oceans. However, the physical processes that are responsible for transporting sediment into the deep sea are still poorly understood. This is simply because the physics and hydrodynamics of these processes are difficult to observe and measure directly in deep-marine environments. This observational impediment has created a great challenge for communicating the mechanics of gravity processes with clarity. Thus a plethora of confusing concepts and related terminologies exist. The primary objective of this article is to bring some clarity to this issue by combining sound principles of fluid mechanics, laboratory experiments, study of modern deep-marine systems, and detailed examination of core and outcrop.

A clear understanding of deep-marine processes is critical because the petroleum industry is increasingly moving exploration into the deep-marine realm to meet the growing demand for oil and gas. Furthermore, mass movements on continental margins constitute major geohazards. Velocities of subaerial debris flows reached up to $500 \, \mathrm{km \, h^{-1}}$. Submarine landslides, triggered by the 1929 Grand Banks earthquake (offshore Newfoundland, Canada), traveled at a speed of $67 \, \mathrm{km \, h^{-1}}$. Such catastrophic events could destroy offshore oil-drilling platforms. Such events could result in oil spills and cost human lives. Therefore, numerous international scientific projects have been carried out to understand continental margins and related mass movements.

Terminology

The term 'deep marine' commonly refers to bathyal environments, occurring seaward of the continental shelf break (>200-m water depth), on the continental slope and the basin (**Figure 1**). The continental rise, which represents that part of the continental margin between continental slope and abyssal plain, is included here under the broad term 'basin'.

Two types of classifications and related terminologies exist for gravity-driven processes: (1) mechanics-based terms and (2) velocity-based terms.

Mechanics-Based Terms

Continental margins provide an ideal setting for slope failure, which is the collapse of slope sediment (**Figure 2**). Following a failure, the failed sediment moves downslope under the pull of gravity. Such gravity-driven processes exhibit extreme variability in mechanics of sediment transport, ranging from mobility of kilometer-size solid blocks on the seafloor to transport of millimeter-size particles in suspension of dilute turbulent flows. Gravity-driven processes are broadly classified into two types based on the physics and hydrodynamics of sediment mobility: (1) mass transport and (2) sediment flows (**Table 1**).

Mass transport is a general term used for the failure, dislodgement, and downslope movement of sediment under the influence of gravity. Mass transport is composed of both slides and slumps. Sediment flow is an abbreviated term for sediment-gravity flow. It is composed of both debris flows and turbidity currents (**Figure 3**). In rare cases, debris flow may be classified both as mass transport and as sediment flow. Thus the term mass transport is used for slide, slump, and debris flow, but not for turbidity current. Mass transport can operate in both subaerial and subaqueous environments, but turbidity currents can operate only in subaqueous environments. The advantage of this classification is that physical features preserved in a deposit directly represent the physics of sediment movement that existed at the final moments of deposition. The link between the deposit and the physics of the depositional process can be established by practicing the principle of process sedimentology, which is detailed bed-by-bed description of sedimentary rocks and their process interpretation.

The terms slide and slump are used for both a process and a deposit. The term debrite refers to the deposit of debris flows, and the term turbidite represents the deposit of turbidity currents. These

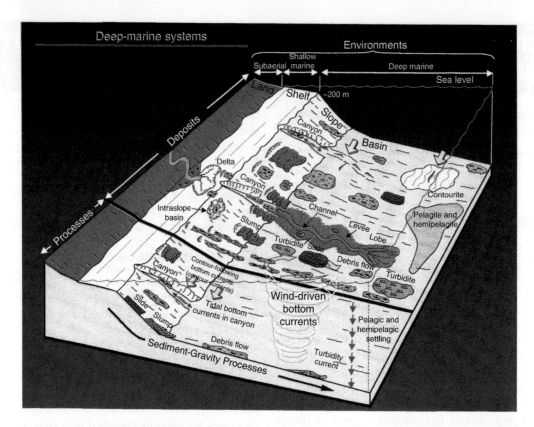

Figure 1 Schematic diagram showing complex deep-marine sedimentary environments occurring at water depths deeper than 200 m (shelf-slope break). In general, sediment transport in shallow-marine (shelf) environments is characterized by tides and waves, whereas sediment transport in deep-marine (slope and basin) environments is characterized by mass transport (i.e., slides, slumps, and debris flows), and sediment flows (i.e., debris flows and turbidity currents). Internal waves and up and down tidal bottom currents in submarine canyons (opposing arrows), and along-slope movement of contour-following bottom currents are important processes in deep-marine environments. Circular motion of wind-driven bottom currents can be explained by eddies induced by the Loop Current in the Gulf of Mexico or by the Gulf Stream Gyre in the North Atlantic. However, such bottom currents are not the focus of this article. From Shanmugam G (2003) Deep-marine tidal bottom currents and their reworked sands in modern and ancient submarine canyons. *Marine and Petroleum Geology* 20: 471–491.

Figure 2 Multibeam bathymetric image of the US Pacific margin, offshore Los Angeles (California), showing well-developed shelf, slope, and basin settings. The canyon head of the Redondo Canyon is at a water depth of 10 m near the shoreline. DB, debris blocks of debris flows; MT, mass-transport deposits. Vertical exaggeration is 6×. Modified from USGS (2007) US Geological Survey: Perspective view of Los Angeles Margin. http://wrgis.wr.usgs.gov/dds/dds-55/pacmaps/la_pers2.htm (accessed 11 May 2008).

Table 1 A classification of subaqueous gravity-driven processes

Major type	Nature of moving material	Nature of movement	Sediment concentration (vol. %)	Fluid rheology and flow state	Depositional process
Mass transport (also known as mass movement, mass wasting, or landslide)	Coherent Mass without internal deformation	Translational motion between stable ground and moving mass	Not applicable	Not applicable	Slide
	Coherent Mass with internal deformation	Rotational motion between stable ground and moving mass			Slump
Sediment flow (in cases, mass transport)	Incoherent Body (sediment–water slurry)	Movement of sediment–water slurry *en masse*	High (25–95)	Plastic rheology and laminar state	Debris flow (mass flow)
Sediment flow	Incoherent Body (water-supported particles in suspension)	Movement of individual particles within the flow	Low (1–23)	Newtonian rheology and turbulent state	Turbidity current

Reproduced from Shanmugam G (2006). *Deep-Water Processes and Facies Models: Implications for Sandstone Petroleum Reservoirs.* Amsterdam: Elsevier, with permission from Elsevier.

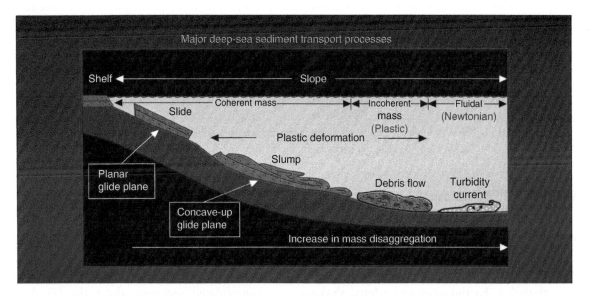

Figure 3 Schematic diagram showing four common types of gravity-driven downslope processes that transport sediment into deep-marine environments. A slide represents a coherent translational mass transport of a block or strata on a planar glide plane (shear surface) without internal deformation. A slide may be transformed into a slump, which represents a coherent rotational mass transport of a block or strata on a concave-up glide plane (shear surface) with internal deformation. Upon addition of fluid during downslope movement, slumped material may transform into a debris flow, which transports sediment as an incoherent body in which intergranular movements predominate over shear-surface movements. A debris flow behaves as a plastic laminar flow with strength. As fluid content increases in debris flow, the flow may evolve into Newtonian turbidity current. Not all turbidity currents, however, evolve from debris flows. Some turbidity currents may evolve directly from sediment failures. Turbidity currents can develop near the shelf edge, on the slope, or in distal basinal settings. From Shanmugam G, Lehtonen LR, Straume T, Syvertsen SE, Hodgkinson RJ, and Skibeli M (1994) Slump and debris flow dominated upper slope facies in the Cretaceous of the Norwegian and northern North Seas (61°–67° N): Implications for sand distribution. *American Association of Petroleum Geologists Bulletin* 78: 910–937.

process-product terms are also applicable to deep-water lakes. For example, turbidites have been recognized in the world's deepest lake (1637 m deep), Lake Baikal in south-central Siberia.

There are several synonymous terms in use:

1. rotational slide = slump;
2. mass transport = mass movement = mass wasting = landslide;
3. muddy debris flow = mud flow = slurry flow;
4. sandy debris flow = granular flow.

Velocity-Based Terms

Examples of velocity-based terms are:

1. rock fall: free falling of detached rock mass that increases in velocity with fall;
2. flow slide: fast-moving mass transport;
3. rock slide: fast-moving mass transport;
4. debris avalanche: fast-moving mass transport;
5. mud flow: fast-moving debris flow;
6. sturzstrom: fast-moving mass transport;
7. debris slide: slow-moving mass transport;
8. creep: slow-moving mass transport.

These velocity-based terms are subjective because the distinction between a fast-moving and a slow-moving mass has not been defined using a precise velocity value. It is also difficult to measure velocities because of common destruction of velocity meters by catastrophic mass transport events. Furthermore, there are no scientific criteria to determine the absolute velocities of sediment movement in the ancient rock record. This is because sedimentary features preserved in the deposit do not reflect absolute velocities. Therefore, velocity-based terms, although popular, are not practical.

Initiation of Movement

The initiation of mass movements and sediment flows is closely related to shelf-edge sediment failures. These failures along continental margins are triggered by one or more of the following causes:

1. *Earthquakes.* Earthquakes and related shaking can cause an increase in shear stress, which is the downslope component of the total stress. Submarine mass movements have been attributed to the 1929 Grand Banks earthquake off the US East Coast and Canada.
2. *Tectonic oversteepening.* Tectonic compression has elevated the northern flank of the Santa Barbara Basin and overturned the slope in southern California along the US Pacific margin.

Uplift (thrusting) along these slopes has led to oversteepening and sediment failure.
3. *Depositional oversteepening.* Mass movements have been attributed to oversteepening near the mouth of the Magdalena River, Colombia.
4. *Depositional loading.* Delta-front mass movements have been attributed to rapid sedimentation in the Mississippi River Delta, Gulf of Mexico.
5. *Hydrostatic loading.* Slope-failure deposits originated during the late Quaternary sea level rise on the eastern Tyrrhenian margin. These slope failures were attributed to rapid drowning of unconsolidated sediment, which resulted in increased hydrostatic loading.
6. *Glacial loading.* Submarine mass movements have been attributed to glacial loading and unloading along the Scotian margin in the North Atlantic.
7. *Eustatic changes in sea level.* Shelf-edge sediment failures and related gravity-driven processes have been attributed to worldwide fall in sea level, close to the shelf edge.
8. *Tsunamis.* The advancing wave front from a tsunami is capable of generating large hydrodynamic pressures on the seafloor that would produce soil movements and slope instabilities.
9. *Tropical cyclones.* Category 5 hurricane Katrina (2005) induced sediment failures and mudflows near the shelf edge in the Gulf of Mexico.
10. *Submarine volcanic activity.* Submarine mass movements have been attributed to volcanic activity in Hawaiian Islands.
11. *Salt movements.* Submarine mass movements along the flanks of intraslope basins have been attributed to mobilization of underlying salt masses in the Gulf of Mexico.
12. *Biologic erosion.* Submarine mass movements have been associated with erosion of the walls and floors of submarine canyons by invertebrates and fishes and boring by animals in offshore California.
13. *Gas hydrate decomposition.* Submarine mass movements have been associated with seepage of methane hydrate and related collapse of unconsolidated sediment along the US Atlantic Margin.

Mass Transport

Various methods are used to recognize mass transport deposits (slides and slumps) in modern submarine environments and their deposits in the ancient geologic record. These methods are (1) direct observations; (2) indirect velocity calculations; (3) remote-sensing technology (e.g., seismic profiling); and (4) examination of the rock (see Glossary).

Slides

A slide is a coherent mass of sediment that moves along a planar glide plane and shows no internal deformation (**Figure 3**). Slides represent translational shear-surface movements. Submarine slides can travel hundreds of kilometers on continental slopes (**Figure 4**). Long-runout distances of up to 800 km for slides have been documented (**Table 2**). Submarine slides are common in fiords because the submerged sides of glacial valleys are steep and because the rate of sedimentation is high due to sediment-laden rivers that drain glaciers into fiords. Submarine canyon walls are also prone to generate slides because of their steep gradients.

Multibeam bathymetric data show that the northern flank of the Santa Barbara Channel (Southern California) has experienced massive slope failures that resulted in the large (130 km²) Goleta landslide complex (**Figure 5(a)**). Approximately 1.75 km³ has been displaced by this slide during the Holocene (approximately the last 10 000 years). This complex has an upslope zone of evacuation and a downslope zone of accumulation (**Figure 5(b)**). It measures 14.6 km long, extending from a depth of 90 m to nearly 574 m, and is 10.5 km wide. It contains both surficial slump blocks and muddy debris flows in three distinct segments (**Figure 5(b)**).

Slides are capable of transporting gravel and coarse-grained sand because of their inherent strength. General characteristics of slides are:

- gravel to mud lithofacies;
- upslope areas with tensional faults;
- area of evacuation near the shelf edge (**Figures 4** and **5(a)**);
- occur commonly on slopes of 1–4°;
- long-runout distances of up to 800 km (**Table 2**);
- transported sandy slide blocks encased in deep-water muddy matrix (**Figure 6**);
- primary basal glide plane or décollement (core and outcrop) (**Figure 7(a)**);
- basal shear zone (core and outcrop) (**Figure 7(b)**);
- secondary internal glide planes (core and outcrop) (**Figure 7(a)**);
- associated slumps (core and outcrop) (**Figure 6**);
- transformation of slides into debris flows in frontal zone;
- associated clastic injections (core and outcrop) (**Figure 7(a)**);
- sheet-like geometry (seismic and outcrop) (**Figure 6**);
- common in areas of tectonic activity, earthquakes, steep gradients, salt movements, and rapid sedimentation.

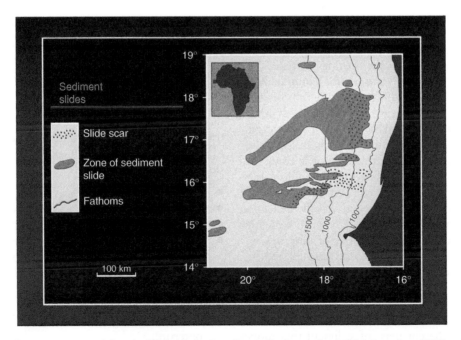

Figure 4 Long-distance transport of detached slide blocks from the shelf edge at about 100 fathom (i.e., 600 ft or 183 m) contour, offshore northwestern Africa. Note that two slide blocks near the 15° latitude marker have traveled nearly 300 km from the shelf edge (i.e., 100 fathom contour). Three contour lines (100, 1000, 1500 fathom) show increasing water depths to the west. From Jacobi RD (1976) Sediment slides on the northwestern continental margin of Africa. *Marine Geology* 22: 157–173.

Table 2 Long-runout distances of modern mass transport processes

Name and location	Runout distance (km)	Data	Process
Storegga, Norway	800	Seismic and GLORIA side-scan sonar images	Slide, slump, and debris flow
Agulhas, S Africa	750	Seismic	Slide and slump
Hatteras, US Atlantic margin	~500	Seismic and core	Slump and debris flow
Saharan NW African margin	>400	Seismic and core	Slump and debris flow
Mauritania–Senegal, NW African margin	~300	Seismic and core	Slump and debris flow
Nuuanu, NE Oahu (Hawaii)	235	GLORIA side-scan sonar images	Mass transport
Wailau, N Molakai (Hawaii)	<195	GLORIA side-scan sonar images	Mass transport
Rockall, NE Atlantic	160	Seismic	Mass transport
Clark, SW Maui (Hawaii)	150	GLORIA side-scan sonar images	Mass transport
N Kauai, N Kauai (Hawaii)	140	GLORIA side-scan sonar images	Mass transport
East Breaks (West), Gulf of Mexico	110	Seismic and core	Slump and debris flow
Grand Banks, Newfoundland	>100	Seismic and core	Mass transport
Ruatoria, New Zealand	100	Seismic	Mass transport
Alika-2, W Hawaii (Hawaii)	95	GLORIA side-scan sonar images	Mass transport
Kaena, NE Oahu (Hawaii)	80	GLORIA side-scan sonar images	Mass transport
Bassein, Bay of Bengal	55	Seismic	Slide and debris flow
Kidnappers, New Zealand	45	Seismic	Slump and slide
Munson–Nygren, New England	45	Seismic	Slump and debris flow
Ranger, Baja California	35	Seismic	Mass transport

On the modern Norwegian continental margin, large mass transport deposits occur on the northern and southern flanks of the Voring Plateau. The Storegga slide (offshore Norway), for example, has a maximum thickness of 430 m and a length of more than 800 km. The Storegga slide on the southern flank of the plateau exhibits mounded seismic patterns in sparker profiles. Although it is called a slide in publications, the core of this deposit is composed primarily of slumps and debrites. Unlike kilometers-wide modern slides that can be mapped using multibeam mapping systems and seismic reflection profiles, huge ancient slides are difficult to recognize in outcrops because of limited sizes of outcrops.

A summary of width:thickness ratio of modern and ancient slides is given in **Table 3**.

Slumps

A slump is a coherent mass of sediment that moves on a concave-up glide plane and undergoes rotational movements causing internal deformation (**Figure 3**). Slumps represent rotational shear-surface movements. In multibeam bathymetric data, distinguishing slides from slumps may be difficult because internal deformation cannot be resolved. In seismic profiles, however, slumps may be recognized because of their chaotic reflections. Therefore, a general term mass transport is preferred when interpreting bathymetric images. Slumps are capable of transporting gravel and coarse-grained sand because of their inherent strength. General characteristics of slumps are:

- gravel to mud lithofacies;
- basal zone of shearing (core and outcrop);
- upslope areas with tensional faults (**Figure 8**);
- downslope edges with compressional folding or thrusting (i.e., toe thrusts) (**Figure 8**);
- slump folds interbedded with undeformed layers (core and outcrop) (**Figure 9**);
- irregular upper contact (core and outcrop);
- chaotic bedding in heterolithic facies (core and outcrop);
- steeply dipping and truncated layers (core and outcrop) (**Figure 10**);
- associated slides (core and outcrop) (**Figure 6**);

(a)

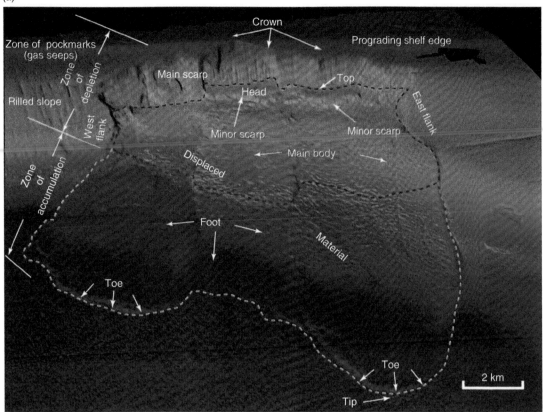

(b)

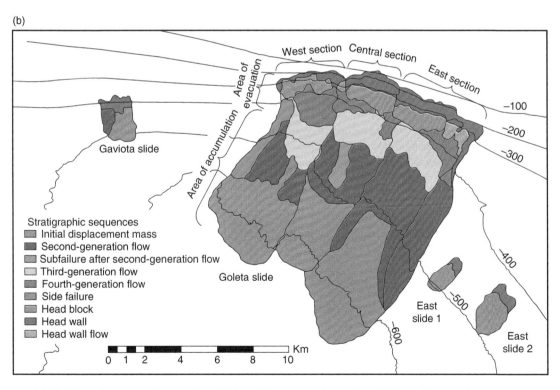

Figure 5 (a) Multibeam bathymetric image of the Goleta slide complex in the Santa Barbara Channel, Southern California. Note lobe-like (dashed line) distribution of displaced material that was apparently detached from the main scarp near the shelf edge. This mass transport complex is composed of multiple segments of failed material. (b) Sketch of the Goleta mass transport complex in the Santa Barbara Channel, Southern California (a) showing three distinct segments (i.e., west, central, and east). Contour intervals (−100, −200, −300, −400, −500, and −600) are in meters. (a) From Greene HG, Murai LY, Watts P, *et al.* (2006) Submarine landslides in the Santa Barbara Channel as potential tsunami sources. *Natural Hazards and Earth System Sciences* 6: 63–88. (b) From Greene HG, Murai LY, Watts P, *et al.* (2006) Submarine landslides in the Santa Barbara Channel as potential tsunami sources. *Natural Hazards and Earth System Sciences* 6: 63–88.

Figure 6 Outcrop photograph showing sheet-like geometry of an ancient sandy submarine slide (1000 m long and 50 m thick) encased in deep-water mudstone facies. Note the large sandstone sheet with rotated/slumped edge (left). Person (arrow): 1.8 m tall. Ablation Point Formation, Kimmeridgian (Jurassic), Alexander Island, Antarctica. Photo courtesy of D.J.M. Macdonald.

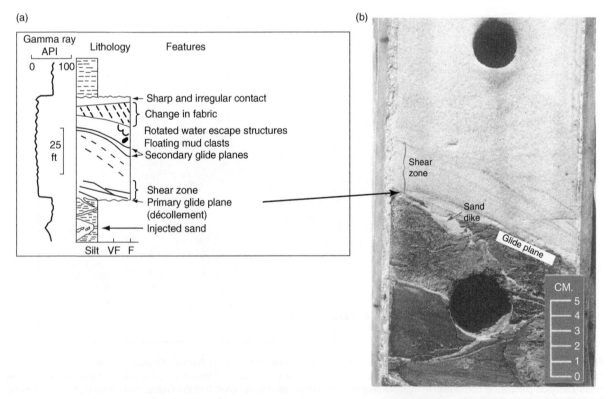

Figure 7 (a) Sketch of a cored interval of a sandy slide/slump unit showing blocky wireline log motif (left) and sedimentological details (right). (b) Core photograph showing the basal contact of the sandy slide (arrow). Note a sand dike (i.e., injectite) at the base of shear zone. Eocene, North Sea. Compare this small-scale slide (15 m thick) with a large-scale slide (50 m thick) in **Figure 6**. From Shanmugam G (2006) *Deep-Water Processes and Facies Models: Implications for Sandstone Petroleum Reservoirs*. Amsterdam: Elsevier.

Table 3 Dimensions of modern and ancient mass transport deposits

Example	Width:thickness ratio (observed dimensions)
Slide, Lower Carboniferous, England	7:1 (100 m wide/long, 15 m thick)
Slide, Cambrian–Ordovician, Nevada	30:1 (30 m wide/long, 1 m thick)
Slide, Jurassic, Antarctica	45:1 (20 km wide/long, 440 m thick)
Slide, Modern, US Atlantic margin	40–80:1 (2–4 km wide/long, 50 m thick)
Slide, Modern, Gulf of Alaska	130:1 (15 km wide/long, 115 m thick)
Slide, Middle Pliocene, Gulf of Mexico	250:1 (150 km wide/long, 600 m thick)
Slide/slump/debris flow/turbidite 5000–8000 BP, Norwegian continental margin	675:1 (290 km wide/long, 430 m thick)
Slump, Cambrian–Ordovician, Nevada	10:1 (100 m wide/long, 10 m thick)
Slump, Aptian–Albian, Antarctica	10:1 (3.5 km wide/long, 350 m thick)
Slump/slide/debris flow, Lower Eocene, Gryphon Field, UK	21:1 (2.6 km wide/long, 120 m thick)
Slump/slide/debris flow, Paleocene, Faeroe Basin, north of Shetland Islands	28:1 (7 km wide/long, 245 m thick)
Slump, Modern, SE Africa	171:1 (64 km wide/long, 374 m thick)
Slump, Carboniferous, England	500:1 (5 km wide/long, 10 m thick)
Slump, Lower Eocene, Spain	900–3600:1 (18 km wide/long, 5–20 m thick)
Debrite, Modern, British Columbia	12:1 (50 m wide/long, 4 m thick)
Debrite, Cambrian–Ordovician, Nevada	30:1 (300 m wide/long, 10 m thick)
Debrite, Modern, US Atlantic margin	500–5000:1 (10–100 km wide/long, 20 m thick)
Debrite, Quaternary, Baffin Bay	1250:1 (75 km wide/long, 60 m thick)
Turbidite (depositional lobe) Cretaceous, California	167:1 (10 km wide/long, 60 m thick)
Turbidite (depositional lobe) Lower Pliocene, Italy	1200:1 (30 km wide/long, 25 m thick)
Turbidite (basin plain) Miocene, Italy	11400:1 (57 km wide/long, 5 m thick)
Turbidite (basin plain) 16 000 BP Hatteras Abyssal Plain	125 000:1 (500 km wide/long, 4 m thick)

Reproduced from Shanmugam G (2006) *Deep-Water Processes and Facies Models: Implications for Sandstone Petroleum Reservoirs.* Amsterdam: Elsevier, with permission from Elsevier.

- associated sand injections (core and outcrop) (**Figure 10**);
- lenticular to sheet-like geometry with irregular thickness (seismic and outcrop);
- contorted bedding has been recognized in Formation MicroImager (FMI);
- chaotic facies on high-resolution seismic profiles.

In Hawaiian Islands, the Waianae Volcano comprises the western half of O'ahu Island. Several submersible dives and multibeam bathymetric imaging have confirmed the timing of seabed failure that formed the Waianae mass transport (slump?) complex. Multiple collapses and deformation events resulted in compound mass wasting features on the volcano's southwest flank. This complex is the largest in Hawaii, covering an area of about 5500 km².

Sediment Flows

Sediment flows (i.e., sediment-gravity flows) are composed of four types: (1) debris flow, (2) turbidity current, (3) fluidized sediment, and (4) grain flow. In this article, the focus is on debris flows and turbidity currents because of their importance. These two processes are distinguished from one another on the

basis of fluid rheology and flow state. The rheology of fluids can be expressed as a relationship between applied shear stress and rate of shear strain (**Figure 11**). Newtonian fluids (i.e., fluids with no inherent strength), like water, will begin to deform the moment shear stress is applied, and the deformation is linearly proportional to stress. In contrast, some naturally occurring materials (i.e., fluids with strength) will not deform until their yield stress has been exceeded (**Figure 11**); once their yield stress is exceeded, deformation is linear. Such materials with strength (e.g., wet concrete) are considered to be Bingham plastics (**Figure 11**). For flows that exhibit plastic rheology, the term plastic flow is appropriate. Using rheology as the basis, deep-water sediment flows are divided into two broad groups, namely (1) Newtonian flows that represent turbidity currents and (2) plastic flows that represent debris flows.

In addition to fluid rheology, flow state is used in distinguishing laminar debris flows from turbulent turbidity currents. The difference between laminar and turbulent flows was demonstrated in 1883 by Osborne Reynolds, an Irish engineer, by injecting a thin stream of dye into the flow of water through a glass tube. At low rates of flow, the dye stream traveled in a straight path. This regular motion of

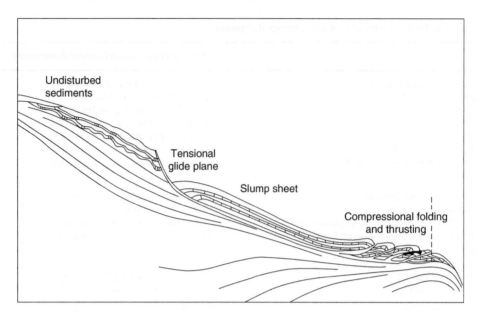

Figure 8   Sketch of a submarine slump sheet showing tensional glide plane in the updip detachment area and compressional folding and thrusting in the downdip frontal zone. From Lewis KB (1971) Slumping on a continental slope inclined at 1°–4°. *Sedimentology* 16: 97–110.

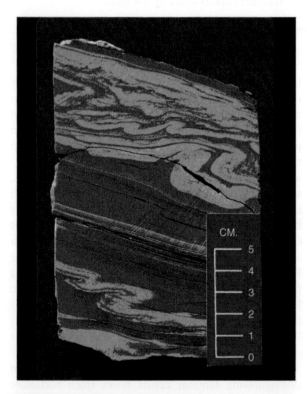

Figure 9   Core photograph showing alternation of contorted and uncontorted siltstone (light color) and claystone (dark color) layers of slump origin. This feature is called slump folding. Paleocene, North Sea. Reproduced from Shanmugam (2006). *Deep-Water Processes and Facies Models: Implications for Sandstone Petroleum Reservoirs.* Amsterdam: Elsevier, with permission from Elsevier.

fluid in parallel layers, without macroscopic mixing across the layers, is called a laminar flow. At higher flow rates, the dye stream broke up into chaotic eddies. Such an irregular fluid motion, with macroscopic mixing across the layers, is called a turbulent flow. The change from laminar to turbulent flow occurs at a critical Reynolds number (the ratio between inertia and viscous forces) of *c.* 2000 (**Figure 11**).

Debris Flows

A debris flow is a sediment flow with plastic rheology and laminar state from which deposition occurs through 'freezing' *en masse.* The terms debris flow and mass flow are used interchangeably because each exhibits plastic flow behavior with shear stress distributed throughout the mass. In debris flows, intergranular movements predominate over shear-surface movements. Although most debris flows move as incoherent material, some plastic flows may be transitional in behavior between coherent mass movements and incoherent sediment flows (**Table 1**). Debris flows may be mud-rich (i.e., muddy debris flows), sand-rich (i.e., sandy debris flows), or mixed types. Sandy debrites comprise important petroleum reservoirs in the North Sea, Norwegian Sea, Nigeria, Equatorial Guinea, Gabon, Gulf of Mexico, Brazil, and India. In multibeam bathymetric data, recognition of debrites is possible.

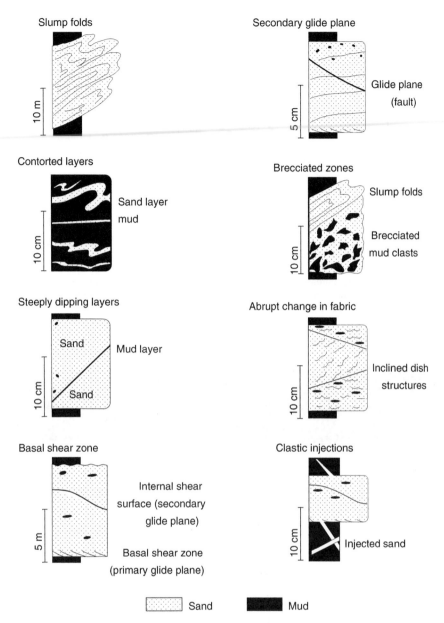

Figure 10 Summary of features associated with slump deposits observed in core and outcrop. Slump fold, an intraformational fold produced by deformation of soft sediment; contorted layer, deformed sediment layer; basal shear zone, the basal part of a rock unit that has been crushed and brecciated by many subparallel fractures due to shear strain; glide plane, slip surface along which major displacement occurs; brecciated zone, an interval that contains angular fragments caused by crushing of the rock; dish structures, concave – up (like a dish) structures caused by upward-escaping fluids in the sediment; clastic injections, natural injection of clastic (transported) sedimentary material (usually sand) into a host rock (usually mud). From Shanmugam G (2006) *Deep-Water Processes and Facies Models: Implications for Sandstone Petroleum Reservoirs.* Amsterdam: Elsevier.

Debris flows are capable of transporting gravel and coarse-grained sand because of their inherent strength. General characteristics of muddy and sandy debrites are:

- gravel to mud lithofacies;
- lobe-like distribution (map view) in the Gulf of Mexico (**Figure 12**);

- tongue-like distribution (map view) in the North Atlantic (**Figure 13**);
- floating or rafted mudstone clasts near the tops of sandy beds (core and outcrop) (**Figure 14**);
- projected clasts (core and outcrop);
- planar clast fabric (core and outcrop) (**Figure 14**);
- brecciated mudstone clasts in sandy matrix (core and outcrop);

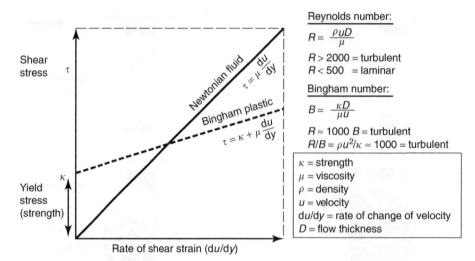

Figure 11 Graph showing rheology (stress–strain relationships) of Newtonian fluids and Bingham plastics. Note that the fundamental rheological difference between debris flows (Bingham plastics) and turbidity currents (Newtonian fluids) is that debris flows exhibit strength, whereas turbidity currents do not. Reynolds number is used for determining whether a flow is turbulent (turbidity current) or laminar (debris flow) in state. From Shanmugam G (1997) The Bouma sequence and the turbidite mind set. *Earth-Science Reviews* 42: 201–229.

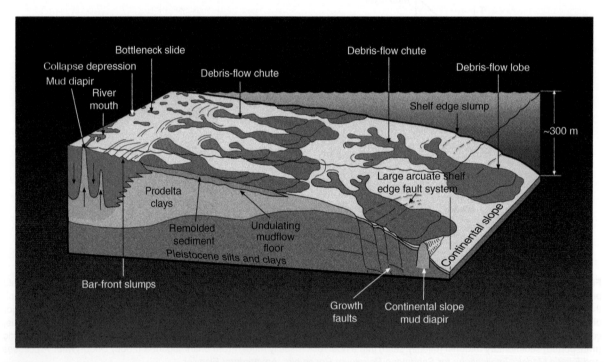

Figure 12 Schematic diagram showing lobe-like distribution of submarine debris flows in front of the Mississippi River Delta, Gulf of Mexico. This shelf-edge deltaic setting, associated with high sedimentation rate, is prone to develop ubiquitous mud diapers and contemporary faults. Mud diaper, intrusion of mud into overlying sediment causing dome-shaped structure; chute, channel. From Coleman JM and Prior DB (1982) Deltaic environments. In: Scholle PA and Spearing D (eds.) *American Association of Petroleum Geologists Memoir 31: Sandstone Depositional Environments*, pp. 139–178. Tulsa, OK: American Association of Petroleum Geologists.

- inverse grading of rock fragments (core and outcrop);
- inverse grading, normal grading, inverse to normal grading, and no grading of matrix (core and outcrop);

- floating quartz granules (core and outcrop);
- inverse grading of granules in sandy matrix (core and outcrop);
- pockets of gravels (core and outcrop);
- irregular, sharp upper contacts (core and outcrop);

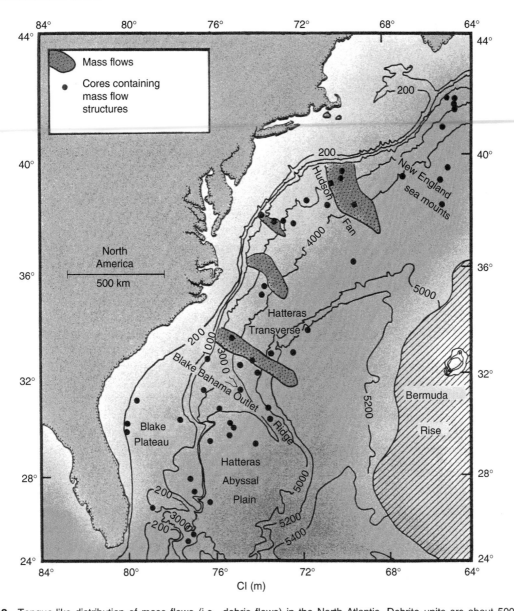

Figure 13 Tongue-like distribution of mass flows (i.e., debris flows) in the North Atlantic. Debrite units are about 500 km long, 10–100 km wide, and 20 m thick. Note a debrite tongue has traveled to a depth of about 5250 m water depth. CI, contour intervals. From Embley RW (1980) The role of mass transport in the distribution and character of deep-ocean sediments with special reference to the North Atlantic. *Marine Geology* 38: 23–50.

- side-by-side occurrence of garnet granules (density: 3.5–4.3) and quartz granules (density: 2.65) (core and outcrop);
- lenticular geometry (**Figure 15**).

The modern Amazon submarine channel has two major debrite deposits (east and west). The western debrite unit is about 250 km long, 100 km wide, and 125 m thick. In the US Atlantic margin, debrite units are about 500 km long, 10–100 km wide, and 20 m thick (**Figure 13**). In offshore northwest Africa, the Canary debrite is about 600 km long, 60–100 km wide, and 5–20 m thick. Submarine debris flows and

their flow-transformation induced turbidity currents have been reported to travel over 1500 km from their triggering point on the northwest African margin.

Turbidity Currents

A turbidity current is a sediment flow with Newtonian rheology and turbulent state in which sediment is supported by turbulence and from which deposition occurs through suspension settling. Turbidity currents exhibit unsteady and nonuniform flow behavior (**Figure 16**). Turbidity currents are surge-type waning flows. As they flow downslope,

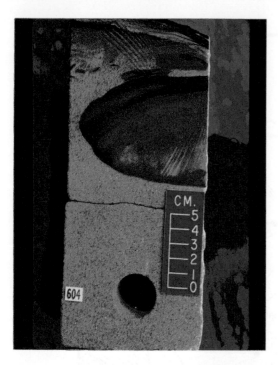

Figure 14 Core photograph of massive fine-grained sandstone showing floating mudstone clasts (above the scale) of different sizes. Note planar clast fabric (i.e., long axis of clast is aligned parallel to bedding surface). Note sharp and irregular upper bedding contact (top of photo). Paleocene, North Sea. Reproduced from Shanmugam G (2006). *Deep-Water Processes and Facies Models: Implications for Sandstone Petroleum Reservoirs.* Amsterdam: Elsevier, with permission from Elsevier.

turbidity currents invariably entrain ambient fluid (seawater) in their frontal head portion due to turbulent mixing (**Figure 16**).

With increasing fluid content, plastic debris flows may tend to become Newtonian turbidity currents (**Figure 3**). However, not all turbidity currents evolve from debris flows. Some turbidity currents may evolve directly from sediment failures. Although turbidity currents may constitute a distal end member in basinal areas, they can occur in any part of the system (i.e., shelf edge, slope, and basin). In seismic profiles and multibeam bathymetric images, it is impossible to recognize turbidites.

Turbidity currents cannot transport gravel and coarse-grained sand in suspension because they do not possess the strength. General characteristics of turbidites are:

- fine-grained sand to mud;
- normal grading (core and outcrop) (**Figure 17**);
- sharp or erosional basal contact (core and outcrop) (**Figure 17**);
- gradational upper contact (core and outcrop) (**Figure 17**);
- thin layers, commonly centimeters thick (core and outcrop) (**Figure 18**);
- sheet-like geometry in basinal settings (outcrop) (**Figure 18**);
- lenticular geometry that may develop in channel-fill settings.

Figure 15 Outcrop photograph showing lenticular geometry (dashed line) of debrite with floating clasts (arrow heads). Cretaceous, Tourmaline Beach, California. Reproduced from Shanmugam G (2006). *Deep-Water Processes and Facies Models: Implications for Sandstone Petroleum Reservoirs.* Amsterdam: Elsevier, with permission from Elsevier.

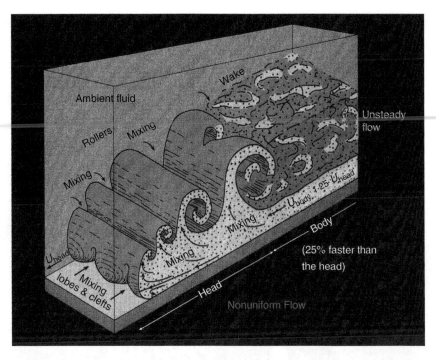

Figure 16 Schematic illustration showing the leading head portion of an unsteady, nonuniform, and turbulent turbidity current. Due to turbulent mixing, turbidity currents invariably entrain ambient fluid (seawater) at their head regions. Modified from Allen JRL (1985) Loose-boundary hydraulics and fluid mechanics: Selected advances since 1961. In: Brenchley PJ and Williams BPJ (eds.) *Sedimentology: Recent Developments and Applied Aspects*, pp. 7–28. Oxford, UK: Blackwell.

In the Hatteras Abyssal Plain (North Atlantic), turbidites have been estimated to be 500 km wide and up to 4 m thick (**Table 3**).

Selective emphasis of turbidity currents Turbidity currents have received a skewed emphasis in the literature. This may be attributed to existing myths about turbidity currents and turbidites. Myth no. 1: Some turbidity currents may be nonturbulent (i.e., laminar) flows. Reality: All turbidity currents are turbulent flows. Myth no. 2: Some turbidity currents may be waxing flows. Reality: All turbidity currents are waning flows. Myth no. 3: Some turbidity currents may be plastic in rheology. Reality: All turbidity currents are Newtonian in rheology. Myth no. 4: Some turbidity currents may be high in sediment concentration (25–95% by volume). Reality: All turbidity currents are low in sediment concentration (1–23% by volume). Because high sediment concentration damps turbulence, high-concentration (i.e., high-density) turbidity currents cannot exist. Myth no. 5: All turbidity currents are high-velocity flows and therefore they elude documentation. Reality: Turbidity currents can operate under a wide range of velocity conditions. Myth no. 6: Turbidity currents are believed to operate in modern oceans.

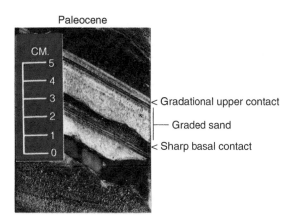

Figure 17 Core photograph showing a sandy unit with normal grading (i.e., grain size decreases upward), sharp basal contact, and gradational upper contact. These features are interpreted to be deposition from a waning turbidity current. Dark intervals are mudstone. Paleocene, North Sea.

Reality: No one has ever documented turbidity currents in modern oceans based on the physics of the flow. Myth no. 7: Some turbidites are deposits of debris flows. Reality: All turbidites are the exclusive deposits of turbidity currents. Myth no. 8: Some turbidites exhibit inverse grading. Reality: All turbidites develop normal grading. Myth no. 9: Cross-bedding is a product of turbidity currents. Reality:

Figure 18 Outcrop photograph showing tilted thin-bedded turbidite sandstone beds with sheet-like geometry, Lower Eocene, Zumaya, northern Spain. Reproduced from Shanmugam G (2006). *Deep-Water Processes and Facies Models: Implications for Sandstone Petroleum Reservoirs.* Amsterdam: Elsevier, with permission from Elsevier.

Cross-bedding is a product of bottom currents. Myth no. 10: Turbidite beds can be recognized in seismic reflection profiles. Reality: Normally graded turbidite units, centimeters in thickness, cannot be resolved on seismic profiles. As a consequence, many debris flows and their deposits have been misclassified as turbidity currents and turbidites.

Sediment Transport via Submarine Canyons

Submarine canyons, which are steep-sided valleys incised into the continental shelf and slope, serve as major conduits for sediment transport from land and the shelf to the deep-sea environment. Although downslope sediment transport occurs both inside and outside of submarine canyons, canyons play a critical role because steep canyon walls are prone to slope failures. Submarine canyons are prominent erosional features along both the US Pacific (**Figure 2**) and the Atlantic margins. Many submarine canyons in the US Atlantic margin commence at a depth of about 200 m near the shelf edge, but heads of California canyons in the US Pacific margin

begin at an average depth of about 35 m. The Redondo Canyon, for example, commences at a depth of 10 m near the shoreline (**Figure 2**). Such a scenario would allow for a quick transfer of sediment from shallow-marine into deep-marine environments. In the San Pedro Sea Valley, large debris blocks have been recognized as submarine landslides. Some researchers have proposed that these submarine landslides may have triggered local tsunamis. The significance of this relationship is that tsunamis can trigger submarine landslides, which in turn can trigger tsunamis. Such mutual triggering mechanisms can result in frequent sediment failures in deep-marine environments.

Tsunamis and tropical cyclones are important factors in transferring sediment into deep-marine environments via submarine canyons. For example, a rapid ($190\,\mathrm{cm\,s^{-1}}$) sediment flow was recorded in the Scripps Submarine Canyon (La Jolla, California) during the passage of a storm front on 24 November 1968 over La Jolla. In the Scripps Canyon, a large slump mass of about $105\,\mathrm{m^3}$ in size was triggered by the May 1975 storm.

Hurricane Hugo, which passed over St. Croix in the US Virgin Islands on 17 September 1989, had generated winds in excess of 110 knots (204 km h^{-1}, category 3 in the Saffir–Simpson scale) and waves 6–7 m in height. In the Salt River submarine canyon (>100 m deep), offshore St. Croix, a current meter measured net downcanyon currents reaching velocities of 2 m s^{-1} and oscillatory flows up to 4 m s^{-1}. Hugo had caused erosion of 2 m of sand in the Salt River Canyon at a depth of about 30 m. A minimum of 2 million kg of sediment were flushed down the Salt River Canyon into deep water. The transport rate associated with hurricane Hugo was 11 orders of magnitude greater than the rate measured during a fair-weather period. In the Salt River Canyon, much of the soft reef cover (e.g., sponges) had been eroded away by the power of the hurricane. Debris composed of palm fronds, trash, and pieces of boats found in the canyon were the evidence for storm-generated debris flows. Storm-induced sediment flows have also been reported in a submarine canyon off Bangladesh, in the Capbreton Canyon, Bay of Biscay in SW France, and in the Eel Canyon, Northern California, among others. In short, sediment transport in submarine canyons is accelerated by tropical cyclones and tsunamis in the world's oceans.

Glossary

Abyssal plain The deepest and flat part of the ocean floor that occupies depths between 2000 and 6000 m (6560 and 19 680 ft).

Ancient The term refers to deep-marine systems that are older than the Quaternary period, which began approximately 1.8 Ma.

Basal shear zone The basal part of a rock unit that has been crushed and brecciated by many subparallel fractures due to shear strain.

Bathyal Ocean floor that occupies depths between 200 (shelf edge) and 4000 m (656 and 13 120 ft). Note that abyssal plains may occur at bathyal depths.

Bathymetry The measurement of seafloor depth and the charting of seafloor topography.

Brecciated clasts Angular mudstone clasts in a rock due to crushing or other deformation.

Brecciated zone An interval that contains angular fragments caused by crushing or breakage of the rock.

Clastic sediment Solid fragmental material (unconsolidated) that originates from weathering and is transported and deposited by air, water, ice, or other processes (e.g., mass movements).

Continental margin The ocean floor that occupies between the shoreline and the abyssal plain. It consists of shelf, slope, and basin (**Figure 2**).

Contorted bedding Extremely disorganized, crumpled, convoluted, twisted, or folded bedding. Synonym: chaotic bedding.

Core A cylindrical sample of a rock type extracted from underground or seabed. It is obtained by drilling into the subsurface with a hollow steel tube called a corer. During the downward drilling and coring, the sample is pushed upward into the tube. After coring, the rock-filled tube is brought to the surface. In the laboratory, the core is slabbed perpendicular to bedding. Finally, the slabbed flat surface of the core is examined for geological bedding contacts, sedimentary structures, grain-size variations, deformation, fossil content, etc.

Dish structures Concave-up (like a dish) structures caused by upward-escaping water in the sediment.

Floating mud clasts Occurrence of mud clasts at some distance above the basal bedding contact of a rock unit.

Flow Continuous, irreversible deformation of sediment–water mixture that occurs in response to applied stress (**Figure 11**).

Fluid A material that flows.

Fluid dynamics A branch of fluid mechanics that deals with the study of fluids (liquids and gases) in motion.

Fluid mechanics Study of the properties and behaviors of fluids.

Geohazards Natural disasters (hazards), such as earthquakes, landslides, tsunamis, tropical cyclones, rogue (freak) waves, floods, volcanic events, sea level rise, karst-related subsidence (sink holes), geomagnetic storms; coastal upwelling; deep-ocean currents, etc.

Heterolithic facies Thinly interbedded (millimeter- to decimeter-scale) sandstones and mudstones.

Hydrodynamics A branch of fluid dynamics that deals with the study of liquids in motion.

Injectite Injected material (usually sand) into a host rock (usually mudstone). Injections are common in igneous rocks.

Inverse grading Upward increase in average grain size from the basal contact to the upper contact within a single depositional unit.

Lithofacies A rock unit that is distinguished from adjacent rock units based on its lithologic (i.e., physical, chemical, and biological) properties (see Rock).

Lobe A rounded, protruded, wide frontal part of a deposit in map view.

Methodology Four methods are in use for recognizing slides, slumps, debris flows, and turbidity currents and their deposits. *Method 1*: Direct observations – Deep-sea diving by a diver allows direct observations of submarine mass movements. The technique has limitations in terms of diving depth and diving time. These constraints can be overcome by using a remotely operated deep submergence vehicle, which would allow observations at greater depths and for longer time. Both remotely operated vehicles (ROVs) and manned submersibles are used for underwater photographic and video documentation of submarine processes. *Method 2*: Indirect velocity calculations – A standard practice has been to calculate velocity of catastrophic submarine events based on the timing of submarine cable breaks. The best example of this method is the 1929 Grand Banks earthquake (Canada) and related cable breaks. This method is not useful for recognizing individual type of mass movement (e.g., slide vs. slump). *Method 3*: Remote sensing technology – In the 1950s, conventional echo sounding was used to construct seafloor profiles. This was done by emitting sound pulses from a ship and by recording return echos from the sea bottom. Today, several types of seismic profiling techniques are available depending on the desired degree of resolutions. Although popular in the petroleum industry and academia, seismic profiles cannot resolve subtle sedimentological features that are required to distinguish turbidites from debrites. In the 1970s, the most significant progress in mapping the seafloor was made by adopting multibeam side-scan sonar survey. The Sea MARC 1 (Seafloor Mapping and Remote Characterization) system uses up to 5-km-broad swath of the seafloor. The GLORIA (Geological Long Range Inclined Asdic) system uses up to 45-km-broad swath of the seafloor. The advantage of GLORIA is that it can map an area of $27\,700\,km^2\,day^{-1}$. In the 1990s, multibeam mapping systems were adopted to map the seafloor. This system utilizes hull-mounted sonar arrays that collect bathymetric soundings. The ship's position is determined by Global Positioning System (GPS). Because the transducer arrays are hull mounted rather than towed in a vehicle behind the ship, the data are gathered with navigational accuracy of about 1 m and depth resolution of 50 cm. Two of the types of data collected are bathymetry (seafloor depth) and backscatter (data that can provide insight into the geologic makeup of the seafloor). An example is a bathymetric image of the US Pacific margin with mass transport deposits (**Figure 2**). The US National Geophysical Data Center (NGDC) maintains a website of bathymetric images of continental margins. Although morphological features seen on bathymetric images are useful for recognizing mass transport as a general mechanism, these images may not be useful for distinguishing slides from slumps. Such a distinction requires direct examination of the rock in detail. *Method 4*: Examination of the rock – Direct examination of core and outcrop is the most reliable method for recognizing individual deposits of slide, slump, debris flow, and turbidity current. This method known as process sedimentology, is the foundation for reconstructing ancient depositional environments and for understanding sandstone petroleum reservoirs.

Modern The term refers to present-day deep-marine systems that are still active or that have been active since the Quaternary period that began approximately 1.8 Ma.

Nonuniform flow Spatial changes in velocity at a moment in time.

Normal grading Upward decrease in average grain size from the basal contact to the upper contact within a single depositional unit composed of a single rock type. It should not contain any floating mudstone clasts or outsized quartz granules. In turbidity currents, waning flows deposit successively finer and finer sediment, resulting in a normal grading (see waning flows).

Outcrop A natural exposure of the bedrock without soil capping (e.g., along river-cut subaerial canyon walls or submarine canyon walls) or an artificial exposure of the bedrock due to excavation for roads, tunnels, or quarries.

Planar clast fabric Alignment of long axis of clasts parallel to bedding (i.e., horizontal). This fabric implies laminar flow at the time of deposition.

Primary basal glide plane (or décollement) The basal slip surface along which major displacement occurs.

Projected clasts Upward projection of mudstone clasts above the bedding surface of host rock (e.g., sand). This feature implies freezing from a laminar flow at the time of deposition.

Rock The term is used for (1) an aggregate of one or more minerals (e.g., sandstone); (2) a body of undifferentiated mineral matter (e.g., obsidian); and (3) solid organic matter (e.g., coal).

Scarp A relatively straight, cliff-like face or slope of considerable linear extent, breaking the continuity of the land by failure or faulting. Scarp is an abbreviated form of the term escarpment.

Secondary glide plane Internal slip surface within the rock unit along which minor displacement occurs.

Sediment flows They represent sediment-gravity flows. They are classified into four types based on sediment-support mechanisms: (1) turbidity current with turbulence; (2) fluidized sediment flow with upward moving intergranular flow; (3) grain flow with grain interaction (i.e., dispersive pressure); and (4) debris flow with matrix strength. Although all turbidity currents are turbulent in state, not all turbulent flows are turbidity currents. For example, subaerial river currents are turbulent, but they are not turbidity currents. River currents are fluid-gravity flows in which fluid is directly driven by gravity. In sediment-gravity flows, however, the interstitial fluid is driven by the grains moving down slope under the influence of gravity. Thus turbidity currents cannot operate without their entrained sediment, whereas river currents can do so. River currents are subaerial flows, whereas turbidity currents are subaqueous flows.

Sediment flux (1) A flowing sediment–water mixture. (2) Transfer of sediment.

Sedimentology Scientific study of sediments (unconsolidated) and sedimentary rocks (consolidated) in terms of their description, classification, origin, and diagenesis. It is concerned with physical, chemical, and biological processes and products. This article deals with physical sedimentology and its branch, process sedimentology.

Submarine canyon A steep-sided valley that incises into the continental shelf and slope. Canyons serve as major conduits for sediment transport from land and the shelf to the deep-sea environment. Smaller erosional features on the continental slope are commonly termed gullies; however, there are no standardized criteria to distinguish canyons from gullies. Similarly, the distinction between submarine canyons and submarine erosional channels is not straightforward. Thus, alternative terms, such as gullies, channels, troughs, trenches, fault valleys, and sea valleys, are in use for submarine canyons in the published literature.

Tropical cyclone It is a meteorological phenomenon characterized by a closed circulation system around a center of low pressure, driven by heat energy released as moist air drawn in over warm ocean waters rises and condenses. Structurally, it is a large, rotating system of clouds, wind, and thunderstorms. The name underscores their origin in the Tropics and their cyclonic nature. Worldwide, formation of tropical cyclones peaks in late summer months when water temperatures are warmest. In the Bay of Bengal, tropical cyclone activity has double peaks; one in April and May before the onset of the monsoon, and another in October and November just after.

Cyclone is a broader category that includes both storms and hurricanes as members. Cyclones in the Northern Hemisphere represent closed counterclockwise circulation. They are classified based on maximum sustained wind velocity as follows:

- tropical depression: 37–61 km h^{-1};
- tropical storm: 62–119 km h^{-1};
- tropical hurricane (Atlantic Ocean): >119 km h^{-1};
- tropical typhoon (Pacific or Indian Ocean): >119 km h^{-1}.

The Saffir–Simpson hurricane scale:

- category 1: 119–153 km h^{-1};
- category 2: 154–177 km h^{-1};
- category 3: 178–209 km h^{-1};
- category 4: 210–249 km h^{-1};
- category 5: >249 km h^{-1}.

Tsunami Oceanographic phenomena that are characterized by a water wave or series of waves with long wavelengths and long periods. They are caused by an impulsive vertical displacement of the body of water by earthquakes, landslides, volcanic explosions, or extraterrestrial (meteorite) impacts. The link between tsunamis and sediment flux in the world's oceans involves four stages: (1) triggering stage, (2) tsunami stage, (3) transformation stage, and (4) depositional stage. During the triggering stage, earthquakes, volcanic explosions, undersea landslides, and meteorite impacts can trigger displacement of the sea surface, causing tsunami waves. During the tsunami stage, tsunami waves carry energy traveling through the water, but these waves do not move the water. The incoming wave is depleted in entrained sediment. This stage is one of energy transfer, and it does not involve sediment transport. During the transformation stage, the incoming tsunami waves tend to erode and incorporate sediment into waves near the coast. This sediment-entrainment process transforms sediment-depleted waves into outgoing mass transport processes and sediment flows. During the depositional stage, deposition from slides, slumps, debris flows, and turbidity currents would occur.

Unsteady flow Temporal changes in velocity through a fixed point in space.

Waning flow Unsteady flow in which velocity becomes slower and slower at a fixed point through time. As a result, waning flows would deposit successively finer and finer sediment, resulting in a normal grading.

Waxing flow Unsteady flow in which velocity becomes faster and faster at a fixed point through time.

See also

Methane Hydrate and Submarine Slides. Ocean Margin Sediments. Tsunami.

Further Reading

Allen JRL (1985) Loose-boundary hydraulics and fluid mechanics: Selected advances since 1961. In: Brenchley PJ and Williams BPJ (eds.) *Sedimentology: Recent Developments and Applied Aspects*, pp. 7–28. Oxford, UK: Blackwell.

Coleman JM and Prior DB (1982) Deltaic environments. In: Scholle PA and Spearing D (eds.) *American Association of Petroleum Geologists Memoir 31: Sandstone Depositional Environments*, pp. 139–178. Tulsa, OK: American Association of Petroleum Geologists.

Dingle RV (1977) The anatomy of a large submarine slump on a sheared continental margin (SE Africa). *Journal of Geological Society of London* 134: 293–310.

Dott RH Jr. (1963) Dynamics of subaqueous gravity depositional processes. *American Association of Petroleum Geologists Bulletin* 47: 104–128.

Embley RW (1980) The role of mass transport in the distribution and character of deep-ocean sediments with special reference to the North Atlantic. *Marine Geology* 38: 23–50.

Greene HG, Murai LY, Watts P, *et al.* (2006) Submarine landslides in the Santa Barbara Channel as potential tsunami sources. *Natural Hazards and Earth System Sciences* 6: 63–88.

Hampton MA, Lee HJ, and Locat J (1996) Submarine landslides. *Reviews of Geophysics* 34: 33–59.

Hubbard DK (1992) Hurricane-induced sediment transport in open shelf tropical systems – an example from St. Croix, US Virgin Islands. *Journal of Sedimentary Petrology* 62: 946–960.

Jacobi RD (1976) Sediment slides on the northwestern continental margin of Africa. *Marine Geology* 22: 157–173.

Lewis KB (1971) Slumping on a continental slope inclined at 1°–4°. *Sedimentology* 16: 97–110.

Locat J and Mienert J (eds.) (2003) *Submarine Mass Movements and Their Consequences*. Dordrecht: Kluwer.

Middleton GV and Hampton MA (1973) Sediment gravity flows: Mechanics of flow and deposition. In: Middleton GV and Bouma AH (eds.) *Turbidites and Deep-Water Sedimentation*, pp. 1–38. Los Angeles, CA: Pacific Section Society of Economic Paleontologists and Mineralogists.

Sanders JE (1965) Primary sedimentary structures formed by turbidity currents and related resedimentation mechanisms. In: Middleton GV (ed.) *Society of Economic Paleontologists and Mineralogists Special Publiation 12: Primary Sedimentary Structures and Their Hydrodynamic Interpretation*, 192–219.

Schwab WC, Lee HJ, and Twichell DC, (eds.) (1993) Submarine Landslides: Selected Studies in the US Exclusive Economic Zone. *US Geological Survey Bulletin 2002*.

Shanmugam G (1996) High-density turbidity currents: Are they sandy debris flows? *Journal of Sedimentary Research* 66: 2–10.

Shanmugam G (1997) The Bouma sequence and the turbidite mind set. *Earth-Science Reviews* 42: 201–229.

Shanmugam G (2002) Ten turbidite myths. *Earth-Science Reviews* 58: 311–341.

Shanmugam G (2003) Deep-marine tidal bottom currents and their reworked sands in modern and ancient submarine canyons. *Marine and Petroleum Geology* 20: 471–491.

Shanmugam G (2006) *Deep-Water Processes and Facies Models: Implications for Sandstone Petroleum Reservoirs*. Amsterdam: Elsevier.

Shanmugam G (2006) The tsunamite problem. *Journal of Sedimentary Research* 76: 718–730.

Shanmugam G (2008) The constructive functions of tropical cyclones and tsunamis on deep-water sand deposition during sea level highstand: Implications for petroleum exploration. *American Association of Petroleum Geologists Bulletin* 92: 443–471.

Shanmugam G, Lehtonen LR, Straume T, Syvertsen SE, Hodgkinson RJ, and Skibeli M (1994) Slump and debris flow dominated upper slope facies in the Cretaceous of the Norwegian and northern North Seas (61°–67° N): Implications for sand distribution. *American Association of Petroleum Geologists Bulletin* 78: 910–937.

Shepard FP and Dill RF (1966) *Submarine Canyons and Other Sea Valleys*. Chicago: Rand McNally.

Talling PJ, Wynn RB, Masson DG, *et al.* (2007) Onset of submarine debris flow deposition far from original giant landslide. *Nature* 450: 541–544.

USGS (2007) US Geological Survey: Perspective view of Los Angeles Margin. http://wrgis.wr.usgs.gov/dds/dds-55/pacmaps/la_pers2.htm (accesed May 11, 2008).

Varnes DJ (1978) Slope movement types and processes. In: Schuster RL and Krizek RJ (eds.) *Transportation Research Board Special Report 176: Landslides: Analysis and Control*, pp. 11–33. Washington, DC: National Academy of Science.

Relevant Websites

http://www.ngdc.noaa.gov
  – NGDC Coastal Relief Model (images and data), NGDC (National Geophysical Data Center).
http://www.mbari.org
  – Submarine Volcanism: Hawaiian Landslides, MBARI (Monterey Bay Aquarium Research Institute)
http://www.nhc.noaa.gov
  – The Saffir–Simpson Hurricane Scale, NOAA/National Weather Service.

CORAL REEFS

J. W. McManus, University of Miami, Miami, FL, USA

Introduction

Coral reefs are highly diverse ecosystems that provide food, income, and coastal protection for hundreds of millions of coastal dwellers. They are found in a diverse range of geomorphologies, from small coral communities of little or no relief, to calcareous structures hundreds of kilometers across. The most diverse coral reefs occur in the waters around southeast Asia. This is primarily because of extinctions that have occurred in other regions due to the gradual restriction of global oceanic circulation associated with continental drift. However, rates of speciation of coral reef organisms in this region may also have been high within the last 50 million years.

Human activities have caused the degradation of coral reefs to varying degrees in all areas of the world. A major focus of present research is on the resilience of coral reefs to disturbances such as storms, diseases of reef species, bleaching (the expulsion of endosymbiotic photosynthetic zooxanthellae) and harvesting to local extinction. Along many coastlines, a combination of increased eutrophication due to coastal runoff and the extraction of herbivorous fish and invertebrates appears to favor the replacement of corals with macroalgae following disturbances. Another important aspect of resilience is the degree to which a depleted population of a given species on one reef can be replenished from other reefs. Most coral reef organisms undergo periods of free-swimming (pelagic) life ranging from a few hours to a few months. Genetic studies show evidence of broad dispersal of the progeny of some species among reefs, but most of the replenishment of a given population from year to year is believed to be from the same or nearby coral reefs.

Coral reefs are complex biophysical systems that are generally linked to similarly complex socioeconomic systems. Their proper management calls for system-level approaches such as integrated coastal management.

General

Types of Coral Reefs

The term 'coral reef' commonly refers to a marine ecosystem in which a prominent ecological functional role is played by scleractinian corals. A 'structural coral reef' differs from a 'nonstructural coral community' in being associated with a geomorphologically significant calcium carbonate (limestone) structure of meters to hundreds of meters height above the surrounding substrate, deposited by components of a coral reef ecosystem. The term 'coral reef' is often applied to both structural and nonstructural coral ecosystems or their fossil remains, although many scientists, especially geomorphologists, reserve the term for structural coral reefs and their underlying limestone. Both types of ecosystem occur within a wide range of tropical and subtropical marine environments, although structural development tends to be greater in waters of lower silt or mud concentration and oceanic salinity. Many reefs survive well amid open ocean waters with low nutrients, aided by efficient 'combing' of waters for plankton, high levels of nitrogen fixation and fast and thorough nutrient cycling. However, extensive coral reefs also occur in coastal waters of much higher nutrient concentrations.

Scleractinian (stony) corals grow as colonies or solitary polyps on a wide variety of substrates, including fallen trees, metal wreckage, rubber tires and rocks. Rates of settlement are often enhanced by the presence of calcareous encrusting algae. Soft sand, silt and mud tend to inhibit the settlement of stony corals, and so few coral ecosystems occur in modern or ancient deltaic deposits. However, coral can grow very near the mouths of small rivers and steams, and fresh or brackish groundwater often percolates through reef structures or emerges periodically through tunnels and caves. Vertical caves are often called 'blue holes'.

Nonstructural coral communities are common on rocky outcrops in shallow seas in many tropical and subtropical regions. They can range from a few clumps of coral to very substantial communities covering many square kilometers of wave-cut shelves near deeper areas.

Structural coral reefs come in many shapes and sizes, from less than a kilometer to many tens of kilometers in linear dimension. It is helpful to differentiate individual coral reefs from systems of coral reefs, such as the misnamed 'Great Barrier Reef' of

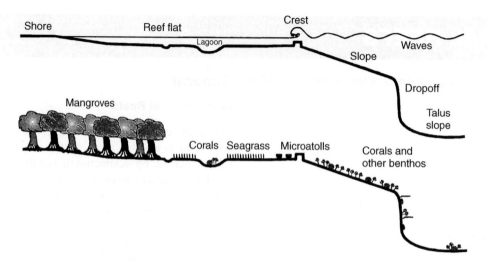

Figure 1 Profiles of a hypothetical fringing reef showing geomorphological and ecological zonation relative to wave action.

Australia, which actually consists of thousands of densely packed coral reefs. Common types of structural coral reefs include fringing reefs, barrier reefs, knoll reefs, pinnacle reefs, platform reefs, ribbon reefs, crescent reefs, and atolls. The term 'patch reef' may refer either to a patch of coral and limestone a few meters across within a structural coral reef (typically in a lagoon or on a reef flat), or to a platform or knoll reef. Thus, the term is best avoided. There is a wide range of structures intermediate between crescent reefs, platform reefs, and atolls in areas such as the Great Barrier Reef System.

A fringing reef is, by definition, always found adjacent to a land mass. Most fringing reefs include a wave-breaking reef crest, one or more meters above the rest of the reef, forming a thin strip offshore (**Figure 1**). Between the crest and the land, there is usually a relatively level area, broken by channels, called a 'reef flat' (**Figure 2**). Fringing reefs differ from barrier reefs, in that the latter are separated

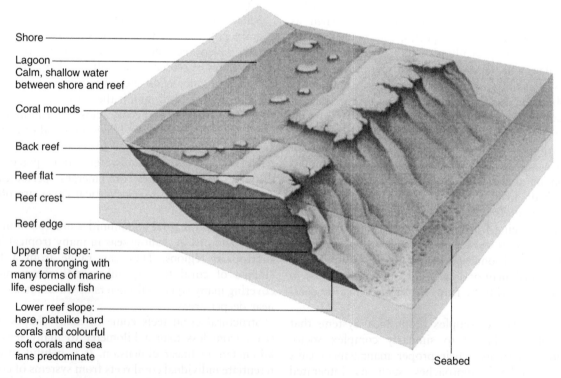

Figure 2 Geomorphology of a typical fringing reef. (Adapted from Holliday, 1989.)

from land by a 'navigable' water body (lagoon). It is useful to differentiate a lagoon from a reef flat in terms of depth; a lagoon is at least 2 m deep at mean tide. Fringing reefs often include lagoons, but the separation of a reef crest and slope from land is more complete in a barrier reef (**Figure 3**).

Structural reefs such as knoll, pinnacle, platform, ribbon, or crescent reefs, are arbitrarily labeled based on their shape. Atolls are donut-shaped structures which, although often supporting islands along the outer rim, do not have an island in the central portion (**Figure 4**). The large Apo Reef, east of the Philippine island of Mindoro, is a double atoll, with two lagoons within adjacent triangular rims, the whole reef being roughly diamond-shaped. Atolls can be quite large, such as the North Male Atoll, which houses the capital of the Maldives (**Figure 5**).

Although one commonly thinks of structural coral reefs as reaching to the sea surface, most of the coral reefs of the world do not. For example, there is a system of atolls and other reefs to approximately 50 km off the north-west of Palawan Island (Philippines) that closely resembles portions of the Australian Great Barrier Reef. However, very few reefs of the Palawan 'barrier system' come within 10 m of the sea surface. Some estimates of coral reef area are based on reefs at or near the sea surface, partly because the larger examples of such reefs tend to be well-charted, whereas most other coral reefs are poorly known in terms of location and characteristics.

Importance

Coral reefs support the highest known biodiversity of marine life, and constitute the largest biologically generated structures on Earth. Coral reefs are of substantial social, cultural, and economic importance. Coral reef systems in Florida, Hawaii, the Philippines, and Australia each account for more than $1 billion in tourist-related income each year. Coral reefs provide food and livelihoods for several tens of millions of fishers and their families, most of whom live in developing countries on low incomes, and have limited occupational mobility. Coral reefs protect coastal developments and farm lands from erosion. In many countries, particularly among Pacific islands, coral reefs are culturally very important, as they are involved in social structuring and interaction, and in religion. Despite poor taxonomic understanding and increasingly strict controls on bioprospecting, coral reef species are yielding substantial numbers of important drugs and other products.

Zonation

The ecological zonation of most coral reefs depends on physical factors, including depth, exposure to

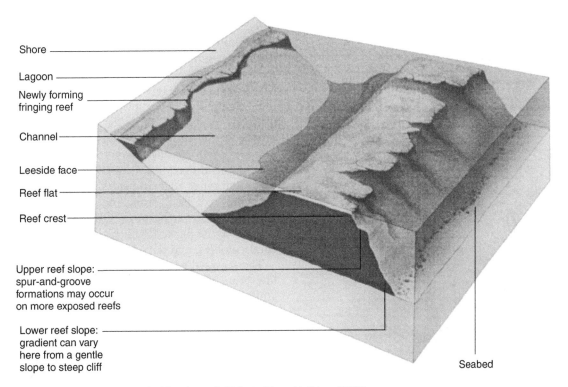

Figure 3 Geomorphology of a typical barrier reef. (Adapted from Holliday, 1989.)

Shore

Lagoon

Newly forming fringing reef

Channel

Leeside face

Reef flat

Reef crest

Upper reef slope: spur-and-groove formations may occur on more exposed reefs

Lower reef slope: gradient can vary here from a gentle slope to steep cliff

Seabed

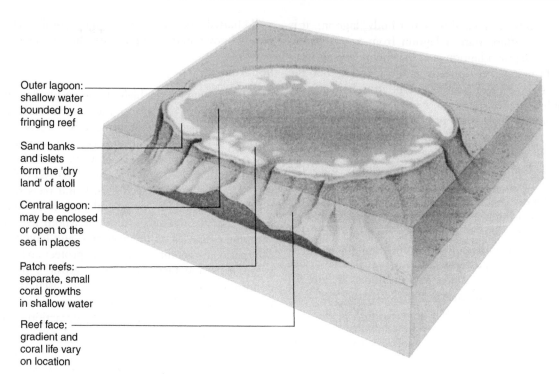

Figure 4   Geomorphology of a typical atoll. (Adapted from Holliday, 1989.)

Outer lagoon: shallow water bounded by a fringing reef

Sand banks and islets form the 'dry land' of atoll

Central lagoon: may be enclosed or open to the sea in places

Patch reefs: separate, small coral growths in shallow water

Reef face: gradient and coral life vary on location

waves and currents, and oxygen limitation. A fringing reef is basically a large slab of limestone jutting out from land. The landward margin may support extensive mangrove forests. A sandy channel may separate these from extensive seagrass beds on a reef flat. Other channels, depressions, and basins may be predominated by sandy bottoms, studded with clumps of coral. Clumps in deeper waters, rising 2 m or more from the bottom and consisting of several species of coral are known as 'bommies.' Some lagoons hold large numbers of 'pillars,' tall shafts of limestone, similarly supporting corals. Throughout the reef flat, there are, typically, low clumps of coral. Macroalgae and seagrass are often kept away from these clumps by the feeding action of the herbivorous fish and sea urchins, particularly active at night and resting during the day in the clumps of coral. Low massive coral colonies may form 'microatolls,' in which central portions of the colony are dead, while the raised outer edge and sides continue to flourish. On some reef flats, branching corals form patches which extend over hectares. Other reef flats and lagoons may be packed with high densities of diverse coral colonies, and have little or no seagrass.

The seaward edge of the reef flat often leads into low branching or massive corals, including microatolls, which become increasingly dense to seaward. A thin band of macroalgae, such as *Turbinaria*, may be present, just before the clumps of coral and hard substrate rise to form a reef crest or 'pavement.' In some areas, the crest may be made of living coral. In areas protected from heavy wave action, the crest may consist of fingerlike *Montipora* or *Porites*

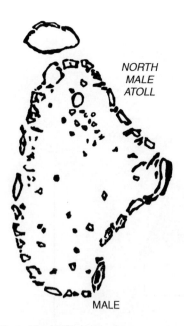

NORTH MALE ATOLL

MALE

Figure 5   North Male Atoll of the Maldives. Large atolls often support substantial human population that will be threatened as sealevels rise. (Adapted from Holliday, 1989.)

forming a band several meters wide, broken periodically by whorl-shaped colonies of leafy corals, all apparently formed to efficiently comb breaking waves for zooplankton. On higher-energy coasts, the corals may be primarily dense growths of wave-adapted *Pocillopora*. Other crests may be covered with various species and growth forms of the brown alga *Sargassum*. The edible and commercially important alga *Caulerpa* may also be abundant. Other reef crests may be densely packed with small clumps of articulated calcareous algae such as *Halimeda* and *Amphiroa*. Still others consist of a pavement of calcareous material or sheared-off ancient corals, and are relatively devoid of all macrobenthos except tiny clumps of algal turf or encrusting algae, sometimes forming a white or pink algal crest. The height of a reef crest above the reef flat is often determined by the height of local tides.

Most reef crests are broken by channels of varying width and depth – exit routes for water piled up behind the reef crest by the breaking waves. These may be studded with corals or smoothed by scouring sand and rock carried along with the exiting water. They are particularly important as breeding grounds for a variety of reef fish.

Beyond the reef crest, on the upper reaches of the 'reef slope,' one often finds small thick clumps of branching *Pocillopora* coral or various species of *Acropora* colonies in similar, wave-resistant growth forms. Encrusting and low lump-like (submassive) colonies may be common. More of the brown algae *Turbinaria* and *Sargassum* may be present. Along a rounded or gentle upper reef slope in the Indo-Pacific, there may be table-shaped *Acropora* colonies of gradually larger size as one proceeds to deeper waters. On many reefs, the reef slope is the most active area of coral growth, because of the oxygen, nutrients, and plankton brought in on the waves and currents. Coral cover may exceed 100%, as colonies overgrow colonies, all competing for light. Soft corals dominate some reef slopes, *Sargassum* others, and on some, the profusion of corals, sponges, algae, and other benthic organisms may prevent the identification of a dominant group. The mean and median stony coral cover (the percentage of the substrate covered with coral) on a reef slope globally are both approximately 40%, with a broad variance.

On most reefs, small channels on the upper slope consolidate into increasingly wider and deeper channels along lower reaches of the slope. Some may be steepened or converted into tunnels by coral growth. The channels may occur fairly regularly at distances of tens of meters, resulting in a 'ridge and rift' or 'spur (or buttress) and groove' structure resembling the toes of giant feet. The bottoms are generally filled with a mix of sand and debris from the reefs, which makes its way downwards, particularly during storms. Sand may drop continuously from escarpments, in flows resembling waterfalls, and spectacular columns of limestone cut away from the reef proper may border deeper rifts.

Many reef slopes lead to a steep 'wall' or 'drop-off,' often beginning about 10–20 m depth. On shelf areas, the drop-off may end at 20–30 m depth, followed by a 'talus' slope of deposited reef materials, often 30°–60°, leading into the more gradual shelf slope. Typically, this shelf will be interrupted by outcrops of limestone and bommies for considerable distances from the reef itself. On reefs jutting into deep waters, the drop-off may extend downwards for hundreds of thousands of meters. Corals and a myriad of other organisms generally cover the slopes, leaving little or no bare substrate.

On some reefs, such as some small fringing reefs along the Sinai, the upper portion consists almost entirely of live massive, platy or occasionally thick branching corals extending from land or from a small reef flat. The corals are joined together tangentially, leaving large spaces of water between. There may be no identifiable reef crest, and the mass simply juts out over a steep dropoff to hundreds of meters depth.

The zonation of atolls and other surface reefs is generally similar. However, as one proceeds across a lagoon basin from the windward to the leeward side of an atoll, one encounters a 'backreef' area (note that the term is also applied by some to the coral-dominated area behind a reef crest). The leeward side of the atoll is subject to less energetic waves and currents. It tends to have less well-defined zonation, less of a distinct crest, and often broader, more gradual slopes.

Storms are very important in determining features of the reef. It is common to find large chunks of limestone on upper slopes, crests and reef flats, representing masses of coral and substrate pulled up from the lower slope and deposited during a storm. Pieces of this jetsam may weight several tonnes each. Smaller pieces of coral and substrate, and masses of sand may pile up during storms, forming islands. Processes of calcification in intertidal areas may glue together pieces of coral and reef substrate thrown up by storms, forming 'beach rock,' which helps to prevent erosion at the edge of low islands. Other islands may be formed on portions of a reef uplifted by tectonic processes. Islands developed by storm action or uplift often have very little relief, but may support human communities. Entire nations, as with the Maldives or some Pacific island nations, may consist of such low islands on atoll reefs.

Geology

Calcification

Carbon dioxide and water combine to form carbonic acid, which can then dissociate into hydrogen ions (H^+) and (HCO_3^-) bicarbonate, or carbonate (CO_3^{2-}) ions as follows:

$$CO_2 + H_2O \Leftrightarrow H^+ + HO_3^- \Leftrightarrow 2H^+ + CO_3{}^{2-}$$

Colder water can hold more CO_2 in solution than warmer water. Calcium carbonate reacts with CO_2 and water as follows:

$$CO_2 + H_2O + CaCO_3 \Leftrightarrow Ca^{2+} + 2HCO_3{}^-$$

The dissolution of $CaCO_3$ depends directly on the amount of dissolved CO_2 present in the water. Thus, in colder waters the higher levels of CO_2 inhibit calcification. Structural coral reefs are primarily limited to a circumglobal band stretching from the Line Islands in Hawaii to Perth, Australia, with nonstructural coral communities being increasingly dominant along the fringes. The distribution band varies, such that cold waters (e.g., Peru) narrow the range of coral reefs closer to the equator, and warm currents facilitate higher latitude development (e.g., Bermuda).

Calcification in stony corals is greatly enhanced by the presence of zooxanthellae in the tissues. Zooxanthellae are nonmotile stages of a dinoflagellate algae. They are contained within specialized coral cells, growing on excess nutrients and trace metals from the carnivorous corals, producing sugars for utilization by the host, and using up excess CO_2 to enhance calcification by the coral. Zooxanthellae are found in other reef organisms, including giant clams, and the tiny foraminifera, amoeba-like organisms (Order Sarcodina) with calcareous skeletons that create substantial amounts of the sand on coral reefs and nearby white-sand beaches. Much of the biology and ecology of zooxanthellae is poorly known.

Paleoecology

Structural coral reefs are forms of 'biogenic reefs,' distinct geomorphological structures constructed by living organisms. Biogenic reefs have existed in various forms since approximately 3.5 billion years ago, at which time cyanobacteria began building stromatolites (**Figure 6**). Bioherms, biogenic reefs constructed from limestone produced by shelled animals, became prominent by 570 million years ago. During the mid-Triassic, the Jurassic and early Cretaceous periods (roughly 200–100 million years ago), scleractinian corals had become significant

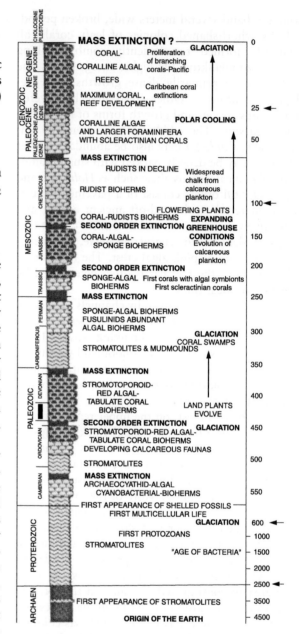

Figure 6 Geological time-scale (in thousand years) highlighting biogenic reef development. Arrows denote changes in scale. (Adapted from Hallock, 1989.)

components of coral–algal–sponge bioherms. Some of these corals may have included zooxanthellae. However, by the mid-Cretaceous, rudist bivalves dominated shallow-water reefs, and scleractinian corals were mainly restricted to deeper shelf-slope environments. This may have been due to seawater chemistry or competition with rudists or both. The massive extinction at the end of the Cretaceous, which ended the reign of dinosaurs on land, also led to the extinction of rudists. Many scleractinian corals survived in their deeper habitats, and evolved into

EARLY PLIOCENE

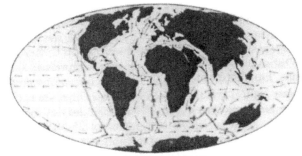

MIDDLE OLIGOCENE

MIDDLE PALEOCENE

Figure 7 Major tectonic movements and inferred patterns of major currents since the Paleocene. Gradually, the circumglobal waterway of the Paleocene, the Tethys Sea, was closed off, resulting in regional extinction of coral reef organisms. This process explains many of the differences in biodiversity now seen among the world's coral reefs. (Adapted from Hallock, 1997.)

most of the modern genera by the Eocene. However, CO_2 concentrations in the atmosphere were high at the time, apparently inhibiting reef development by aragonite-producing corals. Instead, major limestone deposits of the time were formed by calcite-producing organisms such as larger foraminifera (including the limestones from which the Egyptian pyramids were later to be built) and red coralline algae. Calcite is less stable over time than aragonite, but can form in waters of higher CO_2 concentration. Major reef-building by stony corals and their associates did not occur until the middle to late Oligocene. By then, CO_2 concentrations were falling and sea water was warming in tropical areas, whereas at high-latitudes it became cooler. By the late Oligocene, Caribbean coral reefs achieved their greatest development, and by the early Miocene, coral reefs globally had extended to beyond 10° north and south.

During the late Eocene and early Oligocene, tropical oceans were openly connected, in a system of equatorial waterways known as the Tethys Sea. During this period, most scleractinian corals were cosmopolitan. The upward movements of Africa and India restricted circulation, particularly with the closure of the Qatar arch in the Middle East around the time of the Oligocene–Miocene boundary (**Figure 7**). The Central American passageway became restricted, but did not close until the middle Pliocene. However, nearly half of the Caribbean coral genera became extinct at the Oligocene–Miocene boundary, and many more disappeared during the Miocene. Larger foraminifera suffered similar extinctions. Far less dramatic extinctions occurred in the Indo-Pacific.

Reef Geomorphology

Charles Darwin proposed that atolls were formed by a process involving the sinking of islands (**Figure 8**). He noted that most high islands in the Pacific were

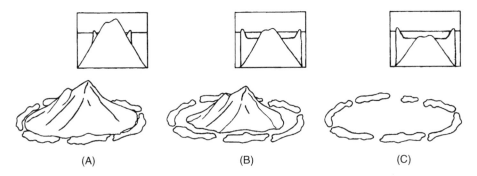

(A) (B) (C)

Figure 8 Stages in the development of an atoll as envisioned by Darwin. Fringing reefs surrounding a volcano (A) become barrier reefs as the volcano sinks (B) and finally form a donut-shaped atoll (C) Sinking islands and/or rising sea levels do help to explain the structures of many reefs, but other factors have been more important in the case of other reefs. (Adapted from Slafford-Deitsch, 1991.)

the tops of volcanic cones. A rocky, volcanic island would tend to form fringing reefs around its shores. Should the island begin to sink, the highly oxygenated outer edges would be able to maintain themselves at the sea surface, while the less actively growing reef flats would tend to sink, forming lagoons defining barrier reefs. Eventually, a sinking cone would disappear altogether, leaving a lagoon in the midst of an atoll. The feasibility of this explanation was confirmed by the mid-twentieth century when drilling on Enewetak and Bikini reefs yielded signs that the atolls were perched over islands that had gradually subsided over thousands of years.

Modern researchers have identified a variety of explanations for the morphology of various reefs. Variations in relative sea level have had profound influences on most structural coral reefs. Few coral reefs are believed to have exhibited continuous growth prior to 8000 years ago, and most are considerably younger. Coral reefs located on extensive continental shelf areas, such as the Florida reef tract and the vast shelf areas off the Yucatan and peninsular south-east Asia, grow on substrates that were generally far inland during the previous ice age.

The three-dimensional shapes of many coral reefs have been highly influenced by underlying topography, often an ancient reef. Wind, storms, waves, and currents have helped to shape many reefs, resulting in tear-drop shapes with broad, raised, steep edges facing predominant winds and currents. The shapes of some reefs are believed to have been particularly dependent on the erosion of much larger blocks of reef limestone during low stands of sea level. It has been demonstrated in the laboratory that small blocks of limestone subjected to rain-like erosion can form many features now identifiable on coral reefs, including lagoons, rifts and ridges, and channels. Evidence has also been found that the ridge and rift structures and some other features of coral reefs can result from collective ecological growth processes as the biota accommodates effects of waves, channeled water, and scouring by sand and other reef matter as calcifying benthic organisms compete for exposure to sunlight. It has been suggested that the double lagoon structure of the Atoll reef off Mindoro, Philippines, resulted from coral reef growth around the subsurface rim of a volcanic crater. Drilling on some reefs has demonstrated continuous growth for 30 m or more. On other reefs, little or no substantial calcification has occurred, the extreme case being nonstructural coral communities.

The Hawaiian and Line Islands are part of a vast chain of islands formed as a tectonic plate has drifted north and west across the Pacific. Volcanic lava has tended to erupt in a fixed 'hot-spot', forming volcanic islands one after the other. As the islands have drifted north, they have subsided gradually into deeper waters until coral growth has ceased. This process has had a strong influence over the structure of coral reefs along these islands and seamounts. However, at smaller scales, coral reefs along the coasts of the Hawaiian islands also show the effects of heavy storm action, exhibiting mound-like features at the sea surface often dominated by calcareous algae, and sometimes broad platforms which produce the famous surfing waves of the islands.

Role in Oil Exploration

Coral reefs are generally highly porous, filled with holes, tunnels, and caves of all sizes. The most famous 'blue-holes' of the Bahamas are the vertical seaward entrances of huge cave and tunnel systems formed by erosion at low stands of sea level, through which sea water flows beneath the islands as water through a sponge. Much of the porosity of coral reefs results from incomplete filling of spaces between coral heads during the active growth of the reef.

Much of the world's crude oil comes from ancient coral reefs that have been subsequently overlain with terrigenous (land-based) sediments. Dense, horizontally layered sediments have often trapped oil within the mounds of porous coral reef limestone. In some areas, subsequent tectonic activity has produced faults, and much of the oil has been lost. In other areas, however, the ancient reefs have yielded vast amounts of oil.

In some parts of the world, modern coral reefs have grown in areas where ancient reefs once occurred and were subsequently covered by dense sediments that trapped oil. Thus, it is common to find oil drilling platforms on modern reefs. This has frequently raised environmental concerns, centering on the damage done to modern corals by the construction and associated activities, the occasional spillages of drilling muds of various compositions and the leakage of oil.

Biology

Biogeography

The highest species diversity on coral reefs occurs in the seas of south-east Asia, often referred to as the Central Indo-West Pacific region. In the Caribbean or Hawaii, there are 70–80 species of reef-building corals. In south-east Asia, there are more than 70 genera and 400 or more species of reef-building corals (**Figure 9**). The diversities of most other groups of reef-associated species, including fish, tend to be higher in south-east Asia than in other regions.

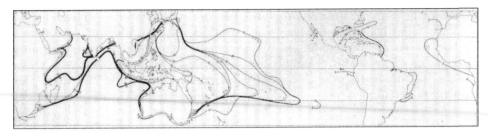

Figure 9 Patterns of generic diversity is scleractinian corals. Contours were estimated from generic patterns and existing data on species distributions. (From Veron 1993; reprinted with permission of the author.)

Diversity begins to drop off gradually to the south below Indonesia, and north above the Philippines. The decline eastward is very dramatic, such that only a few species of corals live in Tahiti and the Galapagos. Westward, the decline is only slight; at least 50 genera of corals live along the coasts of East Africa and the Red Sea. There are only a few species of reef-building corals off the coast of West Africa. Species in south-east Asia tend to have very broad ranges, and endemism is higher in peripheral areas such as the central Pacific.

Much of the global pattern of diversity, particularly at the level of genera and families, can be explained in terms of selective regional extinctions following the gradual breakup of the circumglobal Tethys Sea during the Cenozoic. This is well supported in the fossil records of corals, seagrasses and mangrove trees. Although fish fossils are uncommon, there is a highly diverse assemblage of fossil coral reef fish in Italy, a remnant from the Tethys Sea.

However, there is some reason to believe that rates of speciation have been higher in south-east Asia than elsewhere. The evidence is best known for mollusks, and include unusually high ratios of species to genera and particularly highly evolved armament (such as the spines on many murex shells) in this region. There is a myriad of explanations for this potentially higher speciation rate. Two popular explanations, the Pacific island vicariance hypothesis and the basin isolation hypothesis, both involve changes in sea level. In the former hypothesis, high levels of the sea tend to isolate groups of Pacific islands, facilitating speciation. At low sea levels, the species spread to the heterogeneous refugia habitat of south-east Asia and are gradually lost on the less-heterogeneous Pacific islands due to competition from other species. In the latter hypothesis, low stands of sea level, and, more importantly, periods of mountain building, have isolated particular marine basins within south-east Asia in the past, possibly permitting rapid speciation among whole biotas.

Further understanding of these processes must await increased efforts in taxonomy and systematics.

The vast majority of coral reef organisms, particularly in the Indo-West Pacific region, have never been formally identified. The ranges of known organisms are poorly known. Furthermore, recent decades have seen a rapid decline in interest in and support for taxonomy and systematics. It is likely that human impacts will result in vast changes to the distributions and abundances of coral reef species long before existing biogeography and ecology have been well understood.

Coral Reef Ecosystem Health

Resilience and phase shifts Most coral reefs are subjected to periodic disturbances, such as storms. The ability of an ecosystem to maintain constancy in terms of ecological functions and in the abundances and distributions of organisms is ecological resistance. The capacity of an ecosystem to revert to a previous state (or near to it) in terms of these characteristics is ecological resilience. Terms such as ecosystem health, integrity, and stability usually infer degrees of resistance and resilience. Because profound changes often occur to a coral reef following a perturbation such as a storm, there is increasing focus on the resilience of a coral reef.

Many coral reefs have been known to undergo losses of coral cover from greater than 50% to less than 10%, and then to recover to the former level within 4–10 years. Naturally, in reefs in which colonies may have been decades or centuries old, the age structure of the corals present have often been disrupted. This, in turn, may affect resilience to future perturbations, as certain corals are believed to exhibit higher fecundities in older colonies.

Some coral reefs do not seem to return to high levels of coral cover after a perturbation, especially if they have been subsequently overgrown by fleshy macroalgae. Human intervention, in the form of increased eutrophication and the removal of herbivorous fish and invertebrates, are suspected to favor the growth of macroalgae following perturbations.

Reefs around Jamaica have shown little recovery in more than a decade following coral damage by a hurricane, and both forms of intervention are suspected of being causes of this loss of resilience. Widespread losses of resilience are a concern throughout most coral reef areas.

Reef interconnectivity   Most coral reef organisms undergo pelagic life stages before settling into a reef community. Most corals live as planulae for a few hours to a few days before setting, although longer periods have been recorded. The average benthic invertebrate spends roughly two weeks in waterborne stages, but some survive for months. The average coral reef fish appears to require nearly a month before settling, and many require two months.

The pelagic stages are by no means passive, and although the sizes are very small, the organisms may be adapted to swim into currents and eddies that facilitate their retention or return to particular reefs or groups of reefs. Their success in doing so is believed to depend on factors such as reef geomorphology and the nature and predictability of local oceanography. Analyses indicate that some fish populations regularly exchange genetic material over thousands of kilometers. Although most coral-reef populations are believed to be replenished each year from local progeny, a severely depleted reef may be restocked to some degree from other reefs. This process is crucial to the problem of resilience, and is an active area of controversy and research. The results of this work may have profound implications for the design of marine protected areas, for international agreements on the coordination of management schemes, and for the regulation of harvesting on coral reefs. Furthermore, climate change is likely to alter local oceanographic processes, and has important implications for reef management as such disruptions occur.

Coral Reef Management

Disturbances

Various types of perturbations have affected coral reefs with increasing frequency within recent decades. Some of these are clearly directly related to human activities, whereas others are suspected to be the indirect results of human interventions.

The majority of coral reefs are located in developing countries. In many of these, crowded low-income human populations increasingly overfish, often reducing herbivorous fish and invertebrates, thereby decreasing reef resilience. In a process known as Malthusian overfishing, social norms break down and fishers turn to destructive fishing methods such as the use of poisons and explosives to capture fish.

Organic pollution from coastal habitations and agricultural fertilizing activities are common along coastlines. Runoff from deforested hillsides, mining operations, and coastal construction often contains materials that favor macroalgal growth, and silt and mud that restrict light to zooxanthellae, abrade reef benthos or bury portions of coral reefs entirely.

Increasingly common outbreaks of the coral-eating crown-of-thorns starfish, *Acanthaster planci*, may be related to reductions in predators, such as lethrinid fishes. During 1997–98, a worldwide epidemic of coral bleaching occurred, in which the zooxanthellae of many colonies were expelled and high rates of coral mortality resulted. The cause was unusually warm patches of sea water associated with a strong El Niño event, which some believe to be related to increasing levels of atmospheric CO_2 and global warming. A more controversial suggestion is that the increasing CO_2 levels will cause acidification of the oceans sufficient to result in the net erosion of some coral reefs, especially at high latitudes. Although there have recently been major epidemics of diseases killing corals and associated organisms in western Atlantic reefs, only certain coral diseases appear to be directly linked to stress from human activities.

Assessments

Efforts are underway to gather empirical information on coral reefs via the quantification of benthic organisms and fish by divers. These efforts are being supplemented by the use of remotely sensed data from satellites, space shuttles, aircraft, ships, manned and unmanned underwater vehicles. The usefulness of these approaches ranges from identifying or mapping reefs, to quantifying bleaching and disease. A global database, ReefBase, has been developed by the International Center for Living Aquatic Resources Management (ICLARM) to gather together existing information about the world's reefs and to make it widely available via CD-ROM and the Internet.

Integrated Coastal Management

Management is a process of modifying human behavior. Biophysical scientists can provide advice and predictions concerning factors such as levels of fishing pressure, siltation, and pollution. However, the management decisions must account for social, cultural, political, and economic considerations. Furthermore, almost all management interventions will have both positive and negative effects on various aspects of the ecosystem and the societies that impact it and depend upon it. For example, diverting fishers into forestry may lead to increased deforestation, siltation, and further reef degradation.

It is increasingly recognized that effective management is achieved only through approaches that integrate biophysical considerations with socio-economic and related factors. Balanced stakeholder involvement is generally a prerequisite for compliance with management decisions. The field of integrated coastal management is rapidly evolving, as is the set of scientific paradigms on which it is based. To a large degree, the future of the world's coral reefs is directly linked to this evolution.

See Also

Beaches, Physical Processes Affecting. Geomorphology. History of Ocean Sciences. Manned Submersibles, Shallow Water. Rocky Shores. Satellite Oceanography, History and Introductory Concepts. Sea Level Variations Over Geologic Time.

Further Reading

Birkeland C (ed.) (1997) *Life and Death of Coral Reefs.* New York: Chapman and Hall.

Bryant D, Burke L, McManus J, and Spalding M (1998) *Reefs at Risk: a Map-based Indicator of the Threats to the World's Coral Reefs.* Washington, DC: World Resources Institute.

Davidson OG (1998) *The Enchanted Braid: Coming to Terms with Nature on the Coral Reef.* New York: John Wiley.

Holliday L (1989) *Coral Reefs: a Global View by a Diver and Aquarist.* London: Salamander Press.

McManus JW, Ablan MCA, Vergara SG, et al. (1997) *ReefBase Aquanaut Survey Manual.* Manila, Philippines: International Center for Living Aquatic Resources Management.

McManus JW and Vergara SG (eds.) (1998) *ReefBase: a Global Database of Coral Reefs and Their Resources. Version 3.0 (Manual and CD-ROM).* Manila, Philippines: International Center for Living Aquatic Resource Management.

Polunin NVC and Roberts C (eds.) (1996) *Reef Fisheries.* New York: Chapman and Hall.

Sale PF (ed.) (1991) *The Ecology of Fishes on Coral Reefs.* New York: Academic Press.

Stafford-Deitsch J (1993) *Reef: a Safari Through the Coral World.* San Francisco: Sierra Club Books.

Veron JEN (1993) *A Biogeographic Database of Hermatypic Corals. Species of the Central Indo-Pacific, Genera of the World. Monograph Series 10.* Townsville, Australia: Australian Institute of Marine Science.

MANNED SUBMERSIBLES, SHALLOW WATER

T. Askew, Harbor Branch Oceanographic
Institute, Ft Pierce, FL, USA

Introduction

Early man's insatiable curiosity to look beneath the surface of the sea in search of natural treasures that were useful in a primitive, comfortless mode of living were a true test of his limits of endurance. The fragile vehicle of the human body quickly discovered that most of the sea's depths were unapproachable without some form of protection against the destructive hostilities of the ocean.

Modern technology has paved the way for man to conquer the hostile marine environment by creating a host of manned undersea vehicles. Called submersibles, these small engineering marvels carry out missions of science, exploration, and engineering. The ability to conduct science and other operations under the sea rather than from the surface has stimulated the submersible builder/operator to further develop the specialized tools and instruments which provide humans with the opportunity to be present and perform tasks in relative comfort in ocean bottom locations that would otherwise be destructive to human life.

Over the past fifty years a depth of 1000 m has surfaced as the transition point for shallow vs deep water manned submersibles. During the prior 100 years any device that enabled man to explore the ocean depths beyond breath-holding capabilities would have been considered deep.

History

While it is difficult to pinpoint the advent of the first submersible it is thought that in 1620 Cornelius van Drebel constructed a vehicle under contract to King James I of England. It was operated by 12 rowers with leather sleeves, waterproofing the oar ports. It is said that the craft navigated the Thames River for several hours at a depth of 4 m and carried a secret substance that purified the air, perhaps soda lime?

In 1707, Dr Edmund Halley built a diving bell with a 'lock-out' capability. It had glass ports above to provide light, provisions for replenishing its air, and crude umbilical-supplied diving helmets which permitted divers to walk around outside. In 1776, Dr David Bushnell built and navigated the first submarine employed in war-like operations. Bushnell's Turtle was built of wood, egg-shaped with a conning tower on top and propelled horizontally and vertically by a primitive form of screw propeller after flooding a tank which allowed it to submerge.

In the early 1800s, Robert Fulton, inventor of the steamship, built two iron-formed copper-clad submarines, *Nautilus* and *Mute*. Both vehicles carried out successful tests, but were never used operationally.

The first 'modern submersible' was Simon Lake's *Argonaut I*, a small vehicle with wheels and a bottom hatch that could be opened after the interior was pressurized to ambient. While there are numerous other early submarines, the manned submersible did not emerge as a useful and functional means of accomplishing underwater work until the early 1960s.

It was during this same period of time that the French-built *Soucoupe*, sometimes referred to as the 'diving saucer', came into being. Made famous by Jacques Cousteau on his weekly television series, the *Soucoupe* is credited with introducing the general population to underwater science. Launched in 1959, the diving saucer was able to dive to 350 m.

The *USS Thresher* tragedy in 1963 appears to have spurred a movement among several large corporations such as General Motors (Deep Ocean Work Boat, DOWB), General Dynamics (*Star I, II, III, Sea Cliff*, and *Turtle*), Westinghouse (*Deepstar 2000, 4000*), and General Mills (*Alvin*), along with numerous other start-up companies formed solely to manufacture submersibles. Perry Submarine Builders, a Florida-based company, started manufacturing small shallow-water, three-person submersibles in 1962, and continued until 1980 (**Figures 1** and **2**). International Hydrodynamics Ltd (based in Vancouver, BC, Canada) commenced building the Pisces series of submersibles in 1962. The pressure hull material in the 1960s was for the most part steel with one or more view ports. The operating depth ranged from 30 m to 600 m, which was considered very deep for a free-swimming, untethered vehicle.

The US Navy began design work on *Deep Jeep* in 1960 and after 4 years of trials and tribulations it was commissioned with a design depth of 609 m. A two-person vehicle, *Deep Jeep* included many features incorporated in today's submersibles, such as a dropable battery pod, electric propulsion motors that operate in silicone oil-filled housings, and shaped resin blocks filled with glass micro-balloons

Figure 1 Perry-Link *Deep Diver*, 1967. Owned by Harbor Branch Oceanographic Institution. Length 6.7 m, beam 1.5 m, height 2.6 m, weight 7485 kg, crew 1, observers 2, duration 3–5 hours.

used to create buoyancy. *Deep Jeep* was eventually transferred to the Scripps Institution of Oceanography in 1966 after a stint searching for a lost 'H' bomb off Palomares, Spain. Unfortunately, *Deep Jeep* was never placed into service as a scientific submersible due to a lack of funding. The missing bomb was actually found by another vehicle, *Alvin*. *Alvin* did get funding and proved to be useful as a scientific tool.

The Nekton series of small two-person submersibles appeared in 1968, 1970, and 1971. The *Alpha*, *Beta*, and *Gamma* were the brainchildren of Doug Privitt, who started building small submersibles for recreation in the 1950s. The Nektons had a depth capability of 304 m. The tiny submersibles conducted hundreds of dives for scientific purposes as well as for military and oilfield customers. In 1982, the *Nekton Delta*, a slightly larger submersible with a depth rating of 365 m was

unveiled and is still operating today with well over 3000 dives logged.

A few of the submersibles were designed with a diver 'lock-out' capability. The first modern vehicle was the Perry-Link 'Deep Diver' built in 1967 and able to dive to 366 m. This feature enabled a separate compartment carrying divers to be pressurized internally to the same depth as outside, thus allowing the occupants to open a hatch and exit where they could perform various tasks while under the supervision of the pilots. Once the work was completed, the divers would re-enter the diving compartment, closing the outer and inner hatches; thereby maintaining the bottom depth until reaching the surface, where they could decompress either by remaining in the compartment or transferring into a larger, more comfortable decompression chamber via a transfer trunk.

Acrylic plastic was tested for the first time as a new material for pressure hulls in 1966 by the US Navy.

Figure 2 Perry PC 1204 *Clelia*, 1976. Owned and operated by Harbor Branch Oceanographic Institution. Length 6.7 m, beam 2.4 m, height 2.4 m, weight 8160 kg, crew 1, observers 2, duration 3–5 hours.

The *Hikino*, a unique submersible that incorporated a 142 cm diameter and a 0.635 cm thick hull, was only able to dive to 6 m. This experimental vehicle was used to gain experience with plastic hulls, which eventually led to the development of *Kumukahi*, *Nemo* (Naval Experimental Manned Observatory), *Makakai*, and *Johnson-Sea-Link*.

The *Kumukahi*, launched in 1969, incorporated a unique 135 cm acrylic plastic sphere formed in four sections. It was 3.175 cm thick and could dive to 92 m.

The *Nemo*, launched in 1970, and *Makakai*, launched in 1971, both utilized spheres made of 12 curved pentagons formed from a 6.35 cm flat sheet of Plexiglas™. The pentagons were bonded together to make one large sphere capable of diving to 183 m.

The *Johnson-Sea-Link*, designed by Edwin Link and built by Aluminum Company of America (ALCOA), utilized a *Nemo*-style Plexiglas™ sphere, 167.64 cm in diameter, 10.16 cm thick, and made of 12 curved pentagons formed from flat sheet and bonded together. This new thicker hull had an operational depth of 304 m.

Present Day Submersibles

The submersibles currently in use today are for the most part classified as either shallow-water or deep-water vehicles, the discriminating depth being approximately 1000 m. This is where the practicality of using compressed gases for ballasting becomes impractical. The deeper diving vehicles utilize various drop weight methods; most use two sets of weights (usually scrap steel cut into uniform blocks). One set of weights is released upon reaching the bottom, allowing the vehicle to maneuver, travel, and perform tasks in a neutral condition. The other set of weights is dropped to make the vehicle buoyant, which carries it back to the surface once the dive is complete.

The shallow vehicles use their thrusters and/or water ballast to descend to the bottom and some of the more sophisticated submersibles have variable ballast systems which allow the pilot to achieve a neutral condition by varying the water level in a pressure tank.

One advantage of the shallow vehicles is that the view ports (commonly called windows) can be much larger both in size and numbers, and where an acrylic plastic sphere is used the entire hull becomes a window.

Since the late 1960s and early 1970s acrylic plastic pressure hulls have emerged as an ideal engineering solution to create a strong, transparent, corrosion-resistant, nonmagnetic pressure hull. The limiting factor of the acrylic sphere is its ability to resist implosion from external pressure at great depths. Its strength comes from the shape and wall thickness. Therefore, the greater the depth the operator aims to reach, the thicker the sphere must be, which results in a hull that is much too heavy to be practical for use on a small submersible designed to go deeper than 1000 m.

These shallow-water submersibles, once quite numerous because of their usefulness in the offshore oilfield industry, are now limited to a few operators and mostly used for scientific investigations.

The *Johnson-Sea-Links* (J-S-Ls) stand out as two of the most advanced manned submersibles (**Figure 3**). *J-S-L I*, commissioned by the Smithsonian Institution in January 1971, was named for designer and donors Edwin A. Link and J. Seward Johnson.

Edwin Link, responsible for the submersible's unique design and noted for his inventions in the aviation field, turned his energies to solving the problems of undersea diving, a technology then still in its infancy. One of his objectives was to carry out scientific work under water for lengthy periods.

The *Johnson-Sea-Link* was the most sophisticated diving craft he had created for this purpose, and it promised to be one of the most effective of the new generation of small submersible vehicles that were being built to penetrate the shallow depths of the continental shelf (183 m or 100 fathoms).

Originally designed for a depth of 304 m, the vessel's unique features include a two-person transparent acrylic sphere, 1.82 m in diameter and 10.16 cm thick, that provides panoramic underwater visibility to a pilot and a scientist/observer. Behind the sphere, there is a separate 2.4 m long cylindrical, welded, aluminum alloy lock-out/lock-in compartment that will enable scientists to exit from its bottom and collect specimens of undersea flora and fauna. The acrylic sphere and the aluminum cylinder are enclosed within a simple jointed aluminum tubular frame, a configuration that makes the vessel resemble a helicopter rather than a conventionally shaped submarine. Attached to the frame are the vessel's ballast tanks, thrusters, compressed air, mixed gas flasks, and battery pod.

The aluminum alloy parts of the submersible, lightweight and strong, along with the acrylic capsule which was patterned after the prototype used by the US Navy on the *Nemo*, had extraordinary advantages over traditional materials like steel. They were most of all immune to the corrosive effects of sea water.

The emphasis in engineering of the submersible was on safety. Switches, connectors, and all operating gear were especially designed to avoid possible safety hazards. The rear diving compartment allows one diver to exit for scientific collections while tethered for communications and breathing air supply, while the other diver/tender remains inside as a safety backup. Once the dive is completed and the submersible is recovered by a special deck-mounted crane on the support ship, the divers can transfer into a larger decompression chamber via a transfer trunk which is bolted to the lock-out/lock-in compartment.

Now, 30 years later, the *Johnson-Sea-Links* with a 904 m depth rating, remain state of the art underwater vehicles. Sophisticated hydraulic manipulators work in conjunction with a rotating bin collection platform which allow 12 separate locations to be sampled and simultaneously documented by digital color video cameras mounted on electric pan and tilt mechanisms and aimed with lasers. Illumination is provided by a variety of underwater lighting systems utilizing zenon arc lamps, metal halide, and halogen bulbs. Acoustic beacons provide real time position

Figure 3 *Johnson-Sea-Link I* and *II*, 1971 and 1976. Owned and operated by Harbor Branch Oceanographic Institution. Length 7.2 m, beam 2.5 m, height 3.1 m, weight 10400 kg, crew 2, observers 2, duration 3–5 hours.

and depth information to shipboard computer tracking systems that not only show the submersible's position on the bottom, but also its relationship to the ship in latitude and longitude via the satellite-based global positioning system (GPS). The lock-out/lock-in compartment is now utilized as an observation and instrumentation compartment, which remains at one atmosphere.

Today's shallow-water submersibles (average dive 3–5 h) require a support vessel to provide the necessities that are not available due to their relatively small size. The batteries must be charged, compressed air and oxygen flasks must be replenished. Carbon dioxide removal material, usually soda lime or lithium hydroxide, is also replenished so as to provide maximum life support in case of trouble. Most submersibles today carry 5 days of life support, which allows time to effect a rescue should it become necessary.

The support vessel also must have a launch/recovery system capable of safely handling the submersible in all sea conditions. Over the last 30 years, the highly trained crews that operate the ship's handling systems and the submersibles, have virtually made the shallow-water submersibles an everyday scientific tool where the laboratory becomes the ocean bottom.

Operations

The two *Johnson-Sea-Links* have accumulated over 8000 dives for science, engineering, archaeology, and training purposes since 1971. They have developed into highly sophisticated science tools. Literally thousands of new species of marine life have been photographed, documented by video camera and collected without disturbing the surrounding habitat. Behavioral studies of fish, marine mammals and invertebrates as well as sampling of the water column and bottom areas for chemical analysis and geological studies are everyday tasks for the submersibles. In addition, numerous historical shipwrecks from galleons to warships like the *USS Monitor* have been explored and documented, preserving their legacy for future generations.

Johnson-Sea-Links I and II (J-S-Ls) were pressed into service to assist in locating, identifying, and ultimately recovering many key pieces of the ill-fated Space Shuttle Challenger. This disaster, viewed by the world via television, added a new dimension to the J-S-Ls' capabilities. Previously only known for their pioneering efforts in marine science, they proved to be valuable assets in the search and recovery operation. The J-S-Ls completed a total of 109 dives, including mapping a large area of the right solid rocket booster debris at a depth of 365 m. The vehicles proved their worth throughout the operation by consistently performing beyond expectations. They were launched and recovered easily and quickly. They could work on several contacts per day, taking NASA engineers to the wreckage for first-hand detailed examination of debris while video cameras recorded what was being seen and said. Significant pieces were rigged with lifting bridles for recovery. The autonomous operation of the J-S-Ls, a dedicated support vessel, and highly trained operations personnel made for a successful conclusion to an operation that had a significant impact on the future of the US Space Program.

Summary

There is no question that the manned submersible has earned its place in history. Much of what are now cataloged as new species were discovered in the last 30 years with the aid of submersibles. The ability to conduct marine science experiments *in situ* led to the development of intricate precision instruments, sampling devices for delicate invertebrates and gelatinous organisms that previously were only seen in blobs or pieces due to the primitive methods used to collect them.

While some suggest that remotely operated vehicles (ROVs) could, and have replaced the manned submersible, in reality they are complementary. There is no substitute for the autonomous, highly maneuverable submersible that can approach and collect without contact delicate zooplankton, while observing behavior and measuring the levels of bioluminescence, or probing brine pools and cold seep regions in the Gulf of Mexico for specialized collections of biological, geological, and geochemical samples. Tubeworms are routinely marked for growth rate studies and collected individually, along with other biological species that thrive in these chemosynthetic communities. Sediments and methane ice (gas hydrates) are also selectively retrieved for later analysis.

Some new vehicles are still being produced, but have limited payloads, which restricts them to specific tasks such as underwater camera platforms or observation. Some are easily transportable but are small and restricted to one occupant; they can be carried by smaller support vessels and are more economical to operate. Man's desire to explore the lakes, oceans, and seas has not diminished. New technology will only enable, not reduce, the need for man's presence in these hostile environments.

See also

Deep Submergence, Science of. Manned Submersibles, Deep Water.

Further Reading

Askew TM (1980) *JOHNSON-SEA-LINK Operations Manual*. Fort Pierce: Harbor Branch Foundation.

Busby F (1976, 1981) *Undersea Vechicles Directory*. Arlington: Busby & Associates.

Forman WR (1968) *KUMUKAHI Design and Operations Manual*. Makapuu, HI: Oceanic Institute.

Forman W (1999) *The History of American Deep Submersible Operations*. Flagstaff, AZ: Best Publishing Co.

Link MC (1973) *Windows in the Sea*. Washington, DC: Smithsonian Institute Press.

Stachiw JD (1986) *The Origins of Acrylic Plastic Submersibles*. American Society of Mechanical Engineers, Asme Paper 86-WA/HH-5.

Van Hoek S and Link MC (1993) *From Sky to Sea; A Story of Ed Link*. Flagstaff, AZ: Best Publishing Co.

HUMAN INVOLVEMENT

MARITIME ARCHAEOLOGY

R. D. Ballard, Institute for Exploration, Mystic, CT, USA

Introduction

For thousands of years, ancient mariners have traversed the waters of our planet. During this long period of time, many of their ships have been lost along the way, carrying their precious cargo and the history it represents to the bottom of the sea. Although it is difficult to know with any degree of precision, some estimate that there are hundreds of thousands of undiscovered sunken ships littering the floor of the world's oceans.

For hundreds of years, attempts have been made to recover their contents. In *Architettura Militare* by Francesco de Marchi (1490–1574), for example, a device best described as a diving bell was used in a series of attempts to raise a fleet of 'pleasure galleys' from the floor of Lake Nemi, Italy in 1531. In *Treatise on Artillery* by Diego Ufano in the mid-1600s, a diver wearing a crude hood and airhose of cowhide was shown lifting a cannon from the ocean floor. Clearly, these early attempts at recovering lost cargo were done for economic, not archaeological, reasons and were very crude and destructive.

The field of maritime archaeology, on the other hand, is a relatively young discipline, emerging as a recognizable study in the later 1800s. Not to be confused with nautical archaeology which deals solely with the study of maritime technology, maritime archaeology is much broader in scope, concerning itself with all aspects of marine-related culture including social, religious, political, and economic elements of ancient societies.

Early History

Sunken ships offer an excellent opportunity to learn about those ancient civilizations. Archaeological sites on land can commonly span hundreds to even thousands of years with successive structures being built upon the ruins of older ones. Correlating a find in one area to a similar stratigraphic find in another can introduce errors that potentially represent long periods of time. A shipwreck, on the other hand, represents a 'time capsule', the result of a momentary event where the totality of the artifact assemblage comes from one distinct point in time.

It is important to point out that maritime archaeology's first major field efforts were not conducted underwater but, in fact, were the excavation of boats that are now located in land-locked sites. In 1867, the owners of a farm began to cart away the soil from a large mound some 86 m in length only to discover the timbers of a large Viking ship, the Tune ship, complete with the charred bones of a man and a horse revealing it to be a burial chamber. In 1880, the Gokstad burial ship was discovered in a flat plain on the west side of the Oslo fiord. It was buried in blue clay which resulted in a high state of preservation. Contained within the grave was a Viking king, his weapons, twelve horses, six dogs and various artifacts.

Since the late 1890s, the excavation of boats and harbor installations in terrestrial settings continues to this day, following more or less traditional land excavation protocol. One of the most famous discoveries took place in 1954 near the Great Pyramid of Egypt. While building a new road around the Pyramid, a series of large limestone blocks were encountered beneath which was found an open pit containing the oldest intact ship ever discovered. Dating to 2600 BC, the ship measures 43 m in length, weighs 40 tons, and stands 7.6 m from the keel line.

Although ships found in terrestrial settings provide valuable insight into the culture of their period, many have a more religious significance than one reflecting the economics of the period. Burial ships were commonly modified for this unique purpose and were not engaged in maritime activity at the time of their burial.

Prior to the advent of SCUBA diving technology invented by Captain Jacques-Yves Cousteau and Emile Gagnan in 1942, archaeology conducted underwater by trained archaeologists was extremely rare. In fact, owing to the dangerous aspects of pre-SCUBA diving techniques which included the use of 'hard hats' employed by commercial sponge divers, archaeologists relied upon these commercial divers instead of doing the underwater work themselves.

Even after SCUBA technology became available to the archaeological community it was the commercial or recreational divers by their numbers who tended to discover an ancient shipwreck site. As a result, it was not uncommon for a site to be stripped of its small, unique and easily recovered artifacts before the 'authorities' were notified of a wreck's location.

Even after learning of these sites, many of the early efforts in underwater archaeology were conducted by divers lacking archaeological training.

The first was the famous *Antikythera* wreck off Greece which contained a cargo of bronze and marble sculptures. The site was initially raided in 1900 by sponge divers wearing copper helmets, lead weights, and steel-soled shoes with air supplied to them from the surface by a hand-cranked compressor.

Fortunately, the location of the wreck was soon discovered by the Greek government which with the help of their navy, mounted a follow-up expedition under supervision from the surface by Professor George Byzantinos, their Director of Antiquities, which resulted in the recovery of more of its valuable cargo. The wreck turned out to be a first century Roman ship carrying its Greek treasures back to Rome including the famous bronze statue *Youth*, thought to be done by the last great classical Greek sculptor, Lysippos.

A few years later, another Roman argosy was discovered off Mahdia, Tunisia which was also initially plundered for its Greek statuary. Again, the local government and their Department of Antiquities took control of the salvage operation which continued from 1907 to 1913 enriching the world with its bronze artwork.

Unfortunately, not all of these early shipwreck discoveries were reported to the local government. Their fate was far less fortunate than those just mentioned including a first century BC wreck lost off Albenga, Italy which was torn apart by the Italian salvage ship *Artiglio II* using bucket grabs to penetrate its interior holds.

One of the first ancient shipwrecks to be excavated under some semblance of archaeological control was carried out by Captain Cousteau in 1951 along with Professor Fernand Benoit, Director of Antiquities for Provence off Grand Congloue near Marseilles, France. It, too, was initially discovered by a salvage diver. Although no site plan was ever published after three seasons of diving, the ship was thought to be Roman from 230 BC, 31 m long, at a depth of 35 m, and carrying 10 000 amphora and 15 000 pieces of Campanian pottery bowls, pots, and 40 types of dishes with an estimated 20 tons of lead aboard.

At the same time, a colleague of Cousteau, Commander Philippe Taillez of the French Navy organized a similar excavation of a first century BC wreck on the Titan reef off the French coast but in the absence of an archaeologist failed to actually document the site. Without the presence of trained archaeologists working underwater, there was little hope that acceptable techniques would be developed that met archaeological standards.

In 1958, Peter Throckmorton who was the Assistant Curator of the San Francisco Maritime Museum, went to Turkey hoping to locate an ancient shipwreck site. Like many before him, he learned from local sponge divers the location of a wreck off Cape Gelidonya on the southern coast of Turkey from which a series of broken objects had been recovered. During the following year, Throckmorton's team made a number of dives on the site and recovered a series of bronze tools and ingots revealing the ship to be a late Bronze Age wreck from around 1200 BC but no major effort was mounted.

Throckmorton and veteran diver Frederick Dumas, however, returned in 1960 along with a young archaeologist eager to make his first openocean dives, George Bass. With Bass in charge of the archaeological aspects of the diving program, they began to establish for the first time true archaeological mapping and sampling protocol. Although primitive by today's standard, they established a traditional grid system using anchored lines followed by the use of an airlift system to carefully remove the overburden covering the wreck. Working in 28 m of water, the ship proved to be 11 m in length, covered with a coralline concretion some 20 cm thick. Its more than one ton of cargo consisted of four-handled ingots of copper and 'bun' ingots of bronze as well as a large quantity of broken and unfinished tools, including both commercial and personal goods. These assemblages of artifacts suggested the captain was a Syrian scrap-metal dealer who was also a tinker making his way from Cyprus along the coast of Turkey to the Aegean Sea.

Maritime archaeology truly came of age with the excavation of the Yassi Adav wreck from 1961 to 1964 off the south coast of Turkey. Running offshore near the small barren island of Yassi Adav is a reef which is as shallow as 2 m that has claimed countless ships over the years. Its surface is strewn with Ottoman cannon balls from various wrecks of that period. Several of the ships that have run aground on its unforgiving coral outcrops slide off its rampart coming to rest on a sandy bottom. One such ship, lying in 31 m of water was the focus of an intense effort carried out by the University Museum of Pennsylvania under the direction of George Bass.

Prior to this effort, no ancient ship had ever been recovered in its entirety; this became the objective of this project. With the help of fifteen specialists and thousands of hours underwater, the team carefully mapped the site. The techniques developed during this excavation effort became the new emerging standards for this young field of research and continue to be followed today by research teams around the world.

Bass' team first cleared the upper surface of cargo of its encrustation of weed. Its 900 amphora were then mapped, cataloged, and removed. One hundred were taken to the surface for subsequent conservation and preservation whereas the remaining 800 were stored on the bottom at an off-site location. Simple triangulation was first used to initially delineate the wreck site followed by the establishment of a complex series of wire grids. Each object was given a plastic tag and artists hovered over the grid system making numerous sketches of the site before any object was recovered.

Copper and gold coins recovered from the site revealed the ship to be of Byzantine age sinking during the reign of Emperor Heraclius from AD. 610 to 641. Following the recovery of the ship's contents, the now-exposed hull provided marine archaeologists with their first opportunity to develop techniques needed to document and recover the ship's timbers.

This effort was extremely time consuming but the resulting insight proved worth the effort, providing the archaeological community with a transitional method of ship construction between the classical 'shell' technique to the later evolved 'skeleton' technique. Its length to width ratio of 3.6 to 1 further supported earlier suggestions that ships of this period would have to be built with more streamlined hulls to outrun and outmaneuver hostile or piratical adversaries.

The Growth of Maritime Archaeology

Following the excavation of the Yassi Adav wreck, members of Bass' team then conducted the extensive excavation of a fourth century BC wreck near Kyrenia in Cyprus directed by Michael Katzev. This effort mirrored the Yassi Adav project and resulted in the raising and conservation of the ship's preserved hull structure.

Since these early pioneering efforts, numerous maritime archaeology programs have emerged around the world. Off Western Europe and in the Mediterranean maritime archaeology remains a strong focus of activities. Bass and his Institute for Nautical Archaeology (Texas A&M University in College Station, Texas) continue to carry out a growing number of underwater research projects off the southern coast of Turkey centered at their research facility in Bodrum.

His excavation of the late Bronze Age Ulu Burun wreck off the southern Turkish coast led to the recovery of thousands of artifacts that provided valuable insight in a period of time marked by the reign of Egypt's Tutankhamun and the fall of Troy. His Byzantine 'Glass Wreck' found north of the island of Rhodes and dating from the twelfth to thirteenth century AD, continues to generate important information about this particular period of maritime trade.

In addition to the well-known work with the Wasa, Swedish archaeologists have conducted excellent work in the worm-free waters of the Baltic. In Holland, Dutch archaeologists have drained large sections of shallow water areas which contain a rich history of maritime trade dating from the twelfth to nineteenth centuries.

Another major program directed by Margaret Rule took place in England with the recovery and preservation of the Mary Rose, a large warship lost in July 1545 during the reign of King Henry VIII. From this project, a great deal was learned about the long-term preservation of wooden timbers which is being incorporated in other similar conservation programs.

Research efforts in America span the length of its human habitation. Recent archaeological research suggests that humans arrived in North America more than 12 000 years ago when a southern route was first thought to have opened in the glacial icesheet covering the continent. Some scientists now suggest that early humans may have circumvented this barrier by way of water or overland surfaces now submerged on the continental shelf. New research programs are now being designed to work on the continental shelf looking for early evidence of Paleoindian settlements.

For years Indian canoes, rafts, dugouts, and reed boats have been discovered in freshwater lakes, and sinkholes in the limestone terrains of North and Central America have attracted researchers for many years in search of human sacrifices and other religious artifacts associated with native American cultures.

Ships associated with early explorers, including Columbus, French explorer Rene La Salle, the British, and countless Spanish explorers, have been the focus of research efforts in the Gulf of Mexico, the Northwest Passage, and the Caribbean while shipwrecks from the Revolutionary War and War of 1812 have been discovered in the Great Lakes and Lake Champlain. Warships associated with the American Civil War have received renewed interest including the Monitor lost off Cape Hatteras, the submarine Huntley and numerous other recent finds in the coastal waters of the US east coast.

Within the last two decades, deep water search systems developed by the oceanographic community have been used to successfully locate the remains of the RMS *Titanic*, the German Battleship *Bismarck*, fourteen warships lost during the Battle of Guadalcanal, and the US aircraft carrier *Yorktown* lost during the Battle of Midway.

Ships associated with World War II have been carefully documented including the *Arizona* in Pearl Harbor and numerous ships sunk during nuclear bomb testing in the atolls of the Pacific. Maritime archaeology is not limited to European or American investigators. A large number of underwater sites too numerous to mention have been investigated off the coasts of Africa, the Philippines, the Persian Gulf, South America, China, Japan, and elsewhere around the world as this young field begins to experience an explosive growth.

Marine Methodologies

As was previously noted, early underwater archaeological sites were not discovered by professional archaeologists; they were found, instead, by commercial or recreational divers. It wasn't until the mid-1960s that archaeologists, notably George Bass, began to devise their own search strategies. Being divers, their early attempts tended to favor visual techniques from towing divers behind their boats, to towed camera systems, and finally small manned submersibles (**Figure 1**).

It was not until the introduction of side-scan sonars that major new wreck sites were found. Operating at a frequency of 100 kHz, such sonar systems are able to search a swath-width of ocean floor 400 m wide, moving through the water at a typical speed of 3–5 knots (5.5–9 km h^{-1}). Today, numerous companies build side-scan sonars each offering a variety of options ranging from higher frequencies (i.e. 500 kHz) to improved signal processing, recording, and display.

Various magnetic sensors have been used effectively over the years in locating sunken shipwreck sites having a ferrous signature. This is particularly true for warships with large cannons aboard. Magnetic sensors have also proved effective in locating buried objects in extremely shallow water, on beaches, and beneath coastal dunes.

Over the years, a variety of changes have taken place with regard to the actual documentation of a

Figure 1 Archaeological mapping techniques pioneered by Dr George Bass of Texas A&M University for shallow water archaeology. These techniques were heavily dependent on the use of divers and were limited to less than 50 m water depths.

wreck site. Beginning in the early 1960s, various stereophotogrammetry techniques were used. More recently the SHARPS (sonic high accuracy ranging and positioning system) acoustic positioning system has proved extremely rapid and cost effective in accurately mapping submerged sites. This tracking technique coupled with electronic imaging sensors, has produced spectacular photomosaics.

More recently, remotely operated vehicles have begun to enter this field of research. In 1990, the JASON vehicle from the Woods Hole Oceanographic Institution was used to map the *Hamilton* and *Scourge*, two ships lost during the War of 1812 in Lake Ontario. Using a SHARPS tracking system, the vehicle was placed in closed-loop control and made a series of closely spaced survey lines across and along the starboard and port sides of the ships. Mounted on the remotely operated vehicle (ROV) was a pencil-beam sonar and electronic still camera which resulted in volumetric models of the ships as well as electronic mosaics.

Deep-Water Archaeology

The shallow waters of the world's oceans where the vast majority of maritime archaeology has been done represent less than 5% of its total surface area. For years, archaeologists have argued that the remaining 95% is unimportant since the ancient mariner stayed close to land and it was there that their ships sank.

This premise was challenged in 1988 when an ancient deep-water trade route was first discovered between the Roman seaport of Ostia and ancient Carthage. The discovery site was situated more than 100 nautical miles (185 km) off Carthage in approximately 1000 m of water. Over a nine-year period from 1988 to 1997, a series of expeditions resulted in the discovery of the largest concentration of ancient ships ever found in the deep sea. In all, eight ships were located in an area of $210 \, km^2$, including five of the Roman era spanning a period of time from 100 BC to AD 400. The project involved the use of highly sophisticated deep submergence technologies including towed acoustic and visual search vehicles, a nuclear research submarine, and an advanced remotely operated vehicle. Precision navigation and control, similar to that first used in Lake Ontario in 1990, permitted rapid yet careful visual and acoustic mapping of each site with a degree of precision never attained before utilizing advanced robotics, the archaeological team recovered selected objects from each site for subsequent analysis ashore without intrusive excavation.

Deep-water wreck sites offer numerous advantages over shallow water sites. Ships lost in shallow water commonly strike an underwater obstacle such as rocks or reefs severely damaging themselves in the process. Each winter more storms continue to damage the site as encrustation begins to form. Commonly, the site is located by nonarchaeologists who frequently retrieve artifacts before reporting the wreck's location.

Ships that sink in deep water, however, tend to be swamped. As a result, they sink intact, falling at a slow speed toward the bottom where they come to rest standing upright in the soft bottom ooze. When they are located, they have not been looted. Sedimentation rates in the deep sea are extremely slow, commonly averaging 1 cm per 1000 years. That coupled with cold bottom temperatures, total darkness, and high pressures result in conditions favoring preservation. Although wood-boring organisms remove exposed wooden surfaces, deep sea muds encase the lower portions of the wreck in an oxygen free environment. When deep-sea excavation techniques are developed in the near future, these wrecks may provide highly preserved organic material normally lost in shallow-water sites.

The Roman shipwrecks located off Carthage were found within a much larger area of isolated artifacts spanning a longer period of time. The isolated artifacts appear to have been discarded from passing ships overhead. Given the slow sedimentation rates in the deep sea, it might be possible to easily delineate ancient trade routes by looking for these debris trails, thus learning a great deal about ancient maritime trading practices.

Since this new field of deep-water archaeology has grown out of the oceanographic community, it brings with it a strong expertise in deep submergence technology. The newly developed ROVs possess the latest in advanced imaging, robotics, and control technologies. Using this technology, archaeologists are able to map underwater sites far faster than their shallow water counterparts (**Figure 2**).

Most recently, a second deep-water archaeological expedition resulted in the discovery of two Phoenician ships lost some 2700 years ago. Located in 450 m of water about 30 nautical miles (55 km) off the coast of Egypt, these two ships are lying upright. Due to local bottom currents, both ships are completely exposed resting in two-meter deep elongated depressions.

Ethics

As pointed out earlier, the salvaging of cargo from lost ships goes back much farther in time than marine archaeological research. As a result, this long

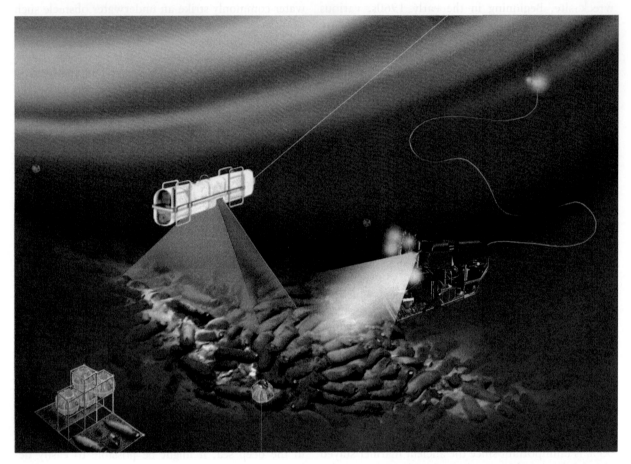

Figure 2 Archaeological techniques pioneered at the Woods Hole Oceanographic Institution and the Institute for Exploration rely exclusively on remotely operated vehicle systems with operating depths down to 6000 m.

history of maritime salvage, rooted in international law, has led to a quasipublic acceptance of salvage operations, making it difficult for the archaeological community to garner moral and legal public support to protect and preserve truly important underwater archaeological sites. 'Finders keepers' remains rooted in the public's mind as a logical policy governing lost ships. Further blurring the boundary between these two extremes in the early years, was the fact that marine archaeologists relied upon the very community that was removing artifacts from underwater sites to tell them where they were located.

This uneasy marriage between the diving community and the archaeological community has, in many ways, stifled the growth and acceptance of the field. Its lack of development of systematization which arises from its immaturity and lack of a large database has further hindered its acceptance into mainstream archaeology.

Today's marine salvagers commonly employ individuals with archaeological experience to participate in their operations. In some cases, this results in

important documentation of the site as was the case with the salvage of the *Central America*. In other cases, however, they are being used to create a false impression that archaeological standards are being followed when they are not.

American salvagers have, in large part, concentrated their attention on lost ships of the Spanish Main beginning with search efforts off the coast of Florida where a large number of silver- and gold-bearing ships were lost in hurricanes between 1715 and 1733. A famous shipwreck in this area, the *Atocha*, was exploited by salvager Mel Fisher.

On the Silver Bank off the Dominican Republic in the Caribbean the richly laden *Nuestra Senora de la Pura y Limpia Concepcion* sank in October 1641. Salvage efforts seeking to retrieve its valuable cargo began almost immediately including one by the British in 1687. Lost from memory, the *Concepcion* was relocated in 1978 by American treasure hunters who continue their recovery efforts to this day.

Fortunately, more and more countries are beginning to enact laws to protect offshore cultural sites, but with the emergence of deep-water archaeology which is conducted on the high seas, the majority of the world's oceans and the human history contained within them are not protected.

One logical step is to add human history to the present Law of the Sea Convention that governs the exploitation of natural resources. Although this would not protect all future underwater sites, it would serve as an important first step.

See also

Sonar Systems.

Further Reading

Babits LE and Tilburg HV (1998) *Maritime Archaeology*. New York: Plenum Press.

Ballard RD (1993) The MEDEA/JASON remotely operated vehicle system. *Deep-Sea Research* 40(8): 1673–1687.

Ballard RD, McCann AM, Yoerger D, Whitcomb L, Mindell D, Oleson J, Singh H, Foley B, Adams J, Piechota D, and Giangrande C (2000) The discovery of ancient history in the deep sea using advanced deep submergence technology. *Deep Sea Research, Part I* 47: 1591–1620.

Bass GF (1972) *A History of Seafaring Based on Underwater Archaeology*. New York: Walker.

Bass GF (1975) *Archaeology Beneath the Sea*. New York: Walker.

Bass GF (1988) *Ships and Shipwrecks of the Americas*. New York: Thames and Hudson.

Cockrell WA (1981) Some moral, ethical, and legal considerations in archaeology. In: Cockrell WA (ed.) *Realms of Gold*. Proceedings of the Tenth Conference on Underwater Archaeology, Fathom Eight San Marino, California. pp. 215–220.

Dean M and Ferrari B (1992) *Archaeology Underwater: The NAS Guide to Principles and Practice*. London: Nautical Archaeology Society.

Greene J (1990) *Maritime Archaeology*. London: Academic Press.

Muckelroy K (1978) *Maritime Archaeology*. New York: Cambridge University Press.

NESCO (1972) *Underwater Archaeology – A Nascent Discipline*. Paris: UNESCO.

MINERAL EXTRACTION, AUTHIGENIC MINERALS

J. C. Wiltshire, University of Hawaii, Manoa, Honolulu, HA, USA

Introduction

The extraction of marine mineral resources represents a worldwide industry of just under two billion dollars per year. There are approximately a dozen general types of marine mineral commodities (depending on how they are classified), about half of which are presently being extracted successfully from the ocean. Those being extracted include sand, coral, gravel and shell for aggregate, cement manufacture and beach replenishment; magnesium for chemicals and metal; salt; sulfur largely for sulfuric acid; placer deposits for diamonds, tin, gold, and heavy minerals. Deposits which have generated continuing interest because of their potential economic interest but which are not presently mined include manganese nodules and crusts, polymetallic sulfides, phosphorites, and methane hydrates.

Authigenic minerals are those formed in place by chemical and biochemical processes. This contrasts with detrital minerals which have been fragmented from an existing rock or geologic formation and accumulated in their present position usually by erosion and sediment transport. The detrital minerals – sand, gravel, clay, shell, diamonds, placer gold, and heavy mineral beach sands – are presently extracted commercially in shallow water. The economically interesting authigenic mineral deposits tend to be found in more than 1000 m of water and have not yet been commercially extracted. Nonetheless, between 1970 and 2000 on the order of one billion US dollars was spent collectively worldwide on studies and tests to recover five authigenic minerals. These minerals are: manganese nodules, manganese crusts, metalliferous sulfide muds, massive consolidated sulfides, and phosphorites. This article will focus on the extraction of these five mineral types.

Descriptions of the formation, geology, geochemistry, and associated microbiology of these deposits are presented elsewhere in this work. As a brief generalization, manganese nodules are black, golf ball to potato sized concretions of ferromanganese oxide sitting on the deep seafloor at depths of 4000–6000 m. They contain potentially economic concentrations of copper, nickel, cobalt, and manganese, and lower concentrations of titanium and molybdenum. Manganese crusts are a flat layered version of manganese oxides found on the tops and sides of seamounts with the highest metal concentrations in water depths of 800–2400 m. They are a potential source of cobalt, nickel, manganese, rare earth elements and perhaps platinum and phosphate. Polymetallic sulfides also come in two forms: metalliferous muds and massive consolidated sulfides. The sulfides contain potentially economic concentrations of gold, silver, copper, lead, zinc, and lesser amounts of cobalt and cadmium. The metalliferous muds are unconsolidated sediments (muds) found at seafloor spreading centers and volcanically active seafloor sites. The metals have been concentrated in the muds by hydrothermal processes operating at and below the seafloor. The best explored site of these metalliferous muds is in the hydrothermally active springs in the central deeps of the Red Sea. By contrast, the massive consolidated sulfides are associated with chimney and mound deposits found at the sites of 'black smokers', hydrothermal vents on the seafloor. The sulfides deposited at these sites have been concentrated by hot seawater percolating through the seafloor and being expelled onto the seabottom at the vents. When in contact with the cold ambient sea water, the hydrothermally heated water drops its mineral riches as sulfide-rich precipitates, forming the sulfide chimneys and mounds. The final authigenic mineral of economic interest is phosphorite, used primarily for phosphate fertilizer. The seabed phosphorites are found as nodules, crusts, irregular masses, pellets, and conglomerates on continental shelves and on the tops of seamounts and rises. Their distribution is widespread throughout the tropical and temperate oceans, although they are preferentially found in areas of oceanic upwelling and high organic productivity.

Whichever of these five mineral types is the target of a mining operation, following regional exploration the process of extracting the commodity of value has seven distinct steps. These steps are: (1) survey the mine site, (2) lease, (3) pick up the mineral, (4) lift and transport, (5) process, (6) refine and sell the metal, and (7) remediate the environmental damage. These steps will be considered in turn (**Figures 1–7**). Naturally, there are differences for each mineral type as well as a range of possible processes that can be used. While many of these processes have been tested and many are used in

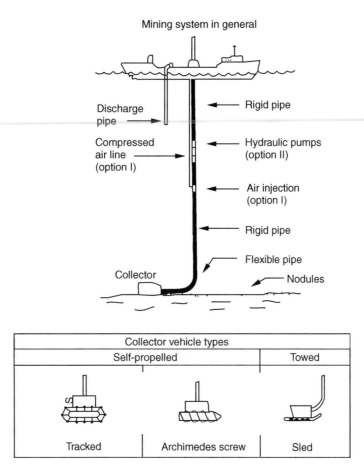

Figure 1 Generalized deep-sea mining system component diagram. This diagram illustrates the proposed collection and lift components which would apply to a wide variety of mineral systems. Note that a considerable variety of collector types are possible on the bottom, including tracked robotic miners, Archimedes screw-driven vehicles, or towed sleds. Both airlift and pumped systems are illustrated as possible lift mechanisms. (Reproduced with permission from Thiel *et al.*, 1998.)

traditional terrestrial minerals operations, to date there is no commercially viable full-scale deep-sea authigenic mineral extraction operation.

Survey

The first step in minerals development is to find an economic mine site. This is found by surveying and mineral sampling. A great deal of mineral sampling has already been done over the last 40 years throughout the world's ocean. These data are available for initial planning purposes. Following a detailed literature review the prospective ocean miner would send out a research vessel to sample extensively in the areas under consideration. New acoustical techniques can be calibrated to show certain kinds of bottom cover, including the density of manganese nodule cover. This is one way to rapidly survey the bottom to highlight areas with potentially economic accumulation of authigenic minerals.

Significant advances in marine electronics, navigation, and autonomous underwater vehicles (AUVs) are being brought together. New 'chirp' sonars which transmit a long pulse of sound in which the frequency of the transmitted pulse changes linearly with time give high resolution and long-range seafloor and sub-bottom imagery. Navigation based on the satellite global positioning system (GPS) can now give accurate underwater positions (≤ 1m) when linked to an acoustic relay. This level of survey equipment is now available on underwater autonomous vehicles, meaning that the cost of a ship is not necessarily an impediment.

Sampling for metal concentrations follows the initial surveys. Sampling may be from a ship, a remotely operated vehicle (ROV), or a submersible. Sampling is likely to begin with dredges, progress to some kind of coring, and finish with carefully oriented drilled samples giving a three-dimensional picture of the ore distribution. These data, after chemical analysis of the contained metals, will give

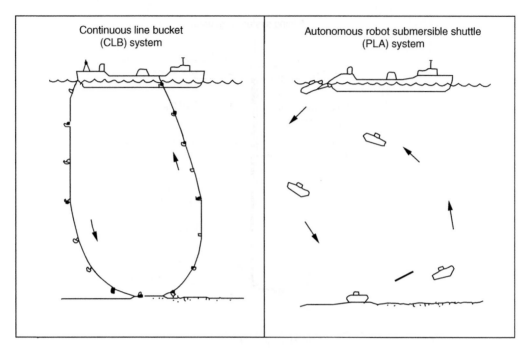

Figure 2 In contrast to the more conventional marine mining systems illustrated in **Figure 1**, two other potential systems are the continuous line bucket system and the autonomous robot submersible shuttle. The continuous line bucket is a series of dredge buckets on a line dragged over the mineral deposit. The submersible shuttle system is a theoretical system using a series of robotic transport submersibles to carry ore from a collection point on the bottom to a waiting surface ship. (Reproduced with permission from Thiel *et al.*, 1998.)

grade and tonnage information. The grade and tonnage estimates of the deposit will be entered into a financial model to determine whether it is economically profitable to mine a given deposit.

In actual fact this process is iterative. A financial model will be used to indicate the type of mineral deposit which must be found. This will narrow the search area. As new data are forthcoming, increasing levels of survey sophistication will follow, assuming that the data continue to indicate an economic mining possibility. In its simplest form the financial model looks at sales of the metals derived from the mine, compared to the total cost of all the operations required to obtain the minerals and contained metals. If the projected sales are greater than the costs by an amount sufficient to allow a profit, typically on the order of a minimum of 20% after taxes, then the mineral deposit is economic. If not, costs must be decreased, sales increased or another more attractive deposit must be found. In the end, a deposit survey will be developed of sufficient detail to allow the production of a mining plan for the development of the property. Normally such a mining plan would need to be approved by the government authority granting mineral leases before actual mining could commence. At some point during the survey and evaluation process, the mining company needs to obtain a lease on the mineral claim in

question. Usually this is done very early in the process, after the first broad area wide surveys.

Lease

Ocean floor minerals are not owned by a mining company. If the minerals are within the exclusive economic zone (EEZ) of a country, they are the property of that country's government. EEZs are normally 200 nautical miles off the coast of a given country, but in the case of a very broad continental shelf may be extended up to 350 miles offshore with recognition of the appropriate United Nations boundary commission. Beyond the EEZ, the ownership of minerals rests with the 'common heritage of mankind' and is administered for that purpose by the United Nations International Seabed Authority in Kingston, Jamaica.

Both national regimes and the International Seabed Authority will lease seabed minerals to *bone fide* mining groups after the payment of fees and the arrangement for filing of mining and exploration plans, environmental impact statements, and remediation plans. The specific details of the requirements vary with the size of the proposed operation, the mineral sought, and the regime under which the application is made. In many countries, the offshore mining laws

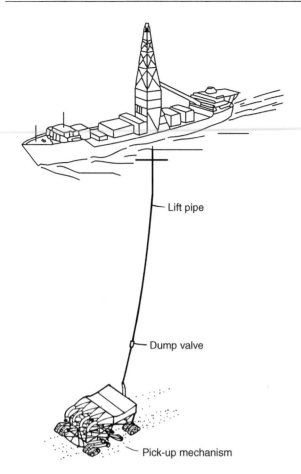

Figure 3 An engineering design for a proposed manganese crust mining system. (Reproduced with permission from State of Hawaii (1987). *Mining Development Scenario for Cobalt-Rich Manganese Crusts in the Exclusive Economic Zones of the Hawaiian Archipelago and Johnston Island.* Honolulu, Hawaii, Ocean Resources Branch.)

Lift pipe

Dump valve

Pick-up mechanism

taken. A typical deep water rental for an oil and gas lease might be $25 per acre per year with a royalty payment equal to 12.5% of the oil extracted. Hard mineral lease rentals and royalties could be expected to be lower, as the commodity is less valuable and the demand for access to offshore mineral resources is much less than for offshore oil.

Mineral Pick Up

Once a deposit has been characterized and leased, a mining plan is drawn up. This plan will depend on the mineral to be mined. In general, several steps must be taken to mine. The first is separating the mineral from the bottom. In the case of polymetallic sulfide, manganese crust or underlying phosphorite, a cutting operation is involved. In the case of phosphorite nodules, manganese nodules, or sulfide muds there is solely a pick-up operation. Cutting requires specialized cutting heads, often these are simply rotating drums with teeth (**Figures 3–5**). Such cutters have been developed for the dredging industry as well as the underground coal industry on land. The size, angle, and spacing of the teeth on a cutter are dependent on the rate of cutting desired and the size to which particles are to be broken. The overall mineral pickup rate is determined by the necessary rate of throughput at the mineral processing plant. This can be worked back to pick-up rate on the bottom. Engineering judgment then dictates whether this is best achieved with larger numbers of smaller cutters or a lower number of larger cutters. This may translate into multiple machines operating on the bottom and feeding one lift system.

When the mineral is broken into sufficiently small pieces, these must then be collected for lifting. This is usually a scooping or vacuuming operation. Scooping is accomplished by blades of various shapes. Vacuuming is usually the result of a powerful airlift or pump farther up the line. One system tested by the Ocean Minerals Co. for picking up manganese nodules involved an Archimedes screw-driven robotic miner, which had two pontoons. A flange in screw-shaped spiral was welded onto each pontoon. The screws both served to drive the vehicle forward as well as pick up the nodules which were sitting in the mud it passed over. There may be a sieving or grinding step between mineral pick-up and lift. This serves two functions; the sieving gets rid of unwanted bottom sediment that may have become entrained in the ore; the grinding ensures that the particle size range going up the lift pipe is in the correct range to get optimum lift without clogging the pipe.

are modeled on the legislation which governs offshore oil development. This is not surprising as the issues are similar and worldwide the annual value of the offshore oil industry is in excess of $100 billion, whereas the offshore mining industry will not be more than a few percent of this value for the foreseeable future.

Leasing arrangements vary from country to country. In the USA, offshore leasing of both hard minerals and oil and gas in the federally controlled waters of the EEZ (normally more than 3 miles offshore) falls under the Outer Continental Shelf Lands Act and is administered by the US Department of the Interior. This act requires a competitive lease sale of offshore tracts of land. The company to whom a lease is awarded is the one which offers the highest payment at a sealed bid auction. In addition to this payment, known as the bonus bid, there is also an annual per acre rental and a royalty payment which would be a percentage of the value of the mineral

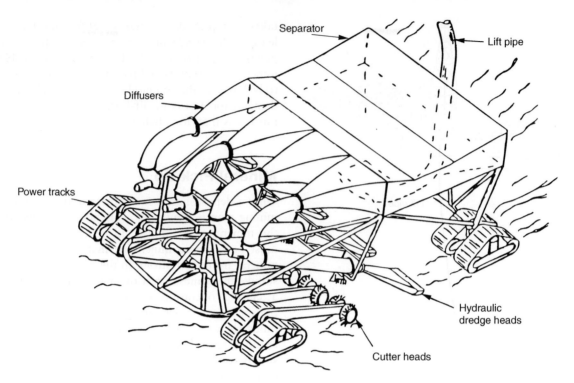

Figure 4 A robotic miner designed to rip up and lift attached flat lying bottom deposits such as manganese crusts. (Reproduced with permission from State of Hawaii (1987) *Mining Development Scenario for Cobalt-Rich Manganese Crusts in the Exclusive Economic Zones of the Hawaiian Archipelago and Johnston Island.* Honolulu, Hawaii, Ocean Resources Branch.)

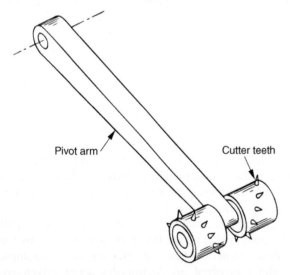

Figure 5 Cutter head design for a bottom miner ripping attached ores such as polymetallic sulfides or manganese crusts. (Reproduced with permission from State of Hawaii (1987) *Mining Development Scenario for Cobalt-Rich Manganese Crusts in the Exclusive Economic Zones of the Hawaiian Archipelago and Johnston Island.* Honolulu, Hawaii, Ocean Resources Branch.)

Lift and Transport

Once the mineral has been cut off or picked up from the bottom it must be lifted to the surface. Over the years a number of systems have been suggested and tested for this purpose. The most successful of these tests have involved bringing the minerals to the surface in a seawater slurry in a steel pipe. This pipe would typically be 30–50 cm in diameter, hence it would be similar if not identical to pipe used in offshore oil drilling operations. Two methods have been tested for bringing the slurry up the pipe, i.e. pumps and airlift. Airlift is a commonly used technique in shallow-water dredging. Compressed air is introduced into the pipe, typically about a third of the way from the top (**Figure 6**). As the air expands moving upward in the pipe it draws the mineral slurry behind it. This works extremely well over lifts of a few hundred meters. It also works over lifts in the deep sea of 6000 m, for example on manganese nodules, except that it is much more difficult to control. Hydraulically driven pump systems along the pipe's length have also been tested to full ocean depth. They also work and although easier to control may ultimately be less efficient lifting systems than an airlift.

Several other lift systems are possible (**Figures 2–5**). The most notable of these is the continuous line bucket, which is a series of buckets on a line which is continuously dragged over the mineral deposit. This system has been successfully tested, although it suffers the disadvantages of lack of control on the bottom, potentially wider spread environmental

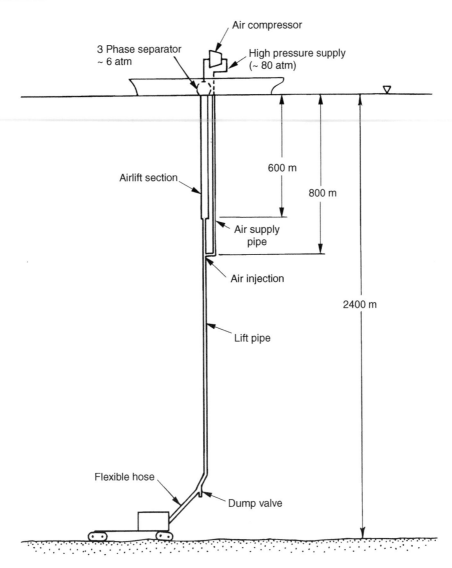

Figure 6 Generalized design of an airlift system to lift ground minerals from the ocean bottom to the surface. (Reproduced with permission from State of Hawaii (1987) *Mining Development Scenario for Cobalt-Rich Manganese Crusts in the Exclusive Economic Zones of the Hawaiian Archipelago and Johnston Island.* Honolulu, Hawaii, Ocean Resources Branch.)

damage, and the possibility of the rope entangling on itself. A system proposed but not tested involves a series of ore-carrying robotic submersible shuttles. These shuttles would use waste mineral tailings as ballast to descend to the bottom, where they would exchange the ballast for a load of ore. Adjusting their buoyancy they would then rise to the surface, off-load the ore onto an ore carrier, reload waste, and return to the bottom. This is potentially a very elegant system in that it handles both tailings waste disposal and mineral lift at the same time, each with the expenditure of very little energy.

Once the ore is on the surface it must be taken to a processing plant. This processing plant can be an existing plant at a site on land, a purpose-built plant near a harbor to process marine minerals, a large offshore floating platform moored near the mine site on which a plant is built, or a ship converted to mineral processing. The latter two could be right at the mine site. The minerals would be processed as soon as they were lifted to the surface. For a shore-based processing plant a fleet of transport ships or barges would be required to move the lifted ore from the mine site to the plant. This could be a tug and barge operation or a series of dedicated ore carrier vessels.

Processing

Once the minerals have been mined and transported to the processing location they must then be treated to remove the metals of interest. There are two basic

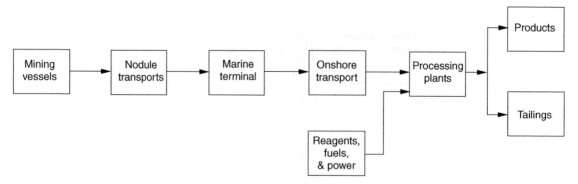

Figure 7 Components involved in a marine minerals extraction operation, each of which has an environmental effect and must be considered in the context of the whole operation. (Reproduced with permission from State of Hawaii (1981) *The Feasibility and Potential Impact of Manganese Nodule Processing in the Puna and Kohala Districts of Hawaii.* Honolulu, Hawaii, Ocean Resources Branch.)

ways to do this: smelting and leaching (pyrometallurgy and hydrometallurgy). Once a basic scenario of either reducing the mineral with acid or melting has been decided upon, there are a number of possible steps to partially separate out unwanted fractions of the mineral. In order to be economically viable, a mineral processing plant needs to process large tonnages of material. In the case of manganese nodules this might be 3 million tonnes y^{-1}. In order to process these large tonnages economically it is important to separate out as much of the waste mineral material as early as possible in the process. This is done most commonly by magnetic separation, froth flotation, or a density separation such as heavy media separation. In an effort to make these processes more efficient the incoming ore would normally be washed to remove salt and ground to decrease the particle size and increase surface area.

Magnetic separation separates magnetic and paramagnetic fractions from nonmagnetic fractions. This technique can be done either wet or dry. Dry magnetic separation involves passing the ground ore through a strong magnetic field underlying a conveyor belt. The magnetic and paramagnetic fraction stay on the conveyor belt while the non-magnetic fractions drop off as the belt goes over a descending roller. Manganese, cobalt, and nickel are paramagnetic, so this is a good separation technique for manganese nodules and crusts. Wet magnetic separation involves passing a slurry of ore over a series of magnetized steel balls. The magnetic and paramagnetic materials stick to the balls and the nonmagnetic fraction flows through. The level of separation depends on the strength of the magnetic field. Usually this is done with an electromagnet so that the field can be turned on and off.

Froth flotation relies on differences in the surface chemistry of the ore and waste grains. Sulfide ores in particular are more readily wet with oil than water.

Grains of clay, sand, and other seabed detritus have the reverse tendency. Typically something like pine oil might be used to wet the mineral grains. The oil-wet ore is then introduced into a tank with an agent that produces bubbles, a commonly used chemical for this is sodium lauryl sulfate. The bubbles tend to stick to the oil-wet grains, which float them to the surface where they can be taken off in the froth at the top of the tank. The waste material is allowed to settle in the tank and removed at the bottom. The separated ore is then washed to remove the chemicals. The chemicals are recycled if possible.

Another technique commonly used in mineral beneficiation is density separation. This can be as simple as panning for gold. In a simple panning operation or more complex sluice boxes or jigging tables, the heavy mineral (the gold) stays at the bottom, whereas the lighter waste materials are washed out in the swirling motion. A more sophisticated version of the same idea is to use a very dense liquid or colloidal suspension to allow less dense material to float on top and more dense material to sink. The density of the medium is adjusted to the density of the particular ore and waste to be separated. In general, heavy media suspensions, such as extremely finely ground ferro-silicon, are superior to traditional heavy liquids such as tetrabromoethane, which have a high toxicity.

Following mineral beneficiation, which removes the non-ore components, the ore itself must be broken down into its component value metals. Marine minerals can contain up to half a dozen economically desirable metals. In order to separate these metals individually, a reduction process must occur to break the chemical framework of the ore. This will either involve smelting or leaching. In a smelting operation the whole mineral structure is melted and the various metals are taken off in fractions of different densities which float on each other. Normally a series of fluxing agents is used. Clean separation in the case of a multi-metal ore can be a very exacting process. For highly complex polymetallic ores such as manganese nodules or crusts many experts favor a leaching approach over smelting in order to facilitate high purity separation of the metals. However, both leaching and smelting processes have been tried on a variety of marine minerals and both types of process have been proved technically viable.

The most common leach agent is sulfuric acid as it is cheap, efficient, and readily available. Other acids, hydrogen peroxide, and even ammonia are also used as leaching agents. To reduce leaching times, temperature and pressure are raised in the leaching vessel. The leaching process breaks up the mineral structure and dissolves the contained metals into an acidic solution known as a 'pregnant liquor'. The metals become positive ions in solution. These individual metal ions must be separated out of the pregnant liquor and plated out as an elemental metal. This is done primarily in three processes: solvent extraction, ion exchange, and electrowinning. These may initially be done with the primary processing of the ore; however, the final phase of these techniques will be done in a dedicated metal refinery to achieve metal purity $>99\%$.

Solvent extraction relies on an organic solvent which is optimized to select one metal ion. This organic extractant is immiscible with the aqueous pregnant liquor. They are stirred together in a mixer forming an emulsion much like oil stirred into water. This emulsion has a very high contact area. The organic extractant takes the desired metal ion out of the aqueous phase and concentrates it in the organic phase. The aqueous and organic phases are then separated by density (usually the organic phase will float on the aqueous phase). The metal ion is then stripped out of the organic phase by an acidic solution and sent to an electrowinning step.

Instead of doing this extraction in a liquid as in the case of solvent extraction, it is possible to run the pregnant liquor over a series of ion-exchange beads in a column. The ionic beads have the property of collecting or releasing a given metallic ion, e.g. nickel, at a given pH. Therefore by adjusting the pH of the liquid flowing over the column it is possible to remove the nickel, for example, from the pregnant liquor, transferring it to its own tank, then remove copper or cobalt, etc. It is possible to use both solvent extraction and ion-exchange column steps in a particularly complex metal separation process or to achieve very high metal purities.

Following solvent extraction or ion exchange, the various metal ions have been separated from each other and each is in its own acidic solution. The standard concentration technique used to remove these metal ions from solution is electrowinning. The metal ions are positively charged in solution. They will be attracted to the negatively charged plate in a cell. A small current is set up between two or more plates in a tank. The positive metal ions plate out on the negatively charged plate. Depending on the purity of the metal plated out and how strongly it is attached, either the metal is scraped off the plate and sold as powdered metal or the entire plate is sold as metal cathode, e.g. cathode cobalt.

Refining and Metal Sales

Each of the metals produced by an offshore mining operation must be refined to meet highly exacting standards. The metals are sold on various exchanges in bulk lots. Silver, copper, nickel, lead, and zinc are largely sold on the London Metal Exchange. Platinum, silver, and gold are sold on the New York Mercantile Exchange. Both of these exchanges maintain informative internet websites with the current details and requirements for metals transactions (www.lme.co.uk and www.nymex.com). Cobalt, manganese and phosphate are sold through brokers or by direct contract between a mining company and end-users (e.g. a steel mill). Metal sales are by contract. The contracts specify the purity of the metal, its form (e.g. powder, pellets or 2 inch squares), the amount, place, and date that the metal must be delivered. Standard contracts allow for metal delivery as much as 27 months in the future (in the case of the London Metal Exchange). This allows a major futures market in metals and considerable speculation. This speculation allows metal producers to lock in a future price of which they are certain. In reality, only a few percent of the contracts written for future metal deliveries result in actual metal deliveries. The vast majority are traded among speculators over the time between the initial contract settlement and the metal delivery date. The price for a metal is highly dependent on its purity. Often metals are sold at several different grades. For example, cobalt is typically sold at a guaranteed purity of 99.8%. It may also be purchased at a discounted price for 99.3% purity and at a premium for 99.95%.

In order to achieve these grades considerable refining takes place. Often this is at a facility which is removed from the original mine site or processing plant. Metal refineries may be associated with the manufacturing of the final consumer endproduct. For example, a copper refinery will often take lower grade copper metal powder or even scrap and produce copper pipe, wire, cookware, or copper plate. Most modern metal fabricators rely on electric furnaces of some form to cast the final metal product.

One of the techniques commonly used at refineries is known as 'zone refining' whereby a small segment of a piece of metal in a tube is progressively melted while progressing through a slowly moving electric furnace coil. The impurities are driven forward in the liquid phase with the zone of melting. The purified metal resolidifies at the trailing edge of the melting zone. This technique has been used to reduce impurities to the parts per billion level.

Remediation of the Mine Site

Once mining has taken place both the mine site and any processing site must be remediated (**Figure 7**). Considerable scientific work took place in the 1980s and 1990s looking at the rate that the ocean bottom recovers after being scraped in a mining operation. While it is clear that recovery does take place and is slow, it is still unclear how many years are involved. A period of several years to several decades appears likely for natural recolonization of an underwater mined site. It also appears that relatively little can be done to enhance this process. Once all mining equipment is removed from the site, nature is best left to her own processes.

An independent yet perhaps even more important issue is the way the waste products are handled after mineral processing. There are both liquid and solid wastes. The most advanced of a number of clean-up scenarios for the discharged liquids is to use some form of artificial ponds or wetlands, most often involving cattails (*Typha*) and peat moss (*Sphagnum*), the two species shown to be most adept at wastewater clean-up. Typically the wastewater will circulate over several limestone beds and through various artificial wetlands rich with these and related species. At the end of the circulation a certain amount of cleaned water is lost to ground water and the rest is usually sufficiently cleaned to dispose in a natural stream, lake, river, or ocean.

The larger and as yet less satisfactorily engineered problem is with solid waste. In fact, recent environmental work on manganese crusts has shown that 75% of the environmental problems associated with marine ferromanganese operations will be with the processing phase of the operation, particularly

tailings disposal. Traditionally, mine tailings are dumped in a tailings pond and left. Current work with manganese tailings has shown them to be a resource of considerable value in their own right. Tailings have applications in a range of building materials as well as in agriculture. Manganese tailings have been shown to be a useful additive as a fine-grained aggregate in concrete, to which they impart higher compressive strength, greater density, and reduced porosity. These tailings serve as an excellent filler for certain classes of resin-cast solid surfaces, tiles, asphalt, rubber, and plastics, as well as having applications in coatings and ceramics. Agricultural experiments extending over 2 years have documented that tailings mixed into the soil can significantly stimulate the growth of commercial hardwood trees and at least half a dozen other plant species. Finding beneficial uses for tailings is an important new direction in the sustainable environmental management of mineral waste.

See also

Hydrothermal Vent Deposits. Mid-Ocean Ridge Geochemistry and Petrology.

Further Reading

Cronan DS (1999) *Handbook of Marine Mineral Deposits.* Boca Raton: CRC Press.

Cronan DS (1980) *Underwater Minerals.* London: Academic Press.

Earney FCF (1990) *Marine Mineral Resources: Ocean Management and Policy.* London: Routledge.

Glasby GP (ed.) (1977) *Marine Manganese Deposits.* Amsterdam: Elsevier.

Nawab Z (1984) Red Sea mining: a new era. *Deep-Sea Research* 31A: 813–822.

Thiel H, Angel M, Foell E, Rice A, and Schriever G (1998) *Environmental Risks from Large-Scale Ecological Research in the Deep Sea – A Desk Study.* Luxembourg: European Commission, Office for Official Publications.

Wiltshire J. (2000) Marine Mineral Resouces – State of Technology Report *Marine Technology Society Journal* 34: no. 2, p. 56–59

OFFSHORE SAND AND GRAVEL MINING

E. Garel, CIACOMAR, Algarve University,
Faro, Portugal
W. Bonne, Federal Public Service Health, Food Chain
Safety and Environment, Brussels, Belgium
M. B. Collins, National Oceanography Centre,
Southampton, UK

Introduction

Offshore sand and gravel extraction involves the abstraction of sediments from a bed which is always covered with seawater. This activity started in the early twentieth century (in the mid-1920s, in the UK), but did not reach a significant scale until the 1960s and 1970s, when markets for marine sand and gravel expanded and dredging technology improved (**Figure 1**). In the mining industry, the term 'aggregates' describes a variety of diverse particulate materials, which are provided in bulk, that is, sand and gravel, together with crushed rocks. The distinction between sand and gravel varies, on the basis of the classification adopted. Hence, the nomenclature may change between the end-users or countries (or even regions within a particular country). For example, in the UK industry, 'sand' refers to (noncohesive) minerals with a mean grain size lying between 0.63 and 5 mm, and 'gravel' is reserved for coarser material; in Belgium, the industry considers as gravel the sediment with a mean grain size larger than 4 mm. For comparison, within the scientific literature, the limit between sand and gravel is generally at 2 mm.

Resource Origin

Marine aggregate deposits from the continental shelf can be either relict or modern. Relict deposits have been formed under hydrodynamic and sedimentological regimes that no longer exist. They consist of river or coastal bank deposits, formed during the lower seawater stands, induced by the glacial periods affecting the Earth over the past 2 My. At that time, the rivers extended widely across the continental

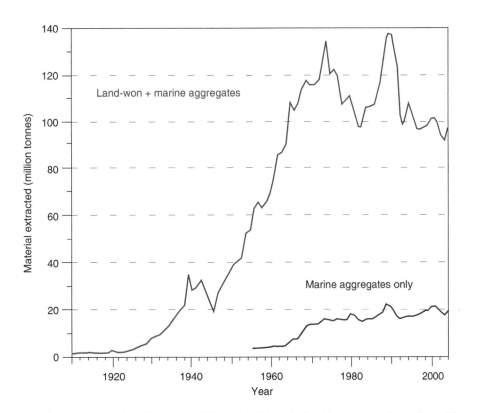

Figure 1 Production of aggregates in the UK between 1900 and 2004 (excluding fill contract and beach nourishment, for marine aggregates). Note the rapid growth of extraction during the late 1960s and the early 1970s, due to the boom in construction in southern England. Sources: Crown-Estate, British Geological Survey.

shelf, transporting and depositing large amounts of sand and gravel in their valleys and coastal waters. This material has been submerged and possibly reworked, especially the mobile sandy deposits, during subsequent warmer interglacial periods, due to the retreating and melting of the ice. Concentration of deposits has resulted from the numerous repetitions of this warm–cold cycle. These sedimentary bodies have remained undisturbed since their formation and are immobile within the context of prevailing hydrodynamic conditions. Relict sand and gravel deposits include mostly thin sheets, banks, drowned beaches and spits, and infilled channel systems, such as river courses and terraces. In contrast, modern deposits have been formed and are controlled by the present prevailing hydrodynamic and sedimentological regime. Therefore, they are related to present-day coastal sediment budgets and sediment dynamics on the seabed. These deposits include, exclusively, sandbanks and sand bedform fields.

The composition, roughness, and grain size of marine aggregate particles vary considerably, depending upon their mode of formation, depositional processes, and history of reworking. Although dependent upon the intensity of post-erosional transport, the grains consist usually of fragments derived from rocks of the surrounding emerged regions. Due to their glacial origin, relict deposits have a composition which is similar to those quarried on land (which are their upstream equivalent), having a low content of shell fragments. In comparison, modern deposits may include high concentration of shelly sediments, which affect the commercial quality of the deposit. In addition, marine aggregates tend to have an higher chloride content. As such, some objections have been raised on their use, in relation to potential alkali–silica reactions, which could affect concrete, or chloride attack on steel reinforcement. However, experience and some experiments have shown that marine aggregates are as suitable as on-land quarried aggregates, for construction purposes.

Production and Usage

Marine aggregates are used mostly in the construction industry, but also for beach replenishment and shore protection, for land reclamation, and in other fill-related uses (such as drainage and capping material). The quality required is governed generally by the usage of the product. High-quality marine aggregates (i.e., well sorted and free from impurities such as clay) are used for mixing with a cementing agent, to produce hard building material (e.g., concrete, mortar, and plaster), or are coated with bitumen for road surfacing. For other usages, such as base material under foundations, roads, and railways, or as the construction of embankments, lower-grade materials can be used. The quality of the material used for beach nourishment may also show strong variability, with location and local/regional policies. Nonetheless, not only should the grain size of the nourished material be similar to (or greater than) that eroded, but it should also not contain fresh biogenic material or contaminants.

Production and usage of marine aggregates vary considerably between countries, with the largest global producers being Japan, the Netherlands, Hong Kong, Korea, and the UK (**Figure 2**). Clusters of extraction areas lie in the vicinity of these countries (e.g., in the North Sea, the Channel, and the Baltic Sea), although aggregate extraction takes place

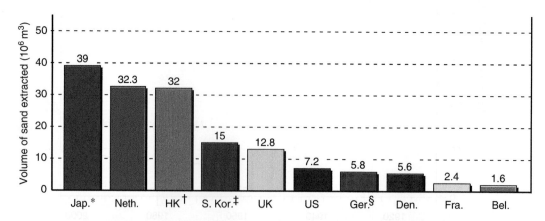

Figure 2 Overview of sand extracted from coastal waters around the world (2002). Annual volumes in million cubic meters per year, for different countries (Japan, the Netherlands, Hong Kong, Korea, United Kingdom, United States, Germany, Denmark, France, Belgium). Key: * data from 1998; † Hong Kong, averaged over 1990–98; ‡ averaged over 1993–95; and § from 2000. Source: Roos PC (2004) *Seabed Pattern Dynamics and Offshore Sand Extraction*, 166pp. PhD Thesis, University of Twente, The Netherlands (ISBN 90-365-2067-3).

in shallow continental seas all over the world. The largest producers present several – if not all – of the following characteristics: limited, or depleted, on-land aggregate resources, increasing environmental pressure on quarries, and large consumption of material. Besides, the yearly national productions may rise drastically as a result of major public works involving, temporarily, a huge volume of extracted material (e.g., Belgium, in 1997, for the construction of pipelines; and the Netherlands, in 2001, for major land reclamation schemes). Countries producing large quantities of marine aggregates extract different types of material, and for distinct purposes (e.g., concrete, filling, or beach recharge). In addition, they may import and export material to and from other countries. For example, in 1998, almost 90% of the marine aggregates production in the UK was for concrete (gravel and coarse sand), but a third of this (c. 7 Mt) was exported to other northwestern European countries. In Spain, by contrast, the dredging of marine aggregates is authorized only for beach replenishment.

Prospecting

The identification of marine aggregate resources is based upon both desk studies and offshore prospecting surveys. The purpose of the desk study is to locate suitable deposits, based upon a scientific literature review and other considerations. Prospecting surveys involve geophysical data acquisition and sediment sampling. In addition to an accurate vessel positioning system (e.g., GPS), the geophysical instrumentation includes usually: an echo sounder, to investigate the seafloor bathymetry; a high-resolution seismic system (e.g., Boomer, Pinger, and Sparker), to produce reflection profiles of the first (c. 10) few meters of the seabed layers; and, a side scan sonar, to characterize the seabed hardness and roughness (**Figure 3**). The collected data set provides information about the thickness, horizontal distribution, and surficial character of the deposit. Vertical and horizontal ground-truth information is provided by core and grab samples, respectively. In addition, grab sampling is useful to examine the surficial sediment grain-size distribution (and also to identify the benthic biological communities present within the area). The selection of the sampling equipment depends upon the difficulty of penetration into the various sediment layers. Large hydraulic grabs can be used to obtain surficial sediment samples, while vibrocorers (up to 10-m long) are capable of coring and recovering coarse-grained material. The level of accuracy of deposit recognition, dependent upon the resolution of both the instruments and the sampling spatial coverage, is governed by the cost of the surveys.

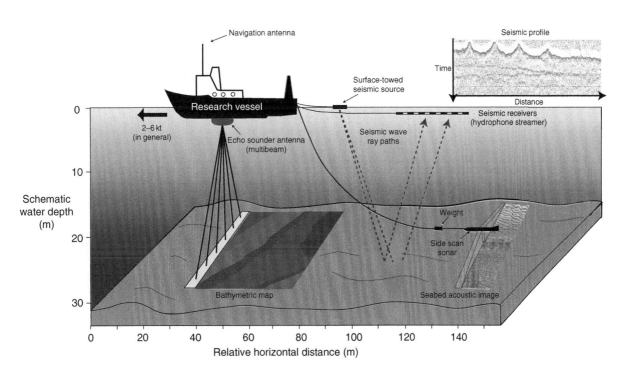

Figure 3 Schematic representation of a survey vessel undertaking geophysical prospecting operations: (multibeam) echo sounder mapping; side scan sonar imaging; and seismic reflection survey.

Extraction Process

Offshore sand and gravel extraction is performed by anchor (or static) suction dredging, and trailer suction dredging. Both techniques utilize powerful centrifugal pumps to draw up the seabed material into the hopper dredger, through pipes of up to 1 m diameter. The dredged material displaces seawater within the hold of the ship, loaded previously as ballast. Anchor suction dredging involves a vessel anchoring or remaining stationary over a deposit, together with forward suction of the bed material through the pipe. As such, the technique is effective for small spatially restricted or, locally, thick deposits. This abstraction procedure creates crater-like pits, or a series of pits, up to 25 m deep and 200 m

in diameter, with a slope of about 5° (**Figure 4**). In contrast, trailer suction dredging requires the dredger to drag the lower end of its rear-facing pipe(s) slowly along the seabed, while the ship is underway. This technique permits relatively large areas with thin and more evenly distributed deposits to be worked. It is used also for thicker deposits, such as sandbanks, to limit the harmful environmental impacts to the superficial layers. At the bottom, the head of the pipe creates linear furrows, of 1–3 m in width and up to 50 cm in depth (**Figure 5**). However, repeated trailer suction hopper dredging over an area can lead to large dredged depressions (**Figure 6**).

The total dredged cargo contains generally a mix of different grain sizes. In some cases, the dredgers

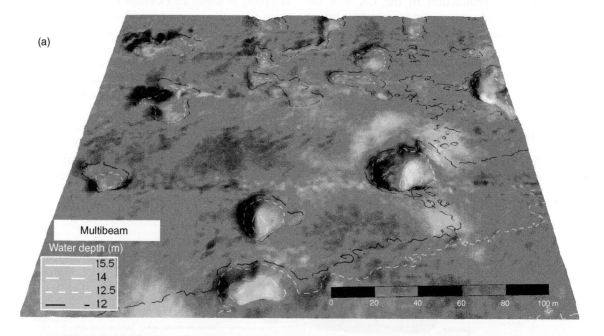

(a)

Multibeam

Water depth (m)

———— 15.5
- - - - 14
- - - - 12.5
———— 12

0   20   40   60   80   100 m

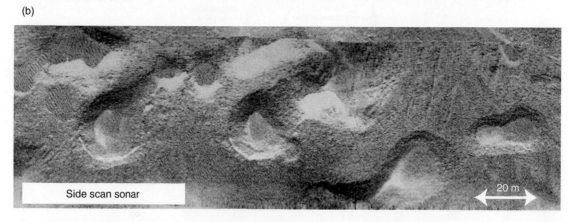

(b)

Side scan sonar

← 20 m →

Figure 4 Examples of gravel pits created by anchor hopper dredging (Tromper Wiek Area, Baltic Sea): (a) bathymetry draped with backscatter (illuminated from the west), obtained from a multibeam system; (b) side scan sonar. Source: Manso F, Diesing M, Schwarzer K, and Garel E (2004) Monitoring of dredged areas by combination of multibeam echosounder and sidescan sonar data. *Conference Littoral 2004*, Aberdeen, UK, 20–22 September 2004.

(a)

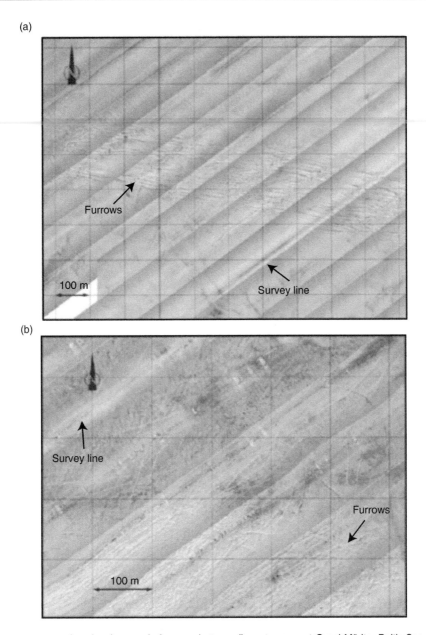

(b)

Figure 5 Side scan sonar mosaics showing sandy furrows, in two adjacent areas, at Graal Müritz, Baltic Sea, in December 2000 (with the survey lines in a NE–SW direction): (a) furrows are observed in the center of the image, with a subequatorial direction (note the reverse in direction of the operating dredger at the western tip of the tracks); (b) the furrows are oriented more randomly, although predominantly NE–SW in the lower part of the image. From Diesing M (2003) *Die Regeneration von Materialentnahmestellen in der südwestlichen Ostsee unter besonderer Berücksichtigung der rezenten Sedimentdynamik*, 158pp. PhD Thesis, Christian-Albrechts-Universität, Kiel.

retain all the dredged material on board, for discharge and processing ashore. However, dredging operations often target only a specific type of aggregate, defined by market demand. Since it is more cost-effective to load and transport only a cargo of the required type, dredgers often have the facility to screen onboard the dredged material, spilling back to the sea the unwanted fines. This screening process spills an important amount of material, up to 3–4 times the retained cargo load, generating sediment

plumes that settle down and spread over the extraction zone (making subsequent extractions sometimes more difficult).

Nowadays, some dredgers with cargo capacities of up to 8500 t operate with pumps capable of drawing aggregates at 2600 t of material per hour, in water depths up to 50 m (even 90 m, for Japanese vessels). Discharging on the wharf is performed by a range of machinery, such as bucket wheels, scrapers, or wire-hoisted grabs, which place the aggregates on a

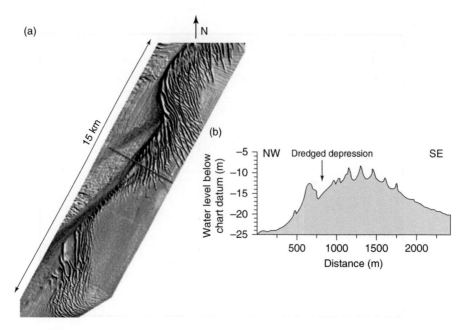

Figure 6 Example of depressions created by repeated trailer suction hopper dredging on a sandbank: the Kwintebank, offshore Belgium. (a) Bathymetric map of the Kwintebank, where the depression is located with the red dotted ellipse; the red straight line indicates the location of the cross section (b); the bank is 15 km in length, 1–2 km in width, and 10–20 m in height. (b) Cross section perpendicular to the bank long-axis, including the dredged depression, 300 m in width, 1 km in length, and up to 5 m deep. (a) Source: Belgian Fund for Sand Extraction.

conveyor belt. Hydraulic discharge is also used, mainly for beach-nourishment schemes.

Environmental Impacts

The environmental effects of sand and gravel extraction include physical effects, such as the modification of the sea bottom topography, the creation of turbidity plumes, substrate alteration, changes in the local wave and current patterns, and impacts on the coast. Biological effects include changes in the density, diversity, biomass, and community structure of the benthos or fish populations, as a consequence of the physical effects. One dredging activity may not have a significant direct or indirect environmental impact, but the cumulative effect of several (adjacent) dredging sites may induce significant changes.

Physical Impacts

The estimation of the regeneration time of dredged features is still very difficult and depends upon the extracted material, the geometry of the excavation, the water depth, and the hydrodynamic regime of the system. Typical timescales for the regeneration of dredged furrows in sandy dynamic substrates lie within the range of months. In very energetic shallow sandy areas, such as those found in estuaries, they may recover after just one (or a few) tidal cycle. In contrast, in deep and low-energy areas of fine sand (e.g., the German Baltic Sea), recovery could take decades, even for small features such as furrows. In waters of the Netherlands, southern North Sea, dredged tracks in a gravelly substrate exposed to long-period waves were found to disappear completely within 8 months, at 38-m water depth. In contrast, in waters of SE England, dredged tracks in gravelly sediments can take from 3 to more than 7 years to recover. Dredged depressions or pits created by static dredging have also been reported to remain, as recognizable seabed features, for several years and up to decades. In some cases, pits have been observed to migrate slowly in the direction of the dominant current.

Surface turbid plumes are generated by the screening process and by the overflow of material from the hopper, during dredging. A further source of turbidity, with a far lesser quantity of suspended material, results from the mechanical disturbance of the bed sediment, by the head of the pipe. Large increases in suspended solid concentrations tend to be short-lived and localized, close to the operating dredger. However, turbid plumes with low suspended sediment concentrations can affect much larger areas of the seabed, over extended time periods (several days instead of several hours), especially when dredging is occurring simultaneously in adjacent extraction areas.

Changes in sediment composition, as a direct result of dredging, range from minor alterations of superficial granulometry to a large increase in the proportion of fines (sand or silt), or to an increase of gravel through the exposure of coarser sediments. In addition, dredged depressions or furrows, in gravel and coarse sand, can trap gravelly sand remobilized by storms, or finer sediment mobilized by tidal currents. This process may reduce significantly the volume of sediment bypassing in an area. In the case of relatively steep and deep dredged trough, a significant increase in silt and clay material, at the bottom, is sometimes observed following dredging, with long-term consequences (i.e., alteration) upon the *in situ* infaunal assemblage.

Local hydrodynamic changes (i.e., in wave heights and current patterns) may result from the modification of the bathymetry by dredging. In turn, the local erosional and depositional patterns may be affected (near-field effects). In general, the dimensions of dredged features are, in relative terms, so small, that there is little influence of the deepened area on the macroscale hydrodynamics. However, local hydrodynamic changes induced by dredging may result in wider environmental effects (far-field effects). The main issue concerns possible erosion at the coastline, according to various scenarios. Beach erosion may occur, as a result of offshore dredging, due to a reduction of the sediment supply to the coast. This effect is induced either by the direct extraction of material which would normally supply the coast, or by trapping these sediments within the dredged depressions. Any 'beach drawdown' is a particular example of this case: if the area of extraction is not located far enough from the shore, that is, on the subtidal part of a beach, beach sediment may slump down into the excavated area (e.g., during winter storms) and not be able to return subsequently to the beach (in the summer). Coastal erosion may take place also if the protection of the coast is reduced. In particular, dredging may lower significantly the crest heights of (shallow) banks that normally reduce wave breaking and hence, when removed, would lead to larger wave heights at the coast. In addition, if the alterations in the sea bottom topography affect the wave refraction patterns, erosion may result from modification of the wave directions at the coastline.

A substantial list of parameters has to be considered to evaluate whether dredging leads to erosion on the shore: water depth; the distance from the shore; the distribution of the banks; the degree of exposure; the direction of the prevailing waves and tidal currents; the frequency, direction, and severity of storms; the wave reflection and refraction pattern; and seabed type and sediment mobility. Predictions on the changes are performed generally using mathematical models in which input data are provided through a number of sources, including surveys, charts, and offshore buoys. The modeling results have often a wide error range due to the relative uncertainties in the estimation of processes, especially for sediment transport calculations.

Biological Impacts

The most obvious harmful direct effect of marine aggregate extraction results from the direct removal of substrata and associated benthic epifauna and inflora. Available evidence indicates that dredging causes an initial reduction in the abundance, species diversity, and biomass of a benthic community. Other impacts on benthic communities (and referred to as indirect effects) arise from the modification of abiotic factors, which determine the broad-scale benthic community patterns. These modifications concern the nature and stability of the sediment, through the exposition of underlying strata, and the tidal current strength, through the alteration of the local topography. The creation of relatively deep depressions with anoxic conditions is another source of concern. Sediment plumes arising from the dredging activities may also have adverse biological effects, by affecting water quality, increasing the sediment deposition, and modifying the sediment type. In the case of clean mobile sands, increased sedimentation and resuspension, as a consequence of dredging, are considered generally to be of less concern, as the fauna inhabiting such areas tend to be adapted to naturally high levels of suspended sediments, caused by wave and tidal current action. The effects of sediment deposition and resuspension may be more significant in gravelly habitats dominated by encrusting epifaunal taxa, due to the abrasive impacts of suspended sediments. In addition, turbidity plumes may have indirect impacts on fisheries, due to avoidance behavior by some species, the interruption of migratory pathways (e.g., lobsters), and the loss of access to traditional fishing grounds.

The importance of the impacts that affect a benthic community is investigated commonly on the basis of the number of species, abundance, biomass, and the age of the population, compared to an undisturbed similar baseline. The existence of negligible or significant impacts depends largely upon the spatial and temporal intensity of dredging and, likewise, upon the nature of the benthos. While the recolonization of a dredged site by benthic species is generally rapid, the complete recovery (also 'restoration' or 'readjustment') of a benthic community can take many years. The recovery rate, after dredging, depends primarily upon several factors: the community diversity and richness of the area prior to dredging; the life cycle and growth rate of species; the nature, magnitude, and duration of the dredging operations; the nature of the new sediment which is exposed, or subsequently accumulated at the extraction site; the larval and adult pool of potential new colonizers; the hydrodynamics and associated bed load transport processes; and the nature and intensity of the stresses (caused, for example, by storms) which the community normally withstands. Thus, among a number of studies addressing the consequences of long-term dredging operations on the recolonization of biota and the composition of sediment following cessation, some disparity appears in the findings. These results range from minimal effects of disturbance after dredging to significant changes in community structure, persisting over many years.

Supply and Demand: The Future of Offshore Sand and Gravel Mining

Offshore sand and gravel constitute a limited and nonrenewable resource. The notion of supply is linked strongly to the usage of the material. As such, a distinction has to be made between the (reserves of) primary aggregates, suitable for use in construction, and the (resources of) secondary aggregates, which may include contaminants, and used typically as in-filling material. For example, in-filling sand is available in relatively large quantities, whereas sand for construction, whose demand for quality is comparatively higher, is not so abundant. Gravel, which is also considered generally as a high-quality material, is in short supply. In the UK waters, the industry estimates that marine aggregate availability will last for 50 years, at present levels of extraction. However, such estimations are subject to important fluctuations, due to limited knowledge of the seafloor surface and subsurface. Furthermore, the level of accuracy of the identification of the resource varies greatly, from country to country (in general, data are far more complete in countries that have a greater reliance on marine sand and gravel). Another frequent difficulty toward a reliable assessment of the supply comes from the dispersal of the data relevant to the resource among governmental organizations, hydrographic departments, the dredging industry, and other commercial companies. In order to tackle this issue, a number of countries are undertaking seabed mapping programs, dedicated to the recognition of marine aggregates deposits. These programs range from detailed resource assessment, to reconnaissance level of the resource.

The precise estimation of the future demand for aggregates in general, including marine sand and gravel, is not an easy task. Difficulties arise from the highly variable and changing character of a number of parameters, such as the construction market and other economic factors, political strategies, environmental considerations, etc. In addition, the estimates can be distorted by extensive public work projects, involving large amounts of material. Moreover, the predictions should consider the exports and imports of material, and, therefore, take into account the demand at a much broader scale than the national one alone. Few countries have published estimates for the future demand of aggregate (only the UK and the Netherlands, in Europe).

However, in a number of countries, on-land aggregate mining takes place within the context of increasing environmental pressure and the depletion of the resource. At the same time, large parts of the coasts, on a worldwide scale, experience increasing coastal erosion. Therefore, it is expected that the future marine sand and gravel extraction is bound to increase, in order to supply high-quality material and to provide the large quantities of aggregates required for the realization of large-scale infrastructure projects designed to manage coastal retreat and accommodate the development of the coastal zones (i.e., beach replenishment, dune restoration, foreshore nourishment, land reclamation, and other coastal defense schemes). Hence, in order to find these new resources, it is expected that the activity will move farther offshore, within the next decade.

See also

Beaches, Physical Processes Affecting. Mid-Ocean Ridge Seismicity. Sediment Chronologies. Seismic Structure. Sonar Systems.

Further Reading

Bellamy AG (1998) The UK marine sand and gravel dredging industry: An application of quaternary geology. In: Latham J-P (ed.) *Engineering Geology Special Publications 13: Advances in Aggregates and Armourstone Evaluation*, pp. 33–45. London: Geological Society.

Boyd SE, Limpenny DS, Rees HL, and Cooper KM (2005) The effects of marine sand and gravel extraction on the macrobenthos at a commercial dredging site (results 6 years post-dredging). *ICES Journal of Marine Science* 62: 145–162.

Byrnes MR, Hammer RM, Thibaut TD, and Snyder DB (2004) Physical and biological effects of sand mining offshore Alabama, USA. *Journal of Coastal Research* 20(1): 6–24.

Desprez M (2000) Physical and biological impact of marine aggregate extraction along the French coast of the Eastern English Channel. Short and long-term post-dredging restoration. *ICES Journal of Marine Science* 57: 1428–1438.

Diesing M (2003) Die Regeneration von Material-entnahmestellen in der Südwestlichen Ostsee unter Besonderer Berücksichtigung der Rezenten Sediment-dynamik, 158pp. PhD Thesis, Christian-Albrechts-Universität, Kiel.

Diesing M, Schwarzer K, Zeiler M, and Kelon H (2006) *Special Issue: Comparison of Marine Sediment Extraction Sites by Mean of Shoreface Zonation. Journal of Coastal Research* 39: 783–788.

ICES (1992) Effects of extraction of marine sediments on fisheries. *Cooperative Research Report, No. 182.* Copenhagen: International Council for the Exploration of the Sea.

ICES (2001) Report of the ICES Working Group on the effects of extraction of marine sediments on the marine ecosystem. *Cooperative Research Report, No. 247.* Copenhagen: International Council for the Exploration of the Sea.

Kenny AJ and Rees HL (1996) The effects of marine gravel extraction on the macrobenthos: Results 2 years post-dredging. *Marine Pollution Bulletin* 32(8/9): 615–622.

Manso F, Diesing M, Schwarzer K, and Garel E (2004) Monitoring of dredged areas by combination of multibeam echosounder and sidescan sonar data. *Conference Littoral 2004*, Aberdeen, UK, 20–22 September 2004.

Newell RC, Seiderer LJ, and Hitchcock DR (1998) The impact of dredging works in coastal waters: A review of the sensitivity to disturbance and subsequent recovery of biological resources on the sea bed. *Oceanography and Marine Biology* 36: 127–178.

Roos PC (2004) *Seabed Pattern Dynamics and Offshore Sand Extraction*, 166pp. PhD Thesis, University of Twente, The Netherlands (ISBN 90-365-2067-3).

van Dalfsen JA, Essink K, Toxvig madsen H, *et al.* (2000) Differential response of macrozoobenthos to marine sand extraction in the North Sea and the Western Mediterranean. *ICES Journal of Marine Science* 57(5): 1439–1445.

van der Veer HW, Bergman MJN, and Beukema JJ (1985) Dredging activities in the Dutch Waddensea: Effects on macrobenthic infauna. *Netherlands Journal of Sea Research* 19: 183–190.

van Moorsel GWNM (1994) The Klaverbank (North Sea), geomorphology, macrobenthic ecology and the effect of gravel extraction. *Report No. 94.24.* Culemborg: Bureau Waardenburg BV.

Van Rijn LC, Soulsby RL, Hoekstra P, and Davies AG (eds.) (2005) *SANDPIT: Sand Transport and Morphology of Offshore Sand Mining Pits Process Knowledge and Guidelines for Coastal Management.* Amsterdam: Aqua Publications.

Relevant Websites

http://www.bmapa.org
– British Marine Aggregate Producers Association (BMAPA).

http://www.dredging.org
– Central Dredging Association (CEDA): Serving Africa, Europe, and the Middle East.

http://www.ciria.org
– Construction Industry Research and Information Association (CIRIA)

http://www.ciria.org
– European Marine Sand and Gravel Group (EMSAGG).

http://www.dvz.be
– ICES Working Group on the Effect of Extraction of Marine Sediments on the Marine Ecosystem (WGEXT).

http://www.iadc-dredging.com
– International Association of Dredging Companies (IADC).

http://www.ices.dk
– International Council for the Exploration of the Sea (ICES).

http://www.sandandgravel.com
– Marine Sand and Gravel Information Service (MAGIS).

http://www.westerndredging.org
– Western Dredging Association (WEDA): Serving the Americas.

http://www.woda.org
– World Organisation of Dredging Associations (WODA).

APPENDICES

APPENDIX 1. SI UNITS AND SOME EQUIVALENCES

Wherever possible the units used are those of the International System of Units (SI). Other "conventional" units (such as the liter or calorie) are frequently used, especially in reporting data from earlier work. Recommendations on standardized scientific terminology and units are published periodically by international committees, but adherence to these remains poor in practice. Conversion between units often requires great care.

The base SI units

Quantity	Unit	Symbol
Length	meter	m
Mass	kilogram	kg
Time	second	s
Electric current	ampere	A
Thermodynamic temperature	kelvin	K
Amount of substance	mole	mol
Luminous intensity	candela	cd

Some SI derived and supplementary units

Quantity	Unit	Symbol	Unit expressed in base or other derived units
Frequency	hertz	Hz	s^{-1}
Force	newton	N	$kg \, m \, s^{-2}$
Pressure, stress	pascal	Pa	$N \, m^{-2}$
Energy, work, quantity of heat	joule	J	$N \, m$
Power	watt	W	$J \, s^{-1}$
Electric charge, quantity of electricity	coulomb	C	$A \, s$
Electric potential, potential difference, electromotive force	volt	V	$J \, C^{-1}$
Electric capacitance	farad	F	$C \, V^{-1}$
Electric resistance	ohm	ohm (Ω)	$V \, A^{-1}$
Electric conductance	Siemens	S	Ω^{-1}
Magnetic flux	weber	Wb	$V \, s$
Magnetic flux density	tesla	T	$Wb \, m^{-2}$
Inductance	henry	H	$Wb \, A^{-1}$
Luminous flux	lumen	lm	$cd \, sr$
Illuminance	lux	lx	$lm \, m^{-2}$
Activity (of a radionuclide)	becquerel	Bq	s^{-1}
Absorbed dose, specific energy	gray	Gy	$J \, kg^{-1}$
Dose equivalent	sievert	Sv*	$J \, kg^{-1}$
Plane angle	radian	rad	
Solid angle	steradian	sr	

*Not to be confused with Sverdrup conventionally used in oceanography: see SI Equivalences of Other Units.

SI base units and derived units may be used with multiplying prefixes (with the exception of kg, though prefixes may be applied to gram $= 10^{-3}$kg; for example, $1 \text{ Mg} = 10^6 \text{ g} = 10^6$kg)

Prefixes used with SI units

Prefix	Symbol	Factor
yotta	Y	10^{24}
zetta	Z	10^{21}
exa	E	10^{18}
peta	P	10^{15}
tera	T	10^{12}
giga	G	10^{9}
mega	M	10^{6}
kilo	k	10^{3}
hecto	h	10^{2}
deca	da	10
deci	d	10^{-1}
centi	c	10^{-2}
milli	m	10^{-3}
micro	μ	10^{-6}
nano	n	10^{-9}
pico	p	10^{-12}
femto	f	10^{-15}
atto	a	10^{-18}
zepto	z	10^{-21}
yocto	y	10^{-24}

SI Equivalences of Other Units

Physical quantity	Unit	Equivalent	Reciprocal
Length	nautical mile (nm)	1.85318 km	km $= 0.5396$ nm
Mass	tonne (t)	10^3 kg $= 1$ Mg	
Time	min	60 s	
	h	3600 s	
	day or d	86 400 s	s $= 1.1574 \times 10^{-5}$ day
	y	3.1558×10^7 s	s $= 3.1688 \times 10^{-8}$ y
Temperature	°C	°C $=$ K $- 273.15$	
Velocity	knot (1 nm h^{-1})	0.51477 m s^{-1}	m s$^{-1} = 1.9426$ knot
		44.5 km d^{-1}	
		$16\ 234 \text{ km y}^{-1}$	
Density	gm cm^{-3}	tonne m$^{-3} = 10^3$ kg m^{-3}	
Force	dyn	10^{-5} N	
Pressure	dyn cm^{-2}	10^{-1} N m$^{-2} = 10^{-1}$ Pa	
	bar	10^5 N m$^{-2} = 10^5$ Pa	
	atm (standard atmosphere)	101 325 N m^{-2} $= 101.325$ kPa	
Energy	erg	10^{-7} J	
	cal (I.T.)	4.1868 J	
	cal (15°C)	4.1855 J	
	cal (thermochemical)	4.184 J	J $= 0.239$ cal

(*Note*: The last value is the one used for subsequent conversions involving calories.)

Energy flux	langley (ly) min^{-1} = cal cm^{-2} min^{-1}	697 W m^{-2}	W m^{-2} = 1.434×10^{-3} ly min^{-1}
	ly h^{-1}	11.6 W m^{-2}	W m^{-2} = 0.0860 ly h^{-1}
	ly d^{-1}	0.484 W m^{-2}	W m^{-2} = 2.065 ly d^{-1}
	kcal cm^{-2} y^{-1}	1.326 W m^{-2}	W m^{-2} = 0.754 kly y^{-1}
Volume flux	Sverdrup	10^6 m^3 s^{-1} 3.6 km^3 h^{-1}	
Latent heat	cal g^{-1}	4184 J kg^{-1}	J kg^{-1} = 2.39×10^{-4} cal g^{-1}
Irradiance	Einstein m^{-2} s^{-1} (mol photons m^{-2} s^{-1})		

*Most values are taken from or derived from *The Royal Society Conference of Editors Metrication in Scientific Journals*, 1968, The Royal Society, London.

The SI units for pressure is the pascal (1 Pa = 1 N m^{-2}). Although the bar (1 bar = 10^5 Pa) is also retained for the time being, it does not belong to the SI system. Various texts and scientific papers still refer to gas pressure in units of the torr (symbol: Torr), the bar, the conventional millimetre of mercury (symbol: mmHg), atmospheres (symbol: atm), and pounds per square inch (symbol: psi) – although these units will gradually disappear (see Conversions between Pressure Units).

Irradiance is also measured in W m^{-2}. Note: 1 mol photons = 6.02×10^{23} photons.

The SI unit used for the amount of substance is the mole (symbol: mol), and for volume the SI unit is the cubic metre (symbol: m^3). It is technically correct, therefore, to refer to concentration in units of mol m^3. However, because of the volumetric change that sea water experiences with depth, marine chemists prefer to express sea water concentrations in molal units, mol kg^{-1}.

Conversions between Pressure Units

	Pa	kPa	bar	atm	Torr	psi
1 Pa =	1	10^{-3}	10^{-5}	$9.869\,23 \times 10^{-6}$	$7.500\,62 \times 10^{-3}$	$1.450\,38 \times 10^{-4}$
1 kPa =	10^3	1	10^{-2}	$9.869\,23 \times 10^{-3}$	7.500 62	0.145 038
1 bar =	10^5	10^2	1	0.986 923	750.062	145.038
1 atm =	101 325	101.325	1.013 25	1	760	14.6959
1 Torr =	133.322	0.133 322	$1.333\,22 \times 10^{-3}$	$1.315\,79 \times 10^{-3}$	1	$1.933\,67 \times 10^{-2}$
1 psi	6894.76	6.894 76	$6.894\,76 \times 10^{-2}$	$6.804\,60 \times 10^{-2}$	51.715 07	1

psi = pounds force per square inch.
1 mmHg = 1 Torr to better than 2×10^{-7} Torr.

APPENDIX 2. USEFUL VALUES

Molecular mass of dry air, $m_a = 28.966$

Molecular mass of water, $m_w = 18.016$

Universal gas constant, $R = 8.31436 \, \mathrm{J\,mol^{-1}K^{-1}}$

Gas constant for dry air, $R_a = R/m_a = 287.04 \, \mathrm{J\,kg^{-1}K^{-1}}$

Gas constant for water vapor, $R_v = R/m_w = 461.50 \, \mathrm{J\,kg^{-1}K^{-1}}$

Molecular weight ratio $\varepsilon \equiv m_w/m_a = R_a/R_v = 0.62197$

Stefan's constant $\sigma = 5.67 \times 10^{-8} \, \mathrm{W\,m^{-2}K^{-4}}$

Acceleration due to gravity, $g \, (\mathrm{m\,s^{-2}})$ as a function of latitude φ and height $z \, (\mathrm{m})$

$$g = (9.78032 + 0.005172 \sin^2 \varphi - 0.00006 \sin^2 2\varphi)(1 + z/a)^{-2}$$

Mean surface value, $\bar{g} = \int_0^{\pi/2} g \cos \varphi \, \mathrm{d}\varphi = 9.7976$

Radius of sphere having the same volume as the Earth, $a = 6371 \, \mathrm{km}$ (equatorial radius $= 6378 \, \mathrm{km}$, polar radius $= 6357 \, \mathrm{km}$)

Rotation rate of earth, $\Omega = 7.292 \times 10^{-5} \, \mathrm{s^{-1}}$

Mass of earth $= 5.977 \times 10^{24} \, \mathrm{kg}$

Mass of atmosphere $= 5.3 \times 10^{18} \, \mathrm{kg}$

Mass of ocean $= 1400 \times 10^{18} \, \mathrm{kg}$

Mass of ground water $= 15.3 \times 10^{18} \, \mathrm{kg}$

Mass of ice caps and glaciers $= 43.4 \times 10^{18} \, \mathrm{kg}$

Mass of water in lakes and rivers $= 0.1267 \times 10^{18} \, \mathrm{kg}$

Mass of water vapor in atmosphere $= 0.0155 \times 10^{18} \, \mathrm{kg}$

Area of earth $= 5.10 \times 10^{14} \, \mathrm{m^2}$

Area of ocean $= 3.61 \times 10^{14} \, \mathrm{m^2}$

Area of land $= 1.49 \times 10^{14} \, \mathrm{m^2}$

Area of ice sheets and glaciers $= 1.62 \times 10^{13} \, \mathrm{m^2}$

Area of sea ice $= 1.9 \times 10^{13} \, \mathrm{m^2}$ in March and $2.9 \times 10^{13} \, \mathrm{m^2}$ in September (averaged between 1979 and 1987)

APPENDIX 3. PERIODIC TABLE OF THE ELEMENTS

Atomic number
Element symbol
Atomic mass

1	2	3	4	5	6	7	8	9	10	11	12	13	14	15	16	17	18
1 H 1.00794																	2 He 4.00260
3 Li 6.941	4 Be 9.01218											5 B 10.811	6 C 12.011	7 N 14.0067	8 O 15.9994	9 F 18.9984	10 Ne 20.1797
11 Na 22.9898	12 Mg 24.3050											13 Al 26.9815	14 Si 28.0855	15 P 30.9738	16 S 32.066	17 Cl 35.4527	18 Ar 39.948
19 K 39.0983	20 Ca 40.078	21 Sc 44.9559	22 Ti 47.88	23 V 50.9415	24 Cr 51.9961	25 Mn 54.9380	26 Fe 55.847	27 Co 58.9332	28 Ni 58.69	29 Cu 63.546	30 Zn 65.39	31 Ga 69.723	32 Ge 72.61	33 As 74.9216	34 Se 78.96	35 Br 79.904	36 Kr 83.80
37 Rb 85.4678	38 Sr 87.62	39 Y 88.9059	40 Zr 91.224	41 Nb 92.9064	42 Mo 95.94	43 Tc (98)	44 Ru 101.07	45 Rh 102.906	46 Pd 106.42	47 Ag 107.868	48 Cd 112.411	49 In 114.82	50 Sn 118.710	51 Sb 121.75	52 Te 127.60	53 I 126.905	54 Xe 131.29
55 Cs 132.905	56 Ba 137.327	57 La 138.906 ★	72 Hf 178.49	73 Ta 180.948	74 W 183.85	75 Re 186.207	76 Os 190.2	77 Ir 192.22	78 Pt 195.08	79 Au 196.967	80 Hg 200.59	81 Tl 204.383	82 Pb 207.2	83 Bi 208.980	84 Po (209)	85 At (210)	86 Rn (222)
87 Fr (223)	88 Ra 226.025	89 Ac 227.028 ◄	104 (261)	105 (262)	106 (263)	107 (262)	108 (265)	109 (267)									

★ Lanthanides

58 Ce 140.115	59 Pr 140.908	60 Nd 144.24	61 Pm (145)	62 Sm 150.36	63 Eu 151.965	64 Gd 157.25	65 Tb 158.925	66 Dy 162.50	67 Ho 164.930	68 Er 167.26	69 Tm 168.934	70 Yb 173.04	71 Lu 174.967

◄ Actinides

90 Th 232.038	91 Pa 231.036	92 U 238.029	93 Np 237.048	94 Pu (244)	95 Am (243)	96 Cm (247)	97 Bk (247)	98 Cf (251)	99 Es (252)	100 Fm (257)	101 Md (258)	102 No (259)	103 Lr (260)

APPENDIX 4. THE GEOLOGIC TIME SCALE

Eon	Era	Period	Epoch	Millions of Years Ago
Phanerozoic	Cenozoic	(Quaternary)	Holocene	
			Pleistocene	0.011
				1.82
		(Tertiary)	Pliocene	
				5.32
			Miocene	
				23
			Oligocene	
				33.7
			Eocene	
				55
			Paleocene	
				65
	Mesozoic	Cretaceous		
				144
		Jurassic		
				200
		Triassic		
				250
	Paleozoic	Permian		
				295
		Carboniferous Pennsylvanian		
				320
		Mississippian		
				355
		Devonian		
				410
		Silurian		
				440
		Ordovician		
				500
		Cambrian		
				540
Proterozoic				
				2500
Archean		Oldest Rock Age of the Solar System		4400 4550

INDEX

Notes

Cross-reference terms in italics are general cross-references, or refer to subentry terms within the main entry (the main entry is not repeated to save space). Readers are also advised to refer to the end of each article for additional cross-references - not all of these cross-references have been included in the index cross-references.

The index is arranged in set-out style with a maximum of three levels of heading. Major discussion of a subject is indicated by bold page numbers. Page numbers suffixed by T and F refer to Tables and Figures respectively. vs. indicates a comparison.

This index is in letter-by-letter order, whereby hyphens and spaces within index headings are ignored in the alphabetization. For example, 'oceanography' is alphabetized before 'ocean optics.' Prefixes and terms in parentheses are excluded from the initial alphabetization.

Where index subentries and sub-subentries pertaining to a subject have the same page number, they have been listed to indicate the comprehensiveness of the text.

Abbreviations used in subentries

AUV - autonomous underwater vehicle
$\delta^{18}O$ - oxygen isotope ratio
DIC - dissolved inorganic carbon
DOC - dissolved organic carbon
ENSO - El Niño Southern Oscillation
MOR - mid-ocean ridge
ROV - remotely operated vehicle
SAR - synthetic aperture radar

Additional abbreviations are to be found within the index.

A

AABW *see* Antarctic Bottom Water (AABW)
ABE *see* Autonomous Benthic Explorer
ABE Autonomous Benthic Explorer (AUV), 105F, 120F, 128, 129F
ABS *see* Acoustic backscatter system (ABS)
ABW *see* Arctic Bottom Water (ABW)
Abyssal circulation
　global pattern, 402, 403F
Abyssal hills
　definition, 178–179
　development models, 179F
　　horst/graben model, 179–180, 179F
　　split volcano, 178–179, 179F
　　whole volcano, 178–179, 179F
　East Pacific Rise (EPR), 179–180
　fast-spreading ridges, 179–180, 179F
　lava flows
　　elongate pillows, 179F
　　syntectonic, 179–180, 179F

　magma supply, 178–179
　propagation, 179–180, 180F
　slow-spreading ridges, 178–179
　structure, 179–180
　volcanic growth faults, 179–180, 179F
Abyssal plain, 535
Acanthaster planci (crown-of-thorns starfish), 548
Accelerator mass spectrometry (AMS), 399–400
Accretionary prisms, **133–139**
　blueschists, 135, 135F
　chemosynthetic organisms, 136
　erosion and redeposition, 134–135
　faulting, 134–135, 137
　　decollement, 134, 134F, 137
　　fluid flow paths, 134F, 136
　　fluid pressure-related weakening, 136
　fluids, 136
　　chemical composition, 136
　　expulsion, 134F, 136
　　fluid pressure-related weakening, 136
　　hydrocarbon formation, 134, 136

　　mineral dissolution, transport and precipitation, 136
　　sediment water content, 136
　　seismic reflection, 136
　　wedge theory, 137
　geothermal gradients, 135, 136
　hydrogen sulfide, 136
　material origin and variation, 133–134
　　Cascadia subduction zone, 133–134
　　Marianas subduction zone, 133–134
　mechanics, 136–137
　　critical Coulomb wedge theory, 137, 137F
　　fluid pressure, 137
　melanges, 133, 135, 135F
　methane, 136
　non-accretionary plate boundaries, 137–138, 138F
　　landslides, 138
　　tectonic erosion, 137, 138F
　seafloor sediment
　　accumulation, 134–135
　　offscraped deposits, 134, 134F, 138

Printed and bound by CPI Group (UK) Ltd, Croydon, CR0 4YY

03/10/2024

01040318-0010